BASIC OPTICS AND OPTICAL INSTRUMENTS

Prepared by

Bureau of Naval Personnel

Dover Publications, Inc.

New York

Published in Canada by General Publishing Company, Ltd., 30 Lesmill Road, Don Mills, Toronto, Ontario.

Published in the United Kingdom by Constable and Company, Ltd., 10 Orange Street, London WC 2.

This Dover edition, first published in 1969, is an unabridged republication of the work originally published by the United States Navy Training Publications Center in 1966 as *Opticalman 3 & 2,* Navy Training Course NAVPERS 10205. The illustrations on page 29, originally in color, are reproduced here in black and white.

Standard Book Number: 486-62291-6
Library of Congress Catalog Card Number: 69-17644

Manufactured in the United States of America
Dover Publications, Inc.
180 Varick Street
New York, N. Y. 10014

PREFACE

This training course was prepared for the Bureau of Naval Personnel by the Training Publications Division, Naval Personnel Program Support Activity, Washington, D. C. Content of the course is intended primarily for men of the Navy and the Naval Reserve who are studying for Opticalman third class and second class.

Technical assistance in preparing the course was provided by the U. S. Naval School, Opticalmen, Great Lakes, Illinois, and the Bureau of Ships.

CONTENTS

READING LIST

Basic Handtools, NavPers 10085-A
Binoculars 7 x 50, NavShips 250-624-2
Azimuth Telescopes, NavShips 250-624-4
Telescopic Alidades, NavShips 250-624-5
Azimuth and Bearing Circles, NavShips 250-624-7
Sextants (David White and Pioneer), NavShips 250-624-10
Navigational Instructions, NavShips 250-624-12
Antiaircraft Gun Mount Telescopes, OP 582 (Chapters 3, 5, 6, 7, and 8)
Collimators for Optical Instruments, OP 1417
Telescope Mark 97 Mod 1, OP 1857

CREDITS

The illustrations below are included in this edition of Opticalman 3 & 2 through the courtesy of the designated companies, publishers, and associations. Permission to reproduce illustrations and other materials in this publication must be obtained from the source.

Source:	Figures:
South Bend Lathe Works	9-1, 9-2, 9-5, 9-8, 9-9, 9-11, 9-12, 9-13, 9-14, 9-15, 9-16, 9-17, 9-19, 9-22, 9-24, 9-25, 9-26, 9-27, 9-28, 9-29, 9-32, 9-33, 9-34, 9-35
Reed-Prentice Corporation	9-3
Lodge and Shipley Machine Tool Company	9-4
Brown & Sharpe Manufacturing Company	9-36, 9-37, 9-38, 9-40, 9-42, 9-43, 9-46, 9-47
Cincinnati Milling Machine Company	9-41, 9-48
Bausch & Lomb Optical Company	18-1

CHAPTER 1

ADVANCEMENT

This training course is designed to help you meet the technical qualifications for advancement to Opticalman 3 and 2. Information presented is based on the June 1965 edition of the Manual of Qualifications for Advancement in Rating NavPers 18068-B. Changes in the qualifications for Opticalmen after the B edition are therefore not reflected in the discussion.

Chapters 2 through 18 of this course deal with the technical subject matter of the Opticalman rating. Chapter 2 gives information on optical glass—its composition, care required in manufacture, and its importance in lenses and prisms for optical instruments. Chapter 3 is devoted to a discussion of the characteristics of light, with special emphasis on wave lengths, reflection, and refraction. By reflecting and refracting light rays, lenses and prisms in optical instruments can create and erect clear and distinct images of distant objects.

Chapters 4 and 5 of the training course present a detailed discussion of the formation of images by mirrors, lenses, and prisms. An understanding of the contents of these chapters is basic to an understanding of the principle of operation of various optical instruments. Different types of lenses and prisms are discussed from the standpoint of their construction and usage in optical instruments. The explanation tells how images are formed by thin and thick lenses, how to use the lens formula, and how to determine the location of an image formed by an optical instrument.

Chapter 6 is devoted to a brief discussion of basic optical instruments. Chapter 7 explains how optical instruments are constructed—the function and construction of mechanical parts, and the location of optical elements within the instrument. An understanding of basic optical instruments at this point in the course will help you to understand better the detailed discussion of optical instruments in the last nine chapters.

Chapter 8 gives the general maintenance procedures applicable to all optical instruments. Maintenance peculiar to each optical instrument is explained in the chapter pertaining to that instrument. Machining operations are discussed in chapter 9, to provide the information required when work of this type must be accomplished.

Chapters 10 through 18 explain the function and operation of various optical instruments, and then give the details for disassembling, repairing, and reassembling and adjusting them. The discussion of these instruments is of the greatest importance to an Opticalman, because he must understand them before he can perform satisfactory repairs on them. Competence in accomplishing work on optical instruments is gained only through supervised work and study in optical shops.

The remainder of this chapter presents information on the enlisted rating structure, the Opticalman rating, requirements and procedures for advancement in rating, and references which will be helpful in studying for advancement and also in performing duties as an Opticalman. This chapter explains the best procedure for using Navy training courses, and you are therefore urged to study it carefully before engaging in intensive study of the remainder of the training course.

THE ENLISTED RATING STRUCTURE

Two of the types of ratings included in the present enlisted rating structure, established in 1957, are general ratings and service ratings.

GENERAL RATINGS identify broad occupational fields of related duties and functions. Some general ratings include service ratings; others do not. Both Regular Navy and Naval Reserve personnel may hold general ratings.

SERVICE RATINGS identify subdivisions or specialties within a general rating. Although service ratings can exist at any petty officer level, they are most common at the PO3 and PO2 levels. Both Regular Navy and Naval Reserve personnel may hold service ratings.

THE OPTICALMAN RATING

Opticalmen maintain, repair, and overhaul telescopic alidades, azimuth and bearing circles, binoculars, compasses, gunsights, sextants, and other optical instruments. This includes inspection, casualty analysis, disassembly, repair, replacement or manufacture of parts, cleaning, reassembly, collimation, sealing, drying, gassing, and refinishing of surfaces.

The Opticalman rating is a general rating ONLY—there are no service ratings. The work of an Opticalman requires a high degree of intelligence and mechanical aptitude. Optical instruments are technical in nature and expensive; and ALL of them are delicate. For these reasons, just ANYONE cannot perform satisfactorily the work of an Opticalman. Intelligence is required to understand the principles of operation; and knowledge of an instrument is necessary in order to repair and collimate it.

OPTICALMAN BILLETS

Opticalmen generally are assigned duty in optical shops aboard repair ships or tenders. Occasionally, however, they are assigned duty ashore as instructors in Opticalman schools. Some Opticalmen are assigned to recruiting duty; other are assigned to Naval Reserve training units.

One important thing to keep in mind is this: The Opticalman rating is important to the Navy; without it, part of the work of the Navy would not be accomplished. Remember, therefore, that Opticalmen make an important contribution toward the fulfillment of the overall mission of the Navy.

Administrative Responsibilities

At the third or second class level, Opticalmen generally do not have the responsibility for administering an optical shop; but an Opticalman 2 is responsible for preparing casualty analysis inspection sheets for instruments and also for the maintenance of records and logs in the shop. Opticalmen on duty at the 3 or 2 level should therefore observe the work of Opticalmen at the first class and chief levels, and learn as much from them as possible about the work of a shop supervisor. This is the only way to develop to the maximum your usefulness to the Navy as an Opticalman. Be prepared for greater responsibility when it is assigned to you.

Shop safety is something you should always emphasize. When using tools and operating machines, it is easy for one to injure himself. This not only causes personal discomfort but results in a pecuniary loss to the Navy during absence from work. Opticalmen should keep the shop in excellent working shape and hazard-free, and work individually and collectively in a manner which minimizes personal injury.

ADVANCEMENT IN RATING

Some of the rewards of advancement in rating are easy to see. You get more pay. Your job assignments become more interesting and more challenging. You are regarded with greater respect by officers and enlisted personnel. You enjoy the satisfaction of getting ahead in your chosen Navy career.

But the advantages of advancing in rating are not yours alone. The Navy also profits. Highly trained personnel are essential to the functioning of the Navy. By each advancement in rating, you increase your value to the Navy in two ways. First, you become more valuable as a technical specialist in your own rating. And second, you become more valuable as a person who can train others and thus make far-reaching contributions to the entire Navy.

HOW TO QUALIFY FOR ADVANCEMENT

What must you do to qualify for advancement in rating? The requirements may change from time to time, but usually you must:

1. Have a certain amount of time in your present grade.
2. Complete the required military and professional training courses.
3. Demonstrate your ability to perform all the PRACTICAL requirements for advancement by completing the Record of Practical Factors, NavPers 760.
4. Be recommended by your commanding officer, after the petty officers and officers supervising your work have indicated that they consider you capable of performing the duties of the next higher rate.

5. Demonstrate your KNOWLEDGE by passing a written examination on (a) military requirements and (b) professional qualifications.

Some of these general requirements may be modified in certain ways. Figure 1-1 gives a more detailed view of the requirements for advancement of active duty personnel; figure 1-2 gives this information for inactive duty personnel.

Remember that the requirements for advancement can change. Check with your division officer or training officer to be sure that you know the most recent requirements.

Advancement in rating is not automatic. After you have met all the requirements, you are ELIGIBLE for advancement. You will actually be advanced in rating only if you meet all the requirements (including making a high enough score on the written examination) and if the quotas for your rating permit your advancement.

HOW TO PREPARE FOR ADVANCEMENT

What must you do to prepare for advancement in rating? You must study the qualifications for advancement, work on the practical factors, study the required Navy Training Courses, and study other material that is required for advancement in your rating. To prepare for advancement, you will need to be familiar with (1) the Quals Manual, (2) the Record of Practical Factors, NavPers 760, (3) a NavPers publication called Training Publications for Advancement in Rating, NavPers 10052, and (4) applicable Navy Training Courses. Figure 1-3 illustrates these materials; the following sections describe them and give you some practical suggestions on how to use them in preparing for advancement.

The Quals Manual

The Manual of Qualifications for Advancement in Rating, NavPers 18068B (with changes), gives the minimum requirements for advancement to each rate within each rating. This manual is usually called the "Quals Manual," and the qualifications themselves are often called "quals." The qualifications are of two general types: (1) military requirements, and (2) professional or technical qualifications.

MILITARY REQUIREMENTS apply to all ratings rather than to any one particular rating. Military requirements for advancement to third class and second class petty officer rates deal with military conduct, naval organization, military justice, security, watch standing, and other subjects which are required of petty officers in all ratings.

PROFESSIONAL QUALIFICATIONS are technical or professional requirements that are directly related to the work of each rating.

Both the military requirements and the professional qualifications are divided into subject matter groups; then, within each subject matter group, they are divided into PRACTICAL FACTORS and KNOWLEDGE FACTORS. Practical factors are things you must be able to DO. Knowledge factors are things you must KNOW in order to perform the duties of your rating.

The written examination you will take for advancement in rating will contain questions relating to the practical factors and the knowledge factors of both the military requirements and the professional qualifications. If you are working for advancement to second class, remember that you may be examined on third class qualifications as well as on second class qualifications.

The Quals Manual is kept current by means of changes. The professional qualifications for your rating which are covered in this training course were current at the time the course was printed. By the time you are studying this course, however, the quals for your rating may have been changed. Never trust any set of quals until you have checked it against an UP-TO-DATE copy in the Quals Manual.

Record of Practical Factors

Before you can take the servicewide examination for advancement in rating, there must be an entry in your service record to show that you have qualified in the practical factors of both the military requirements and the professional qualifications. A special form known as the RECORD OF PRACTICAL FACTORS, NavPers 760, is used to keep a record of your practical factor qualifications. This form is available for each rating. The form lists all practical factors, both military and professional. As you demonstrate your ability to perform each practical factor, appropriate entries are made in the DATE and INITIALS columns.

Changes are made periodically in the Manual of Qualifications for Advancement in Rating, and revised forms of NavPers 760 are provided when necessary. Extra space is allowed on the Record of Practical Factors for entering additional

ACTIVE DUTY ADVANCEMENT REQUIREMENTS

<table>
<tr><th>REQUIREMENTS *</th><th>E1 to E2</th><th>E2 to E3</th><th>E3 to E4</th><th>E4 to E5</th><th>E5 to E6</th><th>† E6 to E7</th><th>† E7 to E8</th><th>† E8 to E9</th></tr>
<tr><td>SERVICE</td><td>4 mos. service— or completion of recruit training.</td><td>6 mos. as E-2.</td><td>6 mos. as E-3.</td><td>12 mos. as E-4.</td><td>24 mos. as E-5.</td><td>36 mos. as E-6.</td><td>48 mos. as E-7. 8 of 11 years total service must be enlisted.</td><td>24 mos. as E-8. 10 of 13 years total service must be enlisted.</td></tr>
<tr><td>SCHOOL</td><td>Recruit Training.</td><td></td><td>Class A for PR3, DT3, PT3, AME 3, HM 3</td><td></td><td></td><td>Class B for AGCA, MUCA, MNCA.</td><td>Must be permanent appointment.</td><td></td></tr>
<tr><td>PRACTICAL FACTORS</td><td>Locally prepared check-offs.</td><td colspan="7">Records of Practical Factors, NavPers 760, must be completed for E-3 and all PO advancements.</td></tr>
<tr><td>PERFORMANCE TEST</td><td colspan="2"></td><td colspan="4">Specified ratings must complete applicable performance tests before taking examinations.</td><td></td><td></td></tr>
<tr><td>ENLISTED PERFORMANCE EVALUATION</td><td colspan="2">As used by CO when approving advancement.</td><td colspan="6">Counts toward performance factor credit in advancement multiple.</td></tr>
<tr><td>EXAMINATIONS</td><td colspan="2">Locally prepared tests.</td><td colspan="4">Navy-wide examinations required for all PO advancements.</td><td colspan="2">Navy-wide, selection board, and physical.</td></tr>
<tr><td>NAVY TRAINING COURSE (INCLUDING MILITARY REQUIREMENTS)</td><td></td><td colspan="5">Required for E-3 and all PO advancements unless waived because of school completion, but need not be repeated if identical course has already been completed. See NavPers 10052 (current edition).</td><td colspan="2">Correspondence courses and recommended reading. See NavPers 10052 (current edition).</td></tr>
<tr><td rowspan="2">AUTHORIZATION</td><td colspan="2">Commanding Officer</td><td colspan="3">U.S. Naval Examining Center</td><td colspan="3">Bureau of Naval Personnel</td></tr>
<tr><td colspan="8">TARS attached to the air program are advanced to fill vacancies and must be approved by CNARESTRA.</td></tr>
</table>

* All advancements require commanding officer's recommendation.

† 2 years obligated service required.

Figure 1-1. —Active duty advancement requirements.

INACTIVE DUTY ADVANCEMENT REQUIREMENTS

<table>
<tr><th colspan="2">REQUIREMENTS *</th><th>E1 to E2</th><th>E2 to E3</th><th>E3 to E4</th><th>E4 to E5</th><th>E5 to E6</th><th>E6 to E7</th><th>E8</th><th>E9</th></tr>
<tr><td></td><td>FOR THESE DRILLS PER YEAR</td><td></td><td></td><td></td><td></td><td></td><td></td><td></td><td></td></tr>
<tr><td rowspan="3">TOTAL TIME IN GRADE</td><td>48</td><td>6 mos.</td><td>6 mos.</td><td>15 mos.</td><td>18 mos.</td><td>24 mos.</td><td>36 mos.</td><td>48 mos.</td><td>24 mos.</td></tr>
<tr><td>24</td><td>9 mos.</td><td>9 mos.</td><td>15 mos.</td><td>18 mos.</td><td>24 mos.</td><td>36 mos.</td><td>48 mos.</td><td>24 mos.</td></tr>
<tr><td>NON-DRILLING</td><td>12 mos.</td><td>24 mos.</td><td>24 mos.</td><td>36 mos.</td><td>48 mos.</td><td>48 mos.</td><td></td><td></td></tr>
<tr><td rowspan="2">DRILLS ATTENDED IN GRADE †</td><td>48</td><td>18</td><td>18</td><td>45</td><td>54</td><td>72</td><td>108</td><td>144</td><td>72</td></tr>
<tr><td>24</td><td>16</td><td>16</td><td>27</td><td>32</td><td>42</td><td>64</td><td>85</td><td>32</td></tr>
<tr><td rowspan="3">TOTAL TRAINING DUTY IN GRADE †</td><td>48</td><td>14 days</td><td>14 days</td><td>14 days</td><td>14 days</td><td>28 days</td><td>42 days</td><td>56 days</td><td>28 days</td></tr>
<tr><td>24</td><td>14 days</td><td>14 days</td><td>14 days</td><td>14 days</td><td>28 days</td><td>42 days</td><td>56 days</td><td>28 days</td></tr>
<tr><td>NON-DRILLING</td><td>None</td><td>None</td><td>14 days</td><td>14 days</td><td>28 days</td><td>28 days</td><td></td><td></td></tr>
<tr><td colspan="2">PERFORMANCE TESTS</td><td colspan="2"></td><td colspan="6">Specified ratings must complete applicable performance tests before taking examination.</td></tr>
<tr><td colspan="2">PRACTICAL FACTORS (INCLUDING MILITARY REQUIREMENTS)</td><td colspan="8">Record of Practical Factors, NavPers 760, must be completed for all advancements.</td></tr>
<tr><td colspan="2">NAVY TRAINING COURSE (INCLUDING MILITARY REQUIREMENTS)</td><td colspan="8">Completion of applicable course or courses must be entered in service record.</td></tr>
<tr><td colspan="2">EXAMINATION</td><td colspan="6">Standard exams are used where available, otherwise locally prepared exams are used.</td><td colspan="2">Standard EXAM, Selection Board, and Physical.</td></tr>
<tr><td colspan="2">AUTHORIZATION</td><td colspan="5">District commandant or CNARESTRA</td><td colspan="3">Bureau of Naval Personnel</td></tr>
</table>

* Recommendation by commanding officer required for all advancements.
† Active duty periods may be substituted for drills and training duty.

Figure 1-2.—Inactive duty advancement requirements.

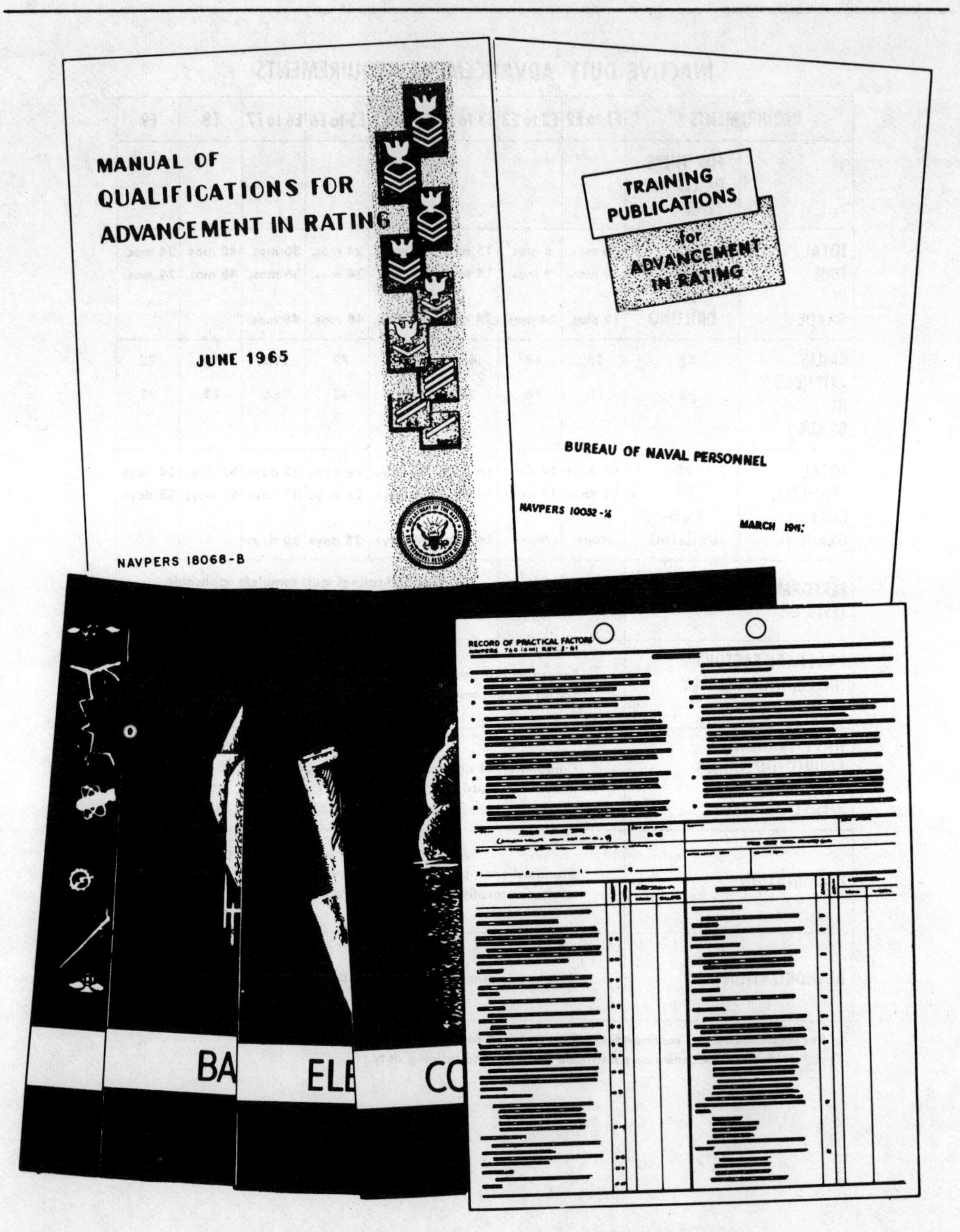

Figure 1-3. —Materials used in preparing for advancement.

practical factors as they are published in the changes to the Quals Manual. The Record of Practical Factors also provides space for recording demonstrated proficiency in skills which are within the general scope of the rating but which are not identified as minimum qualifications for advancement.

If you are transferred before you can qualify in all practical factors, the NavPers 760 form should be forwarded with your service record to your next duty station. You can save yourself a lot of trouble by making sure that this form is actually inserted in your service record before you are transferred. If the form is not in your service record, you may be required to start all over again and requalify in the practical factors which have already been checked off.

NavPers 10052

Training Publications for Advancement in Rating, NavPers 10052 (revised), is a very important publication for anyone preparing for advancement in rating. This bibliography lists required and recommended Navy Training Courses and other reference material to be used by personnel working for advancement in rating. NavPers 10052 is revised and issued once each year by the Bureau of Naval Personnel. Each revised edition is identified by a letter following the NavPers number. When using this publication, be SURE that you have the most recent edition.

If extensive changes in qualifications occur in any rating between the annual revisions of NavPers 10052, a supplementary list of sutdy material may be issued in the form of a BuPers Notice. When you are preparing for advancement, check to see whether changes have been made in the qualifications for your rating. If changes have been made, see if a BuPers Notice has been issued to supplement NavPers 10052 for your rating.

The required and recommended references are listed by rate level in NavPers 10052. If you are working for advancement to third class, study the material that is listed for third class. If you are working for advancement to second class, study the material that is listed for second class; but remember that you are also responsible for the references listed at the third class level.

In using NavPers 10052, you will notice that some Navy Training Courses are marked with an asterisk (*). Any course marked in this way is MANDATORY—that is, it must be completed at the indicated rate level before you can be eligible to take the servicewide examination for advancement in rating. Each mandatory course may be completed by (1) passing the appropriate enlisted correspondence course that is based on the mandatory training course; (2) passing locally prepared tests based on the information given in the training course; or (3) in some cases, successfully completing an appropriate Class A school.

Do not overlook the sections of NavPers 10052 which lists the required and recommended references relating to the military requirements for advancement. Personnel of ALL ratings must complete the mandatory military requirements training course for the appropriate rate level before they can be eligible to advance in rating.

The references in NavPers 10052 which are recommended but not mandatory should also be studied carefully. ALL references listed in NavPers 10052 may be used as source material for the written examinations, at the appropriate rate levels.

Navy Training Courses

There are two general types of Navy Training Courses. RATING COURSES (such as this one) are prepared for most enlisted ratings. A rating training course gives information that is directly related to the professional qualifications of ONE rating. SUBJECT MATTER COURSES or BASIC COURSES give information that applies to more than one rating.

Navy Training Courses are revised from time to time to keep them up to date technically. The revision of a Navy Training Course is identified by a letter following the NavPers number. You can tell whether any particular copy of a Navy Training Course is the latest edition by checking the NavPers number and the letter following this number in the most recent edition of List of Training Manuals and Correspondence Courses, NavPers 10061. (NavPers 10061 is actually a catalog that lists all current training courses and correspondence courses; you will find this catalog useful in planning your study program.)

Navy Training Courses are designed to help you prepare for advancement in rating. The following suggestions may help you to make the best use of this course and other Navy training publications when you are preparing for advancement in rating.

1. Study the military requirements and the professional qualifications for your rating before

you study the training course, and refer to the quals frequently as you study. Remember, you are studying the training course primarily in order to meet these quals.

2. Set up a regular study plan. It will probably be easier for you to stick to a schedule if you can plan to study at the same time each day. If possible, schedule your studying for a time of day when you will not have too many interruptions or distractions.

3. Before you begin to study any part of the training course intensively, become familiar with the entire book. Read the preface and the table of contents. Check through the index. Look at the appendixes. Thumb through the book without any particular plan, looking at the illustrations and reading bits here and there are you see things that interest you.

4. Look at the training course in more detail, to see how it is organized. Look at the table of contents again. Then, chapter by chapter, read the introduction, the headings, and the subheadings. This will give you a pretty clear picture of the scope and content of the book. As you look through the book in this way, ask yourself some questions: What do I need to learn about this? What do I already know about this? How is this information related to information given in other chapters? How is this information related to the qualifications for advancement in rating?

5. When you have a general idea of what is in the training course and how it is organized, fill in the details by intensive study. In each study period, try to cover a complete unit—it may be a chapter, a section of a chapter, or a subsection. The amount of material that you can cover at one time will vary. If you know the subject well, or if the material is easy, you can cover quite a lot at one time. Difficult or unfamiliar material will require more study time.

6. In studying any one unit—chapter, section, or subsection—write down the questions that occur to you. Many people find it helpful to make a written outline of the unit as they study, or at least to write down the most important ideas.

7. As you study, relate the information in the training course to the knowledge you already have. When you read about a process, a skill, or a situation, try to see how this information ties in with your own past experience.

8. When you have finished studying a unit, take time out to see what you have learned. Look back over your notes and questions. Maybe some of your questions have been answered, but perhaps you still have some that are not answered. Without looking at the training course, write down the main ideas that you have gotten from studying this unit. Don't just quote the book. If you can't give these ideas in your own words, the chances are that you have not really mastered the information.

9. Use Enlisted Correspondence Courses whenever you can. The correspondence courses are based on Navy Training Courses or on other appropriate texts. As mentioned before, completion of a mandatory Navy Training Course can be accomplished by passing an Enlisted Correspondence Course based on the Navy Training Course. You will probably find it helpful to take other correspondence courses, as well as those based on mandatory training courses. Taking a correspondence course helps you to master the information given in the training course, and also helps you see how much you have learned.

10. Think of your future as you study Navy Training Courses. You are working for advancement to third class or second class right now, but someday you will be working toward higher rates. Anything extra that you can learn now will help you both now and later.

SOURCES OF INFORMATION

One of the most useful things you can learn about a subject is how to find out more about it. No single publication can give you all the information you need to perform the duties of your rating. You should learn where to look for accurate, authoritative, up-to-date information on all subjects related to the military requirements for advancement and the professional qualifications of your rating.

Some of the publications described here are subject to change or revision from time to time—some at regular intervals, others as the need arises. When using any publication that is subject to change or revision, be sure that you have the latest edition. When using any publication that is kept current by means of changes, be sure you have a copy in which all official changes have been made. Studying canceled or obsolete information will not help you to do your work or to advance in rating; it is likely to be a waste of time, and may even be seriously misleading.

Some helpful publications not included in the reading list for this training course are:

1. Standard First Aid Training Course, NavPers 10081-A, which explains the procedures and methods to follow when administering first aid.

2. Alidade, Telescopic, Marine (Mk 6 Mod 1 & Mk 7 Mod 0), NavShips 324-0654. This publication explains the construction and operation of Mk 6 and Mk 7 telescopic alidades and gives a brief discussion on maintenance.

3. Alidade, Telescopic, Marine (Mk 7 Mod 0), NavShips 0924-001-6000, which provides information on the principles of operation and maintenance of Mk 7 alidades.

A list of training films that may be useful is given in appendix I of this training course. Other films which may be of interest are listed in the United States Navy Film Catalog, NavPers 10000 (Revised).

CHAPTER 2

OPTICAL GLASS

Without glass, there would by NO optical instruments and no Opticalmen in the Navy; because the optical elements of these instruments are made of glass with definite characteristics. For this reason, you should understand the properties and characteristics of lenses and prisms used in optical instruments throughout the Navy. You may wonder why ordinary glass or plastic substances cannot be used for making lenses and prisms. This chapter explains why the use of such elements for this purpose is unsatisfactory.

Because of the high transparency and ease of manufacturing plastic lenses (in almost final shape), glass manufacturers have made plastic lenses. They are impractical for optical instruments, however, for two reasons: (1) they expand and contract more than glass during temperature changes, which means that they change their focal length more than glass lenses and, for this reason, it is impossible to correct aberrations in an optical system composed of plastic lenses; and (2) plastic lenses are so soft in texture that you cannot clean them satisfactorily (even with lens tissue) without damaging them.

NATURE OF GLASS

The only common characteristic of all glass is that its structure is not CRYSTALLINE but AMORPHOUS. Generally speaking, all glasses are hard and brittle; but the DEGREE of hardness and brittleness varies widely—some metals and minerals are harder and more fragile than glass.

Solid bodies have a definite or crystalline structure, but this is not true of such vitreous bodies as glass. The properties of glass are explainable only by assuming that they have the same molecular arrangement as a LIQUID. When a crystalline body passes from the liquid to the solid state, the transition takes place at a definite temperature and is accompanied by considerable heat which temporarily halts solidification. With glass, on the other hand, the transition from the liquid to the solid state is so continuous and gradual that the most delicate instruments have failed to record either evolution of heat or retardation of the solidifying process, which is a GRADUAL STIFFENING WITHOUT CHANGE OF STRUCTURE. All glass, however, assumes a crystalline structure (devitrification) if while in the vitreous state the temperature is maintained too long at the critical state (crystallization point). Crystalline glass gives DOUBLE refraction, and a lens made from it forms TWO SEPARATE IMAGES at the same time.

Glass has NO melting point. When heat is applied to it gradually, it gets soft and can be molded into a thread; when it is red hot, it flows in a thick mass. A temperature of several thousand degrees turns glass into a fluid.

COMPOSITION OF ORDINARY GLASS

In a liquid state, glass is a MIXTURE of certain chemicals in solution. The most common chemicals used for this purpose are the silicates and borates. Under ordinary conditions of cooling, these chemical solutions remain mutually dissolved. Each component of the mixture, however, has its own solidification point; and if the molten glass is maintained SLIGHTLY BELOW one of the critical temperatures of a component, that constituent crystallizes. In some instances, therefore, it is very difficult TO PREVENT devitrification of amorphous glass. Proper cooling of the molten glass—at a RELATIVELY RAPID RATE through the critical range of temperature—is the answer to the problem.

The tendency of amorphous glass to crystallize places a natural limit on the number of bodies which can be obtained in a vitreous state, only a few of which are ordinarily used for making glass.

Ordinary glass is manufactured from MIXED SILICATES of a few bases—alkalies (sodium and potassium), alkaline earths (calcium, magnesium, strontium, and barium), lead oxide, and (in small quantities) iron and aluminum oxides. Metallic elements (lead, potassium, sodium) also form oxides; and when these oxides are melted with silica, they form metallic SILICATES.

AMORPHOUS STATE OF GLASS

Although glass is a liquid, it is also a solid, which scientists generally describe as AMORPHOUS. Solids are characterized by definite shape and volume. Crystalline solids, for example, have a regular arrangement of particles; amorphous solids, on the other hand, have a random arrangement of particles—large, long-chain, entangled molecules.

You perhaps wonder how anything as SOLID as glass can be a LIQUID or in an AMORPHOUS state. The reason for this condition of glass is that the molecules are held together in crystals by VAN DER WAALS FORCES, which means that the electric field of the atoms of one molecule causes a similar variation in the electric field of the atoms of another molecule to generate attraction between them.

You can prove for yourself that GLASS IS AN AMORPHOUS STATE by placing a thin-walled glass tube five feet long (approximately) on two nails driven equidistant from the deck on the bulkhead of your shop and observing the bend in the tube during a five or six months' period. Hold the glass tube against the bulkhead and mark its original position with a pencil, so that you will be able to measure the amount of bend which develops during the period.

One interesting thing about this test is that when you first place the tube of glass on the nails it shows a slight bend, which immediately disappears if you then remove it from the nails. At the end of your test, however, the bend will remain in the tube when you remove it from the nails; because the liquid glass has ACTUALLY FLOWED to its new position.

PROPERTIES OF OPTICAL GLASS

Glass used in optical instruments must have special properties; ordinary glass has too many imperfections which affect light during passage through it. The PURELY OPTICAL PROPERTIES which directly influence light as it passes through glass include: (1) homogeneity, (2) transparency, (3) freedom from color, (4) refraction, and (5) dispersion. These properties are now discussed in the order listed.

HOMOGENEITY

Homogeneity is the most important property of optical glass. If you examine a thick piece of ordinary glass, you will find that the layers of difference densities show clearly in the form of internal irregularities, known as VEINS or STRIAE, little streaks with a higher or lower index of refraction (bending) than the other part of the glass. Many times the striae are also so small that they cannot be detected until the glass is ground (as a lens, for example) and polished. Because these striae affect the sharpness of an image formed by the lens, it cannot be used in an optical instrument.

You can test a lens for striae in the manner illustrated in figure 2-1. If you place a light (S) behind a screen with a hole in it directly in front of the light and then hold a lens (L) with one hand and a knife blade (K) at the point indicated in the other hand, you can look along the optical axis (central point) of the lens and detect the absence or presence of striae. If the lens has no striae, the field appears dark (part B, fig. 2-1); if striae are PRESENT, they show as bright lines (part C, fig. 2-1).

To be homogeneous, a lens must also be free of dust, dirt, and bubbles. A few bubbles in a lens stop the passage of light through the lens at their location, but they do not hurt the quality of the image. The best lenses may have ONE or TWO bubbles, but inspectors of precision lenses reject lenses with more than THREE bubbles; and they also reject a lens with ONE bubble as BIG as half a millimeter in diameter.

TRANSPARENCY AND COLOR

Transparency is a vital property of optical glass, its most important characteristic. No glass, however, is perfectly transparent; and some glass is translucent or opaque. You will also learn in chapter 3 that when light strikes a glass surface (even though it is perfectly polished) part of the light is reflected from the entrance surface, part of it is absorbed, and part of it is reflected from the emergent surface. The amount of absorption of light by the best glass is small if the glass is thin; but a piece of the best glass several inches thick absorbs a noticeable amount of light.

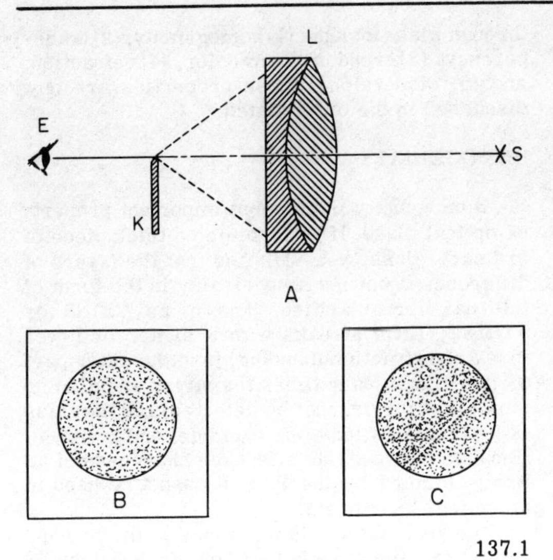

137.1

A. Testing procedure.
B. No striae present.
C. Striae present (white line).
Figure 2-1.—Testing a lens for striae.

The degree of absorption of light by glass varies with the color of the light. NOTE: Optical glass must be free from color. When white light passes through glass, the glass absorbs more of one of its component colors than the other colors, thus causing the emergent light to have a slight color tint. In thick pieces, purest and whitest glasses always show a distinct blue or green tint.

REFRACTION AND DISPERSION

Refraction and dispersion of light are two of the most important properties of any optical element. Refraction is the bending of a ray of light when it enters a lens or prism; dispersion is the separation of light into its component colors as it passes through a prism or lens. This occurs in an uncorrected lens or prism, because the index of refraction (ratio of speed of light in a vacuum to speed of light in a medium) of glass is different for each wave length. (You will learn more about refraction and dispersion in the next few chapters.)

GENERAL REQUIREMENTS

The general requirements of special importance in optical glass are: (1) chemical stability, (2) mechanical hardness, and (3) freedom from internal strains. It is important that you know and remember these qualities of optical glass.

CHEMICAL STABILITY

Chemical stability is an essential feature of optical glass, because the best lenses would soon be useless if they were affected by moisture and traces of chemical fumes in the atmosphere. Condensed water on glass absorbs carbon dioxide and forms carbonic acid, which dissolves glass. Distilled water, for example, must be kept in specially made glass containers or bottles; because it dissolves ordinary glass. High-quality optical glass (HARD CROWN and BORO SILICATE CROWN) resists chemicals and is therefore durable.

Occasionally, the optical qualities listed in the specifications for manufacturing necessitate changes in the composition of the glass which sacrifice chemical stability. Lenses made from unstable glass, however, can be protected by cementing them between two components of stable glass; but this procedure imposes limitations on optical design.

MECHANICAL HARDNESS

ANOTHER IMPORTANT FEATURE OF OPTICAL GLASS IS MECHANICAL HARDNESS (generally accompanied by a low refractive index) because lenses must be HARD ENOUGH to resist the effects of cleaning, which must be accomplished as necessary.

You will learn a little later in this chapter that HARD CROWN GLASS is harder than DENSE FLINT GLASS. This difference in degree of hardness is necessitated by the elements used by the manufacturer in order to get desired OPTICAL QUALITIES in the glass.

FREEDOM FROM INTERNAL STRAINS

Although desirable, ENTIRE FREEDOM from internal strains is essential ONLY for special optical purposes. Manufacturers of glass know from experience the amount of STRAIN PERMISSIBLE in glass intended for various purposes.

Strains in glass result from annealing (cooling and setting). When glass cools, it contracts; and if it cools TOO RAPIDLY, the surface becomes cool while the center is still hot, resulting in strains which usually cause breakage during grinding and polishing.

Strains in optical glass can be detected by polarized light. Perfectly annealed glass, entirely free from internal strains, produces no effect on a beam of polarized light which passes through it. A serious amount of double refraction indicates strain in the glass.

You can test for strain in optical glass by doing the following: (1) mount two polaroid filters in line with a light, one inch apart; (2) look through the filters toward the light and turn one of the filters until the field is dark; and (3) while looking into the dark field, hold the glass you desire to test between the two polaroid filters. If the field remains dark, the glass is free from strains. Strained glass, on the other hand, rotates the plane of polarization, causing you to see RINGS or BANDS of colored light.

MANUFACTURE OF GLASS

The discussion of manufacture of glass is limited to OPTICAL glass, because this is the type of glass in which you are primarily interested. An understanding of the quality of materials used in optical glass and the procedure for manufacturing it will be both interesting and beneficial to you. Why must we use specific amounts of definite ingredients for a particular type of optical glass? Why must so much care be exercised in manufacturing optical glass? The following discussion contains the answers to these questions.

COMPONENTS OF OPTICAL GLASS

In order to secure homogeneity in optical glass, the ingredients used in manufacturing it must be FINELY DIVIDED and THOROUGHLY MIXED.

Excessive furnace temperatures required to fuse certain substances impose a limit on the composition of optical glass. Excessive proportions of silica, lime, and so forth, for example, RAISE the fusing point. When this temperature exceeds 1600° C, glass manufacture in ordinary furnaces is impossible. Pure silica produces excellent glass, but it CANNOT be fused in ordinary regenerative gas furnaces used in making some types of glass.

The CHEMICAL BEHAVIOR of glass during manufacture and during use also places a limit on the procedure for making it. Glass solutions tend toward a state of saturation and, while in a molten state, a glass mixture rich in silica and deficient in bases DISSOLVES and ABSORBS basic materials it contacts. Similarly, a glass rich in bases and poor in silica or other acid constituents ABSORBS acid bodies. Optical glass must therefore be FUSED in FIRE-CLAY CRUCIBLES of such chemical composition as required in order to supply some of the constituents which the glass requires.

Because the transparency of glass enables one to see in the finished products defects of COLOR and QUALITY, raw materials selected for making optical glass must be PRACTICALLY FREE from impurities. Although volatile and combustible substances are usually completely eliminated (by high temperature) during the melting step, all FIXED (stable) substances which compose the mixture appear in the finished glass. The selection of raw materials for optical glass is therefore most important.

Raw materials for optical glass must also be uniform in composition. The materials generally used contain: (1) silica, (2) alkalies, and (3) bases other than alkalies.

Silica

The principal material in SILICA is SAND; and only the purest kind of sand is suitable for optical glass. It must NOT contain more than 0.1 percent of impurities; the remaining material must be PURE SILICA. SILICON (Si) is one of the chemical elements. Silica is one of its oxides (SiO_2). Pure quartz is a crystalline form of silica, and pure sands used for some optical glass are disintegrated quartz.

Silica is also a component of sandstone; but silica obtained in the form of crushed stone is not as satisfactory as sand—nor is it as pure or homogeneous. The same objections are also valid in the use of quartz or flint, with the added objection of the difficulty in crushing them, because of their extreme hardness. Flint glass got its name from ground flint used in its manufacture. FLINT currently denotes a heavy, dense glass of high lead content and refractive power.

Alakalies

The original forms of potassium and sodium chloride contain alkalies. From a manufacturing standpoint, however, it is not possible to use the chlorides because they are volatile at temperatures used for melting glass. They are also affected by hot silica ONLY in the presence of water vapor; consequently, chlorides vaporize before they combine with other ingredients. For this reason, alkalies are used in the form of sulphate of soda (salt cake) and (less frequently) carbonate of soda (soda ash).

Other Basic Materials

Calcium oxide (lime) and lead oxide are the most important basic materials (other than alkalies) used in optical glass. The usual form of calcium oxide used is the carbonate or the hydrate (slacked lime). Lead oxide (generally in the form of red lead) is used extensively in the manufacture of FLINT glass.

Barium oxide (magnesia) is also used for making optical glass; and zinc oxide is included in the ingredients for making a SPECIAL ZINC CROWN glass for certain optical purposes.

With respect to all elements used in a mixture for making optical glass, bear in mind that the manufacturer selects the elements to make a particular type of glass in accordance with specifications, to meet a specific need.

Now that you have learned the components of different types of optical glass, study the following formulas for CROWN and FLINT optical glass. There are quite a few other formulas for making optical glass; these are given as examples.

FORMULA FOR BORO-SILICATE CROWN GLASS

Silica (SiO_2)	68.24%
Boron oxide (B_2O_3)	10.00
Potassium oxide (K_2O)	9.50
Sodium oxide (Na_2O)	10.00
Zinc oxide (ZnO)	2.00
Manganese oxide (Mn_2O_3)	0.06
Arsenic pentoxide (As_2O_5)	0.20

NOTE: The index of refraction of glass produced by this formula for one particular wave length of yellow light is 1.5128.

FORMULA FOR DENSE FLINT GLASS

Silica (SiO_2)	18.00%
Lead oxide (PbO)	82.00

NOTE: the index of refraction of glass made by this formula is 1.9625.

One thing to remember about optical glass components is: CROWN GLASS—fairly low index of refraction and dispersion—contains phosphorus, barium, or boron, but NO lead; FLINT GLASS—higher index of refraction and dispersion than crown glass—may contain a small quantity of barium or boron, but it DOES contain lead—the greater the amount of lead used, the higher the index of refraction of the glass.

FURNACES AND CRUCIBLES

Manufacturers employ two general types of furnaces for making glass: (1) tank, and (2) pot (crucible). They use tank furnaces for producing great quantities of glass, such as rolled plate or sheet glass; but tank furnaces are UNSATISFACTORY for making optical glass.

Pot furnaces are used EXCLUSIVELY for manufacturing optical glass, for the following reasons:

1. The composition of the glass can be accurately regulated.
2. Molten glass can be protected better from contamination by the action of furnace gases or materials from the top of the furnace dropping into the glass.
3. It is possible to FUSE IN POTS materials which cannot be kept together in an open-tank furnace long enough to combine them.

Pot furnaces (crucibles) are gas furnaces similar in principle of operation and construction to furnaces used in making steel. They are lined with some form of FIRE CLAY, because the lining must be able to resist the disintegrating action of heated furnace gases for long periods. Fire clays are affected by the dissolving action of molten glass, so a crucible for optical glass is used ONLY once.

Crucibles employed for making optical glass usually vary between 30 and 50 inches in diameter and are of the covered type in design, as shown in figure 2-2. Furnaces for optical-glass manufacture are made to take ONLY one pot, instead of 10 or 12 generally used for making ordinary glass. The reason for this is that the methods followed for making technically perfect glass cannot be used when they are most desirable, because the raw materials for optical glass must be selected for the chemical composition which gives the desired optical properties. Furnace conditions must therefore be adapted to the materials selected for a particular type of optical glass. Heat (degree and duration) must be adapted to each pot. This control of temperature for melting glass is NOT possible when more than one pot is used in the furnace.

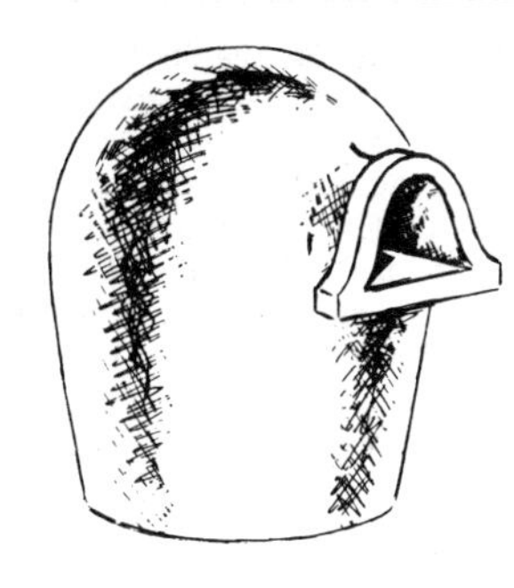

137.2

Figure 2-2.—Covered pot (crucible) for melting optical glass.

MIXING THE INGREDIENTS

Because the quantities of ingredients used for making optical glass are comparatively small, it is best to mix them carefully and thoroughly by hand. A certain amount of CULLET—broken glass from a previous melting—must also be thoroughly mixed with the raw ingredients. Pieces of cullet with striae are generally used for this purpose. Cullet acts as a flux, to cause the other materials to react on each other at a lower temperature than they would without it.

MELTING PROCESS

The empty crucible is placed in a kiln and gradually brought to a red heat—in four or five days. It is then removed from the kiln and transferred as rapidly as possible to the melting furnace, which was also heated to the same temperature as the crucible. NOTE: This is necessary in order to prevent shrinkage and breakage of the pot. The opening to the melting furnace is then carefully sealed temporarily with brick. An opening in the brickwork opposite the mouth of the pot is made with fire-clay slabs and the furnace is then heated to a melting heat.

A quantity of CULLET is next placed in the pot and melted, to form a coating of molten glass on the bottom and on the lower part of the walls of the pot, as protection from violent attack by the melting raw materials. The first charge (part) of the raw materials may be put in the pot as soon as the cullet is melted. When this charge melts completely, a second charge is added, and then a third—in smaller and smaller amounts as the number of charges increases--and so on until the pot IS FILLED with molten glass.

FINING PROCESS

The pot full of molten material is full of bubbles of all sizes. The glass must therefore undergo a process called FINING in order to eliminate the bubbles. This process is accomplished by INCREASING THE FURNACE HEAT, to make the glass more fluid and to increase the size of the bubbles as a result of the expansion of the contained gas. When the bubbles increase in size, they rise to the top of the molten glass. By keeping the glass at this high temperature (sometimes for 30 hours) all bubbles in the glass are eventually eliminated.

It is not difficult to maintain an extreme temperature a long time for lead glass, and other types of easily fusible glass; but for hard-crown glass it is difficult to maintain a high fining temperature without endangering both pot and furnace.

In some valuable optical glass, all bubbles are NOT eliminated; because increased temperature and expansion of gas in the molten glass do NOT force them to the top of the glass. The reason for this is their size; they are very small bubbles. The only optical drawback to bubbles of this size is that they prevent the transmission of a very small percentage of light which falls upon them.

When the molten glass is free from bubbles, the furnace heat is decreased. As the greater part of the impurities are lighter than the molten glass, they float to the top and can be removed by skimming with a ladle.

STIRRING PROCEDURE

Molten glass in the crucible must now be stirred thoroughly and continuously during gradual cooling. A thoroughly burned fire-clay cylinder heated to a RED HEAT is used for this purpose. The stirring process must continue until the molten mass is so stiff that the stirrer CANNOT be moved; it remains in the cooled mass. The time of stirring (usually 4 to 20 hours) varies in accordance with the composition of the glass and the size of the pot.

SETTING AND ANNEALING

When the stirring process is completed, to prevent devitrification, the glass must be cooled as RAPIDLY as safety permits to a temperature where it becomes SET. Once the glass is set, further rapid cooling results in CRACKING and

FRAGMENTATION of the mass. It is therefore necessary that the glass be slowly cooled (annealed) down to natural temperature.

Setting and annealing are accomplished in the following manner:

1. When stirring is completed, the crucible must be withdrawn from the furnace and allowed to cool to annealing temperature, 400° C to 500° C.

2. The pot must then be placed in an annealing kiln at annealing temperature and slowly cooled to natural temperature. This process usually takes from ONE to TWO weeks, in accordance with the amount (quantity) of glass melted.

3. When the glass is cool, the pot is removed from the annealing kiln and broken away from the glass.

Under very favorable conditions, the whole melting may be one big lump of optical glass, perhaps as much as 1,200 pounds. Generally, however, the glass is cracked into a few large lumps and a great number of smaller fragments. The lumps are then inspected and visible defects are removed with a chipping hammer.

The lumps of glass are next heated to a RED heat and placed in RED-HOT FIRE-CLAY molds large enough for each lump. To prevent strain in glass intended for a special purpose, it is allowed to SETTLE into the molds by its own weight—it is NEVER forced into the molds. Final annealing then follows, during which time (6 to 8 days) the glass is cooled down to ambient temperature.

When the glass is cool, it is ready for cutting, grinding, and polishing for the purpose for which it was made. You will learn more about usage of optical glass when you study lenses and prisms, and also when you view Navy films which describe and explain the procedure for grinding and polishing them.

CHAPTER 3

CHARACTERISTICS OF LIGHT

Light is the term commonly used for radiant (heat) energy which affects our eyes and gives us vision. When you study the spectrum later in this chapter, you will learn that the wavelength (light) to which our eyes are most sensitive is 0.5550 micron.

Objects which emit light are luminous, and they are visible ONLY BECAUSE OF THE RADIATED LIGHT. Nonluminous objects do not give off light and are therefore invisible until light from a luminous object strikes them and is reflected to our eyes. When light strikes the surface of an object, some of it is ABSORBED, part of it is REFLECTED; and if the object is transparent or translucent, a portion of it is TRANSMITTED. Objects which emit no light are designated as opaque. If an OPAQUE OBJECT is made thin enough, it will become TRANSLUCENT, or even TRANSPARENT.

We discuss the properties of light in this chapter because you must understand the nature and behavior of light before you can fully comprehend the principle of operation of various optical instruments. Without the properties of light which enable lenses and prisms to change its direction, the construction of optical instruments would be an impossibility.

Another reason why you must understand the characteristics of light is that before you can qualify for Opticalman 3 you must know Huygens' theory of light, the theory of refraction and reflection, and the effect on light rays by convergent and divergent lenses. In addition, you must know what happens to rays of light when they strike prisms and different types of mirrors.

LIGHT THEORIES

Scientists have always been interested in the properties of light: and because of their inquisitive minds and experiments, they developed various theories concerning light. The ancient Greeks, for example, believed that light was generated by streams of particles ejected from the eyes, and then reflected back into the eyes by objects they struck. This theory did not last long, however, because it did not explain why a person could not see as well by night as by day.

Space in this manual does not permit a discussion of all theories of light, but some of them are considered briefly in order to give you an idea concerning their impact on the development of current theories of light.

During the 17th century, Sir Isaac Newton announced his CORPUSCULAR theory of light. He assumed that light was a flight of material particles originating at the light source. He believed that light rays moved with tremendous speed in a state of near vibration and could pass through space, air, and transparent objects.

Newton's theory concurred with the idea that light moves only in a straight line, but it did not explain other characteristics of light. He accidentally discovered a form of light interference which is now known as Newton's rings but he did not realize their significance.

During Newton's era, Christian Huygens announced his concept of light, now known as Huygens' principle, which helps to explain some of the phenomena of optics. Huygens attempted to show that the laws of reflection and refraction (explained later) could be explained by his theory of WAVE MOTION of light.

Although Huygens' wave-motion theory of light appeared to be the logical explanation for some phases of light behavior, it was not accepted for many years. Huygens could explain the passage of waves through water, but he did not know how light waves came from the sun or passed through space. He then proposed that the waves passed through a medium which he called ETHER, the function of which was to serve light rays in the same manner as water serves the familiar waves of water. He assumed that ether occupied all space, even that part already occupied by matter.

About 50 years after Huygens announced his theory of wave motion of light, Thomas Young, Fresnel, and others, supported the wave theory, and Newton's corpuscular theory was virtually abandoned. These three scientists accepted the ETHER theory and assumed that light was waves of energy transmitted by an elastic medium designated by Huygens as ether.

Three other scientists (Boltzmann, Hertz, and Maxwell) conducted experiments which proved that light and electricity are similar in radiation and speed. As a result of their experiments, they developed the ELECTROMAGNETIC theory. They produced alternating electric currents with short wavelengths which were undoubtedly of electromagnetic origin and had all the properties of light waves. This theory (sometimes called the Maxwell theory) held that energy was given off continuously by the radiating body.

For some years after promulgation of the Maxwell theory of light, scientists thought the puzzle of light was definitely solved. In 1900, however, Max Planck rejected the electromagnetic theory. He did not hold the view that energy from a radiating body was given off continuously. His contention was that the radiating body contained a large number of tiny oscillators, possibly resulting from electrical action of atoms in the body. His idea was that the energy given off by the body could be of high frequency and have high energy value, with all possible frequencies represented. Planck argued that the higher the temperature of the radiating body the shorter the wavelength of most energetic radiation would be.

In order to account for the manner in which radiation from a warm, blackbody is distributed among the different wavelengths, Planck found an equation to fit the experimental curves, which were based on lightwaves of different length. He then came to the conclusion that the small particles of radiated energy were GRAINS of energy like grains of sand. He therefore called these units quanta and named his theory the QUANTUM THEORY. He assumed that when quanta were set free they moved from their source in waves.

A few years later, Albert Einstein agreed with Planck relative to his quantum theory, and stated that (when emitted) light quanta retained their original identity as packets of energy.

Through experimentation, R. A. Millikan later proved that the energy caused by motion (kinetic energy) of units of light (photons) behaved in the manner assumed by the quantum theory. In 1921, A. H. Compton learned through experiments that electrons and photons have kinetic energy and momentum, and that they behave like material bodies. This idea was therefore somewhat similar to the old corpuscular theory.

Knowledge gained later by scientists from the study of diffraction, interference, polarization, and velocity (explained later) proved the corpuscular theory of light untenable. More recently, however, phenomena of light have been discovered which are not accounted for by the wave theory, so many scientists now accept Maxwell's electromagnetic theory. Because the phenomena of light propagation can be explained best in terms of the wave theory, this is the theory used to explain the passage of light through optical instruments discussed in this manual.

LIGHT FACTS

We know now that all forms of light obey the same general laws. When light travels in a medium or substance of constant optical density, it travels in waves in straight lines and at a constant speed. When light strikes a different medium from the one in which it is traveling, it is either reflected from or enters the medium. Upon entering a transparent medium, the speed of light is slowed down if the medium is MORE dense, or increased if the medium is LESS dense. Some substances of medium density have abnormal optical properties and, for this reason, they may be designated as optically dense. If the light strikes the medium on an angle, its course is bent (refracted) as it enters the medium. NOTE: Reflection and refraction are discussed fully in this text in the chapters (4 & 5) which deal with lenses and prisms and image formation. When discussing the characteristics of light, however, we must use and explain these and other terms to the extent necessary for you to understand the discussion. As explained previously, without light, lenses and prisms, and some other optical elements, we could not have optical instruments. After you learn the characteristics of light and the types and function of various optical elements, you will then experience less difficulty in understanding image formation—the prime purpose of optical instruments.

SOURCE OF LIGHT

Natural light comes from the sun 93,000,000 miles away and in small amounts from distant stars. Although the moon shines, it merely reflects light from the sun. Such bodies as the sun are called LUMINOUS BODIES.

All light other than natural is called artificial light. It may come from an incandescent bulb, a burning house or forest, glowing coals, neon signs (gas at reduced pressure bombarded with electrons), matches, flashlights, and many other sources.

You perhaps have used an electric exposure meter when you took pictures with your camera. The function of this meter, illustrated in figure 3-1, is to determine the intensity of a light source. All you need do is turn the meter toward the sky and observe the movement of its hand. Although the meter has no batteries, and despite the fact that a spring attached to the hand holds it back, the hand moves when light strikes the face of the meter. The energy which moves the hand comes from the sun, but a light from a match or flashlight also causes the hand to move IN ACCORDANCE WITH the amount of energy provided by the light. This indicates that both natural and artificial light gives off energy.

RADIATION OF WAVES

Take a look now at figure 3-2, which shows in sequential order (A, B, C, D) the radiation of waves generated by a pebble dropped into a tray of milk.

The pictures in illustration 3-2 were taken a fraction of a second apart. Note in part A that the pebble made a dent in the milk and that the surface is recovering its natural position and is rising. Part B shows that the surface of the milk has begun to rise and that the original wave is beginning to spread. Energy is spreading out in the form of little waves from the source of the disturbance of the surface of the milk by the pebble; and the waves are circles which get bigger and bigger as the amount of energy (wave motion) created by the pebble causes them to expand—the bigger the pebble, the greater the size of the waves and circles. When all the energy produced on the milk by the pebble is absorbed by the waves, they stop forming, as illustrated.

TRANSMISSION OF WAVE ENERGY

Visible light is one of many forms of radiant energy transmitted by waves. To illustrate, secure one end of a rope to some object and hold the other end in one hand. Then stretch the rope fairly tight and shake it. A wave motion (pulse) passes along the rope from the end held to the end secured, as shown in figure 3-3.

If you continue to shake the rope, you create a series of waves (a wave train) which passes along the rope. Study illustration 3-4. Note that the different parts of the rope (medium) vibrate successively; each bends back and forth about its own position. The disturbance travels but the medium does not travel. Energy only is carried along the rope.

Imagine now that the light source is a vibrating ball from which a countless number of threads extend in all directions. As the ball vibrates, successive waves are transmitted along the threads. In a similar manner, light radiates from its source.

RADIATION OF LIGHT

Thermal radiation and light waves are of the same nature and exhibit similar properties. Like light waves, thermal radiation normally travels in straight lines and can be reflected from a mirror or polished metal. Thermal radiation is not heat; it is energy in the form of wave motion.

During the latter part of the 18th century, scientists recognized that radiations from hot bodies consisted of electromagnetic waves (not mechanical) of the same fundamental nature as light waves. Luminous light sources such as the sun or the glowing filament of an electric light bulb act as oscillators in radiating energy in the form of light waves, and these waves spread out in all directions from their sources. The sun pours forth radiant energy from its surface at the rate of 70,000 horsepower for every square yard of its surface.

Because light travels outward in all directions from its source, the waves take the form of growing spheres (fig. 3-5), the luminous point of which is the center.

LIGHT RAYS

Single rays of light do not exist; but the term light ray is used throughout this manual for the sake of clarity and convenience in showing the direction of travel of the wave front. Light is indicated by one, two, or more, representative light rays in white lines, with arrow heads to indicate the direction of travel.

Refer now to illustration 3-5 again and observe that light is moving in all directions from the light bulb. Then study figure 3-6, which shows lines with arrow heads to indicate that the direction of travel of the light is along the radii of the sphere of light and at right angles to the fronts of the waves. The light which travels along these radii is designated as light rays.

A wave front which radiates from a light source is curved when it is near the source and the radii of the waves diverge or spread. See illustration 3-7. As these waves move outward, however, the wave front becomes less curved and eventually almost straight, as indicated in figure 3-8. After traveling a distance

137.3

Figure 3-1.—Electric exposure meter.

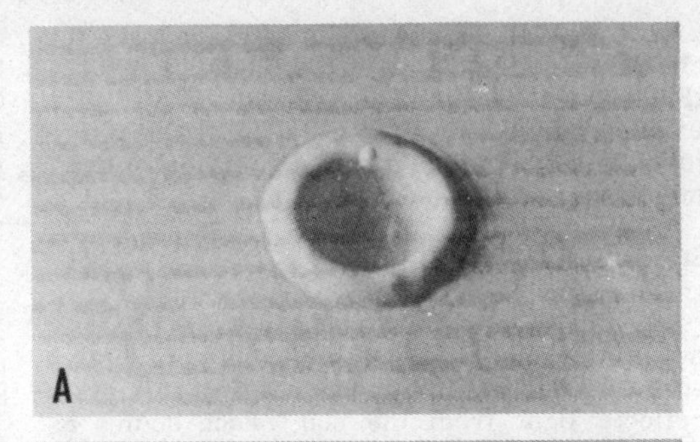

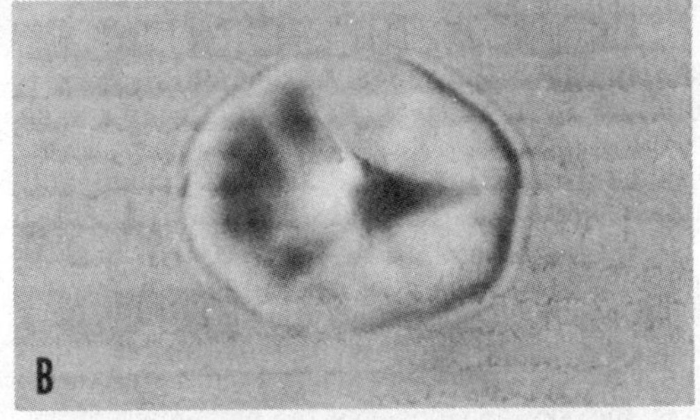

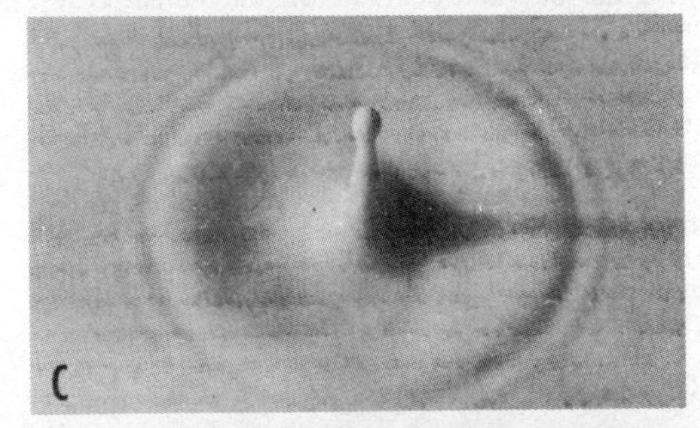

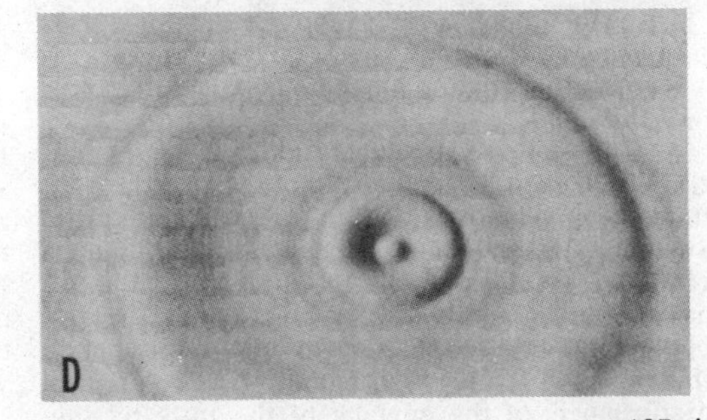

137.4

Figure 3-2.—Creation of waves in a liquid by a dropped pebble.

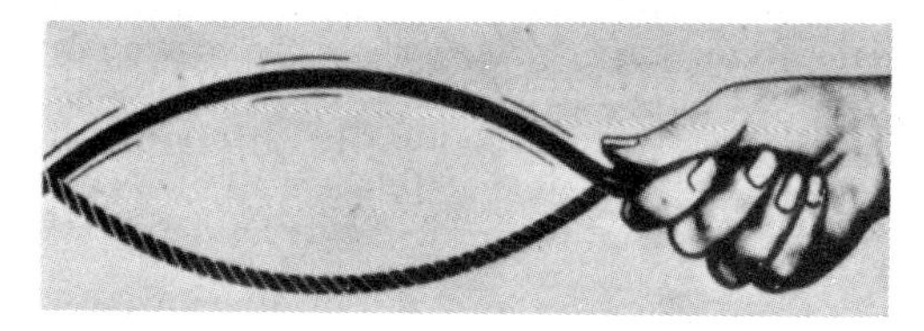

137.5
Figure 3-3.—Wave motion created by a moving rope.

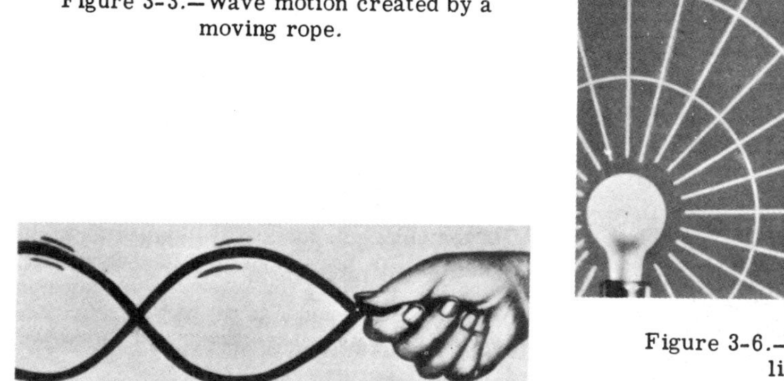

137.6
Figure 3-4.—Wave train created by shaking loose end of a rope.

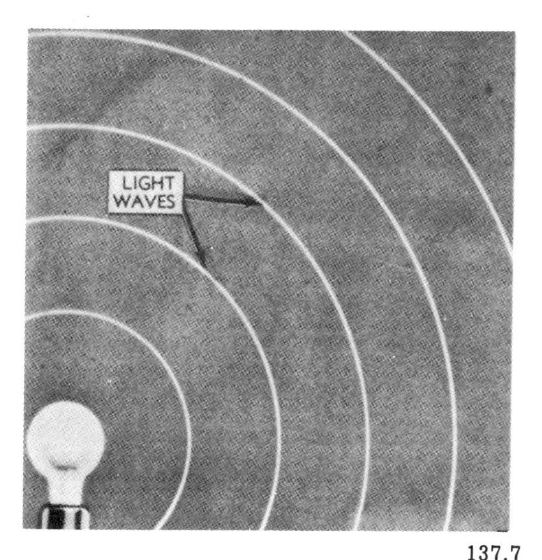

137.7
Figure 3-5.—Light waves created by a light.

137.8
Figure 3-6.—Direction of travel of light waves.

of 2,000 yards from their light sources, wave fronts are considered to be parallel to each other.

A pinhole camera (fig. 3-9) is a good example of the manner in which light travels outward from its source. Such a camera is merely a box with a sheet of film at one end and a tiny pinhole instead of a lens at the other end. Note that the camera is taking a picture of an arrow by light reflected from some luminous source and that each point on the arrow is sending out light rays in a dispersed manner.

One ray of light from each point on the arrow enters the pinhole in the front of the camera and lands upon the film. Since light travels in straight lines, no light reaches a given point on the film except the ray which comes from the corresponding point on the arrow. The rays of light which pass through the pinhole of the camera form an inverted arrow on the film.

ILLUMINATION

The unit used for measuring the luminous intensity of light is called CANDLEPOWER. If a luminous source, for example, gives ten times as much illumination as a standard candle, it has the luminous intensity of 10 candlepower.

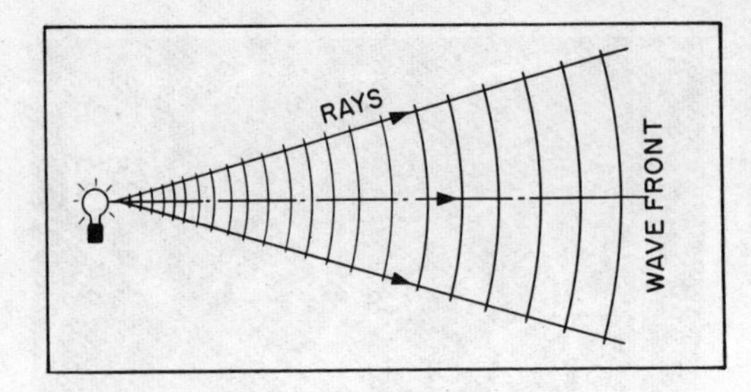

137.9
Figure 3-7.—Light rays and wave front produced by a light.

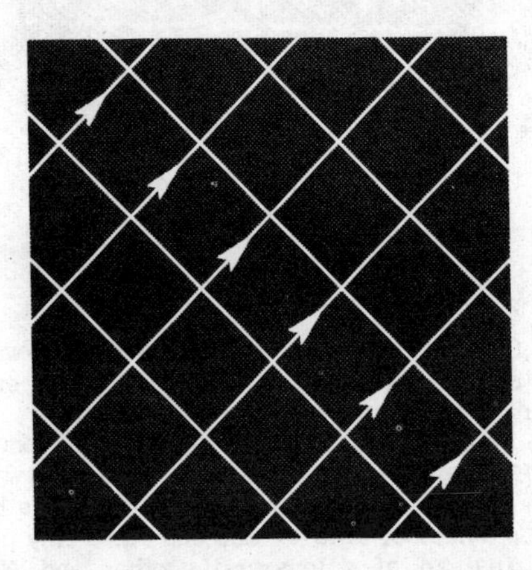

137.10
Figure 3-8.—Waves and radii from a distant light.

Because of the difficulty of getting exact measurements with a standard such as a candle, the National Bureau of Standards maintains a group of incandescent electric lights which fulfill certain conditions as standards of measurement. Secondary standards can be calibrated from these standard lamps by any laboratory. A light bulb of specific size, for example, gives approximately the same amount of candlepower.

The intensity of light which falls on a nonluminous source is generally measured in FOOT-CANDLES. In some instances, however, the intensity of a source of light is measured in terms of LUMINOUS FLUX radiated by the source per unit solid angle. The word luminance is used instead of brightness, though it means the same thing, and measurements are made in lumens per solid angle. The unit of flow of luminous flux through space is called a LUMEN—the amount of light which falls on an area one foot square at a distance of one foot from a standard candle.

The surface of an object is illuminated by one foot-candle when its light source is one candlepower at a distance of one foot. The formula for this is:

$$\text{Foot Candles} = \frac{\text{Candle-power}}{(\text{Distance})^2}$$

Look now at figure 3-10. If the object is two feet from the light source, the light from the candle covers four times the area it covered after traveling one foot. The illumination at this point is ONLY one-fourth of a foot-candle. Illumination provided by a candle is therefore inversely proportional to the SQUARE OF THE DISTANCE between the candle and the object.

SPEED OF LIGHT

The difference in the speed of light through air, glass, and other substances accounts for the bending of light rays. Without this characteristic of light, a glass lens could not bend light rays to a focus, as you will learn later in this text. The length of all waves in the electromagnetic spectrum is also connected to corresponding frequencies and the speed of light.

Because light travels with such high velocity, it was years before any one could measure its

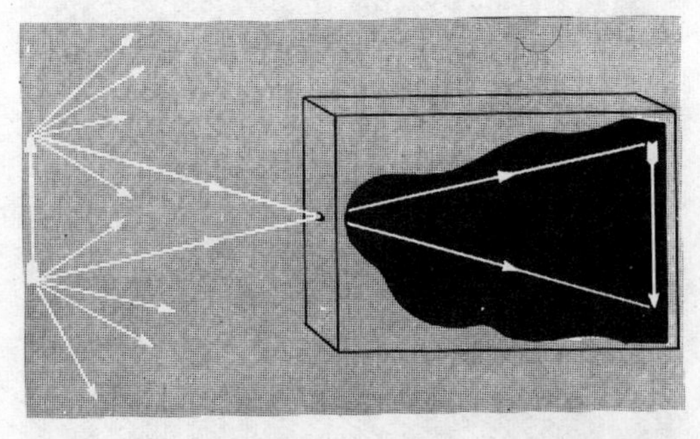

137.11
Figure 3-9.—Light rays creating an image on the film of a pinhole camera.

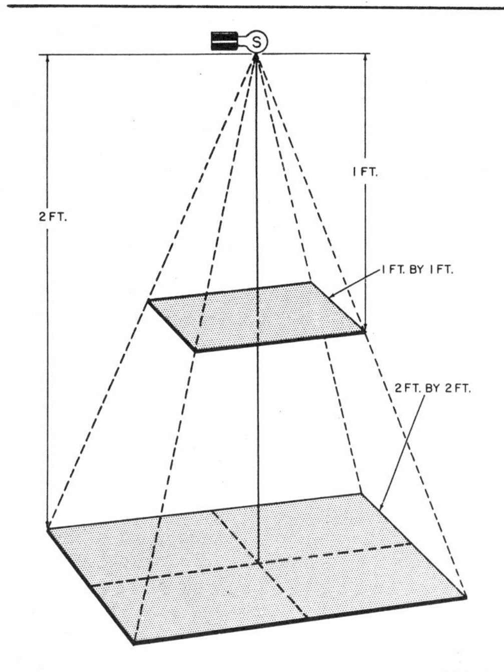

137.12
Figure 3-10.—The inverse square law of light.

speed. Galileo tried to measure it by having two men in towers on hills some distance apart flash lights at each other. Each person flashed his light as soon as he saw the light signal of the other. Galileo reasoned that he could determine the speed of light by dividing the total distance the light traveled by the time required for the transmission of signals. His experiment was not successful; and he concluded that the speed of light was too great to be measured by this method. His final thought relative to the speed of light was that its transmission through space was perhaps instantaneous.

Roemer's Discovery

Olaus Roemer, a Danish astronomer, in 1675 calculated the speed of light by observing the irregularities in the times between successive eclipses of the innermost moon of Jupiter by that planet.

Roemer observed the position of Jupiter's moons revolving around the planet. The moons appeared on one side and then moved across in front of the planet and disappeared behind it. He could calculate accurately when one of the moons would be eclipsed by the planet. When he tried to calculate ahead six months, however, he learned that the moon eclipse occurred about 20 minutes later than he had calculated. He therefore concluded that the light had taken this amount of time to cross the diameter of the earth's orbit, which is approximately 186,000,000 miles. The difficulty was that Roemer did not correctly evaluate the speed of light; later measurements showed that the time was about 1,000 seconds, which gave 186,000 miles per second as the velocity of light.

Michelson's Measurements

The most accurate measurements of the speed of light were made after 1926 by A. A. Michelson, a distinguished American physicist, and his colleagues. Professor Michelson used an octagonal mirror in an apparatus illustrated in figure 3-11. He measured the speed of light in air over the exact distance between Mt. Wilson and Mt. San Antonio, California. The light source (mirror) and the telescope were located on Mt. Wilson and the concave and plane mirrors were located on Mt. San Antonio, about 22 miles distant.

Study the illustration. Mirror M is stationary, and Professor Michelson passed a pencil of light through a slit and a lens to the octagonal mirror. NOTE: A pencil of light is a narrow group of light rays which come from a point source, or converging toward a point. A pinhole opening produces a pencil of light rays.

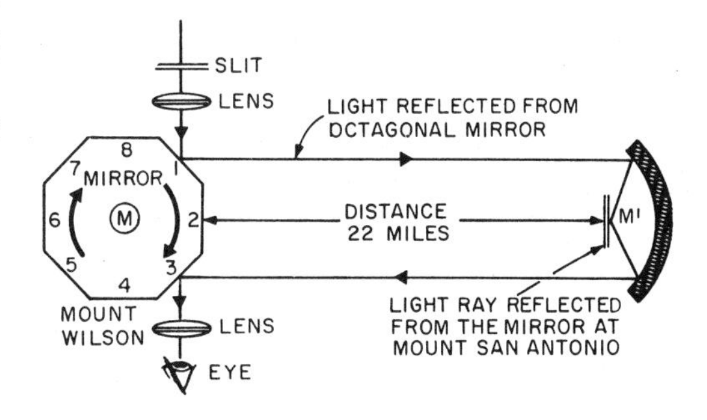

137.13
Figure 3-11.—Michelson's mirror method for measuring the speed of light.

Mirror M then reflected the light from position 1 to Mirror M', which (in turn) reflected the pencil of light back to point 3 on the octagonal mirror. The octagonal mirror was next put into motion and increased in speed enough to move position 2 on the octagonal mirror into the position formerly occupied by position 3 during the time required for the light to travel from position 1 on the octagonal mirror to Mt. San Antonio and return. After several years of observation with his apparatus, Professor Michelson concluded that the speed of light in air was 299,700 kilometers (A kilometer is .6214 mile.) per second.

Sometime later, Professor Michelson used an evacuated tube one mile long to measure the speed of light in a VACUUM. The vacuum tube removed variations in air density and haze from the test, and the experiment showed that the speed of light in a vacuum was slightly higher than in air. The velocity of light in a vacuum is generally accepted as 300,000 kilometers per second, or 186,000 miles per second.

Modern physicists compute the speed of light with great accuracy. Some of their measurements are based on light interference. For all practical purposes, however, the speed of light in air or in a vacuum is considered as 186,000 miles per second. In media more dense than air, the speed of light is slower, as indicated by the speed of yellow light in the following substances:

Quarts.	110,000 miles per second
Ordinary crown glass .	122,691 miles per second
Rock salt . . .	110,000 miles per second
Boro-silicate crown glass .	122,047 miles per second
Carbon disulfide.	114,000 miles per second
Medium flint glass.	114,320 miles per second
Ethyl alcohol	137,000 miles per second
Water	140,000 miles per second
Diamond . . .	77,000 miles per second

NOTE: All colors of light travel at the same speed in air or empty space. In denser media, the velocity of light varies for different colors.

WAVELENGTH AND FREQUENCY

The action of waves on water helps us to understand light as a wave motion. If you measure a distance of 12 feet on the surface of water and drop a pebble at the point where you started to measure, you will find that it takes 4 seconds for the first wave created by the pebble to reach the other end of your measured distance. If you then divide the distance by the time ($12 \div 4$), you get 3, the speed of the waves per second. The formula for this measurement is:

$$\text{speed} = \frac{\text{distance}}{\text{time}}$$

Bear in mind that the distance a wave travels on water in a specified time depends upon the velocity of the wave.

A wavelength is the DISTANCE BETWEEN the crest of one wave and the crest of the next (adjacent) wave, as illustrated in figure 3-12. The best way to measure a wavelength is by the FREQUENCY—the number of waves which pass a point in one (1) second. You can determine this by putting a stake in water and counting the number of waves which pass the stake per second. See figure 3-13.

If waves are moving at a speed of 3 feet per second and have a frequency of 6 waves per

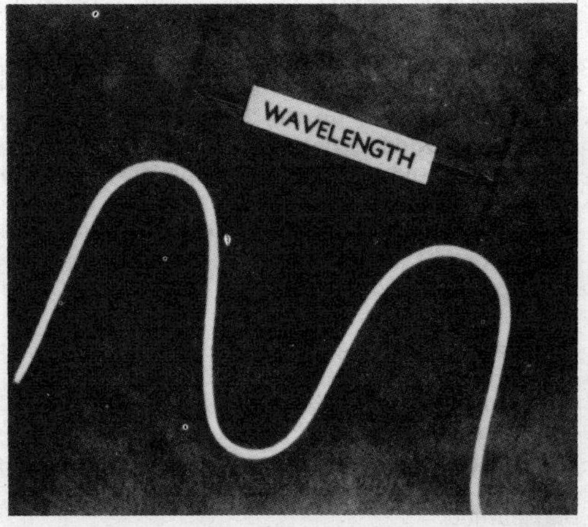

137.14

Figure 3-12.—Measurement of a wavelength.

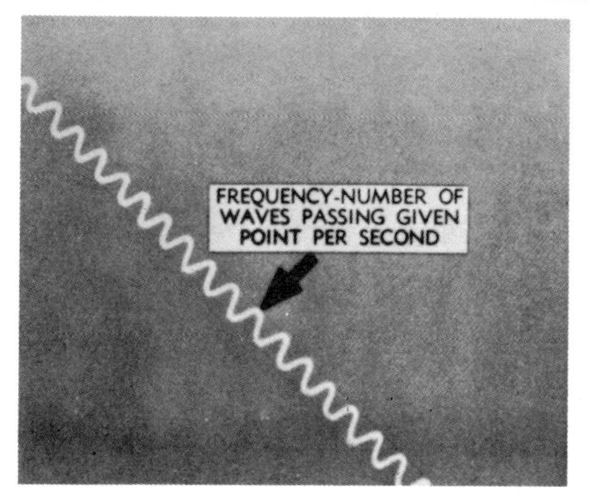

137.15
Figure 3-13.—Determination of wave frequency.

second, you can determine the wavelength by the following formula:

$$WL = \frac{S}{F} = \frac{3}{6} = .5$$

This formula also shows the relation which exists with respect to the speed, frequency, and wavelength of light. As used in physics and optics books, the formula is:

$$c = f\lambda$$

The letter c means the velocity of light in a vacuum, f is the frequency, and λ (wavelength) is the character for LAMBDA in the Greek alphabet. As used in optics, this character always means wavelength.

By using methods based on interference and diffraction, the wavelength of light can be measured accurately. You will learn more about these terms later in this text.

Light waves, in contrast with waves on water, are much too short to be measured in inches or millimeters. (A millimeter is about 1/25th of an inch. A light wavelength is sometimes measured in microns, represented in formulas by μ, the Greek letter m. A micron is one-thousandth of a millimeter.) For measuring a minute wavelength of light, a shorter unit than a micron must be used. This unit is the MILLIMICRON, which represents one one thousandth of a micron and is abbreviated to m mw, or $m\mu$.

Another important unit used for measuring wavelengths is the ANGSTROM UNIT (AU), which is 1/10th of a millimicron, or one ten-millionth of a millimeter. Because these units are still inconveniently long for measuring the shortest electromagnetic waves, the X-ray unit (XU) is used for this purpose. It is one one-thousandth of an Angstrom unit.

ELECTROMAGNETIC SPECTRUM

Heat and radio waves, light waves, and ultraviolet and infrared rays, X-rays, and cosmic rays are forms of radiant energy of different wavelengths and frequencies. Together, all of these form the electromagnetic spectrum, illustrated in figure 3-14. The visible portion of the electromagnetic spectrum consists of wavelengths from .00038 to .00066 millimeters. The different wavelengths represent different colors of light. Note the arrows which point to the wavelengths of the colors of the rainbow in the spectrum. Observe also that the wavelengths in this part of the spectrum (vision and photography) are in millimicrons of wavelengths. Wavelengths in the electromagnetic spectrum (extreme left) are in microns.

Note in illustration 3-14 that the wavelengths we call light are between 400 and 700 millimicrons, each spectral color has its own small range of wavelengths. If light around 60 millimicrons of wavelengths, for example, reaches your eyes, you see RED (sensation of red on the retina). Around 460 millimicrons the wavelengths of light which reach your eyes are BLUE; so the red waves are therefore much longer than the blue waves.

When light with a wavelength of 300 millimicrons reaches your eyes, you receive no sensation of color. Radiation of this wavelength is generally called ULTRAVIOLET LIGHT. Ultraviolet rays (radiation) from the sun cause sunburn and sometimes blisters. CAUTION: All short-wave radiations can do some damage if you get too much of them. A prolonged dose of strong X-rays, for example, causes irreparable damage to the body. Gamma rays are deadly radiation waves given off by atomic bombs.

Note that the infrared light rays are between 1 micron and 100 microns in the electromagnetic spectrum. These rays are called

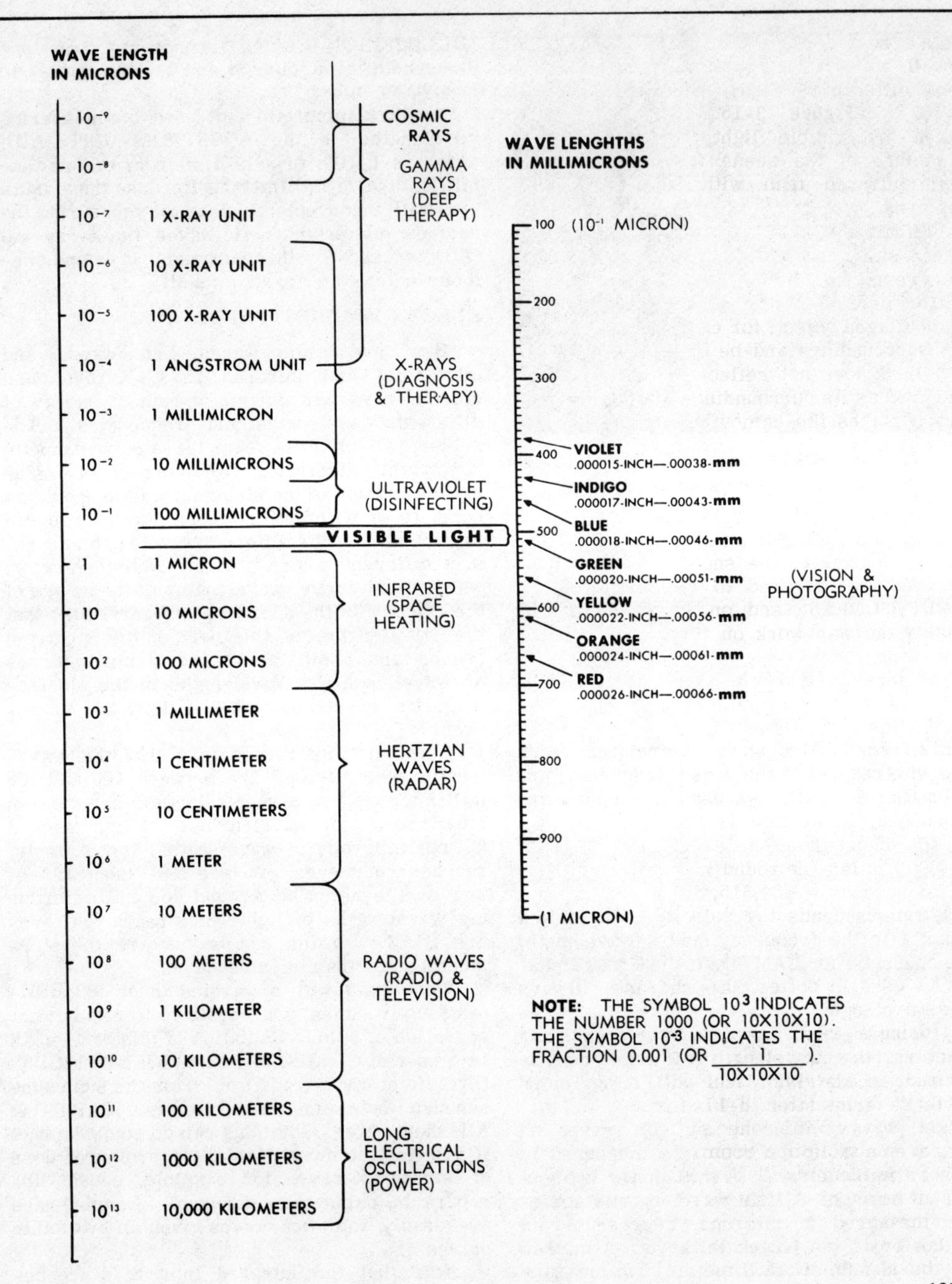

Figure 3-14.—Electromagnetic spectrum.

HEAT rays. We cannot see infrared rays; but if we could see them, everything would look different. Study illustrations 3-15 and 3-16. Figure 3-15 shows a photograph taken by visible light; figure 3-16 shows a picture of the scene in figure 3-15 taken with infrared film with a red filter over the lens.

Infrared light is used also for signaling between ships at night. In aerial reconnaisance, too, we use infrared photography to get more and better details of the area photographed. A camouflaged object, for example, may blend with its surroundings and be invisible from the air; but if it does not reflect the same amount of infrared as its surroundings, an infrared photograph makes the camouflage stand out clearly.

During World War II SNOOPERSCOPES with powerful spotlights which sent forth beams of invisible infrared light were used to watch the enemy at night. When the infrared beams sent out by the spotlight struck an object and reflected it back to the snooperscope, the scope changed the infrared to visible wavelengths. SNIPERSCOPES used on rifles in the Pacific during the war work on the same principle as the snopperscope.

Observe in figure 3-14 that RADAR waves are adjacent to the infrared rays in the electromagnetic spectrum and have wavelengths a little longer than infrared. We know that these wavelengths travel at the same speed as light because they have been sent to the moon and reflected back in about 2.6 seconds. Because the distance of the moon from the earth is approximately 240,000 miles (in round numbers), 2 x 240,000 ÷ 2.6 seconds = 184,615, the speed of radar in miles per seconds.

COLOR OF LIGHT

Because sunlight includes the whole range of wavelengths between 400 mμ and 700 mμ, it is a mixture of all visible colors between red and violet. Illustration 3-17 shows how you can prove this. When the sun is shining, put a prism on a table in a room with one window and cover the window with dark paper or cloth. Then cut a horizontal slit about an inch long and 1/16th on an inch wide in the paper to admit a small quantity of light. Hold the prism close to the slit to ensure passage of sunlight onto one of the long faces of the prism. (Lenses and prisms are discussed in detail in chapter 5.) At the same time, hold a ground glass screen or a sheet of white paper on the other side of the prism, 6 to 8 inches away. When the sunlight passes through the prism, wavelengths of various colors refract at different angles toward the base of the prism and produce the colors of sunlight (the rainbow) on the glass screen or sheet of white paper. See illustration 3-18. This breaking up of white light into its component colors is called DISPERSION.

Selective Reflection and Absorption

If you look at a piece of red paper in the sunlight, you see red; but this does not mean that the paper is making red light. What it does mean is that the paper is reflecting a high percentage of the red light which falls on it and is absorbing a high percentage of all other colors.

When you look through yellow glass, you see yellow; because the glass is transmitting yellow light and is absorbing most of the other colors. Usually, yellow glass absorbs violet, blue, and some green; but it transmits yellow, orange, and red. When yellow, orange, red, and a little green all enter your eye at the same time, however, the color you see is yellow.

Selective absorption of light is what takes place when a color filter is used on an optical instrument. An image may be blurred by haze or fog, but when a yellow filter is put into the line of sight the image becomes sharper. The reason for this is that a thin haze permits most of the light to pass through; but it scatters some of the blue and violet light in all directions. Haze is therefore visible because of the scattered blue and violet colors. The yellow filter absorbs blue and violet and the haze becomes almost invisible.

Color Vision

A pure spectral color is composed of light of one wavelength, or a very narrow band of wavelengths. When this light enters your eyes, it gives a sensation of color; but you cannot judge the wavelength of light from color sensation. Most of the colors you see are not pure spectral colors but mixtures of these colors. The sensations you get from these mixtures are therefore not always what you may expect.

Observe in illustration 3-19 what happens when red, blue, and green lights are mixed. You can perform this experiment yourself by doing the following:

1. Put pieces of red, blue, and green cellophane over the lenses of three flashlights.

2. Take the three flashlights and a piece of white cardboard into a dark room and flash the lights upon the white cardboard. Use two flashlights at first, to produce yellow, sky-blue, and pink, and then mix all three colors to produce white (See illustration 3-19).

By mixing these three colors, red, blue, and green in the correct intensities, you can produce many different colors.

What you just studied concerning the mixing of light colors does not hold true for mixing paints, as you perhaps learned in your science classes in school. If you mix yellow and blue paints, for example, you get green (fig. 3-20). The blue paint reflects some of the spectral violet, a high percentage of the spectral blue, and some of the green. It absorbs all others colors. The yellow paint absorbs the violet

137.17

Figure 3-15.—Photograph of scene in illustration 3-16 taken by visible light.

137.18

Figure 3-16.—Photograph taken by infrared light.

137.19

Figure 3-17.—Dispersion of light into a spectrum by a prism.

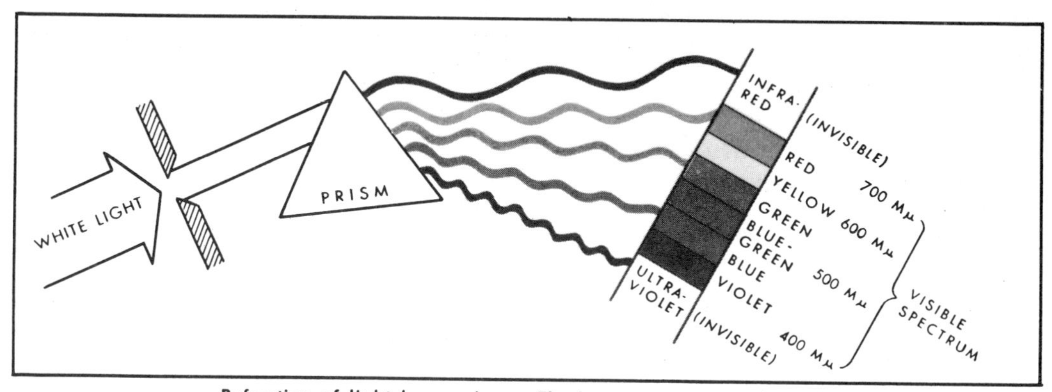

Refraction of light by a prism. The longest rays are infra-red; the shortest, ultra-violet.

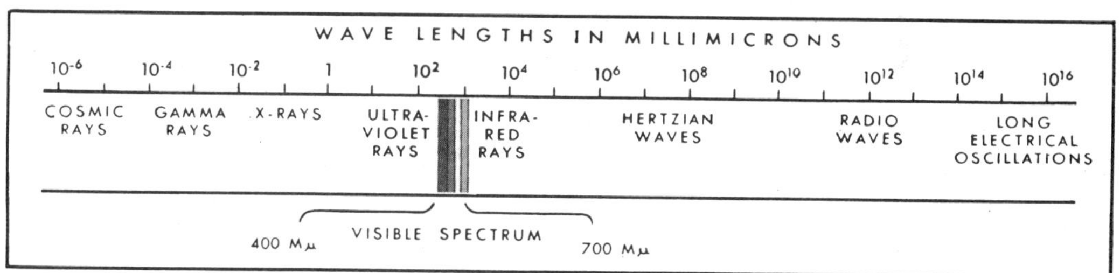

C137.20

Figure 3-18.—Creation of a rainbow by dispersion of light passing through a prism.

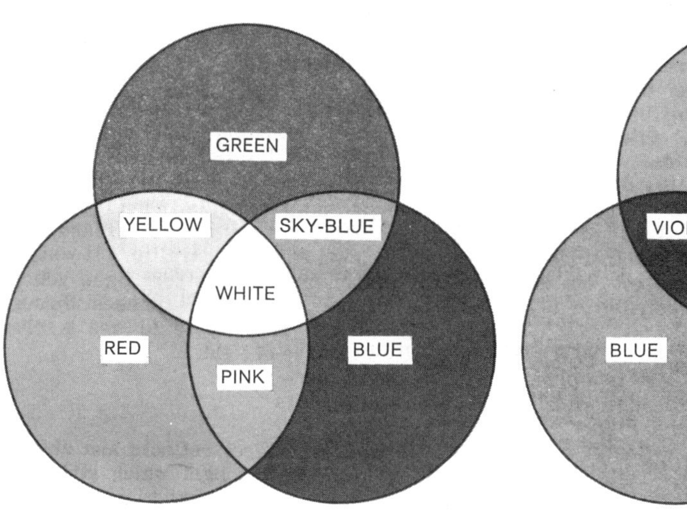

C137.21

Figure 3-19.—Colors created by mixing beams of colored light.

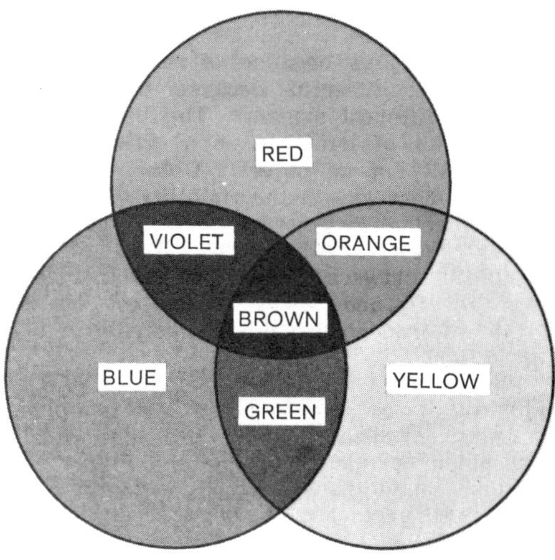

C137.22

Figure 3-20.—Colors produced by mixing paints.

and blue; it reflects some of the green, and nearly all of the yellow, orange, and red. When you mix blue and yellow paint, you combine their absorbing powers. The only color they do not absorb is green.

BEHAVIOR OF LIGHT

In this section we discuss the visibility of objects and also the reflection and refraction properties of light. Without these properties of light, the use of lenses and prisms in optical lenses for the purpose of creating images of objects would be impossible.

When we look at objects, we give little or no thought to the reasons for their visibility. We have previously discussed rays of light, shown by white lines in illustrations; single rays of light, however, do not exist. Study now illustration 3-21, the A part of which shows light as a cone passing through a lens. This is correct; but for the sake of clarity and convenience, the term RAY is used to indicate the direction of travel of wave fronts. Bear in mind, too, as you study the characteristics of light that the illustrations show ONLY height and width; but everything (including light) has three dimensions: height, width, and depth.

VISIBILITY OF OBJECTS

We see things because of reflected light. Objects look different because they reflect light in a different manner. The difference in the intensity of light makes a difference in the visibility of an object. Color, likewise, makes a difference in the visibility of objects. If one object absorbs twice as much color as another object, you have no difficulty in differentiating between them. You can therefore judge the size and shape of an object because of the difference in color or intensity of reflected light.

Refer now to illustration 3-22, one part of which is an egg and the other part is a piece of white cardboard cut to the approximate dimensions of the egg. You can easily distinguish each by the way light is reflected from them. All parts of the cardboard reflect light equally, because all rays of light fall on it at the same angle. Rays of light on the egg, however, strike the shell at different angles; and the amount of light reflected from any surface depends upon the angle of incidence (explained later) with which the rays of light strike the shell.

Another way to tell the difference between the egg and the piece of cardboard is by the shadows cast by the egg. Observe the right side of the egg. Because of the difference in the angles with which the light strikes the egg, you can detect roughness in the shell. This roughness indicates texture, which causes an object to show minute differences in color or shape all over the surface.

For the sake of convenience, we can divide objects into three different classes, according to the reaction of light when it falls upon them: OPAQUE, TRANSLUCENT, AND TRANSPARENT.

Opaque Objects

All the light which falls upon an opaque object is either reflected or absorbed—none of the light passes through. This is important, because most objects are opaque. No object, however, is completely opaque. If it is thin enough, you can see through anything. Even heavy metals such as silver and gold allow some light to pass through them when they are painted in a thin film on glass. When this film is made a little thicker, it permits light to pass through, but you cannot see through the film. It is translucent, not opaque.

Translucent Objects

When light falls upon a translucent object, some of it is absorbed and reflected; but MOST OF THE LIGHT is transmitted through the object and scattered in all directions. This is what happens, for example, when light passes through ground glass plate, stained glass windows, or a thin sheet of paraffin. If you hold these items in front of a strong light, you can see that much of the light passes through, even though you are unable to see a clear image of the source of light.

Transparent Objects

A transparent object reflects and absorbs a small amount of the light which strikes it; but it permits most of the rays to pass through. A window pane is a good example of a transparent object. Air seems transparent when you look through it from a short distance; but

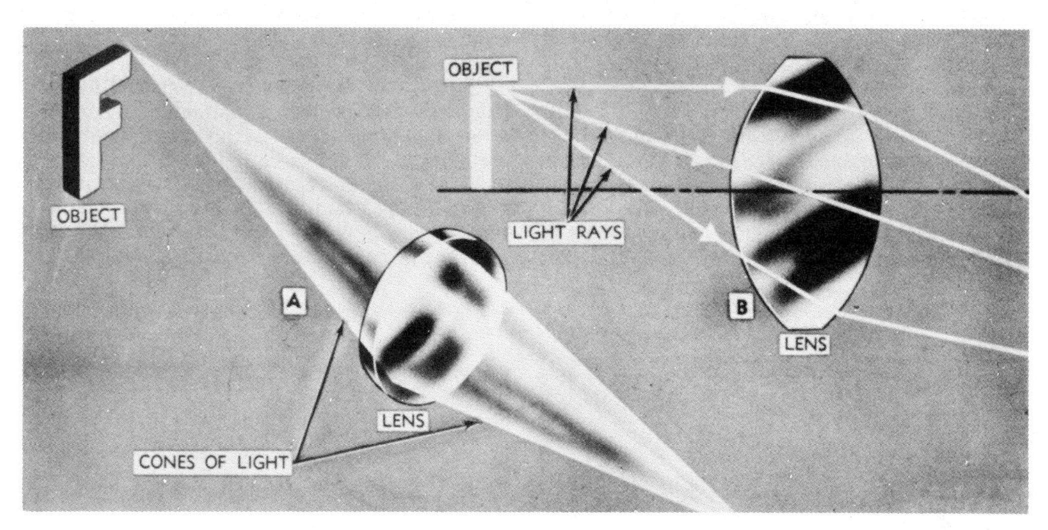

137.23

Figure 3-21.—Cones and rays of light.

137.24

Figure 3-22.—Visual determination of difference between objects.

when you look at objects from a great distance (from an airplane, for example) they look hazy, even on a clear day. The color of the sky (blue) is caused by the scattering of rays of sunlight.

Glass is considered transparent because we generally see it in thin sheets; but the thicker the glass the greater the loss in transparence. Lenses and prisms which you will study in chapter 5 are transparent. As you learned in chapter 2, light absorbed by a lens or a prism is wasted. This is why good optical glass is much more transparent than window glass.

Tubes which hold lenses and prisms in optical systems are opaque, to prevent entrance of light into the system except through the front lens. These tubes are painted a dull or flat-black color inside, so that they will absorb and not reflect light which falls on them. Some optical systems use mirrors, but the reflecting surfaces of the mirrors are opaque.

REFLECTION

You know from experience that a mirror reflects light. If you experiment with a plane mirror with a flat, polished surface in a dark room with a window through which you can admit light, you will find that you can reflect a beam of light to almost any spot in the room. When you hold a mirror at right angles to a beam of light, you can reflect the beam back along the same path by which it entered the room. Figure 3-23 shows how to do this.

If you shift the mirror to an angle from its right-angle position, the reflected beam is snifted at an angle from the incoming beam twice as great as the angle by which you shifted the mirror. Study figure 3-24. If you hold the mirror at a 45° angle with the incoming beam, the reflected beam is projected at an angle of 90° to the incoming beam. Remember this characteristic of light.

The simple experiments you just made illustrate one of the dependable kinds of action of light. You can reflect light precisely to the point where you want it, because any kind of light reflected from a smooth, polished surface acts in the same manner. This property of light is put to use in many types of fire control instruments.

Refer now to figure 3-25, which illustrates how you can control a reflected ray of light from a plane mirror. The ray of light which strikes the mirror is called the INCIDENT ray, and the ray which bounces off the mirror is known as the REFLECTED ray. The imaginary line perpendicular to the mirror at the point where the ray strikes is called the NORMAL

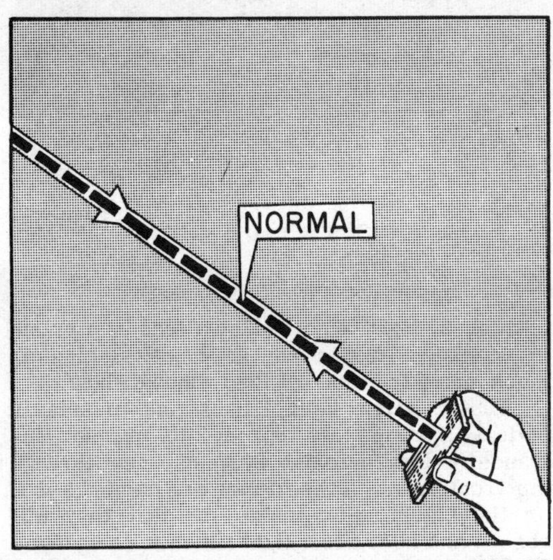

137.25
Figure 3-23.—Reflection of a beam of light back on its normal or perpendicular.

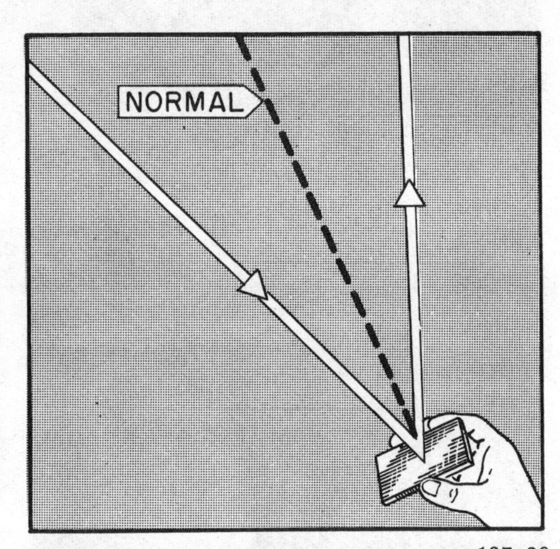

137.26
Figure 3-24.—Reflection of beams of light at different angles.

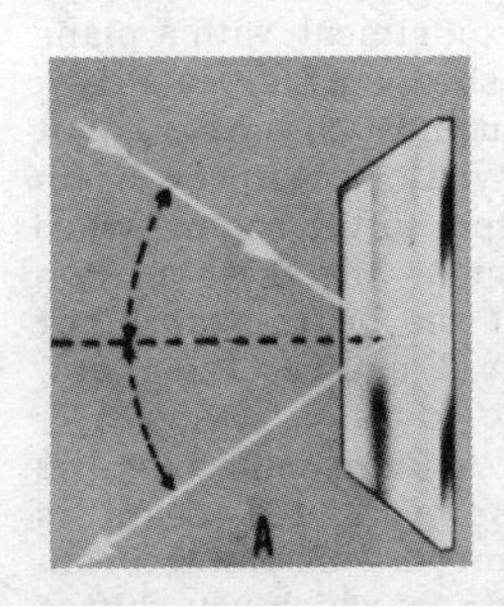

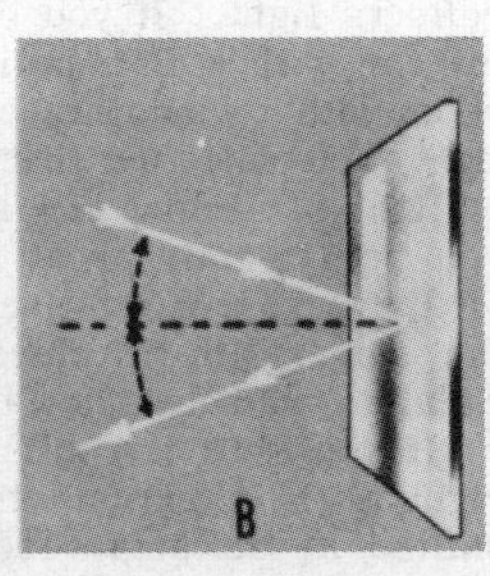

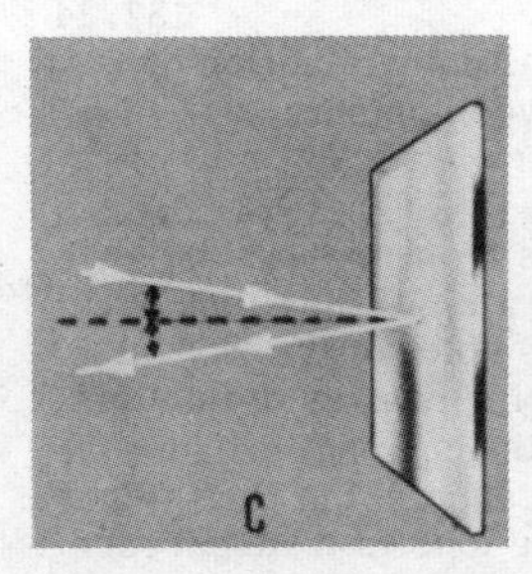

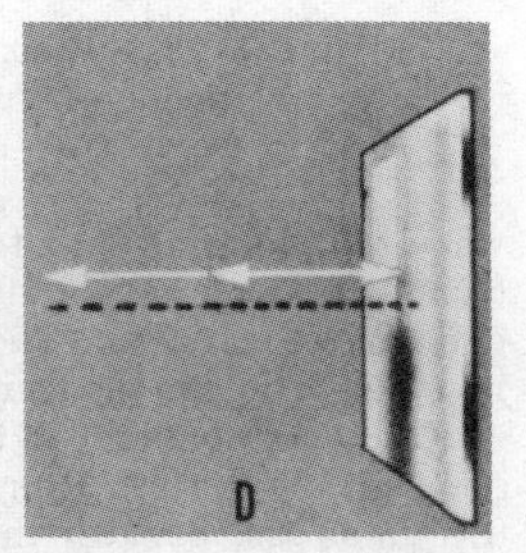

137.27
Figure 3-25.—Control of a reflected ray of light by a plane mirror.

or PERPENDICULAR (fig. 3-26). The angle between the incident ray and the normal is the ANGLE OF INCIDENCE; the angle between the reflected ray and the normal is the ANGLE OF REFLECTION.

Law of Reflection

The law of reflection is as follows: THE ANGLE OF REFLECTION EQUALS THE ANGLE OF INCIDENCE AND LIES ON THE OPPOSITE SIDE OF THE NORMAL. THE INCIDENT RAY, THE REFLECTED RAY, AND THE NORMAL ALL LIE IN THE SAME PLANE. THE INCIDENT RAY AND THE REFLECTED RAY LIE ON OPPOSITE SIDES OF THE NORMAL.

Refer to illustration 3-25 again. Note that the incident and reflected angles in parts A, B, and C become successively smaller. By applying the law of reflection, you can see that in all such cases of reflection the angle of reflection can be plotted as long as the angle of incidence is known, or vice versa. To illustrate, study figure 3-27. In this instance you desire to put the No. 4 ball in the nearest pocket, but your cue ball is behind the 8 ball. If you are an expert pool player, you know where to strike the right side of the pool table with the cue ball in order to have it reflect on a line which will enable it to hit the No. 4 ball and put it in the pocket. Angle b must equal angle a.

Regular Reflection

Regular reflection occurs when light strikes a smooth surface and is reflected in a concentrated manner, as shown in figure 3-28. You can plot the direction in which any single ray of light will be reflected by erecting a perpendicular or normal at the point of impact and by applying the law of reflection.

Light is reflected by regular reflection from nearly every visible object. The glossy finish of a new automobile, for example, reflects light essentially by regular reflection. Surfaces of transparent mediums (optical interfaces) also reflect light. The amount of incident light reflected depends upon the angle of incidence—the greater the angle of incidence the greater the amount of reflected light. When light passes through a piece of glass, about 4 percent of the incident light is lost at the first surface and another 4 percent is lost by reflection at the second surface when the light emerges.

Diffuse Reflection

If a beam of light strikes a rough surface such as a sheet of unglazed paper, the light is not reflected regularly but is scattered in all directions, as shown in figure 3-29. What you see in this illustration is diffuse reflection of light. Any rough surface has many plane surfaces which reflect light in accordance with the angle of incidence. Skin, fur, and dull surfaces all reflect light in an essentially diffused manner. On the other hand, some of the reflected light from a shiny automobile or the bright finish of a polished casting is diffused reflection.

REFRACTION

As you study the meaning of refraction, refer to figure 3-30, which shows what happens to rays of light as they pass through a sheet of glass. Both plane surfaces of this glass plate are parallel and air contacts both surfaces. Glass and air are transparent, but the glass is optically denser than air; so light travels approximately one-third slower in glass than in air.

Observe the dotted lines (N & N') in the illustration. These are the normals erected for the incident and refracted rays. When a light ray (wave front) strikes the surface of the glass at right angles (parallel to the normal), it is not bent as it passes through the glass. This is true because each wave front strikes the surface squarely. The wave front is slowed down when it strikes the surface of the glass, but it continues in the same direction it was going before striking the glass. When it squarely strikes the other surface of the glass, it passes straight through without deviation from its course.

If a wave front strikes the first surface of the glass at an angle, as illustrated in part B of figure 3-30, one edge of the first wave front arrives at the surface an instant before the other edge; and the edge which arrives first is slowed down as it enters the denser medium before the second edge enters. Observe that the second edge continues to travel at the same speed, also, until it strikes the surface of the glass. This slowing down of one edge of the wave front before the other edge slows down causes the front to PIVOT TOWARD THE NORMAL.

The information just given relative to a wave front which strikes glass plate is applicable FOR ANY FREELY MOVING OBJECT. When one side of the object is slowed down as it hits something,

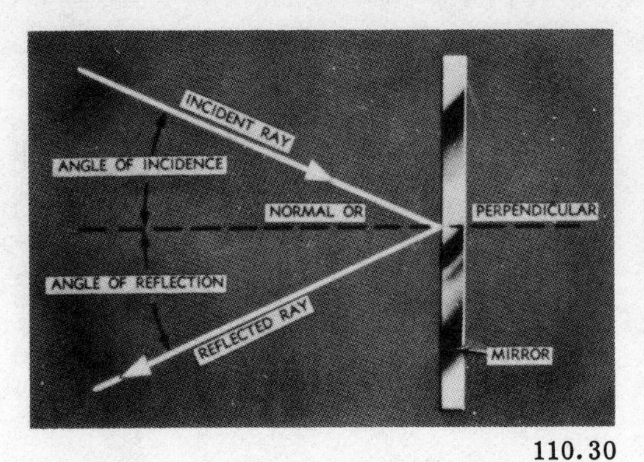

110.30
Figure 3-26.—Terms used for explaining reflected light.

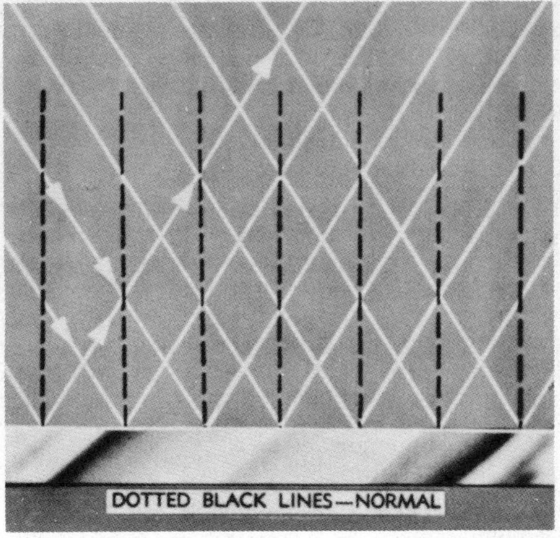

137.30
Figure 3-28.—Regular reflection.

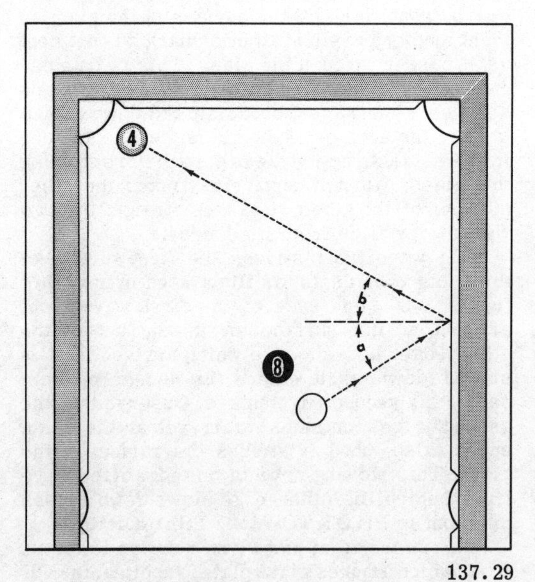

137.29
Figure 3-27.—Application of the law of reflection on a pool table.

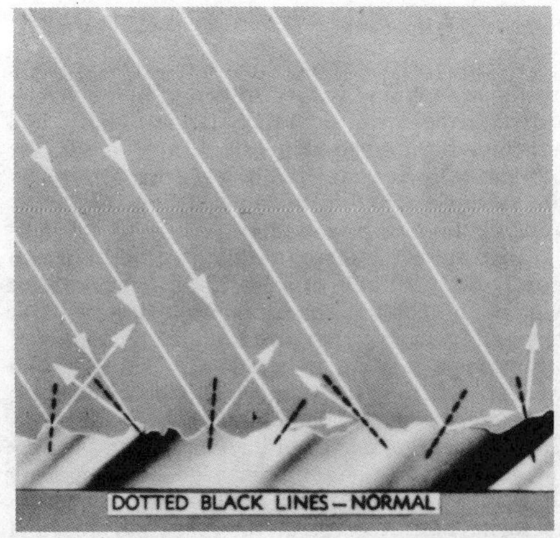

137.31
Figure 3-29.—Diffuse reflection

the other side continues to move at the same speed and direction until it also hits something. This action causes the object to pivot in the direction of the side which hits first and slow down. Pivoting or bending of light rays (wave fronts) as just explained, is called REFRACTION; and the bent (pivoted) rays are labeled REFRACTED RAYS.

If the optical density of a new medium (glass in this case) remains constant, the refracted light rays continue to travel in a straight line, as shown in part B of figure 3-30, until the surface from which they emerge (glass-to-air surface) causes interference. At this point, an opposite effect occurs to a wave front. As one edge of the front reaches the surface (glass-to-air), it leaves the surface and resumes original speed (186,000 miles per second, at which it entered the glass).

Speeding up of one edge of a wave front before the other edge speeds up, causes the front to pivot again; but this time it pivots toward the edge of the front which has not yet reached the surface of the glass. Again, THIS BENDING OR PIVOTING OF THE WAVE FRONT IS CALLED REFRACTION.

If the glass plate has parallel surfaces, the emergent light ray (ray refracted out of the glass) emerges from the second surface at an angle equal to the angle formed by the incident ray as it entered the glass. If you draw a dotted line along the emergent light ray (fig. 3-30), straight back to the apparent source of the ray, you will find that the emergent ray is parallel to the incident ray.

If the optical density of a new medium entered by a light ray (wave front) is constant, the light follows its course in a direct line, as illustrated in part B of illustration 3-30.

Study next illustration 3-31, which show a straight stick in a glass of water. Note that the stick appears bent at the surface of the water. What you see here is an optical illusion created by refraction. When a ray of light passes from air into water, it bends; and when it passes from the water into the air, it also bends. This illustration shows why a fish in water is NOT WHERE HE SEEMS TO BE—he is much deeper.

You can demonstrate refraction visually by placing the straight edge of a sheet of paper at an angle under the edge of a glass plate held vertically (part B, fig. 3-32). Observe that the straight edge of the sheet of paper appears to have a jog in it directly under the edge of the glass plate. The portion of the paper on the other side of the glass appears displaced as a result of refraction. If you move the sheet of paper in order to change the angle of its straight edge, the amount of refraction is increased or decreased.

Laws of Refraction

You should understand thoroughly all laws of refraction. Briefly stated, they are as follows:

1. WHEN LIGHT TRAVELS FROM A MEDIUM OF LESSER DENSITY TO A MEDIUM OF GREATER DENSITY, THE PATH OF THE LIGHT IS BENT TOWARD THE NORMAL.

2. WHEN LIGHT TRAVELS FROM A MEDIUM OF GREATER DENSITY TO A MEDIUM OF LESSER DENSITY, THE PATH OF THE LIGHT IS BENT AWAY FROM THE NORMAL.

3. THE INCIDENT RAY, THE NORMAL, AND THE REFRACTED RAY ALL LIE IN THE SAME PLANE.

4. THE INCIDENT RAY LIES ON THE OPPOSITE SIDE OF THE NORMAL FROM THE REFRACTED RAY.

Study illustration 3-33, and then review carefully all laws of refraction. Note the NORMAL, the ANGLE OF INCIDENCE, and the ANGLE OF REFRACTION.

The angle between the refracted ray of light and a straight extension of the incident ray of light through the medium is called THE ANGLE OF DEVIATION. This is the angle THROUGH WHICH THE REFRACTED RAY IS BENT FROM ITS ORIGINAL PATH BY THE OPTICAL DENSITY OF THE REFRACTING MEDIUM.

The amount of refraction is dependent upon the angle at which light strikes a medium and the density of the new medium—the greater the angle of incidence and the denser the new medium, the greater the angle of refraction. If the faces of the medium are parallel, the bending of light at the two faces is always the same. As illustrated in part B of figure 3-30, the beam which leaves the optically denser medium is parallel to the incident beam. An important thing to keep in mind in this respect, however, is that the emergent beam must emerge from the denser medium into a medium OF THE SAME INDEX OF REFRACTION AS THE ONE IN WHICH IT WAS ORIGINALLY TRAVELING; that is, air to glass to air, NOT air to glass to water (as an example).

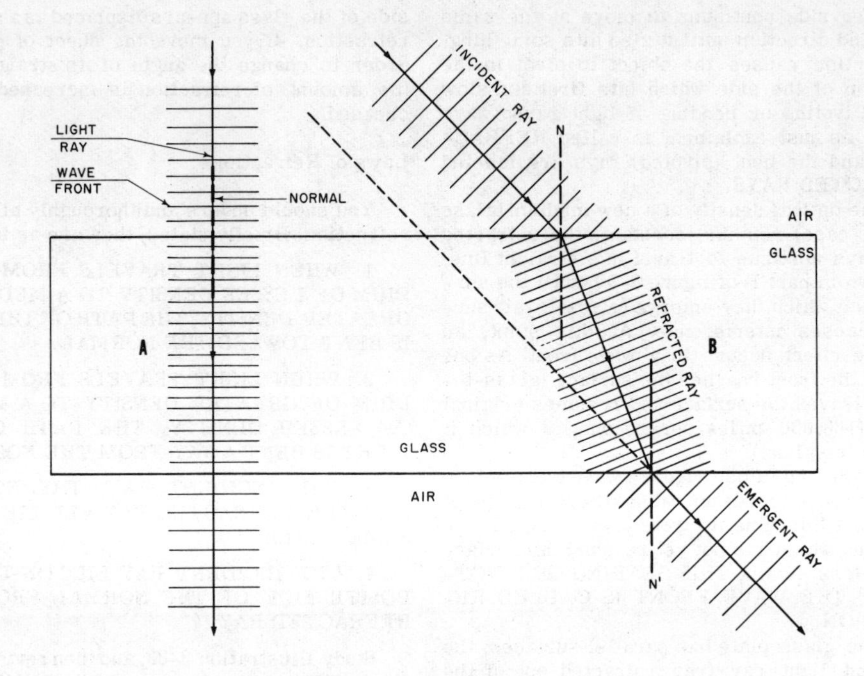

137.32
Figure 3-30.—Refraction of light beams by a sheet of glass.

137.33
Figure 3-31.—Optical illusion caused by refraction.

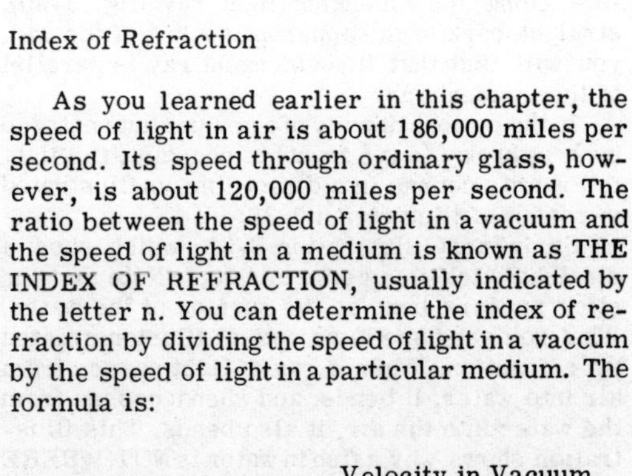

Index of Refraction

As you learned earlier in this chapter, the speed of light in air is about 186,000 miles per second. Its speed through ordinary glass, however, is about 120,000 miles per second. The ratio between the speed of light in a vacuum and the speed of light in a medium is known as THE INDEX OF RFFRACTION, usually indicated by the letter n. You can determine the index of refraction by dividing the speed of light in a vaccum by the speed of light in a particular medium. The formula is:

$$\text{Index of Refraction} = \frac{\text{Velocity in Vacuum}}{\text{Velocity in Medium}}$$

You can determine the index of refraction of a substance by Snell's law, by actually measuring the angles of incidence and refraction in a simple experiment, because air can be used instead of a vacuum for all practical purposes.

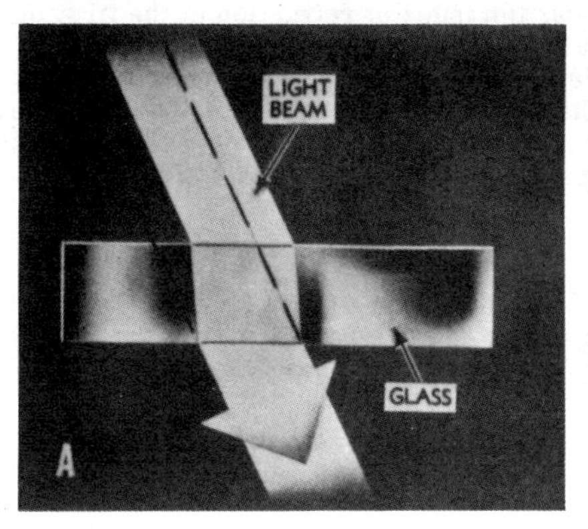

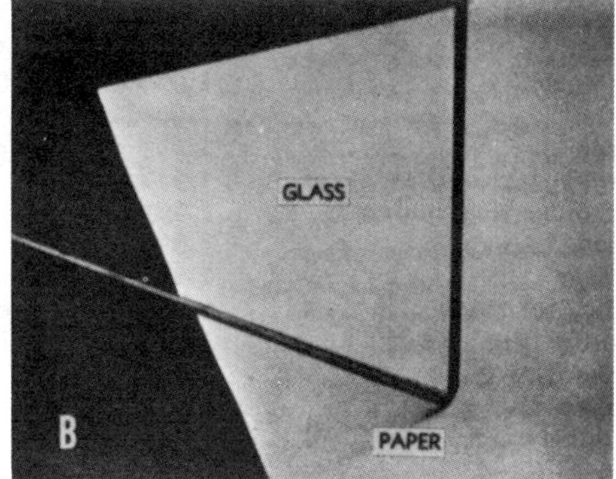

137.34

Figure 3-32.—Effects of refraction.

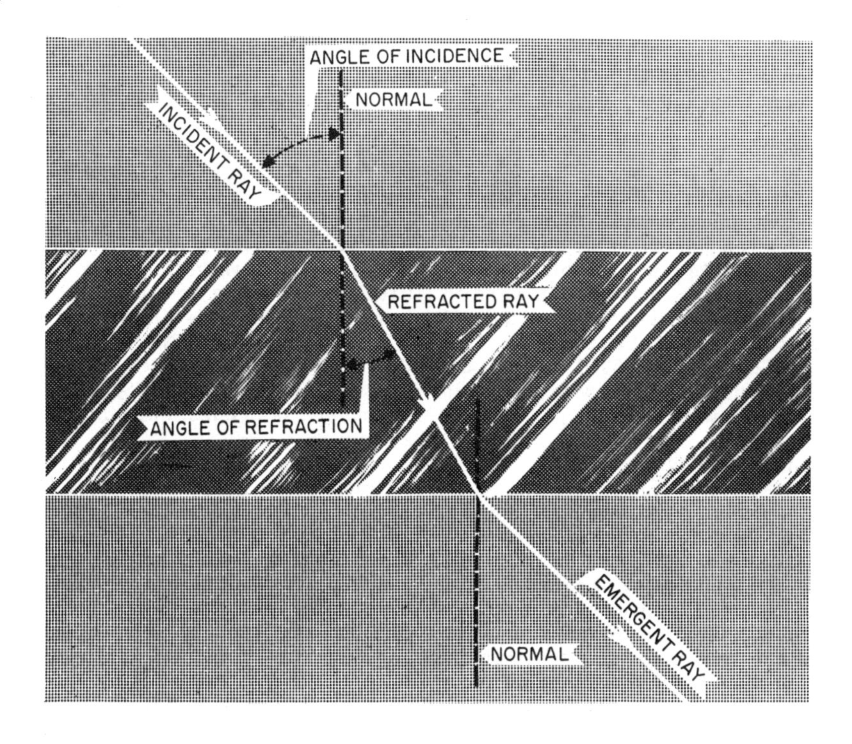

12.233

Figure 3-33.—Terms used for describing refraction.

Following is a list of indices of refraction for some substances:

Vacuum	1.000000
Air	1.000293
Water	1.333
Boro-silicate crown glass	1.517
Thermosetting cement	1.529
Canada balsam	1.530
Gelatin	1.530
Light flint glass	1.588
Medium flint glass	1.617
Dense flint glass	1.649
Densest flint glass	1.963
Diamond	2.416

The index of refraction of a transparent substance of high purity is a constant quantity of the physical properties of a substance. You can therefore determine the identify of substances by measuring their indices of refraction. The refractometer is an instrument which quickly and accurately measures the index of refraction of a substance.

NOTE: For most computations, the index of air is considered to be unity (1.000).

Angle of Refraction

The angle of refraction of light is dependent upon the relative optical density of the medium light enters. You can determine the angle of refraction by using Snell's law. From his observations, Willebrord Snell, a mathematician and an astronomer in the University of Heyden, defined the index of refraction as the SINE OF THE ANGLE OF INCIDENCE TO THE ANGLE OF REFRACTION. You can learn how to apply this law by using the following equation:

$$n \text{ sine } \theta = n' \text{ sine } \theta'$$

In optical formulas, n represents the index of refraction. The character θ is the Greek word for THETA and means ANGLE. The indices of the two media are represented by n for the first and n' for the second. The index of refraction of the first medium multiplied by the sine of the angle of incidence equals the index of refraction of the second medium times the sine of the angle of refraction. Refer to a table of tangents, cosines, and sines.

Assume that the ray of light in figure 3-33 is contacting at a 45° angle a plate of glass whose index of refraction is 1.500. According to Snell's law, the index of refraction of the first medium (air: 1.000) times the sine of the angle of incidence (sine 45°) equals the index of refraction of the second medium (glass: 1.500) times the SINE of the angle of REFRACTION. Enter this information in the equation and you get:

$$1.000 \times \text{sine } 45^\circ = 1.500 \times \text{sine of } \theta$$

Get the value of sine 45° (.707) from the table of sines. Then enter all data in the formula and solve for sine of θ, as follows:

$$\text{sine } \theta' = \frac{1.000 \times .707}{1.500'} = .471$$

Refer to the table again and find the angle with a sine of .471. The answer is 28° 20', the ANGLE OF REFRACTION of the second medium (glass).

If you now reverse the direction of the light ray, its angle of incidence at the surface of the glass is 28° 20', and its angle of refraction is 45 degrees. This may seem strange to you, but use the formula again and you get:

$$1.500 \times \text{sine } 28^\circ 20' = 1.000 \times \text{sine of } \theta$$

When you solve the equation, you get .707 for the sine of θ, which the table shows for a 45° angle, the size of the angle of incidence with which you started.

What you just proved by solving the last equation is known as the LAW OF REVERSIBILITY, something you should remember. The law means that if the direction of a ray of light AT ANY POINT IN AN OPTICAL SYSTEM IS REVERSED, THE RAY RETRACES ITS PATH BACK THROUGH THE SYSTEM, regardless of the number of prisms, mirrors, or lenses in the system.

Refraction in a Prism

You know that when a ray of light passes through plate glass it emerges parallel to the incident ray. This is true because both faces of the glass are parallel. Study figure 3-34 now to find out what happens when light strikes a triangular glass prism, whose faces are not parallel.

Note in part A of this illustration that when the light rays and wave fronts strike the first face of the prism, the rays nearest the base (thickest part) ARE REFRACTED AT THE SAME ANGLE AS THE RAYS NEAR THE APEX OF THE

PRISM. The portion of the light rays and wave front nearest the base, however, are delayed longer in reaching the second face of the prism than the rays nearest the apex. The ray and wave front nearest the apex, therefore, are delayed the least in reaching the second face of the prism; and for this reason, the refraction of light rays and wave front IS ALWAYS TOWARD THE WIDE BASE. A point of interest here is that this refraction of light rays and wave fronts toward the base of the prism is actually caused by the inclined faces of the prism, and that the inclined faces are a necessity in order to have the triangular shape of the prism. Bear in mind that all applicable laws of refraction must be adhered to strictly, and that they may be used in tracing the path of light through a prism.

When a light ray enters a prism, it bends toward the normal; when it leaves the prism, it bends away from the normal. Study illustration 3-34. Note IN PART B that the two normal lines are not parallel, and also that the deviation of the prism is the angle between the incident ray (extended through the prism) and the emergent ray, which is refracted toward the base. You can determine the amount of deviation of any light ray in a prism if you know the angle of incidence, the angle of the prism, and its index of refraction.

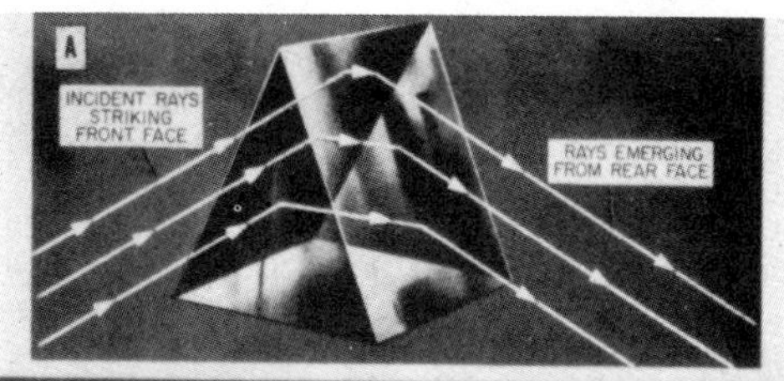

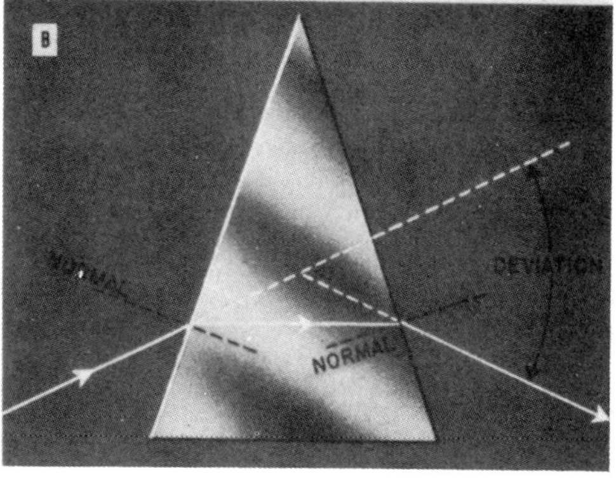

110.32
Figure 3-34.—Passage of light rays through a prism.

Reflection and Refraction Combined

Smooth glass reflects part of the light which falls upon it, about 4 percent (more if the angle of incidence is large); but most of the light which enters the glass is refracted. Figure 3-35 shows a ray of light passing through plate glass. The dotted lines are the normals. The white arrow to the right of the first normal line indicates reflected light. The line of light which extends upward from the second normal represents the amount of light reflected back into the glass when the light strikes the lower surface. This is called INTERNAL REFLECTION. An internally reflected ray of light is refracted at the upper surface of the glass and emerges parallel to the reflection from the incident ray.

Study next illustration 3-36, which shows reflections from both surfaces of a glass plate. Note the two images. If you have several plates of glass in a stack with thin layers of air between the plates on the inside, you can see twice as many reflections as the number of plates of glass. NOTE: You will occasionally find a condition such as this in optical instruments.

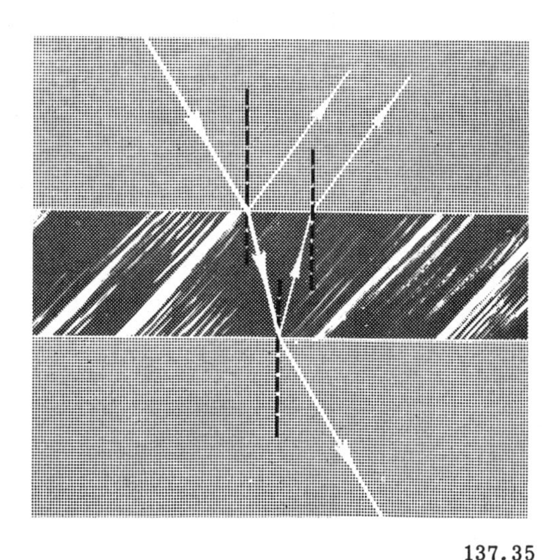

137.35
Figure 3-35.—Reflection and refraction combined.

If you have five lenses in a system, you have ten faces; and each face reflects part of the light. The image you see when you look through the instrument is formed ONLY by the light which passes through the lenses. A complex instrument such as a submarine telescope may have many surfaces which reflect part of the light, and the lenses and prisms must have a coating or film applied to them to eliminate reflection and prevent loss of light in the instrument.

You learned in chapter 2 that optical glass is highly transparent, but it is still visible because of reflected light from its surface. Other glass objects are visible partly because of refraction. You can see part of the background through the glass, but the glass bends the rays from the background before they reach your eyes. In accordance with the angle at which it strikes the surface of the glass, each ray bends at a different angle. The background, therefore, appears distorted when you see it through the glass.

Figure 3-37 shows a piece of white paper behind an empty glass beaker with a glass rod in it. You can see this glass rod as a result of reflection and refraction combined. Light from the white paper is refracted as it passes through the rod, causing the background to look distorted and uneven in brightness.

Reflection can take place ONLY at a surface between two media with different indices of refraction. Because the rod in figure 37 is in air, the difference between the two media is fairly large and the rod is visible. This same rule applies to refraction, as you can prove by Snell's law. If the indices of refraction of the two media are identical, the angle of incidence equals the angle of refraction and there is NO refraction.

Illustration 3-38 is the same as figure 3-37 except that water has been put into the glass beaker, and the appearance of the part of the glass rod IN THE WATER looks different from the part OUT OF THE WATER. The reason for this is that the index of refraction between the two media is now much smaller, so there is less reflection and less refraction.

If the water in the glass beaker is replaced with a solution of the same index of refraction as glass, there is no reflection or refraction and the end of the glass rod in the solution is invisible. See figure 3-39.

Total Internal Reflection and Critical Angle

You have learned that when light passes from one medium to another of lower density (from glass to air, for example) part of the light is reflected internally. When the angle of incidence is small, the surface reflects only a small part of the light; but when the angle of incidence is increased, the percentage of internal reflection increases.

When light which is attempting to leave a more optically dense medium for a less optically dense medium strikes the boundary surface, it may be reflected instead of refracted, even though both media are transparent. When this happens, as it does when the angle of incidence exceeds a certain CRITICAL ANGLE, we get what is known as TOTAL INTERNAL REFLECTION. Study illustration 3-40 carefully.

Observe the critical angle in figure 3-40. This is the angle formed when light about to pass from a medium of greater optical density into one of lesser density is refracted along the surface of the denser medium, as illustrated. The critical angle varies with the INDEX OF REFRACTION of the substance or medium. When the angle is exceeded, the light is reflected back into the denser medium.

Critical angles for various substances (when the external medium is air) are as follows:

Water	48° 36'
Crown glass	41° 18'
Quartz	40° 22'
Flint glass	37° 34'
Diamond	24° 26'

The small critical angle of a diamond accounts for its brilliance, provided it is a well-cut diamond. The brilliance is due to total internal reflection of light; the light is reflected back and forth many times before it emerges to produce bright, multiple reflections.

Rays of light from an underwater source are incident at various angles on the surface which separates the water and air, as shown in figure 3-40. As the angle of incidence of the light rays increases, the deviation of refracted rays becomes proportionately greater. A point is reached where an incident ray is deviated to such an extent that it travels along the surface of

137.36
Figure 3-36.—Reflection from the surfaces of a glass plate.

137.38
Figure 3-38.—Effect on visibility by the reduction of reflection and refraction.

137.37
Figure 3-37.—Visibility resulting from combined reflection and refraction.

137.39
Figure 3-39.—Elimination of visibility by eliminating reflection and refraction.

the water and does not emerge into the air, as illustrated. The angle formed by this INCIDENT RAY AND THE NORMAL is the critical angle of the medium.

All light rays which travel in an optically dense medium and strike the surface with an angle of incidence less than the critical angle of the media are REFRACTED and pass into the optically lighter medium in accordance with the laws of refraction. All such rays which strike at an angle of incidence greater than the critical angle of the media are REFLECTED INWARD in accordance with the law of reflection.

One example of total internal reflection at the surface of water is shown in figure 3-41. Rays of light from the sand and the goldfish strike the upper surface of the water at an angle greater than the critical angle and are reflected downward into the water. The reflected rays, however, strike the end of the aquarium at LESS THAN the critical angle, so they pass through and you can see an image of the fish reflected by the upper surface of the water. The path of each reflected ray is illustrated in figure 3-42.

You can always calculate the critical angle of a substance by the following equation (Snell's law):

$$N \text{ sine } \theta = N' \text{ sine } \theta'$$

In this equation, n represents the index of refraction, i the angle of incidence, and r the angle of refraction. The index of refraction is a number applied to a transparent substance, and it denotes the increased speed of light in a vacuum in comparison to the speed of light in a substance. It also determines the relation between the angle of incidence (n' sine θ) and the angle of refraction (n sine θ) when light passes from one medium to another. The index between two media is called the relative index.

Suppose you desire to calculate the critical angle of a medium when the other medium is air. How can you do this? Use water as one medium, as an example, and air as another; then make proper substitutions in the formula (Snell's law) and solve the equation. The index of refraction of water is 1.333; when the angle of incidence is the critical angle, the angle of refraction is 90 degrees. The procedure for solving the problem follows:

Snell's law: $n \text{ sine } \theta = n' \text{ sine } \theta$

$$1.333 \text{ sine } \theta = 1.000 \text{ sine } 90^\circ$$

$$\text{sine } \theta = \frac{1.000 \times 1.000}{1.333} = \frac{1.000}{1.333}$$

$$\text{sine } \theta = .750817$$

$$\theta = 48^\circ\ 36'$$

$$\text{sine } \theta = \frac{1.1}{333} = .750187$$

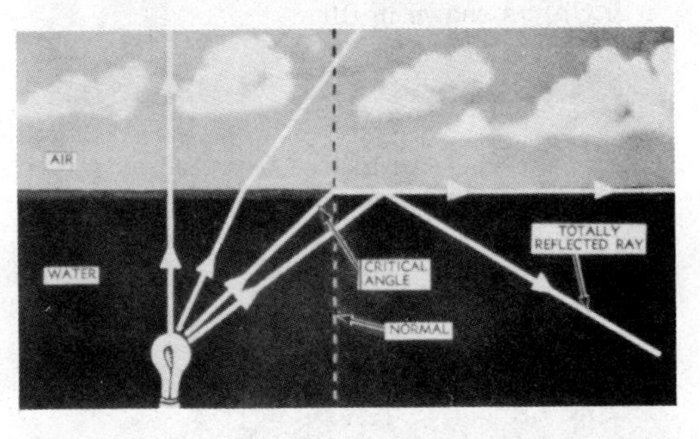

137.40
Figure 3-40.—Angles of light rays from an underwater source.

137.41
Figure 3-41.—Total internal reflection at the surface of water.

Look up .750 in the Table of Tangents, Sines, and Cosines and you will find that the critical angle for water is 48°36'

Atmospheric Refraction

At a surface which separates two media of different indices of refraction, the direction of the path of light changes abruptly when it passes through the surface. If the index of refraction of a single medium changes gradually as the light proceeds from point to point, the path of light also changes gradually and is curved.

Although when air is most dense it has a refractive index of only 1.000292, the index is sufficient to bend light rays from the sun toward the earth when these rays strike the atmosphere at an angle, as shown in illustration 3-43.

The earth's atmosphere is a medium which becomes denser toward the surface of the earth. As a result, a ray of light traveling through the atmosphere toward the earth at an angle does not travel in a straight line but is refracted and follows a curved path. From points near the horizon, in fact, the bending of light is so great that the setting sun is visible even after it is below the horizon (fig. 3-44).

MIRAGES--Over large areas of heated sand or water there are layers of air which differ greatly in temperature and refractive indices. Under such conditions, erect or inverted (sometimes much distorted) images are formed which are visible from great distances. These images are MIRAGES.

Observe the apparent lake of water in a desert in illustration 3-45. This looks like a real lake but it is ONLY a mirage caused by the refraction of light over the hot sand. The sand heats the air directly above it, though the air at a higher level remains comparatively cool. Because cool air is denser than hot air, the index of refracting is fairly low at the surface and gradually increases at higher and higher altitudes.

Study illustration 3-46 to learn what happens to light rays in a mirage. Light rays in cool air do not bend, as shown, but the ray which travels downward toward the hot air curves upward. When an observer looks at an object along the hot air ray, he thinks he sees it along the dotted line in the illustration.

You perhaps have observed mirages on asphalt highways on clean, hot days. When the highway rises in front of you and then flattens out, its surface forms a small angle with your line of sight and you see reflections of the sky. These reflections look like puddles of water in the road. Under proper conditions of the atmosphere and light, you can even see an approaching car reflected in the mirage.

LOOMING.—Looming is the exact opposite of a mirage. Ships, lighthouses, objects, and islands sometimes loom—they appear to hang in the sky above their real locations. On some bodies of water (Gulf of California and Chesapeake Bay, for example) looming is common. Figure 3-47 shows the path of light rays in looming.

The reason for looming is that air warmer than water is cooled at the water's surface and the index of refraction of the air decreases higher up and the rays of light bend downward, as shown in the illustration. This explains why a lighthouse appears to hang in the sky.

HEAT WAVES.—On a hot day the columns of heated air which rise from the earth are optically different from the surrounding air and rays of light are irregularly refracted. The air is turbulent and conditions under which observations are made change constantly. An object viewed through such layers of air therefore appears to be in motion and the air is BOILING, or the image is DANCING BECAUSE OF HEAT WAVES. This condition is particularly bad for using a high-powered telescope, one of more than 20 power. The heat waves are caused by the refraction of light waves at various changing angles, thereby creating a distortion.

RAINBOWS.—A rainbow is formed when sunlight strikes drops of falling water and is refracted, reflected, and dispersed into the atmosphere. You can understand how a rainbow is created by studying figure 3-48 and the discussion which follows.

This illustration shows only three drops of water, but when a rainbow is formed millions of drops of water are involved. Note the rays from the sun which strike these drops of water, and then follow the rays as they pass through and emerge from the drops. Rays of light, of course, strike at many points on the drops of water.

When a ray of sunlight penetrates the surface of a drop of rain, it is refracted as it passes through to the back surface, from which it is reflected by total internal reflection. As it leaves the drop of rain, it is refracted a second time. Because the index of refraction is different for different colors of light, emerging rays of light

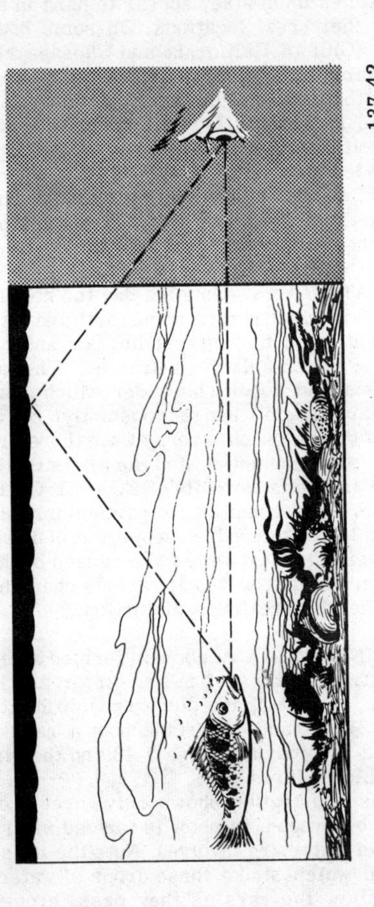

137.42

Figure 3-42.—Effect of total internal reflection on light rays.

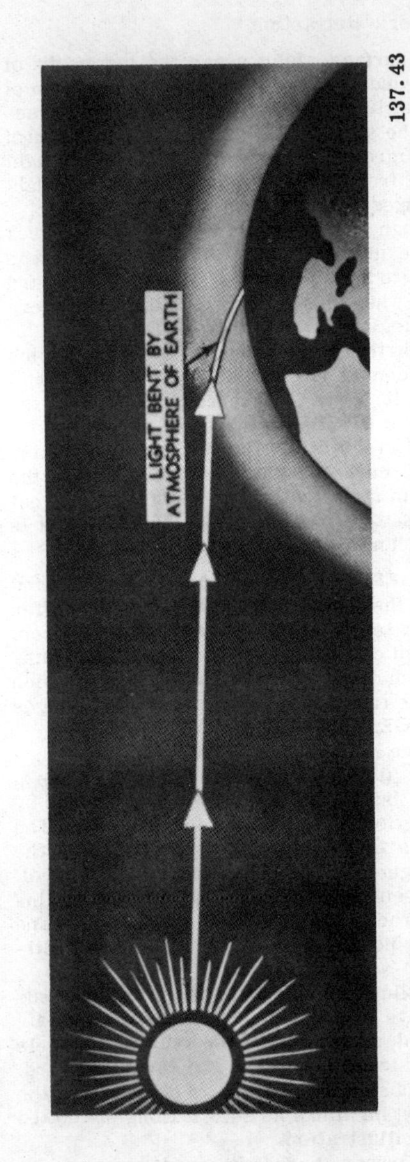

137.43

Figure 3-43.—Bending of the sun's rays by atmospheric refraction.

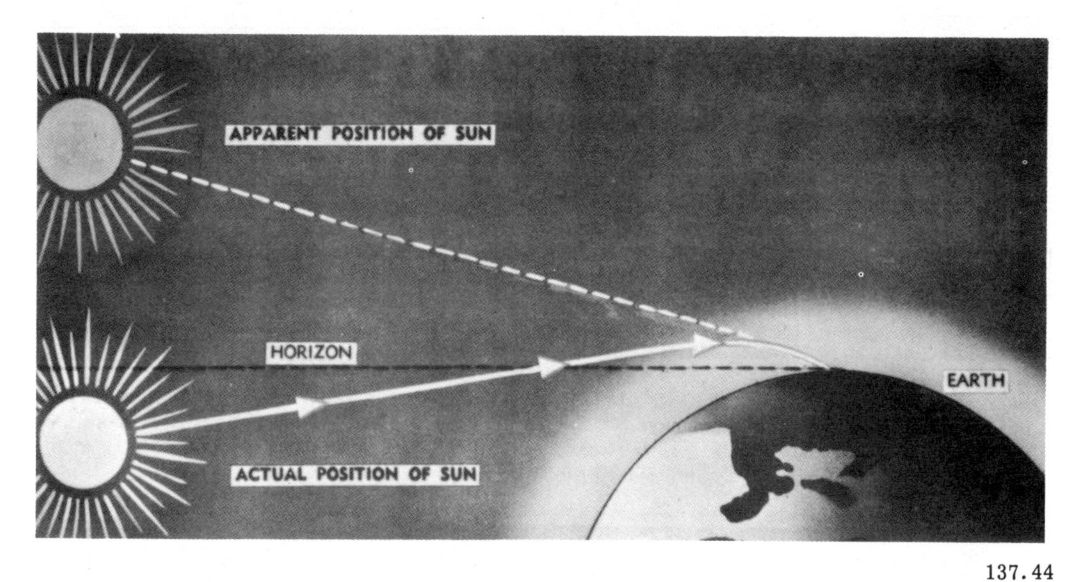

137.44
Figure 3-44.—Visibility of the sun below the horizon as a result of refracted light.

137.45
Figure 3-45.—Picture of a mirage in a desert.

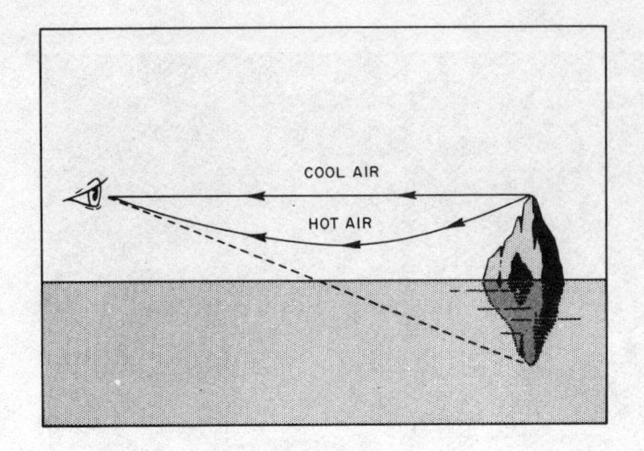

137.46

Figure 3-46.—Path of light rays in a mirage.

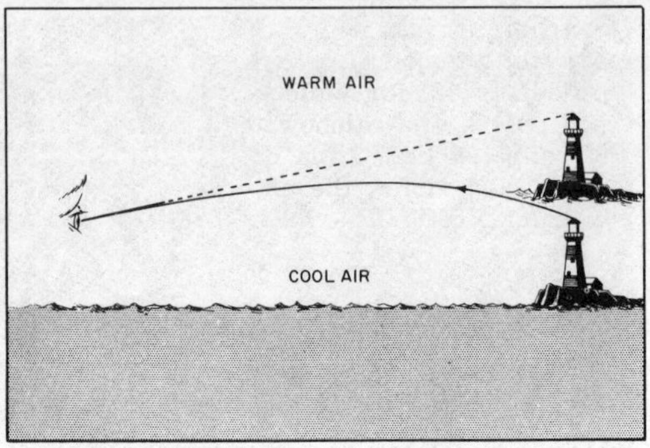

137.47

Figure 3-47.—Path of light rays from a looming object.

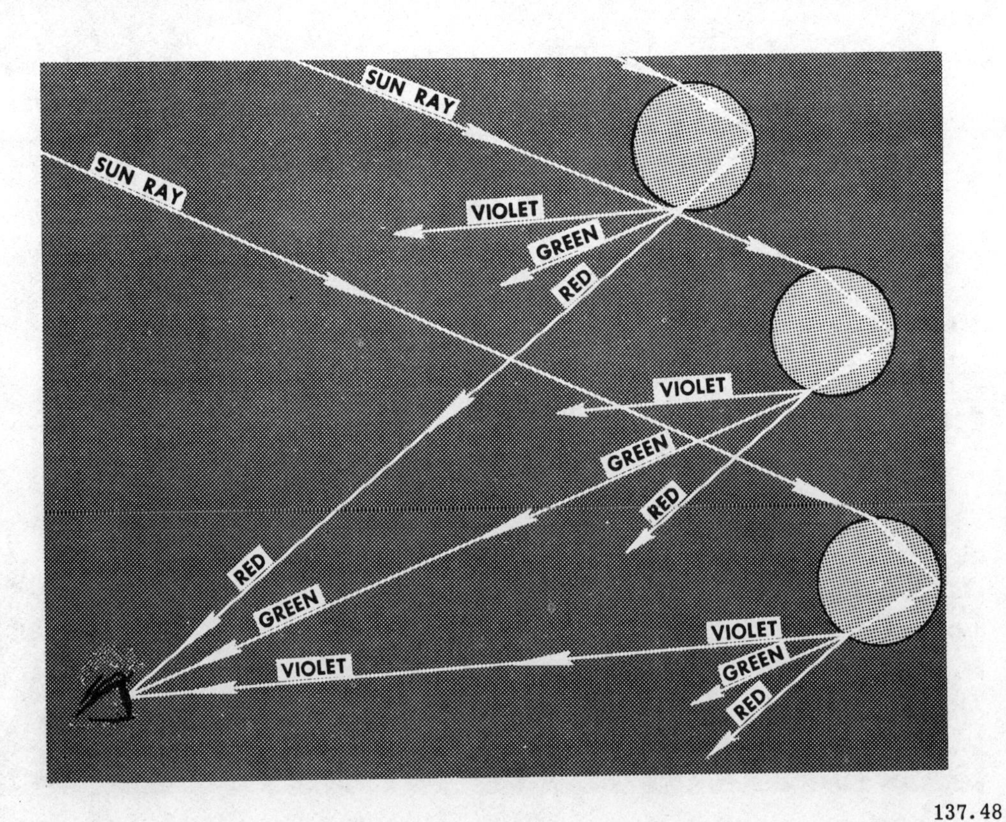

137.48

Figure 3-48.—Formation of a rainbow.

from the drops of water are DISPERSED into a spectrum. The angle which these refracted rays make with the sunlight coming over an observer's shoulder is 40° for violet and 42° for red. All other colors of a rainbow come from refracted and reflected light from drops of rain between the red and violet, the extreme colors in the spectrum: VIBGYOR—violet, indigo, blue, green, yellow, orange, and red. Indigo is sometimes omitted. A rainbow is an arc, because the observer's eye is at the apex of a cone from which he sees colored rays refracted from drops of water at different angles (40° to 42°). Since the sun's rays are parallel, all emerging rays from drops of water are parallel by color—red, green, and so on.

CHAPTER 4

IMAGES AND MIRRORS

This chapter contains information concerning the following items, all of which you must know in order for you to qualify for advancement in rating to Opticalman 3:

1. Formation of images by mirrors.
2. Theory of magnification of images.
3. Effect on light rays when they contact (strike) plane, concave, and convex mirrors.
4. Measuring units and systems used in optics.

Because your job as an Opticalman is concerned with all aspects of repair and overhaul of optical instruments, it is best to describe at this time the general principle of operation of an optical instrument and explain the function it serves. With this knowledge in mind, you can understand better the discussion of optical elements and instruments in the next four chapters.

An optical instrument consists of two major parts: (1) optical, and (2) mechanical. The optical parts are specially manufactured and purified glass made in different shapes and sizes, as required for a specific purpose in an optical instrument. The names of these optical parts are: (1) mirrors, (2) lenses, and (3) prisms.

Smooth surfaces of mirrors and flat surfaces of prisms can change the direction of light by REFLECTION, as you learned in chapter 3; whereas, lenses (with one or two curved surfaces) and prisms (with flat surfaces) change the direction of light by REFRACTION AND/OR REFLECTION. The mechanical structure of optical instruments SERVES MERELY TO CONTAIN, HOLD, AND CONTROL THE LOCATIONS OF OPTICAL PARTS.

An optical instrument enables you to see better by:

1. Enlarging the image of an object you are observing. This enlargement of the image is due to MAGNIFICATION, which makes the object APPEAR LARGER AND CLOSER TO YOU.
2. Increasing your ability to judge the distance of an object from you. This ability to judge distance is called STEREOSCOPIC VISION, dependent upon the difference in view of the same situations by your eyes. If your eyes were farther apart, you could perceive depth better; but this can be accomplished for you through appropriate reflection of light rays by an optical instrument.

Some optical instruments increase your effective interocular or interpupillary distance (effective distance between the eyes with respect to the light entering them). Prism binoculars, for example, increase the effective interocular distance by a SERIES OF REFLECTIONS FROM PRISMS. Ship's binoculars are actually giant binoculars.

3. Optically placing a target at an apparently great distance, thereby determining a fixed line of sight. Such instruments do not need magnification (discussed in chapters 6 and 7). Telescopes and periscopes, on the other hand, are optical instruments constructed for the specific purpose of providing a definite amount of magnification of an object.

As you can understand by studying the preceding discussion of optical instruments, optical elements CONSTITUTE THE HEART of an optical instrument. You will learn more about this when you study basic optical instruments (chapter 6) and the construction of optical instruments (chapter 7).

Before we get into the discussion of images and mirrors, and lenses and prisms (chapter 5), it is best to explain at this point the DIFFERENT SYSTEMS OF MEASUREMENT USED IN OPTICS, in order that you may understand better the discussion in this chapter and in various subsequent chapters.

MEASUREMENT IN OPTICS

Some of the measurement systems which an Opticalman should understand are: (1) metric,

(2) degree, and (3) Navy mil. Each of these systems is discussed briefly in the following pages.

METRIC SYSTEM

Some measuring systems are rather complicated. Units of measurement in the English system, for example, are entirely arbitrary. To illustrate, there are 272 1/2 square feet in a square rod, 57 3/4 cubic inches in a quart, and 31 1/2 gallons in a barrel.

Shortly after the French Revolution, the National Assembly of France decided to appoint a commission for the purpose of developing a more logical measuring system than those current at the time. The product of this commission was the metric system, which has been adopted by most countries of the world. The metric system, in fact, is used almost exclusively for measurements in scientific work.

In your work as an Opticalman, you will most likely use the metric system of measuring more than other measuring systems. The diameter and focal length of lenses are usually stated on optical drawings, for example, in millimeters—not in inches. In addition, with some experience, you will find the metric system much easier to use than the English system.

Decimals are basic in the metric system of measurement, which means that you can easily convert from one unit to another. Suppose you know that an object, for example, is 0.67 meter long and you desire the answer in decimeters. All you need to is multiply by 10 and you get an answer of 6.7 decimeters in length. If you wish the answer in centimeters, multiply by 100, and you get 67 cm. For an answer in millimeters, multiply by 1,000 and you get 670 mm.

Suppose you desire to use the English system of measurement to get in feet an object which is 0.67 yard long. You must multiply by 3 to get the answer in feet, and by 36 to get the answer in inches.

What, then, is the difference in using the English or metric system of measurement? As you can see, in the metric system, all you need do is move the decimal point.

The unit of length in the metric system is the METER, which is equal to 39.37 inches. A meter is divided into 100 equal parts called centimeters; and each centimeter is divided into ten parts called a millimeter, because each part is 1/1,000 part of a meter. All units of linear measurement of the metric system are multiples or fractional parts of a meter in units of 10.

Following is a table of metric units, with their equivalents in inches, yards, and miles:

	. 1 millimeter =	.03937 inch
10 millimeters =	1 centimeter =	.3937 inch
10 centimeters =	1 decimeter =	3.937 inches
10 decimeters =	1 meter =	1.0936 yards
10 meters =	1 dekameter =	10.936 yards
10 dekameters =	1 hectometer =	109.36 yards
10 hectometers =	1 kilometer =	.6214 mile

The names of multiples in the metric system are formed by adding the Greek prefixes: DEKA (ten), HECTO (hundred), KILO (thousand), and MEGA (million). Sub-multiples of the system are formed by adding LATIN PREFIXES: DECI (tenth), CENTI (hundredth), MILLI (thousandth), and MICRO (millionth).

For quick, approximate conversion from inches to the metric system units, or vice versa, refer to a metric unit inch conversion table, which your optical shop will have. For more exact conversion, and for conversion of large units, use the following table:

From	To	Multiply by	
Millimeters	Inches	Millimeters by	.03937
Inches	Millimeters	Inches by	25.4
Meters	Inches	Meters by	39.37
Meters	Yards	Meters by	1.0936
Inches	Meters	Inches by	.0254
Yards	Meters	Yards by	.9144
Kilometers	Miles	Kilometers by	.6214
Miles	Kilometers	Miles by	1.609

The unit of volume in the metric system is the LITER, which is the volume of a cube 1/10th of a meter on a side. A liter is equal to 1,000 cubic centimeters, equivalent to 1.057 quarts.

The unit of mass in the metric system is the GRAM, the weight of one millimeter of distilled water at 4° C. For all practical purposes, a gram may be considered as the weight of one cubic centimeter (cc) of water.

The three standard units of the metric system (meter, liter, and gram) have decimal multiples and sub-multiples which make it easy to use for all purposes. Every unit of length, volume, or mass is exactly 1/10th the size of the next larger unit.

Standard abbreviations for principal metric units are:

Meter	m
Centimeter	cm
Millimeter	mm
Liter	l
Milliliter	ml
Cubic centimeter	cc
Gram	g
Kilogram	kg
Milligram	mg

DEGREE SYSTEM

The degree system is a means of measuring and designating angles or arcs. A degree is 1/360th of the circumference of a circle, or the value of the angle formed by dividing a right angle into 90 equal parts. Each degree is divided into 60 parts called minutes, and each minute is divided into 60 parts called seconds.

NAVY MIL

A Navy mil is a unit of measurement for angles, much smaller than a degree—1/6,400 of the circumference of a circle.

A mil is the value of the acute angle of a triangle whose height is 1,000 times its base. For example, when you look at an object 1,000 meters distant and 1 meter wide, the object intercepts a visual angle of 1 mil. Another way to say this is: A mil is an angle whose sine or tangent is 1/1,000. NOTE: For very small angles, the sine and tangent are practically the same.

Use the following formula when you desire to determine the value of a mil in minutes and seconds:

$$\text{sine } A = \frac{1}{1{,}000}$$

sine A = 0.001

sine A = approximately 3'26.5" (table of sines) (1 mil)

POSITION OF IMAGE

An image is the optical counterpart of an object produced by a lens, mirror, or an optical system (including prisms). Two types of images are produced: (1) real, and (2) virtual.

A REAL IMAGE ACTUALLY EXISTS AND CAN BE THROWN UPON A SCREEN. It is produced by real foci (points of intersection of light rays). An image formed on a photographic plate is a good example of a real image, as is also true of an image formed on a motion picture screen. Study illustration 4-1.

A VIRTUAL IMAGE GETS ITS NAME FROM THE FACT THAT IT HAS NO REAL EXISTENCE. IT IS FORMED BY VIRTUAL FOCI AND CANNOT BE THROWN UPON A SCREEN. Your image in a mirror is a good example of a virtual image. Reflected rays of light which strike your eyes from the mirror seem to extend through the MIRROR TO YOUR IMAGE IN IT. Your image in the mirror, in fact, appears to be on the other side, as shown in figure 4-2; but this is only an optical illusion produced by the plane mirror.

You will learn in this chapter that an optical element produces an image by collecting a beam of light from an object and transforming it into a beam which CONVERGES TOWARD OR DIVERGES FROM ANOTHER POINT. If the beam actually converges to a point, it produces a real image of the object; if the beam diverges from a point, it produces a virtual image of the object.

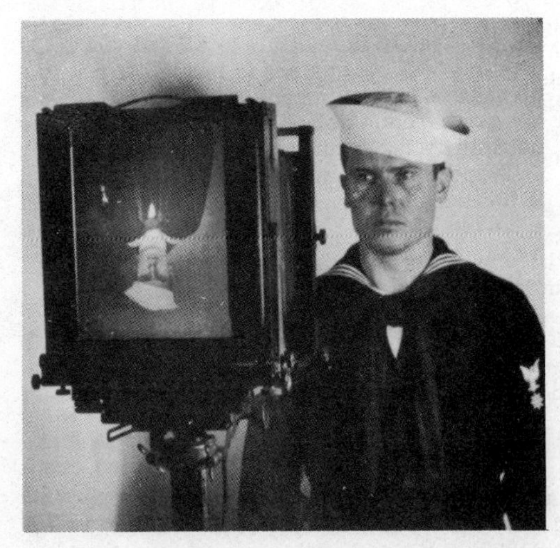

137.49

Figure 4-1.—Real image of a sailor on photographic plate.

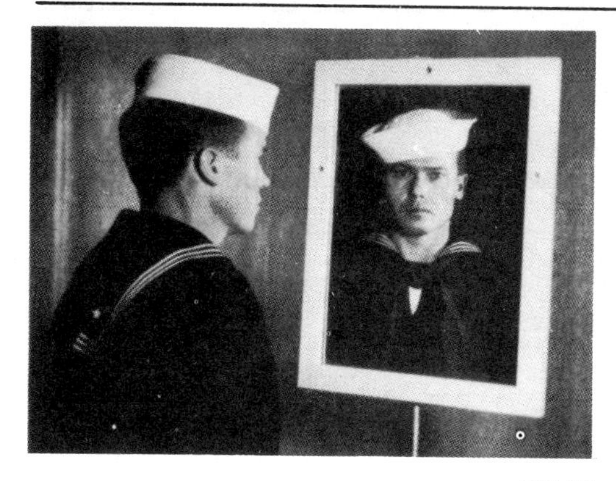

137.50
Figure 4-2.—Virtual image of a sailor formed by a mirror.

Refer again to illustration 4-1. The image of the sailor in this illustration is formed on a photographic plate (ground glass) by the lens of a camera. The image is inverted (upside-down) and reverted (left to right). You know this is true because the sailor is the object, which is erect; and his picture on the ground glass is the image (upside-down). NOTE: ALWAYS COMPARE THE IMAGE WITH THE ACTUAL OBJECT.

You can ascertain whether an image is real by holding a piece of paper or a sheet of ground glass where the image is formed. If you can see the image on the paper or glass plate, IT IS A REAL IMAGE. If you can see the image but CANNOT FORM IT ON PAPER OR GROUND GLASS PLATE, IT IS A VIRTUAL IMAGE, LIKE THE ONE IN ILLUSTRATION 4-2. The sailor in this illustration does not SEE HIMSELF AS OTHERS SEE HIM, because the image is backwards, the optical term for which is REVERTED. The image is erect, however, and ERECT IS THE OPTICAL TERM FOR RIGHT SIDE UP.

When you desire to describe an image—any image with its actual object—you can say that it is:

1. Real or virtual.
2. Erect or inverted.
3. Normal or reverted.
4. Of the same size as the actual object, or larger or smaller than the actual object.

The size of an image is not important at this time, because we are concerned now with the position of an image. The rule of describing an image, in comparison with the object which formed it, is as follows: STAND BETWEEN THE MIRROR AND THE OBJECT AND LOOK AT THE OBJECT. Then stand on the front side of the mirror and LOOK DIRECTLY AT THE IMAGE AND MIRROR, so that you can compare the attitude of the image with the way the object looked.

Take another look at illustration 4-1, in which you see the image as it appears when you look toward the lens which forms it; and you see the object (sailor) as he looks when you stand between him and the lens and observe him from that point.

Now study illustration 4-3, part A of which shows where to stand to view an object itself, and also where to stand to view the image created of that object by a mirror. Note the position of the OBJECT and also the position of the IMAGE, which is seemingly behind the mirror.

Study next part B of illustration 4-3, which shows the proper position to stand for observing an object and the image of the object created by a PORRO PRISM. Observe how the two angular sides of the prism reflect the rays of light.

Part C of figure 4-3 shows the position to stand for viewing an object, and then the position to stand for viewing on plate glass or a screen the image of that object created by a positive lens. (The straight line through the center of the lens is the optical axis; positive lenses converge light rays to a point. Lenses and prisms are discussed in detail in chapter 5.)

Your image in a plane mirror is virtual, erect, and reverted. Your description of the image of another object in a plane mirror, however, depends upon the manner in which you hold the mirror with respect to the object.

To illustrate the position of an image with a plane mirror, do the following: Lay this training course on a table in normal reading position and hold a small mirror (2" x 3") on its edge at the top of A above the letter F in in illustration 4-4. The F in this illustration is like any other F you would see in print, and it is in the normal position (optical term for an image which is not reverted or turned left to right). With the reflecting side of the mirror toward the F, look at the reflection of F in the mirror and compare it with the other image positions in the illustration. It is the F (inverted and normal), as shown below letter C.

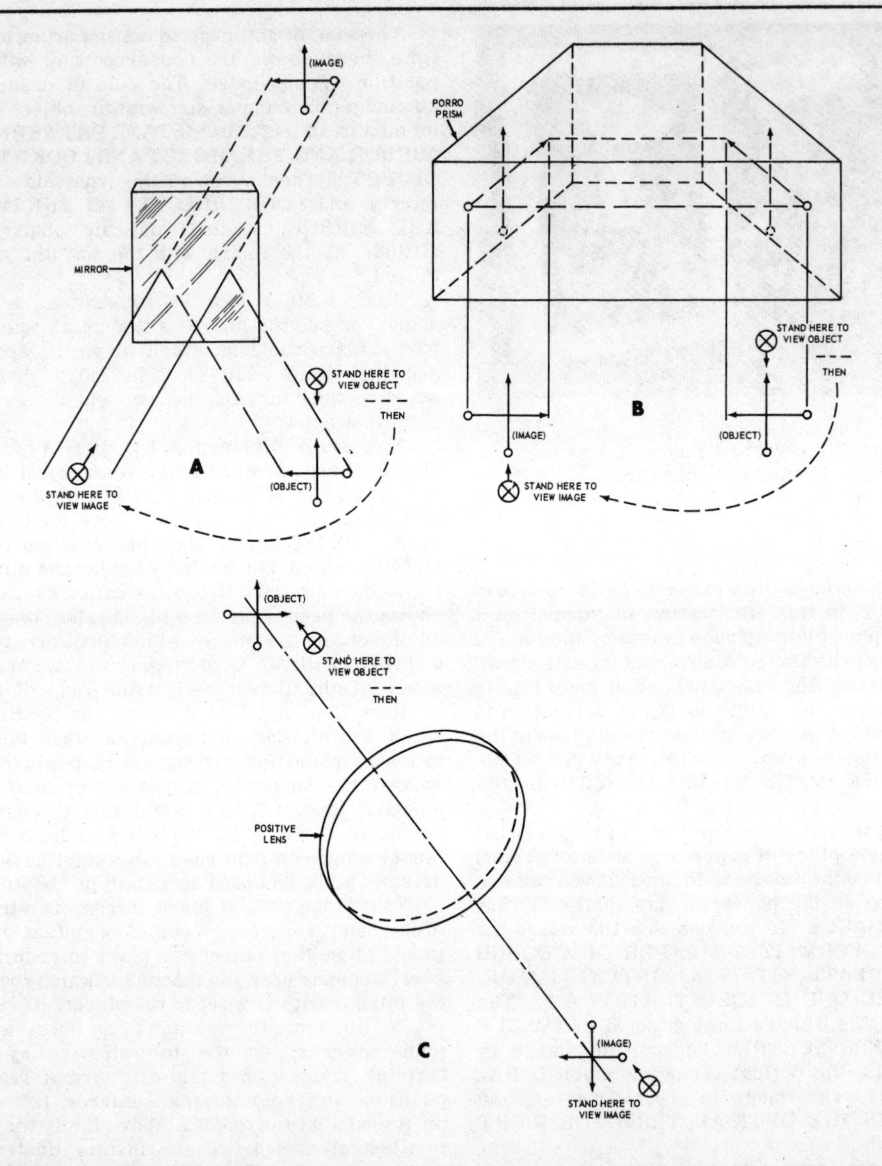

137.51

A. Position to stand for viewing an object itself, and the position to stand for viewing the image created of that object by a mirror.
B. Position to stand for observing an object and the image of the object formed by a Porro prism.
C. Position to stand for viewing an object, and the position to stand for viewing the image of that object created by a positive lens.

Figure 4-3.—Viewing objects and images created of them by optical elements.

The mirror, therefore, inverted F in its natural position but did not revert or turn it left to right.

Now stand the mirror on its edge along the right side of the same F and look at its reflection in the mirror. This time the image looks like the one below B in the illustration. It is erect but reverted. Observe that the mirror did not turn the F upside-down.

Thus far you have learned how to make an inverted or reverted image of the F in the natural position (A, fig. 4-4). How can you make AN INVERTED AND REVERTED IMAGE OF F? Place the mirror at the top of D in the illustration and observe the image (F, inverted and reverted) formed in it.

IMAGES FORMED BY PLANE MIRRORS

When you see the image of an object in a mirror, the image appears to be located at a distance behind the reflecting surface (equal to the distance of the object from the front of the mirror), as indicated in illustration 4-2. The image is erect and reverted, which is true for all single reflections from plane mirrors, held in the vertical plane.

TRACING LIGHT RAYS TO A PLANE MIRROR

In a dark room, the image of a tiny point of light viewed in a mirror appears to be located behind the mirror and on the other side of the room from where it actually is. See figure 4-5. The observer sees along the path of the reflected ray to the point where the incident ray is reflected by the mirror (eye A, fig. 4-5). His line of sight is extended in his mind in a direct line through and beyond the mirror. The apparent position of the point of light in the mirror is located directly across the room from the light source and at the same distance behind the mirror as the light source is in front of the mirror.

As long as the observer can see the reflection of the point of light in the mirror, regardless of his location in the room, its apparent position is unchanged. Observe the line of sight of eye B in illustration 4-5. The source of light (object) is reflected, and the apparent position of its reflection is changed only when the position of the object or the mirror is changed.

If the point of light (source) is replaced by a letter covered with luminous paint (F, fig 4-6), light from every point on the letter sends out incident rays which are reflected by the mirror. Each incident ray and/or reflected ray obeys the laws of reflection and their paths can therefore be plotted accordingly. The entire image formed by a combination of an infinite number of images of individual points of light is consequently reflected to the eye of the observer. As the observer looks along the paths of the reflected rays, he sees the image formed by the points of light (seemingly back of the mirror and in an erect, reverted position).

IMAGE TRANSMISSION BY PLANE MIRRORS

A single mirror can be so mounted that it will reflect light (image) for a practical purpose. An adjustable mirror on a car fender is a good example of such reflection of an image. If the image cannot be reflected satisfactorily with a single mirror, a second mirror can be so placed that it will reflect light from the first mirror and retransmit it. Illustration 4-7 shows how mirrors can be arranged so that they will transmit an image.

The two mirrors shown in the illustration are placed (mounted) together in such manner that the angle they form is 90°, as illustrated. Light from an object (F) strikes the reflecting surface of one mirror, after which light rays from every point on F are reflected by the first mirror to the second mirror, which reflects them again in rays parallel to the original rays (incident rays to the face of the first mirror). The image reflected by the two mirrors is therefore reflected a total of 180 degrees.

Because the two mirrors are so mounted that they form a 90° angle at their point of contact, the image transmitted by them by reflection in the horizontal plane is not inverted or reverted but ERECT AND NORMAL. The illustration shows the rear of the object and the front of the image. When reflection takes place in the vertical plane, the IMAGE IS INVERTED BUT NOT REVERTED.

NOTE: Review the information given earlier in this chapter concerning the comparison of images with their objects.

SPHERICAL MIRRORS

You perhaps have been at an amusement park where a building designated as FUN HOUSE

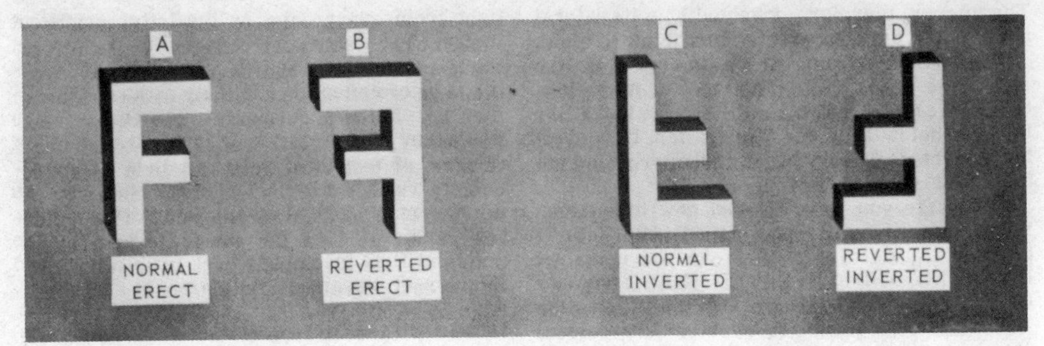

137.52

Figure 4-4.—Positions of images of the letter F created by a small mirror.

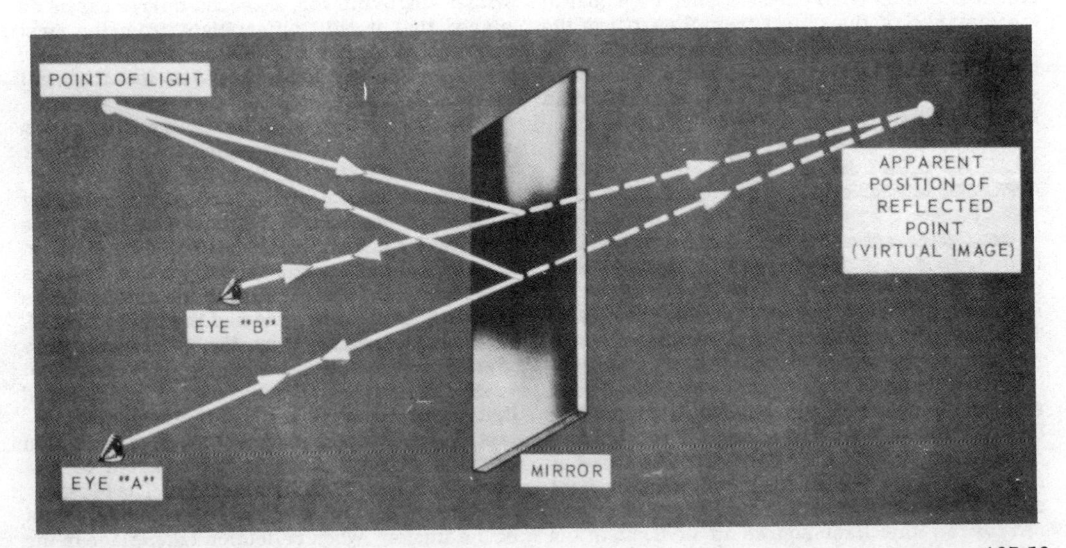

137.53

Figure 4-5.—Apparent position of a virtual image formed by a plane mirror.

had curved mirrors used to make you look ridiculously tall or disgustingly fat. Convex rear-view mirrors are also used on automobiles and trucks to give the drivers a wide view (field of vision).

A curved mirror either increases or decreases a wave front and changes its curvature. Such a mirror is called a SPHERICAL MIRROR (outside, convex mirror; inside, concave mirror).

CONCAVE SPHERICAL MIRRORS

It is important at this time that you learn the procedure for constructing a concave mirror. Refer to illustration 4-8 as frequently as

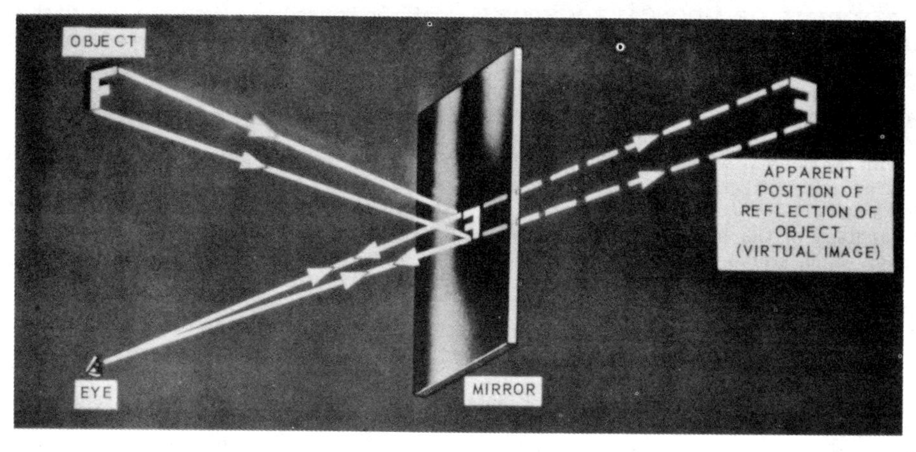

137.54
Figure 4-6.—Apparent position of an object reflected by a plane mirror.

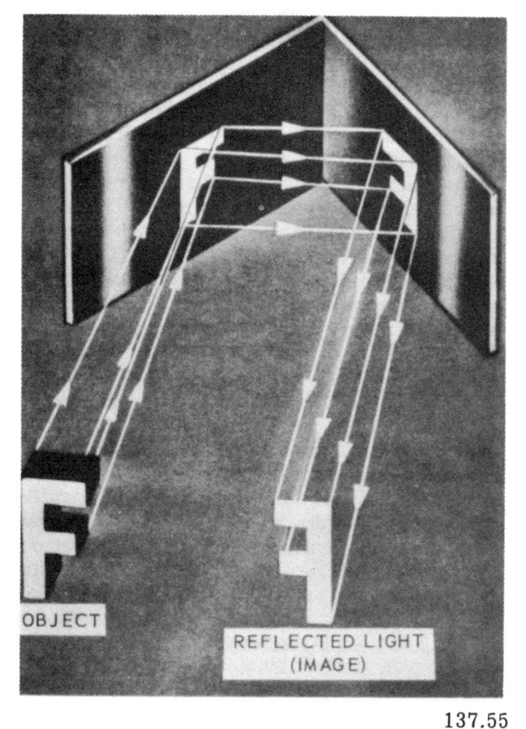

137.55
Figure 4-7.—Image produced by two plane mirrors placed at right angles.

necessary during your study of the following discussion.

Construction of a Concave Mirror

The shape of the curvature of a spherical mirror varies in accordance with the purpose for which it is intended. The procedure for making one must therefore be made accurately in accordance with a specific formula.

Begin the construction by measuring the length of the radius of a circle which will produce the desired curvature of the surface of the mirror. Line CV in figure 4-8 represents the radius of the size of a circle necessary to produce the reflecting surface of the mirror you are constructing. Draw this line after you make the circle with a compass. Point C, where you placed the metal point of the compass, IS THE CENTER OF CURVATURE OF THE SPHERE OF WHICH THE SURFACE OF THE MIRROR IS A PART. Line CV (Radius of the circle) IS THE OPTICAL AXIS OF THE MIRROR.

In order to locate the focal point of the concave mirror you just constructed, bisect line CV, represented by F (focal point) in the illustration. The focal point of a concave mirror is halfway between the center of curvature and the vertex (V) of the mirror. The focal point, or PRINCIPAL FOCUS, IS THE POINT TO WHICH PARALLEL RAYS ARE REFLECTED WHEN THEY STRIKE THE SURFACE OF THE MIRROR.

THE NORMAL OF A CONCAVE MIRROR IS A RADIUS drawn from the CENTER OF CURVATURE to the point of contact OF THE INCIDENT RAY ON THE SURFACE OF THE MIRROR. Observe that the angles between an incident ray of light parallel with the optical axis form an angle with the normal which is equal to the angle formed by the reflected ray and the normal (angles a and a').

Regardless of the number of parallel incident rays which strike the surface of a concave mirror, their reflected rays always converge at the principal focus (focal point). Observe that angle b equals angle b'. As you know, the angle of reflection (b') equals the angle of incidence (b). These angles are measured FROM THE REFLECTED RAY TO THE NORMAL, and FROM THE INCIDENT RAY TO THE NORMAL.

The normal is erected perpendicular to the surface of the mirror by drawing a straight, dotted line from the center of curvature to the point of contact of the incident ray.

Tracing Light Rays to a Concave Mirror

To learn how the law of reflection applies to a concave mirror, study illustration 4-9. The center of curvature of this mirror is in front. Note also the PRINCIPAL FOCAL POINT where the reflected rays converge. If imaginary lines are run from this center to the points of incidence of the incident rays, they indicate the NORMALS of individual light rays. Observe the N's on the edge of the lens. When these lines are drawn, the reflected rays can be so plotted that each forms an angle of reflection equal to the angle of incidence of the corresponding ray.

When diverging rays of light strike a concave mirror, they come together or converge; but the rays of light reflected from a concave mirror are more convergent than the incident rays.

The outer surface of a concave mirror is a part of the arc of a sphere, and the center of this sphere is the CENTER OF CURVATURE OF THE MIRROR. The distance from the center of curvature to the surface of the mirror is the RADIUS OF CURVATURE. Take another look at illustration 4-9. Note that the focal point is exactly halfway between the center of curvature and the surface of the mirror.

If you place a small source of light at the focal point of a concave mirror, the light which strikes the mirror is reflected in a narrow beam of parallel rays. For this reason, a curved mirror is used as a reflector in a flashlight, or in a searchlight which throws an intense beam of light.

Image Formation by a Concave Mirror

Refer now to figure 4-10, which shows how rays of light from an object form an image when they are reflected from the surface of a concave mirror. The object (located between the center of curvature and infinity), arrow AB, actually transmits billions of rays of light in all directions; but for our purpose, a few rays of light are sufficient to give you a general understanding of image formation by a concave mirror.

Ray BH travels parallel to the axis (OV) and strikes the surface of the mirror at H, from which point it is reflected through the focal point (F). Ray BCK, drawn through the center of curvature (C), intersects the reflected ray from ray BH at E. Since ray BCK is drawn through the center of curvature, it coincides with the normal to the mirror and is therefore reflected back in the same direction. Where the reflected rays of ray BH and ray BCK intersect (E) is the location of the image of the top of the arrow. In the same manner, the reflected rays AK and ACH give the location of the bottom of the image at D. Note that this image is located between C and F.

Because this image is formed by an actual intersection of reflected rays of light, it is considered a real image (smaller than the object, normal, and inverted).

The formation of images, by concave mirrors may be grouped by cases as explained next, with the object at varying distances from the surface of a mirror.

OBJECT AT INFINITY.—When an object is at infinity (fig. 4-11), light rays from it are diverging in all directions; but before they arrive at the mirror, they have become so nearly parallel that we may say they are parallel. The surface of the mirror converges the rays of light to the focal point to form a real, normal, and inverted image of the object (diminished in size).

OBJECT BETWEEN INFINITY AND CENTER OF CURVATURE.—When an object is placed at some point between infinity and the center of curvature of the mirror, the image is real, normal, inverted, and diminished in size; and it is located between the center of curvature and the

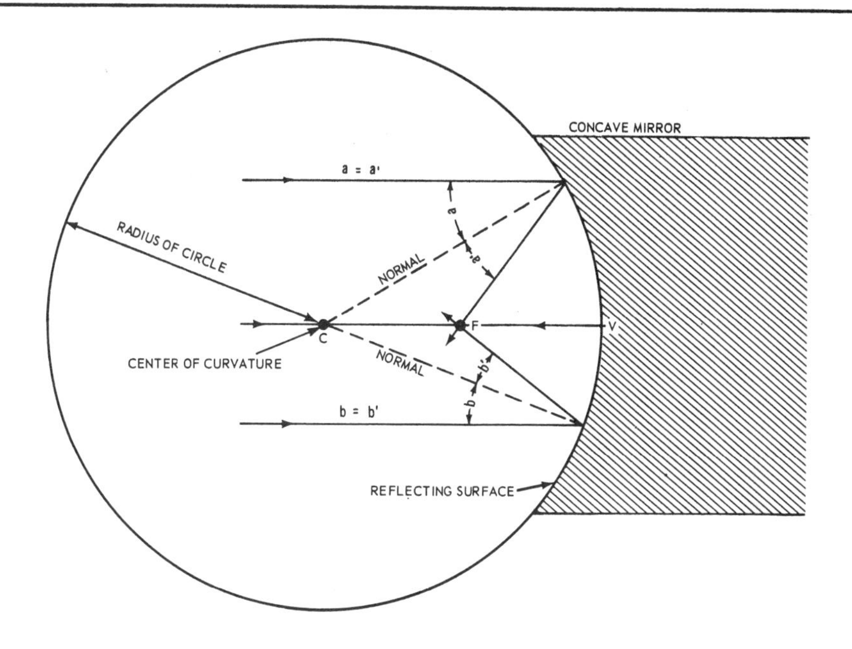

137.56

Figure 4-8.—Construction of a concave mirror.

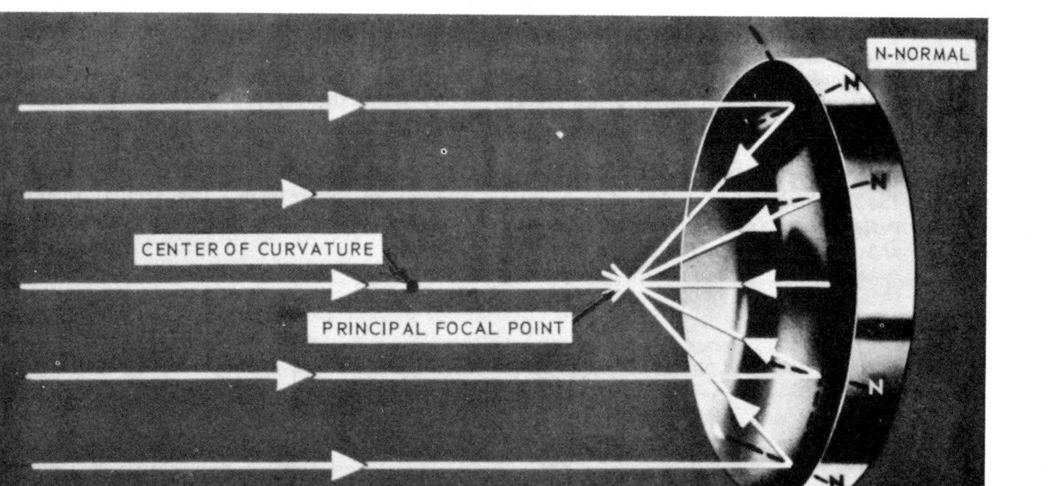

137.57

Figure 4-9.—Reflection of parallel rays of light from a concave mirror.

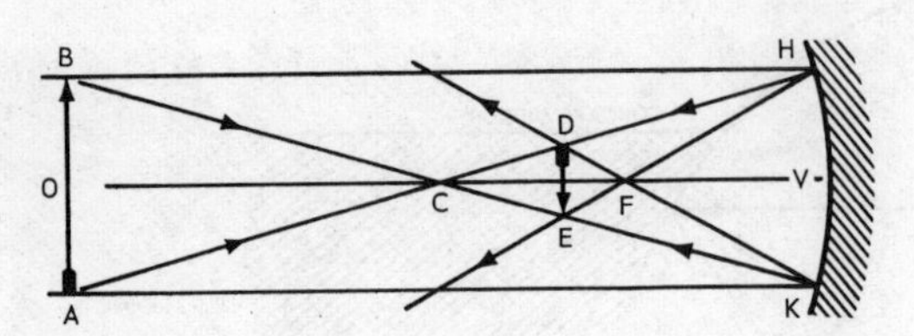

137.58

Figure 4-10.—Image formation of an object by reflected rays of light from a concave mirror.

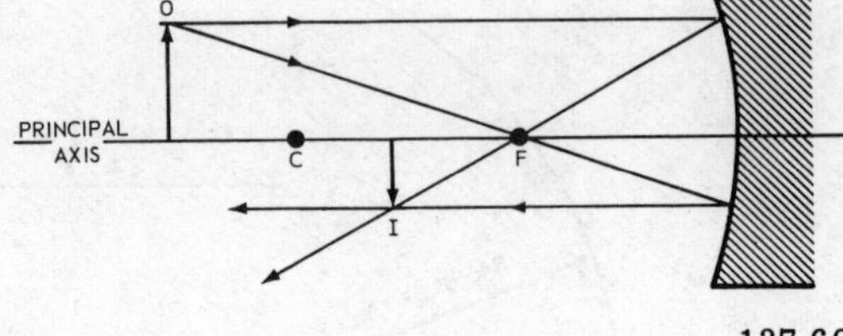

137.60

Figure 4-12.—Position of image formed by a concave mirror when the object is between infinity and center of curvature.

focal point of the mirror, as shown in figure 4-12. NOTE: In this case, the image IS LARGER THAN the image formed in illustration 4-11, but it is still smaller than the actual object.

OBJECT AT CENTER OF CURVATURE.—If an object is located at the center of curvature of a concave mirror, the mirror forms a real, inverted, normal image of the same size as the object, at the center of curvature. See figure 4-13.

OBJECT BETWEEN CENTER OF CURVATURE AND FOCAL POINT.—When an object is placed between the center of curvature and the focal point of a mirror, the image formed by the mirror is real, inverted, normal, and enlarged (larger than the object); and it is located between the center of curvature and infinity, as shown in illustration 4-14.

OBJECT AT FOCAL POINT.—If an object is placed at the focal point (fig. 4-15) of a concave mirror, reflected rays from the mirror are parallel and a real image IS NOT FORMED. If an eye in the front area before the mirror catches the reflected parallel rays, they appear to be coming from infinity behind the mirror; and the eye sees a virtual, erect, reverted, and enlarged image at infinity.

OBJECT BETWEEN FOCAL POINT AND REFLECTING SURFACE.—When an object is placed between the focal point and the reflecting surface of a concave mirror, the reflected rays are divergent. Study illustration 4-16. As seen by an eye in front of the mirror, the rays appear to meet a short distance behind the mirror to form a virtual, erect, reverted, and enlarged image of the object. NOTE: The closer the object is moved toward the mirror, the larger is the image formed; and the image moves farther away from the mirror until the object reaches the principal focus. After passing this point, the image changes from REAL to VIRTUAL and decreases in size as the object approaches the surface of the mirror. The virtual image of a concave mirror is never smaller than the object.

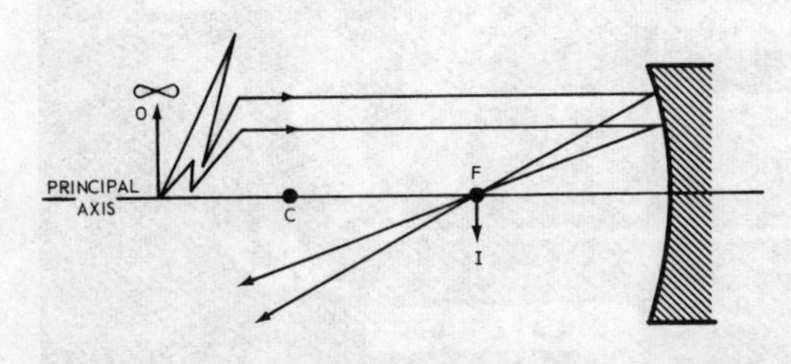

137.59

Figure 4-11.—Position of image formed by a concave mirror when the object is at infinity.

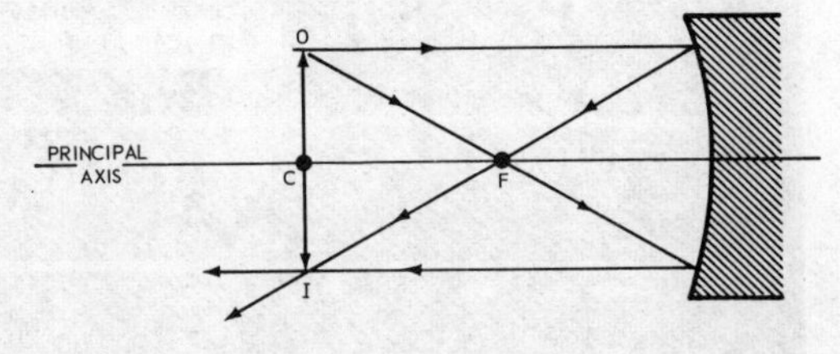

137.61

Figure 4-13.—Position of image formed by a concave mirror when the object is at the center of curvature.

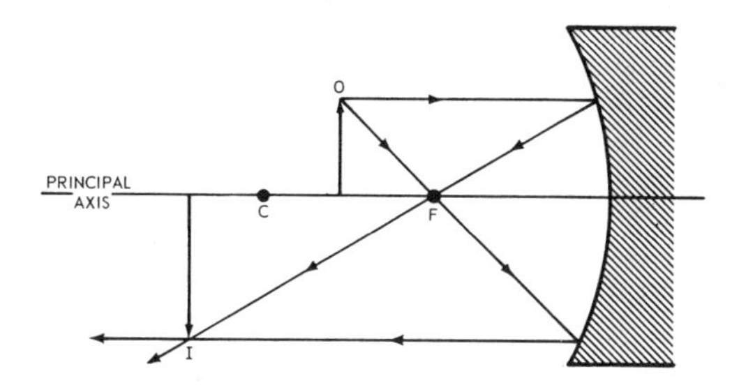

137.62

Figure 4-14.—Position of image formed by a concave mirror when the object is between the center of curvature and the focal point.

Calculation of Image Position

You can determine the position of an image by using the following equation, which is equally applicable for finding the positions of images created by both curved mirrors and lenses:

$$\frac{1}{F} = \frac{1}{D_o} + \frac{1}{D_i}$$

Do, read d-sub-o = distance of object from surface of mirror

D_i, read d-sub-i = distance of image from surface of mirror

F = focal length of the mirror

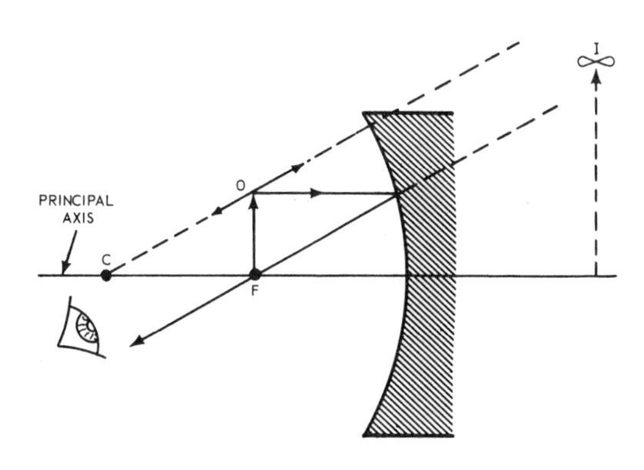

137.63

Figure 4-15.—Position of an image formed by a concave mirror when the object is at the focal point.

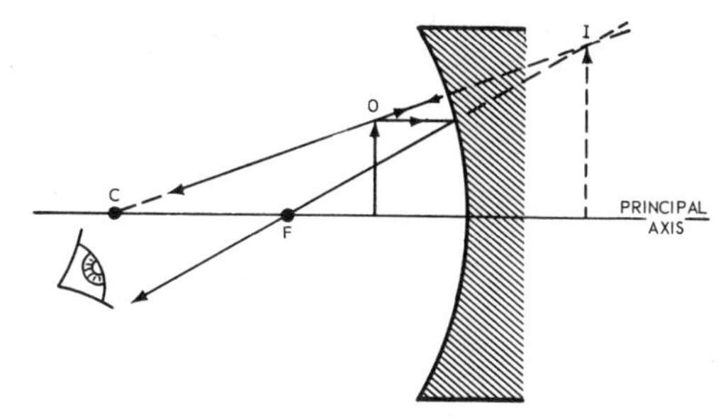

137.64

Figure 4-16.—Position of an image formed by a concave mirror when the object is between the focal point and the reflecting surface.

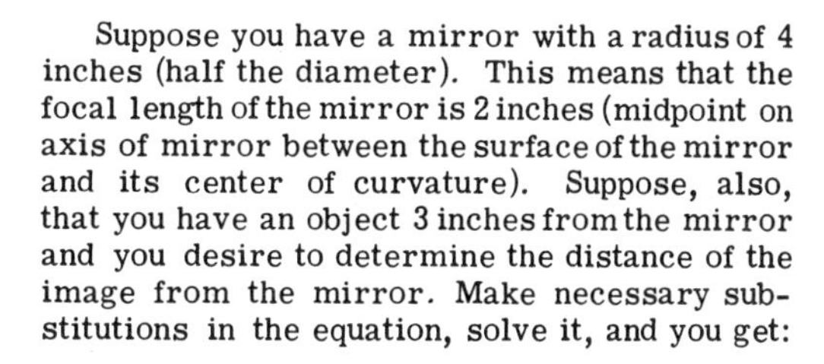

Suppose you have a mirror with a radius of 4 inches (half the diameter). This means that the focal length of the mirror is 2 inches (midpoint on axis of mirror between the surface of the mirror and its center of curvature). Suppose, also, that you have an object 3 inches from the mirror and you desire to determine the distance of the image from the mirror. Make necessary substitutions in the equation, solve it, and you get:

$$\frac{1}{D_o} + \frac{1}{D_i} = \frac{F}{F}$$

$$\frac{1}{F} = \frac{1}{D_o} + \frac{1}{D_i}$$

$$\frac{1}{2} = \frac{1}{3} + \frac{1}{D_i}$$ NOTE: 2 = F, focal length; 3 = D_o, distance of object

The lowest common denominator into which you can divide 3, D_i, and 2 is $6D_i$; so when you divide the denominator by 3, d_i, and 2, respectively, and put the quotient of each in the formula and solve for D_i, you get

$$3D_i = 2D_i + 6$$

$$3D_i - 2D_i = 6$$

D_i = 6 inches, distance of image from mirror

You can also use the formula just explained to ascertain the position of a virtual image. If an object is a distance of 1 inch from a concave mirror with a focal length of 2 inches, for example, you can locate the image by substituting the figures (1 & 2) in the proper places in the formula, as follows:

$$\frac{1}{F} = \frac{1}{D_o} + \frac{1}{D_i}$$

$$\frac{1}{2} = \frac{1}{1} + \frac{1}{D_i}$$

$$D_i = 2D_i + 2$$

$$D_i = 2D_i = 2$$

$$-D_i = 2$$

$$D_i = -2$$

The answer to this equation (-2) is -2 inches in front of the mirror, which means that the image (virtual) is 2 inches BEHIND THE MIRROR.

Magnification of the Image

Magnification is THE SIZE OF THE IMAGE DIVIDED BY THE SIZE OF THE OBJECT, whose distance from the lens is a factor in the amount of magnification produced. When you look through a reading glass, you observe the magnifying power of the glass. If when the glass is close to the eye(s) the image created by the glass looks five times the object's size, the magnifying power of the glass is five. In other words, the lens is five-power (5x).

The lens in this magnifying glass is a five-power lens because the bigger the image, the farther away it is from the lens; and a large object looks small when it is far away, as is true for Mars and the sun. This statement is true, even though the SIZE OF A VIRTUAL IMAGE MAY VARY FROM THE SIZE OF THE OBJECT TO INFINITY. For this reason, when you hold a reading glass close to your eyes and look through it, the size of the image changes only slightly when you move the object.

NOTE: DO NOT CONFUSE MAGNIFICATION WITH MAGNIFYING POWER. Magnification of an image is a variable amount controlled by the distance of the object to the lens and the focal length of the lens. You must know the distance of the object and also the distance of the image before you can determine magnification (size of the object and size of the image). Magnifying power, on the other hand, IS A CONSTANT POWER WHICH NEVER CHANGES.

You can determine the magnification of an image created by a mirror by dividing the size of the image by the size of the object. The formula for determining the amount of magnification of an image is:

$$MAG = \frac{S_i}{S_o} = \frac{D_i}{D_o}$$

This formula shows that the size of the image (S_i) divided by the size of the object (S_o) equals the distance of the image (D_i) divided by the distance of the object (D_o). Magnification of the image, therefore, is equal to:

$$MAG = \frac{S_i}{S_o}, \text{ or } \frac{D_i}{D_o}$$

The following examples illustrate the use of the formula for determining the magnification of an image by a concave mirror.

Suppose the distance of an object from a mirror is 3 inches and the distance of the image from the mirror is 6 inches. By what amount is the image magnified? Substitute 3 and 6 in the formula and solve for MAG (magnification).

$$MAG \; \frac{D_i}{D_o} = \frac{6}{3} = 2, \text{ magnification of image}$$

The answer to the equation (problem) indicates that magnification of the image is TWICE THE SIZE OF THE OBJECT.

Suppose, next, that you have a virtual image 2 inches in front of a mirror and an object 1 inch in front of a mirror and you wish to determine the size of the virtual image. Make proper substitutions in the equation and solve for MAG:

$$MAG = \frac{D_i}{D_o} = \frac{2}{1} = 2, \text{ magnification of image}$$

When an object is at the center of curvature of a mirror, you can plot the distance of the image with the formula in the following manner:

$$\frac{1}{F} = \frac{1}{D_o} + \frac{1}{D_i} \text{ (focal length)}$$

In this instance, the distance of the object is equivalent to the radius of curvature. Because the focal point is halfway between the center of curvature and the surface of the mirror, the radius of curvature is 2F; and the distance of the object is also 2F.

If you now substitute 2F in the equation in the D_o position and solve for D_i, you get:

$$\frac{1}{F} = \frac{1}{2F} + \frac{1}{D_i}$$

$$2D_i = D_i = 2F$$

$$2D_i = D_i = 2F$$

$$D_i = 2F$$

The distance of the image from the mirror is 2F, which means that the image is also at the center of curvature.

To determine the amount of magnification of the image, make proper substitutions in the equation and solve, as follows:

$$MAG = \frac{D_i}{D_o} = \frac{2F}{2F} = 1$$

Your answer shows that the image and the object in this example are of the same size, as they should be.

PARABOLIC MIRRORS

If a very small luminous source is located at the principal point of focus, light rays are almost parallel after they reflect from a mirror—provided the curvature of the mirror is VERY SLIGHT. The rays actually have a slight convergence, particularly those reflected near the edges of the mirror (fig. 4-17). For this reason, a parabolic mirror is used whenever parallel reflected rays are desired. Study illustration 4-18.

A parabolic mirror is a concave mirror with the form of a special geometrical surface—a paraboloid of revolution. Light rays which emanate from a small source at the focal point of a parabolic mirror are parallel after they reflect from its surface.

The source of light (usually a filament or arc) is located in the principal point of focus and the rays diverge, because THERE IS NO TRUE POINT SOURCE. All rays which strike the parabolic mirror (except those which are diffused or scattered) reflect from the mirror toward the focal point and nearly parallel with each other, thereby providing for the formation of a powerful beam of light which diverges only slightly. Most searchlights have parabolic mirrors, as do automobile headlights.

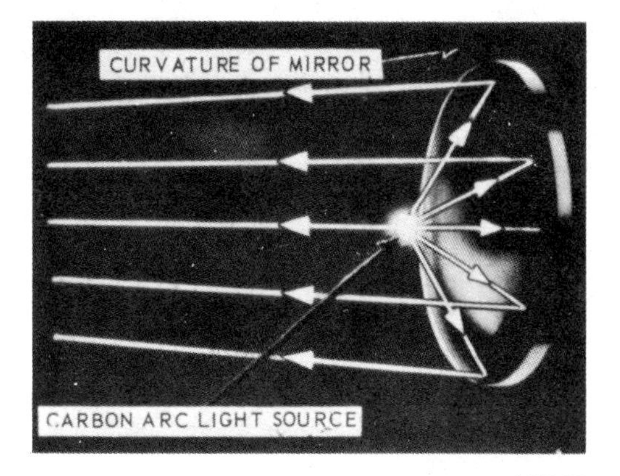

137.65
Figure 4-17.—Reflection of light rays by a spherical mirror.

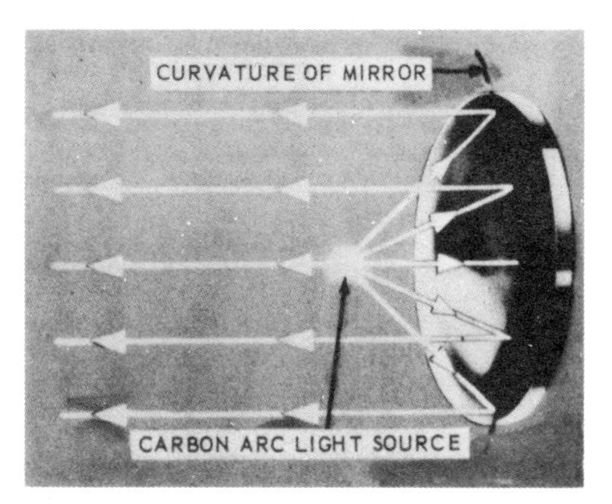

137.66
Figure 4-18.—Reflection of light rays by a parabolic mirror.

Spherical mirrors are generally used for ordinary purposes because the grinding process is easy; but other types of mirrors are used for special purposes. A CYLINDRICAL MIRROR is part of a cylinder—not part of a sphere. When parallel rays reflect from a concave spherical reflecting surface, they form a CONE-SHAPED BEAM which converges to a point. When parallel rays of light reflect from a concave CYLINDRICAL reflecting surface, they form a WEDGE-SHAPED BEAM which converges to a line; and when light converges to a line, it is called ASTIGMATIZED LIGHT. Think of a CYLINDRICAL MIRROR as a silvered portion of the inside of an ordinary tin can. If you silver the inside curved surface of the can and then split the can lengthwise, you have an example of a concave cylindrical mirror.

Construction of a Convex Mirror

Illustration 4-19 shows the procedure for constructing a convex mirror. Note the angles of incidence and the angles of reflection formed by the parallel rays of light which strike the mirror. These angles are equal, as you know; and the normals from the center of curvature of the mirror bisect these two angles. Observe the radius of the circle; all normals to the face of the mirror are actually radii of the circle. THE PRINCIPAL FOCUS OF A CONCAVE MIRROR IS REAL; THE PRINCIPAL FOCUS OF A CONVEX MIRROR IS VIRTUAL.

Tracing Light Rays to a Convex Mirror

Illustration 4-17 shows how rays of light strike and reflect from the surface of a convex mirror. Note the angles formed by reflected rays, the radius to the center of curvature, the normals in relation to the radii, and the position of the virtual image formed by the extensions behind the mirror of the reflected rays of light. Now take another look at illustration 4-17, which shows rays of light striking different portions of the surface of a convex spherical mirror. The principle of reversibility is illustrated by the central ray.

The law of reflection holds true for all surfaces--convex, concave, and plane. The amount of reflected light from curved surfaces depends upon the distance of the light source and the amount of curvature of the reflecting surface.

If light from a distance source such as the sun strikes a convex mirror, the rays are reflected in a convergent manner. The reason for this can be determined by plotting the angles of reflection of individual rays in relation to their angles of incidence and the normals for each light ray. In this case, the normal for each ray is an imaginary line drawn FROM THE CENTER OF CURVATURE OF THE MIRROR TO THE POINT OF INCIDENCE OF THE RAY. The angle of reflection, of course, is equal to the angle of incidence for each ray.

When a light source is close to a mirror, the rays are divergent when they strike the mirror and are also reflected in a divergent manner. In this case, the rays are reflected at different angles from parallel rays of light which strike the mirror, but always equal to the angle of incidence.

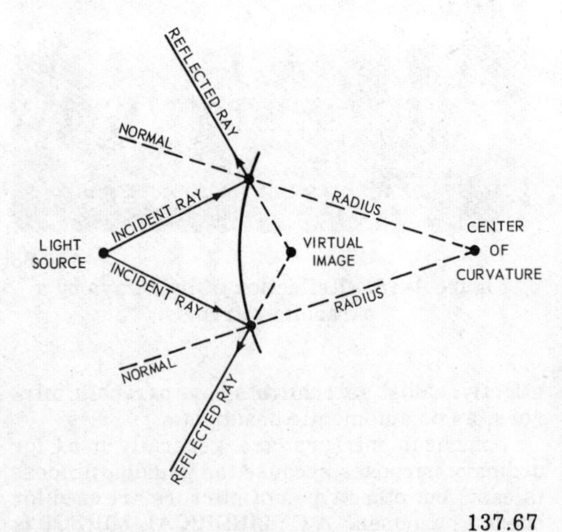

137.67

Figure 4-19.—Procedure for constructing a convex mirror.

Images Formed by Convex Mirrors

Study illustration 4-20, which shows three objects (arrows 0_1, 0_2, and 0_3) of the same size but of different distances from a convex spherical mirror. These arrows are of the same height because they are constructed between a line parallel with the optical axis. The ray of light which passes along the tips of the three arrows strikes the mirror and is reflected in the manner indicated by arrow AF. The dotted extension of this line behind the mirror contacts the optical axis at the focal point.

Rays of light from the three arrow heads to the CENTER OF CURVATURE OF THE MIRROR ARE SECONDARY AXES, and the image formed by each lies between them and the optical axis. (Any straight line which passes through the

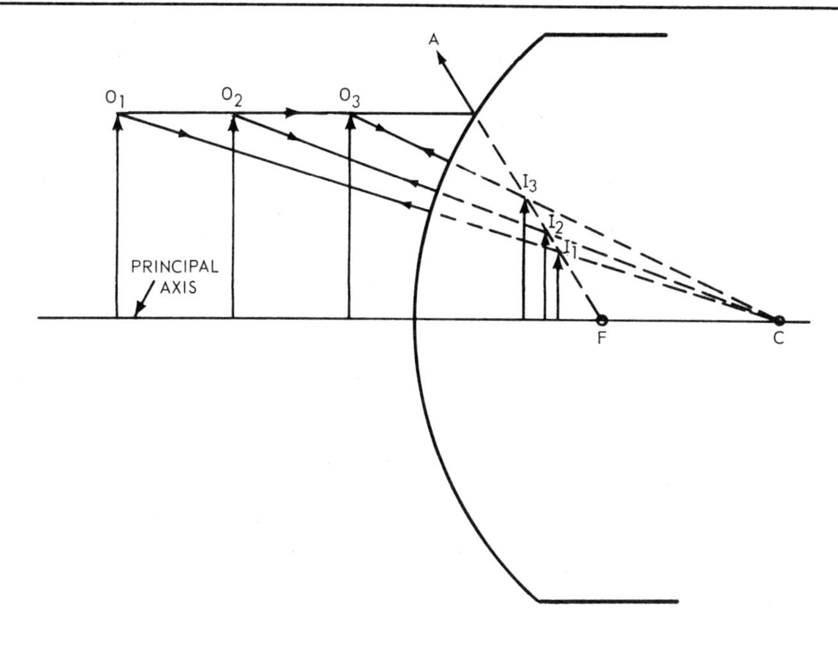

137.68

Figure 4-20.—Image formation by a convex mirror.

center of curvature of a mirror to its surface is called a normal.) Object 0_1 creates I_1, and so forth. Observe that the size of the image is larger when the object which formed it is moved nearer to the mirror, but an image can NEVER BECOME AS LARGE AS ITS OBJECT.

As you can see, these images are virtual, erect, reduced in size, and located behind the mirror between the principal focus and the vertex.

The radius of curvature of a convex mirror is negative and the image is always virtual. The focal length (F) and the image distance (D_i) are therefore negative quantities, as you can determine by using the mirror formula.

Suppose an object 10 cm in front of a convex spherical mirror forms an image 5 cm to the rear of the mirror. How can you find the focal length of the mirror? Solution of the problem is as follows:

$$\frac{1}{F} = \frac{1}{D_o} + \frac{1}{D_i}$$

$$\frac{1}{F} = \frac{1}{10} + \frac{1}{-5}$$

$$10 = IF + (-2F)$$

$$10 = IF$$

$$-10 = F$$

CHAPTER 5

LENSES AND PRISMS

This chapter contains information pertaining to the following, all of which you must know before you can qualify for advancement in rating to Opticalman 3:

1. Effect of light rays produced by divergent and convergent lenses—different types of lenses and their refractive and reflective capabilities.
2. Penta, right-angled, Rhomboid, Porri, and Amici (roof) prisms—how they are constructed and their function in optical instruments.
3. Methods of determining focal lengths, dioptric strength, and magnifying power of positive lenses, using the metric and English systems of measurement.

This chapter also takes into consideration various formulas used for determining the dioptric strength of lenses and prisms, which you must know before you can qualify for advancement in rating to Opticalman 2.

The sequence followed in discussing lenses and prisms in this chapter gives you the knowledge necessary for you to understand current and subsequent discussion of everything pertaining to these optical elements; that is, their function and use in optical instruments.

THIN LENSES

A thin lens is one of the types of lenses (thin, thick, and compound) used in optical instruments, and its thickness is small enough to make it (thickness) UNIMPORTANT IN MEASURING DISTANCES TO IMAGES AND OBJECTS. All lenses discussed in this training course are considered to be thin, UNLESS THEY ARE SPECIFIED AS THICK.

Some types of thin lenses (also called simple lenses) are shown in figure 5-1. Note the sizes and shapes of these lenses, and observe also that there are two groups: (1) CONVERGENT, which converge light rays to a single point; and (2) DIVERGENT, which diverge light rays and does not bring them to a single point.

One important rule to remember about lenses is this: READ THE SURFACES OF LENSES FROM LEFT TO RIGHT. Observe, for example, the plano-convex (bulging face) and the plano-concave (cut-out face) lenses in figure 5-1. In each case, the left face of the lenses is plano or plane (straight).

REFRACTION IN THIN LENSES

Refer now to illustration 5-2, which shows the FOUR PRINCIPAL RAYS of light which pass through any lens, THICK or THIN. When these light rays pass through a lens, they ALWAYS follow the rules which pertain to each. Line XY in the illustration is the optical axis (sometimes called principal axis) of the lens. The optical axis passes through the center of a lens and perpendicular to its principal plane (illustrated). Lens terminology is explained in detail later in this chapter.

LIGHT RAY A.—An incident ray (one entering a medium) passes through the optical center (0, fig. 5-2) of a lens and emerges from the lens without deviation from the path it was following before entering the lens. This is true because the incident ray strikes the surface of the lens parallel to the normal. (The normal of an incident ray at any point on a lens is an imaginary line at right angles to the surface of the lens at the point where the ray enters.) When the ray reaches the second surface of the lens, it is still traveling parallel to the normal. (The normal of an emergent light ray is an imaginary line at right angles to the surface of the lens at the point where the ray emerges from the lens.)

When an "A" light ray passes through a lens at an angle to the optical axis but through the optical center, it is slightly refracted before it reaches the optical center. After it passes through the optical center and strikes the second

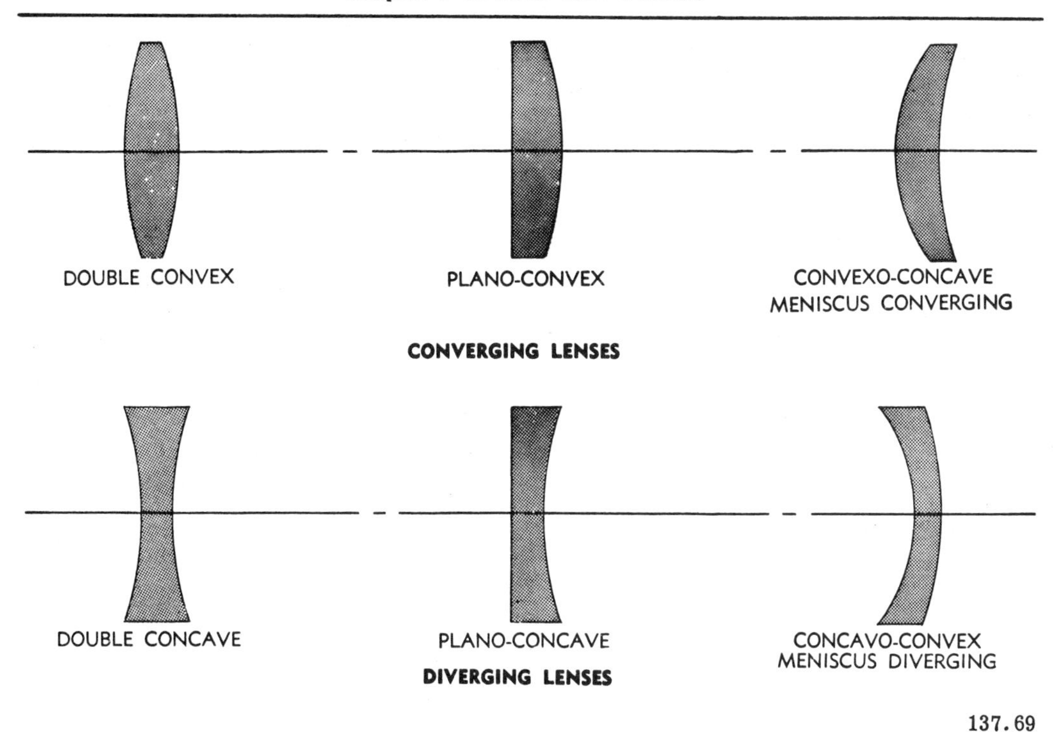

137.69

Figure 5-1.—Types of thin lenses.

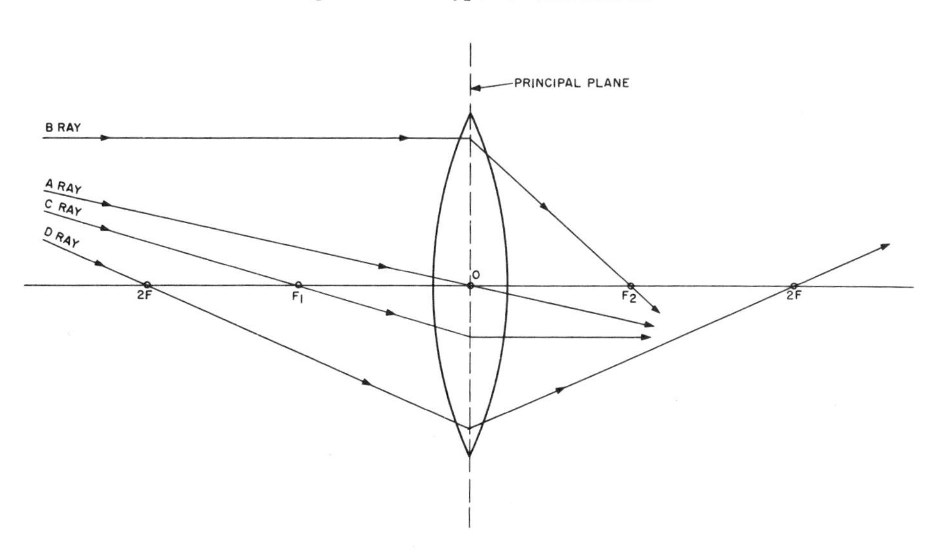

137.70

Figure 5-2.—Principal light rays.

surface, it is slightly refracted again, but at the same angle at which the incident ray struck the first surface. The emergent ray is parallel and offset to the incident ray, but it is offset so slightly that in actual theory the ray is said to have passed directly through the thin lens without refraction or deviation.

LIGHT RAY B.—Any incident light ray which travels parallel to the optical axis of a lens strikes the lens and is refracted to the principal focal point, the one behind the lens.

LIGHT RAY C.—Any ray which passes through the principal focal point and strikes the lens is refracted and emerges parallel to the optical axis. NOTE: The C ray is the opposite of ray B, because it enters the lens from the opposite edge, through the principal focal point, and does not pass through the principal focal point behind the lens, as does ray B.

LIGHT RAY D.—Any ray which passes through a point two focal lengths in front of a lens and strikes the lens is refracted and converges to a point two focal lengths behind the lens. In accordance with the Law of Reversibility, this ray (and all other rays) could be reversed in direction.

NOTE: The four principal light rays just discussed can travel to the lens in any direction or angle, as long as they follow the rules which pertain individually to them.

Observe in illustration 5-2 that refraction appears to take place in the lens at the principal plane (explained later in this chapter), but this is true for illustrative purposes only. A light ray refracts toward the normal as soon as it strikes the surface of the lens, and away from the normal as it leaves the surface of the lens.

Illustration 5-2 is important to you primarily because you can use rays A, B, C, and D to PLOT ANY IMAGE OF AN OBJECT WITH GREAT ACCURACY, provided your measurements are accurate.

Refraction in Convergent Lenses

Refer now to illustration 5-3, which shows light rays passing through two prisms of the same size and shape, placed base-to-base. Observe that the rays of light pass into the prisms and bend toward the bases of the prisms as they pass through. After the light rays emerge from the prisms they cross at the points indicated.

Observe, next, in illustration 5-4 how a convergent lens deviates light rays. When parallel rays of light strike the front surface (left) of a convergent lens, they pass through the lens and CONVERGE AT A SINGLE POINT.

A convergent lens may be thought of as two prisms (fig. 5-3) arranged so that each directs rays of light to the same point. The lens bends light rays in the same manner as a prism; but, unlike a prism, it brings the light rays to a single point. Picture a convergent lens, therefore, as two prisms with surfaces rounded into a curve.

Note that a convergent lens is THICKER IN THE MIDDLE THAN AT THE EDGES. As you observed in illustration 5-1, the two surfaces of a convergent lens may differ in shape. Both surfaces may be convex (double convex), one surface may be plane and the other convex (planoconvex), or one surface may be convex and the other concave (convexoconcave) meniscus converging, as it is termed.

If you apply the law of refraction (chapter 3) to a light ray, you can understand what happens when it passes through a convergent lens. When an incident light ray enters the top of a convergent lens (a medium more dense than air), it bends toward the normal; when the refracted ray (emergent ray) goes back into the air, it bends away from the normal.

Incident light rays which enter the bottom of a convergent lens bend away from the normal. The two sets of light rays (top and bottom) which enter a convergent lens therefore cross AFTER THEY EMERGE from the lens. If the incident rays are parallel when they enter the lens, they cross the optical axis at a single point called the focal point.

Refraction in Divergent Lenses

Take another look at the different types of simple divergent lenses shown in illustration 5-1.

Suppose that we now take two prisms like those shown in figure 5-3 and place them apex-to-apex, in the position illustrated in figure 5-6. What we do here is construct a different type of lens, a divergent lens. When rays of light strike the front surfaces (left face) of the prisms, the rays pass through in the manner illustrated, in accordance with the laws of refraction.

Observe that the light rays in the top prism refract away from the normal; whereas, the light rays which pass through the bottom prism refract toward the base, away from the normal.

If you now assume that the front and rear surfaces of these two prisms have been ground into spherical surfaces, you have a simple divergent lens. Study illustration 5-7.

Divergent lenses are always thinner in the middle than at the edges, just the opposite to convergent simple lenses. The optical center of a divergent simple lens is at the thinnest point of the lens, and the lens causes convergent light to be less converging, parallel rays to diverge, and divergent light to be more diverging.

The two surfaces of a divergent simple lens may differ in shape. Both surfaces may be concave (double concave), one surface may be plane and the other concave (planoconcave), or one surface may be concave and the other convex (concavoconvex) meniscus diverging.

To learn how the law of refraction applies to a divergent lens, study illustration 5-8. Observe the one incident ray used to illustrate the refraction of light as it passes through the top of a divergent simple lens, and the manner in which it is bent on both faces—toward the normal on the first face, away from the normal on the second face.

Light rays which pass through divergent lenses of the optical axis ALWAYS refract toward the thickest part of the lenses.

CONVERGENT LENSES

In this section we discuss the construction of convergent lenses and tracing light rays through them, lens terminology, image formation by convergent lenses, and the procedure for determining the positions of these images.

Construction of a Convergent Lens

One method for constructing a convergent lens follows. Refer to illustration 5-9 as you study the procedure. NOTE: All measurements must be accurate; all steps in the procedure must be correct.

1. Draw a straight line lengthwise through the center of a sheet of paper and parallel with the sides. This will be the optical axis of the lens.
2. Construct line PP' perpendicular with the optical axis. This will be the principal plane.
3. Measure a distance on the optical axis which gives the radius of a circle large enough to draw a lens of the thickness and size you desire, from the optical center of your lens (0) to point B.
4. Draw line AO equal to BO.
5. Find the point on the optical axis which gives you the radius of a circle whose surface passes through point B. This radius is line BC, and the CENTER OF CURVATURE OF THIS LENS' SURFACE IS AT C (center of the circle).
6. With the radius of your compass still adjusted for a circle with a radius equal to BC, locate on the optical axis point C', which is the center of curvature of another circle whose surface (part) passes through point A to form the other side of the convex lens.
7. Draw a ray of light parallel with the optical axis to the surface of the lens. Then, in accordance with the law of refraction, extend the ray through the lens and on to point F on the optical axis (where the refracted ray crosses), as shown in illustration 5-9.
8. Then draw the two normals (NORMAL, and NORMAL'). These are reference lines (at right angles with a surface or other lines) used to determine the angles of incidence and refraction.

Tracing Light Rays through a Convergent Lens

Review at this time illustration 5-5, for it shows how the laws of refraction may be applied to plot the path of any light ray through any types of lens. Then study illustration 5-10, which shows how light rays pass through a convergent lens and converge at a single point.

Millions of light rays may come from every point of light on an object, but we use in illustration 5-10 only three such rays to show how they pass through a convergent lens. As you learned previously in this chapter, the light rays which strike a convergent lens on either side of the optical axis bend toward the thickest part of the lens, and bend again toward the thickest part of the lens when they emerge from its left face. As shown in the illustration, they converge at a single point.

A light ray which passes along the optical axis through a lens does not bend, because it strikes the surfaces of the lens at and parallel to the normal.

Lens Terminology

It is important at this point that you understand the terms used to explain certain parts of lenses. Refer frequently to figure 5-11 as you study lens terminology.

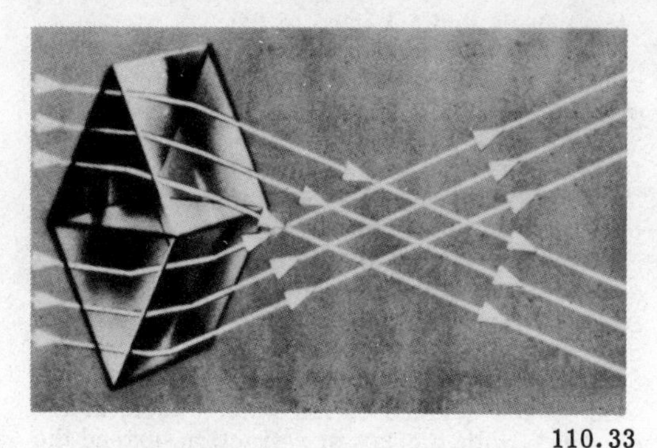

110.33
Figure 5-3.—Deviation of light rays by prisms.

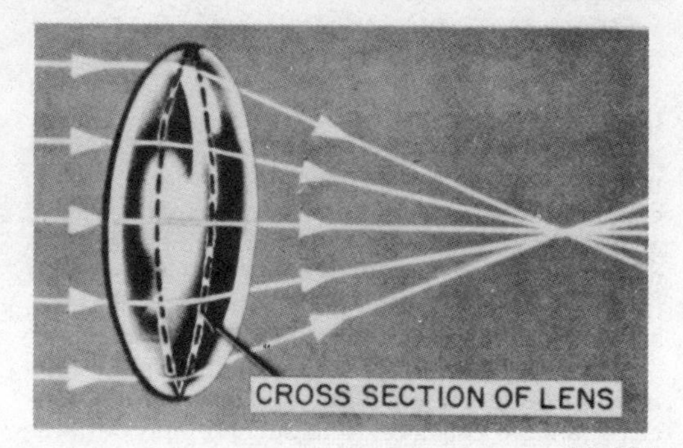

12.234
Figure 5-4.—Deviation of light rays by a convergent lens.

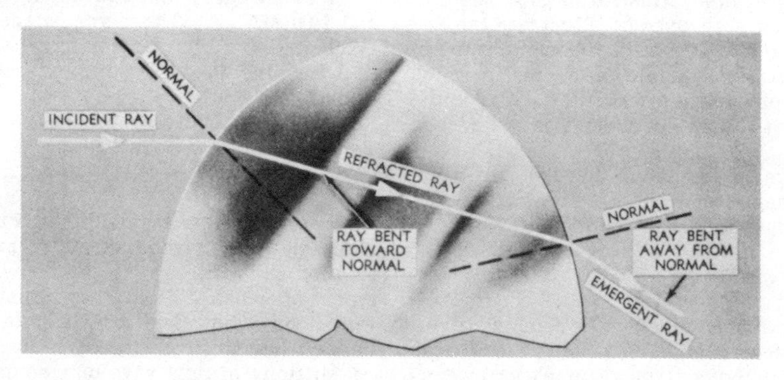

137.71
Figure 5-5.—Refraction of light rays by a convergent lens.

137.72
Figure 5-6.—Deviation of rays by two prisms placed apex-to-apex.

137.73
Figure 5-7.—Deviation of rays by a divergent lens.

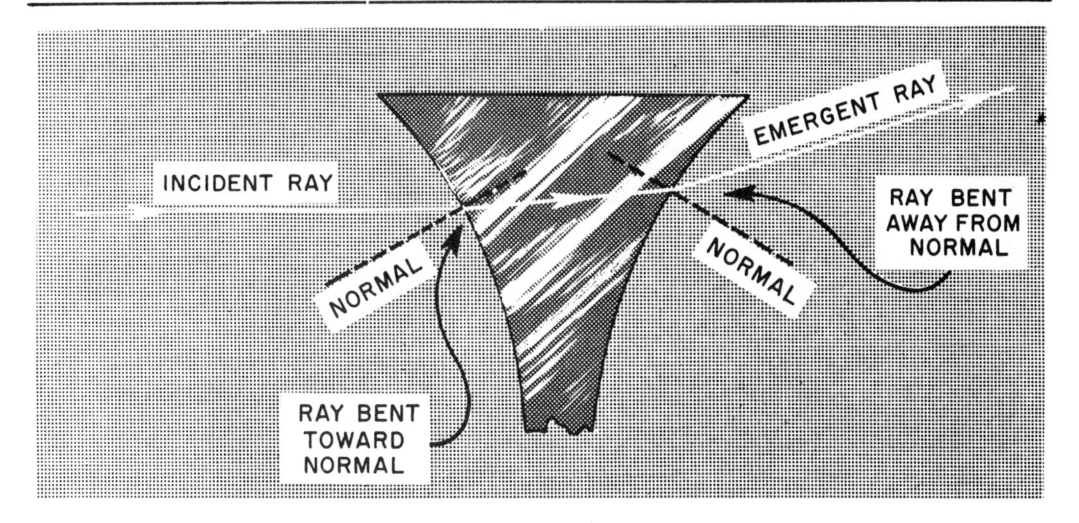

137.74

Figure 5-8.—Application of the law of refraction to a divergent lens.

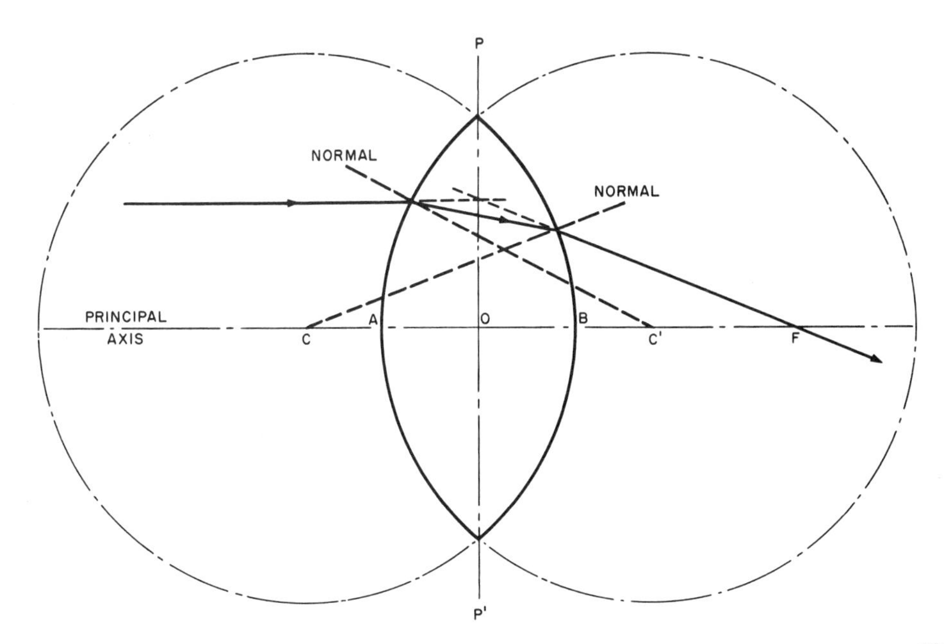

137.75

Figure 5-9.—Construction of a convex lens.

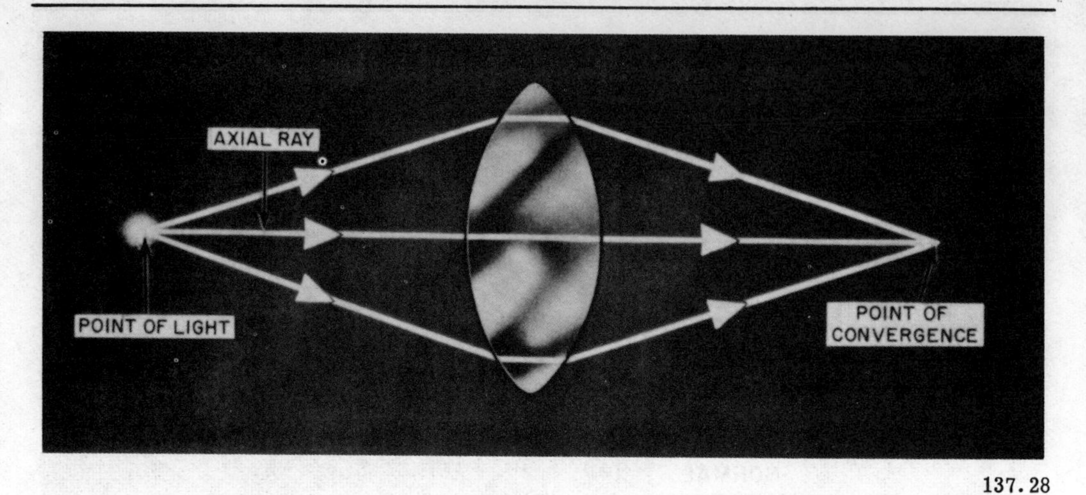

137.28

Figure 5-10.—Tracing light rays through a convergent lens.

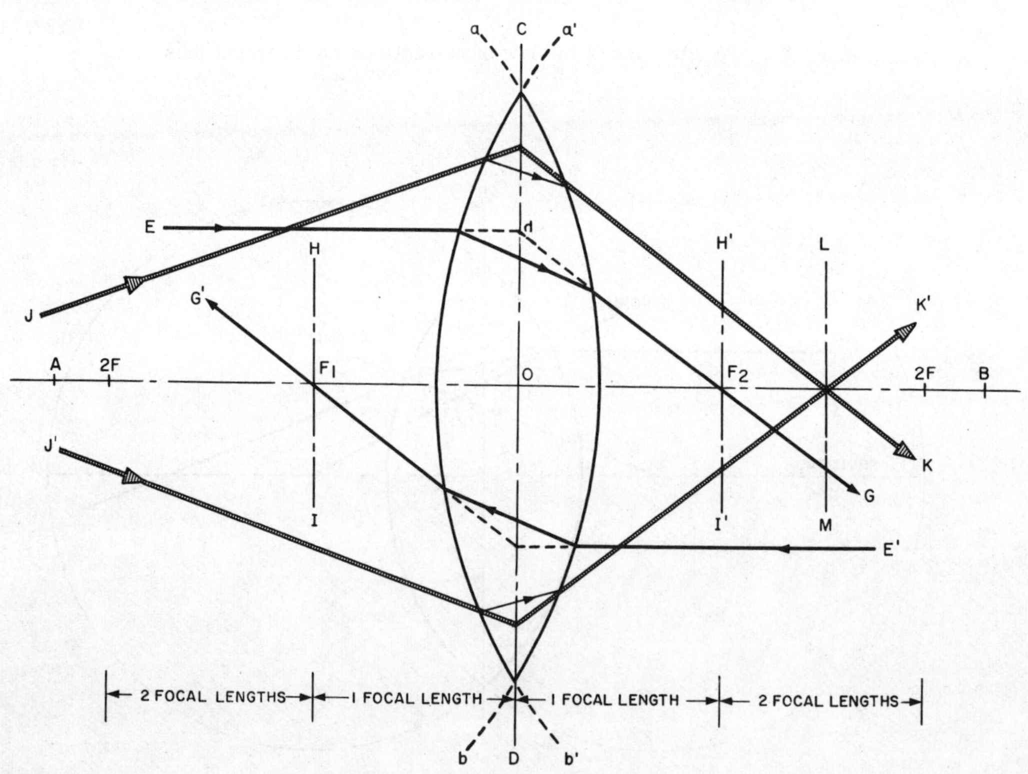

137.76

Figure 5-11.—Lens terminology.

OPTICAL AXIS.—Line AB in illustration 5-11 is the optical axis (principal axis), which is an imaginary straight line passing through the centers of curvature of both surfaces of a lens. Point A is the center of curvature of curve ab; point B is the center of curvature of curve a'b'.

PRINCIPAL PLANE.—Both thin and thick lenses have two principal planes (imaginary planes perpendicular to the optical axis). In a thin lens, the two planes are so close together that they are considered as ONE PRINCIPAL PLANE (CD). Observe that incident ray E, parallel to the optical axis, is refracted upon entering and upon leaving the lens. If both the incident ray and the refracted ray, however, are extended to d on the principal plane of the lens, as indicated by the dotted lines, refraction appears to take place at d.

OPTICAL CENTER.—The point in a lens through which light rays pass without deviation is called the optical center (O). In thin lenses, the optical center is located on the optical axis HALFWAY BETWEEN THE TWO CURVED SURFACES OF THE LENS. The principal plane in a thin lens intersects the optical center.

PRINCIPAL FOCAL POINT (Principal Focus).—The principal focus is the point where parallel incident rays converge after they pass through a convergent lens. Every convergent lens has two points of principal focus, one on each side. The point of principal focus on the left side of the lens is the PRIMARY FOCAL POINT, designated by F_1; the point of principal focus on the right of the lens is the SECONDARY FOCAL POINT (F_2). The incident ray (E') is parallel to the optical axis and, after it is refracted by the lens, passes through the primary focal point (F_1). Ray E is refracted by the lens as it passes through and crosses the optical axis (as ray G at the secondary focal point (F_2). Both F_1 and F_2 are located the same distance from the principal plane of a thin lens, but on opposite sides.

PRINCIPAL FOCAL PLANE.—The principal focal plane is an imaginary line (HI and H'I') perpendicular to the optical axis at the points of principal focus.

PRINCIPAL IMAGE PLANE.—The principal image plane is an imaginary line (LM, fig. 5-11) perpendicular to the optical axis at the point where the image is formed. The principal image plane may be located anywhere along the optical axis of the lens from its focal point to infinity.

FOCAL DISTANCE (Focal Length).—The focal length of all lenses is the distance from the principal focus (F_1 or F_2) to the principal plane (CD). Illustration 5-12 shows the focal lengths of a convergent lens; figure 5-13 gives the focal length of a divergent lens.

You can determine approximately the focal length of a convergent lens by holding the lens as necessary in order to focus the image of an object at infinity on a sheet of paper or ground glass. When the image is CLEAR and SHARP, you have reached the point of principal focus; and if you then measure the distance from the image to the optical center of the lens, you get the focal length.

Image Formation by a Convergent Lens

As you know, light rays in the form of pencils emanate from all points on an object and pass through a lens to a point of convergence behind the lens. This point is called the IMAGE POINT WHEN THE OBJECT IS AT A DISTANCE GREATER THAN THE FOCAL LENGTH OF THE LENS.

Rays of light which pass through the optical center of a lens, and superimposed on the optical axis, do not change their direction; and an image is formed at some point on this ray. Study illustration 5-14 carefully. The central ray in the top portion (A) of this illustration passes through the optical center and does not refract as it continues through the lens. The other light rays (2) refract toward the thickest portion of the lens as they enter it, and as they emerge, and form an INVERTED and REVERTED image (F) at the IMAGE PLANE. The other part of F is on the optical axis.

The image is inverted and reverted because two similar rays from a point at the bottom of the object form a point of the image corresponding to the bottom of the object; and every point on the object forms its point of light on the image in the same manner. Rays from the upper part of the object form points of light on the corresponding image, thereby causing the image to be transposed diametrically and symmetrically across the optical axis from the object. THIS IMAGE IS REAL.

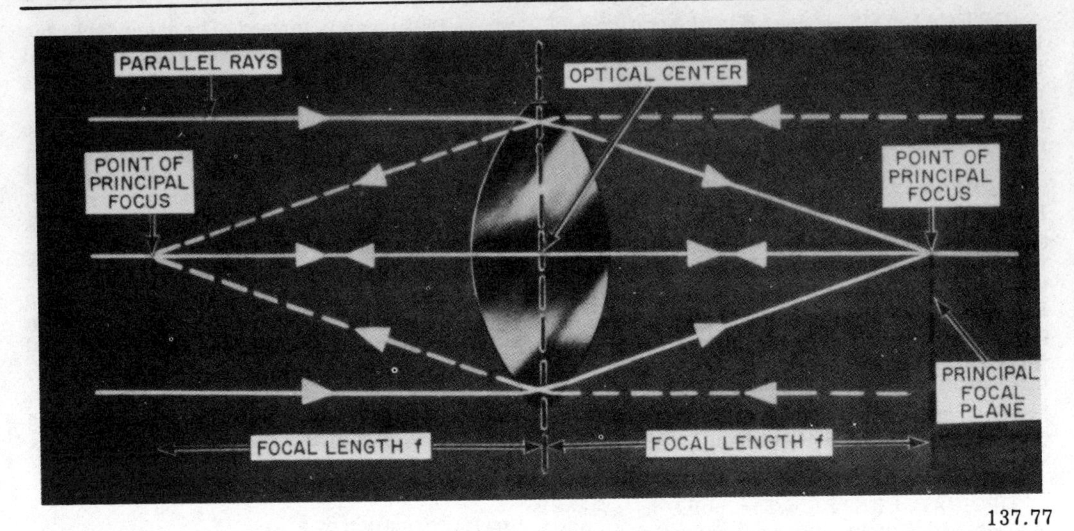

137.77

Figure 5-12.—Focal lengths of a convergent lens.

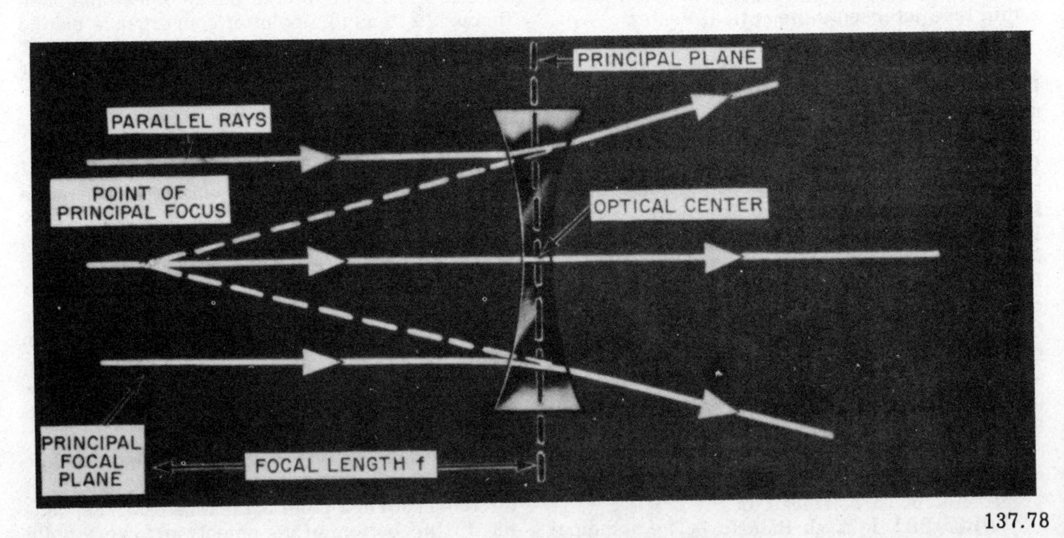

137.78

Figure 5-13.—Focal length of a divergent lens.

Study the bottom portion (B) of illustration 5-14 carefully. Observe the object, the light cone, the focal length, point of principal focus, the focal point, and the image. Note that Y and Y' are equal in length, representing the distance from the optical center to the edges of the lens. The term CONJUGATE FOCI means that two focal points are so related that AN OBJECT AT

ONE POINT IS FORMED AS A REAL IMAGE AT THE OTHER FOCAL POINT, as is true in illustration 5-14.

Calculating the Image Position

When an object is at a great distance (infinity), incident rays of light from it are parallel and the image is real, inverted, reverted, and diminished; and it is formed by the light rays at the secondary focal point, as shown in part A of figure 5-14.

If the object is at a DISTANCE BEYOND TWO FOCAL LENGTHS BUT LESS THAN INFINITY (fig. 5-15), a real, inverted image is formed by light rays from the object between the secondary focal point and 2F on the opposite side of the lens. Note the size of the image in each illustration shown, as compared with the object. When the object is brought closer to the lens, the image formed by it is larger than images formed by the object at greater distances from the lens; but the image is still smaller than the actual object.

In illustration 5-16 you see an object placed at two focal lengths in front of the lens; so the image formed of this object by the lens is real, inverted, reverted, equal in size, and located at 2F on the other side of the lens.

When an object is at a distance between one and two focal lengths from a lens, as illustrated in figure 5-17, 1 1/2F, the IMAGE IS REAL AND LARGER THAN THE OBJECT, inverted, reverted, and at a distance of 3F on the other side of the lens.

Illustration 5-18 shows an object at the principal focus of a lens, in which case the emerging light from the lens is parallel and therefore cannot converge to form an image. The image in the illustration is VIRTUAL, ERECT, NORMAL, ENLARGED, and FORMED AT INFINITY. A searchlight is an example of this type of image formation.

When an object is closer to a lens than the principal focus, divergence of the incident light is so great that the converging power of the lens is insufficient to converge or make it parallel. The emerging light is therefore merely less divergent than the incident light, and the rays appear to come from an object at a great distance than the actual distance of the object. See figure 5-19. These rays thus appear to converge behind the object to produce an ERECT, NORMAL, ENLARGED, and VIRTUAL IMAGE, located on the same side of the lens as the object.

From this discussion of images created by objects, we derive the following conclusion: As you move an object closer to a lens, the image created by the object moves closer to the lens, and it becomes increasingly larger as it moves. When you move the object to the principal focal point of the lens, the image BECOMES VIRTUAL AND IS FORMED AT INFINITY.

IMAGE POSITION.—When you desire the exact location of an image, use the lens law. Suppose, for example, that you wish to find the distance and size of an image from a lens when the object is one inch high and three inches from a lens with a focal length of two inches. Make necessary substitutions in the lens law and solve for D_i (distance of image), as follows:

$$\frac{1}{F} = \frac{1}{Do} + \frac{1}{D_i}$$

$$\frac{1}{2} = \frac{1}{3} + \frac{1}{D_i}$$

$$3D_i = 2D_i + 6$$

$$D_i = 6'', \text{ distance of image from lens}$$

MAGNIFICATION OF THE IMAGE.—Use the following formula to calculate the amount of magnification produced by a convergent lens:

$$\frac{S_i}{S_o} = \frac{D_i}{D_o}$$

NOTE: S_o = size of object
S_i = size of image
D_o = distance of object from optical center
D_i = distance of image from optical center

The procedure for solving this formula is the same as that explained in chapter 4 for determining the magnification of an image created by a mirror.

Thus far you have studied the methods for finding the distance and size of an image formed by a convergent lens, the GRAPHIC and MATHEMATICAL methods; but the objects we have been considering have been a short distance from the

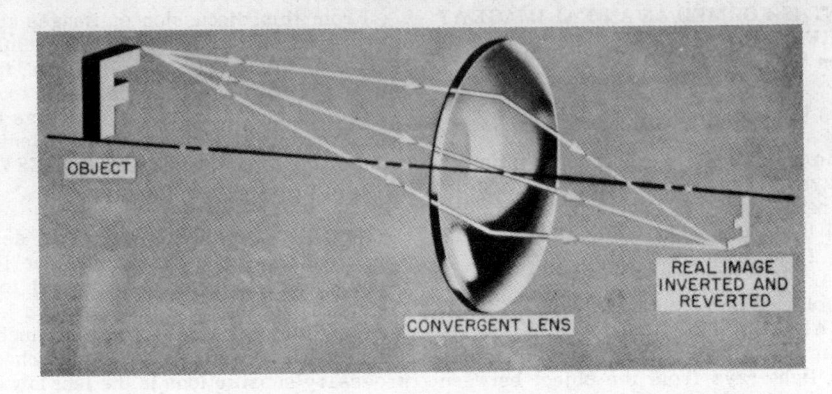

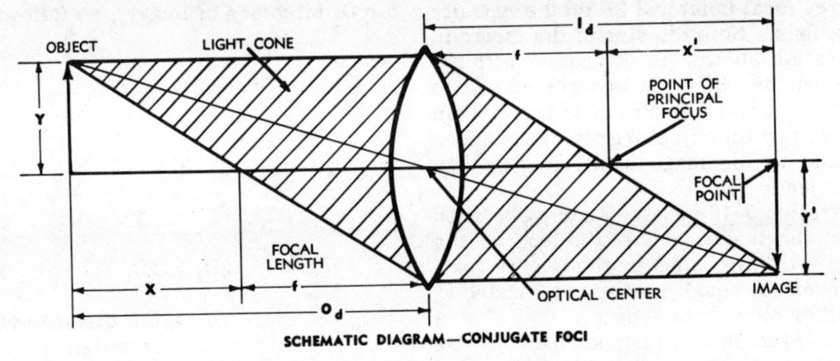

137.79

Figure 5-14.—Image formation by a convergent lens.

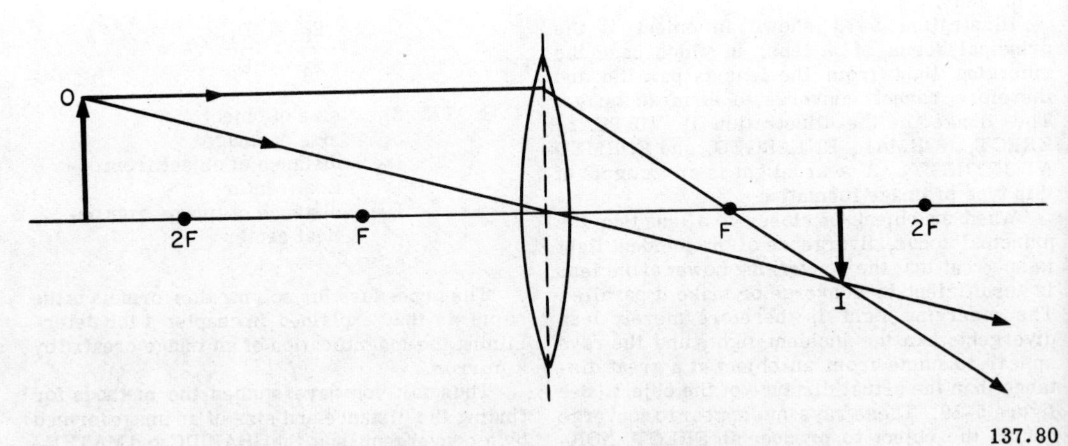

137.80

Figure 5-15.—Position of an image formed by a convex lens when the object is more than two focal lengths distant.

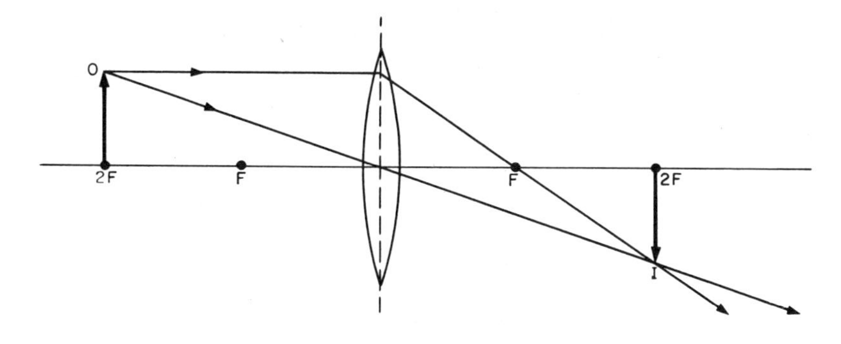

137.81

Figure 5-16.—Position of an image formed by a convergent lens when the object is at a distance equal to twice the focal length.

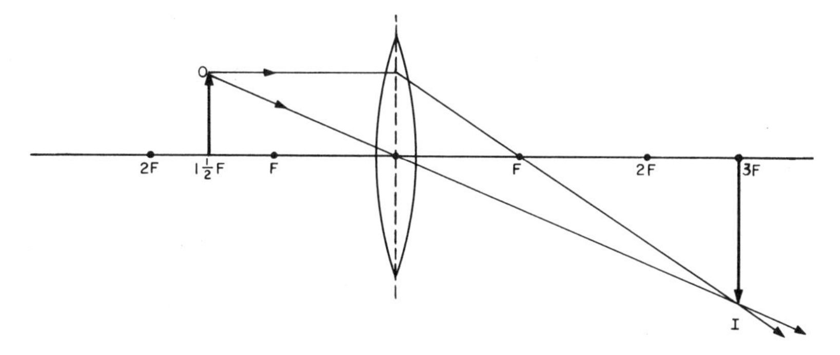

137.82

Figure 5-17.—Position of an image formed by a convex lens when the object is between the first and second focal lengths.

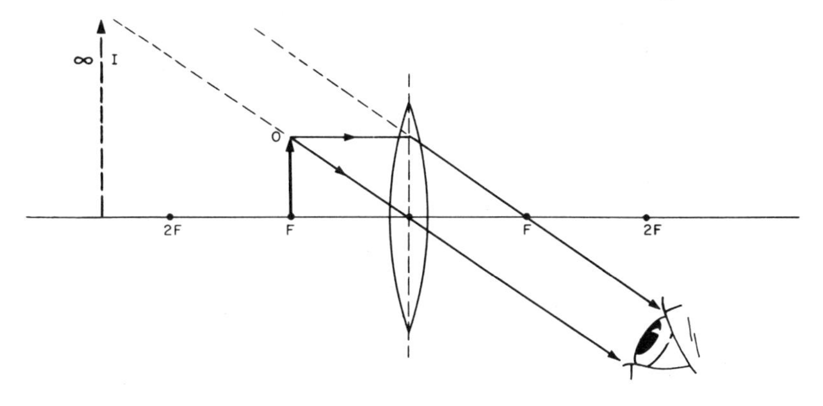

137.83

Figure 5-18.—Lack of image formation by a convergent lens when the object is at the principal focus.

focal point of a lens. What you should learn now about image formation, therefore, is the procedure for solving practical problems similar to those encountered aboard ship.

Suppose, for example, that you are looking at a ship through a telescope with a convergent objective (lens nearest the object) which has a focal length of 10 inches. The ship is 200 yards long and 5,000 yards away, and your problem is to ascertain the length of the image formed by the objective lens and the distance of the image of the ship from the lens. The procedure for solving this problem is as follows:

1. The ship is 200 yards long, or 7,200 inches long (200 x 36). The focal length of the lens is in inches, so change yards to inches in order to have your answer in inches.
2. The range (distance away) of the ship is 5,000 yards, or 180,000 inches (5,000 x 36).
3. The lens law is:

$$1 \frac{1}{F} = \frac{1}{D_o} + \frac{1}{D_i}$$

4. Make proper substitutions in the formula and solve for D_i, as follows:

$$\frac{1}{10} = \frac{1}{180,000} + \frac{1}{F}$$

The lowest common denominator (LCD) which you can use to solve this equation is $180,000D_i$; so when you use it and solve the equation you get:

$$18,000D_i = D_i + 180,000$$

$$18,000D_i - D_i = 180,000$$

$$17,999D_i = 180,000$$

$$D_i = 10.0005 \text{ inches } \left(\frac{180,000}{17,999}\right)$$

Your answer shows that the distance of the image of the ship formed by the objective of the telescope is a little more than 10 inches, which means that the image of a distant object is practically in the principal focal plane of the lens.

You can determine the length of the image formed by the objective by making necessary substitutions in the formula and solving for S_i, by cross multiplication.

$$\frac{S_i}{7,200} = \frac{10}{180,000}$$

$$72,000 = 180,000S_i$$

$$S_i = .4'', \text{ image length } \left(\frac{72,000}{180,000}\right)$$

The lens law also works for determining distances and sizes of virtual images formed by lenses; but the image distances will be negative, denoting that the image is virtual. Disregard the negative sign, however, when you calculate magnification.

MAGNIFYING POWER.—You learned what magnification and magnifying power mean during your study of chapter 3. Bear in mind now as you study magnifying power in lenses that it is constant and never changes.

When an object is inside one focal length of a convergent lens, the image is on the same side of the lens as the object, and it is MAGNIFIED, NORMAL, AND ERECT. The image is now VIRTUAL, which is illustrated in the reading glass in figure 5-20. Study the top part of the illustration first, and then notice in the bottom portion the point of principal focus, the object, and the image. The object is within one focal length.

The size of a virtual image may vary from the SIZE OF THE OBJECT TO INFINITY. For this reason, when you hold a reading glass close to your eyes and look through it, the size of the image changes only slightly when you move the object.

You can determine the magnifying power of a reading glass in the following manner:

The formula is:

$$MP = \frac{10}{4}$$

NOTE: Magnifying power of the lens is the apparent size of the image (10 inches away) divided by the apparent size of the object, also 10 inches from the eyes (distance most people hold a book or paper when they read).

If the lens in a reading glass has a focal length of 2 inches, its magnifying power (MP) is 5 (10 divided by 2, in the formula).

You can find the most comfortable distance for viewing an object with a magnifying glass by adjusting the glass until the virtual image is about -10 inches in front of the lens and then by

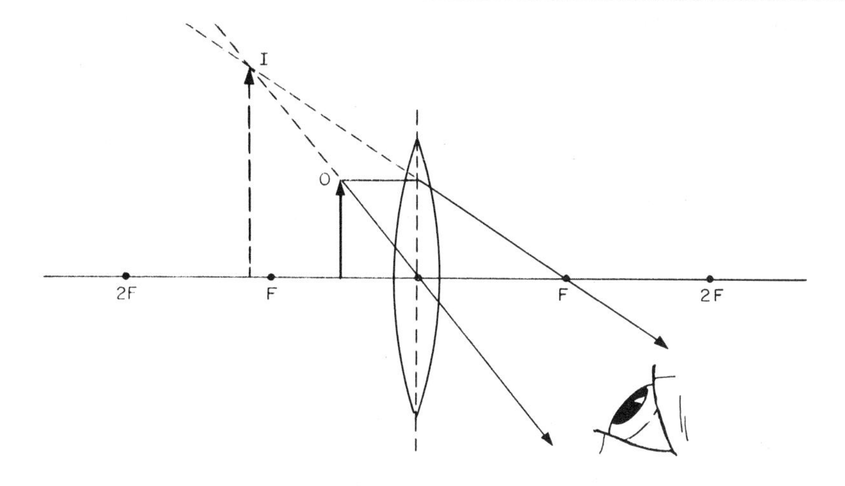

137.84

Figure 5-19.—Formation of a virtual image by a convex lens when the object is closer to the lens than the focal point.

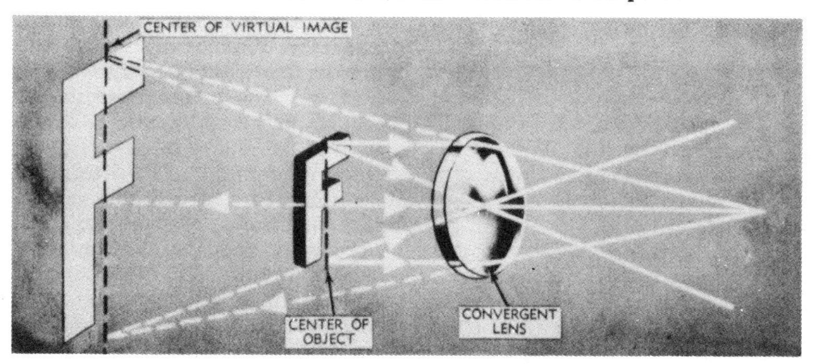

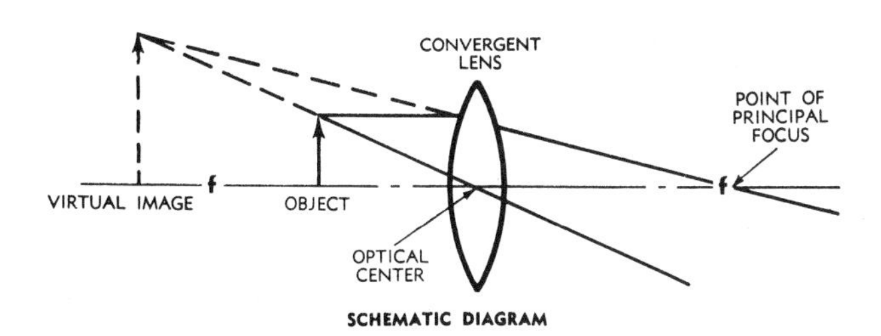

137.85

Figure 5-20.—Magnification by a reading glass when the object is within focal length.

putting this distance and the focal length of the lens in a formula and solving for the distance of the object (D_o), as follows:

1. The formula is:

$$\frac{1}{D_o} + \frac{1}{D_i} = \frac{1}{F}$$

2. The distance of the image (D_i) is -10 inches.

3. In this problem, the focal length of the lens is 7 inches.

4. The lowest common denominator (LCD) which you can use in solving the equation is $-70D_o$; so when you make correct substitutions in the equation and solve it, you get:

$$\frac{1}{D_o} + \frac{1}{-10} = \frac{1}{7}$$

$$-70 + D_o = -10D_o$$

$$17D_o = 70$$

$$D_o = 4.23 \text{ inches}$$

A few simple experiments at this point will enable you to understand better some of the things you just studied about images formed by convergent lenses. Study the discussion carefully. Proceed as follows:

1. Hold a magnifying glass, or a short-focus lens, close to one eye and use a match for an object. Put the head of the match on the lens on the other side and you will see a badly blurred image, because the image is so close to your eye that you cannot focus properly on it. If you slowly move the match head back from the lens, the image becomes sharper as the distance of the object increases; because as the distance of the object increases, the distance of the image also increases. The image distance, however, increases much faster than the object distance.

2. Move the match back until you can comfortably focus your eye on the image—about 10 inches from your eye. The image is then NORMAL AND ERECT, JUST LIKE THE OBJECT BUT BIGGER. As you know, the image is virtual, because it really is not there.

3. Slowly move the match farther away and observe the position of the image. It is at infinity but you can see it clearly. The reason for this is that your eye can focus on anything between approximately 10 inches and infinity. When you move the match farther from the lens, past the focal point, you see only a blur—the virtual image disappears and the image becomes real. Because the image is on the same side of the lens as your eye, you cannot see it clearly.

4. Now use the page of your book as an object and hold it at arm's length with one hand. Use the other hand to put the lens in contact with the page and observe the image created. It is virtual, normal, erect, and slightly magnified. You can see it because your eye is farther away and you can bring it into clear focus on the retina of your eye.

5. Retain the book in the same position and move the lens slowly toward your eye and observe the size of the virtual image. It becomes larger and larger as you move the lens closer to your eye, because you are now comparing the apparent size of the image with the apparent size of the object.

The object looks comparatively small because you are holding it at arm's length. When the lens is close to the page, the image is near the lens and it is also fairly small. As you increase the distance of the object, however, you increase the distance of the image; and when the object and the image are both close to the lens (both at arm's length), they look smaller than they would if the lens were close to your eye. When you increase the object's distance by moving the lens closer to your eye, the apparent size of the image increases, even though the apparent size of the object remains the same.

6. Slowly move the lens toward your eye until the virtual image disappears. At first, the real image looks blurred, because (even though on your side of the lens) it is so far back from the lens that you cannot focus your eyes on it. If you bring the image closer to the lens by moving the lens farther from the book, the image is inverted. You cannot tell at a glance whether it is normal or reverted; but if you turn the book upside down, your lens forms an inverted image of the inverted book. The print is right side up and you can read it, which means that it is backward.

As you move the lens closer to your eye, the image blurs again; because it is too close to your eye and you cannot focus it sharply on the retina. It is rather difficult for the eye to focus on anything closer than 10 inches. You can see a clear image when the distance between your eye and the lens is equal to the image distance plus 10 inches.

LENS DIOPTER (generally called diopter).—A lens diopter is the UNIT OF MEASURE OF THE

REFRACTIVE POWER (dioptric strength) of a lens or a lens system. It is based on the metric system of measurement. All optical diagrams give focal lengths and diameters of lenses in millimeters.

A lens with a focal length of 1 meter has the refractive power of 1 DIOPTER. Study illustration 5-21. The refractive power of a converging lens is POSITIVE; the refractive power of a diverging lens is NEGATIVE.

The refractive power of lenses which do not have focal lengths of 1 meter is the reciprocal of the focal lengths in meters, and it varies inversely as the focal length. This means that a converging lens with a focal length of 20 centimeters (1/5 meter) has a power of +5 diopters; whereas, a diverging lens with a focal length of 50 centimeters (1/2 meter) has a power of -2 diopters. A lens with the shortest focal length has the greatest positive or negative dioptric strength.

A lens with a focal length of 25 centimeters has a positive dioptric strength of 4 diopters. When converted to meters, the 25 centimeters equal .25 meter. The reciprocal of .25 meter equals 4 diopters. The equation for this is as follows:

$$\text{Diopters} = \frac{1}{f \text{ (in meters)}}$$

$$\text{Diopters} = \frac{1}{.25 \text{ meters}}$$

$$\text{Diopters} = 4$$

Another formula for determining the dioptric strength of a lens when its focal length is in millimeters is:

$$\text{Dioptric strength} = \frac{1{,}000 \text{ millimeters (mm)}}{F \text{ (in millimeters, mm)}}$$

If the focal length of a lens is in inches, the formula is:

$$\text{Dioptric strength} = \frac{39.37 \text{ (or 40) inches}}{F \text{ (in inches)}}$$

DIVERGENT LENSES

Refer again to illustration 5-1 and study the types of simple divergent lenses.

Divergent lenses have negative dioptric strength, and they are always thinner in the middle than at the edges. The optical center of a divergent lens is at the thinnest point of the lens, and the lens diverges parallel rays of light.

The point of principal focus, focal points, and focal planes resulting from the nearness of an object or light source to a simple divergent lens are located on the side of the lens toward the light source or object. The point of principal focus and other focal points are located where the emergent rays should intersect on the optical axis if they were extended backward as imaginary lines toward the side of the lens on which the light strikes. Review figure 5-13 for some terminology and focal length of a simple divergent lens.

If you use a page of this book as an object—at arm's length—and look at it through a divergent lens, this is what happens:

1. When the lens is in contact with the page (object), the image you see is erect, normal, and slightly smaller than the object.
2. If you move the lens closer to your eye, the image becomes even smaller.
3. When you have the lens quite close to your eye, you can see only a blur, REGARDLESS OF THE POSITION IN WHICH YOU HOLD THE OBJECT.

You will understand what took place when you held the divergent lens in the positions just described and looked at the page after you study the next few pages, dealing with the construction of a divergent lens and image formation by it.

Construction of a Divergent Lens

Suppose, now, that we construct a divergent lens like the one shown in figure 5-13. Proceed as follows:

1. Sketch the double concave lens on paper.
2. Draw a dotted line through the middle of both ends of the lens, to represent the PRINCIPAL PLANE.
3. Then draw a straight line through the OPTICAL CENTER of the lens, PERPENDICULAR TO THE PRINCIPAL PLANE, to represent the OPTICAL AXIS.
4. Next, draw two lines, to represent rays of light, near the ends of the lens to the left face, through the lens (refraction indicated), and out into space.
5. With your ruler, draw the dotted lines along the straight portion of the emergent light ray to the optical axis. Where the two dotted lines intersect the optical axis is the POINT OF PRINCIPAL FOCUS, as indicated by the terminology and arrow.
6. Draw a dotted line downward from the POINT OF PRINCIPAL FOCUS, and then draw the two arrows in the positions indicated and insert FOCAL LENGTH.

Now sketch another double concave lens on paper (fig. 5-22) and draw a line through the

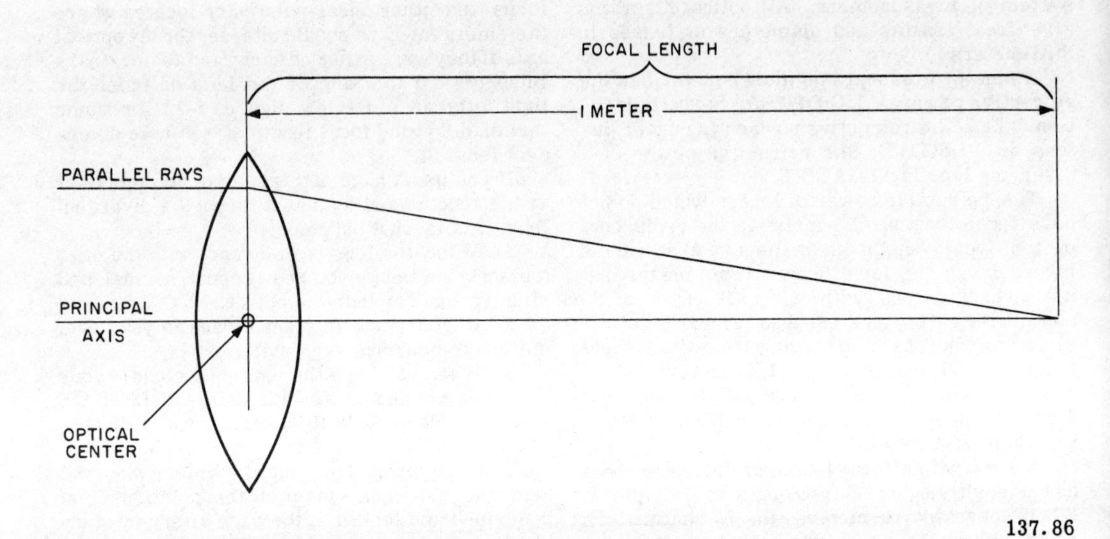

137.86

Figure 5-21. —Lens diopter.

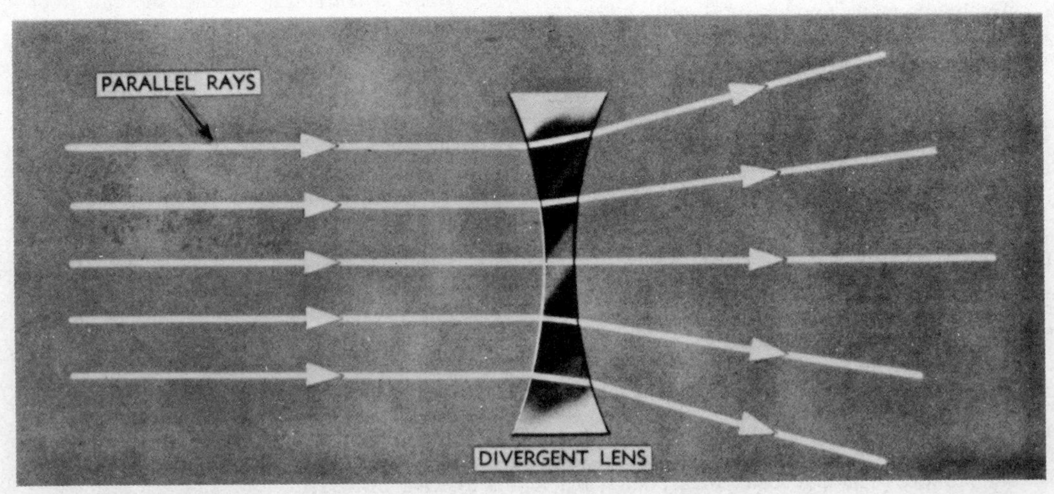

137.87

Figure 5-22. —Effect of parallel rays on a divergent lens.

OPTICAL CENTER, perpendicular with the PRINCIPAL PLANE, to represent the OPTICAL AXIS. Then draw two other lines (parallel) above and below the optical axis, as shown, to represent FOUR LIGHT RAYS.

The light rays you just drew (fig. 5-22) show the PROCEDURE for TRACING LIGHT RAYS THROUGH A DIVERGENT LENS. Rays which pass along the optical axis and through the optical center do not refract (deviate), as you know; rays which pass through the lens at points other than through the optical center (along the optical axis) are deviated in the manner shown in the illustration.

Image Formation by a Divergent Lens

When you look through a divergent lens (fig. 5-22), extensions of the refracted rays of light appear to converge at a point (POINT OF PRINCIPAL FOCUS) on the same side of the lens as the object, as shown in illustration 5-13. In order to learn how an image is created by a divergent lens of this type, draw (sketch) the lens on paper and then do the following:

1. Draw a dotted line through the middle of each end of the lens to represent the principal plane.
2. Draw another dotted line perpendicular to the principal plane and through the optical center to represent the optical axis. See figure 5-23.
3. Using a focal length of 2 inches, put a dot on the optical axis to represent the focal point (F).
4. Next, draw 3 arrows 1 inch high on the optical axis in the positions indicated by O_1, O_2, and O_3. Observe that one arrow is INSIDE THE FOCAL POINT (F), one arrow is ON THE FOCAL POINT, and the third arrow is BEYOND THE FOCAL POINT.
5. Along the tips of the arrow heads, draw a line to the principal plane to represent a parallel ray of light (parallel to the optical axis). Note how this ray diverges up after it contacts the principal plane. If you were to look at this ray from the opposite side of the lens, it would appear to emerge from the first focal point; so extend this line (dotted portion) to the focal point (F).
6. At the point where the ray of light which passes through the optical center from arrow O_1 intersects the dotted extension to the focal point of the refracted ray you drew along the tips of the arrow heads, construct an arrow (erect) between this point and the optical axis. This arrow is designated I_1. Then draw arrows I_2, and I_3 to represent the other images made by the objects (O_1, O_2, and O_3).

SIZE OF IMAGE.—Observe that the images you constructed in illustration 5-23 are erect and normal, between the lens and its focal point, SMALLER THAN THE OBJECTS WHICH CREATED THEM, and VIRTUAL.

Illustration 5-24 shows what is meant by REDUCTION BY DIVERGENT LENSES. The impression you receive that the image you see on the other side of a divergent lens is smaller than the object is known as REDUCTION, in contrast to MAGNIFICATION PRODUCED BY A SIMPLE MAGNIFIER, as you studied earlier in this chapter. Study the schematic diagram at the bottom of the illustration.

CALCULATING IMAGE POSITION.—Now use the lens formula to calculate the positions of the images you constructed in illustration 5-23. The lens formula is:

$$\frac{1}{F} = \frac{1}{D_o} + \frac{1}{D_i}$$

The focal length of a divergent lens is negative, because the image is on the same side of the lens as the object and the image distance is negative.

The focal length of the lens used in the illustration is 2 inches TO THE LEFT of the principal plane of the lens, so the focal length of the lens IS MINUS 2 INCHES.

To find the image distance for arrow O_2 (object) drawn at the focal point some substitutions must be made in the formula (lens law), as follows:

$$\frac{1}{-2} = \frac{1}{2} + \frac{1}{D_i}$$

$$2D_i = -2D_i - 4$$

$$4D_i = -4$$

$D_i = -1$, image distance for arrow O_2

The answer you got by solving the formula means that the image is 1 inch from the principal plane of the lens, but is ON THE SAME SIDE of the lens as the object.

You can calculate the distances of the other images in illustration 5-21 in the manner just described.

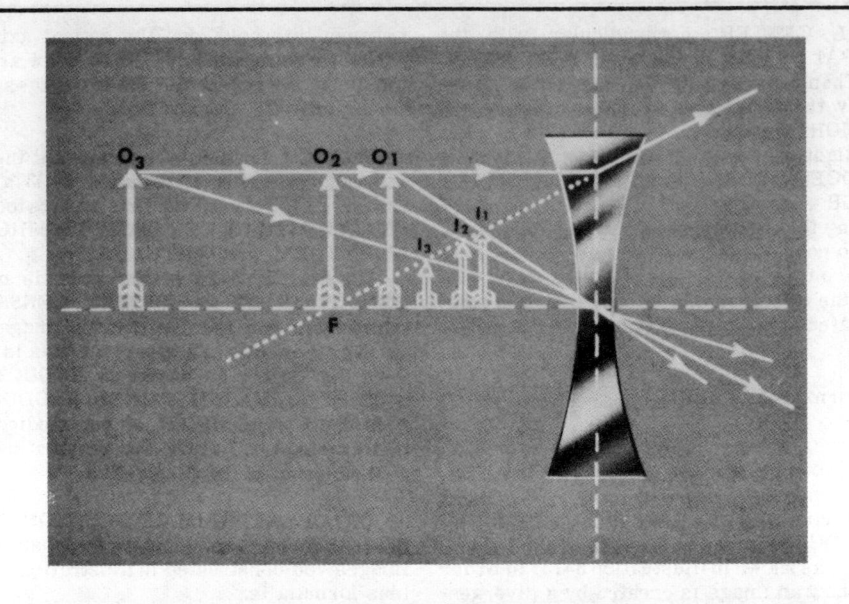

137.88

Figure 5-23. —Image formation by a divergent lens.

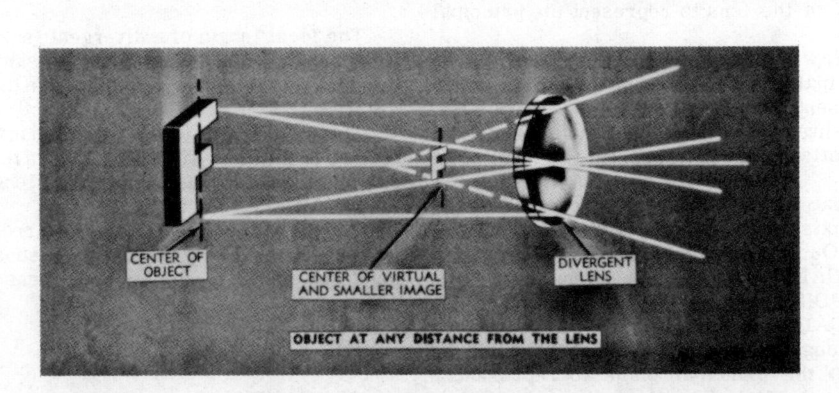

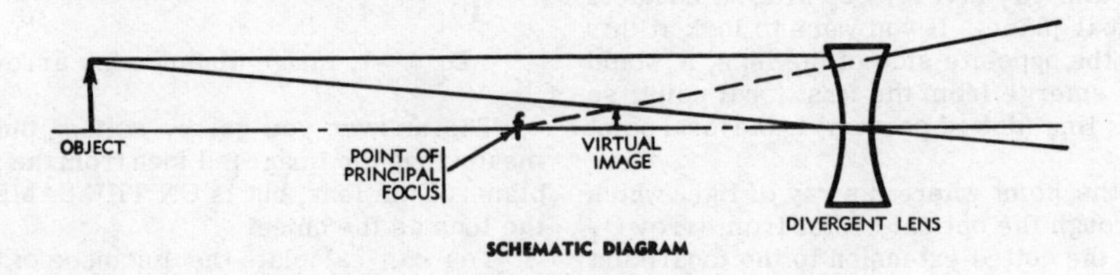

137.89

Figure 5-24. Reduction by a divergent lens.

MAGNIFICATION.—You can use the lens law to determine the size of each image created. Magnification, you will recall, is the size of the image divided by the size of the object. The magnification formula is:

$$M = \frac{S_i}{S_o} = \frac{D_i}{D_o}$$

Suppose that an object's distance is 2 inches and the image distance is 1 inch. What is the amount of magnification produced by the lens? Make necessary substitutions in the formula and solve for S_i, as follows:

$$M = \frac{D_i}{D_o} = \frac{1}{2}\text{, and}$$

$$M = \frac{S_i}{S_o}$$

$$S_i = MS_o$$

$$S_i = \frac{1}{2} \times 1\text{, or .5 inch}$$

The amount of magnification of the images in this illustration is therefore one-half.

NOTE: Negative image distances are considered only in their absolute values in the magnification formula; so the problem just solved in the formula does not give a magnification of -1/2 but merely 1/2. The negative image distance merely means that the image is on the same side of the lens as the object and is virtual. As you learned in chapter 4, the image may even be larger than the object and still be virtual.

THICK LENSES

The types of thick lenses considered in this section are:

1. Single thick lenses.
2. Thick lenses constructed from two thin lenses.
3. Compound lenses.

Lens thickness is one of the factors which control focal lengths in a lens, along with index of refraction and radii of curvature.

You will find from this discussion of thick lenses that they FUNCTION in about the same manner as thin lenses; that is, their effect on light rays is essentially the same.

SINGLE THICK LENSES

Bear in mind that all lenses are made of optical glass in accordance with a specified formula, for binoculars, telescopes, and so forth. All lenses are considered as either thin or thick, and all lenses follow the same laws of refraction.

Thus far, we have been discussing thin lenses; but it is important now that we explain the difference between a thin lens and a thick lens. How thick must a lens be to be considered thick? If the distance between the two principal planes of the lens is great enough to be measured, the lens is said to be THICK. It is actually a matter of judgment of an individual working with lenses, however, to decide when a lens is thick; but in this training manual, we consider any lens whose principal planes are separated by 5 mm to be thick.

Like thin lenses, thick lenses are identified by reading each lens surface; that is, whether they are equi-convex, double convex, planoconvex, and so on.

Tracing Light Rays through a Thick Lens

Two equi-convex lenses are illustrated in figure 5-25. Both lenses have the same index of refraction and radius of curvature; their diameters are equal, but their thicknesses are unequal.

NOTE: A lens is said to be equi-convex or equi-concave if its radii or curvature are equal; that is, the radius of curvature of one surface of the lens must be the same as that of the other surface.

Refer now to the light rays in illustration 5-25. As you know, an A ray is any ray which passes through the optical center of a lens and emerges from the lens parallel to the incident ray without deviation. This rule applies to BOTH THICK AND THIN LENSES; but note the difference in the A rays of the two lenses in illustration 5-25. In the thin lens in part A, the A ray is traveling toward the optical center and passes directly through without refraction or deviation. In a thick lens, however, if an incident light ray (not shown in diagram) traveling toward the optical center of the lens were refracted upon reaching the lens, it would not pass through the optical center. Upon emerging from the lens, the light ray would be slightly deviated from its original path.

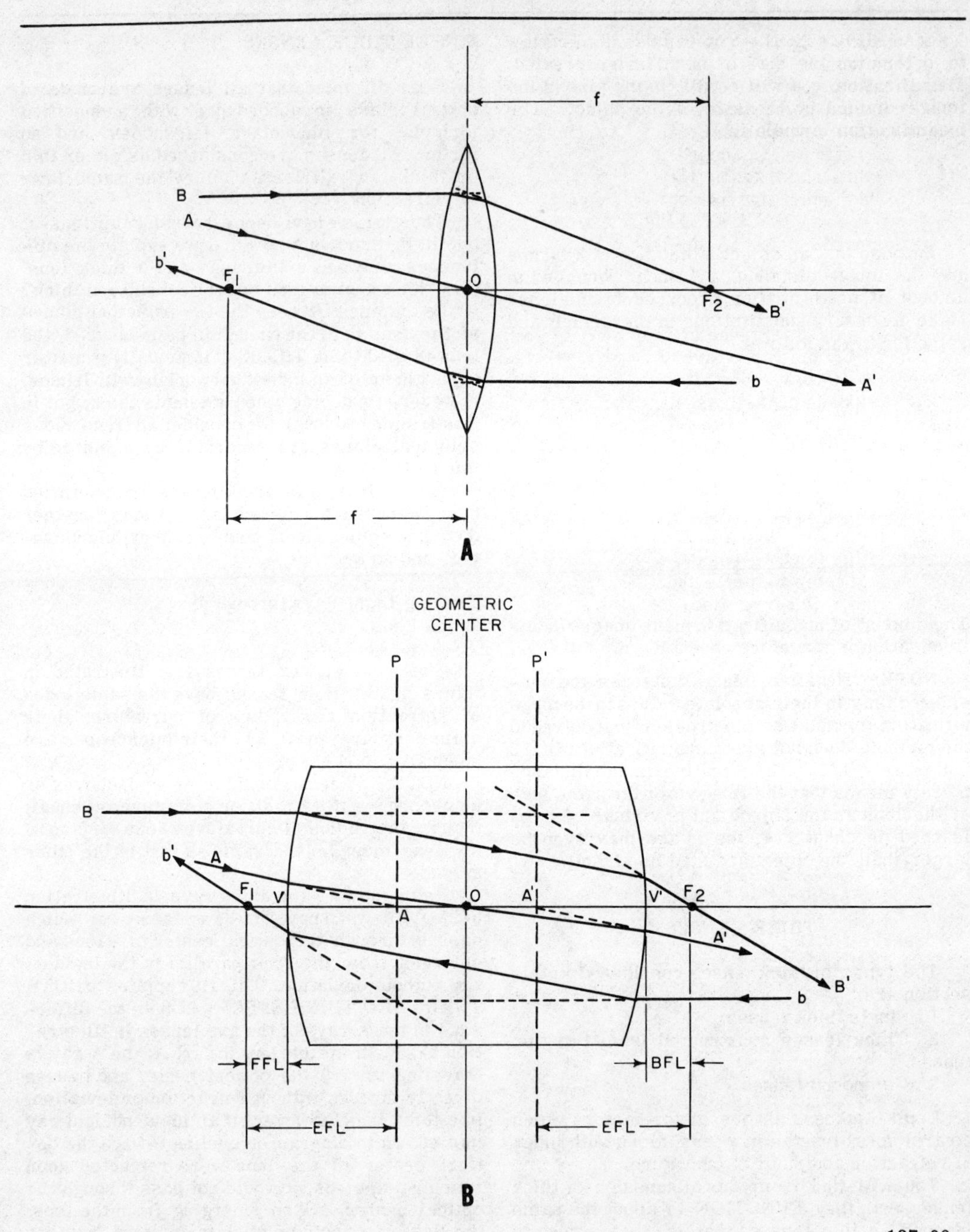

137.90

Figure 5-25.—A thin lens and a single thick lens.

In order for a ray of light to pass through a thick lens without deviation, it must travel along the optical axis, or travel in the direction of the first principal point (where the principal plane intersects the optical axis). When it strikes the lens, it is refracted in accordance with the laws of refraction and passes through the optical center. Upon emerging from the second surface, the emergent ray appears to have come from the second principal point (A') and is parallel to the incident ray, slightly offset (NOT DEVIATED) from its original path.

If a lens is VERY THICK, the degree of convergence or divergence of rays is changed by the DISTANCE the rays travel through the glass. If the light is converging after passing through the first surface of the lens, the degree of convergence is increased before the second surface of the lens is reached. If the lens is thick enough, the light could converge to a focus on the second surface.

Observe that the refraction of the B rays in the two lenses in figure 5-25 is the same; but the ray in the thicker lens converges more and travels a greater distance than the ray in the thinner lens. Observe also that the principal plane (P') of the B ray is now located to the right of the optical center of the thick lens.

Now compare the b rays of the two lenses. The refracted ray in the thin lens appears to be refracted at the same plane where the B ray refracted; but ray b of the thick lens does not appear to be refracted at the same point as the B ray—it traveled a greater distance and is more convergent than in the thin lens. The location of the principal plane (P) for the b ray is to the left of the optical center. Refraction, therefore, DOES NOT TAKE PLACE IN THE EXACT CENTER of the thick lens as it does in thin lenses.

Observe (fig. 5-25) that the A ray deviates slightly as it strikes the face of the lens, passes through the optical center, and then REFRACTS AGAIN as it leaves the left face of the lens. Note, also, that the B ray refracts exactly the same amount as the b ray as it strikes the left face of the lens, as it passes through, and as it leaves the face of the lens. Both of these rays pass through the optical axis of the lens at EXACTLY THE SAME DISTANCE FROM THE LENS.

If you remember the rules just explained for the passage of light rays through a lens, you can trace the rays as necessary in order to locate the image of an object. The important thing to keep in mind about refraction in a lens is that the quality of the glass, its shape, and thickness are all factors which have a bearing on refraction and tracing of light rays through it.

Focal Lengths of Thick Lenses

The focal length of a lens is measured from the principal plane to the principal focus. In thin lenses, the principal plane bisects the optical axis and is in the center of the lens; but in thick lenses the principal planes do not bisect the optical centers. Study parts A and B of figure 5-25. The focal length of a thick lens cannot therefore be measured from the geometric center of the lens.

In thick lenses, the focal length is measured from the principal planes to the principal focus (fig. 5-25), from the principal plane (P') to the principal focus (F_2), or from the principal plane (P) to the principal focus (F_1). In thick lenses, the focal length is labeled EQUIVALENT FOCAL LENGTH (EFL).

If the thickness of a lens is measurable, measure the distances from the principal focus to the surfaces (vertex) of the lens. The distance from the primary principal focus (F_1) to the first vertex (V) is labeled FRONT FOCAL LENGTH (FFL); the distance from the second vertex (V') to the secondary principal focus is labeled BACK FOCAL LENGTH (BFL). ALWAYS READ OPTICAL DRAWINGS FROM LEFT TO RIGHT.

In equiconvex lenses the EFL and the BFL are exactly equal in length; but in double convex lenses with unequal radii of curvatures, the FFL and the BFL are of unequal length. Regardless of whether a lens is double convex or equiconvex, however, the EFL's are ALWAYS EQUAL IN THE SAME LENS.

THICK LENSES CONSTRUCTED FROM THIN LENSES

In single thick lenses, you can locate the EFL, the FFL, the BFL, and the principal planes by following the laws of refraction. Bear in mind that light always follows these laws, regardless of the type, thickness, or shape of glass which refracts the rays.

In the following pages we explain how to construct a thick lens by using two thin lenses, when they are placed reasonably close together.

Construction Procedure

Sketch on a sheet of paper two thin lenses, in the positions shown by lenses X and Y in figure 5-26, and draw all lines (including dotted extensions) required to show the optical axis, light rays B and b, F_1 and F_2, principal plane P', optical axis, and the principal plane of each lens. NOTE: The detailed procedure for making all the drawings of light rays, focal lengths, etc., is not given here because it was explained fully for illustration 5-25.

If you arrange two thin lenses in a predetermined position, they perform as a single thick lens. The two lenses in illustration 5-27 are identical (symmetrical) in every respect.

The light rays (B and b) you drew to the face of lens X would come to a focus at the secondary principal focal point (F_2) if lens Y were not within the focal length of lens X. When these two converging rays enter lens Y, they are refracted and made ever more convergent to the first focal point (F) you put on the optical axis. The extensions of incident ray B and emergent ray B' (dotted lines) provide the top point THROUGH WHICH you drew the principal plane (P'). The dotted extensions of ray b (incident) and emergent ray b' intersect to provide the other point THROUGH WHICH you drew the principal plane (P').

If you were to draw two more parallel rays of light from the right to lens Y, similar to B and b, you would get the points THROUGH WHICH you would construct principal plane P. Review illustration 5-25.

You can get the equivalent focal length (EFL) of two thin lenses used to make a thick lens by using the focal lengths of the two convergent lenses in combination. Study illustration 5-27. The EFL is equal to the distance from the first focal plane to the first principal plane, or the distance from the second principal plane to the second focal plane.

The first focal point of the combination is designated F_1, and a plane which passes through this point perpendicular to the optical axis is the first focal plane. The plane which passes through the second focal point (F_2) is the second focal plane. The front focal length (FFL) is the distance from the first focal plane to the principal plane of lens X. The back focal length (BFL) is the distance from the principal plane of lens Y to the second focal plane.

You can find the first focal point of the combination by measuring from the principal plane of lens X a distance equal to the FFL. Measure to the LEFT if the FFL is positive; measure to the right if FFL is negative. To find the second focal point of the combination, measure the FFL from the center of lens Y—to the right if FFL is positive, to the left if FFL is negative.

You can find the first principal plane of the combination by measuring a distance equal to the EFL from the first focal point (F_1). To find the second principal plane, measure the EFL from the second focal point (F_2). In both cases, measure to the left if EFL is positive; otherwise, measure to the right. NOTE: The first principal plane may lie to the right of the second principal plane, but these planes can be formed most anywhere.

REMEMBER: These lenses are very thin, and the distance between the principal plane and the vertex of each lens is practically negligible. For this reason, measurements for the FFL and the BFL of the lens combination must be from the principal planes of the thin lenses, NOT FROM THE VERTEX OF EACH.

When thin lenses used in combination are identical in optical characteristics, FFL and BFL are equal; but if the focal length of one lens is not equal to that of the other, FFL and BFL are unequal.

Although the two thin lenses used in the combination are not identical in optical characteristics, the EFL ON BOTH SIDES OF THE COMBINATION IS ALWAYS EQUAL. YOU CAN PROVE THIS STATEMENT BY USING THE FORMULA.

The formulas for determining the three types of focal distances just explained are as follows:

$$EFL = \frac{F_1 \times F_2}{F_1 + F_2 - S}$$

$$BFL = \frac{(F_1 \times F_2) - (S \times F_2)}{F_1 + F_2 - S}$$

$$FFL = \frac{(F_1 \times F_2) - (S \times F_1)}{F_1 + F_2 - S}$$

F_1 = focal length of lens A (in combination)

F_2 = focal length of lens B (in combination)

S = separation of the two lenses (X & Y, or left and right) in a combination, measured from their principal planes.

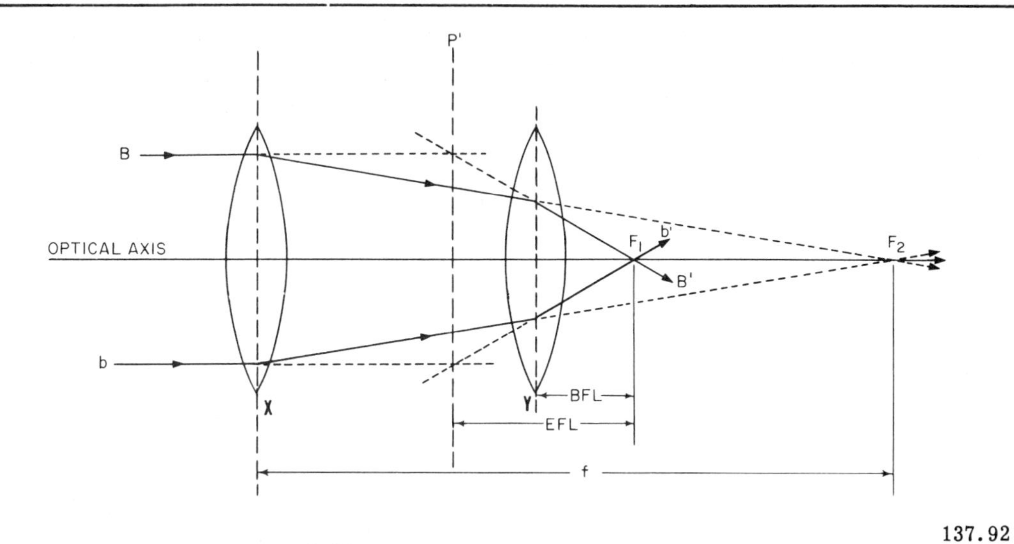

137.92

Figure 5-26.—Two thin lenses used as a thick lens.

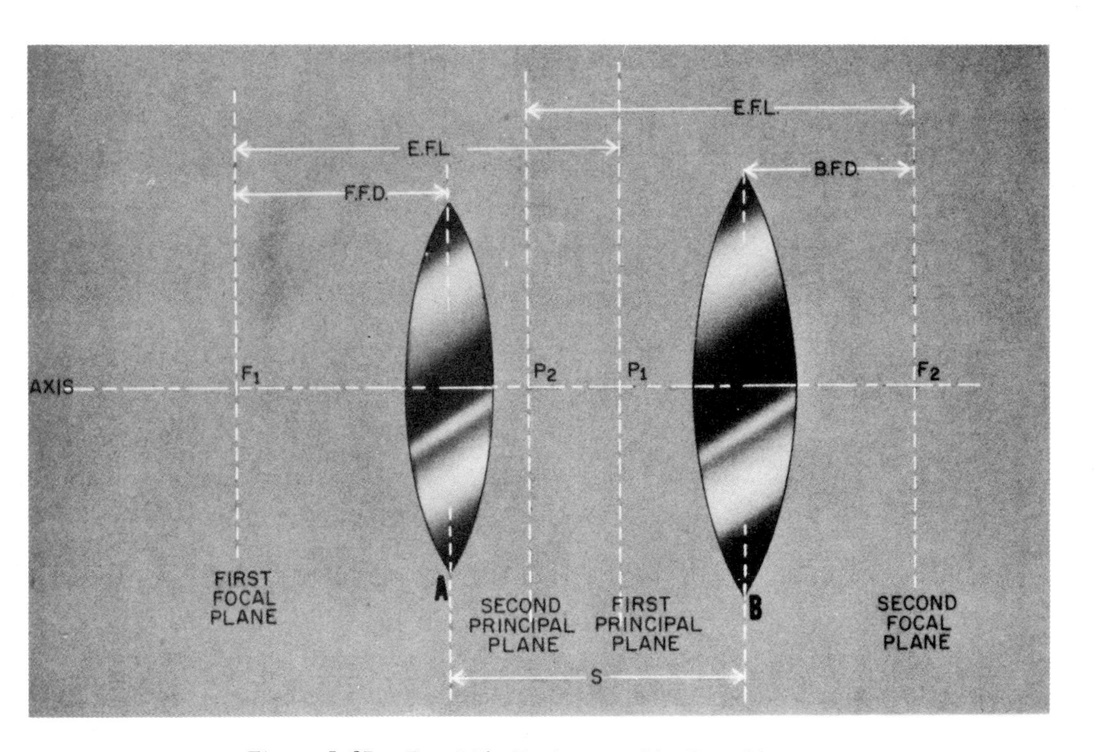

Figure 5-27.—Focal lengths in a combination of lenses.

Suppose you place two lenses together to form a thick lens, with a separation of the lenses at their principal planes of one inch, with a focal length of 3 inches for one lens and a focal length of 2 inches for the other lens, and you desire to know the EFL, the FFL, and the BFL of the combination. Make proper substitutions in the formulas and you get:

$$\text{EFL} = \frac{3 \times 2}{3 + 2 - 1}, \quad \text{EFL} - \frac{6}{4} = 1\,1/2 \text{ inches}$$

$$\text{FFL} = \frac{(3 \times 2) - (1 \times 3)}{3 - 2 - 1}, \quad \frac{6 - 3}{4} = 3/4 \text{ inch}$$

$$\text{BFL} = \frac{(3 \times 2) - (1 \times 2)}{3 + 2 - 1}, \quad \frac{(6) - (2)}{4} = \frac{4}{4} - 1 \text{ inch}$$

Tracing Light Rays through the Combination

The best way to find the paths of light rays through a combination of lenses used as a thick lens is to draw in the two principal planes (after you calculate them accurately by the formula) and mark the principal points where the planes intersect the optical axis. Study illustration 5-28. Next, draw a light ray (AQ) toward the first principal plane (P_1), instead of toward the optical center, and observe that it emerges from lens Y as if it passed through the second principal plane (P_2).

Light ray B, which strikes the surface of lens X parallel to the optical axis, passes through the second focal point (F_2) of the combination; but it travels as though all the refraction took place in the second principal plane. Light ray C, which passes through the first focal plane of the combination, emerges parallel to the optical axis; but it travels as if all the refraction took place in the first principal plane.

If you desire to indicate the paths of light rays inside the lens group, draw a straight line from the point of incidence to the point of emergence for each ray.

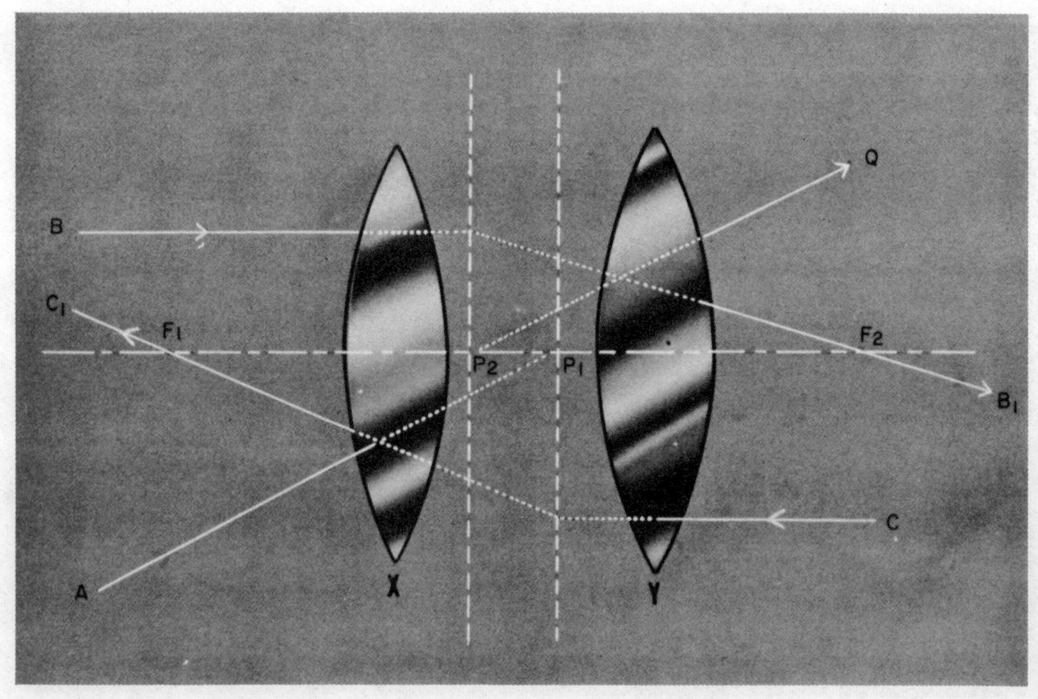

137.94

Figure 5-28.—Tracing light rays through a combination.

COMPOUND LENSES

Because an optically perfect single simple lens cannot be produced, two or more lenses ground from different types of optical glass are frequently combined as a unit to cancel defects present in a single lens.

The refractive power of a compound lens is less than that of the simple convex lens used in its construction. If a double convex lens of crown glass is combined with a concave-plano lens of flint glass, the effective refractive power of the concave-plano lens is reduced about 50 percent; but the latter lens has sufficient power to neutralize the dispersion in the double convex lens. By combining these two lenses to form a compound lens, light which passes through the lens is brought to a practical focus at a point almost double the distance to the point of principal focus of the convex lens alone. Study illustration 5-29.

The elements of compound lenses are frequently cemented together with their optical axes in alignment. Two lenses may be cemented together to form a DOUBLET, three lenses may be cemented together to make a TRIPLET, or each lens of the unit may be mounted together to make a DIALYTE.

NOTE: The lenses of DOUBLETS TOO LARGE in diameter to be cemented together (even if their inner surfaces match) form a lens combination called an AIR-SPACED or UNCEMENTED DOUBLET.

In a dialyte compound lens, the inner surfaces of the two elements do not have the same curvature, which means they cannot be cemented together in order to correct for defects. The two lenses are separated by a thin spacer ring, or tin foil shims, and are secured in a threaded cell or tube.

Cementing the contact surfaces of lenses used in a compound lens is generally considered desirable, because it helps to maintain the two elements in alignment under sharp blows, keeps out dirt, and decreases the loss of light as a result of reflection where the surfaces contact.

NOTE: Cement used for this purpose must have approximately the same index of refraction as that of crown glass.

A triplet compound lens has all three lenses cemented together. A lens of this type has six surfaces which enable a designer of an optical instrument to plan the best corrections for defects.

CYLINDRICAL LENSES

A cylindrical lens is a lens whose surfaces (one or both) are portions of a cylinder. The power of this lens to converge light rays when its axis is in a vertical position is in the horizontal meridian only; no refraction is produced in the VERTICAL PLANE. When the same lens is turned through a 90° angle, the axis is horizontal and its power to converge light rays is exercised ONLY IN THE VERTICAL PLANE. There are two types of cylindrical lenses, positive and negative, or convergent and divergent.

CONVERGENT CYLINDRICAL LENSES

Convergent cylindrical lenses are used rather extensively for magnifying vernier scales on instruments and also for eyeglasses and the azimuth circle in the 90° prism housing.

A convergent cylindrical lens is shown in part X of illustration 5-30. The shaded portions of the illustration represent planes. In planes

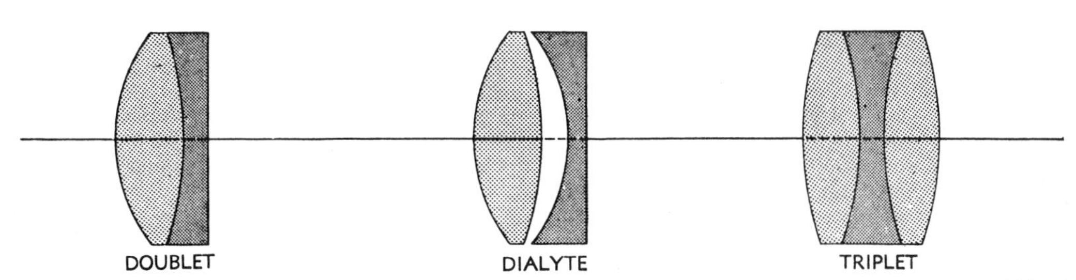

Figure 5-29.—Compound lenses.

which pass through the object point (0) and parallel to the cylindrical surface of the lens, there is NO CONVERGENCE OF LIGHT RAYS. In planes perpendicular to the central plane through the center of the object point (0), light rays are refracted as they pass through the lens and converge at a point beyond the lens with the plane through the middle of the lens.

Observe A, B, and C on the lens. They represent the points at which the planes emerge from the lens. Ray OB in this lens passes straight through the center of the lens and is not refracted; rays OA and OC are refracted as they pass through and converge at I, the focal point. All light rays which come from point O and are refracted by the lens as they pass through it also pass through line I_1I_2, which is a real image of O.

If the refracted light rays were projected back through the lens (dotted lines), they would pass through line I_3I_4 and create a virtual image of object O.

DIVERGENT CYLINDRICAL LENSES

Refer now to part Y of illustration 5-30 to learn what happens to light rays as they pass through a divergent cylindrical lens. Note the object (O), the plane through the lens at B, and also the planes through A and C. Rays of light incident through plane O and B are not refracted. Rays of light incident through points OA and OC are diverged toward the edge of the lens and do not converge to any central point on the central plane, as did the rays through A and C in the convergent lens.

If the rays of light from O through A and C were projected back through the lens, they would pass through the central plane at I_2 and I_1, respectively, and create virtual images where they intersected line I_3I_4.

LENS APERTURE

The aperture of a lens is the largest diameter through which light can enter a lens. The light-gathering ability of a lens is determined by: (1) its aperture, and (2) its focal length.

Take a look now at the lenses in illustration 5-31, both of which have the same diameter but not the same focal length. The arrows on the left, the objects, have the same size; and both lenses receive the same amount of light from the objects, because their apertures are equal.

The bottom lens in the illustration, however, has a longer focal length than the top lens and therefore makes a larger image of the arrow, because the light it receives is spread over a larger area. If the diameters of the two lenses were equal, the lens with the shorter focal length would form a brighter image than the lens with the longer focal length, because the light it receives is concentrated in a smaller area.

Study next illustration 5-32, which shows two lenses with the same focal length but of different diameters. The larger lens at the top therefore forms a brighter image of the object, because it has a greater aperture than the bottom lens and receives more light from the object.

When you compare the light-gathering ability of one lens with another, take into consideration the relative aperture (focal length divided by diameter) of both lenses. To find the relative aperture of a lens, divide its focal length by its diameter. For example, the formula for finding the relative aperture of a lens with a diameter of 2 inches and a focal length of 8 inches is:

$$\text{Relative aperture} = \frac{F}{\text{diameter}} = \frac{8}{2} = 4$$

The relative aperture of this lens is therefore, generally written as f:4.

If you have two lenses with different relative apertures, you can tell which one will form the brighter image by using the formula. Suppose, for example, that you have two lenses with relative apertures of f:4 and f:2, respectively. If both lenses have the same diameter, the focal length of the f:4 lens is twice that of the f:2 lens. Use F_1 for the focal length of the f:2 lens and F_2 for the focal length of the f:4 lens in the formula and solve and you get:

$$\text{Relative aperture} = \frac{F}{\text{diameter}}$$

$$2 = \frac{F_1}{d}, \text{ and } 4 = \frac{F_2}{d}$$

$$F_1 = 2d, \text{ and } F_2 = 4d$$

If the focal lengths of these two lenses were equal, the f:2 lens would be twice the diameter of the f:4 lens. Let d_1 represent the diameter of the f:2 lens and d_2 represent the diameter of the f:4 lens in the formula and solve and you get:

$$2 = \frac{F}{d_1}, \quad \& \quad 4 = \frac{F}{d_2}$$

$$d_1 = \frac{F}{2}, \quad \& \quad d_2 = \frac{F}{4}$$

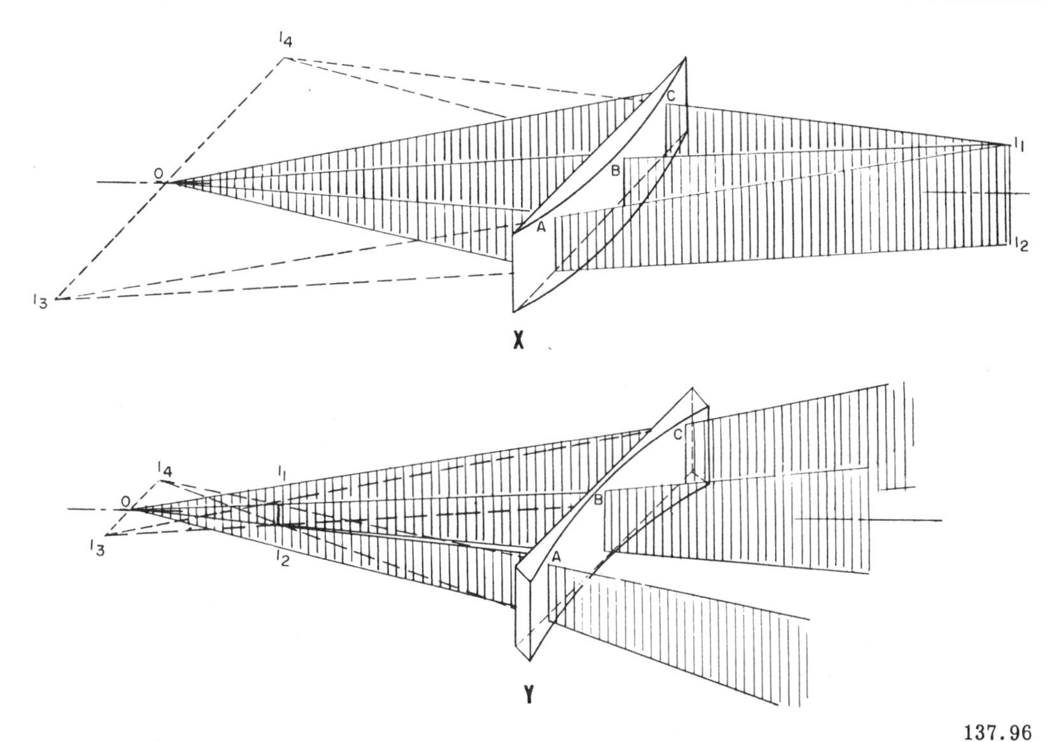

137.96
Figure 5-30.—Cylindrical lenses.

137.97
Figure 5-31.—Passage of light through lens aperture.

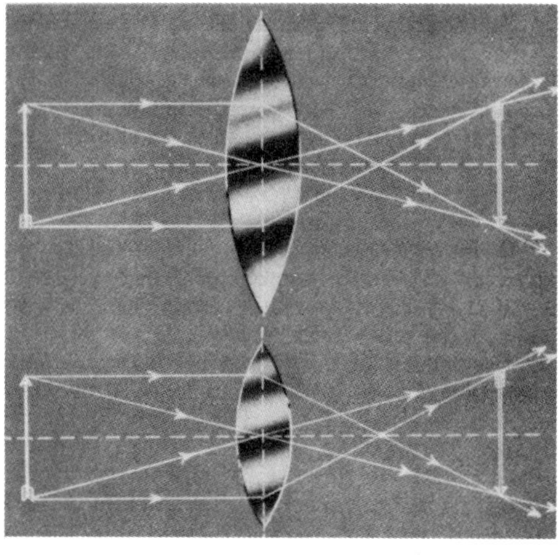
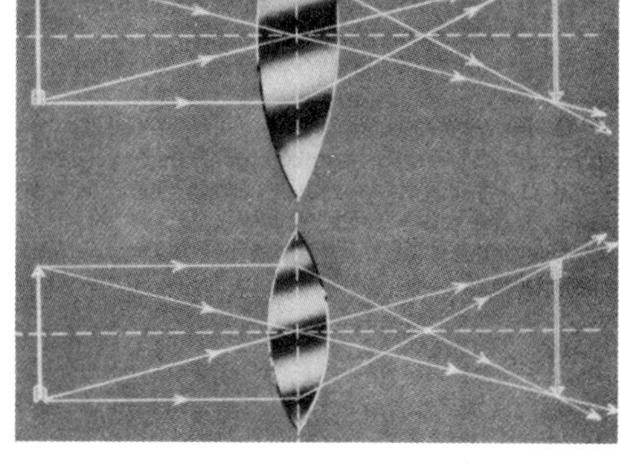

137.98
Figure 5-32.—Image brightness increased by enlarged lens aperture.

In both examples, the f:2 lens forms the brighter image; because BRIGHTNESS OF THE IMAGE is proportional to the light-gathering ability of the lens, and the relative image brightness of two lenses is inversely proportional to the square of their relative apertures.

The relative image brightness of the two lenses just considered (f:2 and f:4) may be determined by using the formula, as follows:

$$\text{Relative image brightness} = \frac{(4)^2}{(2)^2} = \frac{16}{4} = 4$$

This means that the image formed by the f:2 lens is four times as bright as the image formed by the f:4 lens.

LENS ABERRATIONS

Aberration in a lens is an image imperfection which prevents the lens from forming a true reproduction of an object, because the light rays do not converge to a single focus. Aberrations result from a variety of conditions, some of which you studied in chapter 2, Optical Class. The general types of aberration are: (1) chromatic, (2) spherical, (3) astigmatism, (4) coma, (5) curvature of the field, and (6) distortion.

CHROMATIC ABERRATION

You learned in chapter 3 that when white light is refracted through a prism it disperses the light into rays of different wavelengths to form a spectrum. The rays of different colors are refracted to different extents, as illustrated in figure 5-33. Observe that violet rays are refracted most and that red rays are refracted least.

Because a lens may be considered as composed of an infinite number of prisms, as shown in figure 5-34, dispersion also occurs in a lens when light passes through it. Dispersion in a lens produces an optical defect known as chromatic aberration, which is present in every uncorrected single lens. The violet rays focus nearer to the lens than the red rays, and the other rays focus at intermediate points. The lens therefore had different focal lengths for different colors of light and an image created by the lens is fringed with color.

Chromatic aberration may be corrected by proper spacing between lenses, and also by adjusting the curvatures of the lenses. See figure 5-35, part A of which shows how a portion of the aberration can be diminished by equalizing the deviation at the two surfaces of a lens. Part B of this illustration shows how chromatic aberration in a lens can be corrected by a compound lens, one part of which is positive (convergent) and the other part of which is negative (divergent). As you learned previously in this training course, a lens with positive dioptric strength is made of crown glass and a lens with negative dioptric strength is made of flint glass.

Since crown glass is more strongly convergent for blue rays than for red rays, and the flint glass is more strongly divergent for blue rays than for red rays (fig. 5-33), the high color dispersion of the flint divergent lens sufficient to compensate for the lower color dispersion of the crown convergent lens, without complete neutralization of its refractive power. Note in part B of illustration 5-35 that the two rays come to a focus. A compound lens designed in this manner is called an achromatic lens.

SPHERICAL ABERRATION

Spherical aberration is a common fault in all simple lenses. In a convergent lens, refracted light rays through its center do not intersect rays refracted through other portions of the lens at a single point on the optical axis. Study figure 5-36.

The outer rays of light in illustration 5-36 intersect the optical axis closer to the lens; the more central rays intersect the optical axis at a greater distance from the lens. Failure of the refracted rays passing through the lens to intersect the optical axis at a central point causes a blurred image.

Take a look now at illustration 5-37, which shows rays of light passing through a divergent lens and the imaginary extension of the refracted rays. Intersection of outer and inner rays of light on the optical axis of this lens is opposite that of refracted rays from a convergent lens.

The amount of spherical aberration in either a convergent or divergent lens is influenced by: (1) thickness of the lens, and (2) its focal length. A thin lens with a long focal length has less aberration than a thin lens with a short focal length.

One method of reducing spherical aberration, at the expense of light intensity, is to test a lens to find out how much of the area around the optical axis (where the lens is most free of aberration) may be used to form a sharp image, and then to mask out with a field stop all rays which pass through the lens beyond this circle. Study illustration 5-38.

Observe in figure 5-38 the rays blocked by the field stop from passage through the lens. This field stop is a flat ring or diaphragm made of metal (or other suitable opaque material) to mask the outer portion of the lens. The stop prevents rays from striking the lens and thus reduces the amount of light which passes through it.

Spherical aberration in a lens can be minimized also by BENDING THE LENS, which can be accomplished by increasing the curvature of one surface and decreasing the curvature of the other surface. This process retains the same focal length of the lens but reduces the amount of aberration.

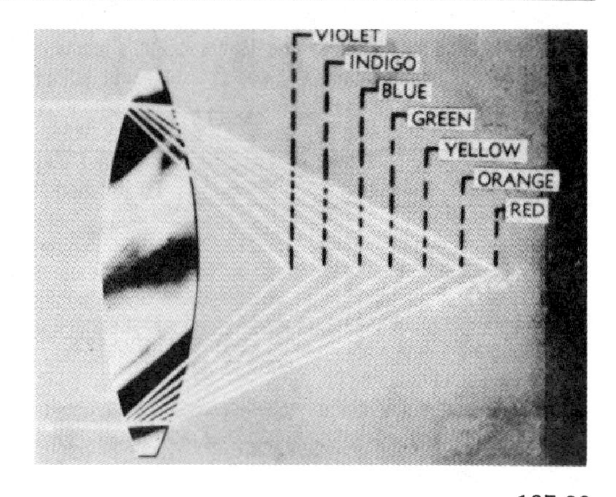

137.99

Figure 5-33.—Chromatic aberration in a lens.

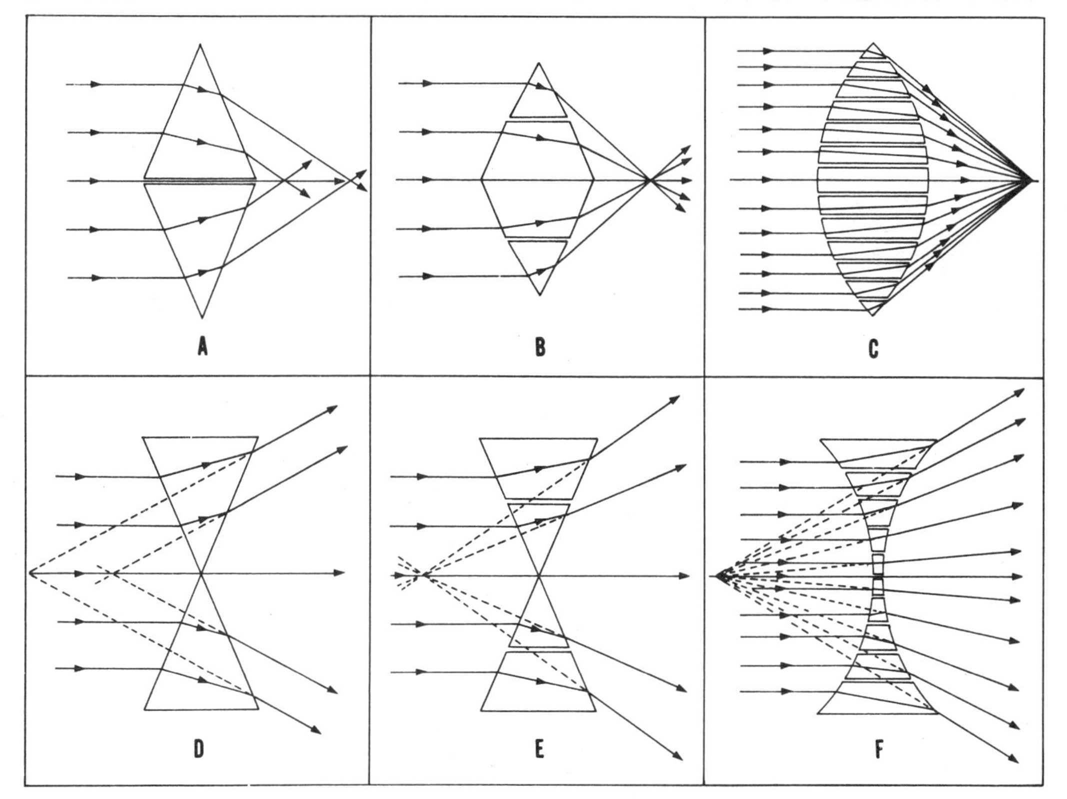

137.100

Figure 5-34.—Lenses constructed from prisms varying in number, size, and shape (principle of refraction shown).

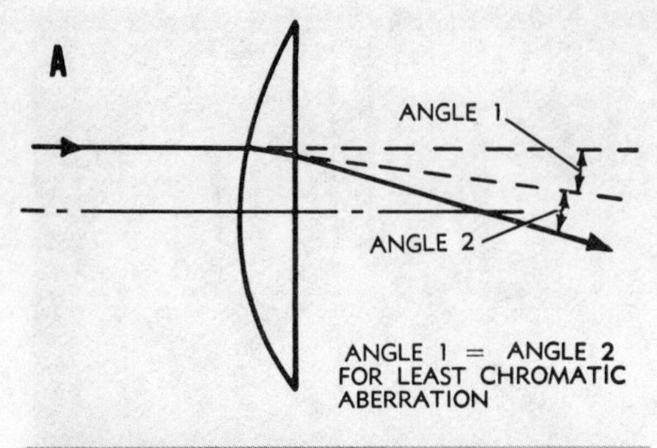

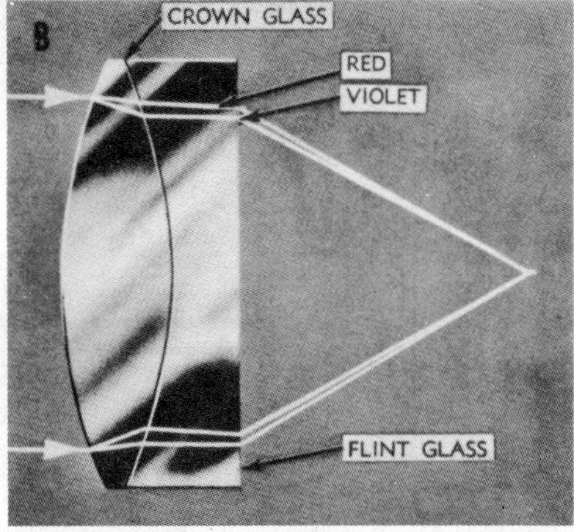

137.101

A. Correction for least chromatic aberration by curvature of the lens.
B. Correction for chromatic aberration by a compound lens.

Figure 5-35.—Correction of chromatic aberration in a lens.

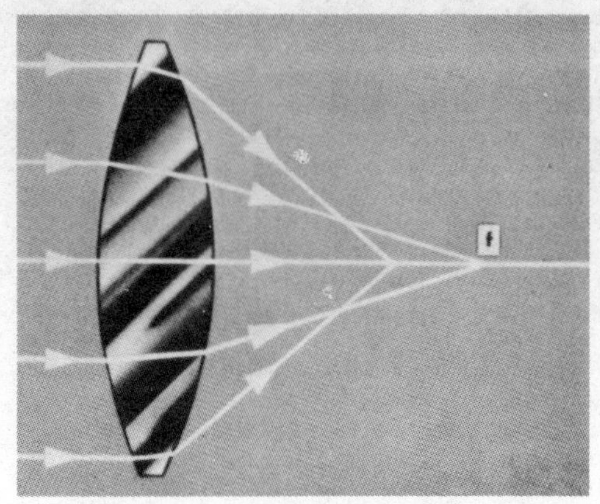

137.102

Figure 5-36.—Spherical aberration in a convergent lens.

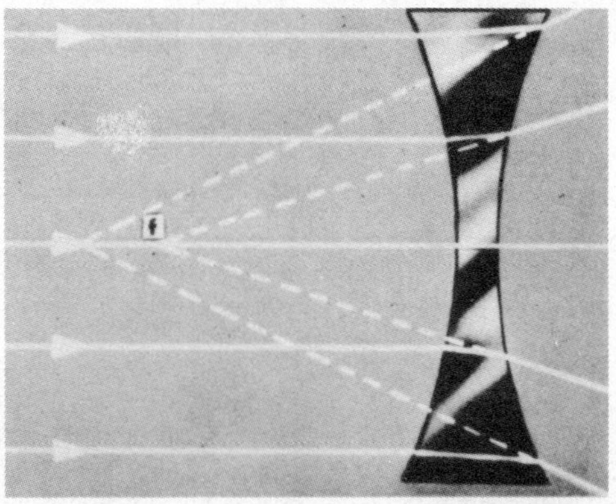

137.103

Figure 5-37.—Spherical aberration in a divergent lens.

In telescopes, spherical aberration is reduced by placing the greater curvature of each lens toward the parallel rays to make the deviation of the rays at each surface nearly equal. In order to reduce the amount of spherical aberration to a minimum, the angle of emergence of a ray (e, fig, 5-38) must equal its angle of incidence (i). In keeping with this rule, telescope objectives are assembled with the crown side facing forward.

Spherical aberration in fire control instruments is generally eliminated by a compound lens (fig. 5-39). The concave curves of the divergent lens neutralize the spherical aberration of the convex curves of the convergent lens. Proper refractive power of the compound lens, however, is retained by selecting two single lenses with correct indices of refraction to form the compound lens.

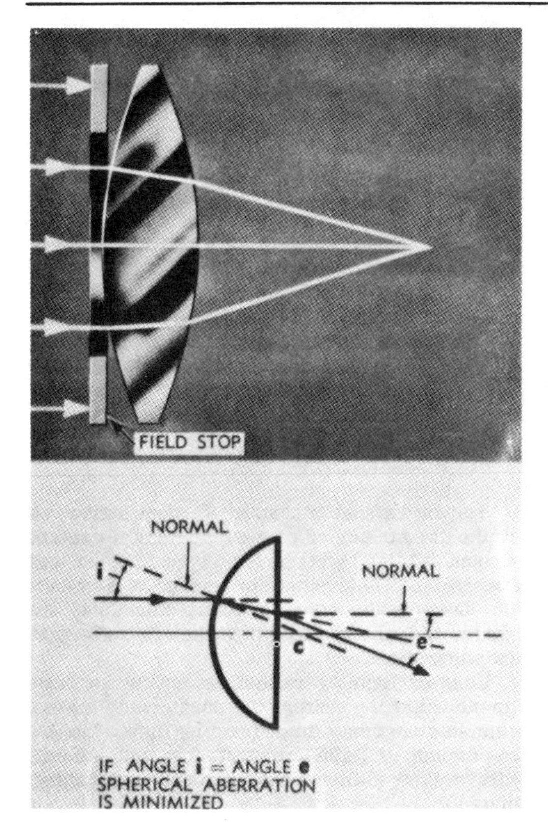

137.104
Figure 5-38.—Reduction of spherical aberration by a field lens.

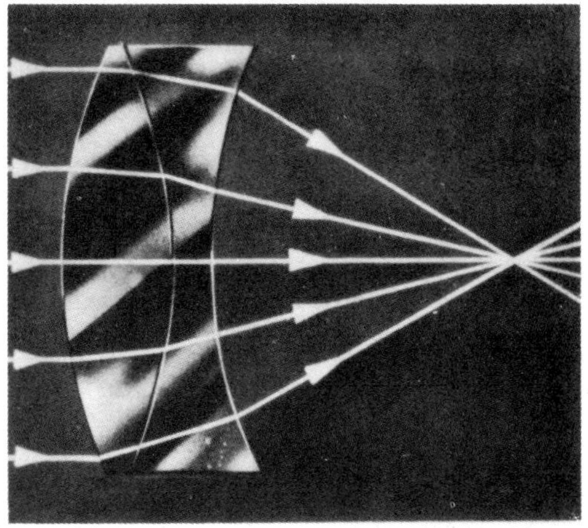

137.105
Figure 5-39. Elimination of spherical aberration by a compound lens.

COMA

Coma is caused by unequal refracting power of concentric ring surfaces or various zones of a lens for rays of light which come from a point a distance off the optical axis. Rays from various surfaces come to a focus at slightly different points, resulting in a lack of superimposition of the rays. Coma appears as blurring of the image for points off the optical axis.

The image of a point of light is formed by a cone of light rays refracted through a relatively wide portion of a lens. In order for them to form a sharply defined point of light, the rays which pass through the concentric circular zones (or rings of varying thickness of the lens) must come to a focus at exactly the same place in the focal plane.

In a lens which is producing coma, rays of light originating at a point located off the optical axis and refracted through the inner zone form a well-defined image of the point. Rays refracted through the next zone, however, form a larger, less-defined image of the point, which is offset slightly from the first. The image formed by each successive zone is larger, less-defined, and farther removed from the initial point of light, as illustrated by part A of figure 5-40. Displacement of the successive images is in a direction TOWARD OR AWAY FROM the center of the lens.

The total image of the point offset from the optical axis may be a blur in any of a wide variety of patterns—egg, pear, or comet. See part B of illustration 5-40. The name COMA COMES FROM the resemblance of the blur to a comet.

When viewed under a microscope, a point of light influenced by coma may have a very fantastic shape, as a result of the effects of all types of aberration upon it. Because coma causes portions of points of light to overlap others, the result is BLURRED IMAGES OF OBJECTS IN THE PORTION OF THE FIELD AFFECTED BY COMA.

Coma can be corrected by compound lenses made of the proper type of glass for each part and with correct curves of the faces. A lens which has been corrected for chromatic and spherical aberration, plus coma, is called an APLANATIC LENS.

ASTIGMATISM

Astigmatism is a lens aberration which makes it impossible to get images of lines equally sharp when the lines run at angles to each other. This optical defect is found in practically all lenses except some relatively complex lenses designed to eliminate this condition.

A perfect lens would refract rays from a point of light to a sharply defined point of light on the image. Rays of light which form the image are refracted as a cone (fig. 5-41). Cross sections of these cones are circular; and successive circles become smaller and smaller until the focal point (illustrated) is reached.

A lens with properly ground spherical or plane faces DOES NOT show astigmatism for points near the optical axis, but it DOES show astigmatism for points at a considerable distance from the axis. The face of the lens is then at an oblique angle to incoming light rays. Cross sections of cones of light refracted by the lens become successively narrow ovals until they are a line in the vertical focal plane. They then are broader ovals and eventually are circular, at which time they again become a line in the horizontal focal plane at right angles to the first line. Study illustration 5-42 carefully. Between the two focal planes (horizontal and vertical) is an area known as the CIRCLE OF LEAST CONFUSION, in which plane the MOST SATISFACTORY IMAGE is formed.

The best way TO REDUCE ASTIGMATISM in a lens is through the use of a combination of several lenses, in the same manner explained for eliminating spherical and chromatic aberrations. When lenses made of optical glasses with different indices of refraction are ground to different curvatures, the various types of aberration CANCEL EACH OTHER.

A lens designer has a difficult task in his endeavors to eliminate aberration in a lens. Anything he does to correct one type of imperfection usually affects other types of aberration. He must consider many variables, including:

1. Index of refraction of different kinds of glass.
2. Difference in dispersion in various types of optical glass.
3. Curvature of refracting surfaces.
4. Thickness of lenses and distance between them.
5. Position of stops along the optical axis.

NEWTON'S RINGS

If convergent and divergent lenses of slightly unequal curvature are pressed against each other, irregular COLORED BANDS or patches of color appear between the surfaces. See figure 5-43. The pattern you see in this illustration is called NEWTON'S RINGS, after Sir Isaac Newton, who first called attention to it. These rings constitute a defect in a compound lens; but the rings can be used advantageously for testing the accuracy of grinding and polishing lenses.

LOSS OF LIGHT

As you learned in chapter 3, when light rays strike the surface of a lens or prism, a certain amount of the light is lost by reflection and absorption. The greater the amount of elements you have in an optical system, therefore, the greater the amount of light you lose by reflection and absorption.

Loss of light by reflection can be reduced considerably by coating the surfaces of optical elements used only for refracting light. The extra amount of light transmitted by instruments with various elements coated produces brighter images.

DISTORTION

Distortion is a form of spherical aberration in which the relative location of the images of different points of the object is incorrect. If a straight line is imaged by a lens with this defect, the line is curved. Distortion of this type in instruments of high magnifying power is a serious defect, because the amount of distortion present is increased in proportion to the power of the instrument.

When an object is held close to a lens, particularly harmful distortion is caused by the refraction of rays of light from different points of the object by dissimilar portions of the lens. When an off-center line located close to the lens extends across the field, rays from the middle of the line strike the lens nearer its center and refract at different angles than do rays refracted from near the margin of the lens, thereby giving a curved appearance to the image.

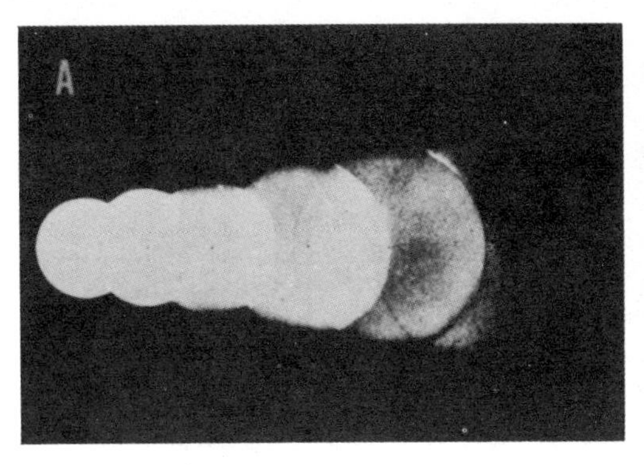

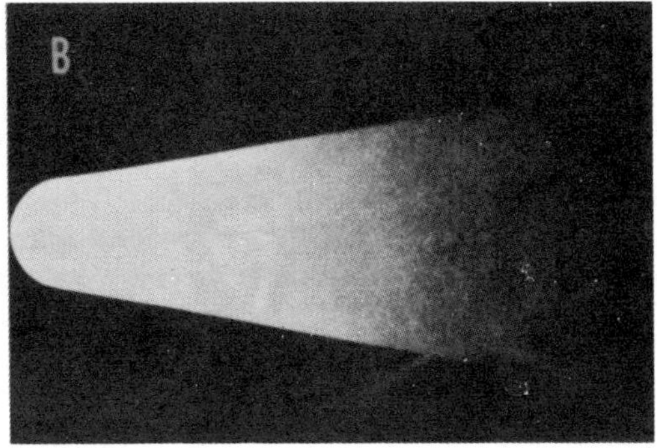

137.106

A. Formation.
B. Appearance after formation.
Figure 5-40.—Coma.

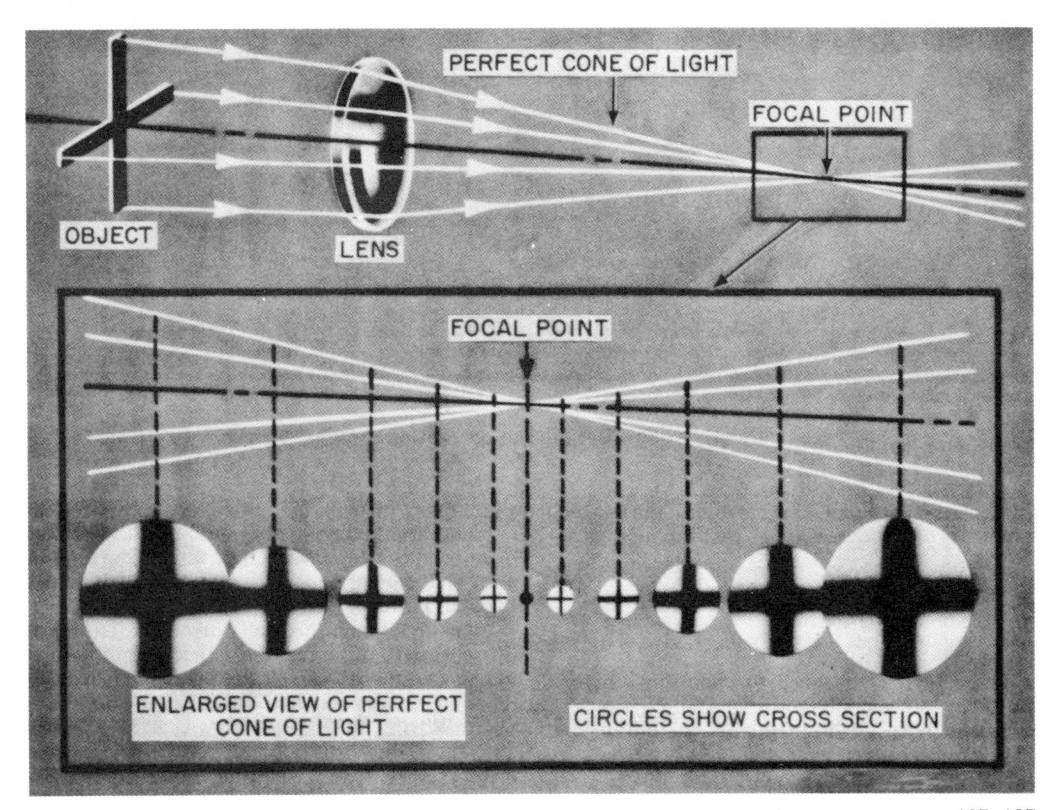

137.107

Figure 5-41.—Refraction of light by a perfect lens.

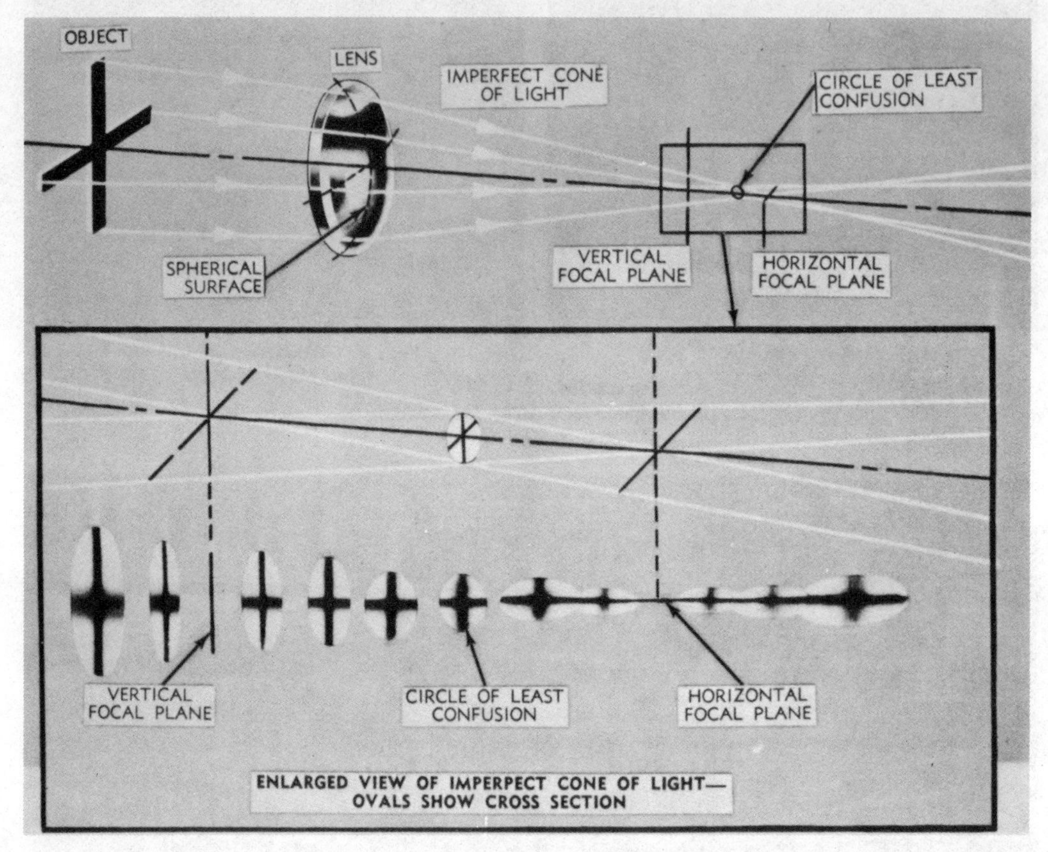

137.108

Figure 5-42.—Astigmatic refraction of light.

The most common types of distortion in a lens system are discussed briefly in the next paragraphs.

Pincushion Distortion

Some terms you need to understand before you study pincushion (hourglass) distortion are discussed now. These terms are the definitions of different types of rays and light pencils.

Marginal rays are rim rays which pass through a lens at points remote from its center. Paraxial rays are rays close to the center of a lens. Oblique centric pencils are cones of light which pass through the optical center of a lens at a considerable angle to the optical axis. Paraxial pencils are pencils of light along the optical axis, and eccentric pencils are those which pass through the lens near the rim.

Paraxial magnification (through a lens near the optical axis) is actually less than the true magnification of a lens. Spherical aberration, therefore, appears as PINCUSHION DISTORTION WHEN A VIRTUAL IMAGE IS VIEWED THROUGH AN UNCORRECTED LENS. Take a look at C and D in illustration 5-44. The extensions of such a virtual image are curved away from the lens, creating curvature of the image (part Y, fig. 5-44).

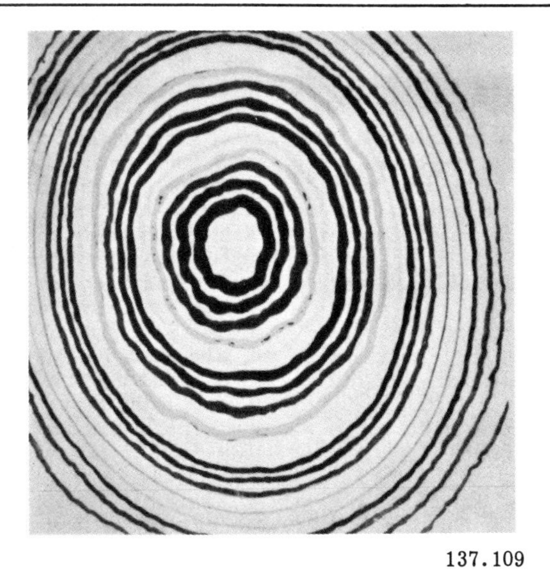

137.109
Figure 5-43.—Newton's rings.

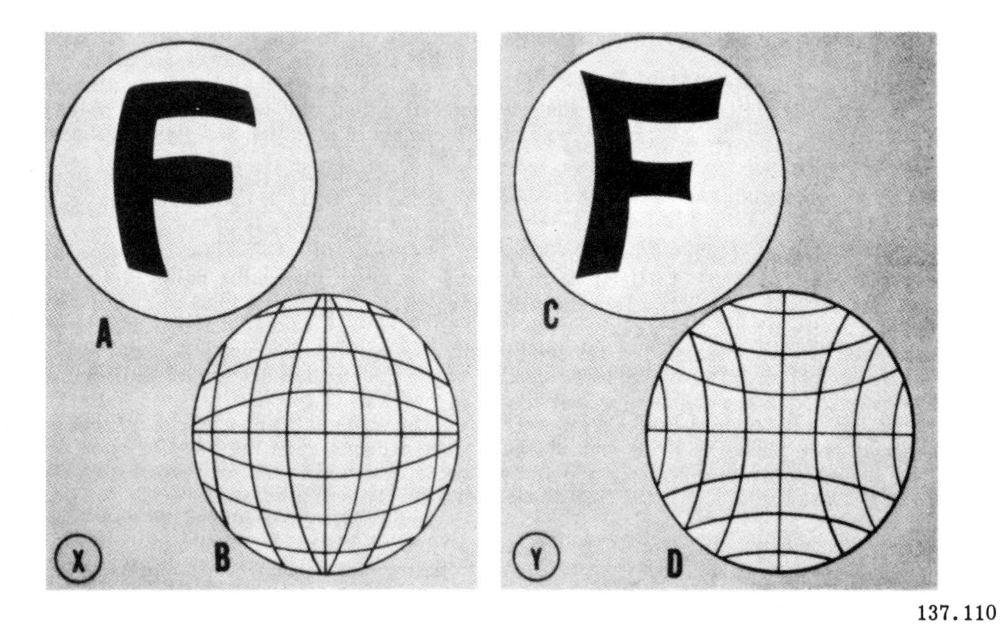

137.110

X. Barrel distortion.
Y. Pincushion distortion.
Figure 5-44.—Distortion.

Barrel Distortion

Marginal light rays are refracted more than paraxial rays and therefore cross the optical axis closer to the lens than paraxial rays. Oblique centric pencils also focus closer to the lens than paraxial pencils. As the real image is formed where the eccentric pencils and the centric pencils come to a focus, such an image formed by an uncorrected lens is curved with the extremities of the image close to the lens (part X, fig. 5-44).

This curvature of image (fig. 5-45) is especially troublesome in wide-angle instruments. On a screen, this image gives barrel distortion (A and B, fig. 5-44); and it is impossible to focus clearly all of the image on the screen at one time. In your study and work as an Opticalman, however, you will learn that if the screen has the same curvature as that of the image, there is NO DISTORTION or OUT-OF-FOCUS effect.

Distortion of the image can be partially corrected by using a compound lens made of convergent and divergent lenses with DIFFERENT TYPES OF DISTORTION. In some cases, it is also possible to USE DIAPHRAGMS OR FIELD STOPS to prevent the passage of undesirable marginal rays.

An eyepiece or a lens system free of distortion is called ORTHOSCOPIC, because it gives a flat field of view and a correct image of normal proportions.

PRISMS

A prism is a piece of glass whose surfaces ARE FLAT BUT AT LEAST TWO OF WHICH ARE NOT PARALLEL. PRISMS are generally made from borosilicate crown glass, because it has high resistance to abrasion and damage by atmospheric elements. Some prisms are used for both refraction and reflection in military optical instruments. Much of your repair work in the optical shop concerns them and, therefore, you should understand fully how prisms ACT IN CONTROLLING THE DIRECTION OF LIGHT.

A prism changes the direction of light rays by:

1. Refraction at two surfaces.
2. Reflection at one or two surfaces, through total internal reflection as a result of the angle of incidence of the light, or by reflection from a silvered surface.

Unlike a lens, a prism is a block of glass bound by plane surfaces, and it can be designed to refract and reflect light in numerous ways. The use of prisms in optical instruments, therefore, permits variations in design which otherwise would be impossible. Plane mirrors, for example, are sometimes used to change the angles of light rays, but the silvered surfaces tarnish and cause loss of light—which becomes more serious as the instrument becomes older. A prism, on the other hand, can be mounted in a simpler and more permanent mount and used for the same purpose.

The surfaces of a prism are not easily disturbed, and it can produce more numerous reflection paths than a mirror. Prisms are used singly or in pairs for changing the direction of light from a few seconds of arc (measuring wedges) to as much as 360 degrees.

There are two general types of prisms, refracting and reflecting, both of which are discussed in some detail in the following pages.

REFRACTING PRISMS

Review illustration 3-34, which shows how light is refracted by a prism. Note that the incident ray of light is bent toward the NORMAL of the front face and away from the normal of the rear face (surface). Observe, also, the angle of refraction, which is a measure of the amount of change in direction of a light ray caused by a prism.

Prisms deviate light rays which strike them because the angle of incidence of the rays is less than the critical angle of the glass.

Prisms with two plane surfaces at slight angles which divert the paths of light through angles by refraction instead of reflection are called optical wedges. Optical wedges are used in fire control instruments; they may be used where the angle of deviation required is a matter of fractions of seconds.

The angle at which a wedge diverts a path of light depends upon the relative slant of its two faces, which are inclined toward each other like the surfaces of a house shingle.

Some wedges employed in fire control instruments appear to be disks or plates of glass with parallel surfaces, because the angle betwen the surfaces is so slight it cannot be detected except by actual measurement.

All wedges cause a certain amount of deviation in the path of light which passes through them. Some instruments which use wedges are therefore designed to create a definite amount of initial deviation of a ray of light when it enters the wedge. This deviation is called CONSTANT DEVIATION, WHICH MAKES IT POSSIBLE FOR THE WEDGE TO NEUTRALIZE the deviation in the path of light, or divert the light at a negative angle.

It is possible to change the path of light passing through a wedge by rotating the wedge. See illustration 5-46. The extent to which a wedge diverts the path of light may also be varied by changing the position of the wedge in relation to the other elements of the optical system, as shown in part X of illustration 5-47.

Another method for changing the path of light by prisms is through the use of pairs of wedges geared to rotate in opposite directions. Two or four elements are used and they are referred to as ROTATING WEDGES OR ROTATING COMPENSATING WEDGES. Part Y of figure 5-47 shows how light is refracted by wedges in three different positions.

The dioptric strength of a prism is a MEASUREMENT OF THE DISTANCE THE REFRACTED RAY OF LIGHT DEVIATES FROM THE PATH OF THE INCIDENT RAY AT ONE METER FROM THE PRISM. Study illustration 5-48. A prism of one diopter bends light to such an extent that when a refracted ray travels one meter beyond the prism it deviates a distance of one cm from the path of the incident ray. If a prism has a power of two diopters, for example, the deviation of the refracted light passing through it is 2 cm at a distance of 1 meter from the prism, and so on.

REFLECTING PRISMS

Most of the prisms used in optical systems are reflecting prisms. Deviation of light by a reflecting prism is brought about by internal, regular reflection. Some of the most common types of reflecting prisms are discussed in the following pages.

Right-Angled Prism

A right-angled prism (fig. 5-49) is a prism whose shape, from a side view, resembles an isosecles right-angled triangle. Prisms with this basic shape are used in many ways in optical instruments.

The name of a right-angle prism implies that it gives reflections of 90° only, but the prism can actually be used to give reflections at a great number of different angles. If a right-angle prism is rigidly mounted and only rays of light parallel to the normal on a side opposite the hypotenuse are permitted to enter it, the rays are not refracted upon entering and leaving the prism—they are merely reflected by the hypotenuse at a true 90° angle.

When a right-angled prism is mounted so that the reflecting hypotenuse can be tilted at various angles, it can be very useful in optical instruments. Most of the light the prism receives is at an angle (by incident rays) with the normal, and any ray which strikes the surface at an angle with the normal is refracted.

When you mount a right-angled prism so that the reflecting hypotenuse may be tilted at various angles to the line of sight, the following statements are applicable:

1. Part of the time, light which strikes the hypotenuse is incident at an angle greater than the CRITICAL angle and total internal reflection occurs.

2. At other times, incident rays of light strike the hypotenuse at an angle LESS THAN the critical angle and total internal reflection is possible only IF THE REFLECTING SURFACE of the hypotenuse is silvered. If you put a coating of silver on the reflecting surface of a right-angled prism, you can change the line of sight from 0° to 180°.

Observe in illustration 5-49 that only one reflection is taking place in the prism, which means that the image is REVERTED when reflection takes place in an horizontal plane and INVERTED when the reflection takes place in a vertical plane. If there is more than one reflection acting upon the line of sight in a right-angled prism, however, the attitude of the final image in relation to the original object depends upon the direction in which the second and subsequent reflections occur. This may seem a little confusing, but keep it in mind as you study next a type of prism which makes use of two reflections.

Porro Prism

A Porro prism is actually a right-angled prism used in a different manner. When the hypotenuse of a right-angled prism is used to receive incident rays of light and exit the same rays after the other two faces of the prism reflects them TWICE, the prism is called a Porro prism. Study illustration 5-50, and observe that the line of sight is reflected a total

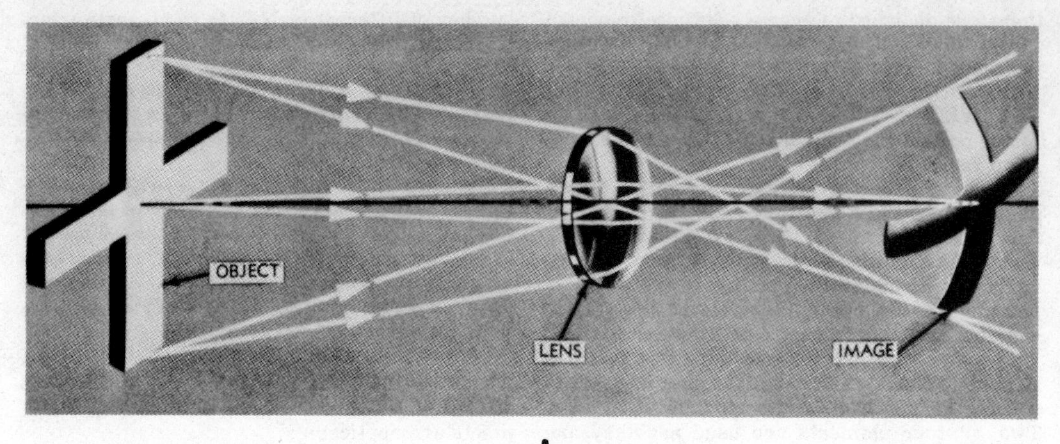

A

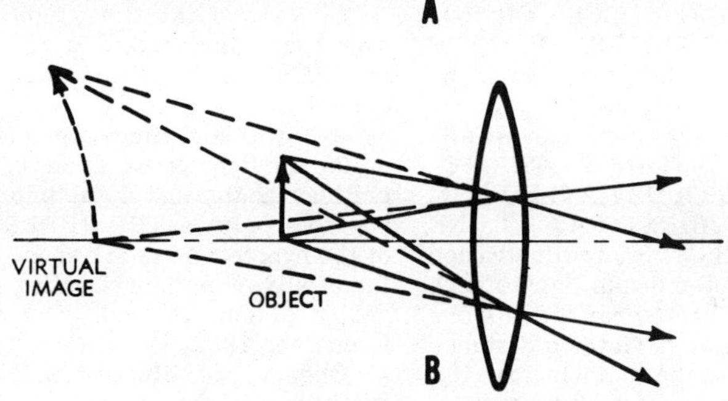

B

A. Curvature of real image. 137.111
B. Curvature of virtual image.
Figure 5-45.—Curvature of the image.

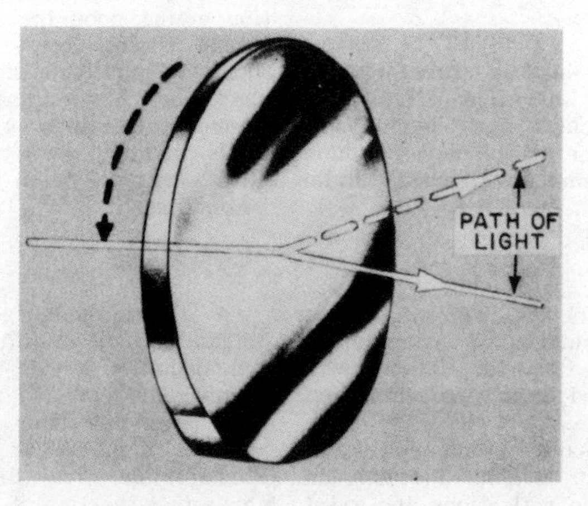

Figure 5-46.—Direction of light changed by a rotating wedge. 137.112

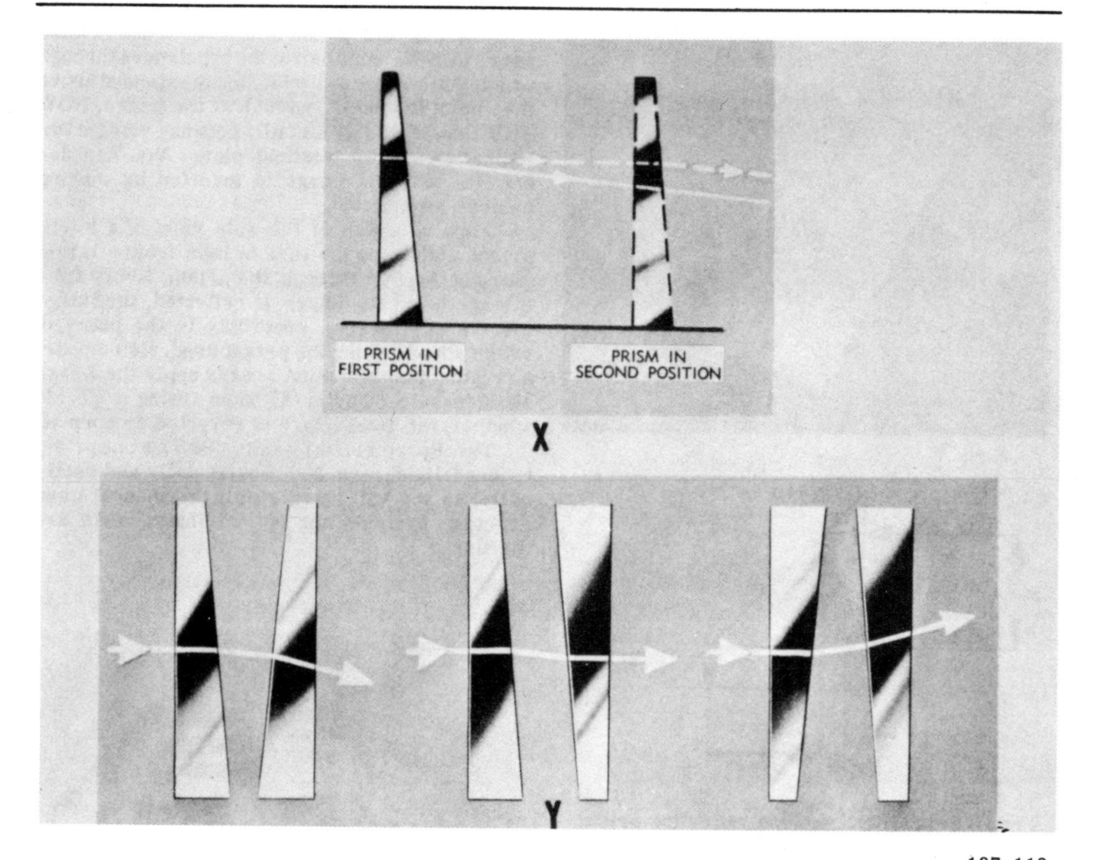

137.113

Figure 5-47.—Path of light changed by pairs of prisms rotating in opposite directions.

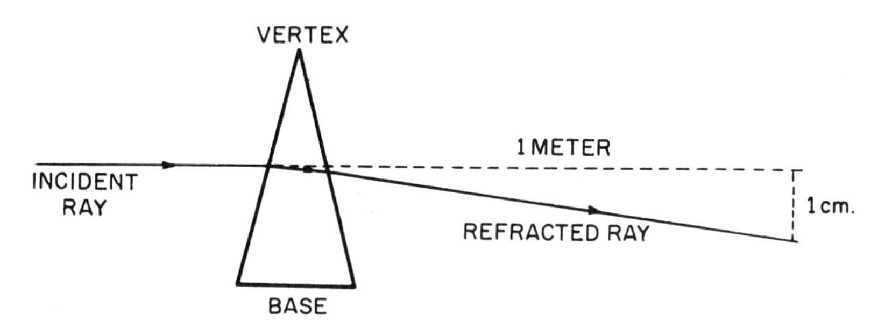

137.114

Figure 5-48.—Prism diopter.

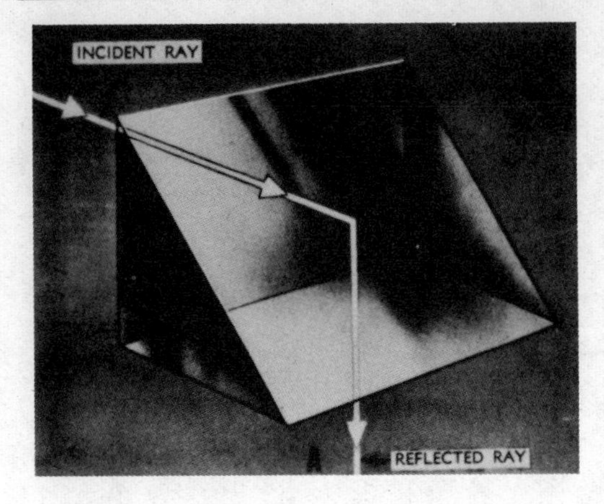

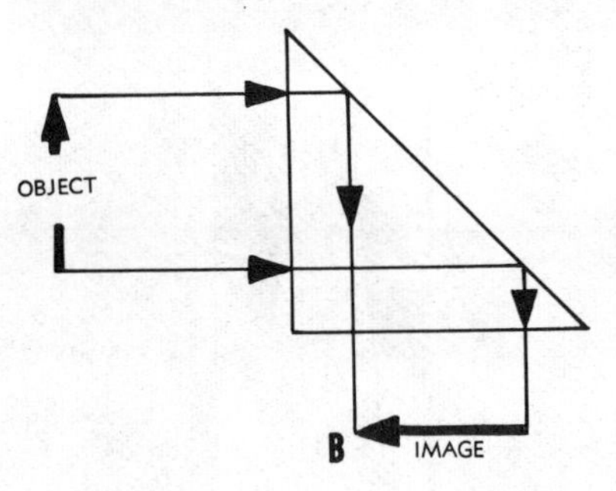

137.115
Figure 5-49.—Right-angled reflecting prism.

of 180°. Note also that the image of F appears reverted; but when we apply the IMAGE ATTITUDE RULE to it, we find that it is NORMAL. You can prove this is true by using the experiment explained next.

Prop this manual up on your table and look at the typed words. The images of the words are normal, because you can read them as they appear in the image. Now take a Porro prism and walk around the table, where you are on the cover side of the book. Then place one-half of the hypotenuse of the prism over the typed page you observed on the other side of the table, and view the typed page in the other half of the prism (hypotenuse) which extends out past the edge of the manual. You can read the words in the image, SO THE IMAGE IS NORMAL.

Now put your prism at the top of the printed page, with the same half of the hypotenuse through which you were viewing the image extended above the top of the page, and look at the image. NOW THE IMAGE IS INVERTED, because you get two reflections in the vertical plane. You can determine why the image is inverted by making another experiment.

Draw a sketch of the side view of a Porro prism and trace the rays of light from a target (upright arrow) through the prism. Every time the arrow or its image is reflected, the image inverts or reverts, according to the plane in which you consider the prism used. Remember, however, that you must always apply the image attitude rule (chapter 4) when trying to decide whether the final image is reverted or normal.

Two Porro prisms can be used as one prism to bend light rays both horizontally and vertically, as you will learn later in this manual when erecting systems for optical instruments are discussed.

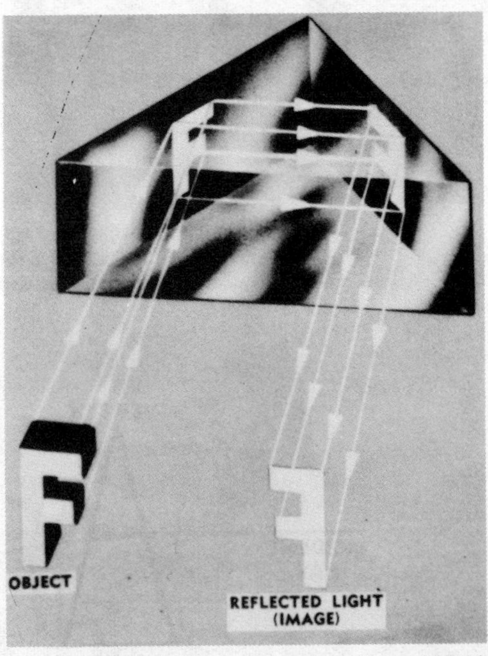

137.116
Figure 5-50.—Image formed by a reflecting prism.

The surfaces of Porro prisms act as plane mirrors and transmit images in practically the same manner as two mirrors placed at right angles, as you learned in chapter 4. The surfaces require a silver coating ONLY when the angle at which light strikes them is less than the critical angle of the material from which the prisms are made.

Dove Prism

A rotating Dove prism (fig. 5-51) resembles a right-angled or Porro prism with its 90° angle sliced off. Light rays which enter one end of the prism are refracted to the longest face and reflected to the opposite face, from which they are refracted out of the prism in the same direction they were traveling before they entered the prism.

The image formed by a Dove prism is inverted or reverted by a single reflection, but not at the same time; the prism can invert or revert the image in accordance with the plane of the prism. Hold up a printed page as an object and observe it through a Dove prism in the vertical plane. THE IMAGE IS INVERTED. Then rotate the prism about your line of sight and observe how the image also rotates. Note, too, that when you rotate the prism through 90° the image rotates through 180 degrees. Observe, also, that the image is then REVERTED AND ERECT, as compared with the object.

RHOMBOID PRISM

A Rhomboid prism consists of two right-angled reflecting prisms built as one piece. You may also consider it as a block of glass with the upper and lower and opposite faces cut at an angle of 45° and parallel to each other. Study illustration 5-52.

A rhomboid prism has two parallel reflecting surfaces which provide two reflections in the same plane and transmit the image unchanged. It does NOT INVERT OR REVERT THE IMAGE OR CHANGE THE DIRECTION OF LIGHT RAYS, but it OFFSETS the light rays from their original direction. This action results from double reflection without reversal of the direction of light.

Regardless of the manner in which you hold or rotate a Rhomboid prism about the line of sight, the image it produces is ALWAYS ERECT AND NORMAL. The only purpose this prism serves is to OFFSET the line of sight, in order to make the new line of sight parallel to the old line of sight.

PENTA PRISM

A penta prism, illustrated in figure 5-53, reflects light rays through an angle of 90° by reflections from two silvered surfaces (part B, fig. 5-50) at a 45° angle to each other. If reflection takes place in the horizontal plane, it does not INVERT or REVERT the image.

A penta prism is useful in rangefinders in which the angles of the light rays must remain constant, and where the rays must be reflected through an angle of 90° and the image must be neither reverted nor inverted.

Amici or Roof-Angle Prism

An Amici prism (fig. 5-54) is a one-piece prism which deviates light through an angle of 90° while (at the same time) it inverts and reverts the image. It may be considered as made up of a right-angled reflecting prism with the hypotenuse face or base replaced by two faces inclined toward each other at an angle of 90 degrees. The last two faces form the ROOF and give the prism its name. When this prism is inserted in the optical system of a telescope, it bends the light rays within the instrument through and angle of 90° and also erects the image. This is a small prism but it transmits a great amount of light.

An Amici prism is used as an erecting prism in an elbow telescope. The light ray enters one short face (fig. 5-54), where it strikes the reflecting surface of one side of the roof and is reflected to the other side of the roof, from which it is reflected outward at right angles to the direction in which it entered the prism.

PRISM ABERRATION

Most prisms used in fire control instruments reflect light; other prisms are used to refract light. When a prism reflects light, there is no chromatic aberration, because the light rays are not dispersed. When a prism refracts light, there is chromatic aberration.

Chromatic aberration in a prism can be corrected by cementing two prisms together, each of which is made of a different kind of glass. See illustration 5-55. The prism in this illustration with the larger refracting angle is made of crown glass and refracts light rays. The prism which has the smaller refracting angle is made of DENSE, flint glass and disperses the colors of light primarily as a result of its greater density.

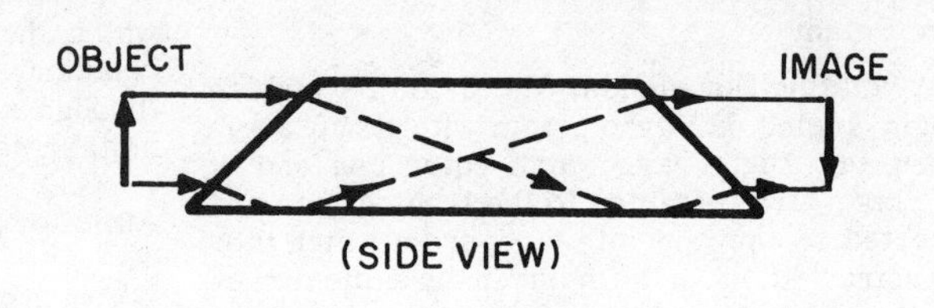

137.117

Figure 5-51.—Rotating Dove prism.

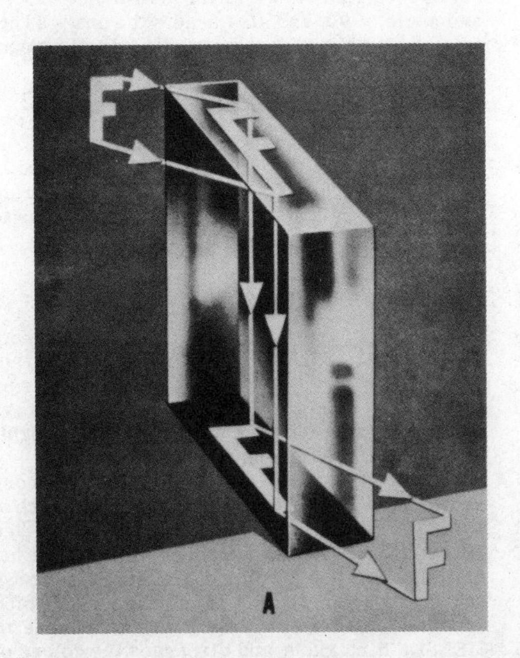

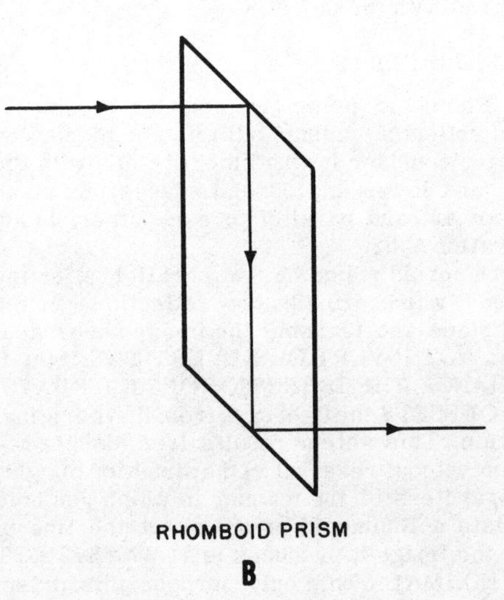

137.118

Figure 5-52.—Rhomboid prism.

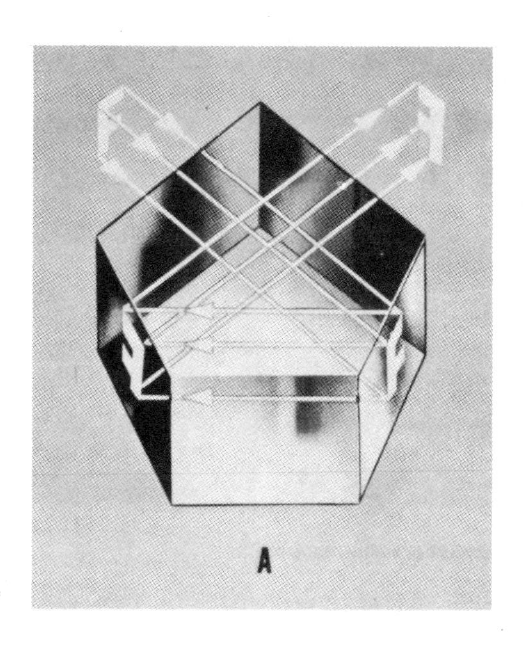

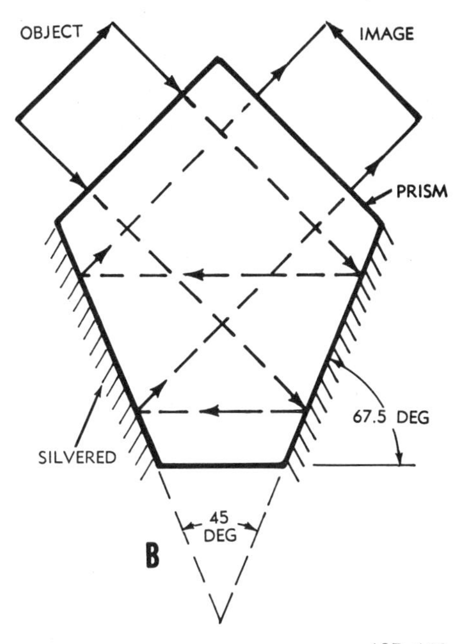

137.119

Figure 5-53.—Penta prism.

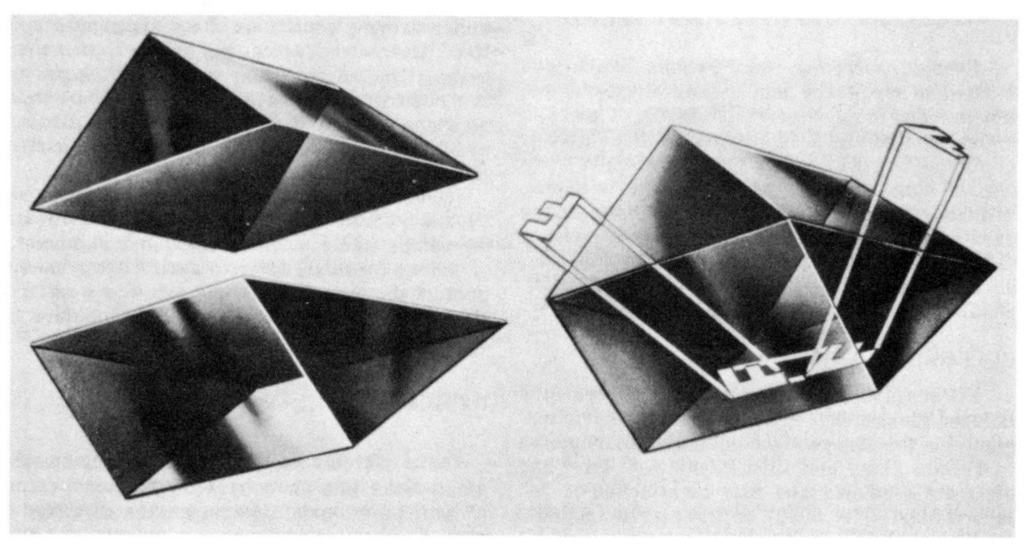

137.120

Figure 5-54.—Amici prism.

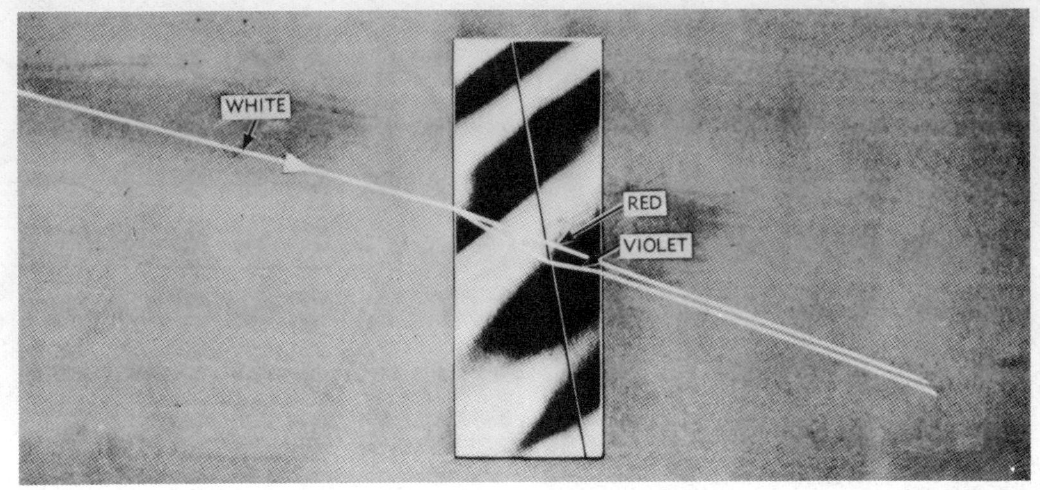

137.121

Figure 5-55.—Correction of chromatic aberration in a prism.

Because of its ability to disperse light, a flint prism neutralizes the dispersion caused by the crown prism without entirely neutralizing the deviation of the path of light.

MISCELLANEOUS OPTICAL ELEMENTS

Lenses, mirrors, and prisms which you studied in chapter 4 and in this chapter form images and hold them in the line of sight in optical instruments. In addition to these optical elements, however, other types of elements are used in some instruments for specific functions, without effect on the line of sight or the image created. These optical elements are known as: (1) color filters, (2) reticles, and (3) windows, each of which is discussed in sufficient detail to enable you to understand its function.

COLORED FILTERS

Filters (sometimes called ray filters) are colored glass disks (with plane parallel surfaces) placed in the line of sight in optical instruments to reduce glare and light intensities. They are separate elements and may be attached or detached (part A, fig. 5-56), or they may be mounted in a manner which makes insertion or removal from an instrument easy, as shown in part C of figure 5-56.

Some of the different types of colored filters employed in optical instruments in order to improve visibility under varying conditions of light and atmosphere are amber, blue, green, red, smoke, and yellow.

Amber and red filters are generally used under varying conditions of fog and ground haze. Red filters are also employed for observing tracer fire. Amber and yellow filters protect the eyes from reflections of sunlight on water and glare from various sources. Blue filters are helpful in determining when objects and/or areas are camouflaged.

Greenish-yellow filters have both green and yellow colors in their composition, and they can serve the same purpose as amber and smoke. A smoke (neutral) filter is a dark filter used to protect the eyes from a bright sun or a searchlight. This type of filter is usually too dark for other purposes.

RETICLES

Most reticles used in optical instruments are glass disks with plane parallel surfaces, on one of which appropriate markings are engraved or etched. In some instances, a planoconvex lens is necessary at the point where a reticle is generally mounted and the markings are therefore

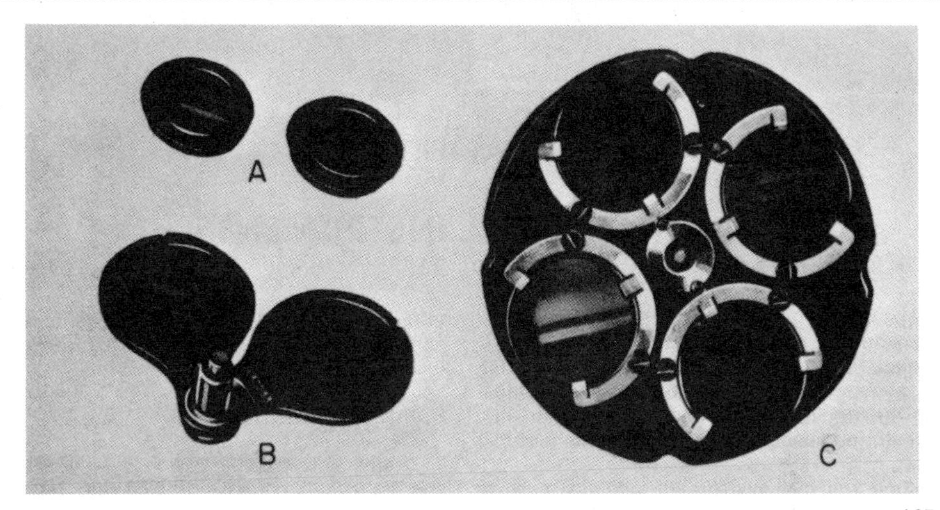

137.122

Figure 5-56.—Color filter mountings.

engraved on it. The function of a reticle is to SUPERIMPOSE reference marks on the view of a target.

Military reticles are made of wire or a filament material, a post (picket), an etching on a plate glass or lens surface, or a punched metal plate. Some are illustrated and explained further in the following chapter on optical instruments.

Crosswires are commonly used in rifle scopes to provide increased transmission of light, by eliminating one piece of glass. A wire reticle does not become dirty. Reticles on glass are generally used in military sights.

WINDOWS

A window is a piece of glass with plane parallel surfaces used to admit light into an optical system and at the same time to prevent the entrance of dirt and moisture. The window glass is actually an optical wedge with very small angles; and it is so mounted that it may be rotated to compensate for the accummulation of errors in the entire system.

CHAPTER 6

BASIC OPTICAL INSTRUMENTS

This chapter pertains to optical elements and the principle of operation of some basic optical instruments. The knowledge you gain by studying these instruments will enable you to understand better the discussion of complicated optical instruments in subsequent chapters of this training manual.

Before you can qualify for advancement in rating to Opticalman 3, you must understand the optical theory of optical instruments; and in order to qualify for advancement to Opticalman 2, you must also know the formulas used to determine the amount of magnification of images by optical instruments. This chapter supplements the information given in chapter 5 on image magnification.

Although we seldom think of them as such, all of us have optical instruments in our bodies. OUR EYES ARE OPTICAL INSTRUMENTS WITH BUILT-IN ADJUSTMENTS WHICH ENABLE US TO SEE OBJECTS CLOSE AT HAND AND AT VARIABLE DISTANCES. An understanding of the functioning of the human eye will therefore help you to comprehend more readily and clearly the operation of optical instruments used in the Navy.

THE EYE AS AN OPTICAL INSTRUMENT

All earthly creatures endowed with sight have either simple or complex eyes, adapted to short and/or long-ranged vision. Animals and birds which inhabit plains and mountains (in the far-open spaces), for example, have such keen visual acuity (sharpness of sight) that they can pick out SMALL OBJECTS AT GREAT DISTANCES. Their ability to do this appears to be associated with small nerve endings in the retinas of their eyes, faster retinal response to motion, and better interpretation by their brains. In other words, vision of these creatures is adjusted to environmental conditions. The other extreme in visual acuity is represented by the short-range, compound eyes of insects and crabs.

STRUCTURE OF THE EYE

Refer now to illustration 6-1, which shows how an eye is constructed and lists its nomenclature. The vitreous humor maintains the bulbous shape of the eye. Refer to this illustration frequently as you study the discussion of the nomenclature.

Coats or Tunics

The three coats or tunics of the eye are: sclera, choroid, and retina, as illustrated. The sclera is the tough, flexible, white portion of the eye. The cornea is the transparent protruding portion of the sclera in the front (center) of the eye.

The middle coat (choroid) is a deep-purple layer composed of veins and blood vessels which provide nourishment. The color is the choroid prevents external light from getting into the eye, except through the cornea. The retina (innermost coat) is a highly-sensitive layer of nerve fibers which transmit visual impressions to the brain.

Refracting Mechanisms

The chief refracting mechanisms of the eye are the cornea and the crystalline lens. The cornea provides the constant part of refraction; the crystalline lens provides variable refraction, which enables the eye to focus on NEAR OR DISTANT OBJECTS. The refracting mechanisms of the eye, therefore, give it the POWER OF ACCOMMODATION, IN ORDER TO SEE OBJECTS FAR AWAY OR CLOSE AT HAND.

The iris (figs. 6-1 & 6-2) is the colored diaphragm in front of the crystalline lens, and ITS

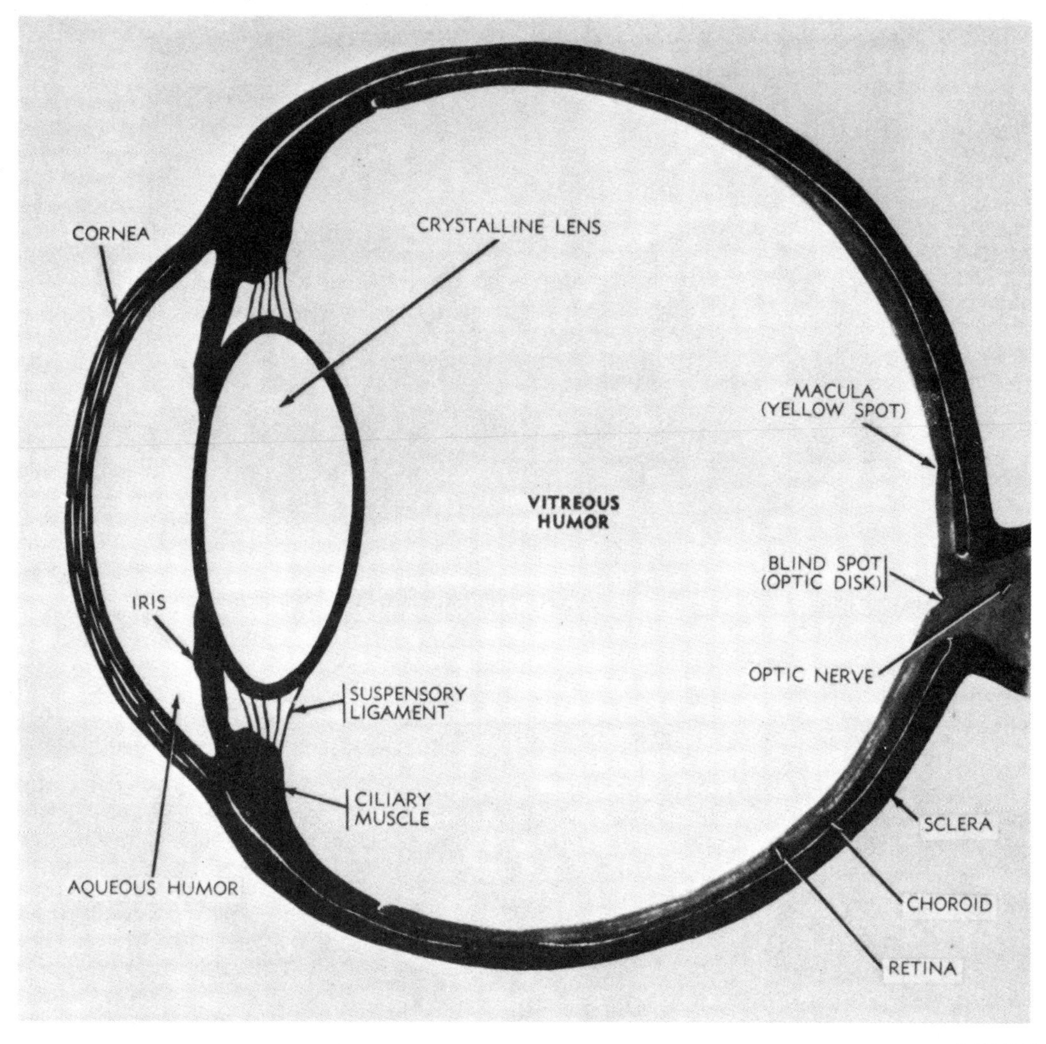

137.123

Figure 6-1.—Construction and nomenclature of the eye.

FUNCTION IS TO CONTRACT AND DILATE AS NECESSARY TO REGULATE THE AMOUNT OF LIGHT WHICH ENTERS THE EYE. In accordance with the amount of illumination, the pupil opening in the center of the iris varies in size from 2 to 7 millimeters—2 mm for intense illumination, 7 mm for faint (night) illumination.

The transparent crystalline lens is suspended by the ligaments and muscles of the ciliary body (part A, fig. 6-3). Observe that the ciliary body encircles the crystalline lens. The front of the lens rests against the aqueous humor and its back rests against the vitreous humor. Ligaments and muscles of the ciliary body COMPRESS AGAINST

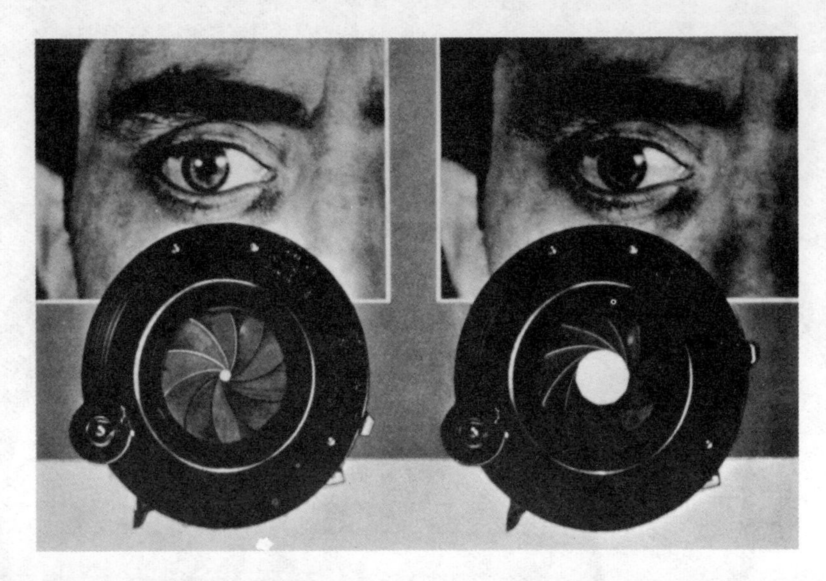

137.124

Figure 6-2.—Comparison of iris with the diaphragm of a camera.

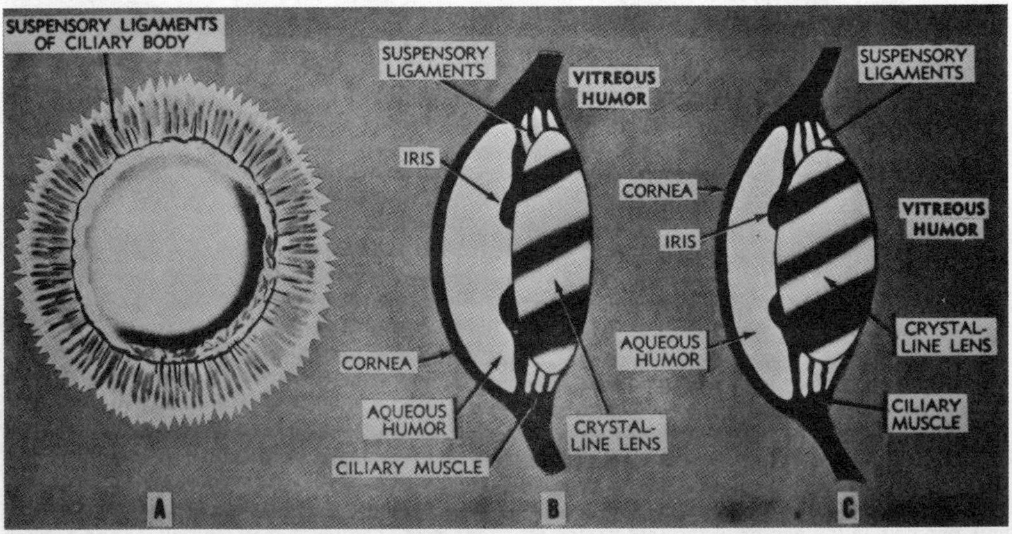

137.125

Figure 6-3.—Suspension and action of crystalline lens.

OR RELEASE THE OUTER EDGES OF THE LENS IN ORDER TO CHANGE ITS REFRACTIVE POWER.

The crystalline lens is double convex, and the front is flatter than the rear (part B, fig. 6-3). When the eye is relaxed, the crystalline lens focuses upon distant objects. Increase in refractive power of the crystalline lens is obtained through compressive action of the muscles of the ciliary body (part C, fig. 6-3), to enable a normal adult eye to view near objects (about 10 inches from the eye, the NEAR POINT OF THE EYE). Because the power of accommodation of the eyes decreases with age, bifocal glasses must be worn to enable the eyes to focus on near objects. Small sections of the glasses are ground to HIGHER REFRACTIVE POWER.

Response Mechanism

The response mechanism of an eye is the area on which images are formed. See figure 6-1. The retina contains a light-sensitive layer of nerve cells (rods and cones) connected to the optic nerve. In a human retina, the light-sensitive layer is at the back of the retina, behind a transparent layer of nerve fibers and blood cells. The macula (spot on the retina) is more sensitive to light than other areas, because it contains mostly cones. The fovea centralis (center spot of the macula) is the most sensitive spot to light because the nerve-fiber and blood-vessel covering of the macula is thinnest at that point, and that area contains cones only. A blind spot occurs at the optic disk because the retina contains neither cones nor rods at the point where it joins the optic nerve.

COMPARISON OF THE EYE AND A CAMERA

Illustration 6-4 shows a human eye superimposed on a camera. Compare each part of the eye with the corresponding part in the camera—lens with lens, diaphragm with iris, retina with film, and sclera with light-proof housing of the camera. As you can see, all parts of a camera must function properly to enable it to form a photograph on sensitized film; all parts of the human eye, likewise, must function properly in order for it to form an image on the retina.

ABERRATIONS OF THE EYE

The optical system of a human eye is afflicted with some of the same aberrations of optical systems which use glass lenses. The refracting surfaces of an eye, however, partially correct spherical aberration, because the surfaces (particularly the front surface of the lens) are not exactly spherical.

A human eye has strong curvature of the field, but this is good because the retina is also curved. Chromatic aberration in the eyes, however, is fairly bad; because when you look at an object, you automatically focus its green and yellow rays on the retina of your eyes. A blue image does not reach the retina; a red image is formed beyond the retina. If available, take a piece of cobalt glass into a dark room and look at a small, bright light through it. The glass transmits red and blue light but absorbs colors in the middle of the spectrum. When you therefore look at a small light through a piece of cobalt glass, you see only out-of-focus red and blue images in the form of halos.

VISION

We can see because the lenses in our eyes project images on the eye retinas, from which the rods and cones in the retinas transmit the images to the optic nerve, which conveys them to the brain for completion of the process. The details of what happens when vision takes place are not fully known; but we do know that the brain converts nerve impulses into pictures. We also know that the visual purple (substance which stimulates rods for keener vision) in the retina is responsible for the brightness or darkness of pictures formed in the brain by impulses sent to it.

NIGHT VISION

Our ability to see in the dark is not fully known. One belief, however, is that the cones on the retina are used for day vision and the rods are used for night vision. Some evidence which supports this theory is that animals which hunt at night and sleep in the day (bats, for example) have retinas composed almost entirely of rods. Such animals as pigeons, on the other hand, go to sleep at dark, and their retinas are composed almost entirely of cones. We have both rods and cones in the retinas of our eyes, and we can see in both dark and light.

As you know, after you spend some time in the dark, your eyes become adjusted to the darkness; and the longer you remain in complete darkness the more acute your night vision becomes. The reason for this is that the visual purple in the rods of your eyes' retinas BUILDS UP TO MAXIMUM CONCENTRATION.

Before lookouts take their positions on the deck of a ship, they generally spend from 20 to 60 minutes in the dark in order to give their eyes a chance to adjust to darkness. Red-colored glasses, however, serve the same purpose, because the rods in the retinas are not sensitive to red light.

COLOR VISION

The manner in which our eyes react to light waves of different lengths so that we can see different colors is not fully known. Thomas Young, an English physician and scientist, and Hermann Helmholtz, a German physicist, promulgated a theory of color vision which is most generally accepted. This theory holds that the retina of the eye has three types of nerve receptors of unequal sensitivity for colors of the visible spectrum. There is one type of receptor for each primary color (red, blue, green), and the greatest amount of sensitivity for each receptor is in a different part of the spectrum.

According to the Young-Helmholts theory, when all three types of receptors are equally stimulated, we see white only. Lack of stimulation gives us the sensation of darkness. If red waves enter our eyes, the waves stimulate the receptors which produce a sensation of red and we therefore see red. When receptors sensitive only to green are stimulated, we see green because the sensation of green was produced on the brain. The reason we see yellow, however, is that the receptors for red and green are stimulated and cause us to see yellow, not red or green singly or in combination. Receptors sensitive to red and blue, when stimulated, cause us to see purple.

Recent experiments on color vision by E. H. Land of the Polaroid Corporation indicate that rays of light do not make a color; instead, they carry information which enables the eye to designate colors for various parts of an image. The color of the image comes from the relative balance of longer and shorter wavelengths, not from the choice of light wavelengths. The real answer to color vision, then, is that the human eye may be some type of electronic computer which differentiates between long and short wavelengths and transforms them into color.

COLOR BLINDNESS

Color blindness is a defect in the eyes of some people which causes them to confuse two or more colors other people can readily distinguish. Less than one percent of women are colorblind; about seven percent of men are colorblind.

Certain diseases, or large doses of some drugs, may produce temporary color blindness; but this defect in the eyes is usually hereditary. Colorblind persons generally cannot distinguish between green, red, and yellow colors, or between blue, blue-green, and violet colors.

RESOLVING POWER OF THE EYES

Ability of an eye to distinguish between extremely fine lines and small angles is called resolving power, which depends upon spacing of the cones in the fovea centralis of the retina. There are no rods in the fovea centralis, but the cones at this spot ARE VERY LARGE. To illustrate, suppose you are looking at two small points of light (fig. 6-5) close together and you desire to know how far apart they must be before your eyes can resolve them—when you can see them as two separate points.

If the image of one point falls on one cone of a fovea centralis and the image of the other point falls on the cone next to it, your eye cannot resolve the two points. The reason for this is that the retina of the eye has no way of distinguishing between the two points and they therefore look like one point. If the two images, however, fall on slightly separated cones, with an unstimulated cone between them, you can see them as two separate points.

If you look at two points 10 inches away, your eyes cannot resolve them until they are approximately .004 inch apart. A normal eye can resolve two points if they are separated by an angle of from 1 to 2 minutes; so we can best express resolving power as an angle. When two adjacent objects become so small in apparent size (part A, fig. 6-5) that further reduction in size results in the eye's failure to separate them, the angle of resolution of the eye has been reached.

VISUAL ACUITY

Acute vision is limited to the fovea centralis area (about .3 mm in diameter) on the macula

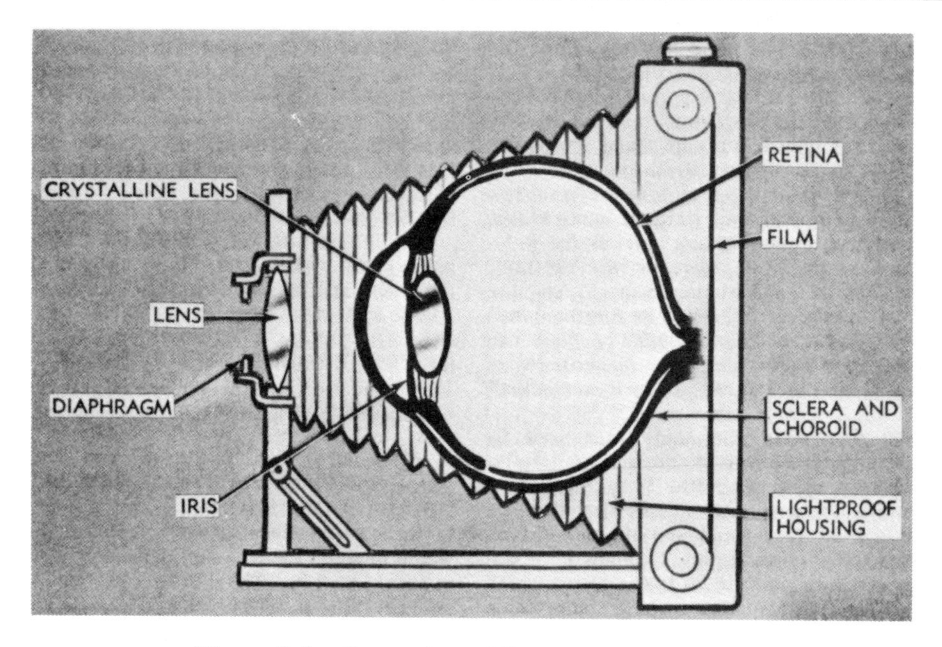

Figure 6-4.—Comparison of the eye with a comera. 137.126

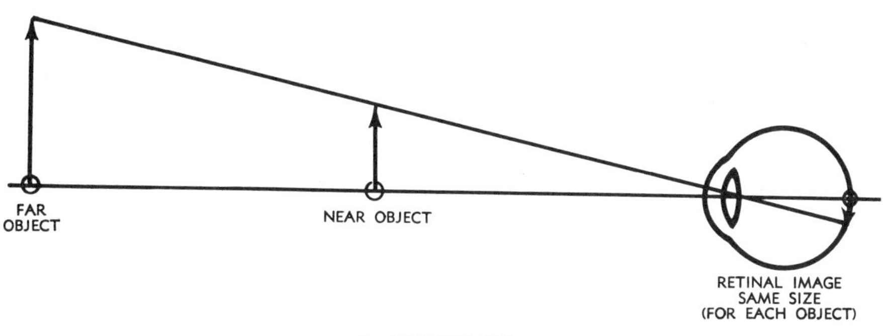

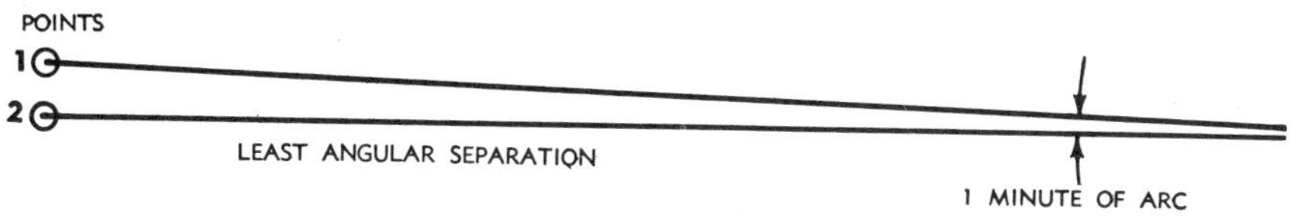

Figure 6-5.—Visual limitations. 137.127

of the eye, which is located on the retina at the visual axis. The cone nerve endings (no rods) at this position have a diameter of .002 millimeters. This area of acute vision covers a true field of view (explained later) generally less than 1° of arc, or a circle about 3.5 mm in diameter at 25 cm distance from the eye.

The least angular separation between any two discernible points in this field of acute vision is normally one minute of arc (part B, fig. 6-5). Coincident readings (on a vernier micrometer), however, can be read closer, because angular displacement of two lines can be distinguished by adjacent cone endings. Some persons can accurately read angular measurement (between two displaced lines) as small as 10 seconds of arc.

Visual acuity is commonly measured by viewing a standard letter of one-minute details.

If you can view the letter E (composed of lines with a width of one minute of angle) at a distance of 20 feet (standard distance), you possess 20/20 vision, which is normal. If you can see at a distance of 20 feet a letter which you should be able to see with perfect vision at 40 feet, your vision is 20/40. Some eyes, on the other hand, have better than average resolving power and can perceive letters which have less than one-minute details. An example of this is 20/15 vision, which means that a person with this type of vision can discern at 20 feet what a normal eye can discern at 15 feet. (Baseball players with this type of vision are usually better hitters.)

BINOCULAR VISION

Both of your eyes are generally identical and operate as a team; the muscles used for adaptation and accommodation dilate, contract, and focus together. Both eyes usually meet with the same light conditions and converge on the same object or field of view; and the images received by them are fused by the brain into a single image. THIS IS COORDINATED, TWO-EYES BINOCULAR VISION.

THE FIELD OF VIEW OF THE EYES IS THE AREA THE EYES CAN OBSERVE WHEN THEY ARE ROTATED IN THEIR SOCKETS. Study illustrations 6-6, which shows the field of view of both eyes and also the binocular field. What the eyes can see without movement IS CALLED THE FIELD OF FIXATION. The field of view of the eyes is 160° horizontally and 70° vertically, as compared with an average telescope with a field of view of about 10 degrees.

When the eyes are stationary (immobile), their field of distant vision is extremely limited; because distinct vision is then limited to a very small central portion of the retina. Portions of the field focused on the remainder of the retina are indistinct; but they help to locate objects of interest, when the eyes turn in their sockets until the image falls upon the fovea and macula to give distinct vision.

As shown in figure 6-6, the field of vision of both eyes includes parts viewed by the right and left eyes, and both eyes together. THE BINOCULAR FIELD EXISTS ONLY IN THAT PORTION OF THE FIELD OF VIEW WHERE THE FIELDS OF THE SEPARATE EYES OVERLAP.

STEREOSCOPIC VISION

Stereoscopic vision is the power of depth perception—THE ABILITY TO SEE IN DEPTH OR THREE DIMENSIONS. This power results from spacing between the pupils of the eyes, which enables them to see objects from slightly different angles. The distance between the eyes (INTERPUPILLARY DISTANCE, IPD) is normally about 64 millimeters.

Your ability to record with each eye a slightly different picture of the image of the same object enables you to see more of one side of a given object with one eye than with the other; but both of your eyes can see farther around the object, thereby enabling you to get a better impression of the object with respect to position, depth, and relation to other objects. This ability is increased by some fire control instruments, because their optical elements increase the virtual distance between the pupils of the eyes. In a military sense, stereoscopic vision implies ability to recognize difference in the range to objects by visual means only. This ability can be developed through training and practice.

You can demonstrate stereoscopic vision by looking at a near object, a small cube, for example. Study illustration 6-7. You can see the left side and front of the cube with your left eye and the front and right side with the right eye. When you look at the cube with both eyes, however, as shown in part C of figure 6-7, you see both sides and the front of the cube, and thus picture it in three dimensions. The brain also fuses the two separate pictures of the eyes into a single image to give the impression of depth.

In like manner, when you observe two objects simultaneously, stereoscopic vision enables you to judge the relative distance of one object from the other, in the direction AWAY FROM YOU.

Your ability to distinguish the relative positions of two objects stereoscopically depends upon the interpupillary distance of your eyes, the distance of the objects from you, and their distance from each other. See illustration 6-8. Other factors of depth perception being equal, the wider your interpupillary distance, the better the appreciation of depth perception you secure through stereovision. In order for you to distinguish the positions of two objects stereoscopically, the distance of the second object from the first object must be approximately equal to the distance of the first object from you.

When you look at two objects and attempt to determine which is farther away, the lines of sight from both eyes converge TO FORM ANGLES OF CONVERGENCE ON BOTH OBJECTS. Study part Y of illustration 6-8. If the angles of convergence to both objects are identical, the objects appear to be the same distance away; but if there is a difference in the angles of convergence to the two objects, one object appears more distant than the other.

Even though the distance between angles of convergence is slight, the brain has the ability to distinguish the difference. Your ability to see stereoscopically, therefore, depends upon your capacity to discern the difference between these angles. Study part A of illustration 6-9, which shows graphically the difference between the angles of convergence shown in figure 6-8.

Angles of convergence become smaller, and the difference between them becomes less discernible, as the objects are moved farther away from you, or as the distance between them is decreased. THIS DIFFERENCE IS KNOWN AS THE DISCERNIBLE DIFFERENCE OF CONVERGENCE ANGLES (part B, fig. 6-9), AND IT IS MEASURED IN FRACTIONS OF MINUTES AND SECONDS OF ARC. This is the procedure followed in determining a person's keenness of stereovision, MEASURED ON A PERCENTAGE BASIS CALLED PERCENT STEREOPSIS. STEREOSCOPIC VISION FOR THE UNAIDED EYE IS EFFECTIVE UP TO 500 YARDS ONLY. This distance, however, can be increased through the use of binoculars or rangefinders, which increase the interpupillary distance between the eyes and therefore increase stereoscopic vision.

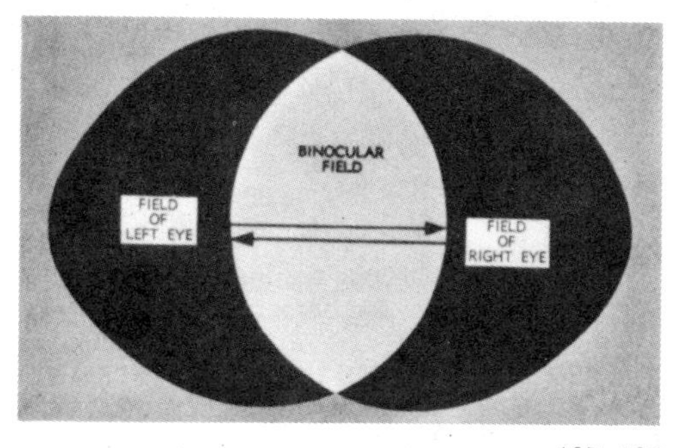

137.128

Figure 6-6.—Field of view of the eyes.

137.129

Figure 6-7.—Stereoscopic vision.

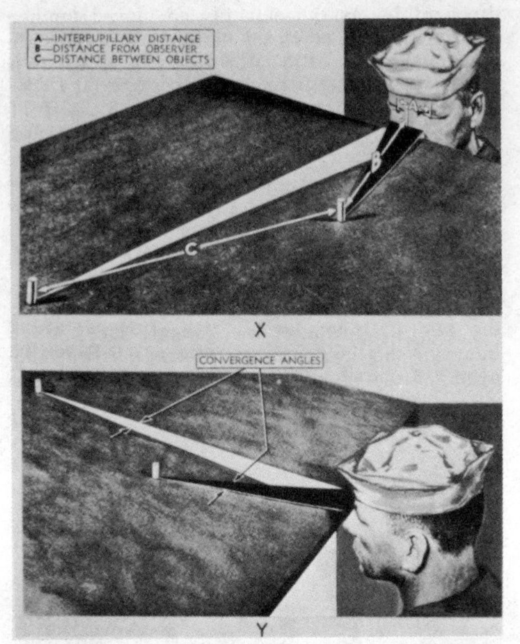

137.130
Figure 6-8.—Distinguishing the distance between objects.

STEREOACUITY, IN CONTRAST WITH VISUAL ACUITY, IS SHARPNESS OF SIGHT IN THREE DIMENSIONS, OR THE ABILITY TO GAGE DISTANCE BY PERCEPTION OF THE SMALLEST DISCERNIBLE DIFFERENCES OF CONVERGENCE ANGLES. The minimum difference which you can discern between two angles of convergence is dependent upon your quality of vision, your training, and conditions which affect visibility.

A well-trained observer can discern an average difference of about 12 seconds of arc; at times, under excellent conditions of observation, this difference may be reduced to 4 seconds of arc for a series of observations. An average, untrained observer should be able to distinguish a minimum difference of 30 seconds of arc between two angles of convergence under normal visibility conditions.

OPTICAL INSTRUMENTS AND THE EYES

Optical fire control instruments may be classified as: (1) monocular, for use by one eye; and (2) binocular, for use by both eyes. Because optical instruments affect functioning of the eyes, certain adjustments must be made to the instruments in order to accommodate them to each eye. A monocular optical instrument, for example, must be so focused (proper positioning of eyepiece, discussed later) that the amount of light which enters the instrument from an object is sufficient to form a distinct image on the retina without undue effort by the muscles of the observer's eyes. The exit pupil (rear opening of eyepiece, illustrated later) must be large enough to admit a maximum amount of light to the pupil of the eye; and stray light must be kept out of the eye.

Interpupillary Adjustment

Adjustment of a binocular optical instrument requires that the two optical systems of the unit be properly aligned with each other and conform to the interpupillary distance of the eyes of the observer. Precise focusing of the instrument changes the position of the eyepiece so that it is in correct relation to the focal plane of the objective and the angles at which the light rays are brought to a focus. The eyepiece of a focusing-type telescope, for example, is generally designed to accommodate the refracting qualities of the eyes of an observer.

Because telescopes with a magnifying power of 4x or less have a sufficiently wide range of accommodation, a single-focus setting is satisfactory. (Eye correction is not extremely large.) These telescopes have fixed-focus eyepiece which cannot be adjusted during operation; hence the name FIXED-FOCUS TELESCOPES, usually with a minus 3/4 to minus 1 dioptric setting.

Eyeshields and Eye Tension

Eyeshields on optical instruments should EXCLUDE STRAY LIGHT FROM THE EYES (particularly at night or during poor conditions of illumination), SO THAT THE PUPIL MAY DILATE AS MUCH AS POSSIBLE. Rubber shields at the eyepiece are considered best. When a monocular instrument is used, the unused eye should also be shielded from light. Light for illumination of the reticle must be held to a bare minimum.

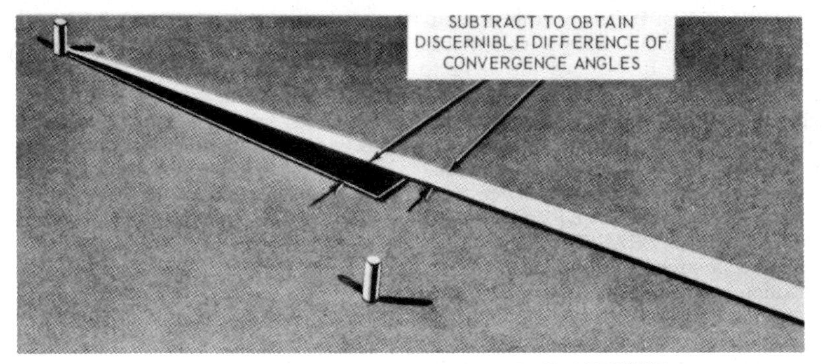

A GRAPHIC VIEW OF DIFFERENCE BETWEEN CONVERGENCE ANGLES SHOWN IN X AND Y, FIGURE 6-8

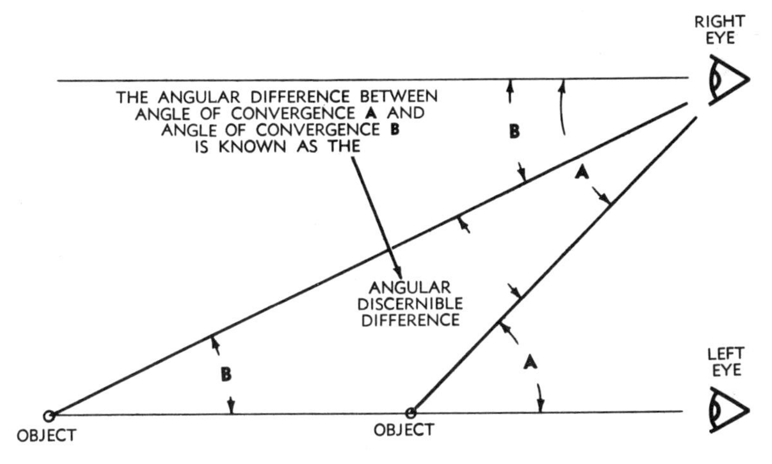

B CONVERGENCE ANGLES IN STEREOSCOPIC VISION

137.131

Figure 6-9.—Angular discernible difference.

Eye tension or fatigue causes the eyes to blink, which is muscular rather than retinal action and is least apparent when the eye is relaxed, as when accommodated for distant objects. In most telescopes, the eyepiece mount is adjustable; and by adjusting the position of the eyepiece you can adapt the instrument to compensate for the inherent refractive errors of the eye. If a person has perfect vision, the light rays which leave the eyepiece of a telescope must be parallel and enter the eye parallel; but the rays which leave the eyepiece must be SLIGHTLY DIVERGENT before entering the eye of a person who is slightly nearsighted. The eyepiece of a telescope for the person must therefore be moved in (toward the objective lens) to bring the final REAL image of the telescope within the focal length of the eyepiece, and moved out (away from the objective) for a farsighted person.

REMEMBER: You CAN SOMETIMES bring the viewed target within focus on your retinas by accommodation of your eyes, as well as by

ADJUSTING THE EYEPIECE of the instrument, which is a serious error repeatedly made by a beginner. He allows his eyes to accommodate on the target first. This is true even though he IS FOCUSING the eyepiece of the instrument, and it causes eyestrain if the instrument is used for long periods. COMPLETELY RELAX YOUR EYES BEFORE YOU ATTEMPT TO FOCUS THE EYEPIECE. THIS TAKES CONSIDERABLE PRACTICE AT FIRST.

The correct procedure for focusing an eyepiece with a diopter scale is as follows:

1. Move the eyepiece to the extreme PLUS diopter position (all the way out).
2. Select an INFINITY target with a background of good contrast.
3. Move the eyepiece slowly in until the image of the target is sharply defined. If you go past the point where the image is sharply defined to a position where the image is blurred, do NOT attempt to focus the eyepiece from this new position; if you do, you will focus from the MINUS to the PLUS position and serious errors in the diopter reading will result. To make proper correction for this error, back the eyepiece out again to the extreme plus diopter position and start over.
4. When you are focusing, do not squint the eye; if you do, errors will result.

Focusing from the PLUS to the MINUS position of the diopter scale prevents the eye from accommodating on the target. If the eye is focused from the minus to the plus position, it readily accommodates to the diverging light rays and causes errors in the diopter reading, and also eye strain.

MICROSCOPES

The two types of microscopes discussed in this section are SIMPLE and COMPOUND. Each is considered in sufficient detail to enable you to understand its importance.

SIMPLE MICROSCOPE

A simple magnifier (microscope) consists of a converging lens located at the first focal plane of the eye, though such positioning is not too important. If an object is viewed by the microscope when it is at or within the focal length of the lens, the eye sees a virtual, erect, and enlarged image. You will recall from previous study of image formation in chapter 5 that the eyes can focus or form an image of parallel rays of light from an object at infinity without accommodation. If the object is at the focal point of the lens (fig. 6-10), the eye can see it without accommodation because the emergent rays are parallel, as indicated. When the object is within the focal length of the lens, however, accommodation by the eye is necessary because the emergent rays are not parallel but divergent, as shown in illustration 6-11. The image formed by this microscope is virtual, erect, and enlarged.

COMPOUND MICROSCOPE

You perhaps used a compound microscope to look at minute plants and animals when you were in high school. Such an optical instrument so magnifies small objects that it increases the usefulness of the eyes at short distances. The eyes by nature are long-range optical instruments of high acuity.

Refer now to illustration 6-12, which shows one of the simplest types of compound microscopes. Study all details and the nomenclature. Note the position of the eye, the eyepiece, the objective, and the object. Then observe the positions of the real and virtual images. NOTE: Objectives and eyepieces will be discussed after microscopes.

Rays of light from the object strike the objective (closest lens to the object) and then details of an object are obtained after the object has been magnified 400 times. Magnifying power of a compound microscope is equal to the magnification of the objective lens ($\frac{D_i}{D_o} = \frac{S_i}{S_o}$) multiplied by the magnifying power of the eyepiece ($MP = \frac{10''}{F \text{ (inches)}}$).

Light waves from an object never focus perfectly at a corresponding point on an image created by them—they form instead a diffused image with a central white spot surrounded by a series of concentric rings of light which fall off rapidly in intensity. THIS IS CALLED A DIFFRACTION PATTERN. See illustration 6-13. Diffraction sets the final limit to the sharpness of the image formed by a lens, resulting from the natural spreading tendency of light waves; and it occurs in images formed by all lenses, regardless of the perfection with which they are constructed. The diffraction pattern (blurred image) created is directly proportional to the wavelength of the light, and inversely proportional to the diameter of the beam of light which enters the optical instrument.

The numerical measure of the ability of an optical system to distinguish fine detail is called resolving power. IN OTHER WORDS, RESOLVING POWER IS THE MEASUREMENT OF THE ABILITY OF A LENS OR OPTICAL SYSTEM TO FORM SEPARATE IMAGES OF TWO POINTS CLOSE TOGETHER. The resolving power of a microscope can measure the shortest distance between two points on an object for which two separate images are distinguishable.

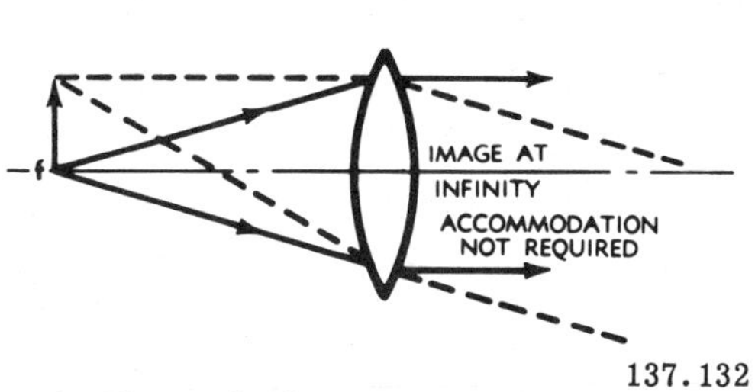

137.132

Figure 6-10—Object at the focal point of a simple magnifier.

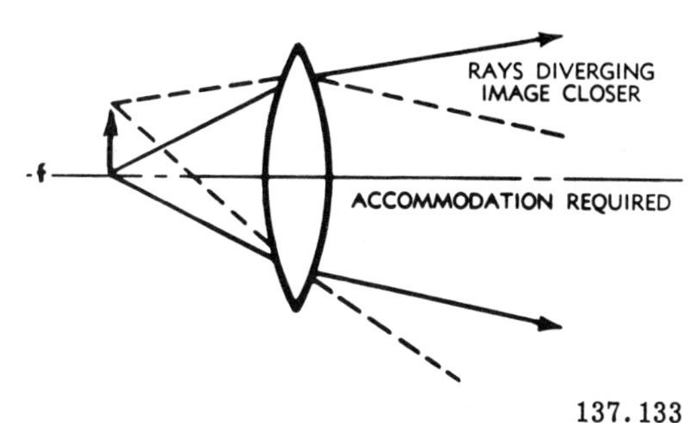

137.133

Figure 6-11.—Object within focal length of a simple magnifier.

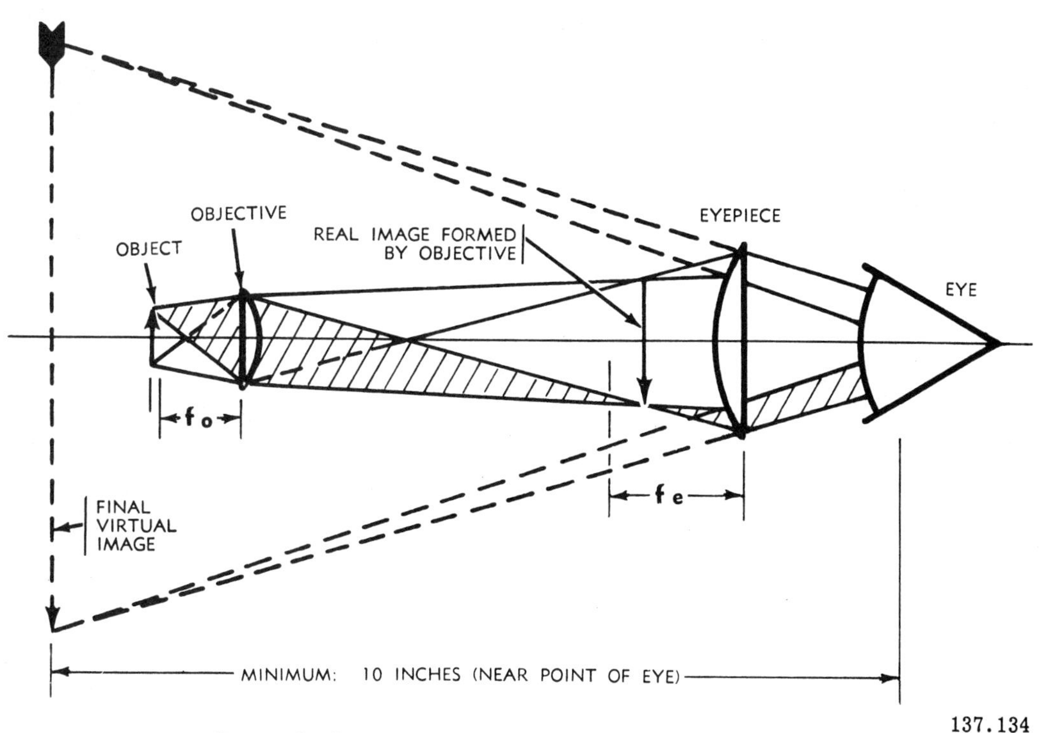

137.134

Figure 6-12.—Image creation by a compound microscope.

Because of their importance in optical instruments, it is best to discuss at this time objectives and eyepieces. If you know the function of these optical elements in image formation, you will understand better the basic optical instruments discussed in the pages which follow.

OBJECTIVES

An objective in a refractive type of optical system is the LENS NEAREST THE OBJECT, and its FUNCTION IS TO GATHER LIGHT FROM THE OBJECT AND FORM A REAL IMAGE OF IT. In order to reduce color and other aberrations to a minimum, objectives are compound lenses WHICH FORM REAL IMAGES IN ALL OPTICAL SYSTEMS.

CONSTRUCTION

Most objectives have two elements: (1) a double convex converging lens of crown glass, and (2) a planoconcave lens, as shown in part A of illustration 6-14.

When the elements of an objective have large diameters, or when the faces of the elements have different curvatures, the elements are not cemented together; they are held in their correct relative positions by retaining and locking rings. Part B of figure 6-14 shows an objective with unsealed elements. This plan of construction allows freedom for proper shaping of the inner surfaces of the elements and also leeway for eliminating aberrations. An objective of this type is generally called a DIALYTE or Gauss objective.

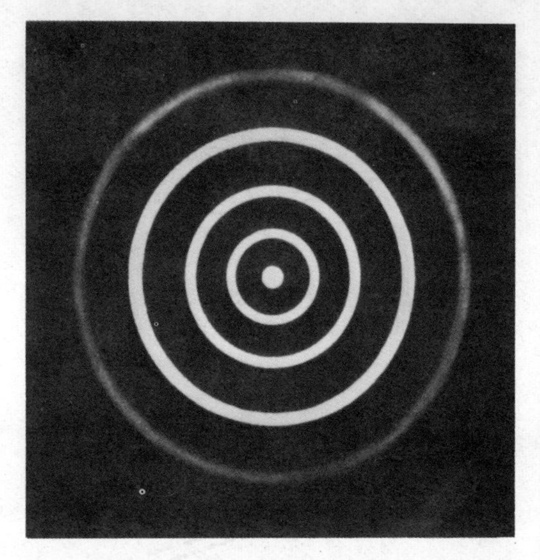

137.135

Figure 6-13.—Diffraction pattern (greatly magnified).

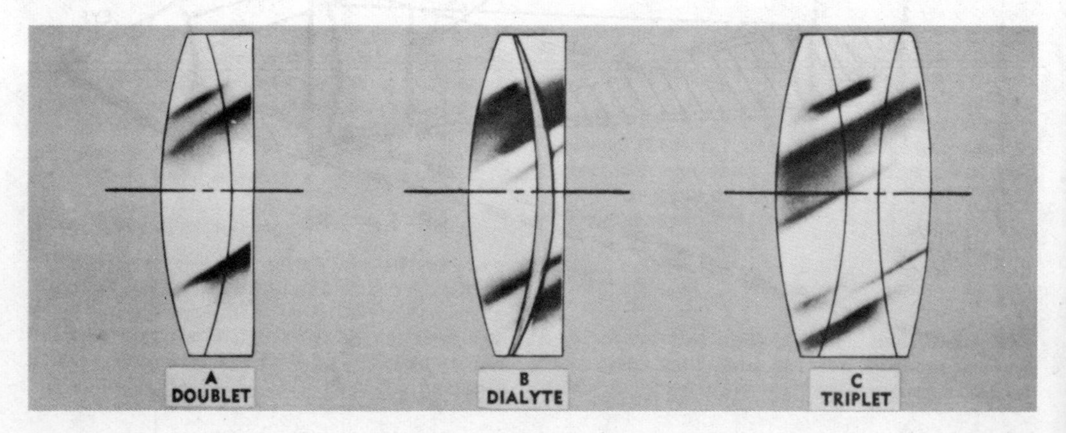

137.136

Figure 6-14.—Types of objectives.

Part C of illustration 6-14 shows a third type of objective, a TRIPLET, which has two positive lenses cemented to a negative lens, though one of the positive lenses can be mounted separately from the negative lens cemented to the other positive lens. An objective of this type provides six surfaces and therefore gives an instrument designer more freedom for correcting aberrations.

RELATION TO OPTICAL SYSTEM

The size of an objective affects the amount of magnification by an instrument, because the amount of light which can pass through a given objective is limited. The focal length of the objective, of course, also affects magnification; and the diameter of the objective affects the RESOLVING POWER of a telescope. The size of the objective also affects brightness of the image.

If greater magnification of the image is desired, more light must be admitted and distributed over a larger image area, RESULTING IN A DIMMER IMAGE. The effective size to which an image may be magnified is therefore GOVERNED TO A GREAT EXTENT BY THE SIZE OF THE OBJECTIVE, AS WELL AS THE LIMITS SET BY DISTORTION RESULTING FROM DIFFRACTION.

An increase in the size of an objective beyond a certain point does not appreciably improve the brightness of an image, because the size of the pupil of the eye imposes a restriction.

Some optical instruments have objectives smaller than their eyepieces, but this arrangement gives the instrument lower magnification.

EYEPIECES (OCULARS)

The function of an eyepiece in an optical instrument is TO ENLARGE THE IMAGE PRODUCED BY THE OBJECTIVE, which means that it is similar to a magnifying glass. Study illustration 6-15.

Eyepieces used in modern fire control instruments generally consist of eyelenses and field lenses, of which one or all may be compound.

CONSTRUCTION

THE EYELENS IS THE LENS (OF THE EYEPIECE) NEAREST THE EYE. It magnifies the image and CAN THUS AFFECT ITS QUALITY, AS SEEN BY THE EYE. The lens in the eyepiece nearest the objective is called the FIELD LENS, which GATHERS LIGHT from the objective and CONVERGES IT INTO THE EYELENS. NOTE: The field lens and the eyelens are used in combination to provide the principle of a thick lens.

Without a field lens, much of the light gathered by the objective would not be brought into the field of the eyelens. Note the marginal rays in parts A and B of illustration 6-15.

RELATION TO OPTICAL SYSTEM

The objective receives parallel rays of light from a distant object, brings them to a focus, and turns them into diverging angular rays. The eyepiece receives these diverging angular rays and directs them to the eye, as a parallel beam when the eyepiece is at zero diopters.

At this point, you need to learn the meaning of another term: EXIT PUPIL. The eyepiece forms an image of the objective lens at the POINT WHERE THE EYE IS PLACED. This image IS THE EXIT PUPIL, WHICH IS DISCUSSED FULLY, and illustrated later in this chapter under MAGNIFICATION IN TELESCOPES.

The axial rays which pass through the eyepiece are so close to the optical axis, and so nearly parallel, that spherical and chromatic aberration in the eyepiece is generally not serious. The field of view must be reasonably flat, and the eyepiece must be corrected for COMA, which automatically gives FLATNESS OF FIELD.

TYPES OF EYEPIECES

General types of eyepieces used in optical fire control instruments are discussed in the following paragraphs. Specific types of eyepieces are usually modified as necessary to meet the requirements for particular optical instruments.

Ramsden Eyepiece

Ramsden eyepiece consists of two plano-convex lenses made of crown glass (with equal focal lengths) and separated by a distance equal to two-thirds of the numerical value of either focal length. The reticle is located in

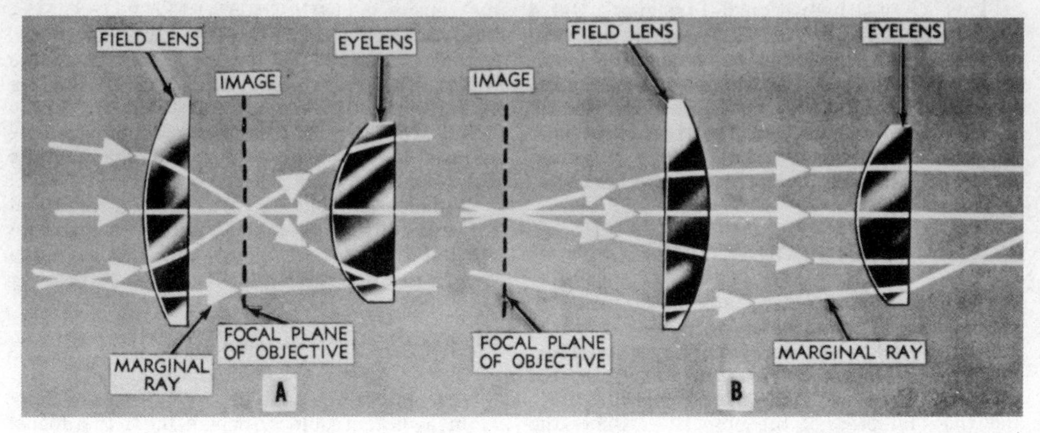

137.137
Figure 6-15.—Typical paths of light through the lenses (field and eye) of eyepieces.

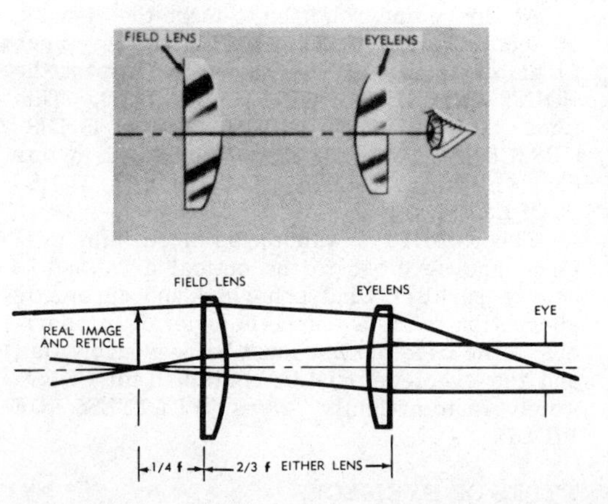

137.138
Figure 6-16.—Ramsden eyepiece.

front of the field lens at a point equal to one-fourth of its focal length, as shown in illustraion 6-16.

The field lens of a Ramsden eyepiece therefore contributes to the amount of magnification and quality of the image produced. Dirt on its principal plane is not in focus and is practically invisible. Dirt on the optical element near the image planes may appear as a blur(s).

A Ramsden eyepiece has one fault: CONSIDERABLE LATERAL COLOR, which can be defined as a DIFFERENCE IN IMAGE SIZE for each color. This type of eyepiece, however, is widely used with reticles, which are marks or patterns placed in the focal plane of the objective of an optical instrument (chapter 7) for the following reasons: (1) to measure angular distance between two points, (2) to determine the center of the field, or (3) to assist in the gaging of distance, determining leads, or measurement.

A reticle may be a pair of crosslines (fig. 6-17) composed of fine wire, or it may be etched on glass plate with plane parallel surfaces. When a reticle is etched on glass, the entire piece of glass is referred to as the reticle.

Because it can be used as a magnifier, a Ramsden eyepiece is called a POSITIVE OCULAR

Kellner Eyepiece

A Kellner eyepiece is a modification of the Ramsden eyepiece. See illustration 6-18. It is a simple type of eyepiece with a field lens and an eyelens. The eyelens is an achromatic DOUBLET with its flat, flint lens toward the eye.

Dense, barium-crown glass and light-flint glass used in a Kellner eyepiece help to reduce aberration; and in order to procure full correction for chromatic aberration, the image MUST LIE in the plane surface of the field lens.

NOTE: If a reticle is desired in an optical system containing a Kellner eyepiece, some

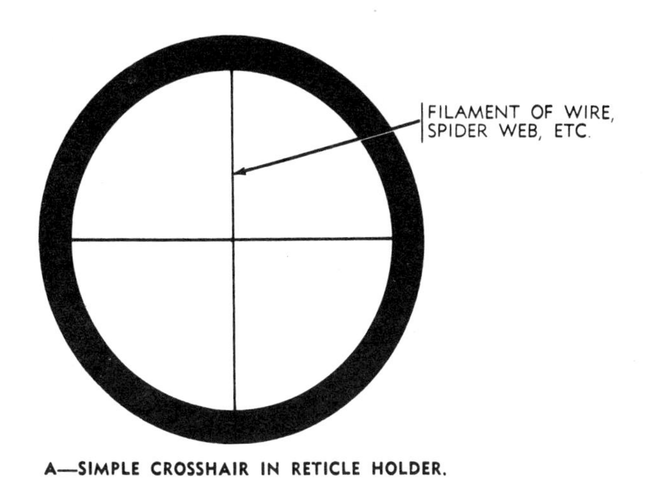

A—SIMPLE CROSSHAIR IN RETICLE HOLDER.

B—RETICLE PATTERN ETCHED ON GLASS.

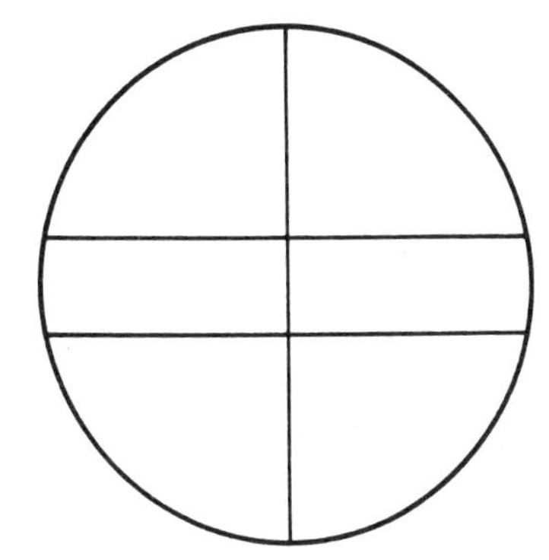

C—STADIA LINES FOR ANGULAR MEASUREMENT.

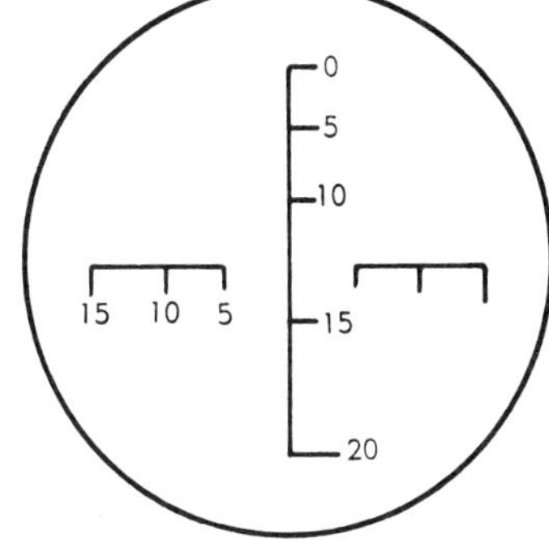

D—ETCHED DESIGN WITH NUMERALS.

137.139

Figure 6-17.—Representative types of reticles.

chromatic aberration is sacrificed in order to use the reticle. This means that all Kellner eyepieces are not oriented to have the image formed on the plane surface of the field lens. When a reticle is desired, the eyepiece is so designed that the image is farther from the field lens, so that the eyepiece may be focused IN or OUT of the instrument without contact of the field lens with the reticle.

A Kellner eyepiece gives an achromatic and orthoscopic field (free of distortion) as large as 50° in some models. Its most serious disadvantage is a pronounced GHOST, resulting from light reflected SUCCESSIVELY FROM THE INNER AND OUTER SURFACES of the lens.

A Kellner eyepiece is commonly used in prism binoculars.

Huygenian Eyepiece

A Huygenian eyepiece, illustrated in figure 6-19, employs a collective (field) lens so moved away from the eyelens that the real image lies between the two lenses. The lenses are separated by a distance EQUAL TO HALF the sum

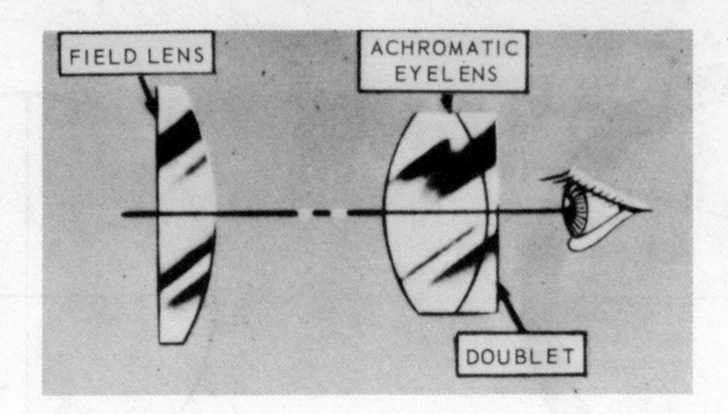

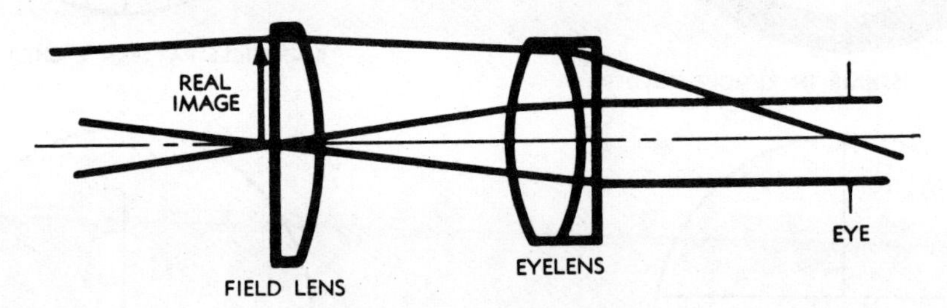

Figure 6-18.—Kellner eyepiece

137.140

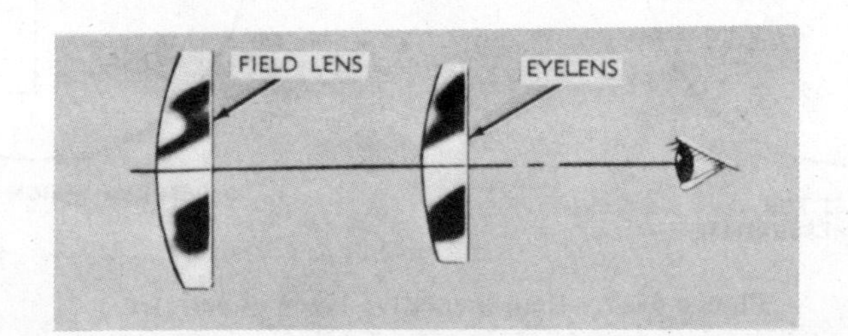

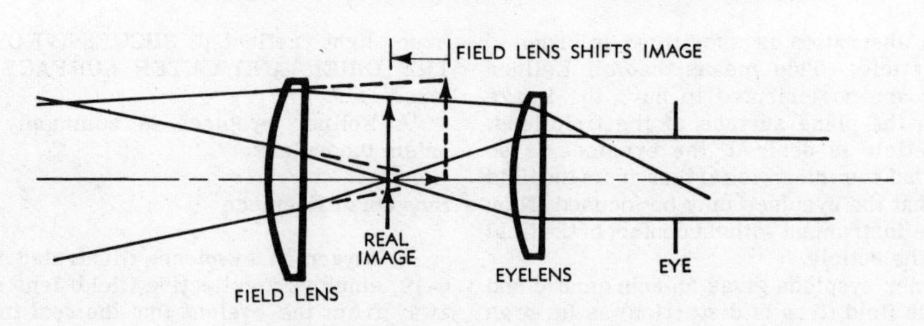

Figure 6-19.—Huygenian eyepiece.

137.141

of their focal lengths. This amount of separation of the lenses corrects chromatic aberration of the eyepiece as a whole, but it does not correct the aberration in the individual lenses.

A Huygenian eyepiece is not suitable for use with a reticle, because the individual components are not corrected for aberration and the reticle is therefore distorted. A Huygenian eyepiece, however, is sometimes used in microscopes with a small reticle in the center of the field; and a reticle of any SIZE MAY BE PLACED IN ANY IMAGE PLANE of the optical instrument, EXCEPT the image plane between the eyepiece lens (ocular).

The focal length generally used in a Huygenian eyepiece is over one inch, which is sufficient to provide adequate eye relief. Both elements of the eyepiece are normally made of the same type of crown glass, with convex surfaces facing forward, as illustrated.

Symmetrical and Two-Doublet Eyepieces

Symmetrical and two-doublet eyepieces are constructed of two cemented, achromatic doublets (fairly close together) with their positive elements facing each other. If the doublets are identical in every respect (diameters, focal lengths, thickness and index of refraction), the eyepiece is symmetrical. If the doublets differ in one respect or another, however, they are considered as a TWO-DOUBLET eyepiece. The eyelens of the two-doublet eyepiece is generally slightly smaller in diameter and has a shorter focal length than its field lens. Doublets in a symmetrical eyepiece, on the other hand, ARE IDENTICAL IN EVERY RESPECT AND CAN BE INTERCHANGED.

Symmetrical and two-doublet eyepieces are often used in fire control instruments which recoil. The eye distance on these instruments must be fairly long, to prevent the eyepiece from striking the gunner's eye. These eyepieces may also be used in terrestial telescopes (not only in gunsight telescopes), or in any other telescope designed to carry them.

A symmetrical eyepiece provides long eye relief, because it has a large exit pupil and low magnification, qualities which ensure eye relief. For this reason, symmetrical eyepieces—along with Kellner—are used extensively in optical instruments, particularly rifle scopes and gunsights.

Orthoscopic Eyepiece

An orthoscopic eyepiece gets its name from the fact that it is free of distortion. It employs a TRIPLET FIELD LENS which may or may not be planoconvex; and a single, planoconvex eyelens, with the curved surface of the field lens facing the curved surface of the eyelens (fig. 6-21).

Because they give a wide field, orthoscopic eyepieces are used extensively in high-powered telescopes. They are also useful in rangefinders, as they permit use of any part of the field.

TELESCOPES

The primary purpose of a telescope is to IMPROVE VISION OF DISTANT OBJECTS. In its simplest form, a telescope consists of two parts: (1) a lens, called the objective (near the object), or an object glass (if a mirror), which ALWAYS FORMS A REAL IMAGE of the field or area the telescope can pick up; and (2) an eyepiece which enables you to view the image.

The objective of a telescope forms a REAL, INVERTED, and REVERTED image of a distant object and therefore serves the same purpose as the lens of a camera. The eyepiece is the optical element which produces an enlarged, virtual image of the real image produced by the objective and thus serves as a simple microscope.

The amount of light which enters a telescope is dependent upon the size of the objective.

In contrast with a microscope, a telescope objective lens has a long focal length suitable for viewing distant objects. The objective of a microscope forms an image relatively far from the lens, but the object is very close to the lens; in a telescope, the objective forms an image relatively near the lens (in the focal plane of the objective lens), but the object is very far from the lens (infinity).

As you study telescopes in the following pages, keep in mind one thing: TELESCOPES ARE THE BASIS OF PRACTICALLY ALL OPTICAL INSTRUMENTS—binoculars, gunsights, periscopes, rangefinders, and so forth.

ASTRONOMICAL TELSCOPES

There are two types of astronomical telescopes: (1) refracting, and (2) reflecting; and

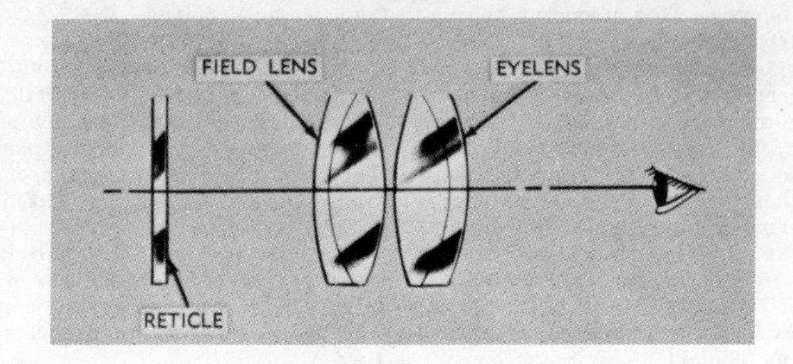

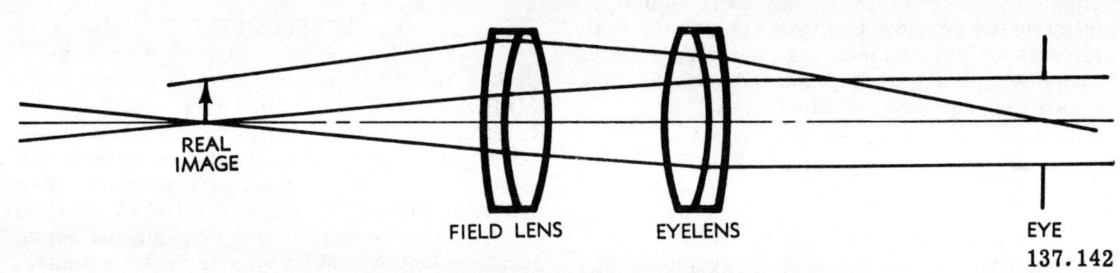

137.142
Figure 6-20. —Symmetrical and two-doublet eyepieces.

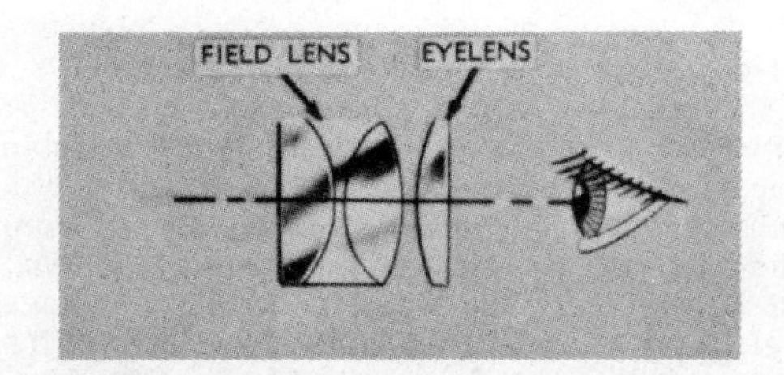

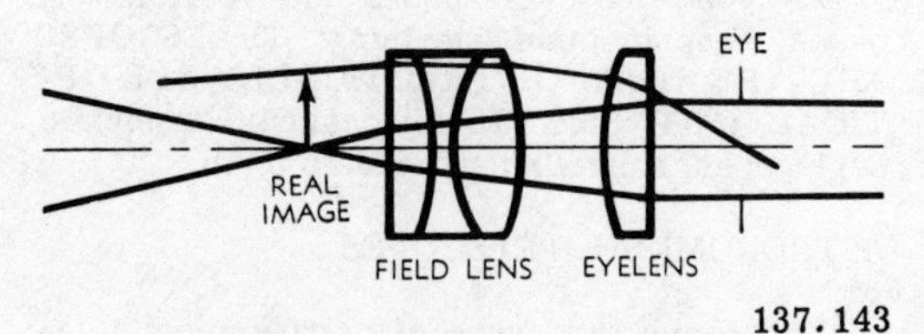

137.143
Figure 6-21. —Orthoscopic eyepiece.

each type is constructed in the manner necessary to have it satisfy a particular requirement. An astronomical telescope used for viewing the heavenly bodies, for example, does not require the image to be erect, because all images of heavenly bodies look round. All astronomical telescopes therefore produce an inverted image.

Refracting Telescopes

A refracting astronomical telescope has high-quality lenses which form images of stars and the sun by refraction. A positive objective lens alone forms only real images of distant objects, but such real images in space cannot be brought to focus by the eye, as shown in illustration 6-22. In order for an eye to bring an image to a focus, the rays of light from the object which enter the eye MUST BE PARALLEL OR ONLY SLIGHTLY DIVERGING, as if from an object no closer than the near point (10 inches) of the eye. If another positive lens, however, is

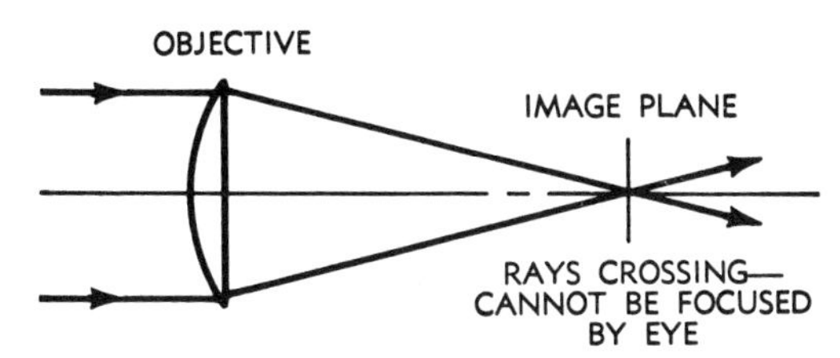

137.144
Figure 6-22.—Passage of refracted rays from an objective through the image plane.

placed between an image and the eye, and the real image is at the primary (first) focal point of the eyepiece, the eye can see without accommodation a virtual image of the object picked up by the objective lens. See figure 6-23. This is known as the KEPLERIAN SYSTEM.

Refer to illustration 6-24, which shows the position of an objective lens in relation to the eyepiece in the Keplerian system of telescope construction. Such an arrangement of optical elements is the simplest form of a refracting astronomical telescope. Observe that the parallel light rays entering the objective lens are refracted and converge to the focal plane of the lens. (The image plane and the focal plane coincide when parallel rays are refracted by any lens.) In the focal plane of the objective lens a real, inverted image of the object is formed. The eyepiece is so placed that the image formed by the objective lens is located on the primary focal point of the eyepiece. The diverging rays, diverging from the real image, enter the eyepiece, are refracted, and emerge parallel to the optical axis of the telescope.

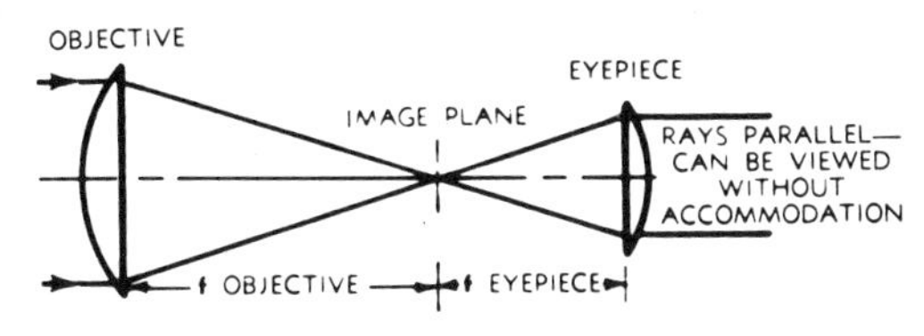

137.145
Figure 6-23.—Objective and eyepiece arrangement in the Keplerian system.

Since the real image formed by the objective lens is located at one focal length (at the primary focal point) of the eyepiece, the eyepiece acts as a magnifying lens to magnify the real image. If you look through the telescope eyepiece, you see a VIRTUAL, INVERTED, ENLARGED image which is formed at infinity.

In an astronomical telescope, in which the focal points of the objective lens and the eyepiece lens coincide, the length of the telescope is the SUM OF THE FOCAL LENGTHS OF THE TWO LENSES.

Observe in illustration 6-24 that the light rays from the object are refracted by the eyepiece and emerge from it in a divergent manner to give a minus value to the emergent rays. If you were to look through the telescope eyepiece in this case, you would see a virtual image which appears to have moved toward the near point of the eye; that is, the image appears to have moved IN from infinity TO or NEAR the near point of the eye (10 inches).

NOTE: In illustration 6-24 the focal point of the objective lens DOES NOT COINCIDE WITH THE FOCAL POINT OF THE EYEPIECE LENS. The image formed by the objective lens is located within the focal length of the eyepiece.

Reflecting Telescopes

Although magnifying power of a telescope is of the greatest importance in astronomical work, most emphasis MUST BE PLACED ON light-gathering ability and resolving power OF THE OBJECTIVE LENS, so that stars (heavenly bodies) may be observed at greater distances.

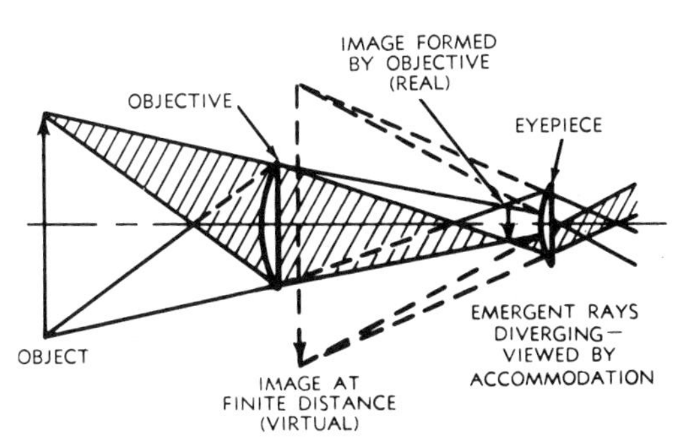

137.146
Figure 6-24.—Refracting astronomical telescope.

Increasing the diameter and the size of an objective lens of a telescope, however, presents many problems; because it is difficult to grind and polish large pieces of optical glass, such as large doublets. For this reason, large, front-surface-aluminized-concave mirrors (The reflecting surface of the concave mirrow is coated with aluminum.) are used as objectives in some astronomical telescopes. The advantages of using a concave mirror instead of a large, converging doublet are as follows:

1. Mirrors do not absorb light and they have no aberration.
2. A mirror has only one surface which requires grinding, as compared to four sides for a doublet.
3. Concave mirrors weigh less than doublets of the same diameter.

You learned in a previous chapter of this manual that when parallel rays of light from an infinity target strike a concave mirror the rays are reflected and made convergent to come to focus at the focal point of the concave mirror. A REAL, INVERTED, DIMINISHED image of the target is formed at the focal point. This image is the same as that formed by a convergent doublet objective lens in a refracting telescope.

Study next illustration 6-25. Before the reflected rays of the concave mirror in this figure are brought to focus, a 90° prism is placed in the converging rays to deviate the rays at an angle of 90°, as shown. The purpose of this deviation is to prevent an observer who is viewing the image from cutting off a large part of the light before it reaches the mirror.

As in the refracting telescope, at the point where the real image is formed an eyepiece is placed to magnify the image. If you look through the eyepiece of a reflecting telescope, you see a VIRTUAL, INVERTED, ENLARGED image formed at infinity.

When there is a need for lengthening the focal length of a concave mirror, or when the design of a telescope is altered (Cassegranian reflecting telescope, for example), a small convex mirror can be used with the concave mirror. See figure 6-26.

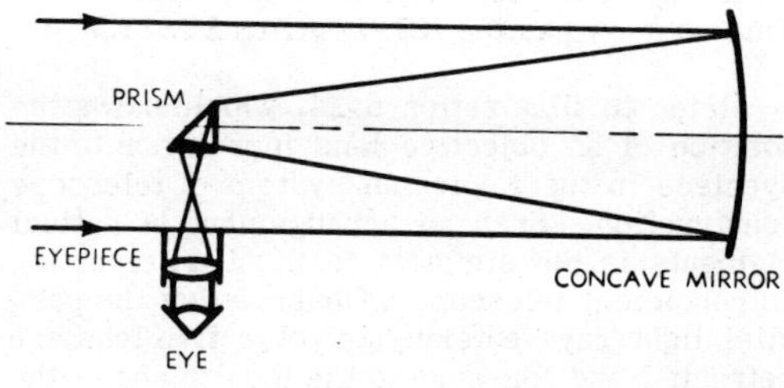

137.147

Figure 6-25.—Reflecting Newtonian telescope.

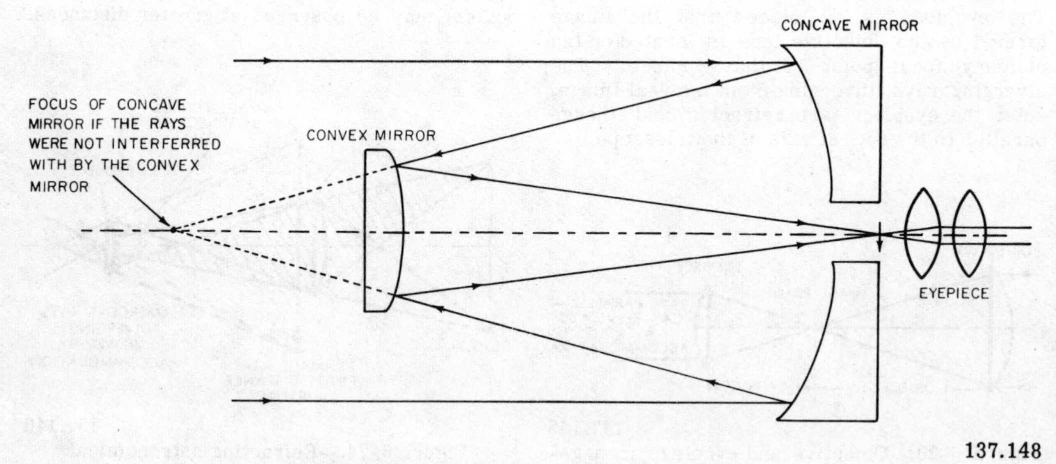

137.148

Figure 6-26.—Cassegranian reflecting telescope.

The concave mirror in a Cassegranian telescope has a small hole ground through the middle (center), and the convex mirror is placed in the converging rays in place of the 90° prism to reflect the rays and make them less divergent and focused at a point greater in distance than the original focal plane of the concave mirror. Study illustration 6-26 carefully.

When the rays are reflected from the convex mirror, they pass directly through the hole in the concave mirror and come to focus to produce an image like the one produced by the doublet lens. The eyepiece is placed behind the reflecting surface of the concave mirror to magnify the real image and give a VIRTUAL, INVERTED, and ENLARGED image of the object at infinity.

Converging mirrors with long focal lengths are used in telescopes as objectives to form real images. The light in this type of mirror is incident on the same side as the center of curvature of the sphere. The focal point is halfway between the center of curvature and the reflecting surface.

A real image created by a convergent mirror used as an objective in a telescope can be viewed through a magnifying eyepiece or photographed by a camera attachment.

TERRESTIAL TELESCOPES

A terrestial telescope gets its name from the Latin word TERRA, which means earth. A terrestial TELESCOPE IS USED TO VIEW OBJECTS AS THEY ACTUALLY APPEAR ON EARTH.

Any astronomical telescope can be converted to a terrestial telescope by inserting a lens or prism erecting system between the eyepiece and the objective to erect the image. See illustration 6-27, which shows the optical elements of the simplest form of terrestial telescopes. Note the positions of the REAL AND VIRTUAL IMAGES.

A lens erecting system requires such positioning of the objective and the eyepiece that the erectors are between the focal point of the objective and the first principal focus of the eyepiece. A prism erecting system, on the other hand, must be placed between the objective and its focal point.

You will learn more details about the use of erecting systems when you study magnification of images in telescopes later in this chapter.

Galilean Telescopes

The first telescope Galileo made had a power of 3, but he later made one with a power of 30. His telescope is based on two major principles:

1. It makes use of an eyepiece consisting of a negative eyelens positioned a distance equal to its focal length (fe, part B, Fig. 6-28) in front of the objective focal point. Such positioning of the negative eyelens makes converging rays from the objective parallel before they converge to form a real image; so no real image exists in this optical system. The light rays do not converge to a point to form a real image; but if you look through the negative lens you see an enlarged, virtual image of the object, which appears to be at a point between 10 inches and infinity.

The virtual image viewed through the negative eyelens is therefore at infinity and can be viewed by the eye without accommodation.

The relation of the optical elements in a Galilean telescope (part B, fig. 6-28) is also referred to as the ZERO DIOPTER SETTING, which means that ALL LIGHT RAYS FROM ANY POINT SOURCE LOCATED AT INFINITY EMERGE FROM THE EYEPIECE PARALLEL. If the eyelens, however, is moved in or out, the emergent light rays converge or diverge and the instrument can therefore be adjusted for farsighted or nearsighted eyes, and also for distance.

The INVERTING EFFECT of the objective lens in a Galilean telescope is canceled by the negative eyelens, because the real image is not allowed to form; that is, the emergent rays from the negative eyepiece are refracted farther away from the axis instead of recrossing it. The virtual image of the object viewed is therefore ERECT.

The principle of the Galilean system is diametrically opposite to that of the Keplerian or astronomical system with a positive lens, which causes the emergent rays from the positive eyelens to recross the axis and form an INVERTED, VIRTUAL IMAGE of the REAL IMAGE formed by the objective lens.

2. A Galilean telescopic system is one in which the diameter of the objective controls the field of view (width of visible area), because the objective is both the field stop and the entrance window, as indicated in figure 6-29. This type of telescopic system is therefore limited to

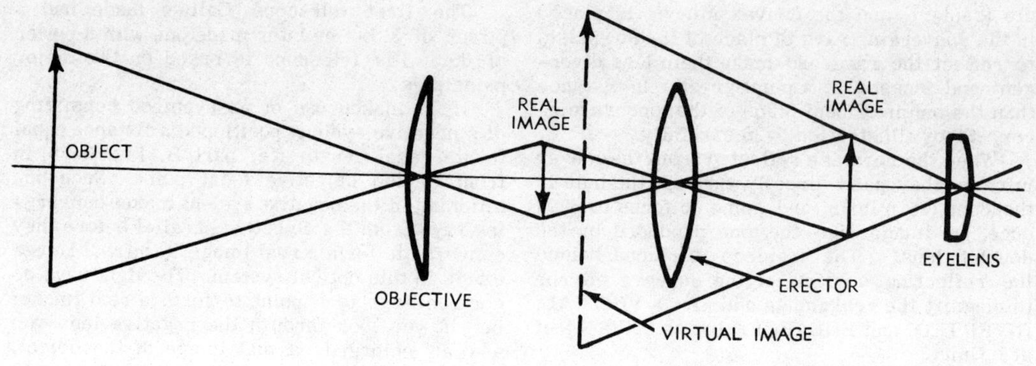

137.149

Figure 6-27. —Terrestial telescope.

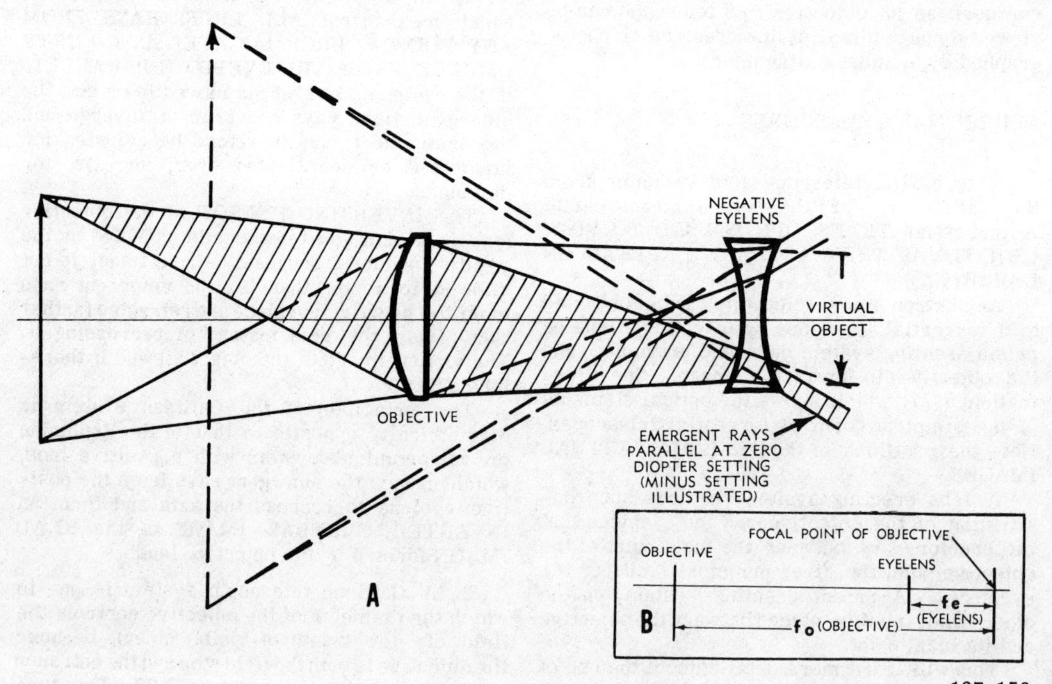

137.150

Figure 6-28. —Galilean telescope.

small fields of view and low magnifying power (2 or 3).

One-Erector Telescope

The arrangement of optical elements in a one-erector lens telescope is illustrated in part A of illustration 6-30. Observe that the parallel rays entering the objective lens from an infinity target are refracted, converged, and form a real, inverted image in the focal plane of the objective lens, as is true for all objective lenses of telescopes.

Rays which leave the real image diverge as though the image itself were an object. When another lens (erector lens) is placed two focal lengths (2F) from the real image, the erector receives the diverging rays and refracts them to two focal lengths (2F) behind the erector. The image produced by the erector is the same size as the image produced by the objective lens. (When an object is placed two focal lengths in front of a positive lens, the image produced by that lens is located two focal lengths behind the object and is the same size as the object but inverted.) An erector lens inverts its image, which in this case is ERECT, in comparison with the target at infinity.

Rays which leave an erect, real image formed by an erector lens are divergent and are received by the eyepiece. The eyepiece is positioned so that the image is in its focal plane (one focal length). For this reason, the refracted rays which leave the eyepiece are parallel to the optical axis of the telescope. If you look through the eyepiece of the telescope, you see a VIRTUAL, ERECT, ENLARGED image formed at infinity.

Part B of illustration 6-30 shows how a single erecting lens diminishes the size of an image. If the erecting lens is placed 3F from the inverted objective image, as shown, the erect image the erecting lens creates is twice as small as the inverted image of the objective lens and is twice as far from the erector lens (3 focal lengths of the erector lens).

A two-power telescope has TWO DIFFERENT MAGNIFYING POWERS (not a magnifying power of 2) which can be selected without changing the eyepiece and without refocusing. All you need do in order to change from one power to the other power in a 2-power telescope is move the erecting lens TO ONE OF TWO DIFFERENT POSITIONS, as illustrated by parts B and C in illustration 6-30 and the C and D positions of the lens in figure 6-31.

Observe in illustration 6-31 the original position of the lens (C), the optical axis of the lens (AB), rays of light (white lines) from point A to the lens, and the refracted rays to point B. As illustrated, the distance the lens is from the object is 1 1/2 F, and the distance of the image from the lens is 3F.

When the lens is moved to position D, the distance of the object (A) from the lens is 3F, and the distance of the image from the lens is 1 1/2 F.

According to the law of reversibility, you know that if the object were at B, its image would be at A. Points A and B are therefore conjugate points, because each is the image of the other. Suppose that the lens is 3 inches from point A and 6 inches from point B and you move the lens to D, 6 inches from point A and 3 inches from point B. POINTS A AND B ARE STILL CONJUGATE POINTS.

This lens, therefore, forms an image of A at B WHEN IT IS AT TWO DIFFERENT POSITIONS. If you place a real object such as an arrow in the plane at A, its image will be in plane B, regardless of whether the lens is at C or D; but WHEN YOU MOVE THE LENS FROM ONE POSITION TO ANOTHER, YOU CHANGE THE SIZE OF THE IMAGE. As you know, the relative size of the object and the image depends upon their relative distances. When the lens is at C, the image is TWICE AS BIG AS THE OBJECT; when the lens is at D, the image is ONLY ONE-HALF THE SIZE OF THE OBJECT.

Two-power terrestial telescopes may be classified as: (1) one-erector, and (2) two-erector (symmetrical). NOTE: In order for a two-erector lens system to be called a symmetrical system, both erectors must be IDENTICAL IN CONSTRUCTION AND HAVE THE SAME FOCAL LENGTH. Both types of terrestial telescopes will be discussed immediately after the following explanation of lens erecting systems.

Lens Erecting Systems

A lens erecting system is employed in an optical instrument to allow changing of magnification of the instrument, because different degrees of magnification can be obtained by changing the RELATIVE POSITIONS of the lenses. This type of erecting system increases the length of an optical system; and it is therefore used in systems where length is a distinct

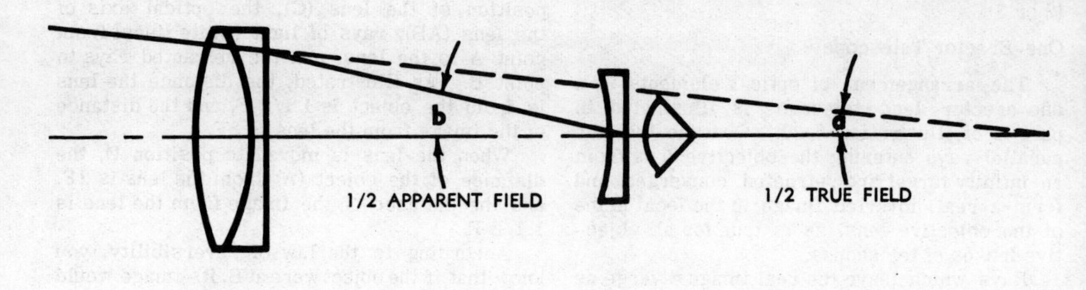

Figure 6-29.—Field of view limitations in a Galilean telescope.

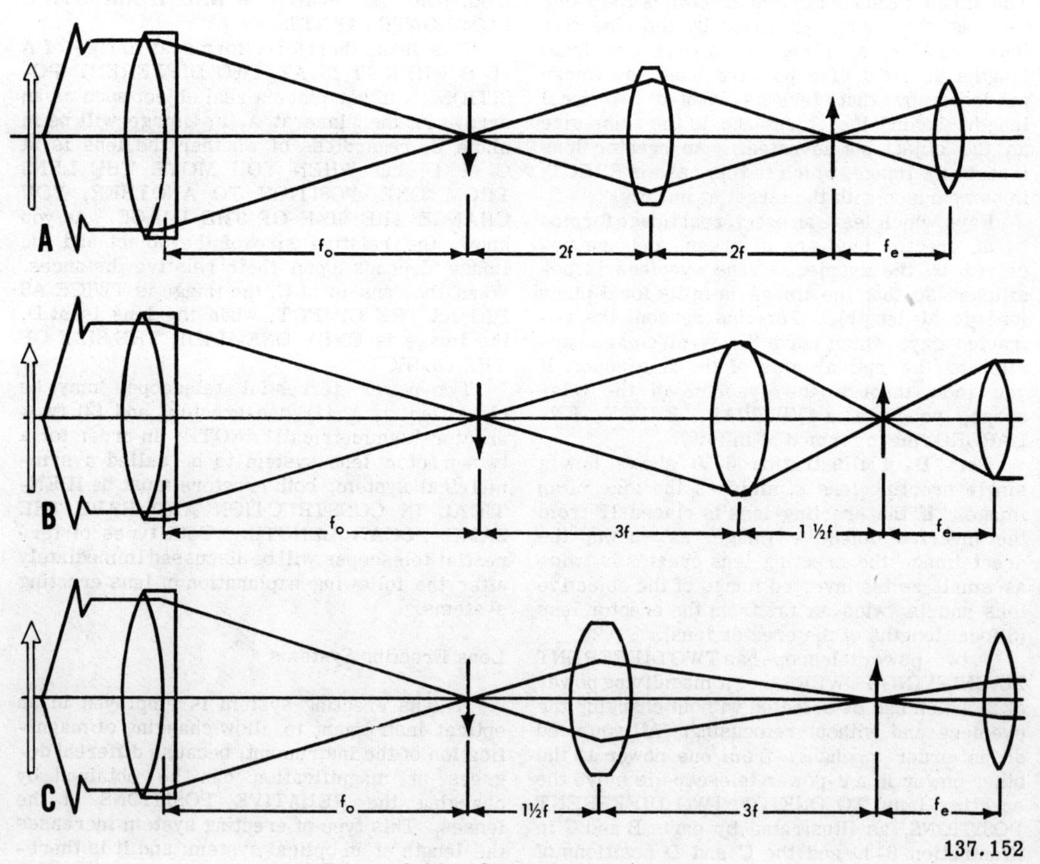

Figure 6-30.—One-erector telescope.

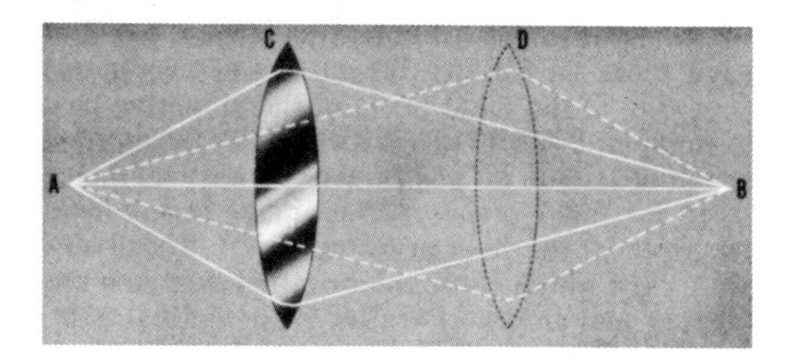

137.153
Figure 6-31.—Conjugate points.

advantage, as is true for periscopes and rangefinders.

At this point, it is best that you learn the distinction between VARIABLE MAGNIFICATION and CHANGE OF MAGNIFICATION. Variable magnification is obtained (produced) in an optical system when the image STEADILY BECOMES LARGER AND LARGER throughout movement of the erecting lenses. Change of magnification in an optical system is obtained ONLY WHEN the instrument is changed FROM ONE POWER TO THE NEXT POWER. Between positions, the image is badly blurred.

Study illustration 6-32, which shows two types of lens erecting systems. The erecting system in the top portion of this illustration is the simplest type, because it uses an erecting lens in the proper position (as indicated) in order to pick up the inverted image formed by the objective and form a second erect image which is magnified by the eyepiece.

Lens erecting systems for fire control instruments generally have two lenses, instead of a single erecting lens, as shown in the bottom part of figure 6-32. Study this system carefully. The complete lens erecting system functions as a single lens.

When two lenses are employed in a lens erecting system, they are compound and achromatic (without color; corrected for aberration). If a third lens is used with an erecting system, it is planoconvex or double convex and serves as a collective lens for collecting and bending light rays to the next optical element.

NOTE: A collective lens ERECTS THE IMAGE. Its ONLY purpose is to collect rays from the optical axis (rays which otherwise may be lost) and send them into the erecting lens, thereby making the final, REAL image brighter, with a larger field of view.

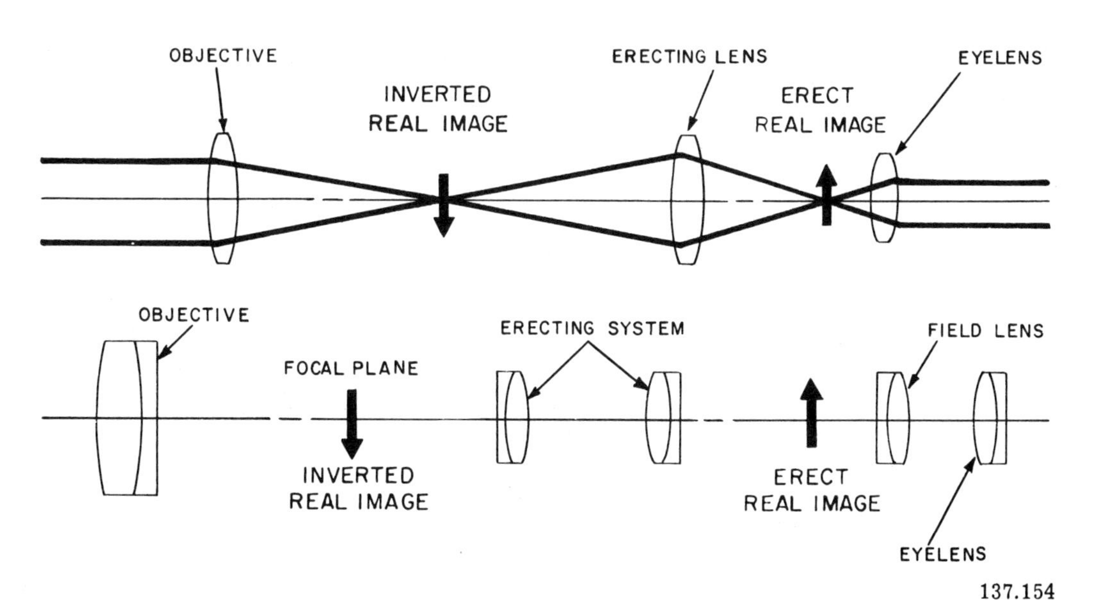

137.154
Figure 6-32.—Lens erecting systems.

Because a collective lens is sometimes placed at the point in the optical system where the image of the objective lens is formed, where a reticle MUST ALSO BE PLACED, the reticle markings are engraved on the flat face of the planoconvex lens used as a collective lens.

A lens erecting system can INCREASE OR DECREASE the magnifying power of an optical instrument; but when the lenses are placed in certain positions, the erecting system DOES NOT AFFECT the magnifying power of the instrument.

If lenses in an erecting system are moved closer to the focal point of the objective lens and farther from the eyepiece, magnification is increased but the field of view is decreased. If the erecting lenses are located at the same distance from the focal points of the objective and the eyepiece, there is no additional magnification of the image. This method of changing the degree of magnification is used in optical instruments which have a change of power.

Two-Erector Telescopes (Symmetrical)

Refer now to illustration 6-33 to see how a terrestial telescope with two erecting lenses is constructed. The erectors (lenses) shown are SYMMETRICAL; that is, they are IDENTICAL in every respect—diameter, thickness, index of refraction, and focal lengths. ASYMMETRICAL erectors (with different focal lengths) may also be used in this type of telescope for design purposes, or to help increase magnifying power, which the objective and eyepiece alone could not do.

As is true for all objective lenses, parallel rays from an infinity target are refracted and converge to the focal plane of the objective lens to form an INVERTED, REAL image. The first erecting lens is so positioned that the real image is in its focal plane. (The image is one focal length from the first erecting lens.) The divergent rays which enter the erecting lens are refracted and emerge parallel to the optical axis.

Since the rays which emerge from the first erecting lens are parallel, the second erecting lens may be placed at any reasonable distance from the first erector, because the rays which enter the second erecting lens are ALWAYS PARALLEL, regardless of the amount of lens separation. Separation of the erectors in fixed-power telescopes is generally THE SUM OF THEIR FOCAL LENGTHS, which is sufficient to ensure good eye relief. As separation of the erectors varies, eye relief of the eyepiece also varies.

Parallel rays which enter the second erecting lens are refracted and converge to the focal plane to form a REAL, ERECT image. If the erectors are SYMMETRICAL, the image produced by the second erector is of the same size as the image produced by the objective lens. If the erectors are ASYMMETRICAL, the size of the image produced by the second erector varies directly in proportion to its focal length—the longer the focal length of the second erecting lens, the larger the image produced by it.

The eyepiece of the telescope is again positioned as necessary in order to have the image of the second erector at its focal plane. When the eyepiece is placed one focal length from the

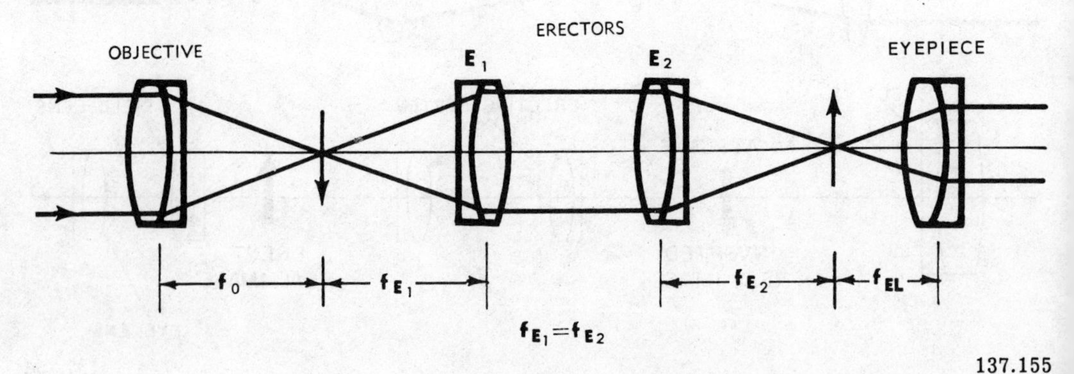

Figure 6-33.—Two-erector telescope (symmetrical).

image, divergent rays from the image are refracted by the eyepiece and emerge parallel to the optical axis. If you look through the eyepiece of the telescope, you see a VIRTUAL, ERECT, ENLARGED image formed at infinity.

A two-erector telescope can also be constructed with a power of two, by moving the erectors together as a unit in the same direction (with their separation fixed). Their distance from the real image formed by the objective lens must be 1 1/2 EFL, or 3 EFL of the erecting lens combination. The two erecting lenses FUNCTION together as a single, thick lens to produce an image in the same manner as the one-erector lens used for the same purpose.

You cannot continuously vary the power in a two-power telescope, because there are ONLY TWO positions of the erecting lens (one-erector lens, or a two-erector lens used as a unit) for which the TWO IMAGE PLANES ARE CONJUGATE.

NOTE: Varying the power means to increase the magnifying power of the telescope steadily and constantly with the image in sharp focus at all times, as is true for a ZOOM LENS on a television camera.

When the erecting lens (or lenses) is in an intermediate position, the image IS OUT OF FOCUS. The only way to keep the two image planes conjugate when the erecting lens is changed to any position, therefore, is to CHANGE ITS FOCAL LENGTH each time you move it; and you can do this by USING THE TWO ERECTING LENSES and VARYING THEIR EQUIVALENT FOCAL LENGTH (EFL) of the combination; that is, by varying the separation between the erectors. By varying the separation of the erectors, simultaneously moving the erectors at different speeds in relation to each other, you can form the image of the second erector in the same plane while its image varies in size.

Variable-magnification erecting systems used in variable-power telescopes provide two to three times as much power in the high-power position as in the low-power position.

Magnification of an erecting system composed of a COMBINATION OF LENSES can be varied by doing the following, simultaneously:

1. Varying the position of the erecting system from its object.
2. Varying the separation between the optical elements of the erecting system.

Now study illustration 6-34, part A of which shows two asymmetrical erectors in the low-power position. When these erectors are shifted toward the object, the position of the image shifts toward the eyepiece (part B, fig. 6-34) and magnification of the telescope is increased. If the distance between the two erectors in the forward position (toward object) is decreased by moving the second lens toward the first lens, magnification is slightly decreased and shifting of the resulting image position is also decreased (part C, fig. 6-34).

The image position CAN BE THE SAME for all possible magnifications produced by the optical system of a variable-power telescope. Separation between the erectors in a variable-magnification telescope (fig. 6-34) is always such that the IMAGE POSITION REMAINS FIXED for any amount of magnification, and the EYEPIECE REMAINS FIXED (part C, fig. 6-34).

GUNSIGHT TELESCOPES

A telescopic sight (military telescope) is generally a terrestial telescope with a reticle in the focal plane of the optical system. A gunsight telescope permits viewing of the reticle and the target in the same optical plane, and it does not require precise alignment of the eye with respect to line of sight. A telescopic sight is universally considered as an instrument with an erecting system.

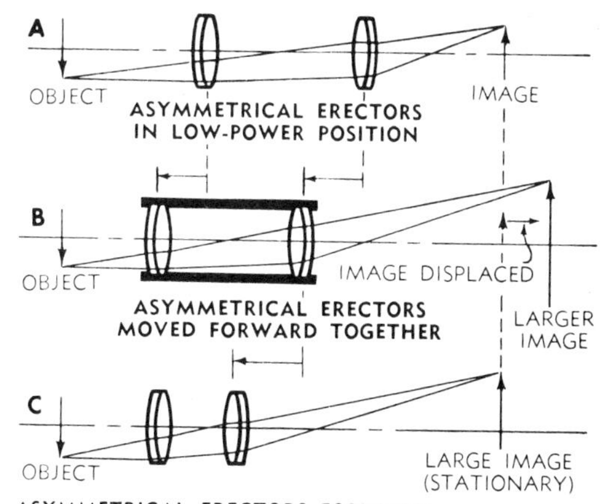

137.156

Figure 6-34.—Two-erector telescope (asymmetrical, variable-magnification).

Reticles are used in fire control instruments for superimposing a predetermined pattern of range and deflection graduations on a target. In its simplest form, a reticle is a POST or PICKET, or two INTERSECTING LINES. Line of sight through the intersection of these two lines is in the center of the field of view. The reticle represents the axis of the bore of the weapon, when it is adjusted for short-range firing or fixed at a definite angle to the bore of the weapon for long-range firing.

A reticle serves as a reference point for sighting a gun, but it may be designed also for measuring the angular distance between two points (grid lines in a gunsight telescope). Since it is in the same focal plane as a real image, it appears superimposed on a target and CAN BE VIEWED BY THE EYE WITH THE SAME ACCOMMODATION REQUIRED FOR VIEWING A TARGET OR FIELD.

In an optical system containing an erecting system, a reticle may be placed in one of two positions, as indicated in illustration 6-35. It may be placed IN THE IMAGE PLANE OF THE OBJECTIVE OR AT THE FOCAL POINT OF THE EYEPIECE (image plane of the erectors).

If the erecting system of the telescope increases its magnification, and the reticle is in the image plane of the objective, reticle lines appear wider on the target than if placed at the focal point of the eyepiece. For this reason, a reticle in a low-powered rifle scope is usually placed in front of the erectors; in a high-powered scope, it is placed at the focal point of the eyepiece.

A reticle in a gunsight telescope may be placed AT THE FRONT OR AT THE REAR OF THE LENS ERECTING SYSTEM OF THE INSTRUMENT, or RETICLES MAY BE PLACED AT BOTH POINTS, IN WHICH CASE THEIR PATTERNS ARE SUPERIMPOSED UPON EACH OTHER. THE PREFERABLE POSITION FOR THE RETICLE IN A TELESCOPIC SIGHT IS BETWEEN THE OBJECTIVE AND THE ERECTING SYSTEM, because the objective and reticle then form a unit and a shift of the erecting system does not disturb the alignment of these elements. When a prism erecting system is used in a telescopic sight, however, the reticle is usually placed between the erecting system and the eyepiece, in the focal plane of the objective lens.

If the objective lens of an optical system is incorrectly located with respect to the reticle, parallax is introduced, because the reticle is not directly in the image plane. Study illustration 6-36. PARALLAX IS APPARENT DISPLACEMENT OF AN OBJECT WHEN AN OBSERVER CHANGES HIS POSITION (part A, fig. 6-36). Parallax in a telescope WHICH HAS A RETICLE IS ANY APPARENT MOVEMENT OF THE RETICLE IN RELATION TO DISTANT OBJECTS IN THE FIELD OF VIEW caused by MOVEMENT OF THE OBSERVER'S HEAD. This condition exists when the image in the telescope lies in one plane and the reticle lies in another, as shown in part B of illustration 6-36. Observe the three positions of the reticle in relation to the image. The middle portion of part B of the illustration shows the reticle

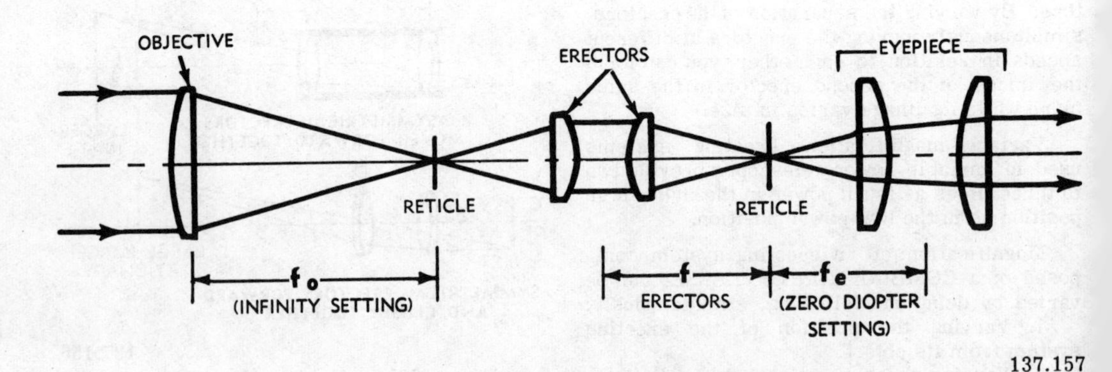

Figure 6-35.—Location of a reticle in a Galilean-type rifle scope.

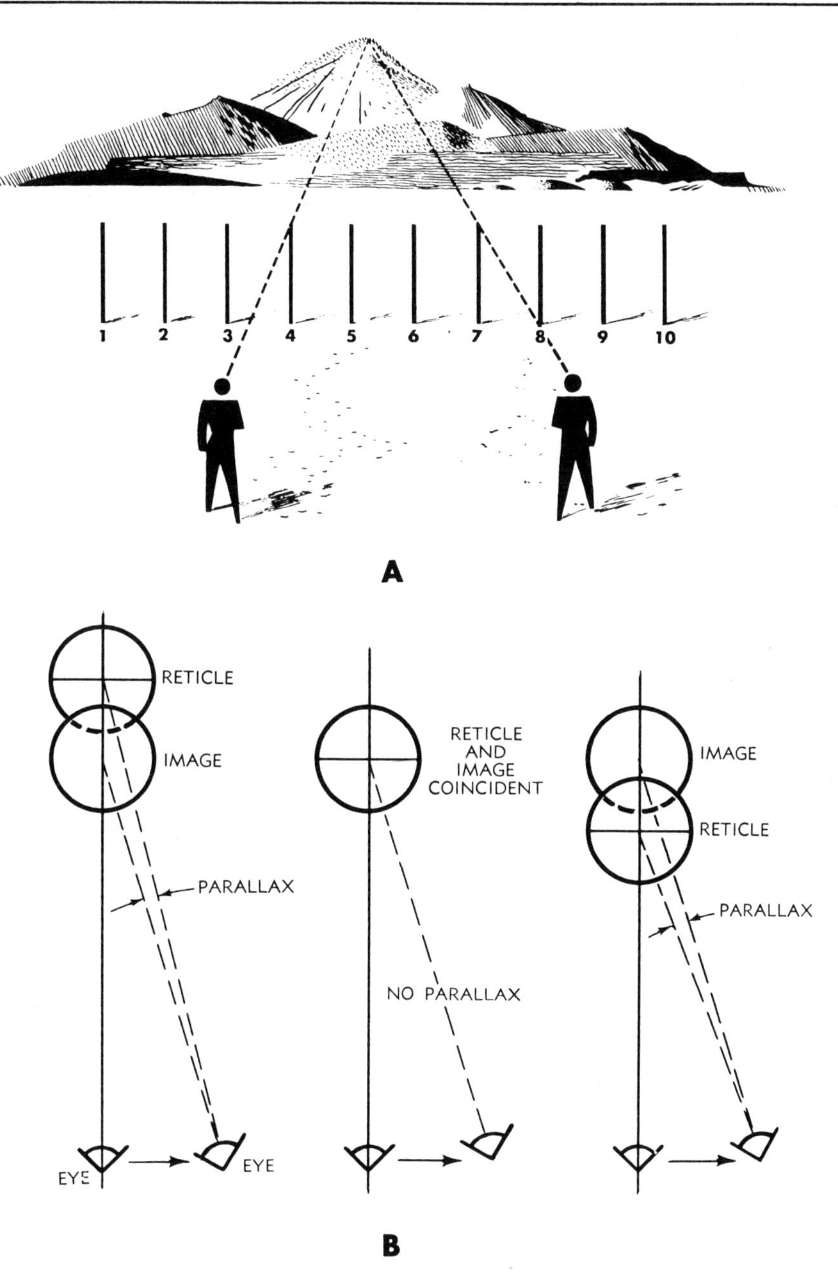

A. Demonstration of parallax.
B. Parallax in a telescopic sight.
Figure 6-36.—Parallax.

137.158

superimposed on the image, in which case parallax does not exist.

Parallax can be removed from the optical system by repositioning the objective lens as necessary to have its image in the same plane as that of the reticle. When the objective lens is moved forward or backward within the instrument, the image moves in the same direction, respectively.

MAGNIFICATION IN TELESCOPES

The formula for computing the magnifying power of an astronomical telescope, which has no erecting system, is:

$$M = \frac{f_o \text{ (focal length of objective)}}{f_e \text{ (focal length of objective)}}$$

This means that you can DETERMINE THE MAGNIFYING POWER of an astronomical telescope by dividing the focal length of the objective by the focal length of the eyepiece, provided the virtual image is at infinity, or the emergent light rays from the object are parallel. Remember the two conditions when this formula can be used for measuring magnifying power in an astronomical telescope. If the image is moved to the near point of the eye (10 inches), it increases slightly in size.

This formula can be used also for determining the amount of magnifying power produced by terrestial telescopes which have PRISM ERECTING SYSTEMS; but it cannot be applied to terrestial telescopes which have LENS ERECTING SYSTEMS, because such erecting systems can (usually do) contribute to the power of the optical system.

Magnifying power in a one-erector optical system for a telescope is equal to the distance of the focal length of the objective divided by the focal length of the eyepiece, multiplied by magnification of the erecting lens.

You learned previously in this chapter that magnification in a variable-power telescope is accomplished by moving the erectors. When the erectors are moved forward as a complete unit to increase magnification, or when the rear erector element only is moved forward, the image formed by the erectors moves back. This is the only case when the image position is not the same for all magnifications produced by the optical system of a variable-power telescope.

Power in an optical instrument is denoted by the letter x; for example, a 7 x 50 binocular is a 7-power instrument with an entrance pupil or objective size of 50 millimeters.

Another method for determining the magnifying power in all types of telescopes is this: DIVIDE THE DIAMETER OF THE ENTRANCE PUPIL BY THE DIAMETER OF THE EXIT PUPIL. The formula to use in doing this is:

$$MP, \text{ or } P = \frac{AP}{EP}$$

AP is the diameter or aperture of the entrance pupil, and EP is the diameter of the exit pupil, as shown in figure 6-37. Observe the position of AP and also the position of EP. You will recall that ENTRANCE PUPIL means the CLEAR APERTURE OF THE OBJECTIVE; and that the EXIT PUPIL is the diameter of the bundle of light which leaves an optical system. The exit pupil is actually AN IMAGE OF THE OBJECTIVE LENS PRODUCED BY THE EYELENS.

You can measure the diameter of the entrance pupil with a transparent metric scale—directly across the objective. This method of measurement is sufficiently accurate for most purposes.

You can determine the diameter of the exit pupil of a telescope by: (1) pointing the instrument toward a light source (out a window, for example), (2) inserting a piece of translucent material in the plane of the exit pupil, and (3) measuring the diameter of the exit pupil on the paper.

The best way to measure the diameter of an exit pupil, however, is with a dynameter. See illustration 6-38. This dynameter is essentially a magnifier or an eyelens with a fixed reticle on a frosted glass plate, both of which move as a unit within the dynameter tube.

To measure the exit pupil with a dynameter, place the dynameter between the eye and the eyepiece of the instrument and focus the dynameter until you have the bright disk of the exit pupil sharply defined on its frosted reticle. Then measure the diameter of the exit pupil on the dynameter reticle (usually graduated in .5 mm) and read the eye distance on the scale on the dynameter tube. This means that in order to keep the image in focus the eyepiece must be moved a distance equal to the amount of shift of the image.

You will learn more details concerning the positioning of elements in the optical system of a telescope (and measuring magnification) when you study chapter 7, which deals with mechanical construction of optical instruments.

BINOCULARS

A binocular is an optical instrument for use by both eyes at the same time. Binoculars used in the Navy consist of two telescopes—one for each eye—hinged together. Most of the first binoculars made consisted of two Galilean telescopes and WERE CALLED FIELD OR NIGHT GLASSES. They were compact in construction, as binoculars should be, but they WERE LIMITED TO EITHER LOW MAGNIFICATION OR A NARROW FIELD OF VIEW.

Navy binoculars in use today consist of an objective, an eyepiece (eyelens and usually a field lens), and Porro prisms (to erect the

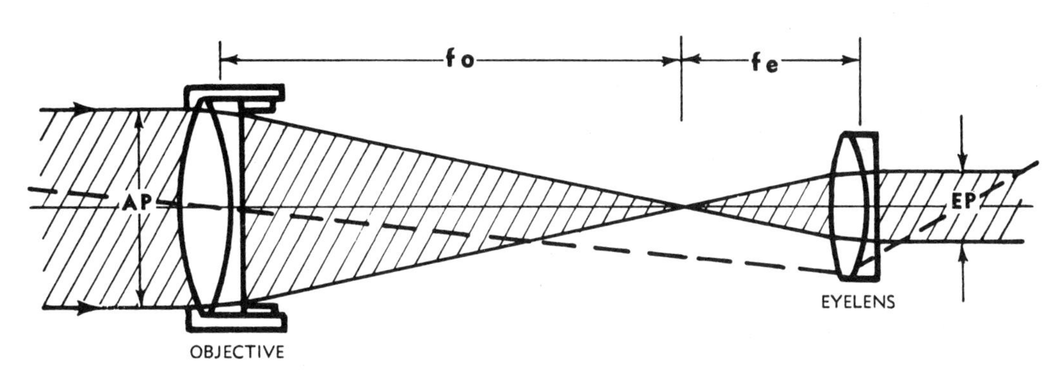

137.159

Figure 6-37.—Entrance and exit pupils.

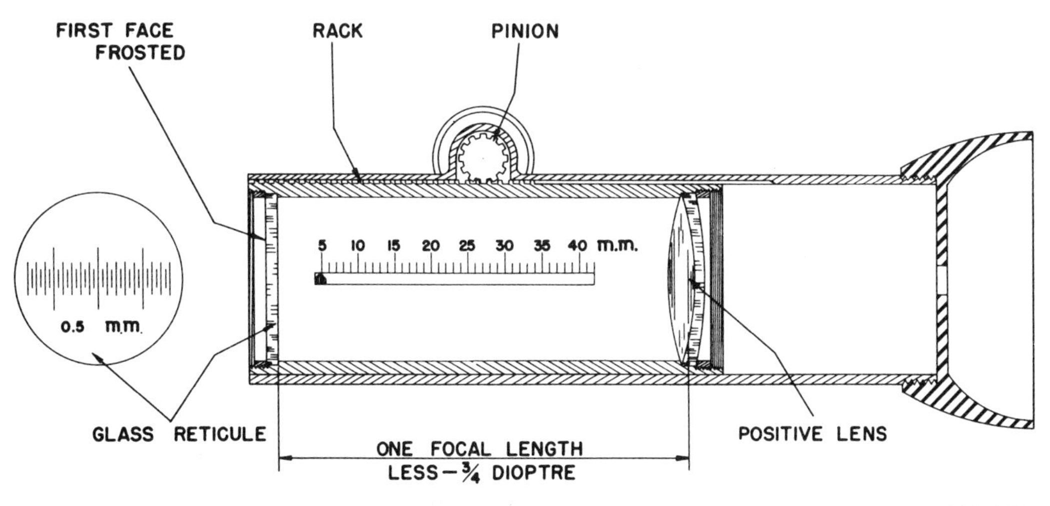

137.160

Figure 6-38.—Dynameter.

image). Prisms are much better erecting systems than lenses in binoculars for three reasons:

1. They enable manufacturers to make binoculars short and compact, to facilitate handling, because no lens erecting system is required. (A lens erecting system must have space equal to four times its focal length.)

2. They double the line of sight back on itself twice. (Four reflections are required to continue the light in an unchanged path.)

3. By increasing an observer's effective interpupillary distance, most prism systems increase the stereoscopic effect.

The left telescope of many models of binoculars contains a reticle. The optical axis of each scope must be parallel to the hinge mechanical axis throughout the entire movement of the interpupillary range; otherwise, each eye sees a different field, resulting in a double image.

Magnification by binoculars depends upon the focal lengths of the objective and eyepiece groups. True field of view varies in accordance with the design of lenses and their power, and the size of the objective determines the brightness of the image. The amount of the exit pupil which can be used is determined by the size of the pupil of the eye of the observer. The size of the pupil of an observer varies from about .1 inch (approximately 2.5 mm) for very brilliant illumination to about .3 inch (about 7 mm) for very faint illumination. Binoculars with large objectives and exit pupils are generally better suited for observation at night and under poor conditions of visibility. The larger exit pupil of the binocular allows the observer to make full use of his larger exit pupil. Binoculars of this type are called NIGHT GLASSES.

Binoculars are usually designated by the power of magnification and diameter of their objectives. A 7 x 50 binocular magnifies 7 diameters and has objectives 50 millimeters in diameter.

The radius of stereoscopic vision is also increased by binoculars, because an observer views an object through them from the two objectives, which are more widely separated than his eyes. Magnification provided by binoculars increases an observer's range of vision. If, for example, the distance between the lines of sight of the observer's eyes is doubled by prism binoculars and a 7-power instrument is used, the radius of stereovision is increased from approximately 500 yards (nonmal) to about 7,000 yards (500 x 2 x 7).

CHAPTER 7

CONSTRUCTION OF OPTICAL INSTRUMENTS

Before you can qualify for advancement to Opticalman 2, you must know the "characteristics pertaining to the construction and assembly of optical instruments," and also the "use and function of drafting machines."

This chapter is therefore limited in scope primarily to a general discussion of optical instruments from the standpoint of basic construction, and the construction and operation of drafting machines. The discussion of instrument body housings, lens mounts, eyepiece arrangements, prism mounts, instrument bearings and gears, and other optical elements will help you to understand better the construction, operation, and maintenance of optical instruments considered in detail in chapters 9 through 17.

BODY HOUSINGS

The housings for optical elements of various optical instruments must be made in accordance with requirements for the location of and the distance between the optical elements of the instruments. You will recall that the function of the mechanical portion of an optical instrument is to hold the optical elements in position, to protect them from foreign substances and damage, and to keep out light where it is not desired. Study illustration 7-1, which shows the body housing for the optical elements of one type of telescope, and you will then understand the importance of the body housing of an optical instrument.

Observe the length of the housing; then note the different optical elements and their positions in the instrument. All of these elements must be so positioned and secured in the housing that they will not move under normal circumstances and impair the effectiveness of the optical system.

Refer now to illustration 7-2 and study the housing and optical elements in a periscope-type telescope. Note the reticle adjustment knob and the monocular eyeguard. The positions of the various optical elements in the instrument are shown, and on the right they are shown in schematic form with the nomenclature listed. The arrow shows the path of light through the instrument. Observe that, in contrast with the optical elements of the telescope illustrated in figure 7-1, the optical elements of the instruments shown in figure 7-2 do not consist of lenses only. It has two right-angled prisms (for bending the light at the elbows) and two reticles, as illustrated.

Refer frequently to appropriate illustrations as you study next some important parts of optical instruments.

EYESHIELDS

Eyeshields (fig. 7-1) are shields of metal, plastic, or rubber fastened over the end of an optical instrument which contains the eyepiece. The function of such shields is (1) to protect the observer's eyes from stray light, wind, and injury from gunfire shock or similar disturbances; and (2) to maintain proper eye distance. Shields made of soft rubber can most effectively meet all of these requirements.

SUNSHADES AND OBJECTIVE CAPS

Sunshades are tubular sections of metal secured in slots around the objective cells of many optical instruments to protect them from rain and direct rays of the sun. See illustration 7-3. The lower part of the shade is generally cut away in the manner illustrated.

Sunshades also eliminate glare produced by sunlight on unprotected objectives (outer faces), and they protect the thermosetting cement used to secure the elements of the objectives.

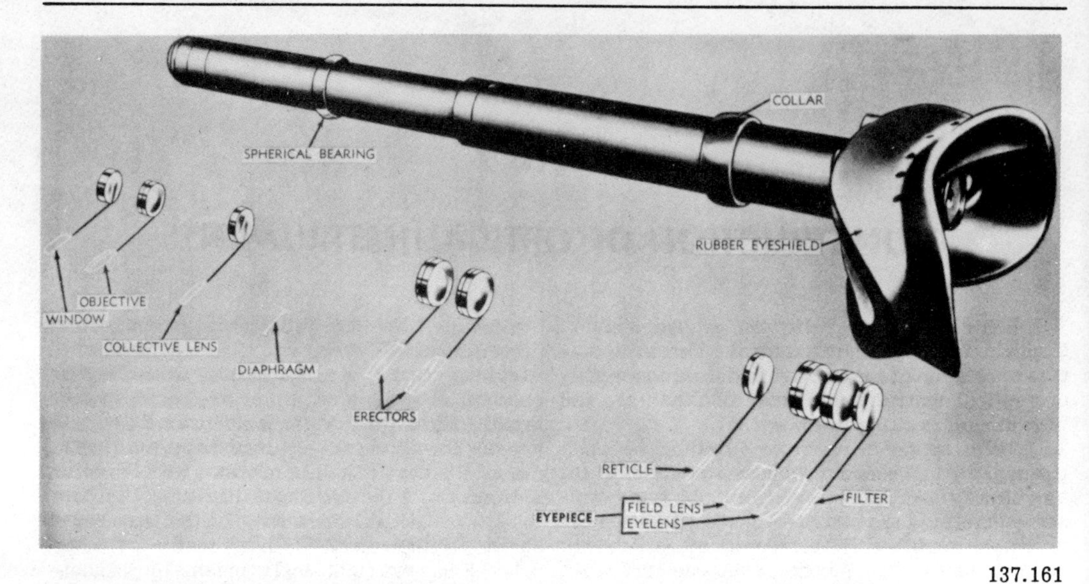

137.161
Figure 7-1.—Body housing and optical elements of a telescope.

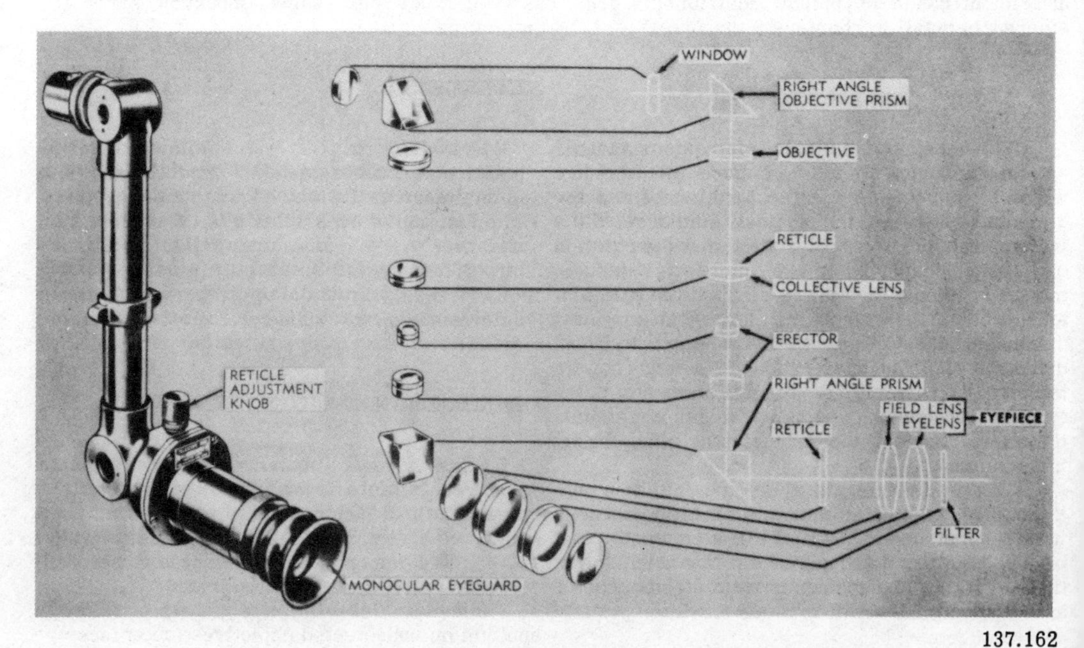

137.162
Figure 7-2.—Housing and optical elements of a periscope-type telescope.

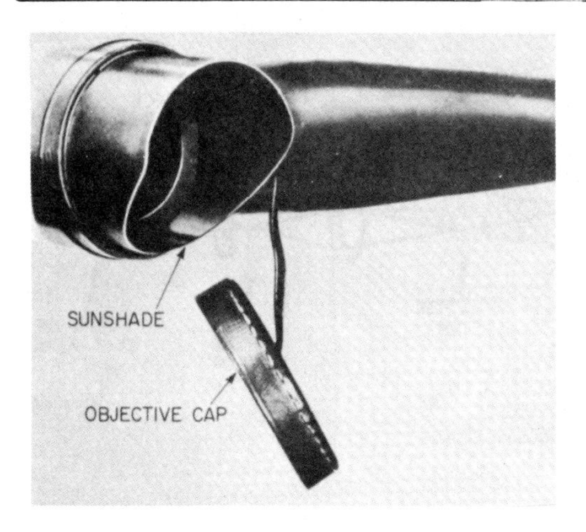

137.163
Figure 7-3.—Sunshade and objective cap.

Objective caps (fig. 7-3) are leather or metal covers which fit over the sunshade and/or objective end of the instrument to protect the objective when the instrument is not in use. Note the strip of leather attached to the objective cap to prevent loss.

DIAPHRAGMS OR FIELD STOPS

Diaphragms (fig. 7-1), sometimes called field stops because they limit the field of view, are rings of opaque material placed in optical systems in such manner that light passes through their centers only. When placed around the edges of lenses, stops prevent rays of light from passing through the margins of the lenses to cause aberrations.

When a diaphragm is placed between the objective and erecting system, between the erecting system and the eyepiece, or between parts of an erecting system, it eliminates marginal rays which would cause glare and haze if they were allowed to reflect from the inside walls. Such a stop in an optical system also prevents ghost images, which result from internal reflection of light rays from curved lens surfaces.

Antiglare stops (fig. 7-4) improve contrast by preventing rays of light exterior to the field of view from bouncing off the interior of the instrument and fogging the field or causing glare (rays 1 & 2, fig. 7-4). Observe that field stops in this illustration are in the image planes. Nonreflecting paint and baffle finish on the inner wall of the instrument housing eliminate most of the undesired light. Light from a brilliant source may still be reflected off a dark wall and cause serious trouble, especially if a glass reticle is used. This is particularly true in wide-angle telescopes and antiaircraft instruments pointed toward the sun.

Stops in straight-tube telescopes (fig. 7-1) are washers or disks with holes in their centers, and placed at intervals along the interior of the telescope body tube to prevent stray rays from reflecting internally, and to control certain aberrations. The prism shelf in binoculars is designed to function as a stop. This is also true of Porro prisms with round corners and grooves in their faces.

Antiglare stops are usually located between the objective and the erectors, or wherever an image of the aperture stop is formed, as shown in figure 7-4. When located between the erectors, they are known as erector stops; and THEY PROVIDE BALANCED ILLUMINATION BY LIMITING THE RAYS OF LIGHT FROM THE CENTER OF THE FIELD TO THE SAME AREA AS THOSE FROM THE EDGE OF THE FIELD.

A field stop limits the field to that area which is fully illuminated and sharply focused, because it eliminates the peripheral region of poor imagery (aberration) and prevents the observer from viewing the inside of the instrument. When a field stop is located in the image plane, as shown, it provides a sharply defined limit to the field. It is designed to admit as much light as is required by the next element in the optical system (erectors or an eyepiece). If a field stop is used in each image plane, as illustrated, the second stop must be slightly larger than the image of the first stop (where the second stop is located), so that inaccuracy in size or positioning of the stop will not conflict with the sharply defined image of the first stop.

LENS CELLS OR MOUNTS

Lens cells or mounts (same meaning) are tubular mountings for optical elements in an optical instrument. See illustration 7-5. This cell is always made of metal, and it holds one or more lenses, generally secured in the cell by a retainer ring or by burnishing (discussed later in chapter). An entire lens assembly may be mounted as a unit in one cell.

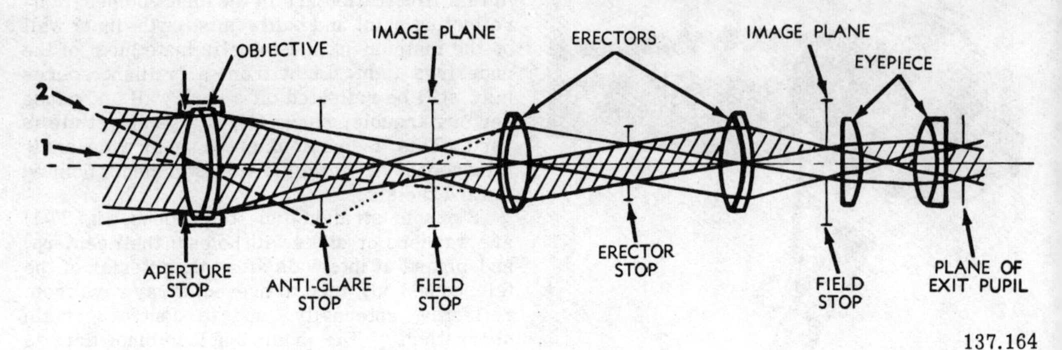

137.164

Figure 7-4.—Antiglare stops.

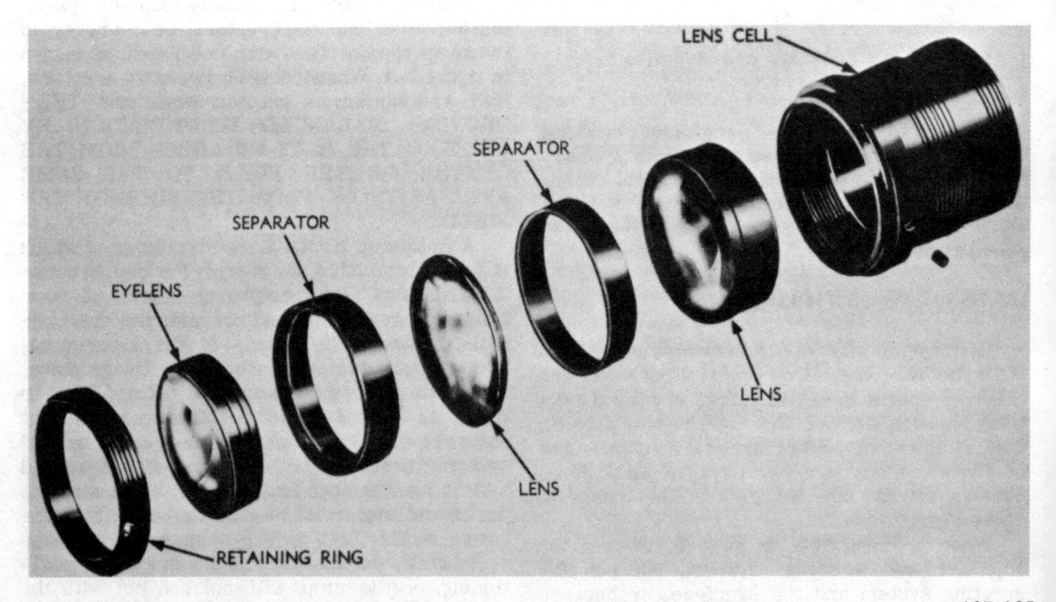

137.165

Figure 7-5.—Lens cells, lenses, separator, and retaining ring.

When two or more lenses are mounted in a cell, they are held in their proper positions by spacers or separators (fig. 7-5). Adjoining faces of these separators are usually beveled to fit snugly on the faces of lenses. As illustrated, the retainer rings are threaded on the outside in order that they may be screwed into lens cells to secure lenses.

Some of the methods employed by the Navy for mounting lenses in optical instruments are considered in the following pages.

RETAINER RING MOUNT

A retainer ring lens mount (fig. 7-6), machined internally, allows a lens to slide into it

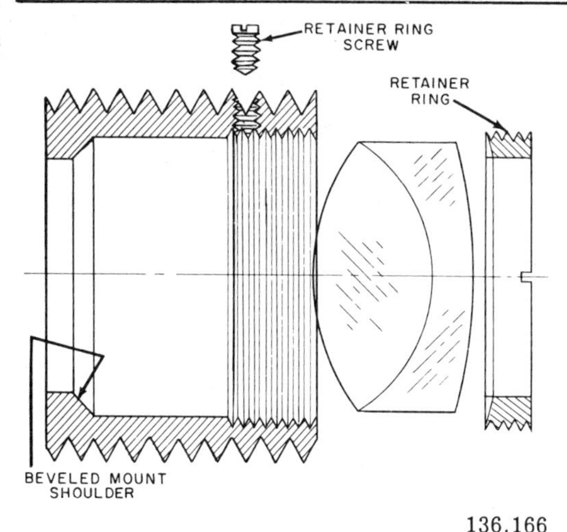

136.166
Figure 7-6.—Retainer ring lens mount.

and fit snugly against a shoulder. An externally threaded ring (retainer ring) screws into the mount and holds the lens in place against the shoulder. Note that the lens cell shown in figure 7-6 is threaded externally, so that it can be screwed into the housing of a telescope.

Some retainer ring mounts are threaded externally so that they may be screwed into the telescope body for axial adjustments. Mounts which do not have an external thread can be slipped into the body of a telescope against a shoulder (as per Navy specifications). Both mounts are then secured in place with another retainer ring which screws into the telescope body.

Another type of retainer ring mount is part of the main body of a telescope (by design), and the lens slides into the body and is held in place with a retainer ring.

A retainer ring mount (sometimes secured by a setscrew, as shown in fig. 7-6), is used for holding compound lenses too large to be cemented together.

CAUTION: To prevent severe strain in the lens, distortion of the lens image, and possible lens breakage, DO NOT TIGHTEN the retainer ring TOO tight, and DO NOT REVERSE the spring ring to such an extent that the three lugs bear against the glass. The positive and negative elements of the lens are therefore separated by three tinfoil shims placed 120° apart around the edges of the lens. The shims, usually .001" or .002" in thickness, are cemented to the concave surface of the negative lens in order to hold them in their proper positions.

The two elements with the shims are then placed in the mount against a shoulder, as in a retainer ring mount. A metal ring (spring), same diameter as the elements, is machined flat on one edge and machined with three lugs 120° apart on the other edge. The spring ring is then placed in the mount with its flat surface against the rear surface of the negative lens, after which a retainer ring is screwed into the mount against the three lugs of the spring ring and tightened, so as the distribute even pressure on the elements and hold them securely in place.

Retainer ring mounts are used for securing optical elements in the housings of optical instruments because they make removal of lenses easy—for cleaning, recementing, or replacing.

BURNISHED MOUNTS

A burnished lens mount is illustrated in figure 7-7. It is constructed in a manner similar to that of a retainer ring mount except that the lens is not secured in a cell by a retaining ring. As shown in the illustration, you must press or form part of the mount over the edges of a lens in order to hold it securely in position.

The difficulty with a burnished lens mount is that NO PROVISION IS MADE for removing the lens for cleaning, recementing, or replacing. If an attempt is made to remove a lens from this type of mount, damage to the lens and mount may result. When necessary, replace both the lens and the mount. Although burnished lens mounts are not used in modern naval optical instruments, some of them are still in use in older naval and foreign-made optical instruments.

ECCENTRIC MOUNTS

An eccentric lens mount (fig. 7-8) is used to mount the objective of an optical instrument in a manner which permits moving of the optical center of the lens laterally to the mechanical axis of the optical instrument. Such movement is necessary to align the optical axis of the lens to the mechanical axis of the instrument. This type of lens mount is used in an optical instrument when extreme exactness in performance of the optical system is required.

Take another look at illustration 7-8. Observe in part A that the objective lens is mounted in an eccentric mount, over which an eccentric ring is placed. By rotating the inner mount as desired in the outer ring, or by rotating the outer ring over the inner mount, you can move the optical axis of the objective to any point in a relatively large area. When you have both rings properly positioned, you can lock them in position.

The outer surface of an eccentric lens mount is concentrically machined to a bearing surface; and the inner surface (which holds the objective lens) is also machined concentrically, but the centers of the concentric surfaces DO NOT COINCIDE. The mount is thicker on one side than on the other side.

An eccentric ring, machined in the same manner as the lens mount, fits over the mount; and both are then ready for placing in the housing of the telescope. A retainer ring is used to secure them. If the ring or mount (or both) is then rotated, the optical axis of the objective rotates in a circle to cause circular movement of the image created by it.

SCREW ADJUSTING MOUNT

A screw adjusting lens mount is employed to position a reticle of a telescope in one of two ways: (1) by adjusting the reticle mount, or (2) by adjusting the entire telescope body. This mount has four adjusting screws placed 90° apart around the mount (fig. 7-9, from Mk 75 boresight) for positioning horizontally and vertically. The ends of these screws (thumbscrew type or slotted-head type) have bearing pads which press against the reticle mount or the telescope body to hold it firmly in position.

PRISM MOUNTS

Prism mounts hold prisms securely in their proper positions in a telescope body. As you know, a prism in an optical system must have correct positioning with respect to all other optical elements in the system. Various types of mounts are used to hold the prisms in correct position, a few of which are explained briefly.

PORRO PRISM MOUNTS

A porro prism mount (fig. 7-10) consists primarily of a flat metal plate shaped to the interior of a telescope body. It is machined to hold one prism on each side of the plate. The hypotenuse surfaces of the prisms are mounted parallel to each other, and they are set over holes machined in the plate to allow light to pass from one prism to the other.

In order to maintain the frosted surfaces of the two mounted prisms at 90° angles with each other, a rectangular, metal adjustment ring (prism collar) is placed snugly around each prism. If the two prisms are NOT AT 90° ANGLES with each other, an effect CALLED LEAN IS CREATED IN THE PRISM CLUSTER, which means that the IMAGE APPEARS TO LEAN AT AN ANGLE IN COMPARISON WITH THE ACTUAL OBJECT.

Each prism is secured to the mount with a spring clip or prism strap, pressed against the apex of the prism. The strap itself is secured to two posts, one on each side of the prism; and the posts, in turn, are screwed into the prism plate. A metal shield placed over each prism under the prism strap prevents stray light from entering the other prism surfaces. These shields must be so placed that they do not touch the reflecting surfaces of the prisms; because if they touch, total internal reflection does not take place and some of the light is refracted through the reflecting surface and absorbed by the light shields.

AMICI PRISM MOUNTS

An Amici (roof-angle) prism mount (part A, fig. 7-11) consists of a right-angled bracket on which the prism sits. Shoulders are ground on the frosted sides of the prisms in order to secure them to the bracket. Two prism straps (one on each side) are placed against the prism shoulder, and then secured by screws to the bracket. The bracket is fastened to the telescope body with four screws which can be loosened when it is necessary to adjust the prism mount. Part B of the illustration shows disassembled parts.

RIGHT-ANGLED PRISM MOUNTS

Mounts for right-angled prisms vary in design in accordance with needs. One mount (fig. 7-12) holds the silvered or reflecting surfaces

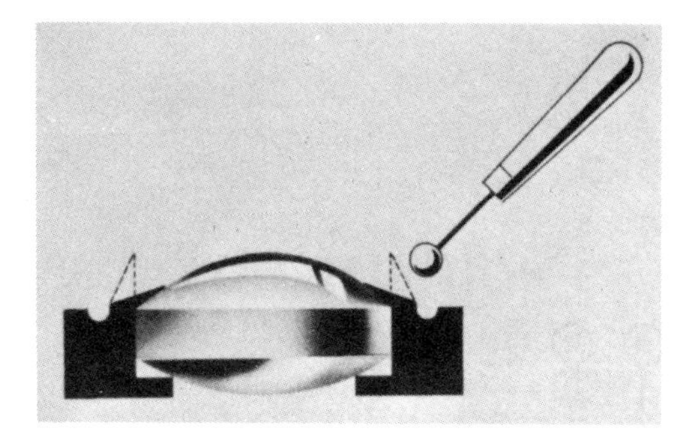

137.167
Figure 7-7.—Burnished lens mount.

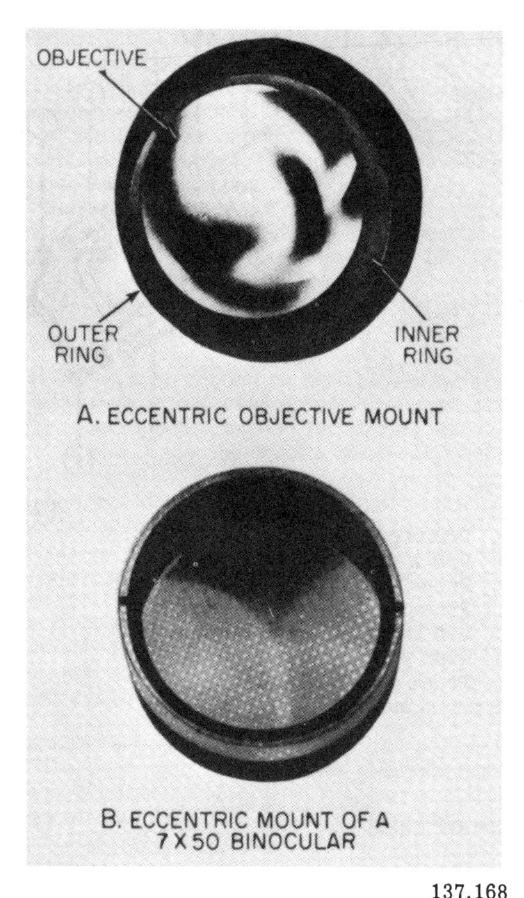

137.168
Figure 7-8.—Eccentric lens mount

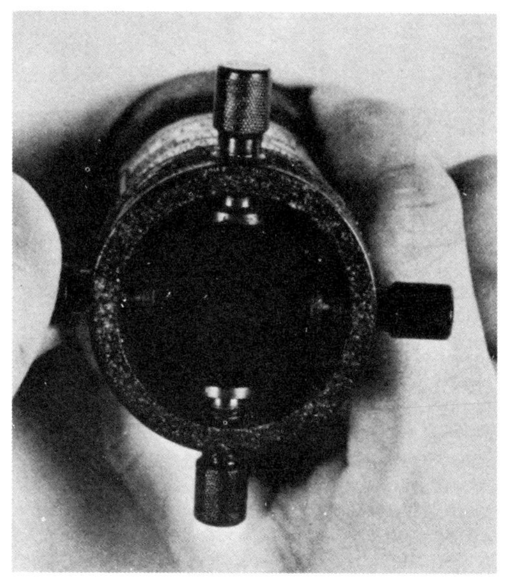

137.169
Figure 7-9.—Screw adjusting lens mount.

of prisms securely in place and properly aligned on bearing pads which prevent the surfaces from touching the base of the mount. Four prism straps, two on each side, hold the prisms in position. The straps also contain bearing pads which help to keep the prisms properly aligned without chipping.

EYEPIECE ASSEMBLIES

Lenses in an eyepiece (figs. 6-15 through 6-21) are secured in a retainer ring-type mount. The field lens and the eyelens may be fastened separately, each with a retainer ring; but they may also be secured together with the same retainer ring, with a separator placed between the field lens and the eyelens to hold both at the correct distance from each other.

The distance between the reticle and the eyepiece in an optical instrument must be so adjusted to the observer's eye that the reticle and image of the object are sharply defined and eye fatigue is eliminated. In order to provide this adjustment, the lenses (2 or more) of the eyepiece are mounted in a single lens

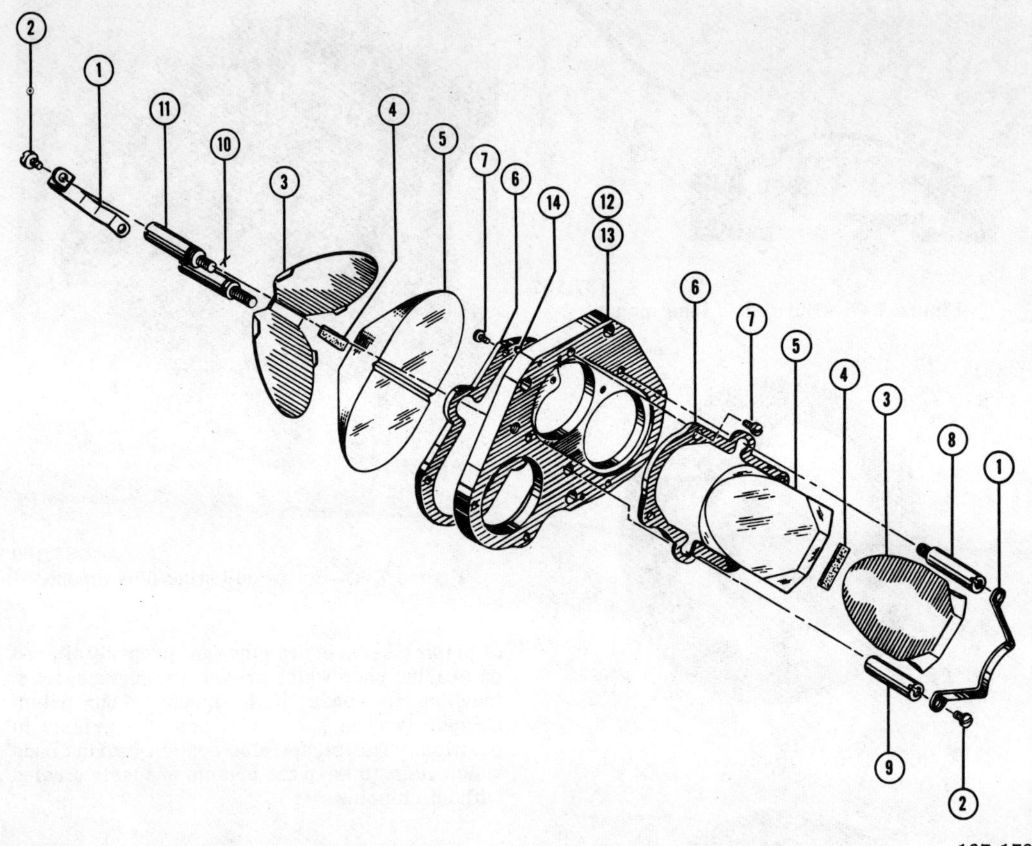

1. Prism clip.
2. Prism clip screw.
3. Prism shield.
4. Prism clip pad.
5. Porro prism.
6. Prism collar.
7. Prism collar screw.
8. Prism post A.
9. Prism post B.
10. Prism post C.
11. Prism post D.
12. Left prism plate and dowel pins.
13. Right prism plate and dowel pins.
14. Prism plate dowel pin.

Figure 7-10.—Porro prism mount.

cell or tube, whose distance from the reticle (also focal plane of the objective) can be adjusted by a rack and pinion, a draw tube, or by rotation of the entire eyepiece during adjustment of the diopter scale.

DRAW TUBE

A draw tube focusing arrangement (fig. 7-13) consists of a metal tube carrying the lenses and their retainer ring. The tube is focused

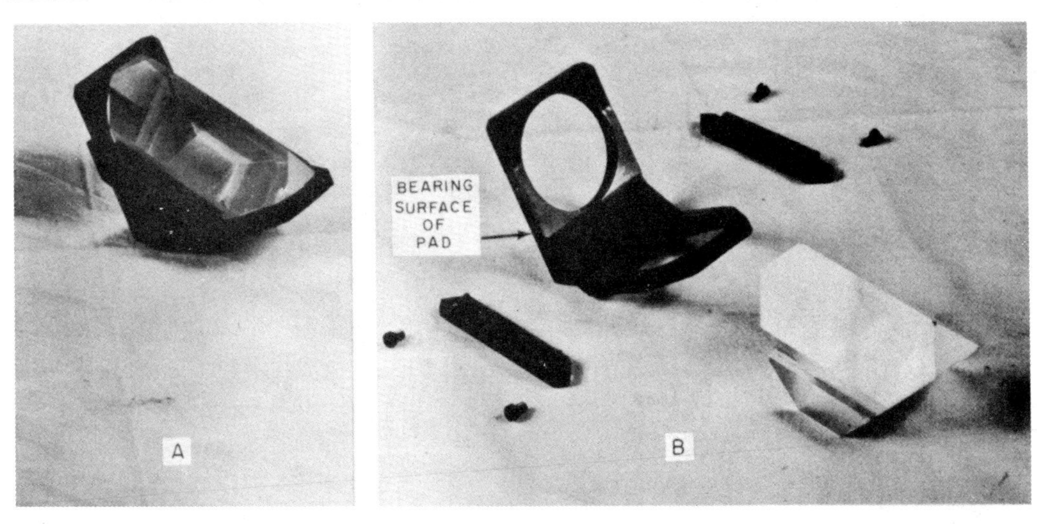

137.171

Figure 7-11.—Amici prism mount.

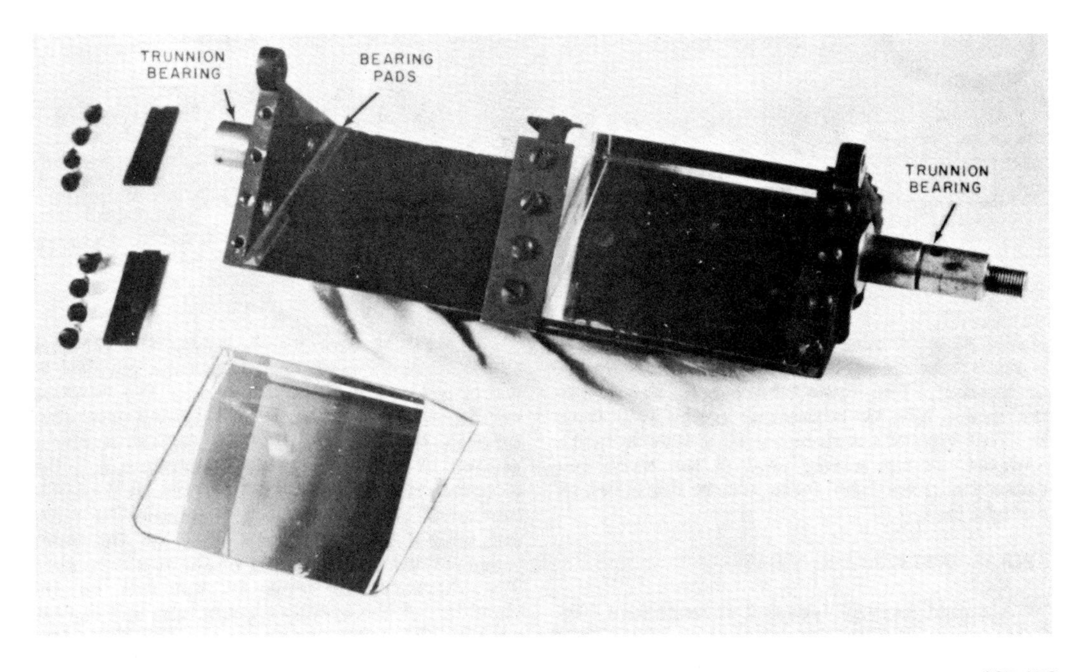

137.172

Figure 7-12.—Right-angled prism mount.

137.173

Figure 7-13.—Drawtube focusing arrangement.

manually by sliding it forward or backward in a guide tube at the rear of the telescope body or housing. The draw tube can be secured to the guide tube or withdrawn completely from it. This type of eyepiece focusing arrangement, however, is not widely used in the Navy, because the draw tube focus can be disturbed by a slight jar.

SPIRAL (HELICAL) KEYWAY

A spiral keyway focusing arrangement (fig. 7-14) is a modification of a draw tube. It is similar in construction to a draw tube, but has the additional following components: (1) a focusing key or shoe, (2) a focusing ring, (3) a retainer ring, and (4) a diopter-ring scale.

A straight slot which guides the focusing key is cut through the guide tube parallel to the optical axis of the telescope. The focusing key is fastened to the draw tube and protrudes through the straight slot to engage a spiral groove or keyway in the focusing ring. The focusing ring is permitted to turn on the guide tube, but it is prevented from moving along the optical axis by a shoulder on the guide tube and the retainer ring on the opposite side. The diopter-scale ring is mounted on the shoulder of the eyepiece guide tube and is read against the index mark on the focusing ring. Study illustration 7-9.

The diopter scale is graduated on either side of 0 DIOPTER TO READ FROM PLUS TO MINUS DIOPTERS. The number of plus or

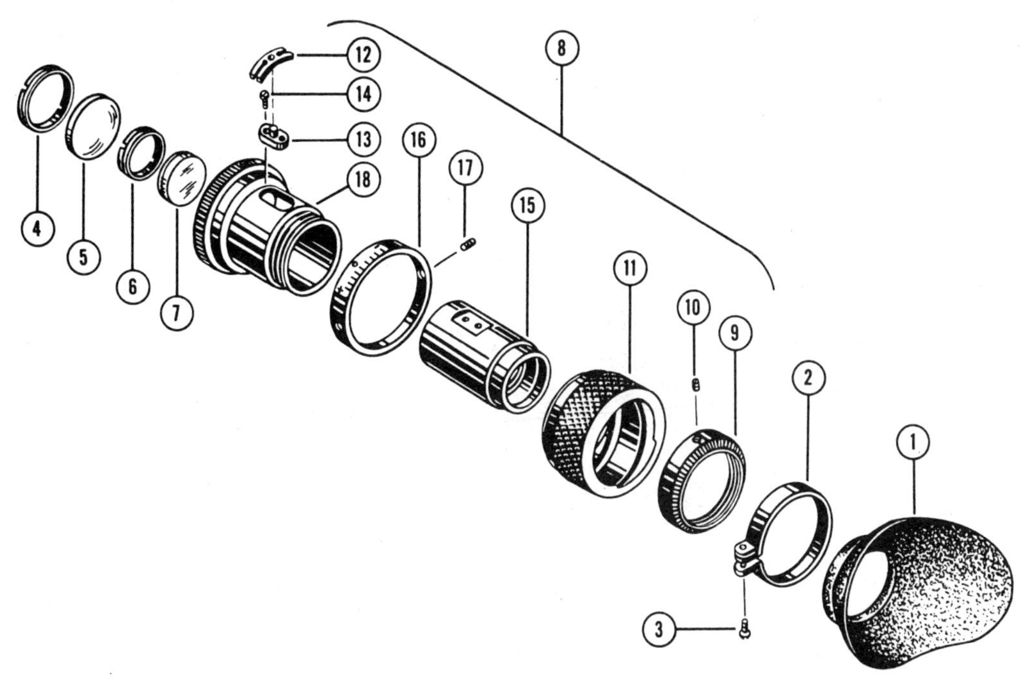

137.174

1. Eyeguard
2. Eyeguard clamp.
3. Eyeguard clamp ring screw.
4. Eyepiece collective retaining ring.
5. Eyepiece collective lens.
6. Eyelens retaining ring.
7. Cemented doublet eyelens.
8. Eyepiece focusing assembly.
9. Focusing ring stop ring.
10. Stop ring lock screw.
11. Knurled focusing ring.
12. Focusing shoe.
13. Focusing key.
14. Focusing key screw.
15. Eyepiece lens mount.
16. Diopter ring.
17. Diopter ring lock screw.
18. Eyepiece diopter.

Figure 7-14.—Spiral keyway focusing arrangement.

minus diopter graduations depends upon the design of the instrument, but it usually runs from +2 to -4 diopters. When the focusing ring is turned either way, the focusing key follows the spiral keyway and moves the draw tube in or out to focus the eyepiece. If an operator focuses the eyepiece to his eye and notes the diopter scale reading, he can save time by adjusting to that reading each time he uses the optical instrument.

THREADED LENS MOUNTS (MULTIPLE-THREAD)

A multiple-thread eyepiece lens mount (fig. 7-15) is a retaining-ring type with external multiple-lead threads, and it screws into a guide tube or eyepiece adapter with multiple-lead threads. When the eyepiece mount is screwed all the way into the adapter, it is stopped by a shoulder in the adapter. A stop ring is

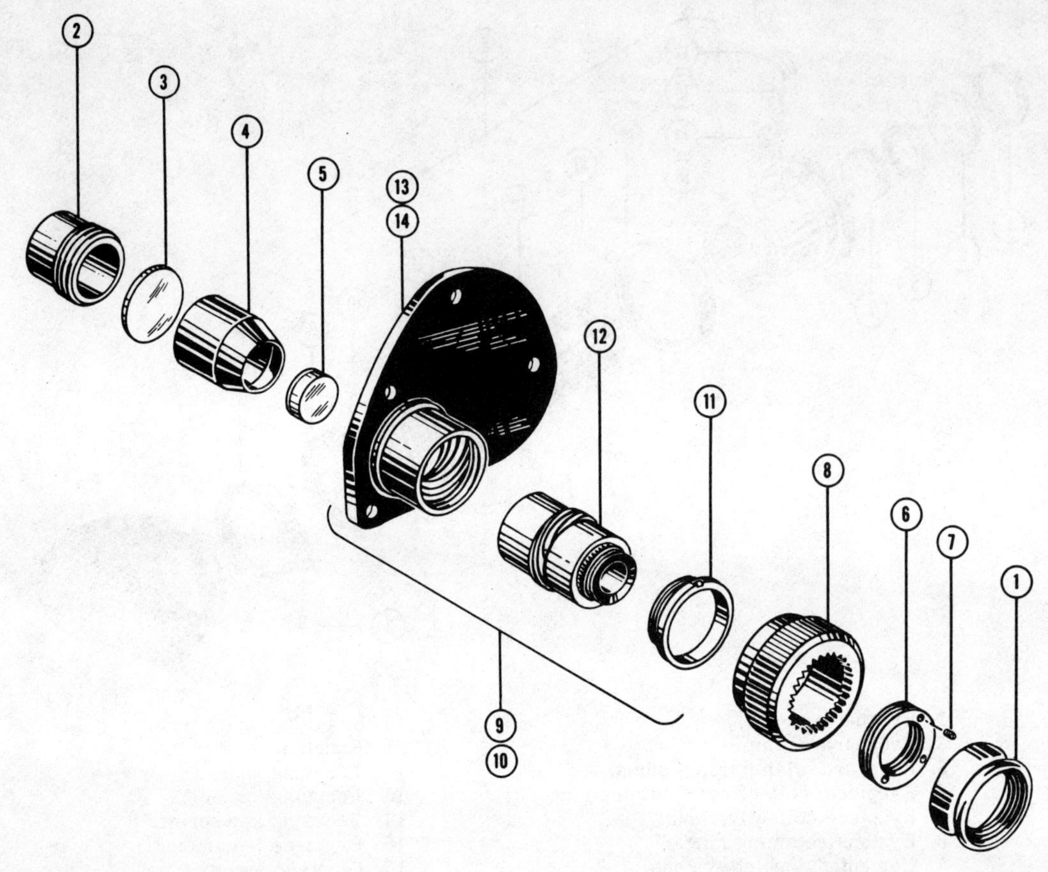

137.175

1. Eyepiece cap.
2. Collective lens retaining ring.
3. Collective lens.
4. Eyepiece lens spacer.
5. Cemented doublet eyelens.
6. Eyepiece clamp ring.
7. Eyepiece clamp ring lock screw.
8. Knurled focusing ring.
9. Left cover and eyepiece mount assembly.
10. Right cover and eyepiece mount assembly.
11. Right eyepiece stop.
12. Eyepiece lens mount.
13. Right cover with eyepiece adapter.

Figure 7-15.—Multiple-thread eyepiece lens mount.

then screwed into the top of the adapter, which prevents extraction of the eyepiece mount when the threads reach the stop ring as the mount is screwed all the way out. A focusing ring with a diopter scale engraved on it is attached to the top of the eyepiece mount.

INTERNAL FOCUSING MOUNT

An internal focusing eyepiece mount shown in figure 7-16, consists of a housing secured and sealed to the rear of the telescope body. The housing contains an eyelens secured by a retaining ring; and in the housing a movable lens mount or cell containing the field lens and an intermediate lens is free to move forward or backward when the focusing knob and shaft are activated. As the focusing knob rotates, it turns the focusing shaft and rotates an eccentrically mounted actuating plate which, in turn, slides the movable lens mount toward or away from the eyelens during focusing for individual eye corrections. The dioptric scale is on the focusing knob and the index mark is on the focusing shaft housing.

A fixed-type eyepiece, (part D, fig. 7-12) as the name implies, is fixed in position and cannot be focused for individual eye correction. The eyepiece mount may consist of a housing secured and sealed at the rear of the telescope body, which contains the eyelens, separator, field lens, and the retainer ring. The eyepiece housing may also be part of the main telescope housing with its component parts. If the eyepiece housing is part of the main telescope housing, the lenses and the spacer slide into the eyepiece housing from the rear and are secured in place with a retaining cap screwed onto the rear of the housing.

Because this eyepiece cannot be focused for individual eye correction, the light rays which leave it are slightly divergent, with a value of -3/4 or -1 1/2 diopters. It is set at this value because the majority of operators set focusing eyepieces slightly on the minus side of the dioptric scale.

Focusing-type eyepieces are mechanically designed to provide fast focusing with minimum turning of the focusing ring or knob. This design permits the eyepiece (when turned completely out) to stop on the plus side of the diopter scale; and to be focused all the way in to the stop on the minus side of the scale, with one rotation (or less) of the focusing ring. Multiple-lead threads of eyepiece mounts, because of their long lead, are responsible for this type of focusing. In internal focusing eyepieces, the eccentric plate slides the lens mounts from maximum to minimum throw with a half turn (or less) of the focusing knob.

The lenses of the spiral keyway and internal focusing eyepieces do not rotate when they are focused, and this is an advantage over a multiple-lead-thread eyepiece. When multiple-lead thread eyepieces are rotated, eccentricity in the lenses or their mounts (if present) causes the image of a target to appear to rotate in a small circle. For this reason, eyepieces with draw tubes which slide in and out without rotating are generally preferred in instruments with reticles. NOTE: The reticle must be superimposed ON THE SAME SPOT OF THE TARGET ALL THE TIME, REGARDLESS OF THE MANNER IN WHICH THE EYEPIECE IS FOCUSED. If the eyepieces or lens mounts rotate with eccentricity in a telescope which has a reticle, the image of the target appears to move under the reticle image in a small circle.

One advantage internal-focusing and fixed eyepieces have over spiral keyway and multiple-lead-thread eyepieces is that they can be sealed well enough to prevent entrance of foreign matter and moisture. Telescopes with these eyepieces can even be submerged under water, because they will not leak. On some optical instruments, fixed-type eyepieces must also be capable of withstanding a hydrostatic pressure test (subjected to water pressures applied externally) prior to approval for service in the fleet.

Spiral keyway and multiple-lead-thread eyepieces cannot be submerged under water, and they also BREATHE DURING FOCUSING; that is, WHEN YOU FOCUS THEM IN, THEY COMPRESS THE AIR WITHIN THE TELESCOPE AND FORCE IT OUT THROUGH THEIR JOINTS AND LOOSE FITTINGS. NOTE: Some telescopes have a small hole near the eyepiece mount which enables the air in them to escape freely. WHEN YOU FOCUS THESE EYEPIECES OUT, THEY DRAW AIR AND DUST INTO THE TELESCOPE. This breathing action can be caused also by changes in atmospheric pressure or temperature changes (day to night, for example). As time passes, dirt and moisture collected in the optical elements of the telescope diminish or obliterate vision through the instrument.

Multiple-lead-thread eyepieces have few mechanical parts and are therefore light in weight. Another advantage they have over some other types of eyepieces is that their threads reduce backlash, hold eccentricity to a tolerable minimum, and provide smooth focusing action.

One disadvantage of a fixed-type eyepiece is that IT DOES NOT PROVIDE MEANS FOR FOCUSING TO THE EYES. This is a fairly serious disadvantage to a slightly farsighted operator who requires that convergent rather than divergent light rays leave the eyepiece. It is also a serious disadvantage for a slightly nearsighted operator who requires that more divergent rays than the eyepiece itself provides leave the eyepiece. Normally, only trained operators with natural 20/20 vision operate gunsights; but both a person of normal vision and another with vision corrected by glasses to normal vision have NEARLY THE SAME diopter setting for focusing an eyepiece. No two individuals, however, with or without eyes corrected by glasses, have the same diopter setting. An individual's setting can change (usually does) from hour to hour during the day, in accordance with the amount of time spent looking through optical instruments.

OPTICAL INSTRUMENT BEARINGS

When a shaft is mounted in a device to hold it during rotation, friction develops at the contact point of the shaft with the device. Friction develops heat, and the amount of friction produced in a shaft housing must therefore be reduced to a minimum in order to obtain satisfactory performance and longer life of the shaft. Devices which reduce the amount of friction produced by shafts in their housings are called BEARINGS. A bearing may also be defined as a device used to guide and support RECIPROCATING and ROTATING elements which may be subject to external loads resolved into components possessing normal, radial, or axial directions, or two-dimensional loads in combination.

Unless it is a simple type such as a single lens reading glass, an optical instrument has many moving parts. Movement of these parts, however, must be so restricted that motion takes place ONLY IN THE DIRECTION DESIRED. Freedom of movement is also essential, and it can be attained by reducing friction between moving parts. Movable parts of an optical instrument must therefore be supported and retained by some suitable means, so that friction-free movement in a specific direction may be obtained.

Before we get into the discussion of different types of bearings, it is a good idea to explain the different types of loads which bearings must carry, as follows:

1. NORMAL LOAD.—A normal load is one applied TOWARD and PERPENDICULAR to the bearing surface.

2. RADIAL LOAD.—A radial load is a load directed AWAY FROM a surface, the opposite of a normal load. Rotation of a wheel or object on an axis is an application of radial load.

3. AXIAL LOAD.—An axial load is one directed along the axis of rotation or surface of an object.

ANGULAR LOAD.—An angular load is a combination of the other loads just described.

Bearings are generally classified as: (1) SLIDING SURFACE, and (2) ROTATIONAL (sometimes called rolling contact bearings).

SLIDING SURFACE BEARINGS

A sliding surface bearing usually has a stationary member which forms the base on which its moving part slides. See figure 7-17. A lathe, for example, has this type of bearing in the holding and guiding of the carriage, and tailstock on the lathe bed. The sliding surfaces are not always flat; they may be square, angular, or circular. The piston and cylinder bore of an internal combustion engine constitute a circular sliding surface bearing.

There are many variations of sliding surface bearings used in optical instruments, some of the more common of which are:

1. Cylindrical
2. Spherical
3. Square (quadrangular)

Cylindrical and spherical sliding surface bearings are used to mount some of the smaller gunsights in order that they may be easily boresighted (aligned with the gun). Refer to illustration 7-18, which shows these two bearings used in an assembly. The cylindrical sliding surface bearing is secured in its mating surface on the gun mount. The function of the spherical sliding surface bearing is to hold the front of the gunsight securely in the cylindrical sliding surface bearing; and at the same time to allow radial movement of the rear end of the gunsight (within certain limitations imposed by the construction of the spherical sliding surface bearing).

The purpose of the SQUARE BEARING (quadrangular) is to move and to hold the rear portion of a gunsight. Study illustration 7-18 again, and then study figure 7-19, which shows the position of a spherical bearing and a quadrangular (square) bearing in an optical instrument. The bearing surfaces in this instance are subjected to NORMAL LOADS by four ADJUSTING SCREWS in an adjusting-screw mount (fig. 7-9). Each adjusting screw exerts pressure on its respective bearing surface; and by loosening and tightening opposing screws, as necessary,

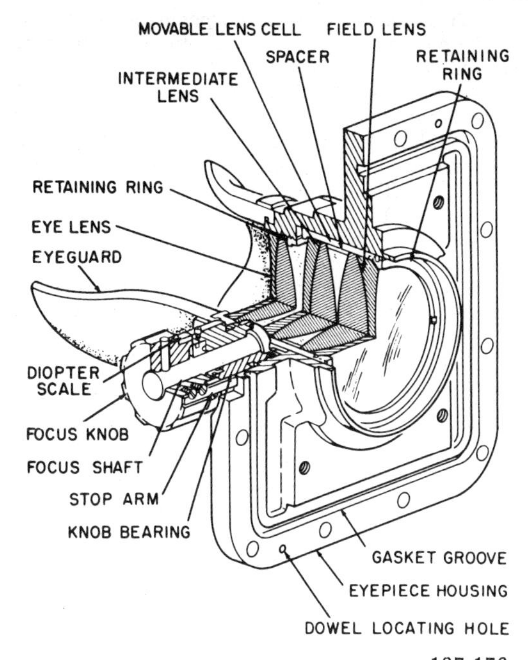

137.176
Figure 7-16.—Internal focusing eyepiece mount.

137.77
Figure 7-17.—Sliding surface bearing.

you can boresight the telescope. Adjusting-screw mounts are also good for holding and adjusting reticle mounts.

Although not a sliding surface bearing, the square bearing is used as a LOCATING BEARING SURFACE, with little sliding motion (if any) exerted upon it. When accurately machined,

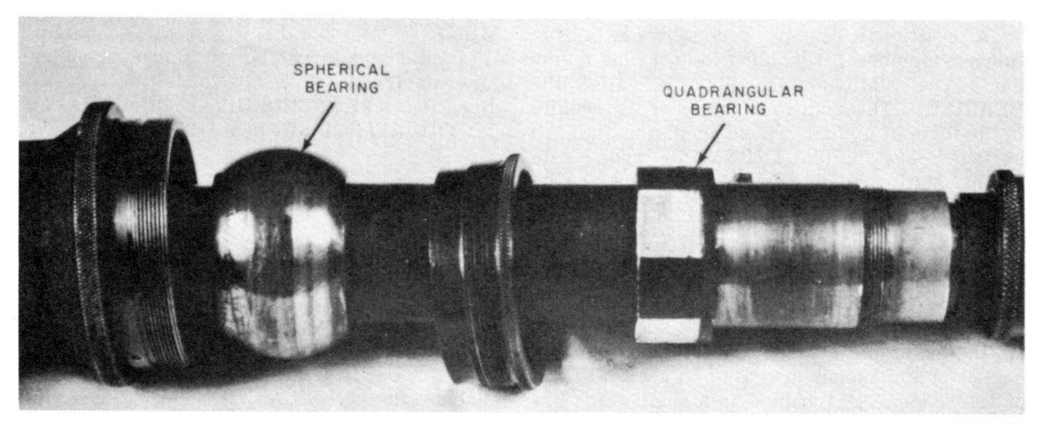

137.178
Figure 7-18.—Cylindrical sliding surface bearing and square bearing in an instrument assembly.

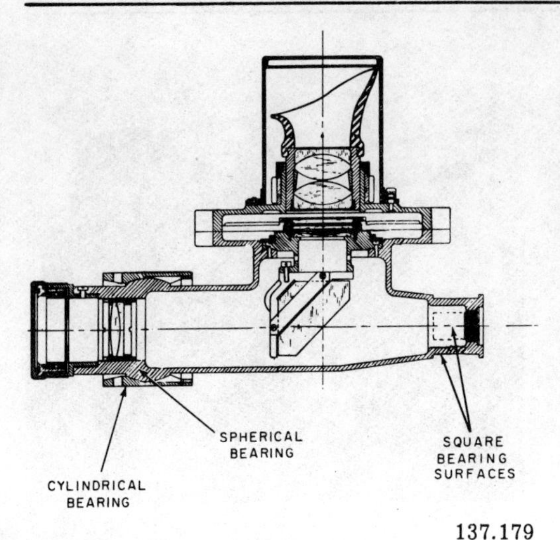

137.179

Figure 7-19.—Positions of sliding surface bearings in a telescope.

a square bearing is used as a bearing pad for holding large gunsights in gun mounts and directors, and for locating and holding parts inside optical instruments. During overhaul of a gunsight telescope, bearing pads become reference surfaces.

ROTATIONAL BEARINGS

A rotational bearing generally has a stationary member for holding the rotating member. The stationary member is called the BEARING. The rotational member is usually in the form of a shaft, whose precision-finished surfaces are called TRUNNIONS and rotate in the stationary member. Trunnions are by necessity circular in cross section, but their profile may be cylindrical, conical, spherical, or an even more complex form. The most common TRUNNION PROFILE in use is cylindrical.

Trunnion bearings (fig. 7-12) such as those on the ends of a Mark 61 telescope, right-angled prism mount are used on MANY KINDS of telescopes. A trunnion is a shaft which rotates around a true horizontal axis in order to keep the optical axis of a telescope or prism mount in a true vertical plane during elevating or depressing operations.

Trunnions are attached permanently to the body casting at the central point of a telescope or mount; but they may be part of the body casting. The trunnions make it possible to rotate a telescope (or mount) during elevation or depression; and if the telescope is stopped at any position, it remains in that position, PROVIDED THE TELESCOPE IS PERFECTLY BALANCED.

Because their resistance to rolling friction is much less than for sliding friction, precision BALL BEARINGS are used extensively in optical instruments. Precision ball bearings in self-contained units are classified in accordance with design. Differences in design in ball bearings are generally not apparent externally. When making a design of these bearings, the OUTER RACE, the INNER RACE, and the STEEL BALLS (which roll between the races) must be taken into consideration.

As you study the most common designs of self-contained precision ball bearings in the following paragraphs, refer to illustrations 7-20 and 7-21 to determine their differences.

Radial and Thrust Ball Bearings

Radial ball bearings (part A, fig. 7-20) are designed to carry loads applied to a plane perpendicular to the axis of rotation in order to prevent movement of the shaft in a RADIAL DIRECTION. Thrust ball bearings (part C, fig. 7-20) are designed to take loads applied in the SAME DIRECTION as the axis of the shaft in order to prevent free ENDWISE MOVEMENT.

Radial and thrust ball bearings are therefore designed to carry loads in a specific direction: PERPENDICULAR OR PARALLEL TO THE AXIS OF SUPPORTED SHAFTS.

Angular Ball Bearings.

An angular ball bearing (part B, fig. 7-20) supports an ANGULAR LOAD—a load which has components of radial and axial thrust—and it is exemplified by the bearing in the front wheel of a bicycle. Angular ball bearings are NORMALLY USED IN PAIRS, in a manner which enables the ANGULAR CONTACT SURFACES of the outer and inner race of ONE BEARING to oppose the ANGULAR CONTACT SURFACES of the OUTER and INNER RACE of the OTHER BEARING. This arrangement of the bearings

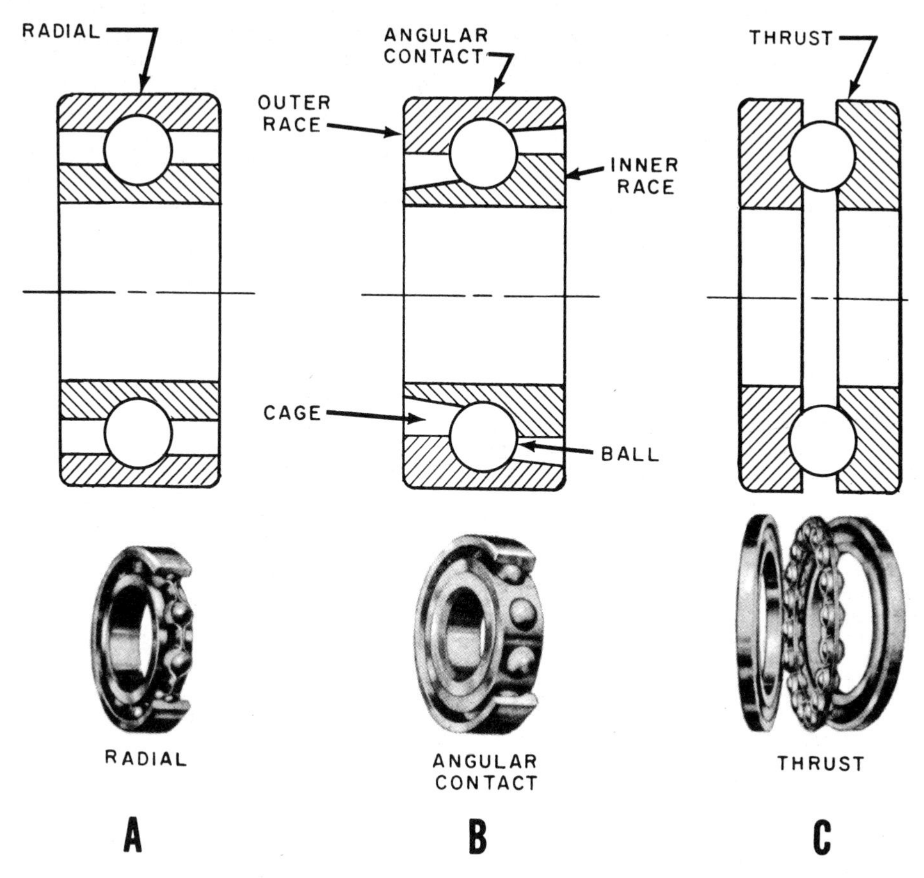

Figure 7-20.—Different types of ball bearings.

provides a technique designated as PRELOADING, which REMOVES what is called GIVE or SOFTNESS before the bearings are subjected to their normal loads.

The principle of PRELOADING is illustrated in figure 7-21. Preloading can be obtained (and normally is) by subjecting the inner races to a STATIC THRUST directed axially TOWARD the angular contact surfaces of the OUTER RACES.

In some cases, individual precision steel balls are used as a bearing between two parts. When this is true, the parts themselves act as the BEARING RACES, with the desired number of steel balls rolling between them. Such a bearing is used between polaroid filter plates in optical instruments, in order to secure SMOOTH and FREE rotation.

The RAY FILTER ASSEMBLY in a ship's telescope uses ONLY ONE precision steel ball as a detent, which starts or stops the movement. The steel ball in this assembly is held against the ray filter plate by a recessed spring and follower. When each glass filter is correctly positioned in the line of sight, the detent ball is thrust into a groove on the plate to hold the desired filter in the line of sight.

CAUTION: Dry metallic surfaces under an appreciable load, though smoothly machined, will not slide over each other without abrasion; so they must be kept covered CONTINUALLY with an approved lubricant, which actually keeps them separated and prevents abrasion and friction. Like metals do NOT RUB TOGETHER WELL unless completely covered with a film of lubricant. If properly lubricated, precision-made ball bearings wear very little. When wear does occur in ball bearings, replace them. Adjustment is impossible.

OPTICAL INSTRUMENT GEARS

An instrument designer must know what types of gears to use for a specific function in order to provide the TYPE OF MOTION and SPEED required. Because you must work with these gears in optical shops, knowledge concerning the basic types will be beneficial to you.

GEAR NOMENCLATURE

In order for you to calculate dimensions of a gear, you need to know the terms used to designate the parts of a gear. The brief discussion which follows provides this informaion.

Refer to illustration 7-22 for terms used in referring to or describing gears and gear teeth. The symbols in parentheses are standard gear nomenclature symbols.

1. Outside diameter (D_O) is the overall diameter of a gear.
2. Pitch diameter (D) represents the diameter of a circle used to calculate the dimensions of a gear. This pitch diameter is less than the outside diameter by an amount equal to twice the addendum.
3. Diametral pitch (P) is a ratio or number of teeth per inch of pitch diameter.
4. Circular pitch (Cp) represents the length of an arc of the pitch circle measured from a point on one tooth to a corresponding point on the next tooth. There are as many circular pitches (of equal length) in a gear as there are teeth in that gear.
5. Addendum (a) is the height of a tooth above the pitch circle along a radial line.

6. Dedendum (b) is the depth of a tooth below the pitch circle along a radial line.
7. Whole depth of tooth (H) represents the total depth of a tooth groove. It consists of one addendum plus (+) one dedendum.
8. Root diameter (D_R) is the outside diameter less TWICE the whole depth.
9. Although not shown in the illustration, center distance (C) is the distance between the axes of a pair of gears correctly meshed.
10. Chordal addendum (a_c) represents the distance (measured on a radial line) from the top of a gear tooth to a chord subtending the intersections of the tooth-thickness arc and the sides of the tooth. Chordal tooth thickness (t_c) represents the length of a chord subtended by the circular-tooth-thickness arc.

Several methods have been devised for checking the accuracy of gear teeth, one of which is checking the thickness of a gear tooth on a straight line through the points at which the pitch circle touches the gear tooth. This is the measurement of CHORDAL THICKNESS.

Various instruments are used for measuring chordal thickness, including a gear-tooth vernier caliper with a horizontal scale and a vertical scale. Tables of chordal thicknesses and corrected addenda are shown in standard engineering handbooks for a range of gears from 10 teeth to 140 teeth (and over), based on a diametrial pitch of 1. For other pitches, divide the values in the tables by the specific diametral pitches.

11. Backlash (B) represents the difference between the tooth thickness and the tooth space of engaged gear teeth at the pitch circle.

SPUR GEARS

Some spur gears are shown in figure 7-23 (from a Mark 74 gunsight), and they are used more than any other type of gear in optical instruments to transmit power from one shaft to another.

Teeth on spur gears vary in size (in accordance with requirements), stated in terms of QUANTITY as PITCH, or DIAMETRAL PITCH (number of teeth per inch of pitch diameter). This means that a spur gear with 16 pitch and a pitch diameter of 1 inch has 16 teeth, and so forth. The FACE OF A GEAR is its thickness, measured across the base of its teeth. A gear with a face of 3/4 inch, for example, is 3/4 inch thick at that point.

Speed ratios between shafts having spur gears is important, and ratio is defined as the RECIPROCAL OF THE RATIO OF THE QUANTITY OF TEETH OF THE TWO GEARS, or reciprocal of the ratio of their pitch diameters.

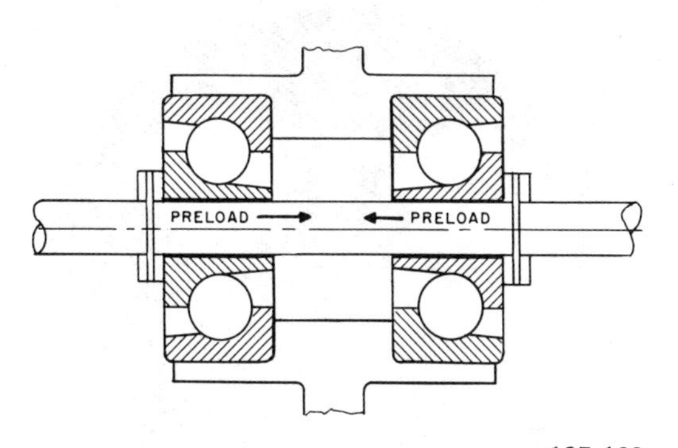

137.180

Figure 7-21.—Preloading produced by pairs of angular ball bearings.

If more than 40 teeth are required on a gear, 14 1/2 degree gears are generally preferable; if 40 teeth are not required, 20 degree gears with full-depth teeth are preferred. When quietness of operation is important, it is best to use gears with at least 20 teeth in a 14 1/2 degree system, and at least 14 teeth should be used on 20 degree gears with full-depth teeth.

Metals generally used in small spur gears are brass and steel; but cast iron is widely used in large spur gears. Spur gears, however, are also made of non-metallic substances.

BEVEL GEARS

Bevel gears used in optical instruments can be put on shafts which intersect at desired angles, provided the angle of the teeth is correct in relation to the shafts.

Bevel gears are made with straight or curved teeth, but they CANNOT BE INTERCHANGED WITH SPUR GEARS. By using the proper type of bevel gear, however, you can get a different speed ratio, as desired. When these gears are used to change the direction of motion 90°, with no change in speed, THEY ARE CALLED MITER GEARS. NOTE: If lapped pairs of bevel gears

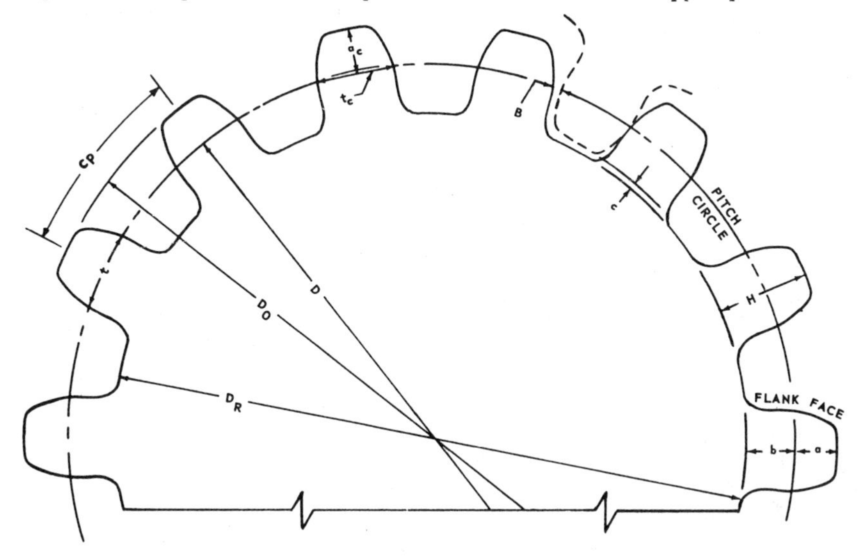

a--addendum	t_c--chordal tooth thickness	H--whole depth of tooth
a_c--chordal addendum	t--arc tooth thickness on pitch circle	D--pitch diameter
b--dedendum	c_p--circular pitch	D_O--outside diameter
c--clearance		D_R--root diameter
		B--backlash

28.259

Figure 7-22.—Gear nomenclature.

are used in an optical instrument (and others), almost perfect quietness of operation is obtained.

The shape of bevel gears, especially those with spiral teeth, causes them to exert much thrust. For this reason, the end of a shaft which contains the gear is generally supported by an angular ball bearing, and the other end has a radial ball bearing.

When one component of a pair of gears which mesh together is bigger than the other (B, C, and D, fig. 7-24) THE BIGGER COMPONENT IS USUALLY CALLED THE GEAR AND THE SMALLER COMPONENT IS CALLED THE PINION.

Spiral bevel gears (part B, fig. 7-13) are used in optical instruments (and others) because they are interchangeable for varying the speed ratio and can be used in different ways, as illustrated. They are cut right- and left-hand, and they are specified like spur gears with reference to face, pitch, and pitch diameter.

Spiral bevel gears which have the same cut (right or left hand) operate at right angles; those which have opposite cuts are used on parallel

5.22.1

Figure 7-23.—Types of spur gears.

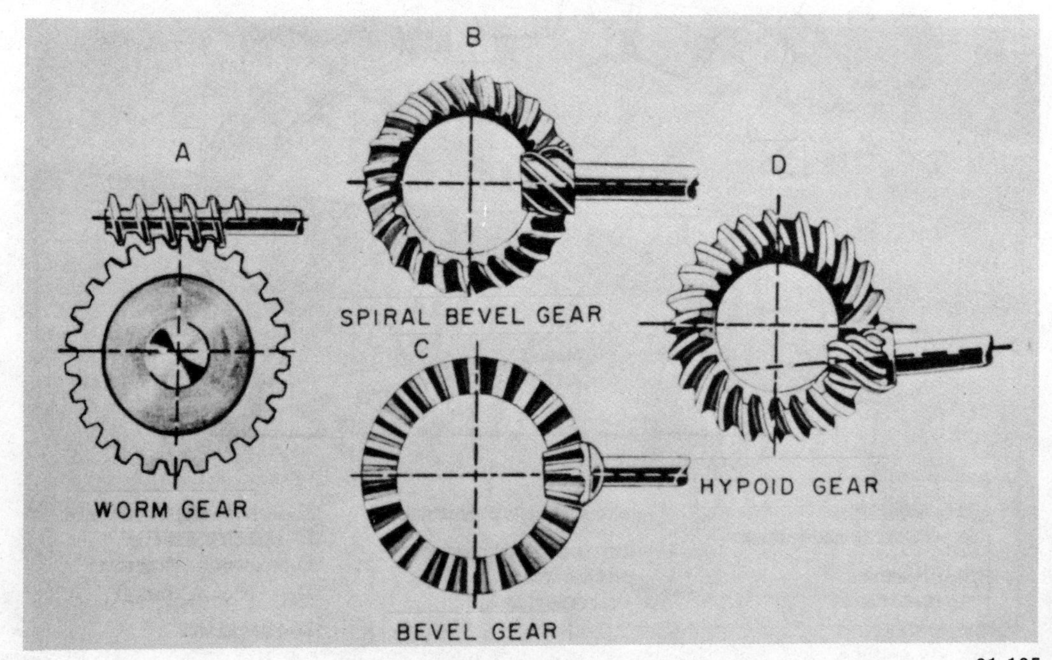

81.195

Figure 7-24.—Types of bevel gears.

shafts. Because the teeth of spiral bevel gears slide over one another, bronze or hardened steel is used in their manufacture to make them more durable.

WORM GEAR AND WORM

Study part A of illustration 7-24, the top part of which is called a WORM, and the bottom part of which is called a WORM GEAR. Worm gears and worms are used extensively in instruments because they provide an effective means for reducing velocity and transmitting power. Study the worm and sector gear from a gunsight elevator shaft in illustration 7-25.

If a worm has ONLY ONE continuous thread, it is called a SINGLE-THREAD worm; but more than one thread may be cut on a worm. Two continuous threads on a worm constitute a DOUBLE-THREAD worm; three continuous threads on a worm are called a TRIPLE-THREAD worm.

Single-thread and double-thread are terms which indicate the TOTAL NUMBER of continuous threads, NOT number of threads per inch. Pitch of a worm means linear distance from a specific point on one thread to a corresponding point on the adjacent thread, measured parallel to the axis of the worm. The pitch can also be determined by dividing one (1) by the number of threads per inch.

Pitch is therefore the RECIPROCAL of the number of threads per inch; and LEAD is the axial distance (parallel to worm axis) moved by the worm thread upon completion of one revolution of the worm. On a worm with a single thread, lead and pitch are therefore equal; but the lead is TWICE the pitch on a double-thread worm and THREE TIMES the pitch on a triple-thread worm.

As is true for spur gears, worms are specified in DIAMETRAL PITCH and PITCH DIAMETER; because they must be so machined that they fit the worm gears with which they mesh. This means that such factors as thread number, threads per inch, face length, and pitch diameter must be taken into consideration.

RACK AND PINION

Some fire control equipment and optical instruments use a rack and pinion such as the one illustrated in figure 7-26. The rack gear moves in a linear motion, as indicated; and it is simply a straight bar into which the gear teeth have been cut. The pinion, of course, moves in a rotary motion.

DRAFTING MACHINES

A person with drafting experience knows the value of a T-square and triangles, used to draw straight parallel lines and angles. T-squares and triangles, however, are not satisfactory for general drafting purposes; but drafting machines (with parallel motion protractors) combine the features of the T-square and triangles.

The drafting machine illustrated in figure 7-27 is one of many different designs. It is constructed with two arms, each with one end connected by and pivoted about a common shaft on ball bearings to give an elbow effect. The free end of ONE ARM is secured to the drafting table by a bracket, shaft, and ball bearings, about which the arm can be rotated. The free end of the OTHER ARM is equipped with a protractor head. There is a pulley at the free end of each arm; and there is an elbow pulley at the center pivot, where the two arms are joined by the same shaft.

Each of the arms shown in figure 7-27 has two hollow bars through which a steel band runs around pulleys at the ends of each arm; and the bands of both arms run around the elbow pulley. Regardless of the manner in which the machine is pivoted or moved about, the arms remain parallel to the surface of the drawing board.

The protractor head is equipped with a protractor and other component parts to enable it to function properly. The protractor is graduated from 0 to 359 degrees and can be rotated and locked in any desired position. A straightedge or scale is attached to other component parts of the head, which may also be rotated and locked in any position.

A drafting machine of this type enables you to select any desired angle with the scale, lock it, and draw that angle at any location on the drafting paper. Regardless of how you move the machine over the paper, provided you lock the protractor head and scale in position, the scale remains locked and parallel to that angle. This is beneficial, because parallel lines (regardless of number chosen) can be drawn by moving the protractor head to the desired position. The steel

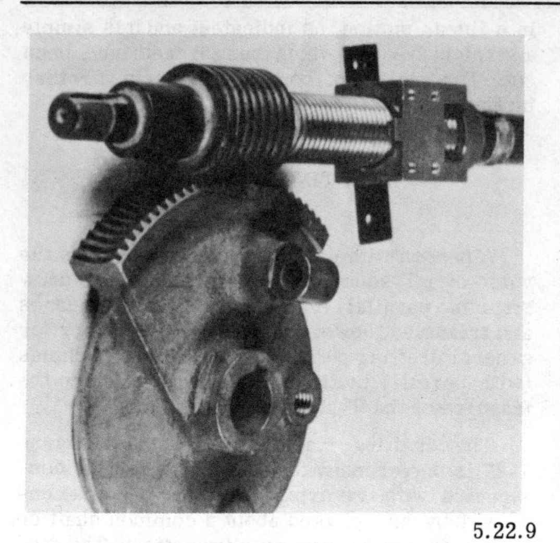

5.22.9

Figure 7-25.—Sector gear and worm.

bands which run around the pulleys make this action possible. When the arms are positioned as desired, the bands and pulleys function together to hold the scale in suspension.

Two scales may be attached to some protractor heads, as shown in figure 7-28. The scales are held at right angles to each other to enable you (the operator) to draw vertical lines and horizontal lines simultaneously. Such drawing is impossible if you have only one scale, because the single scale must be readjusted each time you desire a vertical or horizontal line line.

The procedure for attaching the scales to the protractor head is shown in illustration 7-28. A male dovetail fitting attached to the scale is fitted to the female dovetail fitting of the scale bracket. The scale alignment adjusting screw, which fastens the male dovetail fitting to the scale, permits slight angular adjustments for truing the two scales at a 90° angle to each other. This adjustment is rarely necessary.

5.22.13

Figure 7-26.—Rack and pinion.

45.137
Figure 7-27.—Drafting machine.

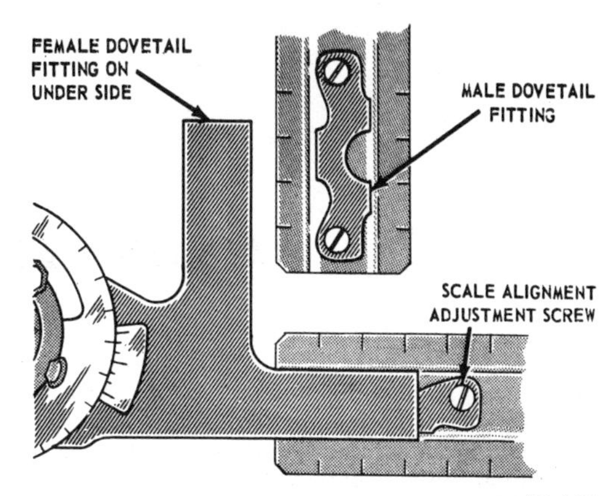

45.137
Figure 7-28.—Scale fittings for a drafting machine.

CHAPTER 8

MAINTENANCE PROCEDURES

The primary purpose of this chapter is to provide repair and maintenance information applicable to all optical instruments. Chapters 10 through 13 contain additional information pertaining to the repair and maintenance of particular types of optical instruments—alidades, binoculars, boresights, collimators, telescopes, and so forth. Other information pertaining to the repair and maintenance of these instruments is available in such publications as: Basic Handtools, NavPers 10085-A; Blueprint Reading and Sketching, NavPers 1007-B; Mathematics, Vol. 1, NavPers 10069-B; and Mathematics, Vol. 2, NavPers 10071-A.

Another objective of this chapter is to impress upon your mind the importance of (1) CARE IN HANDLING and (2) CLEANLINESS OF OPTICAL INSTRUMENTS. These instruments are precision-built and very expensive, and the importance of care in handling and maintaining them CANNOT be overemphasized. You will learn through experience, for example, that even a small amount of dust can impair the efficiency of an optical instrument; and you will also learn from experience that extreme care must be exercised in handling delicate optical elements.

HANDLING AND STOWING

Optical instruments are shipped in specially constructed containers designed for adequate protection during transportation. When you receive optical instruments in the optical shop, check their containers for damage and cleanliness; then, if there is no reason why you should remove the instruments from the containers, stow them in clean storage cabinets or spaces provided for them. CAUTION: When you MUST MOVE AN INSTRUMENT from one location to another, if possible, move it in its container.

HANDLING PROCEDURE

Most containers for optical instruments have catches or locks for securing the instruments in position; so when you put an instrument in its container, place it GENTLY INTO POSITION and carefully close the lid. DO NOT TRY TO FORCE AN INSTRUMENT INTO ITS CONTAINER OR SLAM THE COVER SHUT. The contour of the interior of the case was made in the best manner possible by the manufacturer to hold the instrument snugly in place to prevent damage during handling. If the instrument does not go into its case without difficulty, check for an extended draw tube or something else which is hindering smooth entrance into proper position. CAUTION: Always secure the cover to the container with the catches installed by the manufacturer.

The most important word for you to remember relative to the handling of optical instruments is CARE. If you handle binoculars roughly or drop them, for example, shock may result in misalignment or breakage of optical and mechanical parts. Prisms and reticles usually have adjustable mounts, and a very small amount of shock may knock them out of adjustment. When this happens, you have only one choice: REPAIR. This means that you must unseal the instrument, disassemble parts, repair (as necessary), reassembly, adjust, and collimate. As you know, this adds to much work, all of it resulting (perhaps) from thoughtlessness and negligence in handling.

Gunsight telescopes have bearing surfaces by means of which they are mounted onto other components; and a dent or scratch on the surfaces may make mounting impossible. You will soon find out that it is difficult to repair such bearing surfaces and decide that CARE IN HANDLING CAN PREVENT UNNECESSARY WORK.

STORAGE INFORMATION

The elements which cause most harm and/or damage to optical instruments in storage are: (1) contaminated dust, (2) moisture and chemicals, and (3) heat and light, each of which is discussed in some detail in the next paragraphs.

Contaminated Dust

Much of the dust which settles on objects aboard ship comes from lint-producing bedding, furniture, and clothing. Other dust comes from various sources throughout the ship, including shops and storage spaces. The air surrounding a ship in port, of course, may contain contaminated dust.

Dust alone may cause damage to optical instruments when it settles on them; but dust usually contains mineral particles which are in some instances as hard as glass or steel. When you therefore try to remove the dust, unless you are careful and use proper procedures, you may damage the optical elements or other parts of instruments. If you try to wipe the dust off optical elements, you may simply grind it into their optical surfaces and cause scratches. The procedure for cleaning optical and mechanical parts is outlined later in this chapter.

The best way to protect optical instruments in storage is to keep them in their cases or containers. Before you stow optical parts, WRAP THEM CAREFULLY WITH APPROVED WRAPPING MATERIAL. Should it be necessary for you to remove a part of an instrument for repairs (the objective, for example), cover the opening with a cap of bond paper or masking tape to keep dust out. Secure the paper cap with a rubber band or tape. If an instrument has lens caps, leave the caps in place (except when the instrument is in use).

Moisture and Chemicals

Moisture is always present in the air in minute quantities, and occasionally an optical instrument is covered with spray. Water in itself causes some damage to instruments; but when it contains acid fumes or salt, the damage it inflicts upon optical instruments may be serious.

CAUTION: When an optical instrument receives a spray of salt water, remove the water immediately from optical surfaces with lens paper and dry the housing with a clean cloth.

When the air is humid (in the tropics, for example), protect instrument bearing surfaces with a thin film of petrolatum. Cover unpainted metal surfaces with a thin film of oil; but KEEP OIL AND GREASE OFF OPTICAL SURFACES, and also from locations from which it may be carried to optical surfaces.

The reaction of moisture with acid fumes and dust in contact with optical glass results in rapid deterioration (etching) of the glass. Salt water causes serious corrosion to unprotected metal parts of instruments, as is also true of fresh water mixed with acid fumes. In the presence of moisture, bacteria which settle on glass surfaces may cause a dull-gray film on the glass. Moisture also helps to form stains on lens surfaces.

The reaction of uric acid, acetic acid, and salt with sweat on your fingers causes reflection-reducing films on optical glass and deterioration and etching of the glass itself.

CAUTION: Do NOT put your bare fingers on optical surfaces. Use lens tissue when you pick optical elements up to clean them; and clean them with a swab or clean tissue.

Such mineral acids as sulfuric and nitric do NOT harm most types of glass when other chemicals are absent. You can, in fact, use these acids to clean optical surfaces. Hydrofluoric and phosphoric acids, or their fumes, may quickly ruin optical surfaces. On the other hand, you can use controlled amounts of hydrofluoric acid fumes to etch cross lines on glass to make a reticle. To protect optical instruments and parts from chemical fumes, KEEP THEM AS FAR AS POSSIBLE FROM THE ROOM IN WHICH YOU USE CHEMICALS.

Heat and Light

Optical glass and metal generally are not harmed by moderate amounts of dry heat, but very high temperatures damage them. Even moderate amounts of dry heat, on the other hand, DO HARM CEMENTED FILTERS AND LENSES. High temperature softens the cement, and may cause separation of the elements. When subjected to heat for a certain time, cement used for making compound lenses becomes dry and brittle; and when this is true, slight shock or vibration may crack it, or separate the optical elements.

Strong sunlight hardens cement and often discolors it, thereby affecting the quality of compound lenses or rendering them useless.

For this reason, when available or provided, ALWAYS KEEP LENS HOODS OR SUN SHADES IN PLACE ON OPTICAL INSTRUMENTS EXPOSED TO THE SUN.

INSPECTING OPTICAL INSTRUMENTS

There may be occasions when you will be assigned duty on a large combat vessel and given full responsibility for inspecting all optical instruments aboard the ship. By carefully inspecting the instruments and taking care of little troubles, you will be able to save yourself and your repair activity much work. Make notes on your inspection of each instrument and recommend appropriate remedial action, aboard your ship or at a repair facility.

CAUTION: When you inspect an optical instrument in use aboard ship and follow up with minor repairs, do NOT DISTURB the optical system.

When you are assigned duty in a repair activity afloat or ashore, inspect every instrument sent to the optical shop for repairs. If an instrument is unfamiliar to you, get all information concerning it from Ordnance Pamphlets (OP's), Bureau of Naval Weapons publications, NavShips Manuals, and blue prints. Never attempt to disassemble and repair an instrument until you fully understand it.

During your predisassembly inspection of an instrument, try to locate difficulties. Inspect the physical and mechanical condition of the instrument and also its optical system. Use a casualty analysis inspection sheet and record all your findings on it.

PHYSICAL CONDITION

The defects to look for when inspecting an optical instrument are: dents; cracks; and breaks in the housing, mount, and bearing surfaces. Unless they are on a bearing surface, small breaks are generally not serious, but they still require immediate attention. A crack in the housing (or a loose or broken seal), for example, soon causes condensation of moisture within the instrument.

Temperature changes from day to day cause an optical instrument to BREATHE through a crack; that is, during the day the instrument takes in warm, moist air through the crack, and at night (when the instrument and air cool) the moisture condenses inside the instrument. The next day, and during successive days, this procedure continues until a film of moisture covers an optical surface (perhaps several) and renders the instrument temporarily useless.

Inspect assembly screws for tightness. If retaining rings are exposed at the end of the tube of an instrument, check them also for tightness, by applying light pressure with your fingers.

CAUTION: Do NOT TOUCH the lens with your fingers, and do NOT USE a retainer ring wrench to test the rings for tightness. The setscrew of the ring (or the ring itself) may be sealed with shellac; and if you attempt to turn the ring with a wrench, you may break the seal.

Take a close look at the condition of the paint on exposed metal parts. To prevent corrosion, cover worn or cracked and chipped paint with a thin film of approved oil. As soon as possible, send the instrument to the shop for repainting.

MECHANICAL CONDITION

Carefully examine mechanical adjusting screws, and check knob and gear mechanisms for slack or excessive tightness. If the instrument moves on bearings, test them for binding or looseness.

Try the focusing action of the eyepiece to find out if you can focus it (in and out) without binding or dragging. If binding or dragging exists, the eyepiece adapter or the draw tube is eccentric, which condition is generally caused by dropping or jarring.

Backlash in the focusing action of an eyepiece is usually caused by a loose stop or a retainer ring; but it may be caused by a loose key and its screws in the spiral keyway assembly.

Check the mechanical, 0 diopter setting of the eyepiece to determine whether the index mark points to 0 diopters when the eyepiece draw tube is at mid-throw (halfway in and halfway out). The focusing action should be such that the index mark clears all diopter graduations (plus and minus) during full travel of the drawtube.

If the instrument has turning shafts (ray filter or input), check them by turning the shafts. If rotational action of the ray filter shafts does not turn the color filters in or out of the line of sight, the cause is most likely improper meshing of gears, or detachment of the gear itself from the shaft. If the shaft does not rotate, it is corroded or bent.

All mechanisms must move freely, without binding, slack, backlash, or lost motion. Moving parts should be just tight enough to keep them in proper position.

Check for missing or broken parts—retainer rings, set screws, and so forth. You can locate loose or broken internal parts by shaking the instrument.

If the instrument is gas sealed, CHECK ITS GAS PRESSURE by attaching a pressure gage to the gas inlet fitting. Then crack the valve screw and read the pressure on the gage to find out if it is correct. Correct pressure in most nitrogen-charged optical instruments is approximately two pounds per square inch, or as indicated in the manufacturer's technical manual for a particular instrument. If the gage indicates NO PRESSURE in the instrument, there is a bad gasket, a loose fitting, or a loose screw. Check for all of these defects when you disassemble a gas-filled optical instrument.

Optical System

Because optical elements constitute the HEART of an optical instrument, inspection of the optical system is very important and you should learn to do this phase of your work well. When you first examine an optical system, you may have difficulty in distinguishing one element from another. Through adequate experience, however, you will be able to make this distinction; and you will know where each element belongs in the system and when it is defective.

The best method to follow in inspecting the optical system of an instrument is to point it toward an illuminated area and look for the following:

1. DIRT AND DUST. Dirt and dust show up as dark spots (specks) on the surface of an optical element.

2. CHIPS, SCRATCHES, BREAKS. These defects in an optical element show up as bright, star-like specks, scratches, or areas, when light is reflected from them.

3. GREASE OR OIL. Grease or oil on an optical element in a system is indicated by streaked, clouded, or nebulous areas, with an occasional bright, translucent spot. You may even be able to detect the color by knowing the color of the grease used on the instrument.

4. MOISTURE. Moisture shows up as a sharply defined nebulous area, with brilliant reflection or a diffused, clouded appearance when the area is not illuminated.

5. FUNGUS OR WATER MARKS. Brown or green patches, or stains, indicate the presence of fungus or water marks. Deposits of salt may cause a grainy, milky color similar to that of frosted glass.

6. DETERIORATED BALSAM. Deterioration of Canada balsam used to cement lenses together is indicated by cracking, or a dark or yellow color; and areas between the elements appear milky, colored or opaque, splotched, or net- or thread-like. When the cement just begins to separate, bubbles and areas of splotches shaped like oak leaves appear between the elements. If there are brightly colored bands or rings (Newton's Rings) between the elements, the lenses are under strain in their mounts, or a sudden, sharp blow on the instrument caused the cement to break down.

7. HAZY OR CLOUDED IMAGE. Foreign matter on the objective lens, the erectors, or the prisms of an optical system cause a hazy or clouded image.

You can examine color filters in an optical system, provided they are within the focal length of the eyepiece, by holding one eye a few inches from the eyepiece and turning the ray filter shaft. Defects on a filter show up when it rotates in and out of the line of sight.

If the field of view (true field) is not perfectly round, there is a loose diaphragm within the instrument or the color filter plate is not properly engaged with the detent ball or roller.

Check the anti-reflection (magnesium fluoride) coating on coated optics by holding the instrument under a daylight fluorescent lamp (white light). If the coating is of proper thickness on the optic, its color is light-reddish purple. If the coating shows signs of wear (too thin), it is pale-yellow, straw, copper, or reddish-brown in color.

If the coating on a lens is too thin, the best thing to do is replace the lens. The coating must be of adequate thickness in order for the lens (coating) to reduce reflection properly. If the coating is of satisfactory thickness and color but has scratches, the lens is still usable; for a few scratches do not cause noticeable loss of light.

Optical elements (reticles and collective lenses) placed in or near image planes of an instrument are not coated, because scratches on them, or deterioration of the coating, appear to be superimposed on the image in the field of view. Optical surfaces cemented to other optical surfaces are not coated, as is true for the concave and convex surfaces of a cemented doublet. Cement will NOT adhere to coated surfaces.

Reflecting surfaces of prisms which use the principle of the critical angle and total reflection are NOT coated, for the coating causes too much loss of light.

Inspect prisms and mirrors for signs of wear, peeling, or darkening of the silvered or aluminum surfaces. All of these defects show up as blisters and cracking of the coating or a yellowish color.

Some optical defects are illustrated in parts A through K of illustration 8-1. If available, get some lenses with the defects shown and study them as you read the following discussion of various lens defects.

CHIP. A chip (part A, fig. 8-1) is a break at the edge of a lens or prism caused by uneven pressures or burrs on the lens seats of the lens mounts.

NOTCH. A notch (part B, fig. 8-1) is a ground off surface of a chip on a lens or prism outside the free aperture. A notch, however, cannot be considered a defect in the true sense of the word, because an optical repairman (Opticalman) must place it in the position indicated in order to prevent internal reflections.

SCRATCH AND STRIPE. A scratch (part C, fig. 8-1) remains visible as you rotate a lens or prism through 360 degrees; a stripe, on the other hand, vanishes at some position as you rotate the optical element. You can most easily see scratches and stripes in optical elements when you place the elements against a dark background.

RING. A ring (part D, fig. 8-1), is a circular scratch or stripe around the external edges of a lens, and it is caused by pressure against the lens by the mount seats and the retainer ring. An INTERNAL RING between the elements of the lens may appear at the edges of the lens when lens cleaning fluid dissolves the Canada Balsam.

CRACK. A crack (part E, fig. 8-1) is generally caused by a sudden change of temperature, resulting in sudden contraction or expansion of the outer surface of the glass and fracture of the lens or prism because the center of the optical element does not expand or contract as rapidly as its edge section, which is thicker in convex lenses and some prisms.

BUBBLE. A bubble (part F, fig. 8-1) may result from gases left in the glass during manufacture, or from air which did not escape from the cement when the elements were joined with it.

STRIAE. Striae (part G, fig. 8-1) look like veins or cords running through the glass, and you can see them by looking through the glass at a contrasting light and dark background. This is a manufacturing defect in the optical element.

BLISTER. A blister (part H, fig. 8-1) is an air bubble trapped in the layer of cement between two lenses. If it extends toward the center of the lens, it is called a RUN-IN, generally produced by the dissolving action of a cleaning fluid. A blister, however, may result from uneven mounting during assembly in the instrument, or by dirt between cemented lenses. Blisters can be seen best by reflected light, and they usually increase in size over a period of time.

DIRT FUZZ. Lint, dust, or dirt (part I, fig. 8-1) in the layer of cement between lenses may eventually cause a blister. You can see this type of foreign matter in a lens most easily by transmitted light against a dark background. Dirt fuzz is a manufacturing defect in a lens.

STAIN. A stain (part J, fig. 8-1) is usually brown or green in color and is produced by the evaporation of water or moisture which gets on lenses or prisms and dissolves some of the anti-reflecting coating, thereby causing a very faint deposit (sometimes bacterial in growth).

UNPOLISHED CONDITION. An unpolished state or condition in a glass optic (part K, fig. 8-1) results from the manufacturer's failure to remove grinding pits from it. In some instances, however, this condition is produced on optical surfaces exposed to gases, grit, and particles of all sorts in the atmosphere.

The last step in checking the optical system of an instrument is TESTING FOR PARALLAX, or COLLIMATION of the instrument. Always check the collimation of an instrument before you disassemble it, for the information you thus procure will help you during the making of your casualty analysis.

You can check the collimation of an optical instrument in two ways: (1) by looking through the instrument at an infinity target, or (2) by checking it more accurately with an auxiliary telescope. The first method, however, is generally used when quick results are necessary.

Focus the instrument on a distant target and check for parallax by moving the eye from side to side and up and down. If parallax is present, the reticle (crossline) appears slightly out of focus and seems to move back and forth, up and

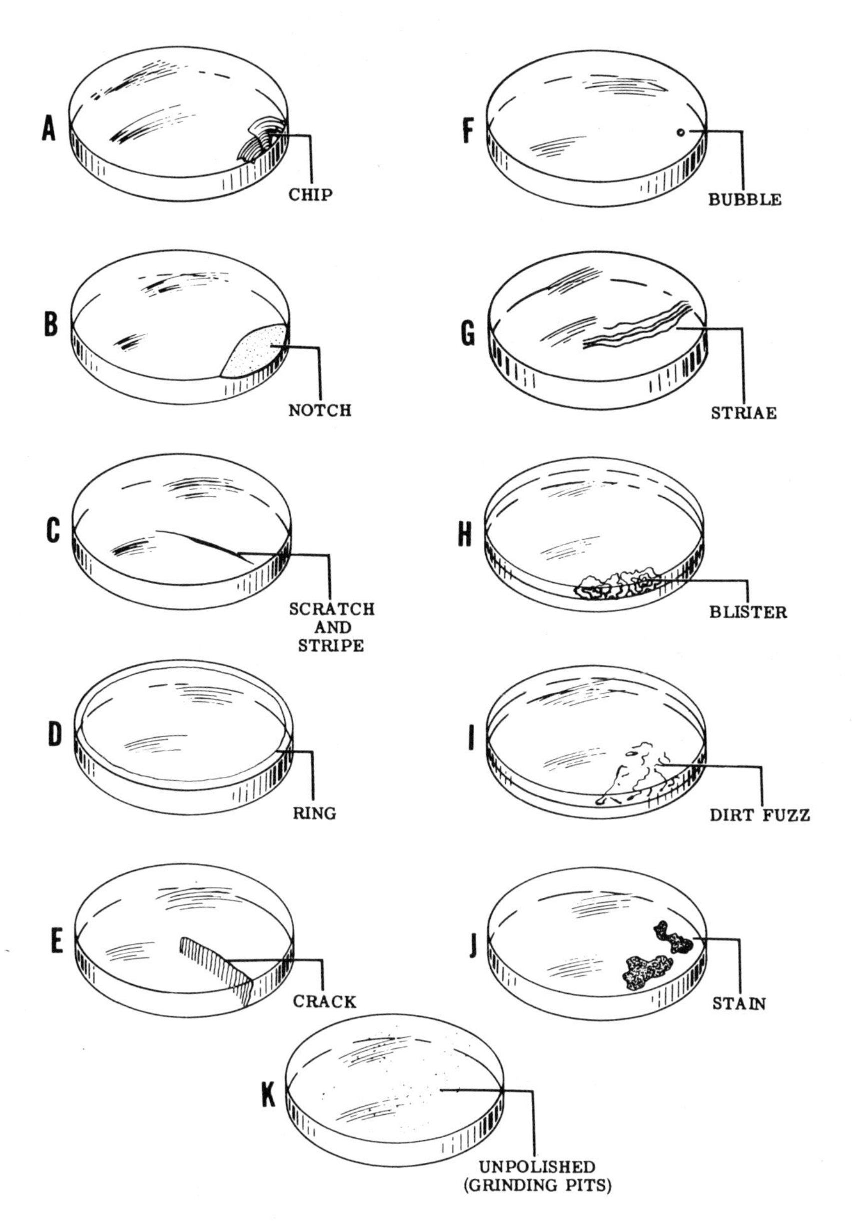

Figure 8-1.—Optical defects.

down, over the target. If parallax is not present, the reticle is in sharp focus with the target and remains superimposed in one spot on the target, regardless of the direction in which you move your eye behind the eyepiece.

Hold the instrument up to your eye in the position in which it is normally used and look at the horizontal wire of the reticle to determine whether it is parallel with the horizon, or square in appearance. The only manner in which you can make an accurate check of the SQUARENESS of the crossline, however, is with a collimator and an auxiliary telescope.

To check the eyepiece diopter setting, focus the instrument on an infinity target and observe the position of the index mark on the diopter scale. If the index mark is not pointing to your personal diopter setting, the 0 diopter is incorrect. At this point, you can determine the number of diopters from the 0 setting your PERSONAL DIOPTER SETTING is off.

If the index mark points to three graduations past your personal setting, for example, the eyepiece diopter setting is off three diopters from 0 diopter to the minus side, provided you focused from plus to minus on the scale. If the index mark points to your personal diopter setting on the diopter scale, the instrument is optically set to 0 diopters, even though the index mark is NOT pointing to 0 diopters.

If the instrument has a porro prism erecting system, check the optical system for LEAN by looking through the instrument with one eye at a vertical target (flag pole or side of building) and by looking directly at the target with the other eye. If the two images are not PERFECTLY parallel, there is leaning (termed LEAN) in the optical system; that is, the image through the instrument appears to LEAN away from the image observed with the naked eye. The reason for this LEAN is that the frosted sides of one porro prism are not at a 90° angle with the frosted sides of the other porro prism.

TESTING OPTICAL QUALITIES OF INSTRUMENTS

The optical qualities of optical instruments include: (1) magnification, (2) size of the field of view, (3) brightness of the image, (4) flares and ghosts, (5) illumination and contrast, and (6) aberrations.

MAGNIFYING POWER

The procedure for determining the magnifying power produced by a telescope, or a similar optical instrument, is summarized in the next paragraphs. Repetition of information given about magnifying power in previous chapters of this manual is held to a minimum.

Ratio of Apparent Field to True Field

You have already learned that the TRUE FIELD of an optical instrument is the width of the target area, or field, which you can see when you look through the eyepiece, expressed as either angular true field or linear true field; and you know the objects you see are greatly magnified.

APPARENT FIELD is the opposite of true field, and it is the width of the target area, or field, which you can see when you look through the objective end of a telescope, expressed in either angular apparent field or linear apparent field. The objects you see through the objective end of a telescope are greatly MINIFIED; that is, they are not as large as they would be when viewed with the naked eye.

The apparent field is always larger than the true field, provided the optical instrument's original purpose was to magnify targets. You can therefore determine the angular magnification of an optical instrument by COMPARING THE RATIO BETWEEN the angular apparent field and the angular true field.

To determine the angular apparent and true field of an optical instrument, you must have some means for measuring these angles directly with the instrument. This you can do by placing the instrument on an angle-measuring instrument such as a bearing circle (chapter 14), or some other instrument by means of which you can measure the angular movement of the instrument when you have it positioned horizontally.

Place the instrument you are checking on the angle-measuring instrument in the normal viewing position (eyepiece toward you), and focus the instrument on a distant object (flag or telephone pole, for example). Then turn the measuring instrument (with the telescope on it as necessary in order to have one side of the pole at one extreme edge of the field of view and take a reading on the measuring instrument. Next, turn the measuring instrument (with pole still in view) until you have the SAME SIDE OF THE

POLE on the OPPOSITE EXTREME EDGE of the field of view and take another reading to find out how many degrees you moved the instrument to get from one reading to the other reading. This measurement is the TRUE FIELD.

Now turn the telescope around on the angle-measuring instrument, with the objective end toward you, and look through the objective lens at the same object. The object is MANY TIMES SMALLER than when you were looking at it through the eyepiece. Finally, turn the measuring instrument and telescope (with it) as necessary to have the same side of the pole on the same extreme edge of the field of view. When you now repeat the same steps you follow in measuring the true field, you will find that your measurement is much larger than the true field. This larger measurement in degrees is the APPARENT FIELD of the telescope.

To determine the magnification of this instrument, divide the apparent field by the true field. The formula follows:

$$\text{MAG} = \frac{\text{Apparent Field}}{\text{True Field}}$$

Now use the formula to determine magnification of an instrument with an apparent field of 50° and a true field of 5 degrees. Substitute the measurement for each field in the formula and solve for MAG and you get 10, which means that the telescope is 10 power.

Ratio of Focal Lengths
(Astronomical Telescope)

You can determine the amount of magnification produced by an astronomical telescope by dividing the focal length of its objective lens by the focal length of its eyepiece.

Ratio of Entrance Pupil to Exit Pupil

You learned in chapter 6 how to use a dynameter to measure the diameter of the exit pupil and the eye distance of an optical instrument. Review that procedure now.

When you know the diameters of the entrance and exit pupils, you can calculate the magnification of an optical instrument by dividing the diameter of the entrance pupil by the diameter of the exit pupil. The formula is:

$$\text{MAG} = \frac{\text{Diameter of Entrance Pupil}}{\text{Diameter of Exit Pupil}}$$

Suppose that the entrance pupil of an instrument is 50 mm and the exit pupil diameter on the dynameter was 10 mm. If you substitute these numbers in the formula and solve for MAG, you get 5, which is the magnfication of the instrument.

In such optical instruments as rangefinders, you cannot measure the diameter of the objective lens; because the objective lens is not at the end of the housing (as generally true for other telescopes) but at a point within the main housing inaccessible without disassembly.

In order to measure the magnification of an instrument of this type, you must have an assistant. Place an outside caliper (opened to the desired measurement) on the entrance window (sometimes several feet from objective) and have your assistant hold the calipers while you place the dynameter to the eyepiece of the instrument and focus it until you get a sharp image of the caliper on the frosted reticle of the dynameter. Then read through the dynameter the opening between the jaws of the caliper.

Use the following formula to determine the magnification of the instrument:

$$\text{MAG} = \frac{\text{Actual Opening between Jaws of Caliper}}{\text{*Measured Opening between Jaws of Caliper}}$$

Suppose the caliper jaw opening is 60 mm and your measured opening between the jaws of the caliper (through dynameter) is 5 mm and you wish to determine the magnification of the instrument. Substitute your actual opening of the jaws of the caliper and the measured opening of the same jaws through the dynameter in the formula and solve for MAG. Your answer is 12, the magnification of the instrument.

IMAGE FIDELITY

Image fidelity is the FAITHFULNESS AND SHARPNESS OF AN IMAGE VIEWED WITH AN OPTICAL INSTRUMENT. As you know, good performance of an optical instrument is obtained only when the images it creates are free of aberrations and distortions.

You can make a rough test of image fidelity in a telescope, or a similar instrument, by doing the following:

1. Find and measure the greatest distance at which you can read clearly a newspaper headline, or any print of comparable size.

*Measurement must be taken through the dynameter

2. Multiply the distance measured by the magnification of the instrument you are testing.

3. Mount the print at the distance you calculated in item 2 and observe it through the telescope. If it is now just as clear and readable as it was with your naked eyes at the closer distance, THE IMAGE IS SHARP (IMAGE FIDELITY IS GOOD.); if the image is fuzzy, IMAGE FIDELITY OF THE INSTRUMENT IS POOR.

You can make a more accurate test for image fidelity by placing a small, glass globe where you can see the reflection of the sun in its surface and focusing your telescope on that reflection. Adjust the eyepiece as necessary to create a small, sharp image of the sun and move the eyepiece in or out from the setting. When you do this, you can see a number of rings around the sun's image. If these rings are circular and concentric, IMAGE FIDELITY (SHARPNESS) OF THE INSTRUMENT IS PERFECT. Any distortion of the rings indicates A LACK OF IMAGE FIDELITY—the greater the distortion of the rings, the poorer the quality of image fidelity.

When you check the image fidelity of an optical instrument, check for two things: (1) CENTRAL ASTIGMATISM, and (2) CENTRAL RESOLUTION. Optical performance is basically a function of design of the instrument and cannot be varied unless the characteristics of the optical elements are changed. There are several possible service defects, however, WHICH CAN CHANGE THE OPTICAL QUALITIES OF ONE OR MORE ELEMENTS. An optical element under strain by mechanical parts, for example, or tilted and improperly positioned elements (faulty mounting), badly matched recemented optics, and even wrong optical parts all cause poor image fidelity.

Tests for central astigmatism and central resolution provide an overall check of both basic optical performance and service defects.

A test for central astigmatism is made ON THE AXIS of the optical system with the test figures in the center of the field of view, which is zero (0) on the axis of an optical system; and it is called CENTRAL OR AXIAL ASTIGMATISM to distinguish it from aberration astigmatism (OFF-AXIAL ASTIGMATISM).

Resolution (resolving power) is the ability of an optical system to distinguish details in an object; so it is therefore a very important characteristic of an instrument, usually stated in terms of the ANGLE SUBTENDED BY THE CLOSEST POINTS ON AN OBJECT THAT THE SYSTEM CAN REVEAL SEPARATED. A strict test of the resolving power on the optical axis—central resolution—is a good check on the overall image fidelity.

Image Fidelity Test Chart

Take a close look now at illustration 8-2, which shows a standard test chart for testing image fidelity in optical instruments. This is a standard test chart available through naval supply channels.

The 56-line-per-inch group at the top of the image fidelity test chart is used for testing central resolution of 7 x 50 binoculars and Mark 1, Mod 0, ship's telescopes. The 28-line-per-inch group is used for testing central astigmatism in these instruments. You can make up a separate chart for each required line spacing by cutting out the groups of lines from two standard charts (fig. 8-2) and pasting them on a white background.

The resolution test requires a test pattern which represents objects at critical distances and spacing. The following paragraphs describe the selection of resolution test patterns. The astigmatism test does not require critical line spacing or distance, but it is convenient to use the same chart and setup for both tests. The group of lines around number 28 is an easily viewed pattern for testing astigmatism in 7 x 50 binoculars and the Mark 1, Mod 0, ship's telescopes.

The width of the black lines on the test chart IS EQUAL TO THE WHITE SPACES BETWEEN THE LINES. Image fidelity test chart values (fig. 8-3) give the reciprocal of the space between the centers of adjacent lines as line per inch, the distance from which you should view the chart and the resolution requirements in terms of ANGLE IN THE FIELD FOR EACH CLASS OF INSTRUMENT. The corresponding patterns for the astigmatism test are also given in terms of the number of lines per inch. These are selected for convenient viewing at the same distance prescribed for the resolution test.

It can be proved that the angular limit of resolution is related inversely to the diameter of the lens. According to Dawes' rule (an approximation): a (in minutes of arc) $= \frac{1}{D_o}$ in fifths of an inch, which means that it is advisable TO HAVE LARGE OBJECTIVES FOR SHARP DEFINITION. A target shooter uses a scope with a 1 1/4" objective (diameter), or even larger. A pair of 7 x 50 binoculars provides

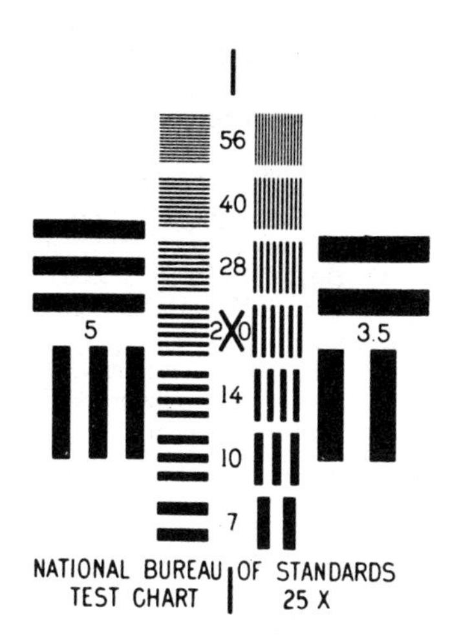

137.182
Figure 8-2.—Image fidelity test chart.

good resolution because of the large size of the objective (50 mm in diameter), and for this reason, it is better in daytime than a pair of 7 x 35 binoculars, with a 35 mm objective.

For best resolution, the power of the eyepiece in an optical system should be ascertained by the following formula:

$$\text{Focal length of Eyepiece} = \text{Diameter of Eye Pupil} \times \frac{\text{Focal Length of Objective}}{\text{Diameter of Objective}}$$

NOTE: All values in this formula are in millimeters. Although the resolving power of the human eye is equal to 1 minute of arc, after long, continuous observation, this is reduced to 2 or 3 minutes of arc by resultant eye fatigue. For continuous operation, therefore, an instrument of greater power is needed to provide the same definition obtainable with a lower-power telescope used for short intervals. Transparent foreign material (grease or fingerprints, for example) on a lens impairs definition (resolving power). Opaque, foreign material on the eyepiece may either impair definition or blot out small portions of the field.

Central Astigmatism

The procedure for testing an optical instrument for astigmatism follows.

1. Use the proper test chart and set it at the distance given in the listing of values. Sight the test pattern for astigmatism on the chart with the instrument to be tested, and line up the center of the astigmatism pattern in the center of the field of view.

Instrument	Θ Resolution Min. Limit in Seconds of Arc	Resolution $\frac{1}{S}$ Lines per inch	D in feet	Astigmatism $\frac{1}{S}$ Lines per inch
7 x 50 Binocular	4	56	77	28
Telescopic Alidade	11	40	39	20
Ship Telescope	4	56	77	28
Azimuth Telescope	8	40	54	20
Sextant Telescope	18	40	24	20

137.183
Figure 8-3.—Image fidelity test chart values.

2. Place an auxiliary telescope to the eyepiece of the instrument undergoing the test and adjust it to bring the horizontal set of lines into sharp focus. Note the diopter reading on the auxiliary telescope. CAUTION: The focusing adjustment on the primary instrument (one undergoing the test) must NOT BE CHANGED after you perform the preceding operation.

3. Check the vertical set of lines for focus. If it is not sharp, ASTIGMATISM IS PRESENT. To put the vertical set of lines in sharp focus, adjust the auxiliary telescope diopter ring. OBSERVE THE DIOPTER READING.

4. The maximum allowable difference in diopters between the horizontal and vertical lines is 0.15 diopters for the primary instrument being tested. Divide the diopter difference found in the auxiliary telescope, steps 2 and 3, by the square of its power to arrive at the corresponding change that would be found in the primary instrument without the auxiliary telescope. For example, the diopter change in the primary instrument equals:

$$\frac{\text{Diopter Change in Auxiliary Telescope (DCA)}}{(\text{Power of Auxiliary Telescope})^2}$$

As you can see, the auxiliary telescope increases the sensitivity of the test BY THE SQUARE OF ITS POWER. The maximum allowable diopter difference for typical auxiliary telescopes is as follows:

Power of Auxiliary Telescope	Maximum Allowable Diopter Difference
3	1.35
4	2.40
5	3.75
6	5.40

5. If the horizontal and vertical lines are in focus within the allowable tolerance, repeat steps 2 and 3 for the diagonal sets of lines. The same tolerance prevails.

NOTE: Excessive astigmatism may be caused by a defective or poorly mounted lens. Check the objective lens first, and then the reflecting surfaces of the prisms (objective prism first). These surfaces must be optically flat to close tolerances.

Central Resolution

If the instrument you are testing passes the astigmatism test, keep the test setup intact and use the following procedure to test for central resolution:

1. Sight the proper chart at the correct distance with the instrument you are testing with an auxiliary telescope and adjust the instrument in order TO BRING THE CENTER OF THE RESOLUTION PATTERN INTO THE CENTER OF THE FIELD OF VIEW.

CAUTION: Be sure you have the pattern centered. The resolving power falls off away from the optical axis. Focus on one set of lines.

Because the minimum resolving power of a well-designed instrument is finer than the eye can observe, always use an auxiliary telescope to make a test for central resolution. The instrument is better than the eye, and the auxiliary power reveals details to the eye.

2. When the black horizontal and vertical lines on the test chart (and the other diagonal sets of lines) APPEAR SHARP AND CLEARLY SEPARATED, resolution in the instrument is satisfactory.

Poor resolution is caused by defective objective lenses and prisms. Always replace the objective lens first. Misplaced, unmatched, and shifted prisms cause trouble because they displace the line of sight. A bad reflecting face on a prism also causes poor resolution.

FLARES AND GHOSTS

To check an optical instrument for flares and ghosts, point it toward a small, bright object against a dark background and focus sharply. If you observe rings or streaks of light, or one or more faint GHOST IMAGES, the image HAS EXCESSIVE INTERNAL REFLECTION. Flares and ghosts in an instrument INDICATE A PROBABLE NEED FOR RECOATING OF LENSES.

ILLUMINATION AND CONTRAST

Illumination of an image depends upon the amount of light received by the objective and the specific intensity (bright daylight or twilight) of these light rays. As you know, the amount of light received is determined by the diameter of the entrance pupil of the objective; and the amount of light which enters the eye is limited by either the exit pupil of the instrument or the pupil of the eye, whichever is smaller.

For maximum illumination at any given light intensity, the exit pupil of an optical instrument must equal the entrance pupil of the eye under the same conditions. With any instrument, furthermore, retinal illumination is never greater than illumination by the unaided eye. Opaque foreign substances—dust or lint, for example—on any optical surface (except one in a real image plane) reduces the amount of illumination in the system.

To test for illumination and contrast in an optical instrument, focus the instrument on a distant object and check the image for brightness. The IMAGE SHOULD BE NEARLY AS BRIGHT AS THE OBJECT APPEARS TO THE NAKED EYE. If the image is dim, the exit pupil may be too large; if the size of the exit pupil is correct (about 0.1" for bright light, to 0.3" for very dim light), look for dirty, stained, or uncoated optical surfaces and darkened mirrors or cement.

Contrast of the image produced by the instrument should be just as good as the contrast of the object seen by the naked eye. If the image is dull and cloudy, look for dirty, oily, or damp optical surfaces.

TESTS FOR OPTICAL DEFECTS

When you test the optical system of an instrument, check for all defects, including aberration (all types), coma, astigmatism, flatness of field, and distortion. All of these defects, singly or in combination, can affect the quality of image formed by an optical instrument; or even render the instrument useless.

Spherical Aberration

To test an optical instrument for spherical aberration, cover the outer half of the objective with a ring of black paper, focus sharply on a distant object, and read the diopter scale. Then remove the ring of paper and cover the inner half of the objective with a black disk. Refocus the instrument and read the diopter scale again. If the amount of movement of the eyepiece for focusing is very small, the instrument is well corrected for spherical aberration.

Chromatic Aberration

You can test an optical instrument for chromatic (color) aberration by doing the following:

1. Set up a white disk against a black background, far enough away to enable you to focus the instrument sharply. When the image is in focus, IT SHOULD HAVE NO COLOR FRINGES.
2. Push the draw tube in a short distance and look for a light-yellow fringe which should be around the image of the disk.
3. Refocus and pull the draw tube out a short distance, AT WHICH POINT THE IMAGE SHOULD BE FRINGED WITH PALE PURPLE.

The two colors (light-yellow and pale-purple) you got by focusing the instrument constitute the SECONDARY SPECTRUM OF THE OPTICAL SYSTEM, and they show that THE SYSTEM IS WELL CORRECTED FOR PRIMARY CHROMATIC ABERRATION (RED AND BLUE).

Coma and Astigmatism

Focus the instrument sharply on a small, round, white object near the edge of the field and study the image produced. If the image is circular and flareless, the INSTRUMENT IS FREE OF COMA.

NOTE: Test for coma at FIVE OR SIX DIFFERENT POINTS around the outer edge of the field.

An optical instrument has excessive astigmatism if one of the cross lines of the reticle shows parallax after you have eliminated parallax for the other cross line. Review illustration 6-30 and the related discussion.

Flatness of Field

Point the instrument being tested toward the horizon and focus sharply on AN OBJECT IN THE CENTER OF THE FIELD. If the edges of the field are IN SHARP FOCUS, THE FIELD IS FLAT; if the edges are not IN FOCUS, REFOCUS THE INSTRUMENT AS NECESSARY in order to create a sharp image of objects at the extreme edge of the field. The change you made in the diopter setting of the eyepiece shows the amount and direction of curvature.

NOTE: If refocusing of the instrument does not sharpen the image of objects at the edge of the field, ASTIGMATISM OR COMA IS RESPONSIBLE.

Distortion

You can test an optical instrument for distortion in the following manner:

1. Rule a pattern of vertical and horizontal lines on a large sheet of cardboard and put it where the pattern nearly fills the field of view of the instrument.

2. Focus the instrument sharply and check the image, which should be composed entirely of straight lines. IF ANY OF THE LINES APPEAR CURVED, THE IMAGE IS DISTORTED.

OVERHAUL AND REPAIR

The discussion of overhaul and repair of optical instruments in this chapter pertains to general procedures for disassembly, repair, reassembly, and adjustment of all basic instruments. A detailed discussion concerning particular optical instruments—disassembly, overhaul and repair, reassembly, collimation, and inspection—is presented in subsequent chapters of this training manual. The procedures for repairing different optical instruments vary in accordance with their variations from the basic design of optical instruments.

REPAIR TOOLS

You cannot accomplish satisfactory work on optical instruments unless you have the tools required for doing a particular task; and you must then acquire skill in using them and keep them in excellent condition.

Many tools used for repairing optical instruments are illustrated and explained in Basic Handtools, NavPers 10085-A. These tools are of the common variety available through normal supply channels. There are some tools, however, which were specially made for use on optical instruments by manufacturers and they are discussed along with the particular instrument for which they were made. There is still a third class of tools which can be used on almost all optical instruments but cannot be procured from a supplier. Your only alternative, therefore, is to draw a sketch or make a blueprint of the tools and make them in the machine shop. The tools considered in the following paragraphs belong in this class.

Grip Wrenches

Illustration 8-4 shows a grip wrench (part A) and the procedure for using it (part B). A grip wrench is made of fiber in sizes at intervals of 1/16 inch until a size of about one inch is reached; and then at 1/8 inch intervals up to sizes of about 3 1/2 to 4 inches.

When you use a grip wrench, select the smallest size which meets a specific need, without forcing it onto the part you must turn.

CAUTION: Grip wrenches have much leverage and you can exert tremendous pressure with them. Most optical parts are by necessity thin and light; so to prevent crushing of parts, try to use the grip wrench over that portion of a tube externally reinforced with a retainer ring or lens mount.

Retainer Ring Wrenches

Study the different types of retainer ring wrenches shown in part A of figure 8-5. These wrenches are also known as SLOT or OPTICAL wrenches. Part B of illustration 8-5 shows an Opticalman using the blade portion of a retainer ring wrench to rotate a slotted retainer ring in a lens mount. A retainer ring may be equipped with two small holes (instead of slots) spaced 180° apart, in which case the pointed tips of the retainer ring wrench are used to turn the retainer ring. This special tool is adjustable, and it can be used to remove or tighten a retainer ring of any size.

CAUTION: Slippage of a retainer ring wrench during use can cause much damage to unprotected optical surfaces, as well as the retainer ring and mount. To prevent such damage, be very careful when you use the wrench; be sure it fits properly in the slots or holes of the retainer ring. Protect optical surfaces with disks—rubber, blotting paper, or clean cardboard.

Geneva Lens Measure

A Geneva lens measure (fig. 8-6) is an instrument designed to measure the dioptric strength of thin lenses, by measuring the amount of curvature of their surfaces.

The dial of a Geneva lens measure is graduated in diopters. The outside red scale is graduated to read clockwise in quarters of a diopter from 0 to -17 diopters; the inner black scale is graduated to read counterclockwise in quarters of a diopter from 0 to +17 diopters.

The index of refraction of the glass for which a Geneva lens measure is designed for measuring dioptric strength is printed on the dial (1.53), and this number is the index of refraction of crown glass. A formula which is provided, however, permits use of the gage to measure types of glass with different indices of refraction.

To use a Geneva lens measure, place the contact points directly on the polished surface of the lens you desire to check for dioptric strength.

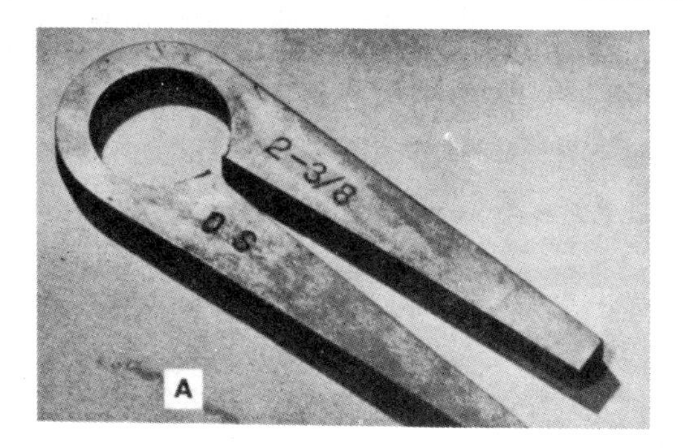

137.184
Figure 8-4.—Grip wrench and usage procedure.

The outer points (2) of the gage are STATIONARY, and the CENTER POINT must be activated until the outer points contact the lens surface. To ensure accurate readings and/or measurements, hold the gage perpendicular to the surface of the lens.

If the dial hand of the lens measure reads 0, the surface of the lens is PLANO or flat. Readings for convex surfaces must be PLUS; readings for concave surfaces must be MINUS. Take the reading in diopters of one lens surface and then measure the other surface. If you ADD THE DIOPTRIC STRENGTH of each lens surface, you get the TOTAL dioptric strength of the lens, provided its index of refraction is 1.53.

When you wish to take a reading of a lens with an index of refraction other than 1.53, use the following formula:

$$\text{True DP of Lens Surface} = \frac{n - 1}{0.53} \times \text{DP reading of lens surface with the gage}$$

(n = index of refraction of lens)

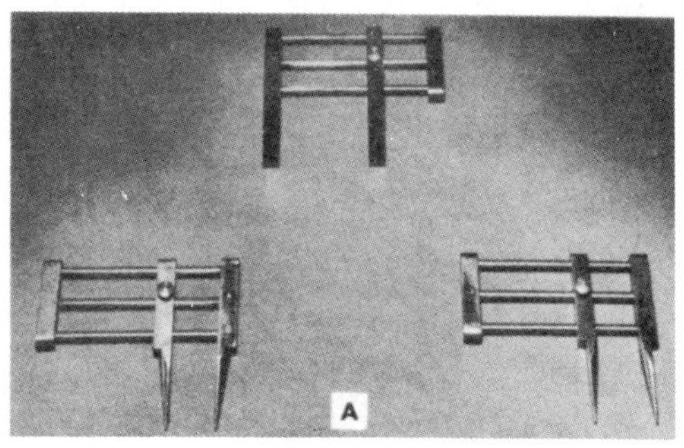

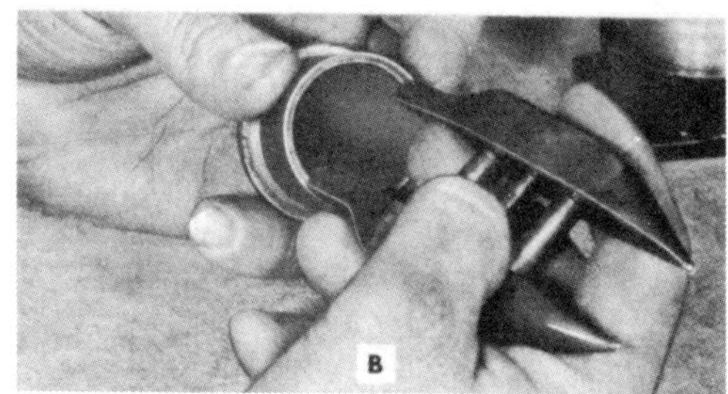

137.185
Figure 8-5.—Retaining ring wrenches.

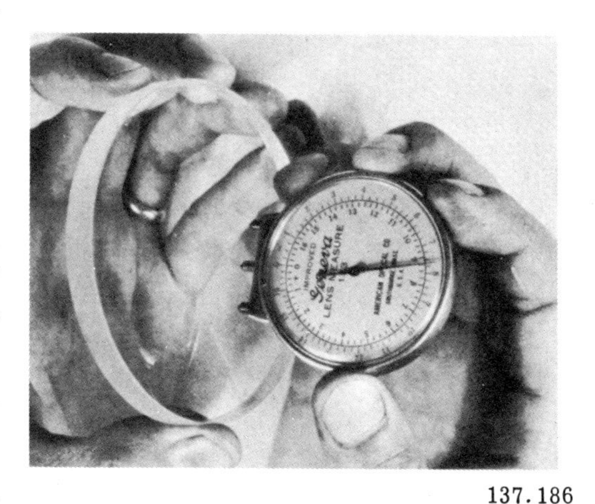

137.186
Figure 8-6.—Geneva lens measure.

To use this formula, take a reading of the first lens surface and transpose its dioptric strength into the formula and obtain a true dioptric strength of the first surface. Then take a reading of the second lens surface, put your results in the formula, and solve it for the true dioptric strength of the second surface. The sum of the two answers you got by solving the formula is the TOTAL dioptric strength of the lens.

Because a compound lens is constructed of a positive and a negative lens of different indices, you cannot use a Geneva lens measure to obtain its dioptric strength; but if the two elements of the lens are separated, you can obtain the dioptric strength of the individual elements and add both results to get the dioptric strength of the combination. NOTE: The lenses must be in contact and used as a unit in order to use the correct dioptric strength of the combination.

Remember that the dioptric strengths of the two lenses have opposite signs; that is, the positive lens has a positive dioptric value and the negative lens has a negative dioptric value. You must remember this when you add the two values.

Another thing to keep in mind concerning a Geneva lens measure is that it is designed to measure only the curvature of a lens' surfaces; so the thicker the lens, the less accurate the results derived through its use. When you are cementing lenses together, use a Geneva lens measure to make certain that the positive lens surface matches the negative lens surface.

Special Telescope Wrenches

Some special wrenches are useful for only one or two purposes and are used on optical instruments with similar design features. A binocular hinge pin puller is an example of such a tool, almost indispensable for repairing binoculars. Part A of illustration 8-7 shows a cross section of a hinge pin puller, with which you can pull and install a tapered binocular hinge pin without damaging other components of the hinge. Part B of figure 8-7 shows a special telescope wrench used for adjusting crossline mounts and tightening adjustable collars.

Take a look now at part A of illustration 8-8 which shows a special wrench used to remove or tighten a retainer ring. Part B of this illustration shows another type of retainer ring wrench which is used frequently. This type of wrench is especially useful for adjusting retainer rings inaccessible to an adjustable retainer ring wrench.

Other special optical instrument tools are illustrated and discussed in later chapters of this manual, as applicable.

PREDISASSEMBLY PROCEDURE

Before you do any repair work on optical instruments, clean your work space and get everything ready and in position. Your working space, your clothes, your tools, your hands, and everything should be almost immaculate before you begin work on an optical instrument, especially on optical elements. Cover your workbench with a sheet of clean, dry paper of light color.

Clean outside metal and painted surfaces with a clean, soft cloth (used for this purpose only). If a solvent is required to remove grease or foreign matter, use benzene or an approved dry cleaning solvent. Clean the outside surfaces of objectives and eyepieces in their mounts. Some particles of dust on an objective do not have a particularly harmful effect on an image produced by the objective, though they do prevent passage of light through the area they cover; but a film of dust on the objective may affect the quality of the image and you must therefore remove it.

If your casualty analysis indicates that the instrument must be partially or completely disassembled in order to effect necessary repairs, follow the procedure discussed next.

DISASSEMBLY

When you disassemble an optical instrument, do not mix non-interchangeable parts of one instrument with non-interchangeable parts of another instrument. In the interest of production and/or competence of performance, experienced optical repairmen work on more than one instrument at a time; and this same statement is true for many Opticalmen, especially when all the instruments require a major overhaul.

If you must work on more than one instrument at a particular time, keep the parts of each instrument in separate containers; and label the parts for double safety and easy identification. One of the surest and best ways to label parts is to scribe each metal part with an identifying mark. When you are giving four pairs of binoculars a general overhaul, for example, you can label the parts of the first binocular #1, the parts of the second binocular

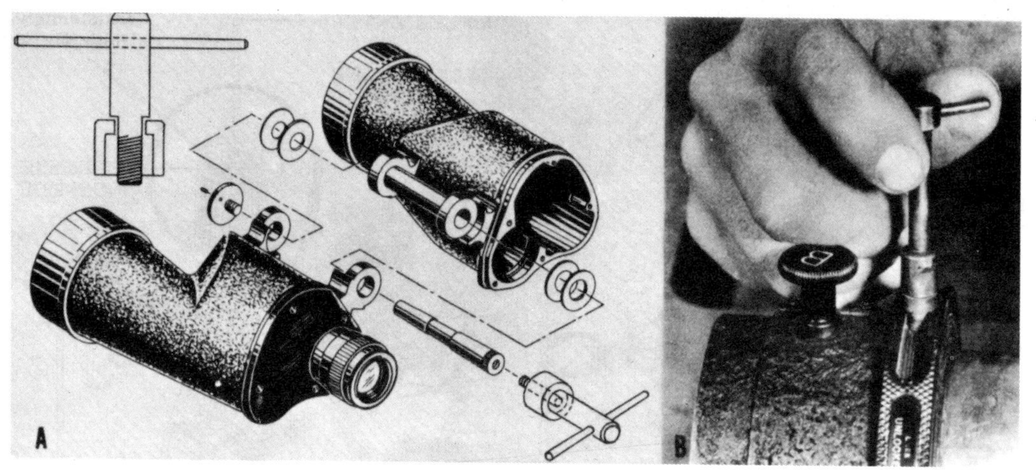

137.187

A. Binocular hinge pin puller
B. Special telescope wrench
Figure 8-7.—Special wrenches for optical instruments.

#2, and so on. In order to identify satisfactorily the parts for right and left barrels, add an R (right) or an L (left), as appropriate. Your markings on the parts for the right and left barrels would then be #1R, or #3L, and so forth. Be careful to scribe these marks where they will not be covered with paint later, and where they will not affect the performance of the instrument.

Other markings which you may be required to make or check during disassembly are ASSEMBLY MARKS. When a manufacturer makes an optical instrument, he fits certain parts by hand; and if there is danger of incorrect assembly of these parts during a later overhaul, he marks them with a small punch mark or a scribe line (on each part of an assembly). When you disassemble an optical instrument, therefore, look for these assembly marks; and if they are missing, make appropriate marks of your own. See figure 8-9, which shows the procedure for marking a part.

Optical elements (glass) require another marking technique, which must meet two requirements: (1) the direction the optic must face when reassembled, and (2) the function the optic serves in the optical system.

You can identify the function of the optic by writing on the frosted portion the following: Obj. (objective lens), #1 Er. (first erector), #2 Er. (second erector), and so on until you mark the last element in the system. The first erector receives the light from the objective and should therefore be numbered first. Use a soft-lead pencil or an instant-drying marking pen.

The accepted method for determining the direction an optical element must face in a system is to mark an arrow on the frosted edge of a lens or prism, the tip of which indicates the direction of light through the instrument.

If you presume a lens in a system is facing the wrong direction, study the diagram for that particular instrument (MARK and MOD) as you remove the lens. You can also use a Geneva lens measure to check the readings of the lens against those listed on the optical diagram.

If you do not fully understand an instrument you must overhaul, obtain and follow a disassembly sheet, or follow the disassembly procedure in the applicable naval publication (NavShips manual; Ordnance Pamphlet, OP; or NavWeps OP). These authentic sources provide information on troublesome areas pertaining to

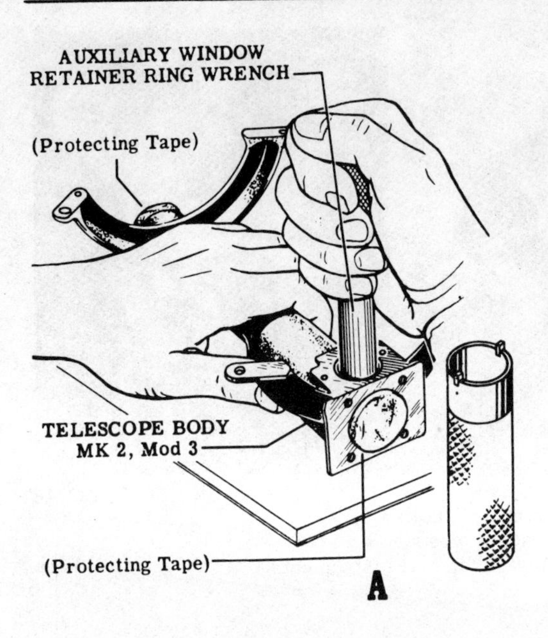

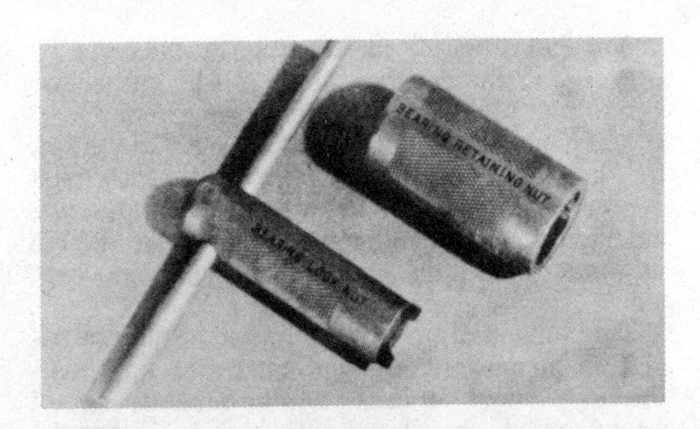

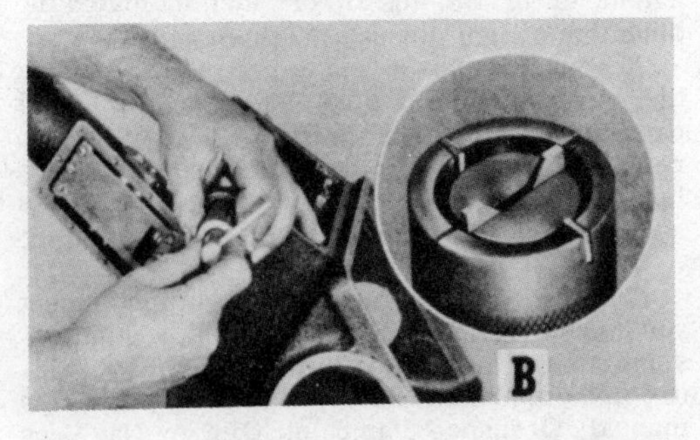

137.187
Figure 8-8.—Special retainer ring wrenches.

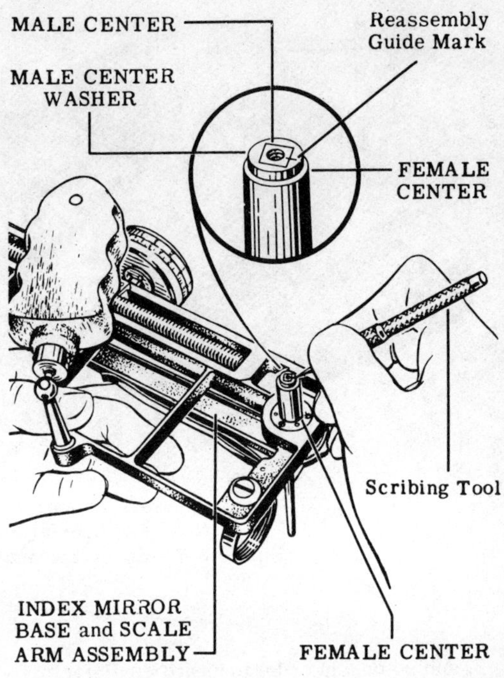

137.188
Figure 8-9.—Scribing an assembly mark.

disassembly, and they also list the precautions you should take.

CAUTION: Before you disassemble any optical instrument, determine whether it is a pressure-tight type. If it is gas filled, release the gas pressure slowly by opening the gas outlet valve. Never remove anything from the instrument until the pressure is fully released.

Start your disassembly of an optical instrument by removing exterior parts which hinder further disassembly, or by removing an exterior retainer ring, cover cap, or access plate (secured by screws). These exterior parts may occasionally be frozen, because they have been exposed to the weather; that is, metal parts in close contact become secured together as a result of corrosion, electrolytic action, or natural affinity for each other. Aluminum-to-aluminum joints have the greatest tendency to freeze (also called seize). Salt-laden atmosphere enhances the tendency of metal parts of navigational instruments to seize together; and if the moisture seal of the instrument was unsatisfactory, salt-laden moisture will most likely be present inside the instrument (breathing process explained previously). If this moisture is present inside the instrument, some of the interior parts may also be frozen.

Removal of Frozen Parts

The procedure for removing frozen parts follows:

1. To prevent damage to parts which come off easily (especially optics), remove them first.
2. Use proper tools, and do not crush parts with wrenches.
3. If you could not remove a lens, cover it with a pad of blotting paper, or a rubber disk of the same size.
4. When time permits, soak frozen joints in penetrating oil.
5. Use shaped wooden blocks to hold a part in a vise. Powdered rosin on the blocks helps to hold a part and prevent it from slipping out of position.
6. If a joint is still frozen after you have soaked it a reasonable time in penetrating oil, proceed as follows:

   a. Wipe off excess penetrating oil and apply heat to the exterior part as you turn it slowly.

   b. If the part breaks free, remove the heat, apply penetrating oil, and carefully separate the parts.

   c. To separate badly frozen parts which can be held solidly, apply penetrating oil and heat. Then apply pressure to the part in the form of an impact, not a steady pressure; for example, put a wrench on a nut or part and strike the wrench (just back of the head over the part) with a fiber mallet. This impact loosens the nut or part, as a general rule.

   d. If the parts are light and springy (body tubes and retainer rings, for example), use a light fiber mallet to tap lightly around the joint as you apply penetrating oil and heat, to help work the penetrating oil into the joint and work the corrosion out.

CAUTION: Use extreme care and patience when you apply heat and pressure to frozen joints, lest you cause distortion (twisting and bending) of metal parts, and breakage of optical elements you could not remove at the outset.

7. If frozen parts do not yield to the procedures just outlined, salvage the most expensive part or parts by carefully cutting, breaking, or machining away the other frozen parts or parts. When a retainer ring is frozen, for example, drill a hole down through it towards the lens; but use care, lest you drill too deeply and ruin the lens with the drill. The diameter of your drill should be slightly less than the thickness of the ring.

   After you weaken the ring by drilling the hole, carefully bend the ring out at that point and remove the free ring and lens. Some retainer rings are kept in place (made vibration proof) by an application of shellac or a similar substance on the threads of the mount and to the edge of the ring. You can soften this compound by repeated applications, as necessary, of acetone or alcohol.

8. To remove screws and set screws with stripped slots or heads twisted off, usually troublesome during disassembly, proceed as follows:

   a. If a screw is frozen in a hole as a result of corrosion, loosen it with penetrating oil and heat. NOTE: Do this before you try to remove the screw.

   b. If the body of a screw protrudes above the surface of a part, file in a new screwdriver slot with a small swiss slotting file and remove the screw with a screwdriver of proper size. You can generally remove some protruding screws with parallel motion pliers.

   c. If a screw is deep in a tapped hole, use a sharp scribe tip and, if possible, make a new slot in the screw. This process is slow and requires patience and care.

   d. If the procedures just described do not work, use one of the following procedures to drill the screw out:

      (1) For very small screws, use a drill slightly smaller in diameter than the minor diameter of the screw and drill through the screw. The outer shell and threads of the screw still remain, and you can run a tap of correct size through the hole to finish the job.

      (2) On screws of larger size, drill a hole of proper size in the screw and remove it with a screw extractor. (Each extractor has a drill of recommended size to use with it.)

Remember that patience and careful, intelligent workmanship are required in order to remove frozen parts from an optical instrument; but do not spend more time on an instrument than it is worth. Consult your shop supervisor whenever you are in doubt.

After you remove all frozen parts, continue with the disassembly. Remember to mark all optical and mechanical parts. Before you turn off a retainer ring or try to unscrew or slide a lens mount, remove the setscrews which secure them. Some of these screws may be hidden under sealing wax, so check for them carefully. Failure to remove these setscrews may cause a part to become seized.

Exercise extreme care when you remove optical elements and geared assemblies through openings in the optical chamber. These parts can be easily damaged by striking other parts and the chamber housing. When you remove a part which exposes the interior of the optical chamber of an instrument, make sure you tape or close it off in some manner in order to exclude foreign matter.

As you remove parts and assemblies from the interior of an instrument, check them for damage not previously noted and write your findings on your casualty analysis sheet for future reference.

Thus far, with few exceptions, our discussion of disassembly of an optical instrument has covered mostly mechanical parts, because this is the proper sequence for disassembling the instrument. As you disassemble an instrument, remove each lens mount and cell and set it aside for disassembly after you complete the disassembly of mechanical components.

Removal of Lenses from Mounts

The techniques discussed at this point for removing lenses from mounts are primarily for lenses mounted with a sealing compound, but they are also applicable for the removal of optics difficult to disassemble. The techniques to follow (and precautions to use) when you disassemble optical elements from their mounts cannot be formulated as step-by-step instructions. The information and/or things which you should keep in mind, however, when doing this work may be classified as follows:

1. Although optical glass is easily chipped or cracked, and easily damaged by shock, steady pressure within limits does not ordinarily crack a lens if the thickness of the glass is sufficient.

NOTE: Removal of the eyepiece and the objective is usually more difficult than removal of other optics, because these two lenses are usually sealed in their mounts with a sealing gasket or compound. Also, these lenses are doublets, which means that excessive or uneven pressure on the lenses can cause damage to the cement used to put them together, or cause the thin planoconcave flint element of the eyepiece to break.

2. Shearing action caused by uneven pressure is the greatest enemy of cement between optical elements; therefore, to force the compound lens out of its mount, press down squarely and evenly over a large part of its area. A device similar to that illustrated in figure 8-10 may be used to support the lens. Note the name of this device, LENS CHUCK AND CLEANING HOLDER, which is a cylindrical brass tube with the edges at one end beveled to match the curvature of the lens. By exerting even pressure on the lens mount, you can break the seal. Observe that the word PRESS in the illustration indicates the point where you should apply pressure.

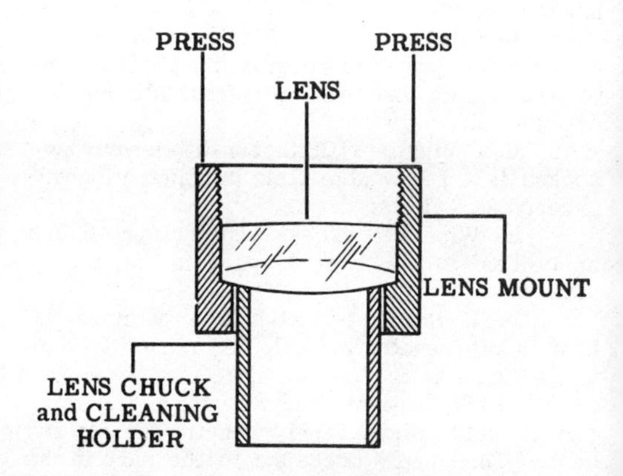

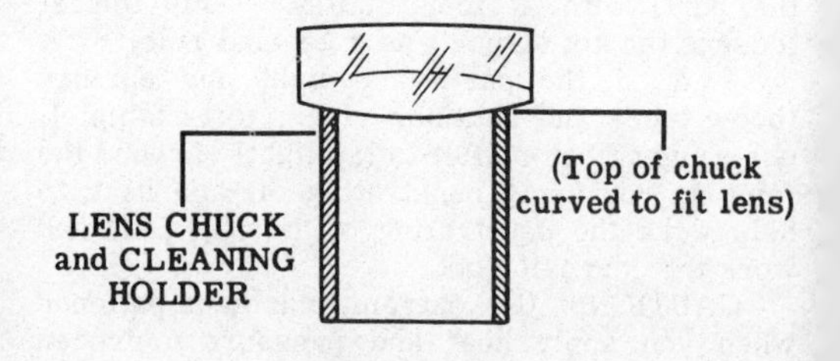

137.189
Figure 8-10.—Lens chuck and cleaning holder.

3. An application of heat to a lens mount helps to loosen it from the lens in two ways:
   a. The metal expands more than the lens.
   b. Most sealing compounds are softened by moderate temperature.

CAUTION: A temperature of 125° F to 140° F softens Canada balsam used to cement the elements of compound lenses together. If a compound lens does not therefore yield to pressure and an application of heat at low temperature, the Canada balsam probably melted previously and ran out between the elements of the lens and the mount and hardened a second time. When this happens, a high temperature is required to soften the cement.

4. When you remove a lens from its mount, protect its surfaces with a clean cloth or tissue paper. DO NOT TOUCH POLISHED GLASS OPTICAL SURFACES WITH YOUR FINGERS. Be sure to mark the path of light through the lens, to make certain that you reassemble it correctly. Then wrap the lens in lens tissue (several thicknesses) and place it where the mechanical metal parts cannot damage it.

5. When you cannot push a lens out from the back, as is sometimes the case, use a small suction cup or piece of masking tape to grip the mount and then ease it out of the mount.

CAUTION: Large thin lenses have a tendency to twist diagonally (COCK) as you try to remove them; so use care in order to prevent sticking. To loosen a COCKED lens, tap LIGHTLY ON THE EDGE OF THE MOUNT, on the side where the lens is stuck. As you tap the mount, so hold it that the lens will eventually drop out into your hand. If you accidentally touch the lens with your fingers, clean it thoroughly at once, to remove salts and acids deposited by your fingers.

REPAIR PROCEDURE

When you start to overhaul and repair an optical instrument, refer to the notations you made on the casualty analysis sheet for it prior to and during disassembly; and use this information as you proceed with the repair process.

Cleaning and Inspecting Parts

The first phase of overhaul of the instrument is cleaning of mechanical parts. Always use approved cleaning solvents, which may be slightly toxic and irritating to your skin and necessitate cleaning in well ventilated spaces only. Avoid prolonged contact of the hands with the solvent. The best policy (safest) is to use solvents only in a space specified for their use.

A cleaning machine of the type shown in figure 8-11 is excellent for cleaning some mechanical parts of optical instruments. An electric motor in the machine revolves a basket of parts a sufficient amount of time in an approved cleaning solvent and thus thoroughly cleans the parts. The second step in this process is to put the clean parts into another basket and rinse them in a container of approved rinsing solution. The final step (usually) in this cleaning process is to wash off the rinsing solution and dry the parts in the machine.

Another type of instrument cleaning machine (not illustrated) agitates solvent around the parts by vibration. The newest types of cleaning machines employ an ultrasonic oscillator to act on an approved liquid cleaning agent and thereby clean the parts.

NOTE: When you use a cleaning machine, follow the instructions listed in the manufacturer's technical manual. If you do not have this manual, consult your shop supervisor.

If your shop does not have a cleaning machine, use a stiff-bristle brush to clean instrument parts in a tank of cleaning solvent. This is one of the best and simplest methods for cleaning some (if not all) instrument parts. Some solvents leave an oily residue on clean parts, and you must remove it by rinsing the parts in an approved degreasing agent. Traces of oil on the interior of an optical instrument may later get on the lenses and affect image formation, or render the instrument useless.

After you clean instrument parts, inspect them for traces of lubricants, grease, sealing compound, or dirt. Scrape off dirt and grease not removed during the cleaning process.

CAUTION: Do NOT scrape bearing surfaces. As you examine each cleaned part, look for defects previously hidden by dirt, wax, or grease; and also check them for corrosion. Replace badly corroded parts.

Place the cleaned parts you intend to use in a suitable, clean container and cover the container to protect the parts from dust and dirt.

Repair Categories

Now that you have cleaned and inspected the parts of the instrument undergoing repair, proceed IMMEDIATELY with the repairs. The

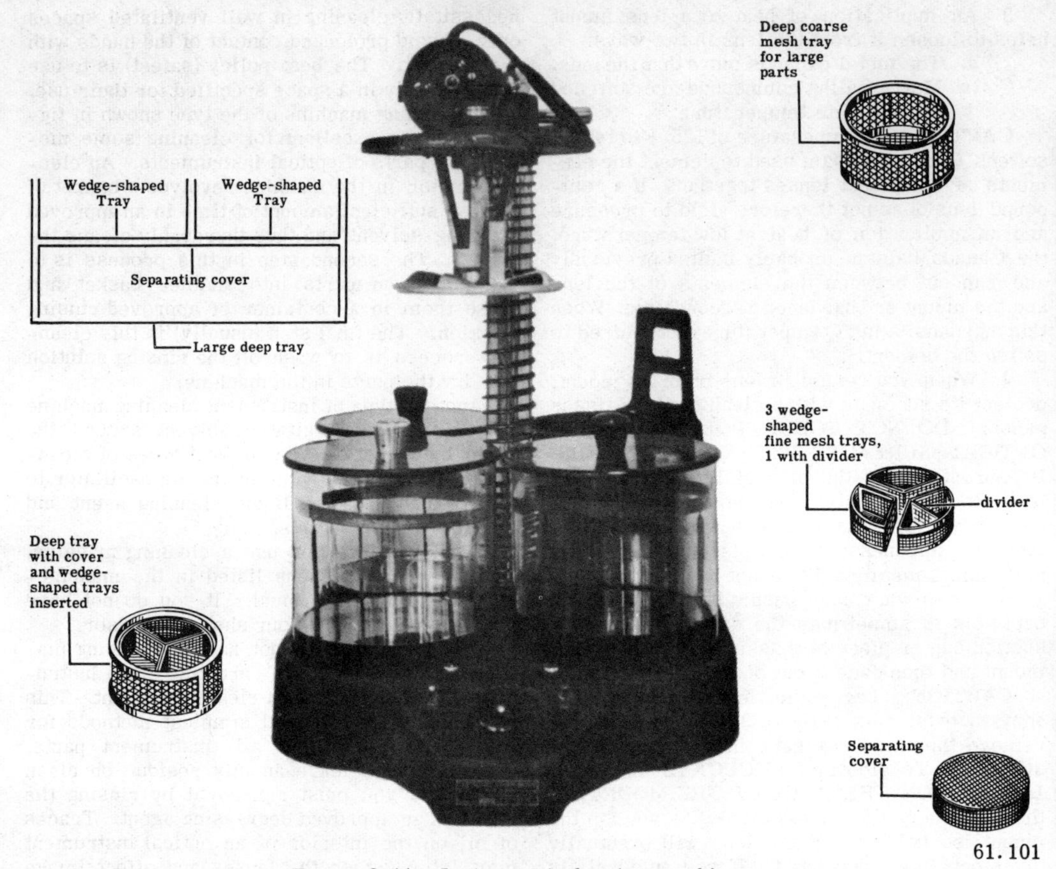

Figure 8-11.—Instrument cleaning machine.

repair process generally consists of three phases or categories: (1) repair and refitting of old parts, (2) using a new part (replacement) from stock, and (3) manufacturing and refitting a new part. Each of these categories is discussed in some detail in the following pages.

REPAIRING OLD PARTS.—Repair reusable old parts, as necessary, and refit them into the instruments from which you removed them. The repairs which you may have to make on a part are discussed next.

If a part must be straightened or reformed to its original shape, strike it carefully at the proper place with a soft-faced hammer. CAUTION: Give the part necessary support before you strike it, lest you inflect further damage upon it.

When a part has stripped or damaged thread (usually in a tapped hole), whenever possible, drill the hole out and retap it for a screw of larger size; but do not go over one or two screw sizes larger than the original size stated on the blueprint. If the screw size must be exactly as stated on the blueprint, proceed as follows:

1. On steel, bronze, and brass parts, drill and tap the hole two or three screw sizes larger than originally and fill the hole with the same material of which the part is made. Use silver solder to secure the plug. NOTE: A screw which fits the larger size makes a good plug.

Then file the plug flush with the surface of the part and drill and retap a hole of correct size.

2. If a larger size screw can be used, repair aluminum parts with stripped threads in the same manner as you repair parts made of other metals. It is difficult to solder aluminum parts, however, and it is best to ask the shop supervisor to have the soldering accomplished in another facility, if possible. When the soldered part is returned to you, dress the soldered area, and redrill and tap the hole to the size specified on the blueprint.

If aluminum welding is not feasible, repair the part by drilling the hole two or three sizes larger, tap it, and then fill it with an approved synthetic metal such as EPOXY. Allow the metal time to dry and then drill and tap a new hole to specified size. The synthetic aluminum metal (when dry) is as tough as most cast aluminum and gives satisfactory results.

3. If a part is broken, repair it by soldering, brazing, or welding. When pieces of a broken part are missing, replace the part.

4. Dress up scratched, burred, and dented parts, in accordance with prescribed shop procedures.

Use much care when you repair parts, to prevent damage to precision bearing surfaces machined on them. Use a stone or a bearing scraper to remove burrs from a bearing surface, and be careful to remove only as much metal as is essential to do a good job. Do NOT file a bearing surface, for filing may completely ruin it.

When you complete repairs on an instrument part, refit the part on the instrument and check its action and/or operation for accuracy. If necessary, scrape off a slight amount of a surface in order to make a part fit properly; and redrill undersized holes and make other necessary changes of your repair job in order to have the part fit correctly. After you fit a part, DO NOT FORGET to make an assembly mark on it to indicate direction of installation.

CAUTION: Reassembly of an instrument containing improperly fitted parts may necessitate unnecessary subsequent disassembly on part or all of the instrument.

REPLACEMENT PARTS.—Sometimes a part is damaged to such an extent that it must be replaced with a new part. One source of replenishment is from stock, for some purposes only.

When you receive a replacement part from stock, try it for proper fit in the instrument or assembly. If it does not fit, take necessary action, including machining. A manufacturer, for example, does not drill dowel pin and screw holes; so you must drill them of correct size wherever required. A manufacturer also makes bearing parts slightly oversized, so that you can fit them properly by hand. Do not forget to make assembly marks on the new parts after you fit them, to ensure correct fitting of them into the instrument later.

MANUFACTURED PARTS.—Occasionally, your shop supervisor can have parts made by submitting an intershop job order; but there will also be times when you will be compelled to manufacture parts. The procedure for doing this is as follows:

1. Use information on the old part, or its name, to locate the blueprint. Use its dimensions to make or procure a new part when the blueprint is unavailable.

2. If the foundry can cast the part, give the old part and the blueprint which covers it to the pattern shop so that it can make an accurate pattern of the part for the foundry.

3. After you receive the manufactured part, machine it as necessary and then fit it by hand to the instrument.

Miscellaneous Repairs

When you gave the instrument on which you are working a pre-disassembly inspection, you perhaps noted undamaged moving parts in the instrument which were dry, tight, grinding, or rough in action. You also perhaps found in some instances a combination of these malfunctions, and even others not mentioned here.

When you effect miscellaneous repairs on an instrument, look for all types of trouble and remedy it, including lack of or dirty lubrication, excessive or insufficient clearances, incorrect alignment, and improper assembly. If the cause of malfunctioning is not readily apparent, proceed as follows:

1. Clean all parts of the bearing assembly.

2. Make a trial assembly, but do not force parts.

3. Check parts for proper clearance in order to determine the cause of binding or excessive lost motion.

When cleaning, lubrication, and proper aligning of parts fail to correct casualties and/or malfunctioning, take the action discussed in the following paragraphs.

INSUFFICIENT CLEARANCE.—If there is an insufficient amount of clearance on such parts as eyepiece draw tubes, tapered sleeve bearings, ball and socket bearings, and multiple-lead thread eyepieces, do this:

1. Make a thin solution of pumice and clock oil (small amount of pumice at first) and put a little portion of the solution on the parts as you reassemble the bearing.
2. Work the parts of the bearing back and forth, or rotate them until their movement is of desired freedom.
3. Disassemble the bearing and wash out all traces of pumice and oil.
4. Reassemble the bearing, lubricate with the proper type of lubricant, and check the motion.

Follow the procedure just described until you obtain the desired fit.

When there is insufficient clearance on a flat, sliding-surface bearing, do the following:

1. Put a thin coat of Prussian blue machinist's dye on a surface plate and rub the oversized portion of the bearing assembly over the Prussian blue.
2. Carefully scrape away the high spots on the bearing indicated by the Prussian blue.

CAUTION: Remove only a small amount of metal at a time, and make a trial assembly after you remove each amount. The important thing here is prevention of the removal of too much metal from the bearing.

Another method for removing excess metal from a sliding-surface bearing is to spread a small portion of a thin mixture of pumice and clock oil over the surface of a flat lap and rub the high part of the bearing over the surface of the coated flat lap. Use a sweeping figure-of-eight motion to ensure uniform removal of the metal. Do NOT remove too much metal.

EXCESSIVE CLEARANCE.—If there is no way of adjusting a bearing by removing excessive clearance with shims, or the bearing does not have some means by which it can be adjusted, replace it with a new one. If there is some way to adjust the bearing, however, adjust it as necessary in order to get a tight fit and then remove high spots in the manner described for obtaining sufficient clearance.

NOTE: Always mark bearing parts to ensure proper assembly after you hand fit them in the manner just described.

CLEANING AND PAINTING PARTS

Having completed all repairs to your instrument, you are now ready to accomplish essential cleaning prior to painting. Reclean all parts on which you made repairs, to remove traces of moisture, dirt, metal chips, and grease from its surfaces. If a part does not require painting, put it in the container with other cleaned parts of the instrument. Those parts which require painting have the old paint on them, and you must remove it before you apply a new finish. The procedure for doing this follows.

Removing Old Paint

There are two types of approved paint remover: (1) BRUSH-ON, and (2) PAINT AND CARBON, each of which is discussed in some detail next.

Brush-on paint remover dissolves synthetic-bristle brushes, so use a natural-bristle brush and brush it on the painted surface of an instrument part. Leave the paint remover on the part as long as necessary for it to dissolve the paint and then wipe it off. Finish the job by rinsing the part in lacquer thinner or benzene to remove wax used in the remover as one of the ingredients.

Because it is difficult to wipe brush-on paint out of holes and corners, you will experience some difficulty in using it.

Paint and carbon removers are available through Navy supply channels and also commercially. They are designated as SUPER cleaners. Besides removing paint, they remove heavy carbon, grease, varnish, and sticky gums.

You will obtain the best results with a paint and carbon remover by putting at least 10 gallons in a stainless steel tank and soaking the parts as long as necessary in it. Then wash each part with hot water, remove the water with a compressed air hose, and bake it briefly in an open. It is then ready for painting.

CAUTION: Paint and carbon removers are not explosive or flammable, and they contain a chemical seal-top layer which prevents evaporation; but use them ONLY in well ventilated spaces and protect your eyes with goggles.

Removing Corrosion

Paint removers do not eliminate corrosion from instrument parts, so it must be removed with an approved corrosion removal compound, available through naval supply channels and commercially. NOTE: Use only Navy approved commercial products.

Always follow the manufacturer's instructions when you use any product, and protect yourself by following safety precautions.

Corrosion generally eats into a part, and the best way to remove it is to soak the part for a sufficient amount of time in a tank (stainless steel) of the compound.

If you do not have an approved corrosion removal compound, you may make some (for different metals) by using the following formulas:

1. To make a corrosion removal compound for CAST IRON AND STEEL, use a 50 percent solution of sulfuric acid and distilled water (about 150° F). Then dip the corroded metal parts in the warm acid for about 5 seconds and wash them immediately in several changes of hot water.

CAUTION: Do not handle chemicals until you understand the safety precautions (end of this chapter) which pertain to them. NEVER USE ACID ON BEARINGS, or GEAR TEETH.

2. You can make a corrosion removal compound for brass by using the following formula:

Water (pure, distilled)	491 cc
Sulfuric acid (concentrated)....	435 cc
Nitric acid (concentrated)	72 cc
Hydrochloric acid (concentrated)	2 cc

If a brass surface is bright in spots, there is probably some clear lacquer on it. Submerge the part in paint remover and then rinse it with hot water. Continue by dipping the part in the correct amount of the corrosion removal solution for 4 or 5 seconds, rinsing it in water, drying thoroughly with an air hose, and applying at least one coat of clear lacquer before the surface oxidizes. NOTE: Do not use lacquer if the part requires paint.

CAUTION: Do not use a brass dip on bearing surfaces.

3. To clean corrosion from aluminum, dip it for 5 to 10 seconds in a 10 percent solution of sodium hydroxide (lye) at a temperature of about 150° F and wash the lye off immediately with hot water.

You can also use some non-chemical methods for removing corrosion and giving a bright, clean finish to metal parts. These methods involve types of abrasives, wire brushes, buffing wheels, and abrasive cloth, listed in the order of discussion.

1. REMOVING CORROSION WITH A WIRE BRUSH. There are two types of wire brushes which you may use to remove corrosion from metal, rotary-power and hand.

CAUTION: To prevent damage to your eyes, wear your goggles to protect them from flying wire. Do not use a wire brush on a bearing surface or an engraved part.

To use a rotary-power wire brush, hold a part against the wheel with enough pressure to force the moving wire bristles into the corrosion and keep the part moving slowly and evenly against the wheel. Run the wheel from the center of the part toward the edges, to ensure thorough cleaning of the edges. Use a hand wire brush, emery paper, or a scraper to remove corrosion from the inside corners of the part.

2. REMOVING CORROSION WITH A BUFFING WHEEL. A buffing wheel gives a part a brighter, polished finish than a wire brush (wheel), but it will not remove heavy corrosion. For this reason, do not use these wheels on large areas, but use them to polish metal parts which must remain bright.

Use a polishing compound with a buffing wheel, and polish a part until you have the desired brightness and polish. Then remove the remains of the polishing compound with a solvent, dry the part thoroughly, and apply at least one coat of clear lacquer.

NOTE: To speed up the buffing process, clean the parts first in a corrosion remover.

3. REMOVING CORROSION WITH ABRASIVE CLOTH. You can remove corrosion from metal with an abrasive cloth in the following manner:

a. Polish flat pieces by hand on crocus cloth (embedded with an oxide of metal) laid on a flat surface.

b. Polish irregular pieces which you cannot buff on a wheel by hand. Use wood or metal in the jaws of a vise to protect these pieces and secure them ONLY as tightly as essential. To polish a piece in the vise, use a strip of fine emery cloth and complete the job with a piece of crocus cloth, to remove grains produced by the emery cloth.

c. Put small, round parts of an instrument in the collet of a lathe and (with the lathe running at high speed) touch the parts lightly with emery cloth or crocus cloth to the extent necessary to obtain the polish desired.

CAUTION: Do not use abrasive cloth on bearing surfaces. When a bearing surface has deep pits caused by corrosion, it is worthless; the bearing is ruined. Use a non-abrasive cleaner to remove light corrosion from the surface of a bearing. Unless there are provisions provided in the construction of a bearing for refitting it, and you can do this in your shop, never remove metal from a bearing surface. When you remove corrosion from a bearing surface, rub it off carefully with crocus cloth or a fine paste of clock oil and pumice, or by scraping.

Painting Procedure

After you remove corrosion from instrument parts, you are then ready to paint those which require paint. There are three reasons for painting metal parts of optical instruments, as follows (in order of importance);

1. To protect the metal from rust and corrosion. This is most important for instruments used aboard ship, where salt spray and damp, salty air quickly corrode unprotected metals.
2. To kill reflections. The glare of bare metal in the sunlight is very annoying to the user of an optical instrument; and under some conditions, a brilliant reflection from a metal surface may reveal the observer's presence to an enemy.
3. To improve appearance. A good-looking, pleasing appearance of an optical instrument creates a good impression on all who see and use the instrument. Inspection of painted surfaces of instruments is part of your mandatory inspection procedure.

Some paint manufacturers make lacquers and enamels which put a fine finish on instruments. Many optical parts have a very smooth, hard finish which appears to be part of the metal itself. This finish is called ANODIZE, applied to the metal by an electrochemical process.

A baking enamel of high quality gives a hard, durable finish, but air-dried enamel is good for touching up or painting an instrument which cannot be subjected to heat in a baking oven. Lacquers have one outstanding characteristic, quick-drying, but they cannot resist chemicals and are therefore not as durable as enamels.

CAUTION: Never cover enamel with lacquer, because the lacquer loosens the enamel from its base and causes it to blister.

Lacquers and enamels which give a dull-flat, black finish are used to cut down surface reflections, and they are also used to kill internal reflections on the inside of optical instruments. You will generally use a dull-black finish paint on most optical instruments.

Paints which give a semigloss, black appearance and a hard, durable finish are used on parts which receive considerable handling, and on such small articles as eyepiece focusing rings, knobs, handles, and pointers.

Always use clear lacquer on parts subject to corrosion but which are not painted, to protect their high polish.

Most paints and their thinners are flammable, and some are explosive; so use a spray booth with an explosion-proof exhaust fan. To prevent spontaneous combustion, put rags used for wiping up paints, oils, thinners, etc., in a container with a self-closing cover and dispose of them completely as soon as practicable. Stow paint materials in a locker which will not tip over, and at a temperature less than 95° F, preferably—never over 95° F.

CAUTION: Permit no smoking in the spray room, and have a CO_2 fire extinguisher available in the room's equipment. Do NOT play with the air hose, or point it toward your own person or any one else.

When you paint with a spray gun (usually the case), mask bearing surfaces, threads, and holes to the interior of the instrument, from which you desire to exclude the paint. Tear off strips of the tape and put them over the surfaces of the bearings, with the edges of each successive strip (one side) slightly overlapping the last strip applied. Then trim off excess tape with a sharp knife or razor blade. Mask off also all points on the instrument you do not wish to paint.

Punch holes in a small box top or piece of cardboard and stick the bodies of screws whose heads you desire to paint into the holes, to keep paint off the bodies of the screws. You can also place on the cardboard top small parts which you intend to paint on one side only. String parts which you desire to paint all over on small pieces of brass wire.

PREPARING THE PAINT.—Prepare both the primer coat and the finish coat in the same manner for use in a spray gun, as follows:

1. Stir the paint thoroughly in order to mix the pigment back into the liquid vehicles used to suspend it. Unless you do this, the paint will

not cover surfaces with uniform thickness and will not have luster and the same color all over.

2. Thin thick paint before you put it into a spray gun; otherwise, it will clog the gun and not go through it. Use your experience and the manufacturer's instructions when you thin the paint to proper consistency. Dip a pencil vertically into the lacquer or enamel and then withdraw it. If the consistency is correct for spraying, the lacquer or enamel will run off the pencil in a smooth, thin stream. When thinning paint, however, do not add much over 20 percent of thinner to the paint, lest you get it so thin that it will not cover material properly. (Total volume of paint and thinner should be about 20% thinner.) After you add thinner, stir the paint thoroughly.

3. When you have the paint at the right consistency for spraying, strain it through several thicknesses of cheesecloth or medical gauze to eliminate lumps of undissolved pigment, dirt, and any other particles which could clog the spray gun and give a poor finish on your work.

APPLYING THE PAINT.—Before you use a spray gun for the first time, seek good information concerning its operation, or closely follow the manufacturer's instructions for its use. Check the spray gun for cleanliness. If it is dirty or has old paint on the inside, disassemble it completely and soak the metal parts in a paint remover. Clean the gaskets in lacquer thinner. CAUTION: Paint remover will ruin the gaskets. When you reassemble the spray gun, lubricate all moving parts.

Fill the cannister of the spray gun with your prepared paint and turn on the air pressure, about 40 to 60 pounds per square inch, or as recommended in the manufacturer's technical manual for the gun. Then so adjust the gun that it delivers a fine spray with enough density to cover surfaces rapidly with a uniform, wet appearance. Then begin your spraying.

Hold the spray gun about 10 inches from your work and keep it moving horizontally, back and forth. Be sure to carry each swing of the gun out past the end of the work before you start back, to prevent piling up of the paint near the edges of the work and subsequent sagging. Start at the top of a surface and work down, back and forth in horizontal motion, and cover the last old lap with about half of your new lap. If you follow this procedure, your paint will be uniformly thick over the entire surface.

After you finish a paint job with a spray gun, spray lacquer thinner over the gun to remove lacquer and enamel from the small openings. At the end of the day, if you use the gun last, completely disassemble the gun and wash all parts in lacquer thinner. Then dry, lubricate, and reassemble it so that there will be no delay of work the next day. The best time to clean a spray gun is while the paint on its surfaces is still wet.

NOTE: Your spray gun should have an air pressure and reducing valve with a water and oil trap (and filter) which should work correctly all the time. Drain this trap regularly. If water and oil get into your spray gun and paint, it will ruin the appearance of your work; and the lacquer or enamel will not dry.

Following is a list of difficulties sometimes experienced with a spray gun, with the reason for each difficulty given.

1. FINISH REFUSES TO DRY. You forgot to remove the oil and grease from the metal surfaces of your work, or from your air supply.

2. FINISH COVERED WITH TINY ROUGH SPOTS. There was too much dust or moisture in the air, or in the paint or spray gun.

3. FINISH HAS SMALL CIRCULAR MARKINGS. There was water in the air hose, or water dripped or condensed on the work before it was completely dry.

4. FINISH SHOWS HORIZONTAL STREAKS. Your spray was too fine and the last lap had started to dry before you applied the next one, or you forgot to cover half of each old lap with the following lap.

5. FINISH IS UNIFORMLY ROUGH. The spray was too fine, or you held the gun too far from the work, and the droplets began to dry before they hit the work.

6. THE FINISH HAS LUMPS OR BLOBS. The spray gun or hose line was dirty, or you forgot to strain the paint.

7. THE FINISH RUNS. The consistency of the paint was too thin.

8. THE FINISH SAGS. You moved the gun too slowly or held it too close to the work.

9. THE FINISH SHOWS ORANGE-PEEL EFFECT. The consistency of the paint was too thick, your spray was too fine, or you held the gun too far from the work.

When you intend to paint and bake instrument parts, remove all masking tape before you put the parts in the oven. If you cannot remove the tape before you bake the parts, remove it immediately upon taking the parts out of the oven.

This is also a good time to apply engraver filler, commonly called MONOFILL, a soft, wax-base compound (generally in crayon form) used to fill in and accentuate engraved index lines and numbers. While the part is hot, the filler flows easily into an engraving. When the part cools, wipe off the excess filler with a soft cloth.

CLEANING GLASS OPTICS AND RECEMENTING LENSES

Clean the lenses and prisms of an instrument you repair while paint on the finished work is drying, and also accomplish necessary lens cementing.

The Navy standard for cleaning glass optical elements is this: OPTICS MUST BE CLEANED TO ABSOLUTE PERFECTION.

Bear in mind that an optical instrument with components of the highest quality arranged in the best design possible is of little or no value if vision through it is obscured by dirty optics. This statement does not mean grime or mud; IT MEANS THE SMALLEST VISIBLE SPECK OF DUST. EVEN A SPECK on a reticle may obscure much detail of an image, and a fingerprint or film of oil will most likely blur the overall image.

For the reasons just given, you must learn the proper technique for cleaning glass optics, and you must then APPLY THEM WITH PATIENCE, CARE, AND THOROUGHNESS. Knowledge of procedure, plus appreciation for quality work, will enable you to attain the absolute-perfection standard required.

Cleaning Equipment

The equipment you need for cleaning optical elements includes a rubber or metal bulb syringe, several camel's-hair brushes (small), alcohol, medically pure acetone, lens tissue (soft, lintless paper), absorbent cotton or silk floss, wooden swab sticks, stoppered containers for alcohol and acetone, and a container to keep the cotton or silk floss absolutely clean. To this list you may also wish to add a special lintless cloth for cleaning optics, the best types of which are SELVYT CLOTH and CAMBRIC NAINSOOK.

You can make a lens cleaning swab of cotton, silk floss, or lens tissue. To make a cotton or silk floss swab, use the end of a wooden swab stick to pick up the top fibers of the material. Thrust the tip of the stick into the material and rotate the stick until some fibers catch on it; then pull the captured fibers loose from the mother material. Repeat this process as often as necessary until you have the swab of desired size. Shape the swab by rotating its tip against a clean cloth or lens tissue.

CAUTION: Do NOT touch the tip of the swab with your fingers or lay it down on the bench top where it will pick up dirt.

Figure 8-12 shows the procedure for making a swab out of lens tissue, step by step. Swabs made in this manner are useful for picking up individual specks of dirt from a lens or reticle, using acetone as a cleaner. Make a supply of lens tissue strips for fabricating swabs by cutting a packet of 4" x 6" lens tissue down the center, lengthwise, so that you can remove the strips one at a time.

The fourth step for making a swab (4, fig. 8-12) shows how to press the tip of the round swab between the cover and the top tissue in order to obtain a flat, chisel-like cleaning tip, as shown in step five (5) of illustration 8-12.

You can make a large, useful lens cleaning pad by folding two thicknesses of 8" x 11" lens cleaning tissue along its length and bringing the two ends together. When you dampen this pad with acetone, you can clean a large area of glass quickly and effectively.

Cleaning Glass Optics.

The recommended procedure for cleaning glass optics is presented by steps in the following paragraphs:

1. Blow all coarse and loose dust from the surface of the lens with a bulb syringe. Then brush the surface of the lens with a camel's-hair brush, using quick, light strokes. Flick the brush after each stroke to dislodge the dust it picked up, and blow off newly loosened particles of dust on the lens (optic) with the bulb syringe.

2. If the lens is large, use several pads of lens tissue dampened with alcohol to remove remaining dirt and/or grease. Change cleaning pads or swabs frequently enough to prevent damage to the optic by the dirt or grit. Use a cotton, silk, or floss swab, or lens tissue on small lenses.

3. Finish the cleaning of the optic by using a pad or swab dampened with a few drops of acetone, to remove traces of film of the alcohol used during precleaning.

CAUTION: If you use a swab or pad moistened with acetone for more than 20 seconds on

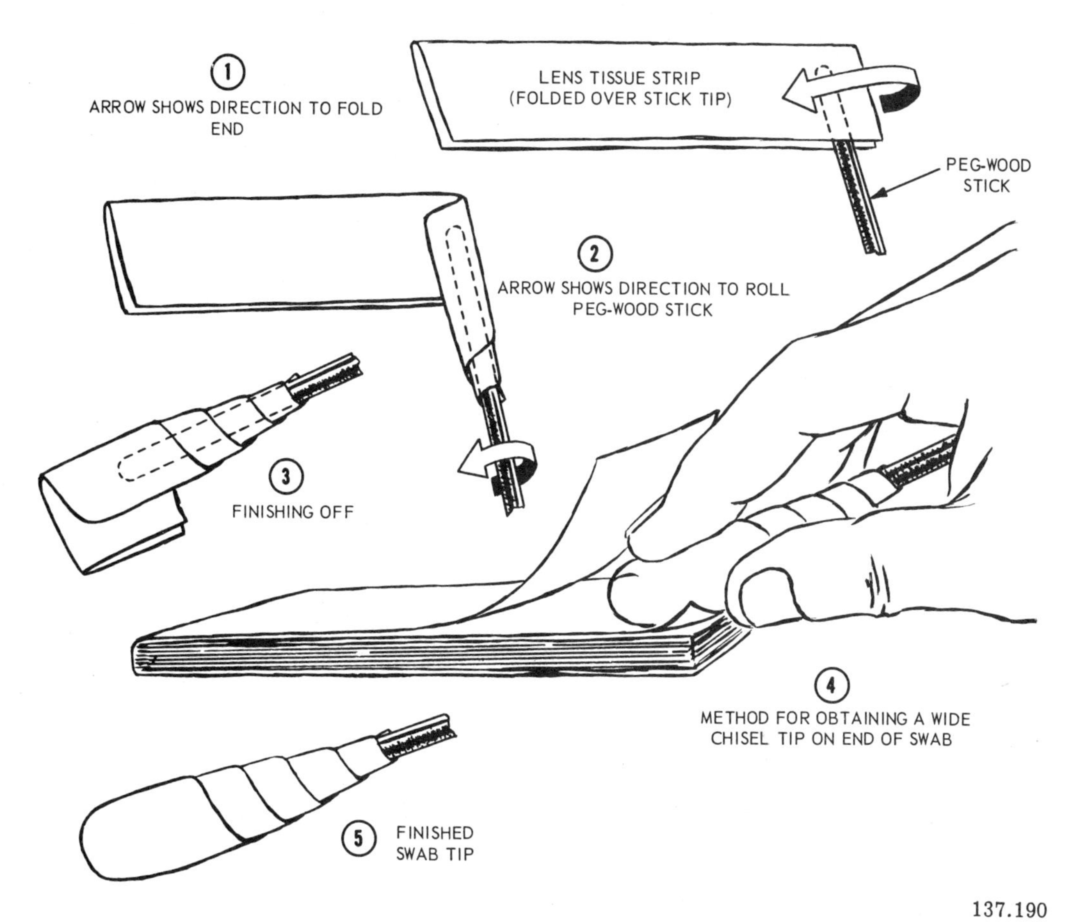

137.190

Figure 8-12.—Procedure for making a lens-tissue swab.

an optic, it leaves a film or water marks on the lens. Acetone evaporates quickly and moisture in the surrounding air condenses in the swab or pad. Medically pure acetone (triple-distilled) leaves an optical surface perfectly clean and free of film when used as described. ACETONE IS HIGHLY FLAMMABLE; KEEP IT AWAY FROM FIRE AND HEAT.

4. As you clean an optic, swab lightly with a rotary motion, working from the center to the edges. Avoid excessive rubbing to prevent damage to the coating of an optic and charging with static electricity. Study figure 8-13 for the correct procedure to follow when you clean a lens with a swab.

5. Under a strong light, examine the lens for dirt, fingerprints, and film which you may have missed. If these are difficult to remove, do the following:

a. Swab the surface with concentrated nitric acid solution and rinse with distilled water. Then reclean with alcohol and acetone.

b. If this procedure does not clean the optic, rub the surface with a damp piece of lens cloth dipped in precipitated chalk. Then clean with alcohol and acetone.

CAUTION: Rub just enough to remove dirt and/or stains, some of which may be in the reflection-reducing magnesium fluoride coating and cannot be removed by rubbing with chalk, for this would ruin the film.

6. If you are satisfied with your cleaning job, wrap the lens in clean lens tissue and put it back in a safe place where it will not become damaged.

Lens Cementing

You must recement a lens when the balsam between the elements deteriorates or breaks down. When you do this, proceed as follows:

1. Separate the cemented elements.
2. Remove old cement and clean the surfaces of the lenses.
3. Recement the elements together.
4. If available, use a lens-centering machine to align the optical axes of the lenses; if you do not have this machine, use V-BLOCKS.
5. Allow the parts adequate time to cool and then clean off excess cement.

The equipment you need for cementing lenses includes:

1. Electric hotplate with controlled heat (LOW, MEDIUM, AND HIGH).
2. Canada balsam or other types of approved lens cement.
3. Sheet asbestos (several pieces) and black paper (few sheets).
4. A rubber-tipped tool and a pair of tongs or brass tweezers for use on the particular optical elements on which you are working.
5. Glass bell jar (small). If you do not have this, use a small, cardboard box.
6. V-BLOCKS or a lens-centering machine.

NOTE: Do all lens cementing in a dust-, dirt-, and draft-free area, very clean, and have all the required tools in place for a particular job.

SEPARATING CEMENTED ELEMENTS.— Turn your electric stove on LOW and place a piece of 3/8'' asbestos on top, over which you now need a piece of the black paper. Put the lens on the paper and cover it with the bell jar or cardboard box. Then watch the black paper for signs of scorching, which shows that the stove is too hot and more asbestos is required over the hotplate.

When the lens is hot enough (between 275° F and 300° F), gently pry the elements of the lens apart with your rubber-tipped tool and allow them to cool slowly. When the temperature of the separated elements is approximately equal to that of the room, remove old balsam from them with alcohol, and then clean them thoroughly with acetone.

RECEMENTING OPTICAL ELEMENTS.— Put the clean lenses on the hotplate, with the surfaces to be cemented together facing upward. Inspect them for dust or dirt which may have fallen on them since they were cleaned, cover with the bell jar, and apply just enough heat to melt balsam.

When the elements are hot enough, put a little balsam on the surfaces to be joined together, pick up the positive element with your tweezers, and join the two cemented surfaces. Then use your rubber-tipped tool to work the top element over the lower one as much as necessary to squeeze OUT all air bubbles. The black paper on the heater makes air bubbles in the elements appear bright.

Use the lens-centering instrument (fig. 8-14) to center (align their optical axes) the elements. This instrument consists of an astronomical telescope with a crossline and a collimator telescope mounted on a tripod, with the objective lens of one instrument facing the objective lens of the other instrument. The crossline mount of the collimator telescope moves in a drawtube, which enables you to bring the image of its crossline into focus with the image of the astronomical telescope. A lens chuck mounted between the two telescopes can be rotated 360°, or more.

Heat the chuck jaws with a small torch or a hot piece of metal and then transfer the hot lens to the chuck. NOTE: Cold chuck jaws may crack one or both elements of the lens. Mount the hot, freshly cemented lens in the warm chuck, which grips ONLY the negative elements of the lens.

Sight through the eyepiece while you rotate the chuck and observe the eccentric movement of the lower crossline. Then move the upper element of the cemented lens over the lower one as necessary to have the crossline intersections coincide.

Allow the lens to cool for a few minutes in the machine and recheck the alignment, remove the asbestos sheet from the hotplate, and place the lens on the asbestos sheet. Then cover the lens with the bell jar (or box) and allow the lens adequate time for cooling. Remove the bell jar and scrape excess balsam from the edge of the lens with a razor blade, after which the lens is ready for final cleaning and inspection.

NOTE: If you do not have a lens-centering machine, use V-BLOCKS in the following manner to align the optical axes of a compound lens: Heat the V-BLOCKS on the hotplate while you are cementing the lens elements; and when you have the elements joined, slide the V-BLOCKS against the edges of the lens from opposite directions. Then turn off the hotplate, cover the lens and V-BLOCKS, and allow the combination to cool simultaneously. NOTE: Lenses whose edges are not concentric when aligned cannot be cemented in this manner.

You will occasionally find a lens doublet (generally from a gunsight, where it is used because it withstands the shock of gun fire) that will not separate when heated. If the elements of a compound lens do not separate at a temperature of 300° F, they were probably cemented together with a thermo-setting plastic, which a manufacturer sometimes uses for two reasons:

1. It resists temperature changes better than balsam.
2. It speeds up lens production.

When you have reason to believe that lens elements have been secured together with a thermo-setting plastic, check the lens with ultraviolet light for FLUORESCENCE. If the cement between the elements is a thermo-setting type, there will be little or no fluorescence; if the cement is balsam, you will see a definite, hazy-white fluorescence. When in doubt about the cement used in lenses, consult your supervisor.

Canada balsam is usually available in prepared form in metal tubes, through Navy supply channels. Use this lens cement on all lenses except very small or very large ones, which can be cemented together better with cements made by specific formulas, as explained next.

1. CEMENT FOR LARGE LENSES. Put three parts of rosin and one part of Canada balsam in a clean cup or bowl in a water bath at a temperature of 130° F. CAUTION: Keep the temperature of the water constant, as determined by a thermometer. Do not get any water in the cement. Stir the cement every 10 or 15 minutes, over a 2 1/2 hour period and then strain it through a piece of clean silk, after which you may use it.

2. CEMENT FOR VERY SMALL LENSES. Mix 4 parts of rcsin with 1 part of refined camphor and follow the procedure just described for large lenses to make the cement.

Most lenses with a diameter over 2 1/2 inches are not cemented together; they are air-spaced. The elements of the lenses are made of constituents with different coefficients of expansion which causes breakage of the cement during expansion and contraction. Some large lenses are also ground with different curvatures on their mating surfaces which make joining by cement impossible.

The reasons for joining the elements of a lens by cement are as follows:

1. Cementing keeps the elements optically aligned.

2. Cementing reduces the number of glass surfaces exposed to the air, which serves the same purpose as a film on optics, to make the image brighter and clearer. Since the index of refraction of Canada balsam is about the same as that of crown glass, there is practically no reflection when two crown glass surfaces are cemented together, and very little reflection when a crown glass surface is cemented to a flint glass surface.

3. Because a soft glass (hydroscopic) has special optical properties, a lens designer may sometimes desire to use it. This type of glass, however, is unstable and quickly deteriorates when used alone; but it can be used satisfactorily when cemented in place between two stable elements.

4. Groups of cemented lenses reduce the number of parts used in an optical instrument.

REASSEMBLY

Now that you have effected essential repairs to instrument parts, performed necessary refinishing, accomplished required cementing of optical elements, and cleaned everything perfectly clean, you are ready to begin the reassembly process.

If you have accomplished your repair and overhaul well, reassembly will be smooth and easy. Unless you know the instrument on which you are working very well, follow a reassembly sheet. Because reassembly is different for each instrument, no set procedure for accomplishing it can be given in this manual. The reassembly tips presented in the next few pages, however, will be helpful.

Reassemble lenses in their cells and tape the ends of the cells to exclude dirt; then proceed as follows: Hold the call so that one end is down, and use a light, fiber mallet (2 or 3 oz.) to tap lightly around the exterior of the cell. This action jars the dirt on the lens (if any) loose and causes it to drop and stick on the tape over the end of the cell. Use the same procedure on body tubes and castings.

After you tap a lens cell in order to jar specks of dirt from it and onto the tape in its end, partially remove the tape and clean the lens with a swab dampened with acetone. Repeat this process until you have the lens perfectly clean, and then keep the tape over the ends of the cell until you wish to reassemble the cell in the body tube or casting.

Use the following steps for reassembling a lens in its mount:

1. Carefully unwrap the clean optic and use the correct tool to reassemble it in the instrument. A lens chuck and cleaning holder (fig. 8-10) is a good tool at this time for cleaning. After you have the lens thoroughly cleaned, install it in its mount.

2. Tighten the retainer ring to seat the lens properly and clean off fingerprints (if any) and dirt. Use a silk floss or lens tissue swab dampened with acetone for cleaning a lens in its mount or cell. See illustration 8-15. To make certain the lens is actually clean, use the cleaning and tapping process described previously in this chapter until you are satisfied that the lens is thoroughly clean.

Some lenses must be sealed in their mounts, and the actual seal is provided by a string of

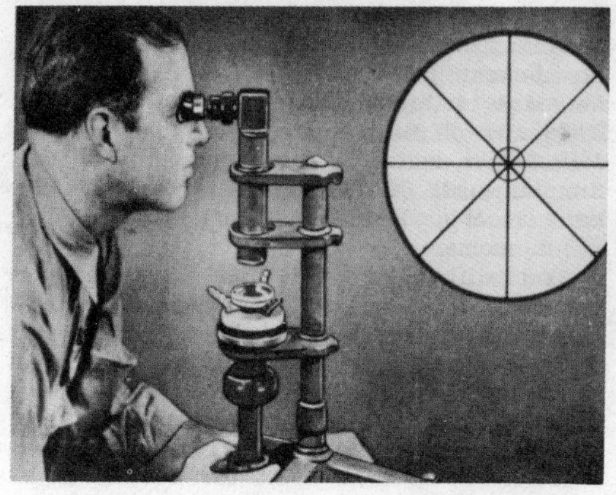

137.192
Figure 8-14.—Lens-centering instrument.

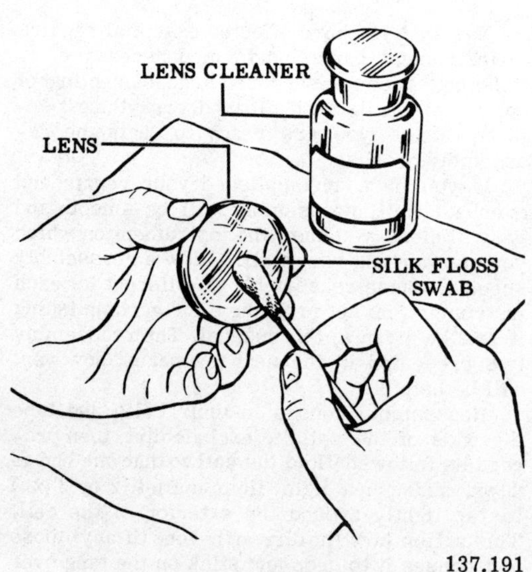

137.191
Figure 8-13.—Cleaning a lens with a silk-floss swab.

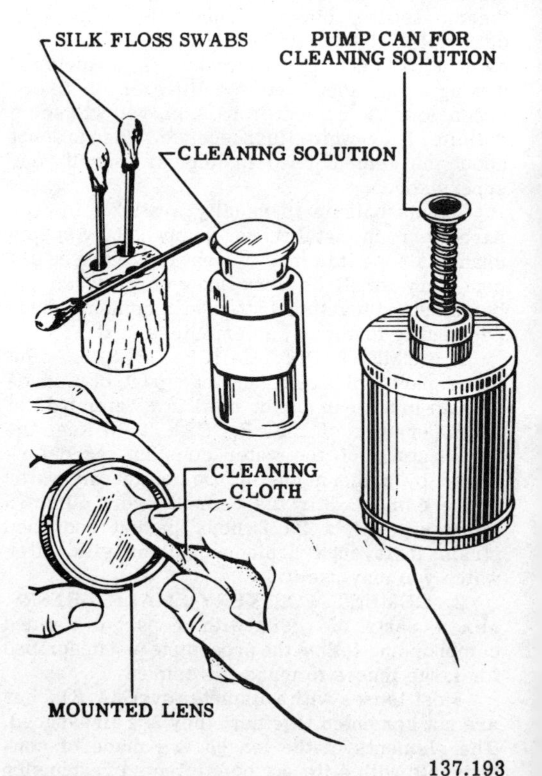

137.193
Figure 8-15.—Cleaning a lens in its mount.

wax about 1/16" in diameter in the form of sealing compound in a space between the lens and its mount, as illustrated in figure 8-16. The procedure for pressing the sealing compound into position in the mount is shown in illustration 8-17. Use a hardwood stick of correct size and shaped like a burnishing tool to press the sealing compound into the recesses provided for it in the mounts.

After you seat a lens in wax, remove the excess wax with a chisel-tipped hardwood stick, as shown in illustration 8-18. Then clean the lens very carefully (fig. 8-15).

Thin gaskets are used to seal lenses in some instruments, in which case you must use the same procedures and observe the same precautions required for sealing lenses in mounts with a compound.

Follow the method illustrated in figure 8-19 to place the lens in its mount and screw the retainer ring snugly against it. A small amount of heat applied carefully to the mount with a torch at this time helps to seat the lens properly. After you apply the heat, screw the retainer ring a bit tighter.

CAUTION: If you make the retainer ring too tight, you may crack the lens, or cause strain which will distort the image. Insufficient pressure, on the other hand, will eventually allow the lens to become loose. It is therefore important that you make certain a lens is actually tight because the retainer ring is snug against it and not because it appears tight only because the compound is holding it in position. If this is true, when the compound dries, the lens will be loose.

To ensure correct pressure on the lens by the retainer ring, tighten the ring snugly against the lens; then turn the retainer ring backward about 1/16 turn to release strain which may be present. Check for strain in the mounted lens by viewing all portions of it in a polariscope, shown in illustration 8-20.

After you assemble all lenses in their cells and mounts, assemble the prism clusters, or prism mounts (if any). Secure the prisms in their mounts by straps and/or collars, which must fit snugly enough to hold the prisms but not so tight that they may cause strain. A collar should fit over a prism with a slight press. If the fit is too tight, strain and breakage usually result; if the fit is too loose, the prism may shift its position and throw the instrument out of adjustment.

When you assemble a prism cluster used as an erector assembly, check the assembly for LEAN before you put it into the instrument. In a prism erecting system, LEAN results when the prisms are not oriented exactly 90° to each other. Illustration 8-21 shows how to correct lean with a prism-squaring fixture. Note that the repairman is adjusting the prism by loosening the prism collar and shifting it slightly. He can also detect LEAN and remove it by using the grids on a sheet of graft paper.

To remove lean from a prism cluster with a prism-squaring fixture, look at the grid through the prism cluster with one eye and at the same time look directly at the grid with the other eye. If there is lean in the prism cluster, the grids will not look parallel. This procedure for detecting LEAN in the cluster is not as easy as it sounds and takes practice in order to attain perfection.

If you assemble a pair of binoculars with LEAN in one or both barrels, the instrument will probably hurt the operators' eyes and will require disassembly and correction of the clusters for LEAN. So check for LEAN in prisms WITHOUT FAILURE before you assemble them in binoculars; and do not forget to check the prism clusters for strain after you assemble them and also prior to installation in the instrument.

When you assemble parts in an instrument, lubricate all moving parts. CAUTION: Use only the lubricant approved for that particular instrument.

As you assemble parts in an instrument, be sure to match all assembly marks; otherwise, you will be compelled to disassemble the instrument, make corrections, and reassemble it.

Check each part as you reassemble it for fragments of foreign matter clinging to it. Each part MUST BE IMMACULATELY CLEAN before you assemble it in the instrument. Keep openings to the interior of the instrument closed with masking tape and remove it only when you must make additional installations. Follow this procedure as you reassemble each part, until you make the final closure.

As you replace components and parts in an instrument, try to work from the top down, to prevent unnecessary work over an optical element, and perhaps damage to it.

Do not force a part into place in an optical instrument; use a light press with the fingers; unless the part must be fitted in position by force in accordance with specifications. If there is a bind, determine the cause.

You can make some adjustment on parts as you assemble them in an instrument. Whenever

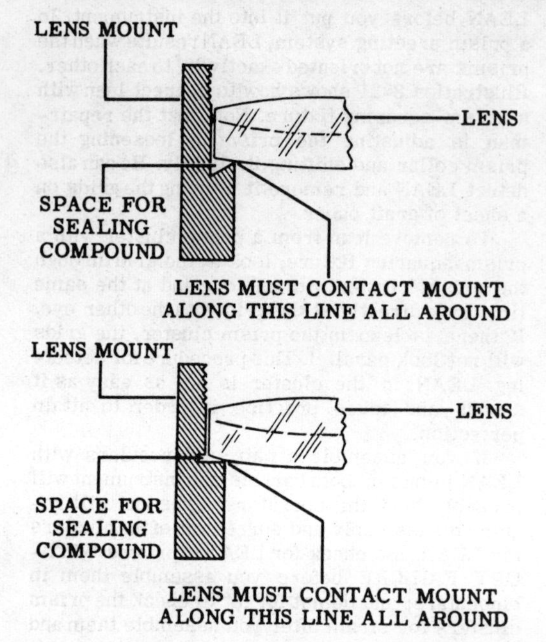

137.194
Figure 8-16.—Space for sealing compound in lens mounts.

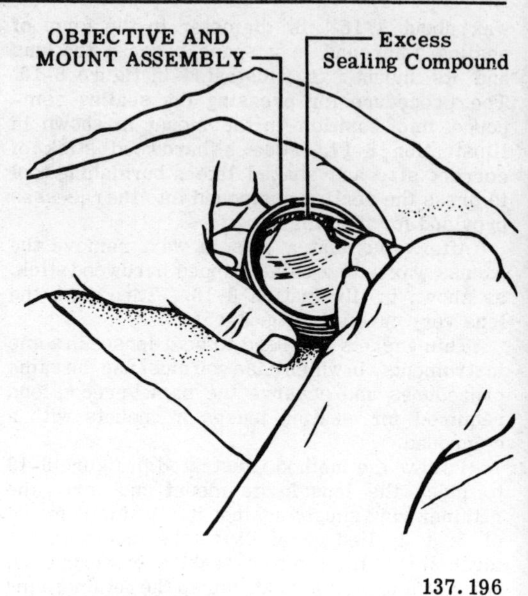

137.196
Figure 8-18.—Removing excess sealing compound from a lens mount.

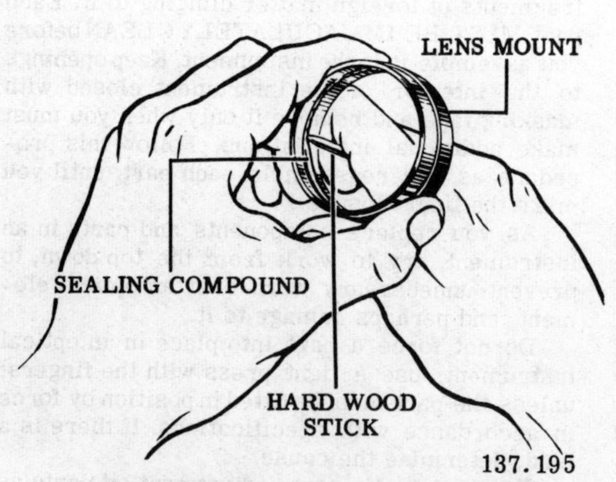

137.195
Figure 8-17.—Pressing sealing compound into the recess of a lens mount.

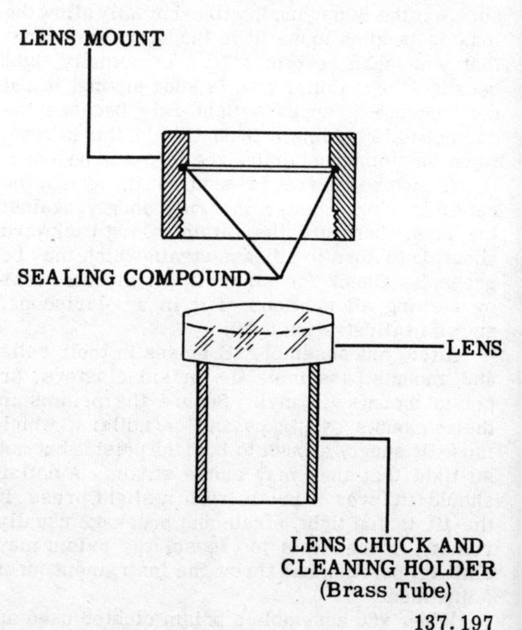

137.197
Figure 8-19.—Placing a lens in its mount.

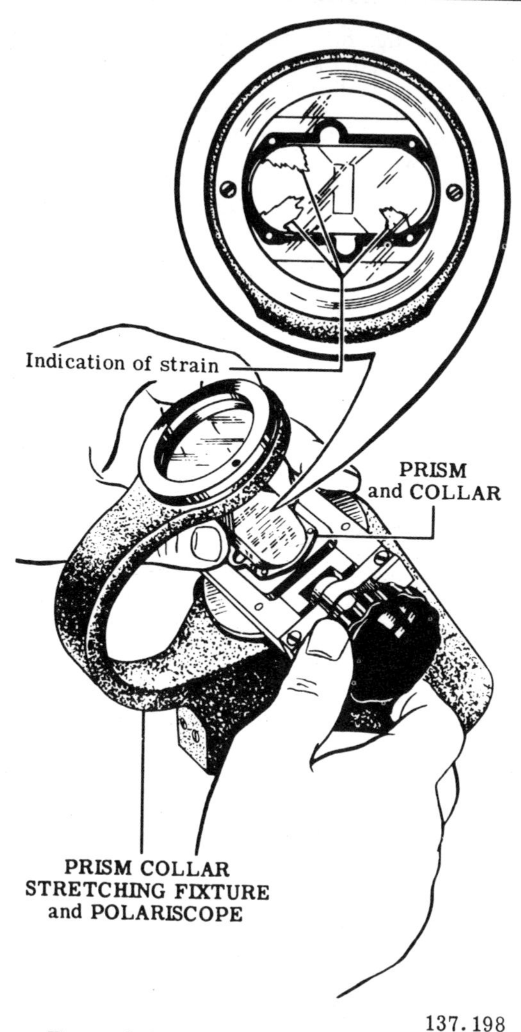

137.198
Figure 8-20.—Checking for strain in a mounted lens with a polariscope.

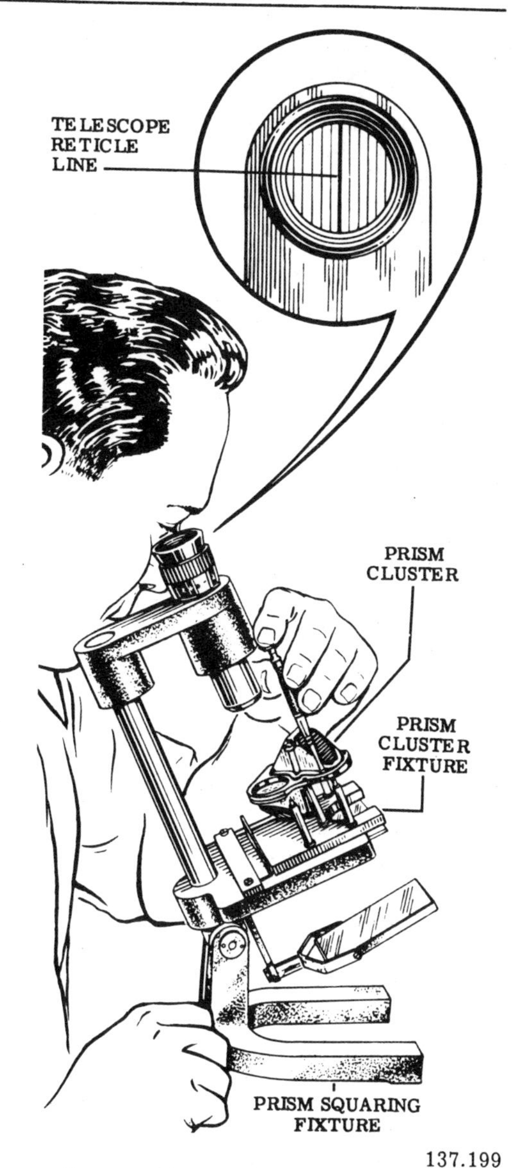

137.199
Figure 8-21.—Prism squaring fixture.

possible, these adjustments should be made during collimation; but in some instances an adjustment is impossible after reassembly because of inaccessibility of parts. The removal of LEAN in an erector prism cluster is a good example of an adjustment which must be made during assembly.

Threads on retaining rings, lens mounts, caps, screws, and setscrews are extremely fine and can be cross threaded easily. When you therefore insert them, turn in a counterclockwise direction until the threads snap into place, and then turn clockwise. NOTE: Always place a small amount of grease or anti-seize compound on threads before you turn them into place.

Seal the final cover to the interior of an optical instrument with sealing wax, gaskets, or packing. The function the instrument must serve determines the method for sealing it, and this is included in design specifications.

Seals used on optical instruments can be placed in one of the following types:

1. Moisture seal.
2. Gas-tight seal.
3. Pressure seal.

After you finish the reassembly process, seal all openings except those you must use when you collimate the instrument. Upon completion of the collimation process, final sealing, drying, and charging of the instrument must be accomplished, as you will learn next.

COLLIMATION

One of the final steps in overhaul and repair of an optical instrument is collimation, which is the ALIGNMENT OF THE OPTICAL AXIS OF THE INSTRUMENT to its mechanical axis. In simpler terms, orientation of all the axes of lenses in an optical system in such manner that they coincide with each other in a straight line and parallel to the mechanical axes of the bearing surfaces (telescope's mounting pads, for example) of the instrument is known as collimation.

Suppose you have an instrument constructed only of a straight tubular housing mounted in two ball bearings like a shaft. It contains no optics: it is only a straight, hollow tube. If you now peer through this tube and rotate it on its bearings like a rotating shaft, you will find that the least amount of rotation is in the center of the tube. THIS CENTER OF LEAST ROTATION IS THE MECHANICAL AXIS OF THE TUBE.

If you point the tube toward an infinity target, you can superimpose this mechanical axis on the object. Regardless of the direction in which you rotate the tube, its mechanical axis REMAINS SUPERIMPOSED on the same spot of the object.

Suppose that you now place in the tube the optical elements required to construct a telescope which will magnify the infinity target and then rotate the tube (telescope) again. If the target (now magnified) appears to revolve around in a circle in the same direction, the optical axes of the lenses are NOT ALIGNED with the mechanical axis of the tube; but you can so position the optical elements of the system (usually laterally) that the target will remain stationary when you rotate the tube. When this is true, you have the optical axes of the lenses aligned with the mechanical axis of the tube. The process you just completed, therefore, is collimation.

Collimation varies for different optical instruments; that is, the procedure for collimating one instrument may be exactly opposite that for collimating another instrument. Some instruments are also collimated on targets at a distance less than infinity (2,000 yards, or more); but most of them are collimated at infinity, because they are used to observe targets at infinity.

If you must collimate an optical instrument on an infinity target, you must have access to such a target every hour of the day and every day of the year, regardless of weather conditions. Because it is difficult or impossible to obtain and maintain an infinity target for long periods of time under ideal weather conditions, you must be able to produce and/or use a suitable artificial target at infinity. Such an infinity target can be produced by an instrument known as a collimator, which is discussed next.

COLLIMATORS

Collimators are precision instruments (with both optical and mechanical elements) which provide an infinity target suitable for use in aligning and adjusting the optical and mechanical components of optical instruments, so that they will perform accurately.

Although collimators may vary in design and/or construction, the optical principle employed in them is the same. Illustration 8-22 shows one type of collimator, but there are many different designs. Observe the nomenclature. This is Mk 4, Mod 0, telescope collimator used to collimate small telescopes, gunsights, and navigational instruments. It has a steel base several feet long with a precision, flat bearing surface machined on its entire top. A keyway is cut down the center of the bearing surface, as shown, for supporting fixtures.

This collimator telescope is secured to the bearing surface with a V-block support, with a highly precisioned bearing surface which slides (rides) on the bearing surface of the base. Other types of V-block supports may be part of the collimator base, and other designs may have keyways cut along the bottoms of their bearing surfaces. A key is then inserted half its thickness into the keyway of the V-block support, with the other half of its thickness in the keyway of the collimator base, to keep the V-block support and collimator telescope aligned parallel with the keyway in the collimator base.

The telescope of the collimator (with its bearing rings) can be secured or rotated on the bearing surfaces of the V's of the V-block support for making adjustments. This telescope consists of a tube with an achromatic doublet objective lens and a crossline reticle mounted internally in the principal focal plane of the lens. Located a short distance behind the reticle is a frosted-glass diffusing plate, and located behind the diffusing plate is either a plain reflecting mirror or a lamp, as illustrated in figure 8-23, which shows the optical principle of the collimator telescope.

When light from the lamp, or reflected light from the mirror, strikes the diffusing plate, the plate diffuses the light evenly over the entire crossline reticle. The reticle then becomes a new light source and emits diverging rays which are received and refracted parallel by the objective lens, as illustrated. If you were to look through the objective lens, the crossline would appear to be at infinity.

AUXILIARY FIXTURES AND EQUIPMENT

Auxiliary fixtures and equipment consist of special attachments, stands, supports, riggings, fixtures, and other optical instruments you must use with a collimator when you collimate various optical instruments.

An auxiliary fixture may be any piece equipment which can be attached to a collimator or its base, or a piece of mechanical or optical equipment, and used during the repair and collimation of an optical instrument. Auxiliary fixtures most generally used are: dynameters, collimator telescopes, checking telescopes, auxiliary telescopes, special support fixtures, and various other fixtures.

Some auxiliary support fixtures, or mounts, are illustrated in figure 8-24. These fixtures securely hold an optical instrument on the base of a collimator during collimation.

Auxiliary support fixtures differ from special support fixtures in that they are used in the collimation of a large number of optical instruments. Special support fixtures are used during the collimation of a limited number of instruments. In some instances, a special support fixture may be used to collimate only one specific instrument.

Checking Telescopes

A checking telescope (fig. 8-25), or a dummy telescope, is a relatively small standard or master instrument used to align collimator components and instrument support fixtures. Design of the telescope varies in accordance with its use, but a checking telescope generally consists of an astronomical telescope with a crossline reticle. Study the nomenclature of the telescope shown in figure 8-23 carefully, noting particularly the position of the optical elements.

Checking telescopes are generally used to collimate collimators employed on many different optical instruments, but one may be designed to collimate a collimater used only on one instrument.

A checking telescope is a master instrument whose delicate components must receive the best care. NOTE: Never attempt to repair a checking telescope. Only its manufacturer has the equipment required to repair it satisfactorily.

Auxiliary Telescopes

An auxiliary telescope is probably used more than any other auxiliary fixture in instrument collimation. It is an astronomical telescope with a Kellner eyepiece, and its main purpose is to compensate for inherent eye errors of a person who is collimating an instrument. Study part A of figure 8-26. Observe the position of all components and their nomenclature.

If individuals who work on collimation of optical instruments have normal vision, no near- or farsightedness, an auxiliary telescope may not be required for doing some phases of collimation. Because most persons have some sort of eye defects, however, an auxiliary telescope must be used to determine what dioptric errors they have in their eyes before they collimate an instrument.

You can determine what the dioptric settings of your eyes are by focusing (from plus to minus on the diopter scale) the auxiliary telescope on an infinity target, or on a collimator telescope crossline until the image is sharply defined. For best results, take five readings and use the reading which appears most during the readings. This is the MEAN.

After you get this dioptric setting, do NOT change the focus until you decide to check your setting again for eye fatigue or strain.

COLLIMATOR TELESCOPE
KEYWAY FOR SUPPORT FIXTURES
SUPPORT FIXTURE
COLLIMATOR BASE

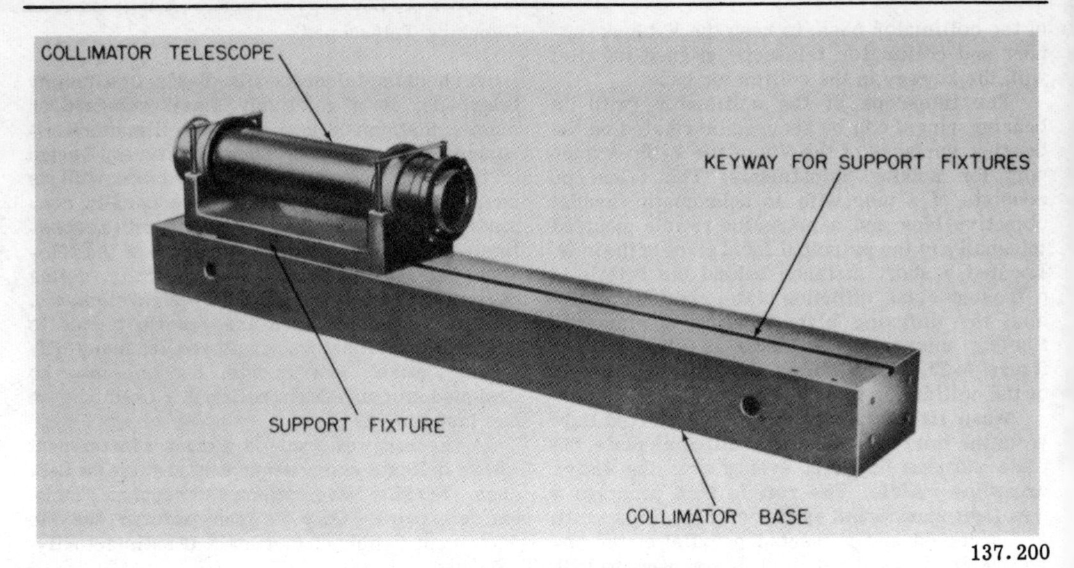

137.200

Figure 8-22.—Telescope collimator.

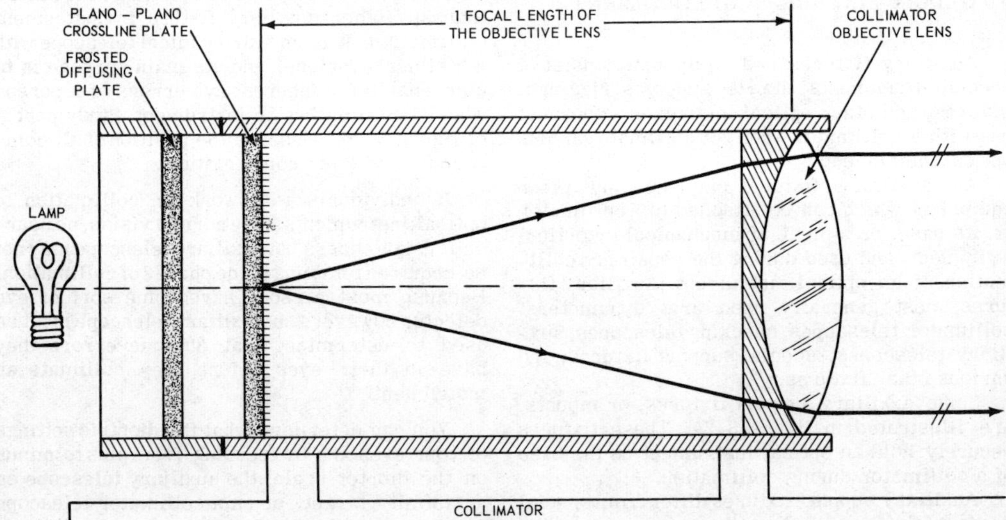

PLANO - PLANO CROSSLINE PLATE
FROSTED DIFFUSING PLATE
1 FOCAL LENGTH OF THE OBJECTIVE LENS
COLLIMATOR OBJECTIVE LENS
LAMP
COLLIMATOR BASE

137.201

Figure 8-23.—Principle of operation of a simple collimator telescope.

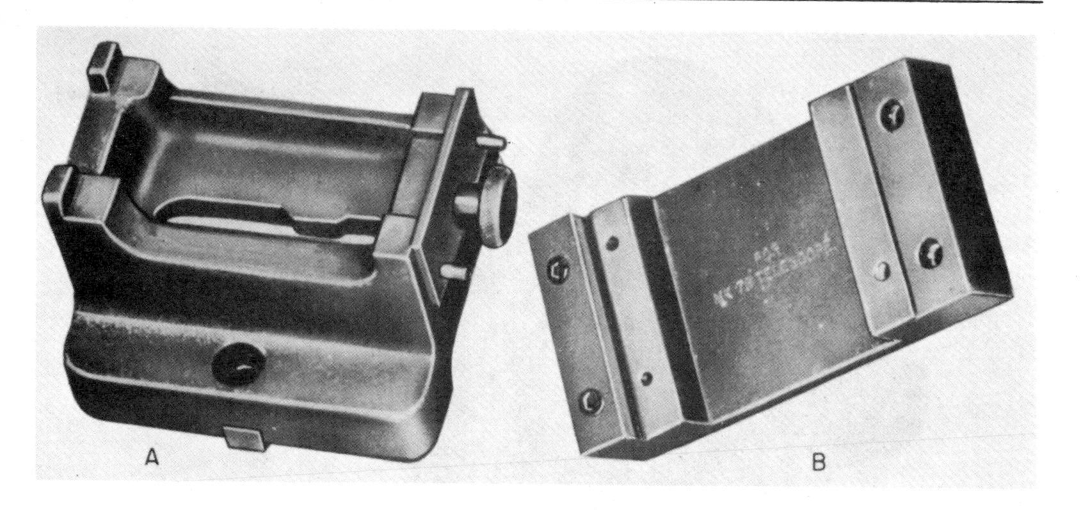

137.202

Figure 8-24.—Auxiliary support fixtures.

You can also use an auxiliary telescope for the following collimating operations:

1. Setting focusing eyepieces to the NORMAL or ZERO diopter setting.
2. Setting fixed-type eyepieces to their required diopter setting.
3. Checking for and aiding in the removal of parallax in an instrument.
4. Increasing magnification of another instrument, by placing the auxiliary telescope to the eyepiece of the instrument. Increase in magnification of the instrument is equal to the combination of the powers of the two telescopes, obtained by multiplying the power of the auxiliary telescope by the power of the instrument (3X x 10X = 30, for example).
5. Collimating hand-held binoculars by means of an auxiliary telescope rhomboid prism attachment, as shown in part B of figure 8-26.

COLLIMATION PROCEDURE

Before you collimate an optical instrument, you must first collimate the collimator; that is, you must adjust and align the optical and mechanical components of the collimator as necessary to have it conform with the specifications of the optical system of the instrument to be collimated.

Collimation of a collimator telescope generally consists of adjusting:

1. The collimator telescope mechanically, so that its optical axis is parallel to the bearing surfaces of the collimator's base and the instrument's auxiliary support fixture bearings.
2. The collimator telescope's crossline, so that its vertical wire is perpendicular to the bearing surface of the collimator base. This is called squaring the collimator crossline.

Collimation of collimators varies in accordance with the design of each collimator and for each instrument to be collimated, and no attempt is made here to establish specific standards or procedures for collimating a collimator. The most common practice is to use the following fixtures (fig. 8-27): auxiliary eyepiece, machinist's square, checking telescope, auxiliary objective lens, and the master instrument.

Auxiliary pieces are most commonly used on collimators when the collimating telescope is not designed for horizontal and vertical adjustment. The collimating telescope is permanently aligned on the V-block support in such manner that its optical axis is parallel to the bearing surface of the collimator's base. The Mk 4, Mod 0, telescope illustrated in figure 8-25 is this type of collimator.

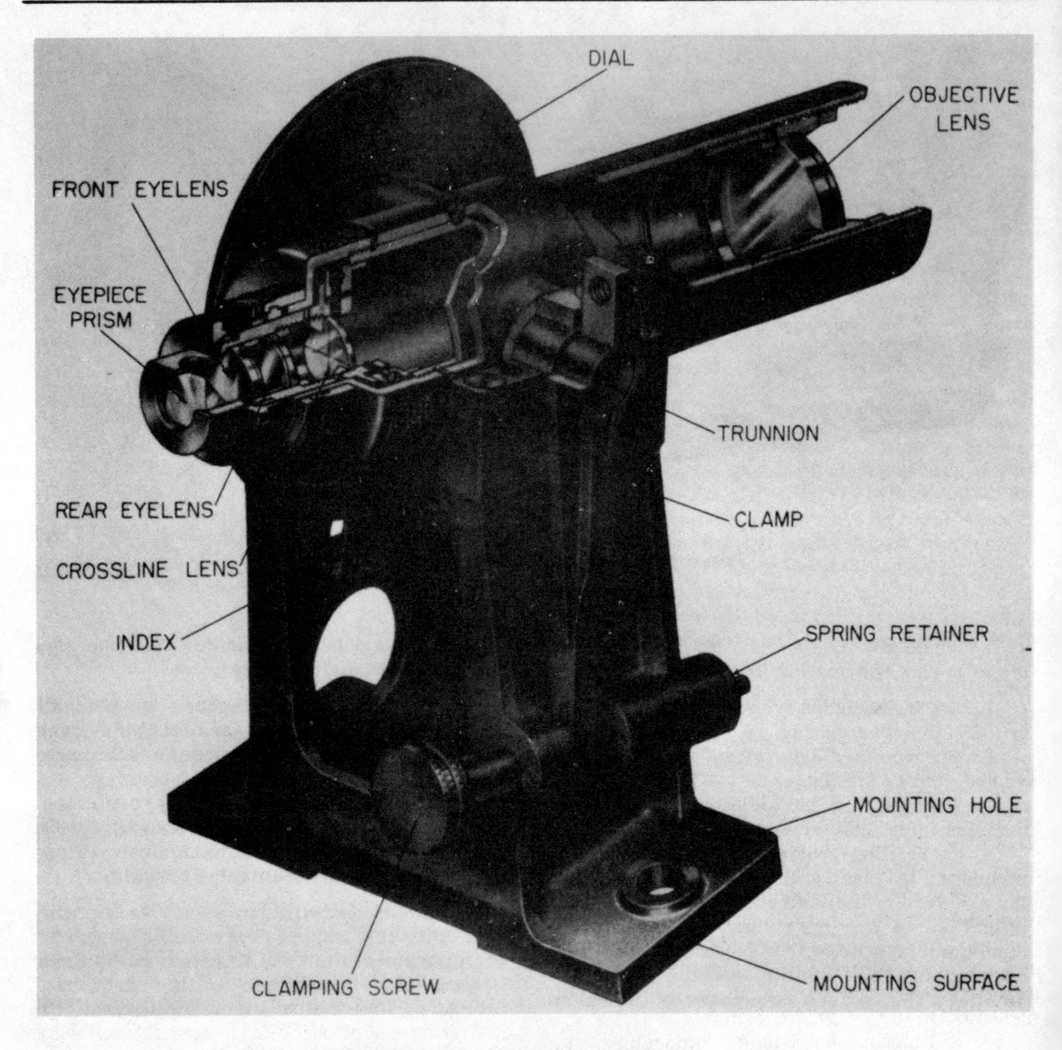

Figure 8-25.—Checking telescope.

The procedure for collimating a collimator with auxiliary pieces is as follows:

1. Level the collimator base with its adjusting screws on the legs.

2. Place the auxiliary objective lens and the eyepiece lens in the V-blocks on top of the collimator bearing surface. NOTE: Any two lenses may be used, but the objective must have a longer focal length than the eyepiece lens. You may place the auxiliary objective lens in front of the collimator objective at a reasonable distance, as desired. Place the eyepiece lens behind the auxiliary objective lens at a distance equal to the sum of their focal lengths in order to construct an astronomical telescope. If you now look through the astronomical telescope you see a magnified image of the collimator's telescope crossline.

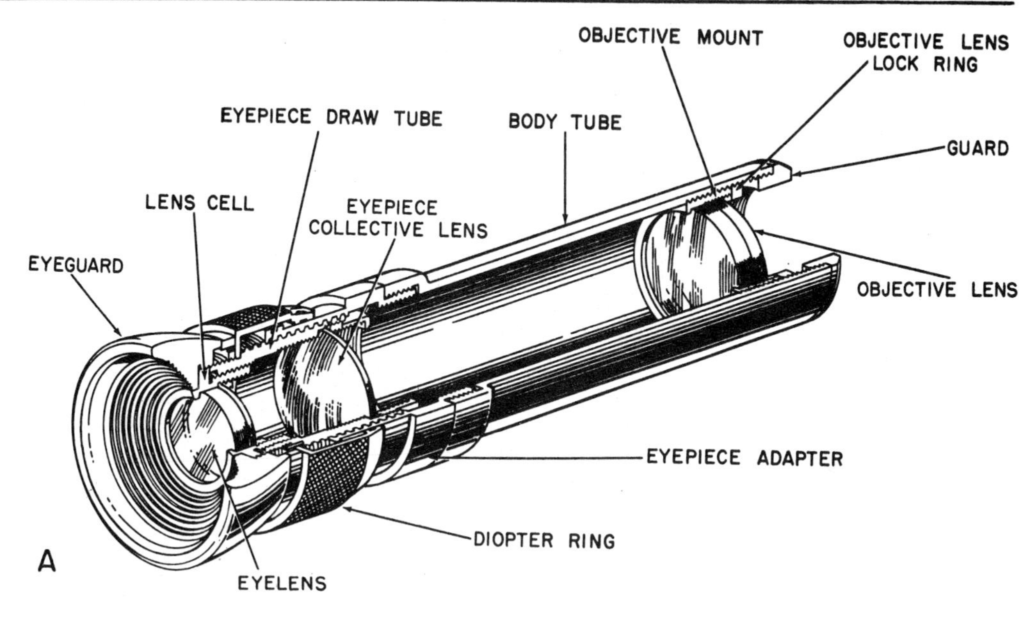

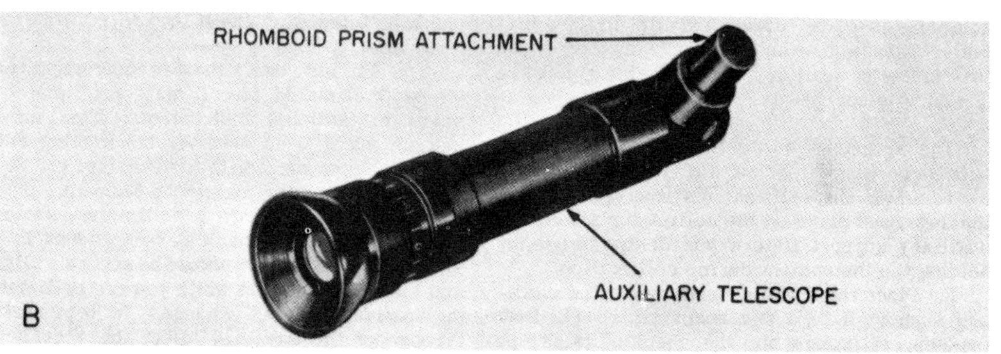

137.205

Figure 8-26.—Auxiliary telescopes.

3. Place the machinist's square on the bearing surface of the collimator base, with its straight edge perpendicular and in the focal plane of the auxiliary objective lens. If you now look through the astronomical telescope you see sharply defined both the crossline and the machinist's square's straight edge.

4. To square the collimator in such manner that the vertical wire of the collimating telescope is perpendicular to the bearing surface of the collimator's base, rotate the collimating telescope until the vertical wire is parallel to the straight edge of the machinist's square. This step should complete the collimating process for the collimator.

For collimators with adjustable collimating telescopes (horizontal and vertical adjustments), use the procedure just described only for squaring the collimating telescope. If the collimator,

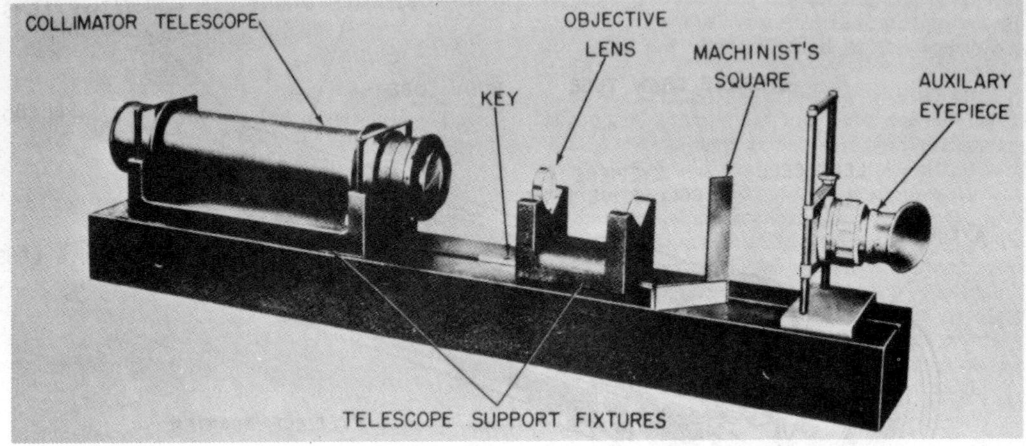

137.206

Figure 8-27.—Squaring collimator and auxiliary pieces.

however, is to be aligned parallel to the collimator's base or the bearing surfaces on an instrument's auxiliary support fixture, use a checking telescope.

The procedure for collimating a collimator with a checking telescope follows:

1. Level the collimator's base by adjusting the legs, and place on the collimator's base the auxiliary support fixture you desire to use for holding the instrument during collimation.

2. Place the checking telescope in the auxiliary support fixture and secure it. NOTE: Both the support fixture and the checking telescope must be placed flat and square in their positions, to ensure proper alignment of the collimator.

3. Peer through the checking telescope and focus it on the collimating telescope's crossline. If this crossline is not square with the crossline of the checking telescope, rotate the collimating telescope until its crossline vertical wire is parallel to the vertical wire in the checking telescope.

4. Superimpose the crossline of the collimating telescope on the crossline of the checking telescope by adjusting the screws under the objective lens of the collimating telescope, to move the collimating telescope horizontally or vertically, as desired. Both crosslines should now appear as one when you look at them through the checking telescope; and if they do, collimation is completed.

You can also use a master instrument, previously collimated, to collimate a collimator for small navigational instruments. This master instrument is used only for collimating collimators. Adjustment on the collimator must be the same as that on the master instrument.

After you collimate a collimator, securely lock all of its mechanical components in position. The collimator should be securely aligned and locked in position while you are collimating an instrument, but changes in temperature throughout the day may affect the accuracy of collimation of the collimator, because its mechanical parts expand and contract in accordance with changes in temperature. For this reason, NEVER assume that a collimator is collimated; check its alignment frequently to make certain that it is collimated.

The procedure for collimating optical instruments varies with different instruments; and for this reason, collimation procedures for a specific type of optical instrument are not listed in this manual. The collimation procedures considered here are general in nature and applicable to all optical instruments. For information relative to collimating procedures for a specific instrument, refer to applicable publications and/or blueprints.

The general steps in collimation of an optical instrument are:

1. Collimate the collimator on the proper telescope support fixture.

2. Put the telescope to be collimated on its support fixture and adjust the auxiliary telescope to your eye correction.

You are now ready to remove parallax, square and superimpose the instrument's crossline, and set the eyepiece diopter setting. The things you generally do to collimate an instrument are explained next.

NOTE: When you use an auxiliary telescope during collimation, do not change the eye correction after you set it properly. When you focus the eyepiece of an instrument, focus from PLUS to MINUS on the diopter scale.

Removal of Parallax

As defined earlier in previous chapters of this manual, parallax is a condition brought about when the reticle of an instrument does not lie in the same plane as one of the image planes, usually the image plane of the objective lens.

To check for parallax in an instrument, place an auxiliary telescope to the eyepiece of the instrument, sight through both, and focus the eyepiece of the instrument until the image of the collimator crossline or the crossline of the instrument (whichever comes into view first) is sharply defined. If parallax is present, one of the two crosslines will come into focus first; if there is no parallax, both crosslines will come into focus at the same time.

The amount of parallax between the two crosslines can be measured in diopters on the diopter scale of the instrument's eyepiece. You can determine the amount of parallax by focusing the eyepiece of the instrument in until the first crossline is sharply defined and by observing the diopter reading to which the index marker points. Then continue to focus until the other crossline is sharply defined and observe where the index mark is pointing on the diopter scale, and also note the number of diopters between the position of clarity of the first crossline and the point of clarity of the second crossline. If the instrument's crossline, for example, came into focus at plus two diopters on the diopter scale and the collimator's crossline came into focus at minus 3 diopters on the diopter scale, the total amount of parallax is 5 diopters.

By knowing which crossline came into focus first, we know the location of the instrument's crossline in relation to the focal plane (image plane) of the objective lens. If the instrument's crossline comes into focus first before the collimator's crossline, the instrument's crossline is farther from the objective lens than its focal plane (image plane). If the collimator's crossline comes into focus before the instrument's crossline, the instrument's crossline is closer to the objective lens than its focal plane.

The problem in collimation now is to place the instrument's crossline in the focal plane of the objective lens, in one of two ways:

1. Move the instrument's crossline forward or aft axially until it is in the focal plane of the objective lens.

2. Move the objective lens until its focal plane is in the same plane as the instrument's crossline. This method is preferred for placing the crossline of an instrument in the focal plane of its objective lens. The objective lens is mounted in an externally threaded mount which can be moved axially along the interior of the instrument. When the objective lens mount is moved any amount, the focal plane and image of the collimator's crossline in the focal plane move in the same direction and the same amount as the objective lens.

In some instruments, spacers or separators are placed in front and at the rear of the objective lens mount (not threaded externally) to allow for axial positioning of the mounts in order to remove parallax.

Removal of parallax by axial adjustment of the instrument's crossline is not preferred over axial adjustment of the objective lens, however, because a position of the telescope's body must usually be disassembled in order to reach the crossline. Instruments which provide for adjustment of the crossline have it mounted in an externally threaded mount which can be adjusted by screwing forward or backward. Some optical instruments also provide for adjusting both the objective lens and the crossline.

After you completely remove parallax from an instrument, both crosslines must come into focus at the same time on the same diopter reading on the diopter scale of the instrument.

NOTE: There is NO tolerance for parallax in any optical instrument.

Squaring and Superimposing the Crossline

You can square and superimpose the crossline in the following manner:

1. Square the crosslines of the instruments; that is, have the vertical line of one parallel

with the vertical line of the other. You can do this by rotating the crossline of the instrument in its mount with a cotton swab or a soft, rubber-tipped eraser—NOT THE FINGERS. You must do this carefully in order to prevent scratches on the glass surface of the crossline, for these defects appear greatly magnified when superimposed on the target.

2. When you have the crossline positioned correctly, tighten its retainer ring. If you find that the crossline rotates with the retainer ring when you tighten it, so position the crossline that it will rotate into correct position (squared) when you tighten its retainer ring.

3. Superimpose the instrument's crossline with the crossline of the collimator, so that both crosslines appear as one when you look at them through the instrument. You can do this in several ways, but the method generally used is to rotate the objective lens' eccentric mount and ring (if provided). When you rotate the mount and ring, or each singly, the optical axis of the objective lens moves laterally and causes the image of the collimator's crossline to move in the same direction and in a circle. So manipulate the eccentric mount and ring that you superimpose the collimator's crossline image on the crossline of the instrument.

In some objective lenses, the optical centers are slightly different from the geometrical centers, which means that you can rotate the objective lens in its mount and give the same effect you get by using an eccentric mount and ring.

Another method for superimposing the crossline is to adjust the crossline and its mount laterally with a screw adjustment mount (described in chapter 7). Adjust the screws as necessary to push the crossline of the instrument horizontally and vertically and superimpose it over the image of the collimator's crossline.

You can superimpose the crossline of instruments containing a prism erecting system by adjusting the erecting prism. An excellent example of this is an Amici prism in a gunsight telescope. The prism is positioned at a definite point between the objective lens and its focal plane; and movement of the prism causes the optical axis and focal plane to move in the desired direction until the image of the collimator's crossline is superimposed with the crossline of the instrument.

Optical Adjustment of the Diopter Setting

The diopter setting of eyepieces varies in accordance with type, focusing or fixed-type, each of which is set to different optical values.

A focusing-type eyepiece is set to a value called O DIOPTERS, which can be accomplished when an infinity target (collimator crossline) image is defined sharply, with parallel rays of light emerging from the rear eyelens, and with the index mark pointing to 0 diopters on the diopter scale of the instrument's eyepiece. NOTE: Parallel rays of light have 0 dioptric value.

You will recall that when you removed parallax from an instrument the images of the crosslines of the instrument and the collimator came into focus with the same reading on the diopter scale, regardless of the reading (plus or minus). If one image, for example, comes into focus at -4 diopters on the diopter scale, the other image must do likewise. This means that the eyepiece focusing mechanism must be focused in from its mid-throw (mechanical 0 diopters) position to allow the images to coincide with the principal focal plane of the eyepiece. (When images or objects are in the principal focal plane of any lens, the rays which leave the images diverge, enter the lens, are refracted, and emerge parallel.) The problem, then, is to move the images OUT toward the observer to the mid-throw position, so that you do not need to focus the draw tube of the eyepiece IN past its mid-throw position.

A condition exactly opposite to that just described may also exist; that is, if both images come into focus on the plus side of 0 diopters, your problem is to move the images IN to the mid-throw position so that you need not move the draw tube OUT in order to have the principal focal plane coincide with the images.

The procedure for moving these images together simultaneously depends upon the type of erecting system in the instrument; that is, a single erector or a two erector (lens) system. The rule to follow for moving a single erector lens is as follows: If the images come into focus on the MINUS side of the 0 diopter graduation on the diopter scale, move the erector lens AWAY from the eyepiece; if the images come into focus on the PLUS side of the 0 diopter graduation, move the erector lens TOWARD the eyepiece. Review the discussion on construction of telescopes in chapter 7 of this manual.

It may appear that the opposite effect occurs to the movement of the image when you move the erector lens; but if you remember the optical theory involved here, you know that the images move in the opposite direction to the movement of the single erector lens. Move the erector lens in the desired direction until the images are in focus, with the index mark pointing to 0 diopters on the eyepiece diopter scale. NOTE: You need an auxiliary telescope for setting the 0 diopter on any telescope eyepiece.

Give the diopter setting a final check by placing the auxiliary telescope to the eyepiece of the instrument and then by focusing from a PLUS to a MINUS position until you have the images sharply defined. The index mark must point to 0 diopters on the diopter scale within a quarter of a diopter tolerance.

To set a two-erector lens erecting system to 0 diopters, move the second lens in the system in the direction in which the images must be moved. The theory involved here is this: The light rays which enter the second erector lens are parallel and the images formed by the lens are in the focal plane of the second erector lens. The images in the focal plane therefore always move in the same direction as the lens. When you have the second erector lens properly positioned, when focused on the images, the eyepiece comes into focus with the index mark pointing to 0 diopters on the diopter scale.

The required diopter setting for a fixed-type eyepiece must be explained only for collimation of a fixed-prism gunsight (Mk 77 & Mk 79, for example), because the mechanical construction of this telescope must be known before you set the diopter setting.

SEALING, DRYING, AND CHARGING

After you collimate an optical instrument, the last step in the repair process involves sealing, drying, and charging, which is discussed next.

Methods used for sealing, drying, and charging differ for the various types, designated for this purpose as: (1) moisture-tight, (2) gas-tight, and (3) pressure-tight.

Optical instruments which are held by hand, or not permanently mounted on a ship's weather decks, must be moisture-tight. These instruments always have focusing-type eyepieces and are sealed:

1. Against the entrance of moisture, and humidity.
2. With black or green wax, or gaskets.
3. From the objective lens to the sealing windows (if provided), or to the rear eyelenses.
4. At atmospheric pressure, with or without gas.

A gas-tight optical instrument is generally mounted on a weather deck and is constantly subjected to the weather. It contains either a focusing or a fixed-type eyepiece, and it is used ONLY on surface ships.

A gas-tight optical instrument should be sealed:

1. Against the entrance of moisture and water.
2. With gaskets and packing only.
3. Between the objective lens (or window) and the sealing plate or crossline (which acts as a sealing window in some instruments) of an instrument with a focusing-type eyepiece.
4. From the objective lens or window to the rear eyelens of an instrument with fixed-type eyepiece.

A pressure-tight optical instrument is mounted on sub-surface craft and must be able to withstand the force of external water pressure and it must be sealed:

1. Against the entrance of high (hydrostatic) water pressure.
2. With gaskets and packing only.
3. From the objective window to the rear eyelens.

The primary purpose of sealing, drying, and charging an optical instrument with gas is to prevent moisture from getting into the instrument and condensing on parts, thereby inflicting damage to them.

A gas-tight instrument may be charged with dry nitrogen or dry helium. A pressure-tight instrument should be charged with dry nitrogen ONLY. Dry nitrogen and dry helium are used to charge instruments because they contain no moisture or oxygen; whereas, dry air contains about 20% oxygen and must NEVER be used as a final charging agent for an optical instrument.

Gas used to charge optical instruments is normally not completely free of moisture and foreign matter and must therefore be cleaned before you use it. This you can do by forcing the gas through an optical instrument dryer, which is actually a gas dryer containing a quantity of silica gel to absorb moisture from the gas as it passes through. See illustration 8-28. The silical gel used on instruments must be impregnated with cobalt chloride, which serves as a moisture indicator. When the silica gel is completely dry, it is deep blue in color. When the silica gel is saturated with 30% of water, its color is lavender; and when it contains 50% of moisture, its

137.207
Figure 8-28.—Optical instrument dryer.

color is pale pink. At a saturation of almost 100% with moisture, silica gel is decidedly pink in color.

A window on the side of the cylinder enables you to observe the color of the silica gel; and when it changes to pink, remove it from the cylinder, place it in a container, and bake it in an oven at a temperature of 300° F to 350° F for a minimum of 4 hours, after which its color should be a deep blue.

All optical instruments except moisture-tight types are equipped with gas inlet and outlet valves, also called plugs; and on most instruments they are located on opposite ends of the instruments. As the gas enters through the inlet valve and circulates throughout the instrument, it becomes saturated with moisture in the instrument and carrier it out through the outlet valve.

After you overhaul a gas-tight instrument, check its gaskets, fittings, and so forth, for air tightness. Use dry air to check for leaks in an instrument, in order to conserve valuable nitrogen or helium. Proceed as follows:

1. Connect a hose to your air supply (oil, moisture, and grease free) outlet valve and connect the other end of the hose to the inlet valve of the instrument dryer.

2. Connect another hose to the outlet valve of the instrument dryer. (This is the hose which you later will connect to the instrument to be gassed.) The free end is brass and male-threaded.

3. Turn on the air to blow dust, moisture, or foreign matter from the hoses.

4. Connect the hose from the outlet valve of the dryer to the inlet valve screw fitting (the small screw by the male-threaded end) on the optical instrument.

5. Open the gas inlet screw (large screw) on the inlet valve.

6. Tighten the gas outlet valve screw on the opposite end of the optical instrument.

7. Turn on the air supply until the pressure gage on the instrument dryer reads approximately five (5) pounds per square inch.

8. While you maintain this pressure, use a liquid soap solution to test for leaks around all fittings, gaskets, screws, the objective window, and the rear eyelens.

9. If you find leaks, mark them with a soft lead pencil, white crayon, or chalk, turn off the air supply, disconnect the hose from the instrument, and then repair the leak(s).

10. After you repair leaks, connect the hose to the instrument and apply the same pressure test and check again for leaks with soap suds.

11. After the instrument passes the soap suds test, maintain the 5 pounds of pressure and close the gas valve screw on the inlet valve.

If the instrument you are working on has a fixed-type or an internal-focusing eyepiece, continue with the following tank test: (NOTE: You can also submerge an instrument with an external-focusing eyepiece, but ONLY up to the eyepiece.)

1. Submerge the instrument in a tank of water.

2. Check for slow rising bubbles which may appear anywhere on the instrument. A few hours may elapse before any bubbles are visible.

3. Mark the leak(s) as soon as you remove the instrument from the tank, and then repair them. Follow up by submerging the instrument in the tank again and make a double check for leaks.

When you are certain there are no leaks in the instrument, remove it from the tank and dry its exterior with a clean, soft cloth. Then recharge it to exactly 5 pounds. Twenty-four hours later, attach a pressure gage to the gas inlet valve of the instrument and check its pressure. If it has dropped, repeat either the soapsuds test or the tank test as often as necessary until you find the leak(s). Then make necessary repairs and dry the instrument.

You now have the instrument ready for charging with nitrogen, which you can do in the following manner:

1. Connect one end of a hose to the outlet valve of a nitrogen bottle and the other end of the hose to the inlet valve of the dryer.
2. Repeat at this time the steps (2 through 5) you used to test for leaks.
3. Remove the screw from the gas outlet valve of the optical instrument.
4. Turn on the nitrogen gas and let it cycle through the entire instrument.
5. Purge the instrument by holding a finger over its outlet valve. When the gage on the dryer shows a pressure up to but not exceeding five pounds, remove your finger from the outlet valve and allow the gas to escape from the instrument. At about five minute intervals during a period of approximately one half hour, repeat the purging operation.
6. When you have the instrument purged (completely free of moisture), replace the outlet valve screw and let the pressure on the dryer build up to approximately two pounds, or as indicated in the overhaul manual for the instrument.
7. When the pressure reaches the specific amount, close and secure the gas valve screw (large one) on the gas inlet valve and disconnect the hose from the optical instrument. Then turn off the nitrogen bottle and replace the small, inlet valve screw.

Some moisture-tight instruments have inlet and outlet screws (not valves) which can be used for drying the instrument only. When you seal a moisture-tight instrument, test it for leaks and dry it; then replace the inlet and outlet screws.

Pressure-tight instruments must withstand a special testing procedure, so check with your instructor or shop supervisor for the instructions and specifications applicable to a particular pressure-tight instrument.

Charge all gas- and pressure-tight instruments with gases and pressures specified for them, at the times stated next:

1. Prior to the conclusion of each ship's overhaul by a tender.
2. When inspection indicates condensation on internal optical surfaces.
3. Immediately after completion of an overhaul of an optical instrument.
4. At the completion of twelve months of service.

Some general rules to follow when you recharge an optical instrument are:

1. NEVER recharge an optical instrument when the temperature is below 32° F.
2. NEVER charge an instrument with nitrogen, or helium after the pressure in the bottle or tank falls below 400 pounds per square inch.
   NOTE: If there is a trace of moisture, oil or grease in the bottle, it starts to come out when the pressure falls below 400 pounds.
3. Recharge each instrument with only the type of gas and pressure specified for that particular instrument. If in doubt this, use nitrogen, and pressurize the instrument to two (2) pounds.
4. When the inlet valve or the area near it is painted ORANGE or YELLOW, always charge the instrument with HELIUM. CAUTION: NEVER use nitrogen. Follow recommended and/or Navy approved instructions for charging an instrument with helium.

BLUEPRINTS

Mechanical drawings of all optical instruments manufactured in accordance with Navy specifications will be available in the repair department of your ship in the form of blueprints or photoprints. They are assembled in sets, and each set covers one item. Photoprints are usually bound in books.

Blueprints of all optical instruments on which you will perform work show the dimensions of each part of the instrument and the material which composes it. They also show the tools to use for disassembling and assembling an instrument, give the assembly procedure, and provide lubrication charts.

NOTE: In the lower, right-hand corner of each blueprint is a number, known as THE DRAWING NUMBER. On each detail pictured in the drawing is a separate, smaller number, known as THE PIECE NUMBER. Occasionally, a letter follows the piece number, and it shows the number of times the original design of that piece has been changed or modified. The first number is the DRAWING NUMBER; the second number is the PIECE NUMBER. The number 240962-2, for example, should be read: Drawing Number 240962, Piece Number 2. Use these numbers when you need new parts, and when you make a report on a particular piece.

SAFETY PRECAUTIONS

You have perhaps learned a great deal about safety precautions in basic naval training courses you studied previously; and they need little or no repetition here UNLESS THEY ARE ESPECIALLY APPLICABLE TO OPTICALMEN. The safety precautions listed and discussed in the next section belong in the category of those important in a particular way to Opticalmen.

HANDLING CHEMICALS

Study the following rules applicable when you work with all kinds of chemicals. If you know them well, you may on occasions be able to prevent extensive harm and/or damage to your body as a result of their contact with it; you may, in fact, even be able to save your life by the knowledge you have about chemicals.

1. DIRECTIONS FOR USE.—Study the directions on the container for using a specific chemical. If you mix chemicals improperly, or in incorrect proportions, they WILL NOT WORK and they MAY BE DANGEROUS. Such mixtures sometimes explode and cause much harm and damage.

CAUTION: NEVER MIX CHEMICALS AT RANDOM, OR PLAYFULLY, JUST TO FIND OUT WHAT HAPPENS. IF YOU DO THIS, YOU MAY NEVER LIVE LONG ENOUGH TO FIND THE ANSWER TO YOUR CURIOSITY.

2. LABELS.—Keep labels on containers and bottles of chemical intact. If you notice that a label is coming loose, glue it back in place. Then coat the label WITH PARAFFIN WAX TO PROTECT IT.

CAUTION: NEVER USE A CHEMICAL FROM AN UNLABELED CONTAINER—GET RID OF IT IN THE PROPER MANNER.

3. WATER AND ACID.—If you must mix water and acid, POUR THE ACID VERY SLOWLY INTO THE WATER.

CAUTION: If you pour water into acid, the MIXTURE WILL BOIL OVER QUICKLY AND BURN YOUR HANDS AND EVERYTHING IT TOUCHES.

4. ACID AND CYANIDE.—The chemical reaction of acid and cyanide generates a deadly poison.

5. CLEANLINESS.—Keep chemicals and their containers clean, as well as all equipment, supplies, and spaces you use when handling chemicals. Even a small amount of dirt or grease, for example, may ruin your work.

6. CHEMICAL POISONING.—Most chemicals are poisonous, and many of them can burn your clothes and hands. CAUTION: WEAR RUBBER GLOVES, A RUBBER APRON, AND GOGGLES WHEN YOU MIX CHEMICALS OR WORK WITH THEM.

Learn by heart the antidotes for poisoning and burning by chemicals. This knowledge may save your life.

Treat acid burns AS QUICKLY AS POSSIBLE. Wash the acid off with an abundance of water and then wash your hands under a spigot, if they were involved. Continue by neutralizing all acid which remains with lime water, a mixture of equal parts of lime water and raw linseed oil, or a paste of baking soda and water. REMEMBER THIS: Baking soda is a base and it neutralizes acids. If acid gets in your eyes, wash it out with cold water and then WASH YOUR EYES WITH WEAK LIME WATER.

WASH ALKALI BURNS WITH PLENTY OF COLD WATER; then neutralize remaining portions of the alkali WITH VINEGAR OR LEMON JUICE. REMEMBER THIS: Acids such as VINEGAR OR LEMON JUICE neutralize bases (alkalies) such as lye.

ANTIDOTES FOR POISON

Some good antidotes for poisons are listed next. Study them carefully; better still, memorize as many as possible.

ACETIC ACID.—Use an emetic to cause vomiting. Magnesia, chalk, whiting in water, soap, oil, mustard, and salt are emetics. A quick method for making a GOOD EMETIC is to stir a TABLESPOONFUL OF SALT OR MUSTARD into a glass of warm water.

HYDROCHLORIC, NITRIC, AND PHOSPHORIC ACID.--Use milk of magnesia, raw egg white, cracked ice, or a MIXTURE OF BAKING SODA AND WATER as an antidote for poisoning by these acids.

CARBOLIC ACID.—Some good antidotes for carbolic acid are: egg white, lime water, olive or castor oil with magnesia suspended in it, zinc sulfate in water, cracked ice, pure alcohol, or about 4 ounces of camphorated oil. Remember particularly: Egg white, lime water, and cracked ice, for they will most likely be readily available.

ALKALIES (sodium or potassium hydroxide).—Good antidotes for poisoning by sodium or potassium hydroxides are: vinegar, lemon juice, orange juice, oil, or milk. You can easily remember these antidotes.

ARSENIC (including rat poison and Paris green).—Use milk, raw eggs, sweet oil, lime water, or flour and water as an ANTIDOTE FOR ARSENIC POISONING.

CYANIDE.—Cyanide poisoning works so rapidly that you can do little to prevent death, which this poison causes in less than a minute. If possible, GIVE HYDROGEN PEROXIDE TO A VICTIM. If breathing stops, apply artificial respiration and let the patient breathe ammonia or chlorine produced by chlorinated water. If the victim is conscious, give him ferrous sulfate in water; then give him emetics and keep him warm.

DENATURED ALCOHOL.—Antidotes for poisoning by denatured alcohol are: emetics, milk, egg white, and flour and water. If breathing stops, give artificial respiration.

IODINE.—Give emetics, or plenty of starch or flour in water (stirred) as an antidote for iodine poisoning.

LEAD ACETATE (sugar of lead).—Use emetics; sodium sulfate, potassium sulfate, or magnesium sulfate in water; milk; or egg whites as antidotes for lead acetate poisoning.

MERCURIC CHLORIDE (corrosive sublimate).—Some good antidotes for mercuric chloride poisoning are: emetics; egg white, milk, table salt, castor oil, and zinc sulfate.

SILVER NITRATE.—For poisoning by silver nitrate, give a solution of table salt and water.

The first thing to do for all types of gas poisoning is this: GET THE VICTIM IN FRESH AIR IMMEDIATELY. IF HE STOPS BREATHING, GIVE HIM ARTIFICIAL RESPIRATION.

Breathe ammonia or amyl nitrite for poisoning by carbon monoxide, illuminating gas, ethylene, or acetylene.

The antidote for poisoning by chloroform and ether is COLD WATER ON THE HEAD AND CHEST.

The treatment given for poisoning by carbon monoxide is also applicable for poisoning by trichloroethylene and perchloroethylene.

NOTE: You may use one of these gases in a gas degreaser, provided you are careful and keep vapors from rising over the edge of the container. If the vapor spills over the edge and the air in the room is quiet, it sinks to the floor. When this happens, an exhaust fan AT FLOOR LEVEL must be used to remove the vapor. A fan in any other position in the room, or natural ventilation, causes the vapor to rise high enough for a person to enhale it.

Both of these gases have strong odors; and if you detect a questionable odor, GET OUT OF THE ROOM IMMEDIATELY. Both gases are also skin irritants; so if you must work with them, wear rubber gloves.

CAUTION: As soon as practicable, have a doctor or Hospital Corpsman treat all persons poisoned by chemicals and/or gases (poisonous).

CHAPTER 9

MACHINING OPERATIONS

In order for you to qualify for advancement to Opticalman 3, you must be able to perform basic operations on lathes, milling machines, bench grinders, and drill presses; and you must know the safety precautions pertaining to them.

In addition to the above, before you can qualify for advancement in rating to Opticalman 2, you must know how to make mathematical calculations for determining taper per foot on a piece of work, and what cutting and surface speeds to use in manufacturing mechanical parts for optical instruments. In order to make these parts, you must be able to operate skillfully and safely the machines mentioned in paragraph one; and you must also know what cutting lubricants to use when operating them.

Information in this chapter pertains only to lathes and milling machines. Refer to Basic Handtools, NavPers 10085-A, for a discussion of bench grinders and drill presses. As you study this chapter, bear in mind one thing: Information presented will enable you to pass the examination for advancement in rating; skill in operating the machines will come only through experience.

LATHES

Although machine shop work is generally performed by men in other ratings, there may be times when you will find a lathe essential for completing a repair job. Different types of lathes are installed in machine shops on various Navy ships, including the engine lathe, horizontal turret lathe, vertical turret lathe, and several variations of the basic engine lathe (bench, toolroom, and gap lathes).

All lathes, except the vertical turret type, have one thing in common—for all usual machining operations, the workpiece is held and rotated about a horizontal axis while being formed to size and shape by a cutting tool. In a vertical turret lathe, the workpiece is rotated about a vertical axis.

The type of lathe you will most likely use is the engine lathe; so this chapter deals only with lathes of this type, and the machining operations you may be required to perform with them.

ENGINE LATHE

An engine lathe such as the one shown in figure 9-1, or one similar to it, is found in every machine shop, however small. It is used principally for turning, boring, facing, and screw cutting; but it may also be used for drilling, reaming, knurling, grinding, spinning, and spring winding. The work held in the engine lathe can be revolved at a number of different speeds, and the cutting tool can be accurately controlled by hand or power for longitudinal and cross feed. (Longitudinal feed is movement of the cutting tool parallel to the axis of the lathe; cross feed is movement of the cutting tool perpendicular to the axis of the lathe.)

Lathe size is determined by two measurements: (1) diameter of work it will swing over the bed, and (2) length of the bed. For example, a 14-inch x 6-foot lathe will swing work up to 14 inches in diameter, and has a bed 6 feet long.

Engine lathes are built in various sizes, ranging from small bench lathes with a swing of 9 inches to very large lathes for turning work of large diameter, such as low pressure turbine rotors. The 16-inch lathe is the average size for general purposes, and it is the size usually installed on ships which have only one lathe.

Principal Parts

In order to learn the operation of a lathe, you must first become familiar with the names and functions of its principal parts. Lathes of different manufacture differ somewhat with respect to details of construction; but all of them

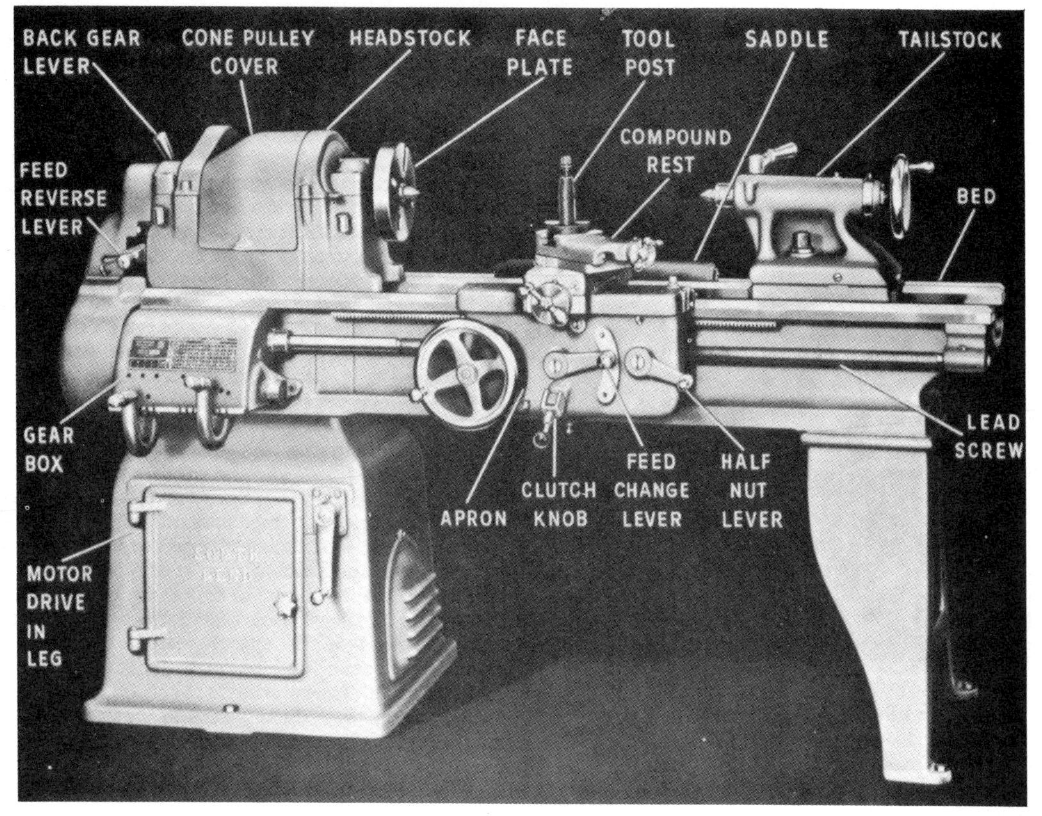

28.69X(75)

Figure 9-1.—An engine lathe.

are built to perform the same general functions. As you read the description of each part, refer to figure 9-1 and others (as applicable) to find its location on the lathe. (For specific details of construction features and operating techniques, refer to the manufacturer's technical manual for the machine you are using.)

The lathe BED is the base or foundation of the working parts of a lathe. The main features of its construction are the ways formed on its upper surface, and which run the full length of the bed. They provide the means for maintaining the tailstock and carriage (which slide on them) in alignment with the headstock, permanently secured by bolts at one end (at operator's left).

The lathe HEADSTOCK carries the headstock spindle and the mechanism for driving it. In the belt-driven type, shown in figure 9-2, the driving mechanism consists merely of a cone pulley which drives the spindle directly, or by means of back gears. When driven directly, the spindle revolves with the cone pulley; when driven by back gear, the spindle revolves more slowly than the cone pulley, which, in this case, turns freely on the spindle. Thus, two speeds are obtainable with each position of the belt on the cone; if the cone pulley has four steps as illustrated, eight spindle speeds can be obtained.

The geared headstock shown in figure 9-3 is more complicated but more convenient to operate, because speed changes are accomplished by merely shifting the gears. It is similar to an automobile transmission except that it has more gear-shift combinations and therefore a greater

28.71X
Figure 9-2.—Belt-driven headstock.

number of speed changes. A speed index plate attached to the headstock indicates the lever positions for obtaining the different spindle speeds.

The primary purpose of the lathe TAILSTOCK is to hold the DEAD center to support one end of work being machined on centers. However, it can also be used to hold tapered shank drills, reamers, and drill chucks. It is movable on the ways along the length of the bed to accommodate work of varying lengths and can be clamped in the desired position by means of the tailstock clamping nut.

Before inserting a dead center, drill, or reamer, carefully clean the tapered shank and wipe out the tapered hole of the spindle. When holding drills or reamers in the tapered hole of a spindle, be sure they are tight enough so that they will not slip. If allowed to slip, they will score the tapered hole and destroy its accuracy.

The function of the CARRIAGE (composed of saddle and apron) is to carry the compound rest, which in turn carries the cutting tool in the tool post. Figure 9-4 shows how the carriage travels along the bed over which it slides on the outboard ways.

The carriage is provided with T-slots or tapped holes for clamping work for boring or milling. When used in this manner the carriage movement feeds the work to the cutting tool, which is revolved by the headstock spindle.

You can lock the carriage in any position on the bed by tightening up on the carriage clamp screw. This is done only when doing such work as facing or cutting-off, for which longitudinal feed is not required. Normally, the carriage clamp should be kept in the released position. Always move the carriage by hand to be sure it is free before applying the automatic feed.

The APRON is attached to the front of the saddle and contains the mechanism for controlling the movement of the carriage for longitudinal feed and thread cutting, and the lateral movement of the cross-slide.

The FEED ROD transmits power to the apron to drive the longitudinal and cross-feed mechanisms. The feed rod is driven by the spindle through a train of gears, and the ratio of its speed to that of the spindle can be varied by changing the gear combination to produce various rates of feed. The rotating feed rod drives gears in the apron, and these gears in turn drive the longitudinal and cross-feed mechanisms through friction clutches.

The LEAD SCREW is used for thread cutting. Along its length are accurately-cut Acme threads which engage the threads of the half-nuts in the apron when the half-nuts are clamped over it. When the lead screw turns in the closed half-nuts, the carriage moves along the ways a distance equal to the lead of the thread in each revolution of the lead screw. Since the lead screw is driven by the spindle through a gear train which connects them (fig. 9-5), the rotation of the lead screw bears a direct relation to the rotation of the spindle. Therefore, when the half-nuts are engaged, the longitudinal movement of the carriage is directly controlled by the spindle rotation, and the cutting tool is consequently moved a definite distance along the work for each revolution it makes.

The COMPOUND REST provides a rigid adjustable mounting for the cutting tool. The compound rest assembly has the following principal parts:

1. Compound rest SWIVEL, which can be swung around to any desired angle and clamped

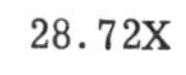
28.72X

Figure 9-3.—Sliding-gear headstock.

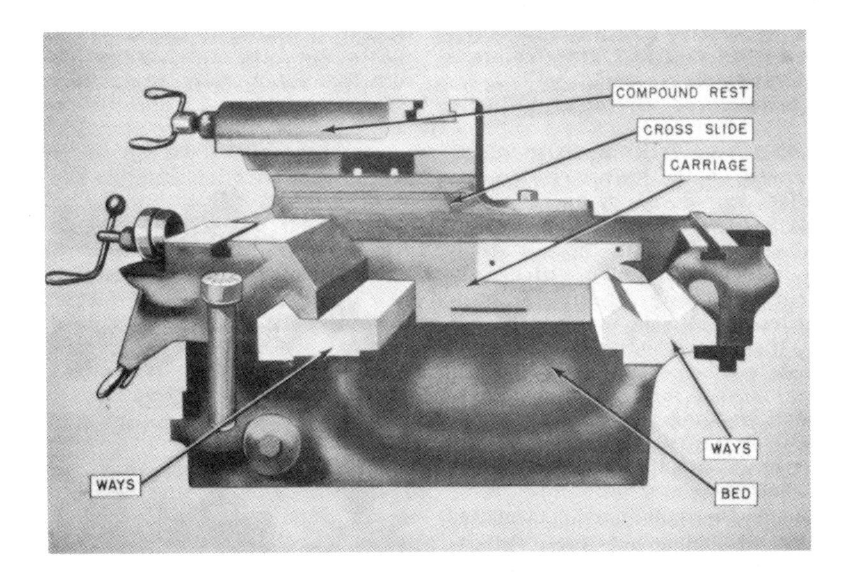

28.76X

Figure 9-4.-Side view of a carriage mounted on the bed.

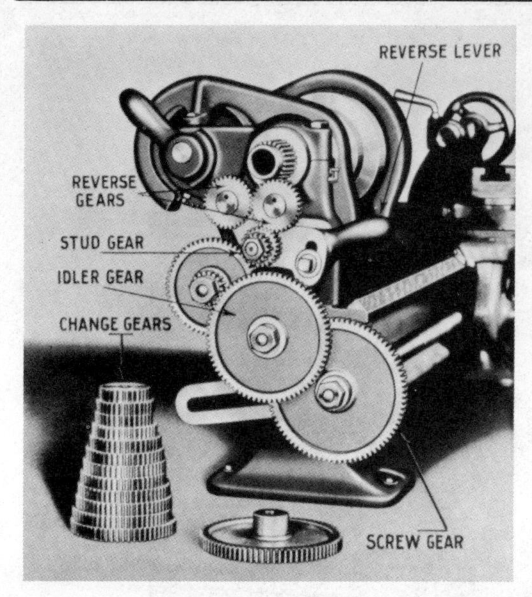

28.80X
Figure 9-5.—Lead screw gear train.

in position. It is graduated over an arc of 90° on each side of its center position to facilitate setting to the angle selected. This feature is used when machining short, steep tapers such as the angle on bevel gears, valve disks, and lathe centers.

2. Compound rest TOP or TOP SLIDE, which is mounted on the swivel section of a dovetailed slide and moved by means of the compound rest feed screw.

This arrangement permits feeding at any angle (determined by the angular setting of the swivel section), while the cross-slide feed provides only for feeding at right angles to the axis of the lathe. The collars on the cross feed and compound rest feed screws are graduated in thousandths of an inch for fine adjustment in regulating the depth of cut.

Attachments and Accessories

Accessories are the tools and equipment used in routine lathe machining operations. Attachments are special fixtures which may be secured to the lathe to extend the versatility of the lathe to include taper cutting, milling, and grinding. Some of the common accessories and attachments used on lathes are described in the following paragraphs.

The sole purpose of the TOOL POST is to provide a rigid support for the cutting tool. The tool post is mounted in the T-slot of the compound rest top, and a forged cutting tool or a toolholder is inserted in the slot in the tool post. By tightening a setscrew, the whole unit can firmly clamp in place with the tool in the desired position.

Commonly used lathe toolholders are illustrated in figure 9-6. Notice the angles of the tool bit in the holder. These angles must be considered with respect to the angles ground in the tools and the angle at which the toolholder is set with respect to the axis of the work.

Figure 9-7 shows the most popular shapes of ground lathe tool cutter bits and their application.

The LATHE CHUCK is a device for holding lathe work, and it is mounted on the nose of the spindle. The work is held by jaws which can be moved in radial slots toward the center to clamp down on the sides of the work. These jaws are moved in and out by screws turned by a chuck wrench applied to the sockets located at the outer ends of the slots.

The 4-jaw independent lathe chuck, part A in figure 9-8, is the most practical for general work. The four jaws are adjusted one at a time, making it possible to hold work of various shapes and to adjust the center of the work to coincide with the center of the lathe. The jaws may be

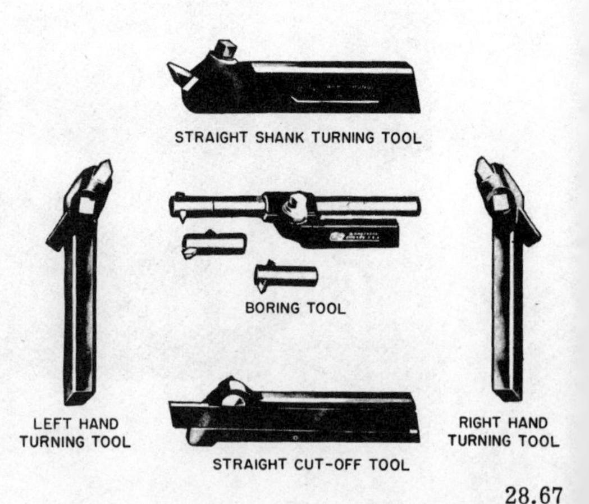

28.67
Figure 9-6.—Common types of tool holders.

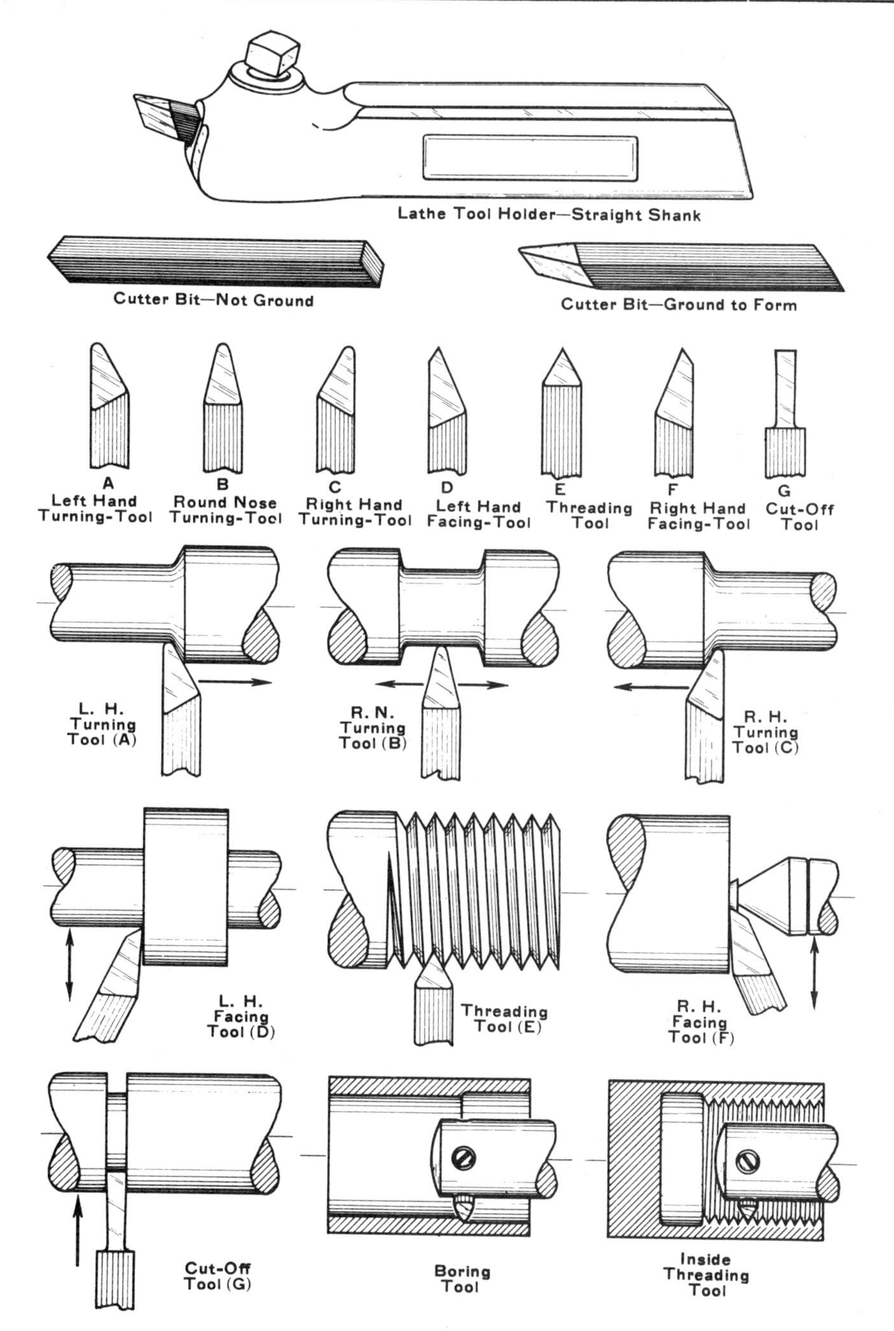

Figure 9-7.—Lathe tools and their application. 28.66

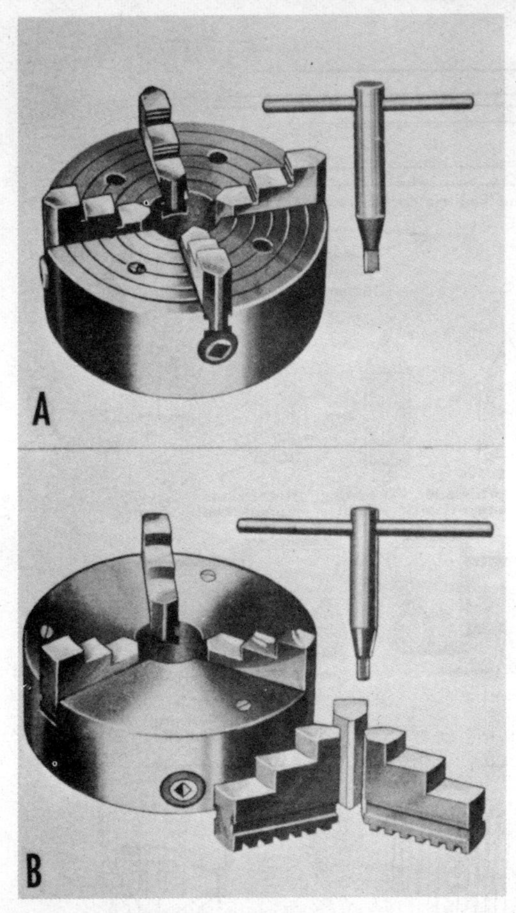

28.90X
Figure 9-8.—A, four-jaw chuck; B, three-jaw chuck.

turned end-for-end to clamp either inside or outside surfaces.

The 3-law universal or scroll chuck, part B in figure 9-8, can be used only for holding round or hexagonal work. All three jaws are moved in and out together in one operation, and they move universally to bring the work on center automatically. This chuck is easier to operate the the four-jaw type; but when its parts become worn its accuracy in centering cannot be relied upon. Proper lubrication and constant care are necessary to ensure reliability of operation.

The DRAW-IN COLLET CHUCK is used to hold small work in the lathe for machining. It is one of the most accurate types of chucks made and is intended for precision work. Figure 9-9 shows the parts assembled in place in the lathe spindle.

The collet chuck which holds the work is a split-cylinder with an outside taper that fits into the tapered closing sleeve and screws into the threaded end of the hollow drawbar. Screwing up on the drawbar by turning the handwheel pulls the collet back into the tapered sleeve, thereby closing it firmly over the work and centering it accurately and quickly. The size of the hole in the collet determines the diameter of the work it can accommodate.

You will use the FACEPLATE for holding work of such shape and dimensions which prevent its being swung on centers or in a chuck. The T-slots and other openings on its surface provide convenient anchors for bolts and clamps used in securing the work to it. The faceplate is mounted on the nose of the spindle.

The driving plate is similar to a small faceplate and is used principally for driving work held between centers. The radial slot receives the bent tail of a lathe dog clamped to the work and thereby transmits rotary motion to the work.

The function of the 60° LATHE CENTERS shown in figure 9-10 is to provide a means for holding the work between points so it can be turned accurately on its axis. The head spindle center is called the LIVE center because it revolves with the work. The tailstock center is called the DEAD center because it does not turn. Both live and dead centers have shanks turned to a Morse taper to fit the tapered holes in the spindles; both have points finished to an angle of 60 degrees. They differ only in that the dead center is hardened and tempered to resist the wearing effect of the work revolving on it. The live center revolves with the work, and it is usually left soft. The dead center and live center must never be interchanged.

NOTE: There is a groove around the hardened tail center to distinguish it from the live center.

The centers fit snugly in the tapered holes of the headstock and tailstock spindles. If chips, dirt, or burrs prevent a perfect fit in the spindles, the centers will not run true.

To remove the headstock center, insert a brass rod through the spindle hole and tap the center to jar it loose; it can then be picked out by hand. To remove the tailstock center, run the spindle back as far as it will go by turning

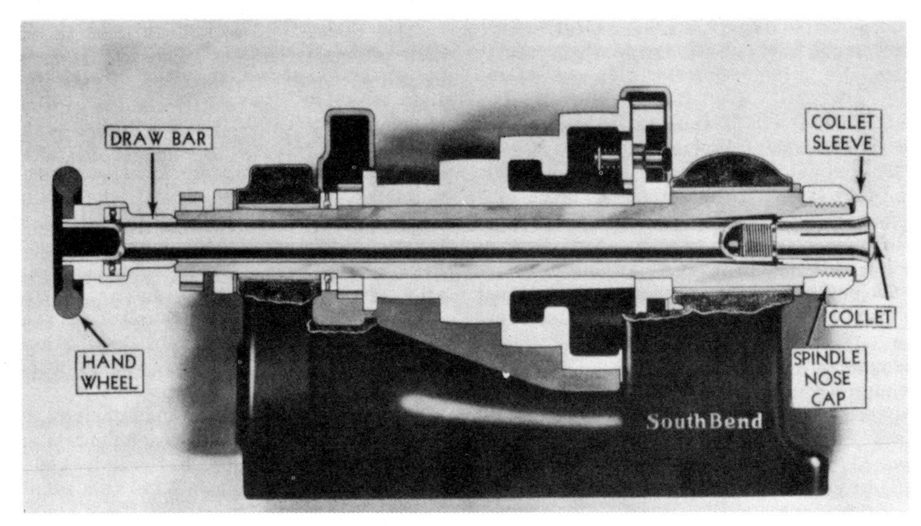

28.91X
Figure 9-9.-Draw-in collet chuck.

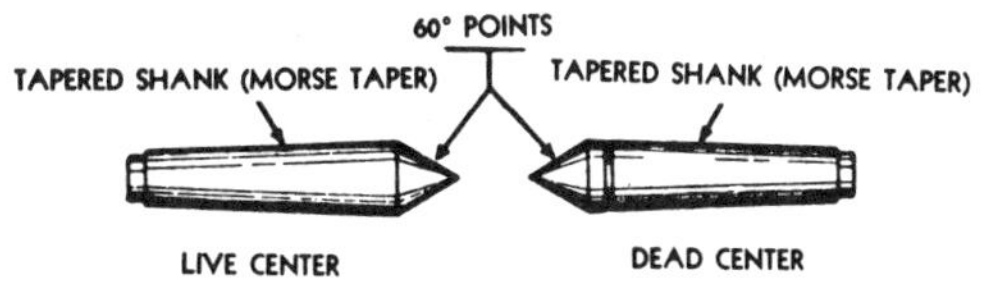

28.93X
Figure 9-10.-Sixty-degree lathe centers.

the handwheel counterclockwise. When the end of the tailstock screw bumps the back of the center, it will force the center out of the tapered hole.

LATHE DOGS are used in conjunction with a driving plate or faceplate to drive work being machined on centers, the frictional contact alone between the live center and the work not being sufficient to drive it.

The common lathe dog shown at the left in figure 9-11 is used for round work or work having a regular section (square, hexagon, octagon). The piece to be turned is held firmly in hole A by setscrew B. The bent tail (C) projects through a slot or hole in the driving plate or faceplate, so that when the latter revolves with the spindle it turns the work with it. The clamp dog illustrated at the right in figure 9-11 may be used

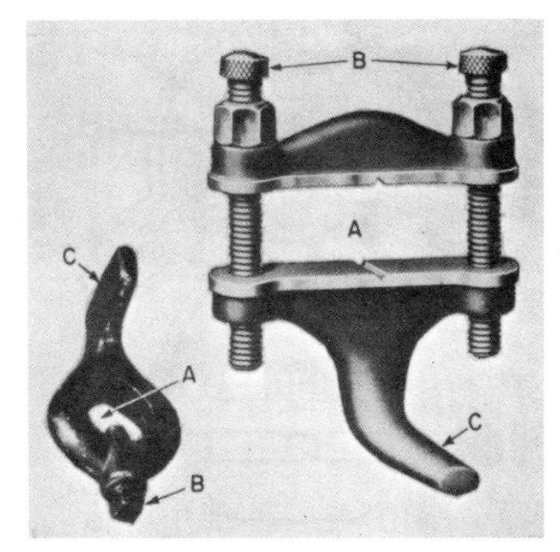

28.95X
Figure 9-11.-Lathe dogs.

for rectangular or irregular shaped work. Such work is clamped between the jaws. Figure 9-35 illustrates the use of a driving plate and clamp dog.

The CENTER REST (also called the steady rest) is used for the following purposes:

1. To provide an intermediate support or rest for long, slender bars or shafts being machined between centers. It prevents them from springing under cut, or sagging as a result of their otherwise unsupported weight.

2. To support and provide a center bearing for one end of work, such as a spindle, being bored or drilled from the end when it is too long to be supported by a chuck alone. The center rest is clamped in the desired position on the bed on which it is properly aligned by the ways, as illustrated in figure 9-12. It is important that the jaws (A) be carefully adjusted to allow the work (B) to turn freely and at the same time keep it accurately centered on the axis of the lathe. The top half of the frame is hinged at C to facilitate placing it in position without removing the work from the centers or changing the position of the jaws.

The FOLLOWER REST is used to back up work of small diameter to keep it from springing under the stress of cutting. It gets its name from the fact that it follows the cutting tool along the work. As shown in figure 9-13, it is attached directly to the saddle by bolts (B). The adjustable jaws bear directly on the finished diameter of the work opposite the cutting tool.

The TAPER ATTACHMENT, illustrated in figure 9-14, is used for turning and boring tapers. It is bolted to the back of the carriage. In operation, it is so connected to the cross-slide that it moves the cross-slide laterally as the carriage moves longitudinally, thereby causing the cutting tool to move at an angle to the axis of the work to produce a taper.

The angle of the taper it is desired to cut is set on the guide bar of the attachment. The guide bar support is clamped to the lathe bed. Since the cross-slide is connected to a shoe that slides on this guide bar, the tool follows along a line parallel to the guide bar and hence at an angle to the work axis corresponding to the desired taper.

Operation and application of the taper attachment will be further explained under the subject of taper work.

The THREAD DIAL indicator, shown in figure 9-15, eliminates the necessity of reversing

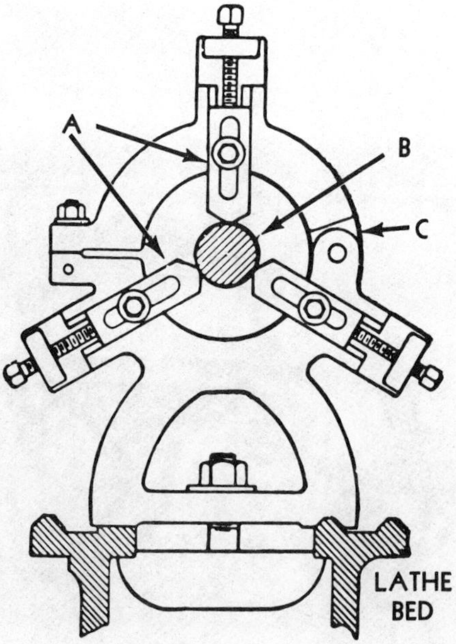

A
B
C
LATHE BED

28.96X

Figure 9-12.—Center rest.

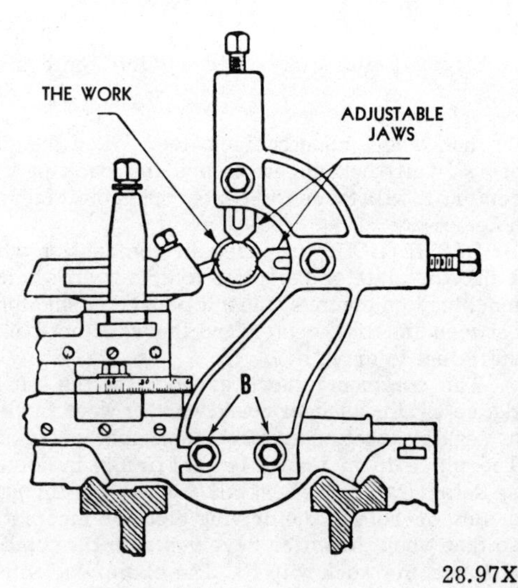

28.97X

Figure 9-13.—Follower rest.

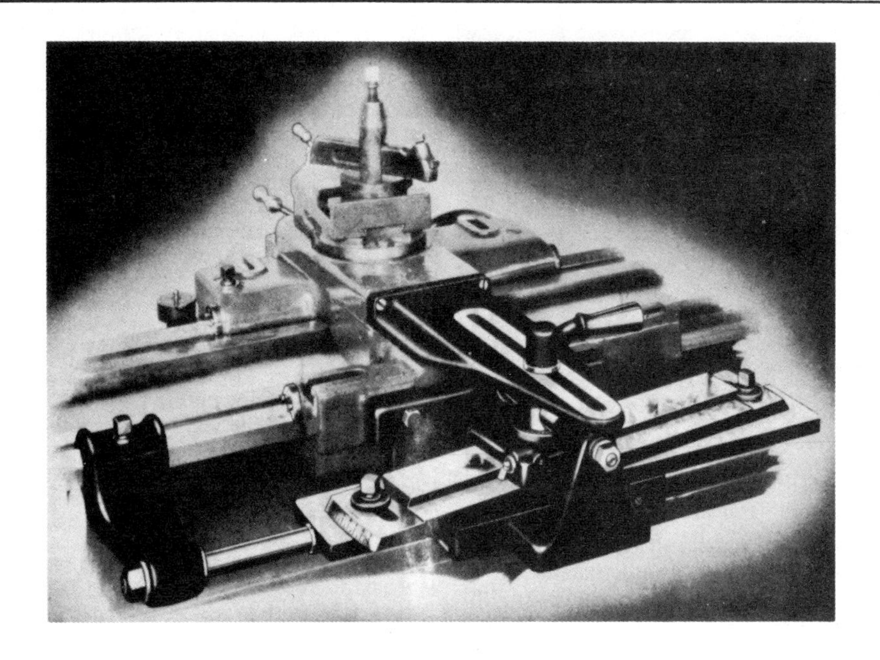

28.98X

Figure 9-14.—A taper attachment.

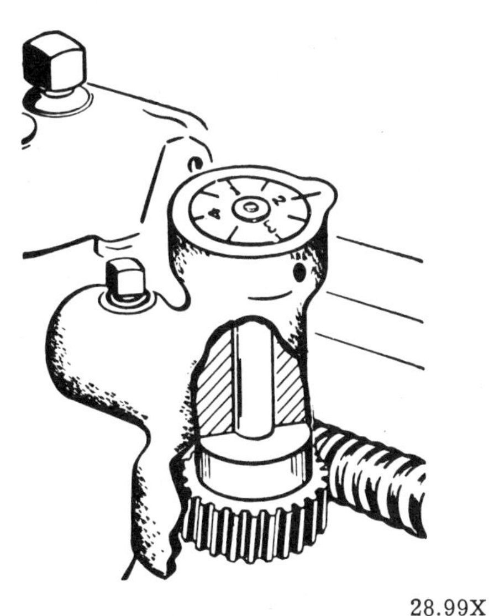

28.99X

Figure 9-15.—Thread dial indicator.

the lathe to return the carriage to the starting point to catch the thread at the beginning of each successive cut taken. The dial, geared to the lead screw, indicates when to clamp the half-nuts on the lead screw for the next cut.

The threading dial consists of a worm wheel attached to the lower end of a shaft and meshed with the lead screw. On the upper end of the shaft is the dial. As the lead screw revolves, the dial is turned and the graduations on the dial indicate points at which the half-nuts may be engaged.

You can attach the CARRIAGE STOP to the bed at any point where you wish to stop the carriage. It is used principally when turning, facing, or boring duplicate parts, as it eliminates the necessity of repeated measurements of the same dimension. In operation, the stop is set at the point where it is desired to stop the feed. Just before reaching this point, you shut off the automatic feed and carefully run the carriage up against the stop. Carriage stops are provided with or without micrometer adjustment. Figure 9-16 shows a micrometer carriage stop. It is clamped on the ways in the approximate position

28.100X
Figure 9-16.—Micrometer carriage stop.

required and then adjusted to the exact setting by means of the micrometer adjustment.

NOTE: Some carriages are equipped with a stop which automatically stops the carriage by disengaging the feed or stopping the lathe. This type of stop is called AUTOMATIC CARRIAGE STOP, and it is usually a built-in feature of the lathe design.

FACTORS RELATED TO THE USE OF A LATHE

Knowledge of many factors is necessary before you can become proficient in performing machine work with a lathe. Some of these factors are considered in the following paragraphs.

Phases of the Operation

Before attempting to operate any lathe, make sure you know how to run it. Read all operating instructions supplied with the machine. Ascertain the location of various controls and learn how to operate them. When you are satisfied that you know how they work, start the motor; but first check to see that the spindle clutch and the power feeds are disengaged. Then become familiar with all phases of operation, as follows:

1. Shift the speed change levers into the various combinations; start and stop the spindle after each change. Get the feel of this operation.

2. With the spindle running at its slowest speed, try out the operation of the power feeds and observe their action. Take care not to run the carriage too near the limits of its travel. Learn how to reverse the direction of feeds and how to disengage them quickly. Before engaging either of the power feeds, operate the hand controls to be sure parts involved are free for running.

3. Try out the operation of engaging the lead screw for thread cutting. Remember that the power feed mechanism must be disengaged before you can close the half-nuts on the lead screw.

4. Practice making changes with the QUICK-CHANGE GEAR MECHANISM by referring to the thread and feed index plate on the lathe you intend to operate. Remember that changes in the gear box are made with the lathe STOPPED; but disengage the clutch and stop the lathe before you shift gears. If a gear does not engage, do NOT force the operating lever—rotate the spindle by hand as you try to engage the gear.

Maintenance

Maintenance is an important part of operational procedure for lathes, and one requisite is PROPER LUBRICATION. Make it a point to oil your lathe daily. Oil the ways daily—not only for lubrication but to protect their scraped bearing surfaces. Oil the lead screw often while it is in use. This is necessary to preserve its accuracy, for a worn lead screw lacks precision in thread cutting. Make sure the headstock is filled up to the oil level; drain it out and replace the oil when it becomes heavy or gummy. If your lathe is equipped with an automatic oiling system for some parts, make sure all parts are getting oil. Check frequently the lubrication of all moving parts. CAUTION: Do NOT oil a lathe when it is running. Always use a high-grade oil of proper viscosity.

Don't treat your machine roughly. When you shift gears for changing speed or feed, remember that you are putting solid gear teeth into mesh with each other; feel the gears into engagement.

Before engaging the longitudinal feed, be certain that the carriage CLAMP SCREW is loose and that the CARRIAGE can be moved by hand. Avoid running the carriage against the headstock or tailstock while under power feed; this puts an unnecessary strain on the lathe and may jam the gears.

Do not neglect the motor just because it may be out of sight; check its LUBRICATION.

If it does not run properly, notify the Electrician's Mate, who is responsible for its care. He will cooperate with you to keep it in good condition. If the lathe is belt-driven, avoid getting oil or grease on the belt when oiling the lathe or motor.

Keep your lathe CLEAN. A clean machine is usually an indication of a good mechanic. Dirt and chips on the ways, on the lead screw, and on the cross-feed screws will cause serious wear and impair accuracy of the machine.

Never put wrenches, files, or other tools on the ways. If you must keep tools on the bed, use a board to protect the finished surfaces of the ways. When you change chucks, protect the ways with a board laid across the ways under the chuck. If a chuck is dropped on the unprotected ways it will do great damage.

Never use the bed or carriage as an anvil; remember that the lathe is a precision machine whose accuracy must not be destroyed.

Cutting Speeds and Feeds

CUTTING SPEED is the rate at which the surface of the work passes the point of the cutting tool, and it is expressed in feet per minute. To find the cutting speed, multiply the circumference of the work (in inches) by the number of revolutions it makes per minute (rpm) and divide by 12 (Circumference = diameter x 3.1416). The result is the peripheral or cutting speed in feet per minute (fpm). For example, a 2-inch diameter piece turning at 100 rpm will produce a cutting speed of

$$\frac{(2 \times 3.1416) \times 100}{12} = 52.36 \text{ fpm}$$

Conversely, the rpm required to obtain a given cutting speed is found by dividing the product of the given cutting speed and 12 by the circumference of the work (in inches).

FEED is the amount the tool advances each revolution. It is usually expressed in thousandths of an inch per revolution of the spindle. The index plate on the quick-change gear box indicates the setup for obtaining the feed desired. The amount of feed to use is best determined from experience.

Cutting speeds and tool feeds are determined by various considerations: hardness and toughness of metal being cut; quality, shape, and sharpness of the cutting tool; depth of the cut; tendency of the work to spring away from the tool; and strength and power of the lathe. Since conditions vary, it is good practice to find out what the tool and work will stand, and then select the most practicable and efficient speed and feed consistent with the finish desired.

If the cutting speed is too slow, the job takes longer than necessary and often the work produced is unsatisfactory. On the other hand, if the speed is too great, the tool edge dulls quickly, and frequent grinding is necessary. Cutting speeds possible are greatly affected by the use of a suitable cutting lubricant. For example, steel which can be rough turned dry at 60 rpm can be turned at about 80 rpm when flooded with a good cutting lubricant.

The accompanying chart gives the approximate, recommended cutting speeds for different metals, using high-speed steel tool bits. Figures indicate feet per minute (fpm).

Type of metal	Roughing cut	Finishing cut	Thread-cutting
Cast iron	60	80	25
Machine steel . .	90	125	35
Tool steel	50	75	20
Brass	150	200	50
Bronze	90	100	25
Aluminum	200	300	50

When ROUGHING parts down to size, use the greatest depth of cut and feed per revoltuion that the work, the machine, and the tool will stand at the highest practicable speed. On many pieces where tool failure is the limiting factor in the size of a roughing cut, it is usually possible to reduce the speed slightly and increase the feed to a point where the metal removed is much greater. This will prolong tool life. Consider an example where the depth of cut is 1/4 inch, the feed 20 thousandths of an inch per revolution, and the speed 80 fpm. If the tool will not permit additional feed at this speed, it is usually possible to drop the speed to 60 fpm and increase the feed to about 40 thousandths of an inch per revolution without having tool trouble. The speed is therefore reduced 25 percent, but the feed increased 100 percent; so the actual time required to complete the work is less with the second setup.

On the FINISH-TURNING OPERATION, a very light cut must be taken, since most of the stock was removed on the roughing cut. A fine feed can usually be used, thereby making it possible to run at a high surface speed. A 50-percent increase in speed over the roughing speed is commonly used. In particular cases, the finishing speed may be twice the roughing speed. In any event, run the work as fast as the tool will withstand to obtain maximum speed in this operation. Use a sharp tool to finish the turning.

Coolants

A cutting coolant serves two main purposes: (1) it cools the tool by absorbing, carrying away, and dissipating a portion of the heat; and (2) it reduces the amount of heat produced in the cutting process by reducing the friction between the tool and the metal being cut. A secondary purpose of a coolant is to keep the cutting edge of the tool flushed clean.

Coolants have some qualities which make them especially desirable for certain applications. The following coolants are the ones most commonly used:

1. LARD OIL. Lard oil is one of the oldest and best coolants, but it is the most expensive. It is especially good for cutting screw threads, drilling deep holes, and reaming. This coolant, however, provides excellent lubrication, increases tool life, prevents rust, and produces a smooth finish on the work.

2. MINERAL LARD OIL MIXTURES. Various types of mixtures of lard oil and mineral oils with a petroleum base are generally used instead of lard oil because they are more fluid, less expensive, and are almost as effective as lard oil.

3. SOLUBLE OILS. Some specially processed mineral oils mix with water to form an emulsion which provides an excellent low cost coolant. Although such emulsions carry away heat better than lard oil or mineral oil, their lubricating qualities are comparatively poor and their use is usually limited to rough turning. Even though these emulsions contain water, they leave a protective film on metal and prevent rust.

4. SODA WATER MIXTURES. Soda water mixtures are the cheapest of all coolants and are very effective coolants, but they have practically no lubricating qualities and they cause steel or iron to rust.

You can make this coolant by mixing one pound of carbonate of soda, one quart of lard oil, one quart of soft soap, and eight to ten gallons of water and by boiling the mixture for one half hour. When it cools, the mixture is ready for use.

Other coolants which may be used separately or in mixtures are: kerosene, turpentine, and white lead. You will find that all the coolants herein listed, with the exception of soda water mixture, are available through the general supply system.

Lubricants generally used for turning the listed metals are:

Metal	Lubricant
Cast iron	Usually worked dry.
Mild steel	Oil or soapy water.
Hard steel	Mineral lard oil.
Monel metal . . .	Dry (or mineral lard oil).
Bronze	Dry (or mineral lard oil).
Brass	Dry (kerosene or turpentine sometimes used on the hard compositions).
Copper	Dry (or mixture of lard oil and turpentine).
Babbitt	Dry (or mixture of lard oil and kerosene).
Aluminum	Dry (or kerosene or mixture of lard oil and kerosene).

For threading, a lubricant is more important than for straight turning. Mineral lard oil is recommended for threading in all steels and cast iron, kerosene mixed with oil for aluminum, white lead mixed with oil (to the consistency of glue) for monel metal, and kerosene or turpentine for brass compositions.

Chatter

If you are unaware of the meaning of the word CHATTER, you will soon learn when you work with machine tools. Briefly, chatter is vibration in either the tool or the work. The finished work surface appears to have a grooved or lined finish instead of a smooth surface. The vibration is set up by a weakness in the work, work support, tool, or tool support; and it is about the most elusive thing to find in the entire field of machine work. As a general rule, strengthening the various parts of the tool support train will help. It is also advisable to support the work by a center rest or follower rest.

Possibly the fault may be in the machine's adjustments. The gibs may be too loose, the bearings may be worn; the tool may be sharpened improperly, and so on. If the machine is in perfect condition, the fault may be in the tool or tool setup. Grind the tool with a point or as near a point as the finish specified permits; avoid a wide, round, leading edge on the tool. Reduce the overhang of the tool as much as possible, and be sure that all the gib and bearing adjustments are properly made. See that the work receives proper support for the cut; and above all, do not try to turn at a surface speed too high. Excessive speed is probably the greatest cause of chatter, and the first thing you should do when chatter occurs is to reduce speed.

Direction of Feed

Regardless of how the work is held in the lathe, the tool should feed toward the headstock, so that most of the pressure of the cut is exerted on the work-holding device and spindle thrust bearings. When it is necessary to feed the cutting tool toward the tailstock, take lighter cuts at reduced feeds. In facing, feed the tool from the center of the workpiece out toward the periphery.

PRELIMINARY PROCEDURES

Before starting a lathe machining operation, always be sure that the machine is set up for the job you are doing. If the work is mounted between centers, check the alignment of the dead center with the line center and make any changes required. Ensure that the tool holder and cutting tool are set at proper height and angle. Check the work-holding accessory to ensure that the workpiece is held securely. Use the center rest or follower for support of long workpieces.

Preparing the Centers

The first step in preparing the centers is to mount them accurately in the headstock and tailstock spindles. The centers and the tapered holes in which they are fitted must be perfectly clean. Chips and dirt left on the contact surfaces will impair accuracy by preventing a perfect fit of the bearing surfaces. Make sure that there are no burrs in the spindle hole. If burrs are found, remove them by careful scraping or reaming with a Morse taper reamer. Burrs produce the same inaccuracies as chips or dirt.

Center points must be accurately finished to an angle of 60°. Figure 9-17 shows the method of checking this angle with a center gage. The large notch of the center gage is intended for this particular purpose. If this test shows the point is not perfect, true it in the lathe by taking a cut over the point with the compound rest set at 30 degrees. The hardened tail-center must be annealed before it can be machined in this manner, or it can be ground without annealing if a grinding attachment is available.

Checking Alignment

To turn a shaft straight and true between centers, the centers must be in a plane parallel to the ways of the lathe. Move the tailstock laterally with the adjusting screws to accomplish this alignment after you release it from the ways. At the rear of the tailstock are two zero lines, and the centers are approximately aligned when these lines coincide. You can check this approximate alignment by moving the tailstock up until the centers almost touch, and observing their relative positions, as shown in figure 9-18. For very accurate work, especially if it is long, make the following test to correct small errors in alignment not otherwise detected.

Mount the work to be turned, or a piece of stock of similar length, on the centers. With a turning tool in the tool post, take a small cut to a depth of a few thousandths of an inch at the headstock end of the work. Then remove

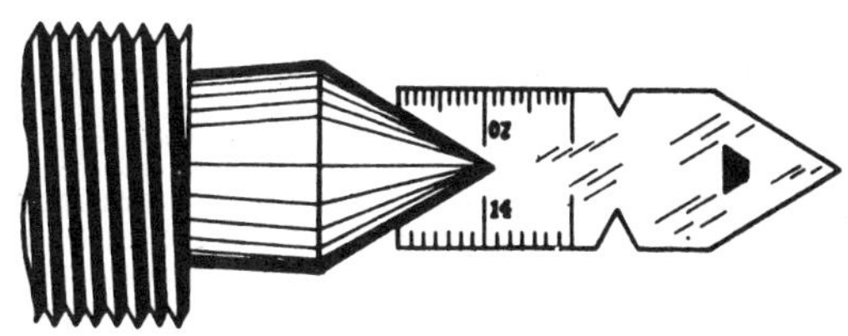

28.105X

Figure 9-17.—Checking center point with center gage.

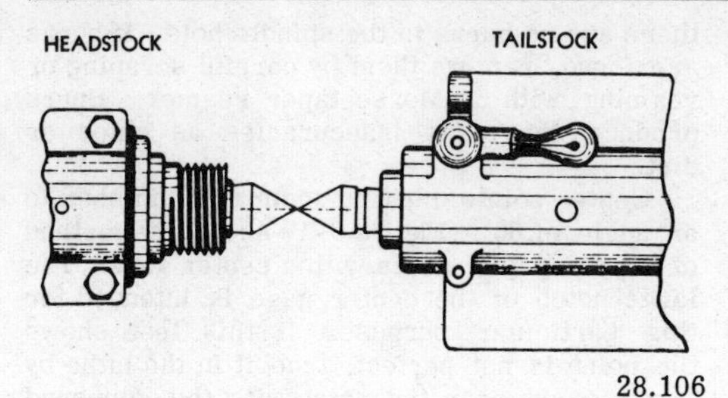

28.106

Figure 9-18.—Aligning lathe centers.

the work from the centers to allow the carriage to be run back to the tailstock without withdrawing the tool. Do not touch the tool setting. Replace the work in the centers; and with the tool set at the previous depth, take another cut coming in from the tailstock end. Compare the diameters over these cuts with a micrometer. If the diameters are exactly the same, the centers are in perfect alignment; if they are different, adjust the tailstock in the direction required by means of the set-over adjusting screws. Repeat the above test and adjustment until a cut at each end has the same diameter.

You can also check positive alignment of the centers by placing a test bar between the centers and bringing both ends of the bar to a zero reading, as indicated by a dial indicator clamped in the tool post. Clamp the tailstock to the ways and properly adjust the test bar between the centers when you take the indicator readings.

Another method which you may use for positive alignment of lathe centers is to take a light cut over the work held between centers. Then measure the work at each end with a micrometer; and if the readings differ, adjust the tailstock accordingly. Repeat the procedure until the alignment is satisfactory.

Setting the Toolholder and Cutting Tool

The first requirement for setting the tool is rigidity. Make sure the tool holder sits squarely in the tool post and that the setscrew is tight. Reduce overhang as much as possible to prevent springing when cutting. If the tool has too much spring, the point of the tool will catch in the work, cause chatter, and damage both the tool and the work. The distances represented by A and B in figure 9-19 show the correct overhang for the tool bit and the tool holder.

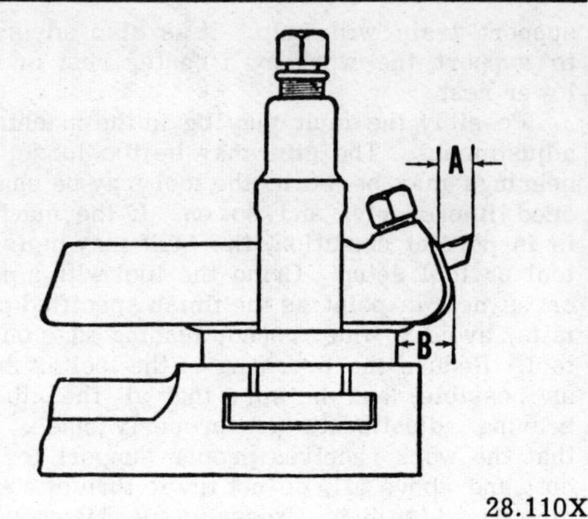

28.110X

Figure 9-19.—Tool overhang.

The point of the tool must be correctly positioned on the work. Place the cutting edge slightly above the center for straight turning of steel and cast iron, and exactly on the center for all other work. To set the tool at the height desired, raise or lower the point of the tool by moving the wedge in or out of the tool post ring. By placing the point opposite the tail center point, you can adjust the setting accurately.

Holding the Work

You cannot perform accurate work if you mount it improperly. Requirements for proper mounting are:

1. The work center line must be accurately centered with the axis of the lathe spindle.
2. The work must be rigidly held while being turned.
3. The work must not be sprung out of shape by the holding device.
4. The work must be adequately supported against any sagging caused by its own weight and against springing caused by action of the cutting tool.

There are four general methods of holding work in the lathe: (1) between centers, (2) on a mandrel, (3) in a chuck, and (4) on a faceplate. Work may also be clamped to the carriage for boring and milling, in which case the boring bar or milling cutter is held and driven by the headstock spindle.

Other methods of holding work to suit special conditions are: (1) one end on the live center or in a chuck and the other end supported in a center rest, and (2) one end in a chuck and the other end on the dead center.

HOLDING WORK BETWEEN CENTERS.—To machine a workpiece between centers, drill center holes in each end to receive the lathe centers. A lathe dog is then secured to the workpiece and the work is mounted between the live and dead centers of the lathe.

To center finished round stock such as drill rod or cold-rolled steel, where the ends are to be turned and must be concentric with the unturned body, the work can be held on the head spindle in a universal chuck or a draw-in collet chuck. If the work is long and too large to be passed through the spindle, a center rest must be used to support one end. The centering tool is held in a drill chuck in the tail spindle and is fed to the work by the tailstock handwheel (fig. 9-20).

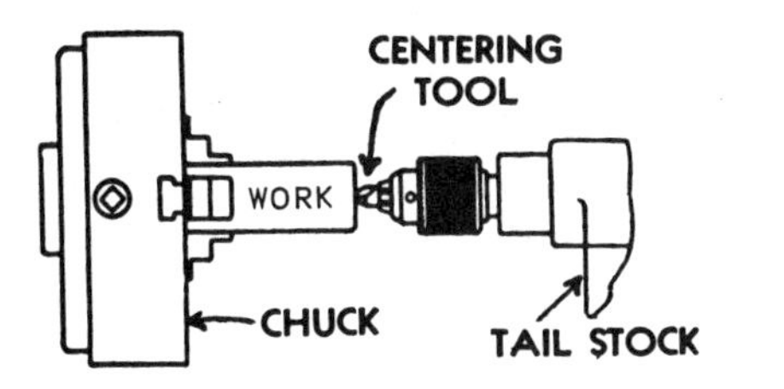

28.111

Figure 9-20.—Drilling center hole.

If a piece must be centered very accurately, the tapered center hole should be bored after center drilling to correct any run-out of the drill. This is done by grinding a tool bit to a center gage at a 60° angle. Then with the tool holder held in the tool post, set the compound rest at 30° with the line of center as shown in figure 9-21. Set the tool exactly on the center for height and adjust to the proper angle with the center gage as shown at A. By feeding the tool as shown at B, any run-out of the center is corrected. The tool bit should be relieved under the cutting edge as shown at C to prevent the tool from dragging or rubbing in the hole.

For center-drilling a workpiece, the combined drill and countersink is the most practical tool. These combined drills and countersinks vary in size and the drill points also vary. Sometimes a drill point on one end will be 1/8 inch in diameter, and the drill point on the opposite end 3/16 inch in diameter. The angle of the center drill is always 60°, so that the countersunk hole will fit the angle of the lathe center point.

If a centerdrill is not available, the work may be centered with a small twist drill. Let the drill enter the work a sufficient length on each end; then follow with a special countersink, the point of which is 60 degrees.

In center-drilling, use a drop or two of oil on the drill. Feed the drill slowly and carefully to prevent breakage of the tip. Extreme care is needed when the work is heavy because it is then more difficult to FEEL the proper feed of work on the centerdrill.

If the centerdrill breaks while you are countersinking and part of the broken drill remains in the work, remove the broken part. Sometimes you can drive it out with a chisel or jar it loose; but it may stick so hard that you cannot remove it. In this case anneal and drill out the broken part of the drill.

The importance of proper center holes in the work and a correct angle on the point of the lathe centers cannot be overemphasized. To do an accurate job between centers on the lathe, countersunk holes must be of proper size and depth, and the points of the lathe centers must be true and accurate.

Figure 9-22 shows correct and incorrect methods of mounting work between centers. In the correct example, the driving dog is attached to the work and rigidly held by the setscrew. The tail of the dog rests in the slot of the faceplate and extends beyond the base of the slot so that the work rests firmly on both the headstock center and tailstock center.

In the incorrect example (fig. 9-22), note that the tail of the dog rests on the bottom of the slot on the faceplate at A, thereby pulling the work away from the center points, as shown at B and C, and causing the work to revolve eccentrically.

When mounting work between centers for machining, there should be no end play between the work and the dead center. However, if held too tightly by the tail center when revolving, the work will heat the center point and destroy both the center and the work. For the same

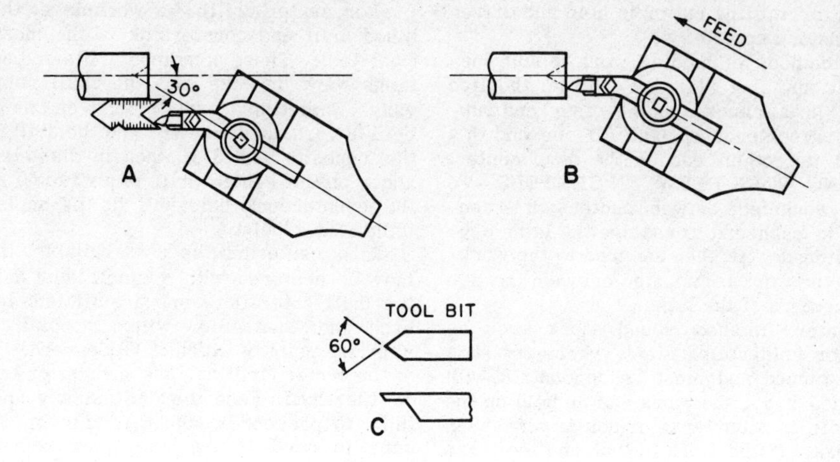

28.112
Figure 9-21.—Boring center hole.

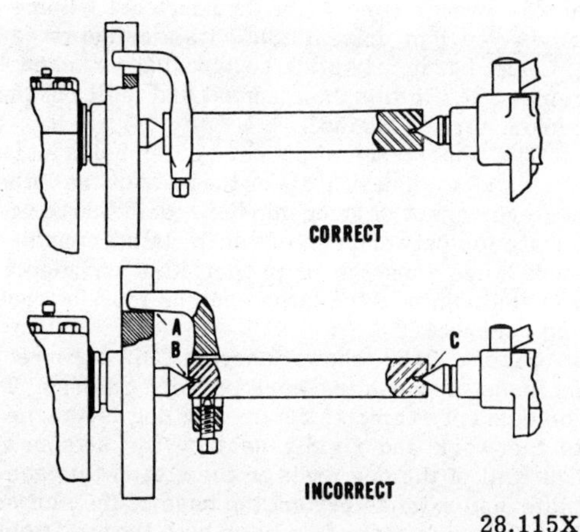

28.115X
Figure 9-22.—Examples of work mounted between centers.

reason, the tail center must be lubricated with a heavy mixture of white lead and oil.

HOLDING WORK ON A MANDREL.—Many parts, such as bushings, gears, collars, and pulleys, require all the finished external surfaces to run true with the hole which extends through them; that is, the outside diameter must be true with the inside diameter or bore. The general practice is to finish the hole to a standard size, within the limit of the accuracy desired. Thus, a 3/4-inch standard hole would ordinarily be held from 0.7505 inch or a tolerance of one-half thousandths of an inch above or below the true standard size of exactly 0.750 inch. First, drill or bore the hole to within a few thousandths of an inch of the finished size; then remove the remainder of the material with a machine reamer, following with a hand reamer if the limits are extremely close.

Then press the piece on a mandrel tight enough so the work will not slip while being machined and clamp a dog on the mandrel being mounted between centers. Since the mandrel surface runs true with respect to the lathe axis, the turned surfaces of the work on the mandrel will be true with respect to the hole in the piece.

A mandrel is simply a round piece of steel (of convenient length) which has been centered and turned true with the centers. Commercial mandrels are made of tool steel, hardened and ground with a slight taper (usually 0.0005 inch per inch). On sizes up to 1 inch the small end is usually one-half thousandth of an inch under the standard size of the mandrel; on larger sizes this dimension is usually one thousandth of an inch under standard. This taper allows the standard hole in the work to vary according to the usual shop practice, and still provides a drive to the work when the mandrel is pressed into the hole. The taper is not great enough to distort the hole in the work. The countersunk centers of the mandrel are lapped for accuracy, and the ends are turned smaller than the body of the mandrel and provided with flats which give a driving surface for the lathe dog.

HOLDING WORK IN CHUCKS.—The independent chuck and universal chuck are more often used than other workholding devices in performing lathe operations. The universal chuck is used for holding relatively true cylindrical work when accurate concentricity of the machined surface and holding power of the chuck is secondary to time required to do the job. When the work is irregular in shape, must be accurately centered, and must be held securely for heavy feeds and depth of cuts, the independent chuck should be used.

Figure 9-23 shows a rough casting mounted in a four-jaw independent lathe chuck on the spindle of the lathe. Before truing the work, determine which part you wish to have turn true. To mount this casting in the chuck, proceed as follows:

1. Adjust the chuck jaws to receive the casting. Each jaw should be concentric with the ring marks indicated on the face of the chuck. If there are no ring marks, be guided by the circumference of the body of the chuck.

2. Fasten the work in the chuck by turning the adjusting screw on jaw No. 1 and jaw No. 3, a pair of jaws opposite each other. Next tighten jaws No. 2 and No. 4. (Brass shim stock may be used to protect the work when it is mounted in the chuck.)

3. At this stage, the work should be held in the jaws just tight enough so it will not fall out of the chuck during truing.

4. Revolve the spindle slowly and, with a piece of chalk, mark the high spot (A, fig. 9-23) on the work while it is revolving. Steady your hand on the tool post while you hold the chalk.

5. Stop the spindle. Locate the high spot on the work and adjust the jaws in the proper direction to true the work by releasing the jaw opposite the chalk mark and tightening the one nearest the mark.

6. Sometimes the high spot on the work will be located between adjacent jaws. In this case, loosen the two opposite jaws and tighten the jaws adjacent to the high spot.

7. To mount the work in a four-jaw chuck so that previously machined surfaces may be centered accurately, use a dial test indicator.

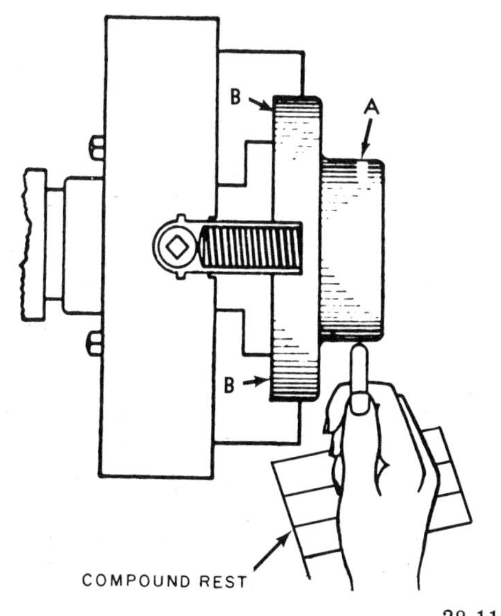

28.119
Figure 9-23.—Work mounted in a 4-jaw chuck.

The three-jaw universal or scroll chuck is so made that all jaws move together or apart in unison. A universal chuck will center almost exactly at the first clamping; but after a period of use it is not uncommon to find inaccuracies of from 2 to 10 thousandths of an inch in centering the work, and consequently the run-out of the work must be corrected. Sometimes this may be accomplished by inserting a piece of paper or thin shim stock between the jaw and the work on the high side.

After you position a piece in a chuck, be sure to tighten all the screws to have each jaw tight against the piece to prevent it from slipping.

When chucking thin sections, be careful not to clamp the work too tightly; as the diameter of the piece is machined when it is in a distorted position, and when pressure of the jaws is released, there will be as many high spots as there are jaws, and the turned surface will not be true.

To preserve a chuck's accuracy, handle it carefully and keep it clean and free from grit. Never force a chuck jaw by using a pipe as an extension on the chuck wrench.

Before mounting a chuck, remove the live center and fill the hole with a rag to prevent chips and dirt from getting into the taper hole of the spindle. Removal of the center is necessary to prevent the possibility of its being ruined when drilling work held in the chuck. (The operator may inadvertently drill right through the center.)

Clean and oil the threads of the chuck and the spindle nose—dirt or chips on the threads will not allow the chuck to run true when it is screwed up to the shoulder. Screw the chuck on carefully, and avoid bringing it up against the shoulder so fast that the chuck comes up with a shock, which strains the spindle and the threads and makes removal difficult. Never use mechanical power in screwing on the chuck. Rotate the spindle with the left hand while holding the chuck in the hollow of the right arm.

To remove a small chuck, place an adjustable jaw wrench on one of the jaws and start it by a smart blow with the hand on the handle of the wrench. To remove a heavy chuck, rotate it against a block of wood held between a jaw and the lathe bed. When mounting or removing a heavy chuck, lay a board across the bed ways to protect them; the board will serve as a support for the chuck as it is put on or taken off.

The above comments on mounting and removing chucks also apply to faceplates.

HOLDING WORK ON A FACEPLATE.—A faceplate is used for mounting work which cannot be chucked or turned between centers. This may be necessary because of the peculiar shape of the work. A faceplate may be used when holes are to be accurately machined in flat work, or when large and irregularly shaped work is to be faced on the lathe.

Work is secured to the faceplate by bolts, clamps, or any suitable clamping means. The holes and slots in the faceplate are used for anchoring the holding bolts. Angle plates may be used to present the work at the desired angle, as shown in figure 9-24. Note the counterweight added for balance.

In order for work to be mounted accurately on a faceplate, the surface of the work in contact with the faceplate must be accurately faced. For very accurate work, the faceplate itself should be refaced by a light cut over its surface. It is good practice to place a piece of paper between the work and the faceplate to prevent slipping.

Before securely clamping the work, move it about on the surface of the faceplate until the

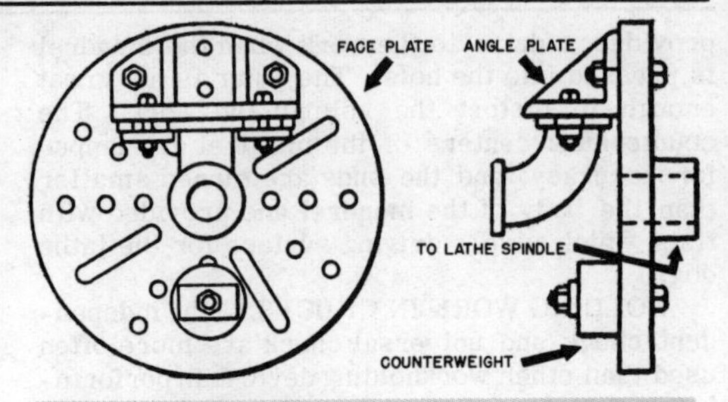

28.124X
Figure 9-24.—Work clamped to an angle plate.

point to be machined is centered accurately over the axis of the lathe. Suppose you wish to bore a hole whose center has been laid out and marked with a prick punch. Proceed as follows: Clamp the work to the approximate position on the faceplate. Then prepare a rod with a countersunk center hole to fit the tailstock center at one end, and with an accurate center point on the other end. Slide the tailstock up and place the rod with the point in the prick punch mark on the work and the other end on the tail center. Then revolve the work slowly. If the punch mark is off center, the point of the rod will describe a small circle (appear to wobble); it it is right on center, the rod will remain stationary. For very accurate centering, a dial indicator held in the tool post and applied to the rod will indicate a very small movement of the rod (to a thousandth of an inch).

USING THE CENTER REST AND FOLLOWER REST.—Although supported at the ends by the lathe centers, long slender work often requires support between ends during turning; otherwise, the work would spring away from the tool and chatter. The center rest is used to support such work so it can be accurately turned with a faster feed and cutting speed than would be possible without it.

The center rest should be placed where it will give the greatest support to the piece to be turned. This is usually at about the middle of its length. Ensure that the center point between the jaws of the center rest coincides exactly with the axis of the lathe spindle. To do this, place a short piece of stock in a chuck and machine it to the diameter of the workpiece to be supported. Without removing the stock from the chuck, clamp the center rest on the ways of the lathe and adjust the jaws to the machined surface. Without changing the jaw settings, slide the center rest into position for supporting the workpiece. Remove the stock used for setting the center rest and set the workpiece in place. Use a dial indicator to true the workpiece at the chuck. Figure 9-25 shows how a chuck and center rest are used when machining the end of a workpiece.

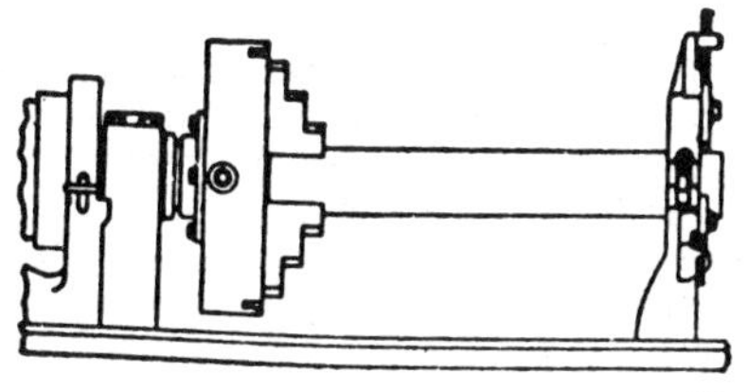

28.126X

Figure 9-25.—Work mounted in a chuck and center rest.

The follower rest differs from the center rest in that it moves with the carriage and provides support against the forces of the cut only. The tool should be set to the diameter selected and a SPOT turned about 5/8 to 3/4 inch wide. Then the follower rest jaws should be adjusted to the finished diameter to follow the tool along the entire length to be turned.

Use a thick mixture of white lead and oil on the jaws of the center rest and follower rest to prevent SEIZING and scoring the workpiece. Check the jaws frequently to see that they do not become hot. The jaws may expand slightly if they get hot, thus pushing the work out of alignment (when using the follower rest) or binding (when using the center rest).

HOLDING WORK IN A DRAW-IN COLLET CHUCK.—The draw-in collet chuck is used for very fine accurate work of small diameter. Long work can be passed through the hollow drawbar, and short work can be placed directly into the collet from the front. The collet is tightened on the work by rotating the drawbar to the right. This draws the collet into the tapered closing sleeve; the opposite operation releases the collet.

Accurate results are obtained when the diameter of the work is exactly the same size as the dimension stamped on the collet. In some cases, the diameter may vary as much as 0.002 inch; that is, the work may be 0.001 inch smaller or larger than the collet size. If the work diameter varies more than this, it will impair the accuracy and efficiency of the collet. This is why a separate collet should be used for each small variation of work diameter, especially if precision is desired.

LATHE MACHINING OPERATIONS

Up to this point you have studied the preliminary steps leading up to the performance of machine work in the lathe. You have learned how to mount the work and the tool, and which tools to use for various purposes. Now, you need to learn how to use proper tools in combination with the lathe in order to perform various machining operations.

Facing

Facing is the machining of the end surfaces and shoulders of a workpiece. In addition to squaring the ends of the work, facing provides a means of accurately cutting the work to length. Generally, in facing the workpiece, only light cuts are required as the work will have been cut to approximate length or rough machined to the shoulder.

Figure 9-26 shows the method of facing a cylindrical piece. The work is placed on centers

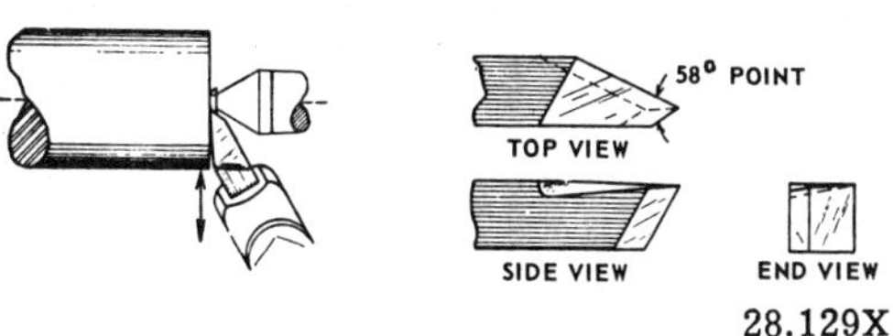

28.129X

Figure 9-26.—Facing a cylindrical piece.

and driven by a dog. A right-hand side tool is used as shown, and a light cut is taken on the end of the work, feeding the tool (by hand cross-feed) from the center toward the outside. One or two cuts are taken to remove sufficient stock to true the work. Then place the dog on the other end of the work and face it to the proper length. A steel rule is used to measure off the length. Another rule or straightedge held on the end that has just been faced provides an accurate base from which to measure. Be sure there is no burr on the edge to keep the straightedge from bearing accurately on the finished end. Use a sharp scriber to mark off the dimension desired.

Figure 9-27 shows the application of a turning tool in finishing a shouldered job with a fillet corner. A finish cut is taken on the small diameter. The fillet is machined with a light cut; then the tool is used to face from the fillet to the outside diameter of the work.

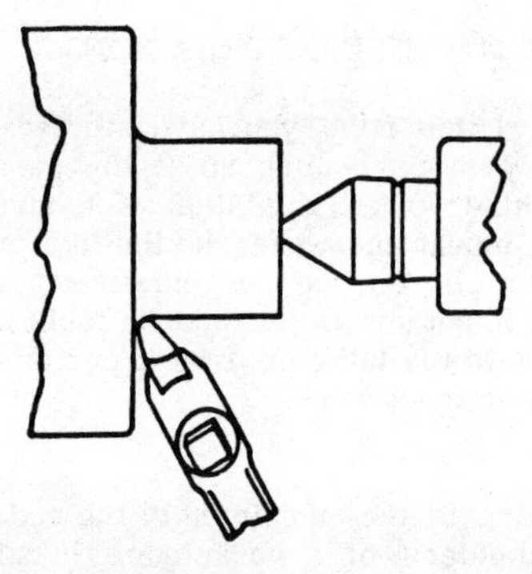

28.130X

Figure 9-27.—Facing a shoulder.

In facing large surfaces, the carriage should be locked in position, since only cross-feed is required to traverse the tool across the work. With the compound rest set at 90° (parallel to the axis of the lathe), the micrometer collar can be used to feed the tool to the proper depth of cut in the face. For greater accuracy in obtaining a given size in finishing a face, the compound rest may be set at 30°. In this position, one-thousandth of an inch movement of the compound rest will move the tool exactly a half of a thousandth of an inch in a direction parallel to the axis of the lathe. (In a 30°-60° right triangle, the length of the side opposite the 30° angle is equal to one-half the length of the hypotenuse.)

Turning

Turning is the machining of excess stock from the periphery of work in order to reduce its diameter. If a considerable amount of the stock must be removed from the periphery, make a series of rough cuts first to remove most of the excess and then use a finishing cut to finish the job.

Select the proper tool for taking a heavy chip. The speed of the work, and the amount of feed of the tool should be as great as the tool and machine will stand.

When taking a roughing cut on steel, cast iron, or any other metal with scale on its surface, be sure to set the tool deep enough to get under the scale during the first cut. Unless you do this, the scale on the metal will dull the point of the tool. Figure 9-28 shows the position of the tool for taking a heavy chip on large work. Set the tool so that it will not dig into the work, but will move in the direction of the arrow—away from the work. Setting the tool in the position shown in the illustration sometimes prevents chatter.

After you turn the work to within about 1/32 inch of the finished size, take a finishing cut. A fine feed, proper lubricant, and a keen-edged tool are necessary to produce a smooth finish. With a micrometer, measure the work carefully to be sure you are maching the work to proper dimensions.

When very close limits must be held, make certain the work is not hot when you take the finishing cut. If you cut it to the exact size while hot, it will be undersized when it cools.

On work to be finished with a cylindrical grinder, leave a limited amount of stock (from

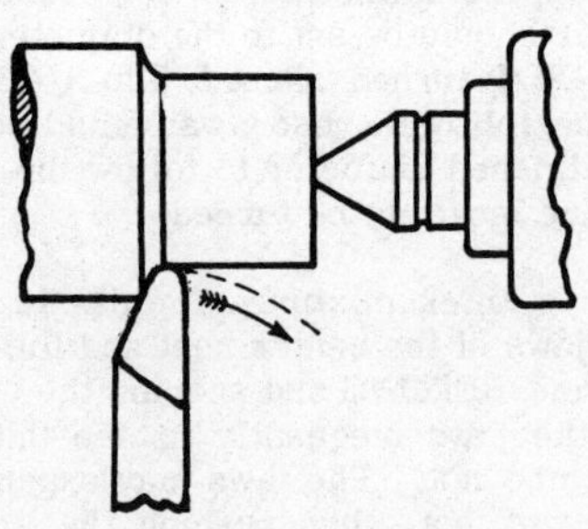

28.132X

Figure 9-28.—Position of tool for heavy cut.

1/64 to 1/32 inch) for grinding to finished dimensions.

Perhaps the most difficult operation for a beginner in machine work is making accurate measurements. So much depends on the accuracy of the work that you should make every effort to become proficient in the use of measuring instruments. A certain FEEL in the application of a micrometer is developed through experience; so do not be discouraged if your first efforts do not produce perfect results. Practice taking micrometer measurements on pieces of known dimensions.

Machining to a shoulder is often accomplished by locating the shoulder with a parting tool. Insert the parting tool about 1/32 inch back of the shoulder line, so that it enters the work within 1/32 inch back of the shoulder line, of the smaller diameter of the work. Then machine the stock by taking heavy chips up to the shoulder thus made. Shouldering eliminates detailed measuring and speeds up production.

Figure 9-29 illustrates the method of shouldering. A parting tool has been used at P and the turning tool is taking a chip. It will be unnecessary to waste time taking measurements. Devote your time to rough machining until the necessary amount of stock is removed. Then finish to accurate measurements.

Boring

Boring is the machining of holes or any interior cylindrical surface. The piece to be bored must have a drilled or cored hole, and the hole must be large enough to insert the tool. The boring process merely enlarges the hole to the desired size or shape. The advantage of boring is that a perfectly true round hole is obtained, and two or more holes of the same or different diameters may be bored at one setting, thus ensuring absolute alignment of the axes of the holes.

The usual practice is to bore a hole to within a few thousandths of an inch of the desired size and finish with a reamer to the exact size.

Work to be bored may be held in a chuck, bolted to the faceplate, or bolted to the carriage. Long pieces must be supported at the free end in a center rest.

When the boring tool is fed into the hole in work being rotated on a chuck or faceplate, the process is called single-point boring. It is the same as turning except that the cutting chip is taken from the inside. The cutting edge of the boring tool resembles that of a turning tool. Boring tools may be of the solid-forged type or the inserted cutter-bit type.

When the work to be bored is clamped to the top of the carriage, a boring bar is held between centers and driven by a dog. The work is fed to the tool by the automatic longitudinal feed of the carriage. Three types of boring bars are shown in figure 9-30.

Note the countersunk center holes at the ends to fit the lathe centers.

Part A of illustration 9-30 shows a boring bar fitted with a fly cutter held by a headless setscrew. The other setscrew, bearing on the end of the cutter, is for adjusting the cutter to the work.

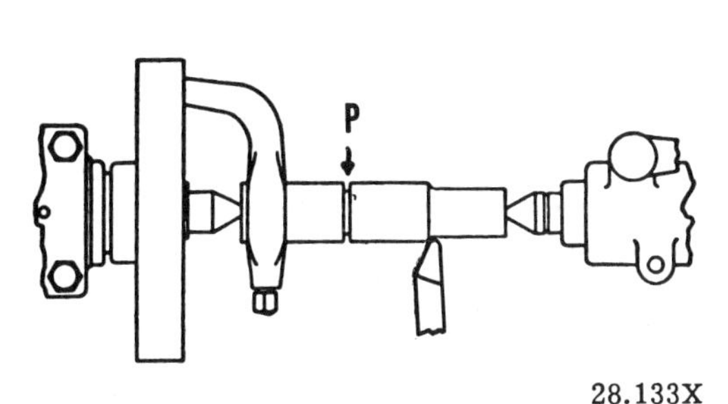

28.133X

Figure 9-29.—Machining to a shoulder.

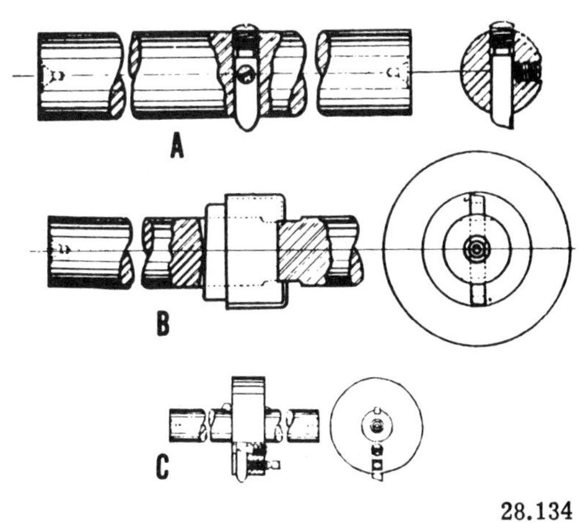

28.134

Figure 9-30.—Boring bars.

Part B of figure 9-30 shows a boring bar fitted with a two-edge cutter held by a taper key. This is more of a finishing or sizing cutter, as it cuts on both sides and is used for production work.

The boring bar shown in part C of figure 9-30 is fitted with a cast-iron head to adapt it for boring work of large diameter. The head is fitted with a fly cutter similar to the one shown in part A of figure 9-30. The setscrew with the tapered point adjusts the cutter to the work.

Tapers

Taper may be defined as gradual lessening of the diameter or thickness of a piece of work toward one end. The amount of taper in any given length of work is found by subtracting the size of the small end from the size of the large end. Taper is usually expressed as the amount of taper per foot of length, or as an angle. Use the following formula and work two examples:

$$\text{Taper per foot} = T \times \frac{12}{L}$$

T = amount of taper
L = length

EXAMPLE 1.—Find the taper per foot of a piece of work 2 inches long, with a diameter at the small end of 1 inch and a diameter at the large end of 2 inches. The amount of the taper is 2 inches minus 1 inch, which equals 1 inch. The length of the taper is given as 2 inches. Therefore, the taper is 1 inch in 2 inches of length. In 12 inches of length it would be 6 inches. See figure 9-31.

EXAMPLE 2.—Find the taper per foot of a piece 6 inches long, with a diameter at the small end of 1 inch and a diameter at the large end of 2 inches. The amount of taper is the same as in problem 1; that is, 1 inch. The length of this taper, however, is 6 inches; hence the taper per foot is 1 inch x 12/6 = 2 inches per foot (fig. 9-31).

From the foregoing discussion, you can see that the length of a tapered piece is very important in computing the taper. If you bear this in mind when machining tapers, you will have no difficulties.

Now let us consider the angle of the taper. In a round piece of work the included angle of the taper is twice the angle the surface makes with

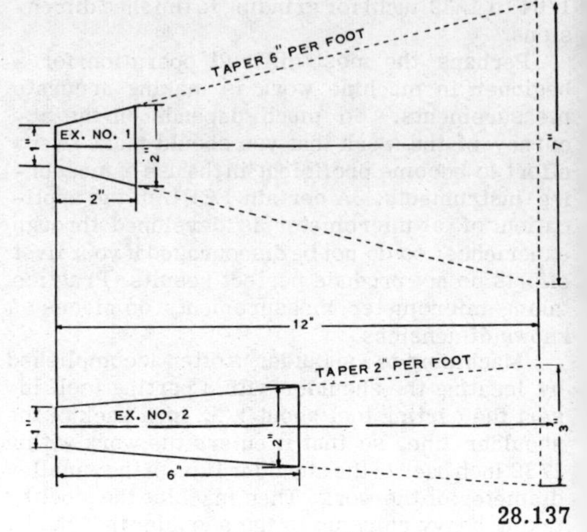

Figure 9-31.—Tapers.

the axis or center line. In straight turning, the diameter of a piece is reduced by twice the depth of the cut taken from its surface. For the same reason, the included angle of the taper is twice the angle the path of the cutting tool makes with the axis or center line of the piece being turned. Tables or charts in machinists' handbooks give the angles for different amounts of taper per foot.

In ordinary straight turning, the cutting tool moves along a line parallel to the axis of the work, causing the finished job to be the same diameter throughout. If in cutting the tool moves at an angle to the axis of the work, a taper is produced. When you therefore turn a taper, so mount the work in the lathe that the axis upon which it turns is at an angle to the axis of the lathe, or causes the cutting tool to move at an angle to the axis of the lathe.

There are three methods in common use for turning tapers:

1. SETTING OVER THE TAILSTOCK. This method moves the dead center away from the axis of the lathe and causes work supported between centers to be at an angle with the axis of the lathe.
2. COMPOUND REST. When set at an angle this rest causes the cutting tool to feed at the desired angle to the axis of the lathe.
3. TAPER ATTACHMENT which also causes the cutting tool to move at an angle to the axis of the lathe.

In the first method, the cutting tool is fed by the longitudinal feed parallel to the lathe axis, but a taper is produced because the work axis is at an angle. In the second and third methods, the work axis coincides with the lathe axis; but a taper is produced because the cutting tool moves at an angle.

The tailstock top may be moved laterally on its base by means of adjusting screws. In straight turning, you will recall that these adjusting screws were used to align the dead center with the tail center by moving the tailstock to bring it on the centerline. For taper turning, we deliberately move the tailstock off center, and the amount we move it determines the taper produced. The amount of setover can be approximately set by means of the zero lines inscribed on the base and top of the tailstock, as shown in figure 9-32. For final adjustment, measure the setover with a scale between center points, as illustrated in figure 9-33.

In turning a taper by this method, the distance between centers is of the utmost importance. To illustrate, figure 9-34 shows two very different tapers produced by the same amount of setover of the tailstock, because in one case the length of the work between centers is greater than in the other. THE CLOSER THE DEAD CENTER IS TO THE LIVE CENTER, THE STEEPER THE TAPER PRODUCED.

Suppose you desire to turn a taper on the full length of a piece 12 inches long with a diameter of 3 inches at one end and a diameter of 2 inches at the other end. The small end is to be 1 inch smaller than the large end; so we set the tailstock over one-half this amount, or 1/2 inch in this case. At one end the cutting tool will therefore be 1/2 inch closer to the center of the work than at the other end; so the diameter of the finished job will be 2 x 1/2 or 1 inch less at the small end. Since the piece is 12 inches long, we have produced a taper of 1 inch per foot. If you wish to produce a taper of 1 inch per foot on a piece only 6 inches long, the small end is only 1/2 inch less in diameter than the large end; so the tailstock must be set over 1/4 inch or one-half of the distance used for the 12-inch length.

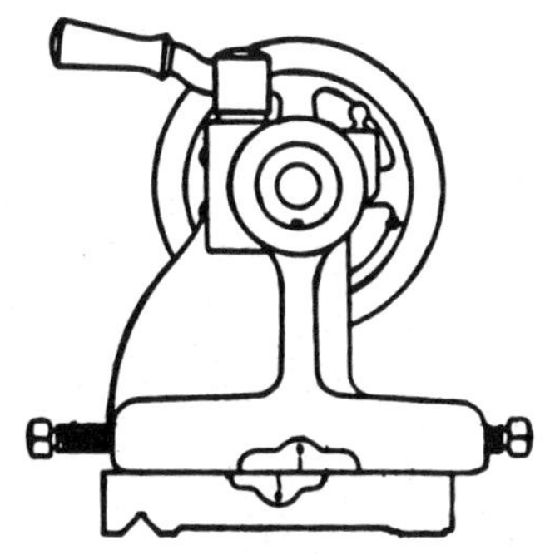

28.139X
Figure 9-32.—Tailstock set-over for taper turning.

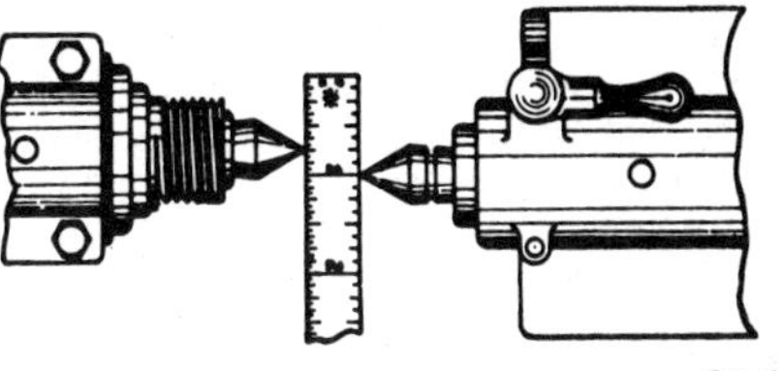

28.140X
Figure 9-33.—Measuring set-over of dead center.

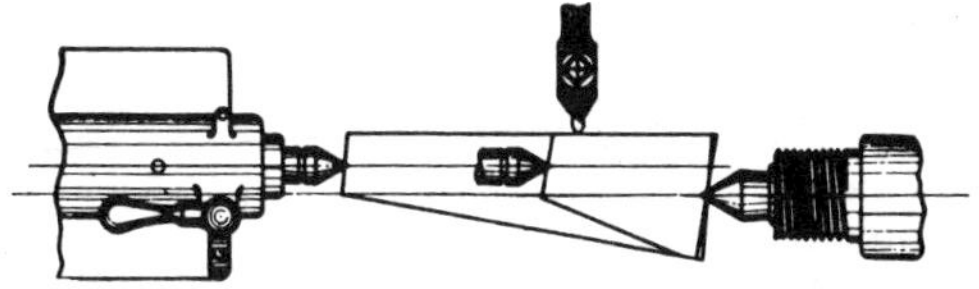

28.141X
Figure 9-34.—Set-over of tailstock showing importance of considering length of work.

From the foregoing discussion, you can see that the setover is proportional to the length between centers and may be computed by the following formula:

$$S = \frac{T}{2} \times \frac{L}{12}$$

S = setover in inches.
T = taper per foot.

$$\frac{L}{12} = \text{length in feet}$$

Remember that L is the length (in inches) of the work from live center to dead center. If the work is on a mandrel, L is the length of the mandrel between centers.

The setover tailstock method cannot be used for steep tapers, because the setover necessary would be too great and the work would not be properly supported by the lathe centers. It is obvious that with setover there is not a true bearing between the work centers and the lathe center points, and that the bearing surface becomes less and less satisfactory as the setover is increased.

After you turn a taper by the tailstock setover method, do not forget to realign the centers for straight turning of your next job.

The compound rest is generally used for short steep tapers. It is set at the angle the taper, is to make with the centerline (half the included angle of the taper). The tool is then fed to the work at this angle by means of the compound rest feed screw. The length of taper which can be machined is necessarily short because of limited travel of the compound rest top.

Truing a lathe center is one example of usage of the compound rest for taper work. Other examples are: refacing of an angle type valve disk, machining the face of a bevel gear, and similar work. Such jobs are often referred to as WORKING TO AN ANGLE rather than as taper work.

The graduations marked on the compound rest provide a quick means for setting to the angle desired. When set at zero, the compound rest is perpendicular to the lathe axis. When set at 90° on either side, the compound rest is parallel to the lathe axis.

On the other hand, when the angle to be cut is measured from the centerline, the setting of the compound rest corresponds to the complement of that angle. (The complement of an angle is that angle which added to it makes a right angle; that is, angle plus complement = 90°.) For example, to machine a 50° included angle (25° angle with centerline), the compound rest must be set at 90°–25°, or 65°.

When a very accurate setting of the compound rest is to be made to a fraction of a degree, run the carriage up to the faceplate and set the compound rest with a vernier bevel protractor set to the required angle. Hold the blade of the protractor on the flat surface of the faceplate, and the stock of the protractor against the finished side of the compound rest.

For turning and boring long tapers with accuracy the taper attachment is indispensable. It is especially useful in duplicating work; identical tapers can be turned and bored with one setting of the taper guide bar.

The guide bar is set at an angle to the lathe axis corresponding to the taper desired. By means of a shoe which slides on the guide bar as the carriage moves longitudinally, the tool cross slide is moved laterally. The resultant movement of the cutting tool is along a line parallel to the guide bar, and therefore a taper whose angular measurement is the same as that set on the guide bar is produced. The guide bar is graduated in degrees at one end, and in inches per foot of taper at the other end to facilitate rapid setting.

When preparing to use the taper attachment, run the carriage up to the approximate position of the work to be tuned. Set the tool on line with the centers of the lathe. Then bolt or clamp the holding bracket to the ways of the bed (the attachment itself is bolted to the back of the carriage saddle) and tighten clamp C, figure 9-35. The taper guide bar now controls the lateral movement of the cross slide. Set the guide bar for the taper desired and the attachment is ready for operation. Make the final adjustment of the tool for size by means of the compound rest feed screen, since the cross feed screw is inoperative.

MILLING MACHINES

A milling machine removes metal by means of a revolving cutting tool called a milling cutter.

28.143X

Figure 9-35.—Turning a taper with a taper attachment.

With various attachments, a milling machine may be used for: (1) boring, broaching, circular milling, dividing, and drilling; (2) cutting of keyways, racks, and gears; and (3) fluting of taps and reamers.

BED TYPE and KNEE AND COLUMN TYPE milling machines are generally found in most Navy machine shops. A bed type milling machine has a vertically adjustable spindle. The horizontal boring mill discussed in this chapter is representative of the bed type. The knee-and-column milling machine has a fixed spindle and a vertically adjustable table. These milling machines have various classes, but only those classes with which you will be concerned are discussed in this chapter.

KNEE-AND-COLUMN MILLING MACHINES

The type of milling machine most commonly used in the Navy is KNEE-AND-COLUMN. Because of its ease of setup and versatility, this machine is more efficient than other types. The main casting consists of an upright column, to which is fastened a bracket, or KNEE, which supports the table. The knee is adjustable on the column, so that the table can be raised or lowered to accommodate work of various size.

Vertical cuts may be taken by feeding the table up or down. The table may be moved in the horizontal plane in two directions at right angles to the axis of the spindle or parallel to the axis of the spindle. Because of this feature, work can be mounted at practically any location on the table. Knee-and-column milling machines are made in three designs: plain, universal, and vertical spindle.

Although knee and column milling machines vary slightly in design, the components labeled in figure 9-36 are common to most milling machines. The column has an accurately machined and scraped vertical dovetail bearing surface. The knee is firmly gibbed to the column dovetail, thus providing a means of vertical movement of the knee on a sliding bearing. The saddle slides on a horizontal dovetail sliding bearing (parallel to the axis of the spindle) on the knee. The swivel table (on universal machines only) is attached to the saddle and can be swiveled approximately 45 degrees (forward to back). The spindle nose has a standard internal taper. Driving keys or lugs are provided on the face of the spindle nose for driving the cutter directly, or for driving an arbor or adapter on which the cutter is mounted. The overarms, yokes, and overarm

28.197X
Figure 9-36.—Universal milling machine.

supports are used to provide accurate alignment and to support arbors. The overarm may be retracted into the column or extended out of the column by the amount necessary to support any length arbor. The overarm supports are extremely beneficial for supporting the cutter when taking heavy cuts and are used in conjunction with the yokes and overarms.

Standard Equipment

Standard Equipment provided with milling machines on Navy ships includes workholding devices, spindle attachments, cutters, arbors, and any special tools needed for setting-up the machine for milling. This equipment permits holding and cutting of many milling jobs encountered in Navy repair work.

VISES commonly used on milling machines are the flanged plain view, the swivel vise, and the toolmakers universal vise (fig. 9-37). The flanged vise provides a rigid workholding setup when the surface to be machined must be parallel to the surface seated in the vise. The swivel vise is used similarly to the flanged vise, but the setup is less rigid and permits the workpiece to be swiveled in a horizontal plane to

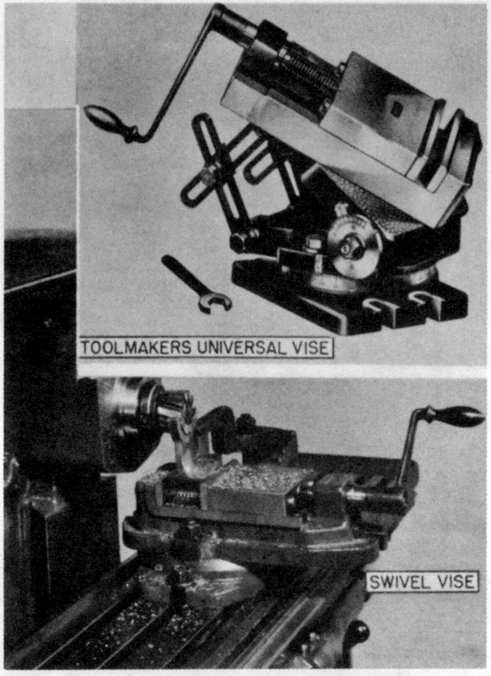

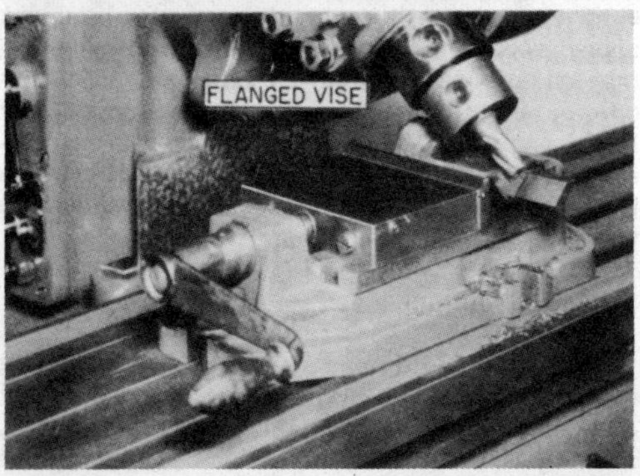

28.199X
Figure 9-37.—Milling machine vises.

any required angle. The toolmakers universal vise is used when the workpiece must be set up at a complex angle in relation to the axis of the spindle and to the table surface. These vises have locating keys or tongues on the underside to ensure that they are positioned correctly in relation to the T-slots on the milling machine table.

INDEXING EQUIPMENT provided with milling machines is illustrated in figure 9-38. Indexing equipment is used to hold the workpiece and to provide a means of turning the workpiece so that a number of accurately spaced cuts may be made (gear teeth for example). The workpiece is held in a chuck, attached to the index head spindle, or between a live center in the index head and a dead center in the footstock. The center rest may be used to provide support for long slender work. The center of the footstock may be raised or lowered as required for setting up tapered workpieces.

The basic components of an index head are shown in figure 9-39. The ratio between the worm and gear is 40 to 1; thus by turning the worm one turn, the spindle is rotated 1/40 of a revolution. The index plate, which has a series of holes in concentric circles, permits accurate gaging of partial turns of the worm shaft and allows the spindle to be turned accurately in amounts smaller than 1/40 of a revolution. The index plate may be secured to the index head housing or the the worm shaft. The crank pin can be adjusted radially for use in any circle of holes. The sector arms can be set to span any number of holes in the index plate to provide a guide for rotating the index crank for partial turns.

The index head spindle can be turned directly by hand, by the index crank through the worm and worm gear, or by the table feed mechansim through a gear train. The first two methods are used for indexing; the third method is used for rotating the workpiece (while it is being cut) to provide a means of making helical cuts. The spindle of the index head is set in a swivel block so that the spindle can be set at any angle from slightly below horizontal to slightly past vertical. An index plate, usually with a 24-hole circle, is provided for placement back of the chuck or center so that the spindle can be indexed rapidly by hand for commonly required divisions.

The CIRCULAR MILLING ATTACHMENT or rotary table shown in figure 9-40 provides a means of setting-up work which must be rotated in a horizontal plane. The worktable is graduated (1/2° to 360°) around its circumference. The table may be turned by hand or by its feed mechanism through a gear train. An 80 to 1 worm and gear drive in the rotary table and index plate arrangement makes this device useful for accurate indexing of horizontal surfaces.

The UNIVERSAL MILLING ATTACHMENT shown in figure 9-40 is clamped to the column of the milling machine. The cutter can be secured

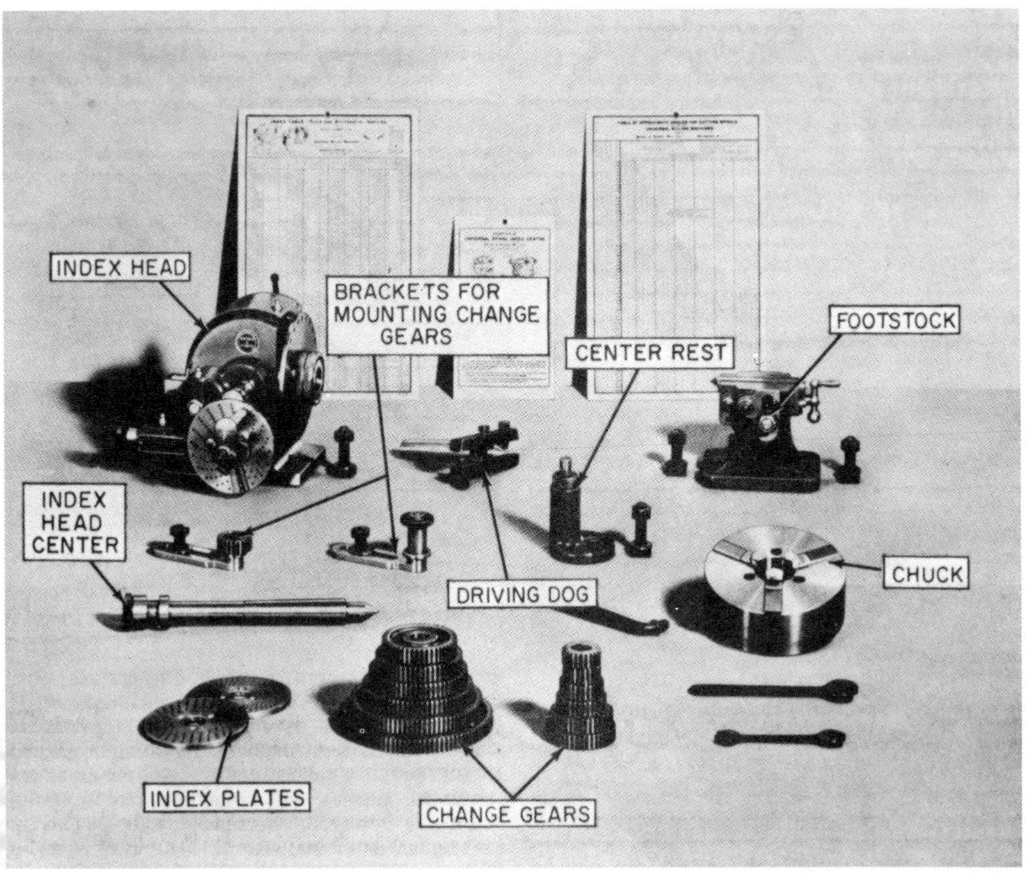

28.200X

Figure 9-38.—Index head with footstock.

in the spindle of the attachment and then—by means of the two rotary swivels—can be set so that the cutter will cut at any angle to the horizontal or the vertical plane. The spindle of the universal milling attachment is driven by gearing connected to the milling machine spindle.

Milling machine cutters are generally classified according to methods of mounting. Arbor cutters are cutters with straight, tapered, or threaded holes for mounting on an arbor. The most common type has a straight hole with a keyway through it or across one end. By means of a key inserted in this keyway, the cutter is prevented from turning on the arbor. Shank cutters have straight or tapered shanks and are mounted in collets or adapters. Facing cutters are attached to either a stub arbor or directly to the milling machine spindle. Figure 9-41 illustrates various arbors, sleeves, and adapters for mounting cutters.

Some milling cutters are illustrated in figure 9-42. These cutters are made from carbon steel, high-speed steel, Stellite, or tool steel with cemented carbide teeth. Some of the more common cutters and the operations to which they are best suited are described in the following paragraphs.

PLAIN-MILLING CUTTERS (numbers 27 and 30, fig. 9-42) are the most widely used of all cutters. They are used for milling flat surfaces parallel to the cutter's axis. The cutter is cylindrical and has teeth cut on the periphery only.

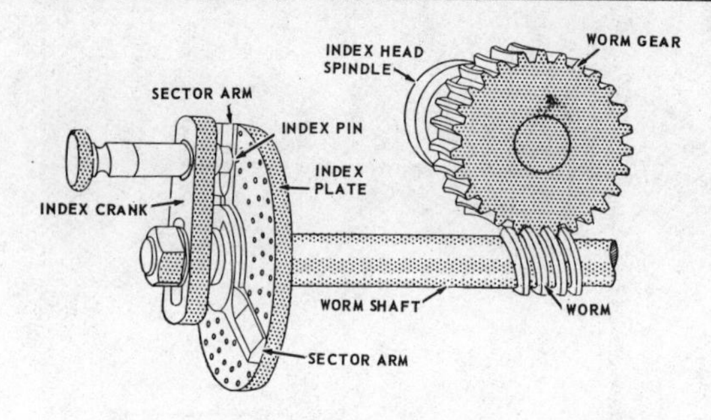

28.201
Figure 9-39.—Index head mechanism.

28.202X
Figure 9-40.—Circular milling attachment and universal head.

Plain milling cutters are made in a variety of diameters and widths. The teeth may be either straight or spiral in shape, but the latter type is generally used when the cutter is more than 3/4 inch wide. A cutter tooth straight or parallel to its axis receives a distinct shock as the tooth starts to cut. To eliminate this shock and thereby

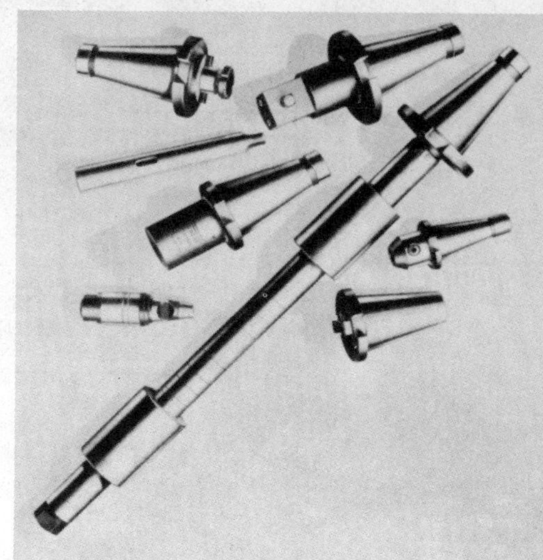

28.204X
Figure 9-41.—Arbors, sleeves, and special adapters.

produce a free cutting action, cutters are often made with helical teeth. A spirally gashed cutter, particularly when used on wide surfaces, gives a much smoother result than a straight gashed cutter. Spirally gashed cutters also require less power to operate and, since the stress on the cutter is relieved, the tendency to chatter is reduced. When the plain milling cutter is made with relatively few teeth and a fairly steep angle of spiral, the cutter is commonly called a coarse tooth cutter. Such cutters are used because of their ability to remove considerable quantities of metal with minimum power. Plain mills with very few teeth and helical milling cutters have a very steep angle of spiral. They are particularly efficient on heavy slabbing cuts. Because of the shearing action of the teeth, they can be used to advantage in removing an uneven amount of stock without gouging. They are made in both HOLE AND ARBOR types for milling forms from solid metal.

SIDE-MILLING CUTTERS (numbers 14, 28, and 29, fig. 9-42) are comparatively narrow milling cutters with teeth on each side as well as on the outer surface. When used in pairs, with an appropriate spacer between them, these cutters can mill parallel sides.

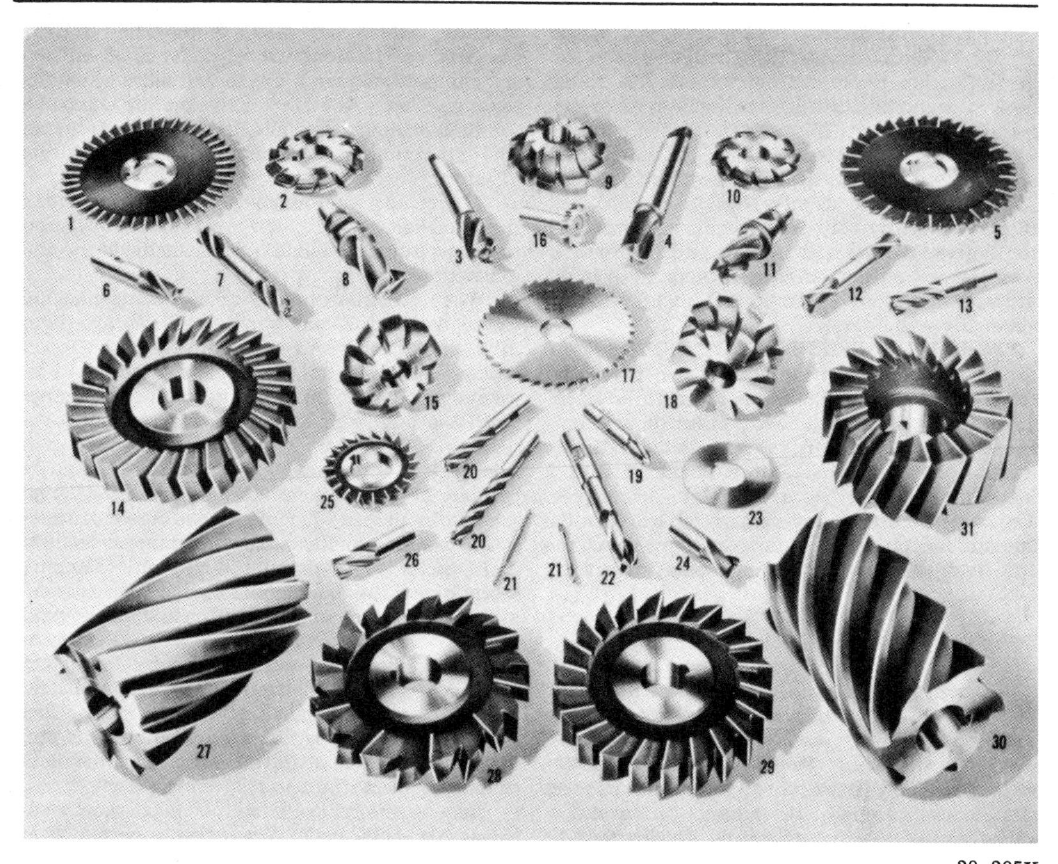

28.205X

1. Metal slitting saw.
2. Involute spur gear cutter (undercut teeth).
3. Spiral end mill, taper shank.
4. Two-lipped spiral mill, taper shank.
5. Metal staggered-tooth slitting saw.
6. Long two-lipped end mill, single end.
7. Long spiral end mills, double end.
8. Two-lipped spiral end mill, double end.
9. Corner rounding cutter.
10. Involute form cutter.
11. Spiral end mill, cam-locking.
12. Long two-lipped spiral end mill, double end.
13. Long spiral end mill, single end.
14. Half side milling cutter.
15. Convex cutter.
16. Woodruff keyseat cutter.
17. Metal slitting saw.
18. Concave cutter.
19. Ball end mills.
20. Long single-end end mill.
21. Double-end end mills.
22. Two-lipped long single-end end mill.
23. Screw slotting cutter.
24. Two-lipped spiral end mill, straight shank.
25. Angular cutter.
26. Spiral end mill, straight shank.
27. Plain heavy duty milling cutter.
28. Staggered tooth side milling cutter.
29. Side milling cutter.
30. Helical plain milling cutter.
31. Shell end miss for use with shell end mill arbor.

Figure 9-42.—Milling machine cutters.

METAL SLITTING SAWS (numbers 1, 5, and 17, fig. 9-42) used on milling machines are essentially thin, plain milling cutters. The thickness of a metal slitting saw tends to decrease toward the center. This is to provide clearance between cutter and work when you are milling deep slots or cutting off thick sections of metal. Slitting saws are usually less than 3/16 inch thick. Generally, slitting saws have more teeth for a given diameter than plain milling cutters. For heavy sawing in steel, staggered tooth slitting saws, from 3/16 to 3/8 inch thick, are generally used.

WOODRUFF KEYSEAT CUTTERS are used for cutting semicylindrical keyways in shafts. Cutters under 1 1/2 inches in diameter are provided with a shank and have teeth on the circumferential surface. Their sides are ground slightly concave for clearance. Cutters larger than 1 1/2 inches in diameter are usually of the arbor type. The larger cutters have staggered teeth on the circumferential surface and on the sides. The side teeth are ground for clearance but not for cutting.

Setting Up the Milling Machine

Before starting a milling operation, ensure that the workpiece and the milling machine are arranged properly; and make certain that the workpiece is firmly secured to the holding device. If indexing is required, select the correct method and calculate the number of turns required in the indexing operation. Ensure that the center is correctly secured to the spindle, positioned properly over the workpiece, and set to rotate in the proper direction. Select the correct cutting speeds and feeds. Consider each job individually, as speeds and feeds for milling often vary considerably even on similar jobs.

HOLDING THE WORK.—An efficient and positive method of holding work on the milling machine is most important if the machine tool is to be used to its best advantage. Regardless of the method used in holding the work, there are certain factors that should be observed in every case. The work must not be sprung in clamping; but it must be secured to prevent its springing or moving away from the cutter, and so aligned that it may be correctly machined.

Milling machine tables are provided with several T-slots used either for clamping and locating the work itself or for mounting the various holding devices and attachments. These T-slots extend the length of the table and are parallel to its longitudinal axis. Most milling machine attachments (vises and index heads, for example) have keys or tongues on the underside of their bases to enable the operator to locate the attachments correctly in relation to the T-slots.

There are various methods of holding work, in accordance with type of work and the operation to be performed. Some of these methods are discussed next.

When work is clamped to the milling machine table, both table and work should be free from dirt and burrs. Work having smoothly machined surfaces may be clamped directly to the table, provided the cutter does not come in contact with the table surface during the machining operation. When work with unfinished surfaces is clamped in this way, the table face should be protected with pieces of soft metal. Clamps should be placed squarely across the work to give a full bearing surface. These clamps are held by bolts inserted in the table's T-slots. Clamping bolts should be placed as near the work as possible so that full advantage of the fulcrum principle may be obtained. When it is necessary to place a clamp on an overhanging part, a support should be provided between the overhang and the table to prevent springing or breakage. When heavy cuts are to be taken, fasten a sturdy stop piece to the table at the tail end of the workpiece to help prevent sliding of the workpiece.

Index centers are used to support work centered on both ends. When the work has been previously reamed or bored, it may be pressed on a mandrel and then mounted between centers.

INDEXING THE WORK.—Indexing may be accomplished by the PLAIN, DIRECT, COMPOUND, or DIFFERENTIAL METHOD. The plain and direct methods are most commonly used; compound and differential methods are used only when the job cannot be done by plain or direct indexing. Only the plain and direct methods are described in this course.

Direct indexing, sometimes called rapid indexing, makes use of the direct index plate mounted just back of the work end of the index head spindle (fig. 9-43). The direct index plate is equipped with 24 evenly spaced holes around its outer edge, which means that only division which divide evenly into 24 can be cut; that is, numbers 1, 2, 3, 4, 6, 8, 12, and 24. To obtain rapid indexing, disengage the worm and worm wheel from the spindle (with the index pin out of

28.209X
Figure 9-43.—Direct index plate.

contact with the direct index plate) and turn the spindle by hand.

Manufacturers supply index heads with various means of disengaging the index head. To divide work into two equal parts, disengage the index pin and revolve the plate and spindle until 11 holes in a 24-hole circle pass the index pin. Then insert the index pin into the 12th hole in the plate to hold the spindle in the proper position. In any indexing operation, always start counting from the hole adjacent to the crankpin. During heavy cutting operations, clamp the spindle by the clamp screw to relieve strain on the index pin.

Plain indexing, accomplished by using the universal index head, is governed by the number of times the index crank must be turned to cause the work to make one revolution. Charts specifying the required number of turns or fractions of a turn and giving the proper index plate for various divisions are furnished by index head manufacturers. If these charts are unavailable, calculate the required number of turns and parts of turns.

The number of turns of the index crank required to index a fractional part of a revolution is determined by dividing 40 by the number of divisions required. For example, if you are required to make 40 divisions on a piece of work, divide 40 by 40, which indicates that one complete turn of the index crank is required for each division. If 10 divisions were required, you would divide 40 by 10, and 4 complete turns of the index crank would be required for each division. Index plates are used to assist in making the division when the quotient of the ratio of the index head and the division desired results in a fraction, thus making it necessary to turn the crank a part of a revolution in indexing. The numerator of the fraction, determined by dividing 40 by the number of divisions required, represents the number of holes in a circle of holes the index crank should be moved for each desired division. The denominator of this fraction represents the number of holes in the correct circle of holes which should be selected on the index plate. For example, the calculation for determining 800 divisions when an index plate with 20 holes is available is as follows:

$$\frac{40}{800} = \frac{1}{20} = \text{1 hole on the 20-hole circle.}$$

When the fraction is such that none of the available index plates contains the number of holes represented by the denominator, multiply both the numerator and denominator by a common multiplier. For example, the calculation for determining 9 divisions when an index plate having a 27-hole circle is available is as follows:

$$\frac{40}{9} \times \frac{3}{3} = \frac{120}{27} = 4\frac{12}{27} = \text{4 complete turns plus 12 holes on the 27-hole circle}$$

If the denominator of the fraction is larger than the number of holes available in an index plate, divide both the numerator and denominator by a common divisor which will give a fraction whose denominator represents the number of holes for which the index plate is available. For example, the calculation for determining 76 divisions when an index plate with a 19-hole circle is available is as follows:

$$\frac{40}{76} \div \frac{4}{4} = \frac{10}{19} = \text{10 holes in the 19-hole circle.}$$

If when reducing the fraction the denominator becomes so small that no available index plate contains the number of holes represented by the denominator, raise the fraction to an available number. For example, the calculation for determining 52 divisions when an index plate with a 39-hole circle is available, is as follows:

$$\frac{40}{52} \div \frac{4}{4} = \frac{10}{13} \times \frac{3}{3} = \frac{30}{39},$$ or 30 holes in a 39-hole circle.

To use the index head sector arms, turn the left-hand arm to the left of the index pin, and insert the pin into the first hole in the circle of holes to be used. Then loosen the set screw and so adjust the right-hand arm of the sector that the correct number of holes is contained between the two arms (fig. 9-39). After you make this adjustment, lock the set screw to hold the arms in position. When setting the arms, count the required number of holes from the one in which the pin is inserted, considering this hole as zero. By subsequent use of the index sector arms, counting the holes for each division is eliminated. When you use the index crank to revolve the spindle, unlock the spindle clamp screw. However, before you cut work held in or on the index head, lock the spindle again to relieve the strain.

MOUNTING THE CUTTER.—To mount a cutter on an arbor (fig. 9-44):

1. Select an arbor having the same diameter as the hole in the cutter.
2. Remove the arbor nut and as many spacers as necessary so that the cutter can be positioned as near the spindle nose as practical. Remember that the cutter must be far enough away from the spindle nose to permit the workpiece to clear the column during the milling cut. NOTE: Watch for left-hand threads on the arbor and arbor nut.
3. Place the cutter on the arbor and align the cutter and arbor keyways and insert a key.
4. Replace the required number of spacers so that tightening the arbor nut will clamp the cutter between the spacers.
5. Screw the arbor nut by hand. NOTE: Do NOT tighten the arbor nut with a wrench at this time.
6. Place the arbor in the milling machine spindle and insert the draw-in bolt through the spindle, and screw the bolt into the arbor by hand as far as possible. Then back the draw-in bolt out of the arbor about one turn.
7. Tighten the draw-in bolt locking nut with a wrench until the arbor is tightly secured in the spindle.
8. Position the overarm and yoke to provide adequate support for the cutter. Then, using a wrench, take up on the arbor nut to clamp the cutter securely.

The procedure for installing an adapter for tapered shank cutters is similar to the procedure for installing the arbor in the mill spindle. Then insert the taper shank cutter in the tapered hole of the adapter. Tap the cutter end lightly with a rawhide mallet to ensure secure seating.

Face mills are usually mounted directly on the spindle nose of the mill. The back of the face mill is counterbored to fit the spindle nose and

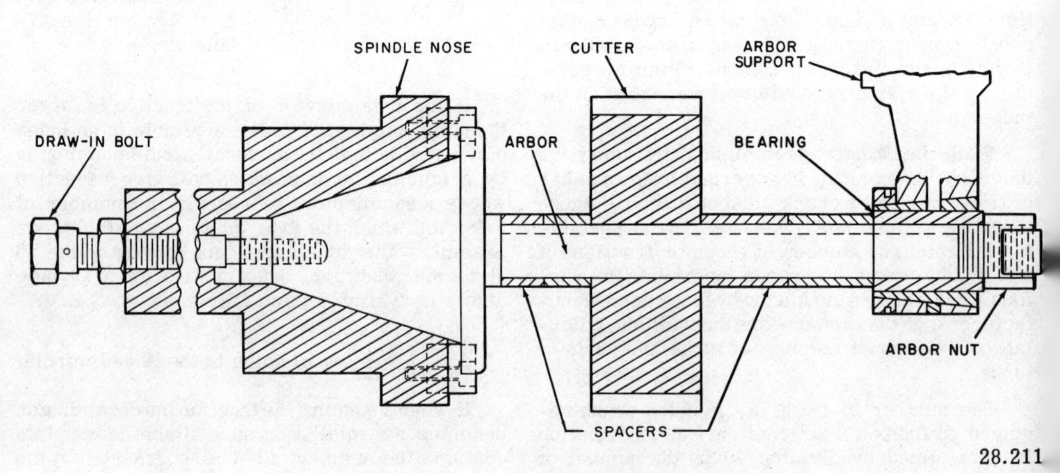

Figure 9-44.—Standard arbor.

has radial slots which fit the driving lugs of the spindle. The cutter is secured to the spindle nose by bolts inserted through the face of the cutter and screwed into the spindle nose.

Before mounting a cutter, always ensure that the cutter, adapter, arbor, and mill spindle are clean and free of burrs and upset edges.

Figure 9-45 shows common methods of positioning a cutter. Methods A, B, and C in the illustration can be used on cylindrical or noncylindrical workpieces. Methods D and E are used when centering the cutter on the axis of cylindrical workpieces; method E is used when the workpiece is mounted between centers.

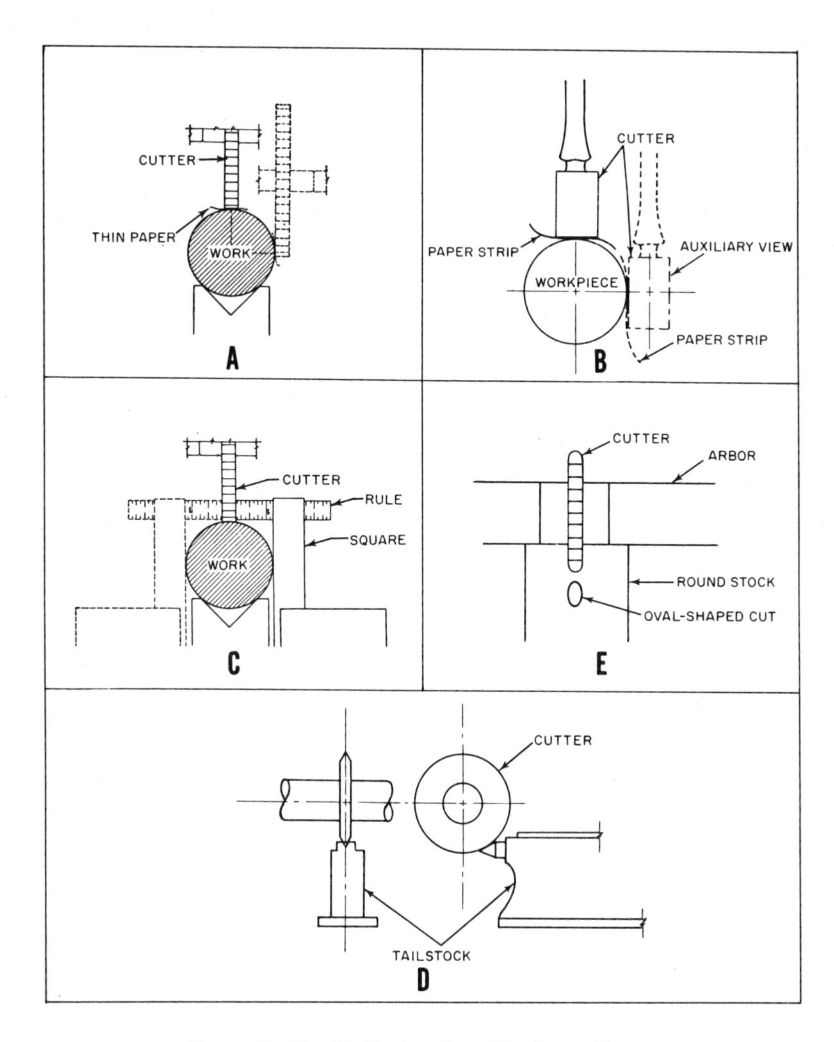

28.212

Figure 9-45.—Methods of positioning cutter.

The methods illustrated in A and B of figure 9-45 are the most accurate methods and should be used when possible. To position a cutter by these methods, do the following:

1. Move the workpiece into position as shown by the auxiliary views in A and B of figure 9-45, with the cutter about 0.010 inch away from the workpiece.

2. Insert a strip of paper (0.003 inch thick) between the cutter and the side of the workpiece and hold it in place.

3. Start the motor and turn the cutter slowly, and feed the workpiece toward the cutter until the cutter tears the paper strip; then feed the table toward the cutter another 0.003 inch (thickness of the paper) to bring the cutter in to very light contact with the workpiece.

4. Lower the workpiece enough for the cutter to clear the top of the workpiece.

5. Set the micrometer collar on the transverse feed handwheel to zero.

6. Move the worktable transversely by an amount equal to one-half the thickness of the cutter plus one-half the diameter of the workpiece (part A, fig. 9-45). The cutter is now centered on the axis of the shaft, and can be set to the proper depth of cut by moving the table upward to the prescribed depth.

The method just described works equally well on cylindrical and noncylindrical workpieces and with end mills as well as arbor type cutters. If the cutter is so small that the arbor or spindle nose touches the workpiece, the cutter can be aligned with some degree of accuracy by using a straightedge placed on the side of arbor type cutters or periphery of end mills for aligning the cutter to a zero point. In moving the workpiece transversely, remember that the thickness (of an arbor cutter) or the diameter (of an end mill) will affect the final transverse position of the cutter. CAUTION: Keep your hands clear of the cutter when using the paper strip.

Part E of figure 9-45 illustrates a method of centering a cutter on the axis of a workpiece used when the tooth profile of the cutter is convex. The work is so adjusted that the cutter is approximately centered over the work. Then the work is moved up until the rotating cutter takes a light depth of cut. If a regular oval-shaped cut appears, the cutter is entered; if the profile of one side of the oval differs from the other side, the workpiece must be adjusted transversely.

When you select the direction of cutter rotation and table travel, make the cutter revolve against the advancing table (fig. 9-46). This is the conventional milling practice, sometimes called the UP METHOD. In milling deep slots, or in cutting off thin stock with a metal slitting cutter, another system known as the CLIMB MILLING process is used. When using this process, feed the work with the rotation of the cutter to make the cutter cut down into the work. The system diminishes the probability of crooked slots produced when the cutter is drawn to one side.

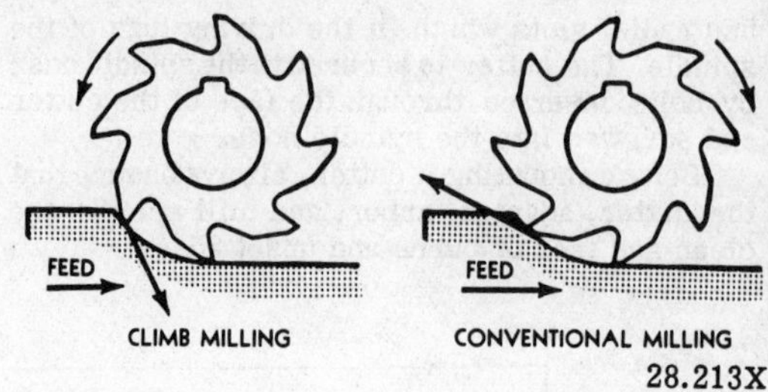

28.213X
Figure 9-46. —Conventional and climb milling.

When the work moves with the cutter, set the table gibs snugly in order to eliminate looseness and/or lost motion in the table. If you fail to eliminate looseness, the cutter teeth may draw the work in and perhaps cause a sprung arbor, a badly damaged cutter, a ruined piece of work, or serious personal injury.

SELECTING FEEDS, SPEEDS, AND COOLANTS.—Milling machines usually have a spindle speed range from 25 to 2000 rpm and a feed range from 1/4 inch to 30 inches per minute (ipm). The feed is independent of the spindle speed, which means that a workpiece can be fed at any rate available in the feed range, regardless of the spindle speed being used. Some of the factors concerning the selection of appropriate feeds and speeds for milling are discussed in the following paragraphs.

Heat generated by friction between the cutter and the work may be regulated by using proper speed, feed, and cutting coolant. Regulation of this heat is very important, because the cutter is dulled or made useless by overheating. It is almost impossible to set down any fixed rules concerning cutting speeds, because conditions vary from job to job. Generally speaking, however, select a cutting speed which gives the best compromise between maximum production and longest life of the cutter. In any particular operation, consider the following factors when dermining proper cutting speed:

1. HARDNESS OF THE MATERIAL BEING CUT. The harder and tougher the metal being cut, the slower should be the cutting speed.

2. DEPTH OF CUT AND DESIRED FINISH. The amount of friction heat produced is directly proportional to the amount of material being removed. Finishing cuts may therefore often be made at a speed 40 to 80 percent higher than the speed used for rough work.

3. CUTTER MATERIAL. High-speed steel cutters may be operated from 50 to 100 percent faster than carbon steel cutters, because they have better heat resistant properties than carbon steel cutters.

4. TYPE OF CUTTER TEETH. Cutters which have undercut teeth cut more freely than cutters which have a radial face; therefore, cutters with undercut teeth may be run at higher speeds.

5. SHARPNESS OF THE CUTTER. A sharp cutter may be run at a much higher speed than a dull cutter.

6. USE OF COOLANT. In most cases, a sufficient amount of coolant will prevent overheating of the cutter, even at relatively high speeds.

Approximate values listed in table 9-1 may be used as a guide when you are selecting cutting speed. If you find that the machine, the cutter, or the work cannot be suitably operated at the suggested speed, make necessary readjustments.

Refer to table 9-2 to determine the cutter revolutions per minute for cutters varying in diameter from 1/2 inch to 8 inches. When you are cutting with a 7.16-inch cutter and a surface speed of 160 feet per minute is required, for example, the cutter revolutions per minute should be 1,398.

If you don't have a table to determine cutter revolutions per minute (RPM), use the one given for the lathe, as follows:

$$RPM = \frac{CFPM \times 12}{\pi \times D}$$

CFPM = Required surface feed in feet per minute.

D = Diameter of cutter in inches.
Circumference of cutter equals 3.1416 (π) x D.

NOTE: Because the cutter revolves and the worm is stationary, the DIAMETER in the formula is for the cutter, NOT the piece of work.

Table 9-1. —Surface Cutting Speeds

	Carbon steel cutters (ft. per min.)		High Speed steel cutters (ft. per min.)	
	Rough	Finish	Rough	Finish
Cast iron:				
Malleable	60	75	90	100
Hard castings	10	12	15	20
Annealed tool steel	25	35	40	50
Low carbon steel	40	50	60	70
Brass	75	95	110	150
Aluminum	460	550	700	900

The rate of feed is the rate of speed at which the workpiece travels past the cutter. When selecting the feed, consider the following factors:

1. Forces are exerted against the work and the cutter, and their holding devices, during the cutting process. The force exerted varies directly with the amount of metal removed and can be regulated by the feed and depth of the cut. The feed and depth of cut are therefore interrelated and (in turn) are dependent upon the rigidity and power of the machine. Machines are limited by the power they can develop to turn the cutter, and also by the amount of vibration they can withstand when coarse feeds and deep cuts are used.

2. Feed and depth of cut also depend upon the type of cutter used. Deep cuts or coarse feeds, for example, should not be attempted with a small diameter end mill, for this action springs or breaks the cutter. Coarse cutters with strong cutting teeth can be fed at a relatively high rate, because the chips are washed out easily by the cutting lubricant.

3. Do not use coarse feeds and deep cuts on a frail piece of work or on work so mounted that the holding device springs or bends.

4. The desired degree of finish affects the amount of feed. When a fast feed is used, metal is removed rapidly and the finish is not very smooth. A slow feed rate and a high cutter speed, however, produce a finer finish. For roughing, use a comparatively low speed and a coarse feed. More mistakes are made by overspeeding the cutter than by overfeeding the work. Overspeeding may be detected by a squeaking, scraping sound. If chatter occurs in the milling machine during the cutting process, reduce the

Table 9-2.—Cutter Speeds in Revolutions Per Minute

28.213.0

Diameter of cutter (in.)	Surface speed (ft. per min.)																
	25	30	35	40	50	55	60	70	75	80	90	100	120	140	160	180	200
	Cutter revolutions per minute																
1/4	382	458	535	611	764	851	917	1,070	1,147	1,222	1,376	1,528	1,834	2,139	2,445	2,750	3,056
5/16	306	367	428	489	611	672	733	856	917	978	1,100	1,222	1,466	1,711	1,955	2,200	2,444
3/8	255	306	357	408	509	560	611	713	764	815	916	1,018	1,222	1,425	1,629	1,832	2,036
7/16	218	262	306	349	437	481	524	611	656	699	786	874	1,049	1,224	1,398	1,573	1,748
1/2	191	229	268	306	382	420	459	535	573	611	688	764	917	1,070	1,222	1,375	1,528
5/8	153	184	214	245	306	337	367	428	459	489	552	612	736	857	979	1,102	1,224
3/4	127	153	178	203	254	279	306	357	381	408	458	508	610	711	813	914	1,016
7/8	109	131	153	175	219	241	262	306	329	349	392	438	526	613	701	788	876
1	95.5	115	134	153	191	210	229	267	287	306	344	382	458	535	611	688	764
1-1/4	76.3	91.8	107	123	153	168	183	214	230	245	274	306	367	428	490	551	612
1-1/2	63.7	76.3	89.2	102	127	140	153	178	191	204	230	254	305	356	406	457	508
1-3/4	54.5	65.5	76.4	87.3	109	120	131	153	164	175	196	218	262	305	349	392	436
2	47.8	57.3	66.9	76.4	95.5	105	115	134	143	153	172	191	229	267	306	344	382
2-1/2	38.2	45.8	53.5	61.2	76.3	84.2	91.7	107	114	122	138	153	184	213	245	275	306
3	31.8	38.2	44.6	51	63.7	69.9	76.4	89.1	95.3	102	114	127	152	178	208	228	254
3-1/2	27.3	32.7	38.2	44.6	54.5	60	65.5	76.4	81.8	87.4	98.1	109	131	153	174	196	218
4	23.9	28.7	33.4	38.2	47.8	52.6	57.3	66.9	71.7	76.4	86	95.6	115	134	153	172	191
5	19.1	22.9	26.7	30.6	38.2	42	45.9	53.5	57.3	61.1	68.8	76.4	91.7	107	122	138	153

speed and increase the feed. Excessive cutter clearance, poorly supported work, or a badly worn machine gear are also common causes of chatter.

The purpose of a cutting coolant is to reduce frictional heat and extend the life of the cutter's edge. A coolant also lubricates the cutter's face and flushes the chips away, thereby reducing the possibility of damage to the finish. Direct the coolant to the point where the cutter strikes the work, and allow it to flow freely on the work and the cutter.

To determine the proper coolant to use for a particular type of metal, refer to the table on coolants given previously in this chapter for use on a lathe. These coolants are also used for milling operations.

Use kerosene as a cutting coolant when you machine aluminum. Machine cast iron dry.

MILLING OPERATIONS

Milling operations may require shifting of the workpiece, changing the cutter and readjusting feeds and speeds before the job is finished. Each change in setup can usually be considered as a separate job and the methods of cutting and typical examples of milling jobs described here provide information that can be applied to almost any milling operation.

Methods of cutting may be classified under four general headings:

FACE MILLING—machining flat surfaces at right angles to the axis of the cutter.

PLAIN OR SLAB MILLING—machining flat surfaces parallel to the axis of the cutter.

ANGULAR MILLING—machining flat surfaces on an inclination to the axis of the cutter.

FORM MILLING—machining surfaces with irregular outlines.

Explanatory names such as sawing, slotting, gear cutting, etc., have been given to special operations. Routing is the term applied to the milling of an irregular outline while you are controlling the work movement by hand. The grooves in reamers and taps are called flutes. Gang milling is the term applied to an operation in which two or more cutters are used together on one arbor. Straddle milling is the term given to an operation in which two or more milling cutters are used to mill two or more sides of a piece of work at the same time.

Face Milling

Use end and side milling cutters for face milling operations, in accordance with size and nature of the work and the type and size of cutter required. In face milling (fig. 9-47), the teeth on the periphery of the cutter do practically all of the cutting. The face teeth actually remove a small amount of stock left from the spring of the work or cutter, thereby producing a finer finish. Be sure all end play of the spindle is eliminated and that the cutter is properly placed.

When face milling, you may clamp the work to the table or an angle plate or hold it in a vise, fixture, or jig. Feed the work against the cutter in such a way that the pressure of the cut is downward, thereby holding the work against the table.

When setting the depth of cut on a flat surface the work should be brought up to the cutter so that a .002-inch feeler gage, held between the work and the cutter, can just be inserted (or a thin piece of paper will just tear when held between the cutter and the work). At this point, the graduated dial on the transverse feed should be locked and used as a guide in determining the depth of cut. When starting the cut, move the

28.214X

Figure 9-47.—Face milling.

work so that the cutter is nearly in contact with the edge of the work; now the automatic feed may be engaged. If a cut is started by hand, avoid pushing the corner of the work between the cutter teeth too quickly as this may cause the cutter tooth to break. The feed trips (if automatic feed is used) should be adjusted to stop the table travel just as the cutter clears the work. This will avoid idle time during the milling operation. Automatic feed trips protrude from the front side of the table and can be adjusted so that the table will advance the work to the milling cutter at a fast rate. Then, as the cut is taken, the feed is at a slower (predetermined) rate. At the end of the cut, the feed lever is again automatically tripped and the table returns to the start at a fast rate and trips to stop TABLE FEED at the END of the cut only.

Plain and Slab Milling

Plain or slab milling is the term generally used to describe the removal of stock from an uninterrupted horizontal surface as shown in figure 9-48. As slabbing or plain milling usually removes a great amount of stock in a short time, it is essential that maximum rigidity of the workpiece and cutter be provided. Cutters with coarse teeth (to withstand heavy cutting pressures) and large helix angles, up to 45 degrees (to maintain continuous tooth contact and an even cutting pressure) are generally used in slab milling. Note in figure 9-48 that the cutter is mounted on an arbor of large diameter and that the distance between the yoke and column is just enough to permit the workpiece to clear as it passes. Notice also that an overarm-support bracket provides additional support for the cutter setup.

28.215X

Figure 9-48.—Slab milling.

Angular Milling

Angular milling is the milling of surfaces at an angle (other than horizontal or vertical) to the reference or base surface. Angular milling may be performed with formed angular cutters (such as dovetail cutters) by mounting the workpiece at an angle to the cutting surface of the cutter, or by setting the cutter at an angle to the base surface of the workpiece, as when using the universal milling attachment.

MANUFACTURING INSTRUMENT GEARS

You will generally be able to get necessary replacement parts for optical instruments through normal supply channels. On occasions, however, this may not be true and you will be compelled to manufacture spur gears and/or other parts. You learned gear nomenclature in chapter 7 of the manual, and the following discussion of gear manufacture, utilization of spur gear formulas, plus some good experience, will enable you to make a spur gear which will perform satisfactorily in an optical instrument.

The two most important things pertaining to the manufacture of a gear are: (1) calculating gear dimensions, and (2) selecting the proper cutter for machining the gear teeth. Review the discussion of gear nomenclature in chapter 7 and then study the following procedure for making a gear.

Gear calculations and measurements were greatly simplified by perfection of the diametral pitch system, which is based on the diameter of the pitch circle—not the circumference. The circumference of a circle is 3.1416 times its diameter, and you must always consider this constant when you calculate measurements based on the pitch circumference; and in order to simplify computations, this constant (3.1416 x diameter) has been BUILT IN, or made a part of, the diametral pitch system.

When you use the diametral pitch system, you need not calculate circular pitch or chordal pitch—indexing devices based on the system accurately space the teeth, and the formed cutter associated with the indexing device forms the teeth within required accuracy. Calculations of teeth depth, center distances, and all other calculations, have been simplified by the diametral pitch system.

Usually the outside diameter (D_O) of a gear and the number of teeth (N) are listed on the blueprint for a gear. By using these factors, and appropriate gear formulas, you can calculate the data you need for making a gear.

Suppose, for example, that you must make a gear with 24 teeth and a diameter of 3.250 inches. The procedure for doing this is:

1. Find the pitch diameter with this formula:

$$D = \frac{ND_O}{N + 2}$$

When you make proper substitutions in this formula and solve for D, you get:

$$D = \frac{24 \times 3.250}{24 + 2} = \frac{78}{26} = 3.000 \text{ inches}$$

2. Find the diametral pitch (P) by solving (with proper substitutions) the following formula:

$$P = \frac{N}{D}, \text{ or } P = \frac{24}{3} = 8$$

3. Make proper substitutions in the following formula and solve for H to get the whole depth of the tooth:

$$H = \frac{2.157}{P}, \text{ or } H = \frac{2.157}{8} = 0.2696 \text{ inch}$$

After you compute the diametral pitch for your gear, select the proper gear cutter to cut 24 teeth on it.

Formed gear cutters are made with eight (8) different forms (numbered from 1 to 8) for each diametral pitch, in accordance with the number of teeth for which the cutter is to be used. The accompanying chart indicates the range of teeth for various cutters.

Number of Cutter	Range of teeth
1	135 to a rack
2	55 to 134
3	35 to 54
4	26 to 34
5	21 to 25
6	17 to 20
7	14 to 16
8	12 to 13

Since the gear in this example must have 24 teeth, you need a number 5 cutter, which cuts gears which have from 21 to 25 teeth. Most cutters are stamped by number, diametral pitch, range, and depth.

After you cut the teeth on your gear, check your dimensional accuracy with a vernier caliper. Find first the arc tooth thickness and the addendum by using the following formulas, respectively:

$$t = \frac{1.5708}{P} = 0.1964 \text{ inch, in your example}$$

$$a = \frac{3.000}{24} = 0.125 \text{ inch}$$

Then adjust the vertical scale of the caliper to the chordal addendum, the formula for calculating which is:

$$a_c = a + \frac{t^2}{4D}, \text{ or } a_c = 0.125 = \frac{(0.1964)^2}{4 \times 3}$$

$$0.125 + \frac{0.0286}{12} \quad 0.128 \text{ inch}$$

MILLING MACHINE PRECAUTIONS

A milling machine operator's first consideration should be for his own safety, and he should attempt nothing that may endanger his life and limb. CARELESSNESS and IGNORANCE are the two great menaces to personal safety. Milling machines are not playthings and must be accorded the respect due any machine tool. For your own safety, observe the following precautions:

1. Never attempt to operate a machine unless you are sure you thoroughly understand it.
2. Do not throw an operating lever without knowing in advance the outcome.
3. Do not play with control levers, or idly turn the handles of a milling machine, even though it is not running.

4. Never lean against or rest your hands upon a moving table. If it is necessary to touch a moving part, be certain you know in advance the direction in which it is moving.

5. Do not take a cut without making sure that the work is secure in the vise or fixture, and that the holding member is rigidly fastened to the machine table.

6. Always remove chips with a brush or other suitable agent—never with the fingers or hands.

7. Before you attempt to operate any milling machine, study its controls thoroughly so that if an emergency arises during operation you can stop it immediately.

8. Above all, you must keep clear of the cutters. Do not touch a cutter, even when it is stationary, unless there is a good reason for doing so; and if you must touch it, be very careful.

If you follow certain safety practices, operation of a milling machine is not dangerous. There is always danger, however, of getting caught in the cutter. CAUTION: Never attempt to remove chips with the fingers at the point of contact of the cutter with the work. There is some danger to the eyes from flying chips and you must always protect your eyes with goggles and keep them out of line of the cutting action.

CHAPTER 10

SHIP TELESCOPES AND SPYGLASSES

Up to this point in this training manual you have studied: theory of light, reflection and refraction, lenses, mirrors, prisms, image formation, basic optical instruments (including construction), and maintenance procedures for optical instruments. The following chapters therefore deal with details of construction and operation of some optical instruments on which you will perform maintenance. This chapter is concerned with ship telescopes and spyglasses (quartermaster and OOD). Subsequent chapters deal with other optical instruments.

SHIP TELESCOPE

A ship telescope (fig. 10-1) is mounted on or near the ship's bridge—usually the signal bridge—where an observer has the best view (in all directions). This instrument is too bulky for a person to hold in his hands, so it is mounted in a yoke which permits 360° of horizontal rotation. See figure 10-1. The signalman uses a ship telescope to read flag and light signals, to determine the location of bearing sightings (lighthouses, etc.), and to identify ships and other objects at great distances.

The optical system of a Mk 1, Mod O, ship telescope consists of an achromatic doublet (uncemented), a pair of Porro prisms (for erecting images), and an eyepiece. The elements of the doublet are separated by three small tinfoil shims (0.1 inch or 0.002 inch in thickness) equally spaced around the edge of the lens. This system of spacing the elements in a large objecive is called THREE-POINT SUSPENSION, and it is used because deterioration of balsam cement is so great for large lenses that its use is impractical.

Observe in illustration 10-2 the position of the objective lens and the Porro prism assembly, whose prisms have their hypotenuse surfaces parallel and are mounted exactly 90° to each other. When you must replace one of the prisms (or both), check your 7 x 50 binocular prisms in stock, some of which are the same as the Porro prisms used in the Mk 1, Mod O, telescope.

The color filter mount (fig. 10-2) usually contains three disks (light smoke, dark smoke, and clear glass). The smoke filters help to REDUCE GLARE FROM THE WATER. The clear glass disk does not affect image brightness, and it is included in the color filter mount to obviate refocusing of the eyepiece when you turn one of the smoke filters into your line of sight. You can select the filter desired by turning the ray filter knob under the eyepiece, and a spring-loaded ball detent beneath the mount (fig. 10-2) holds the filter in position.

A Mk 1, Mod O, ship telescope is sometimes referred to as a change-of-power telescope, because it has four interchangeable eyepieces which give powers of 13x, 21x, 25x, and 32x. The 21x eyepiece is orthoscopic; the others are Kellner eyepieces. The cells of these eyepieces are so constructed that their front focal planes coincide with the real image formed by the objective when you screw them into the eyepiece tube. This telescope is a moisture-tight instrument and is like an astronomical telescope, except that it has a Porro prism cluster to erect the image produced by the objective.

PREDISASSEMBLY INSPECTION

When you receive an optical instrument in the shop for repair and/or overhaul, check the job order for difficulties listed; then prepare (initially) a casualty analysis sheet for the instrument and continue with the following predisassembly inspection:

1. Inspect the general appearance and housing.
2. Check the cleanliness and physical condition of optical parts and record your findings and pertinent information pertaining thereto

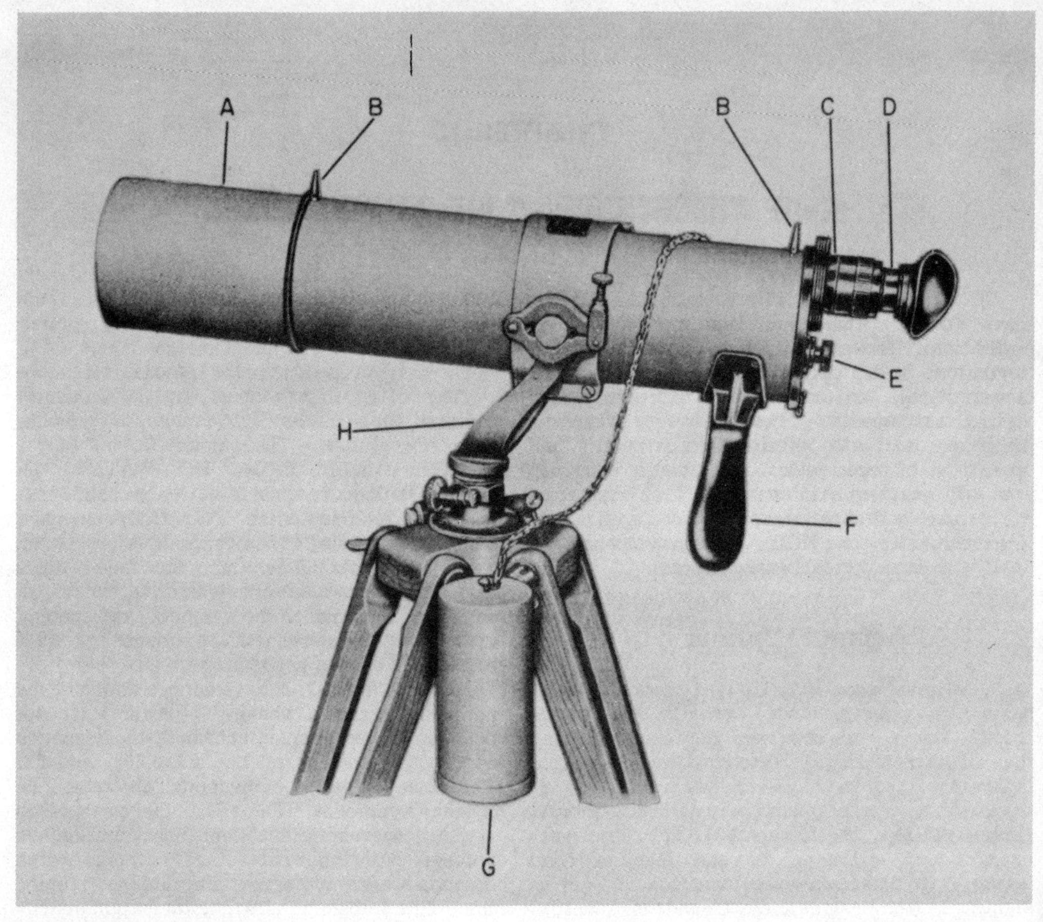

37.2

A. Sunshade assembly.
B. Sighting vane.
C. Diopter scale.
D. Eyepiece assembly.
E. Filter shaft knob.
F. Grip handle.
G. Eyepiece cover.
H. Yoke assembly.

Figure 10-1.–Mark 1, Mod 0, ship telescope.

on your casualty analysis sheet. Test the eyepieces separately. Rotate the filter shaft knob and check for freedom of movement and action. The filters should snap into place and not interfere with the field of view, and the clear filter should be invisible.

3. Inspect the mechanical components, including the yoke assembly and the carrying case, for appearance and condition of finish. Put the telescope on its yoke assembly and check the action. It should balance and have freedom of movement 360° in rotation and 90° above to 25° below the horizontal, and also vertical rotation. Check the eyepiece case for broken or defective parts, and the canvas cover for serviceability.

4. Check the action of the eyepiece focusing ring for smoothness of action. It should turn with smooth, even motion over the full diopter

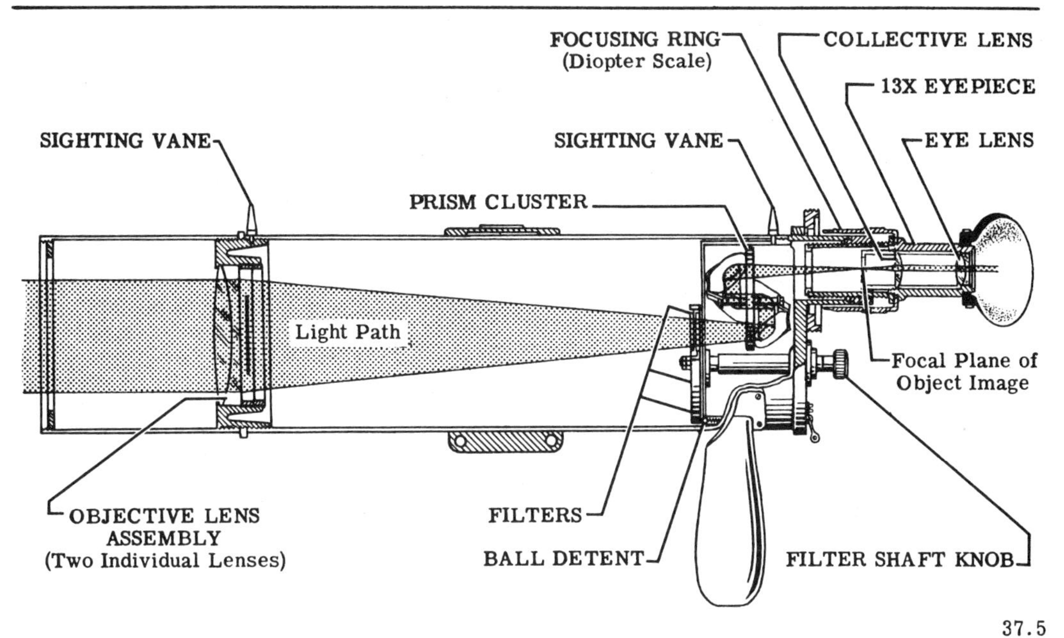

Figure 10-2.—Optical system of a Mk 1, Mod 0, ship telescope.

scale range. Ease of this action indicates the condition of the lubricant in the focusing thread; and it may also indicate whether the eyepiece tube and/or eyepiece adapter is eccentric, causing them to bind during focusing of the eyepiece.

5. Mount the telescope on a collimator and test the diopter scale reading and parfocalization of the eyepieces. Assemble one of the eyepieces. Then adjust an auxiliary telescope to your eye and sight through the ship telescope at the collimator target. Adjust the ship telescope focusing ring to get a sharp image and record the diopter reading and power of the eyepiece. Repeat this operation for the other eyepiece assemblies.

If all four eyepieces give readings within ± 1/4 of zero diopters, as indicated by the diopter scale of the ship telescope, the eyepieces are PARFOCALIZED AND THE DIOPTER SETTING IS CORRECT. If the eyepieces give readings within 1/2 diopter of each other, they are also parfocalized; but the readings may be about some common reference other than zero (one diopter, for example), in which case the diopter scale setting is incorrect and must be adjusted. When the readings are not within 1/2 diopter of each other, the eyepieces are not parfocalized and they and the diopter scale setting must be adjusted.

When you complete the predisassembly inspection, review all notations you made on the casualty analysis sheet for the instrument and then make your recommendations concerning disposition; that is, whether it is feasibly economical to overhaul and repair the instrument or whether it should be salvaged or disposed of otherwise. Submit your findings and recommendations to the shop supervisor, who will make the final decision relative to action to be taken on the instrument.

DISASSEMBLY

If the shop supervisor decides that the ship telescope should be overhauled and repaired, perform only the amount of disassembly required to do the work. Follow your notations made during predisassembly, and also the recommendations of your supervisor.

Before you start to disassemble the telescope, make certain your work space is clean and all tools are in readiness. Then follow these general instructions:

1. Study the disassembly instructions listed by the manufacturer of the instrument.
2. Do not damage parts by forcing them.
3. Work carefully, and as slowly as necessary.
4. Put disassembled parts, by assembly, in parts trays.
5. Mark or tag matched parts with the serial number of the instrument. Many parts are not interchangeable.
6. For information pertaining to disassembly of frozen parts, review the discussion of the subject in chapter 8.

The procedure for disassembling a Mk 1, Mod O, ship telescope is given by steps for assemblies and subassemblies.

Eyepiece Cover Assembly and Objective Cap

To disassemble the eyepiece cover assembly and objective cap, do the following:

1. Put the telescope on a V-block stand, as illustrated in figure 10-3, and remove the chain retaining screw from the face of the prism box, filter, and eyepiece focusing assembly.
2. Unscrew the eyepiece cover assembly from the eyepiece cover adapter (fig. 10-4) and unscrew the objective cap from the end of the sunshade.

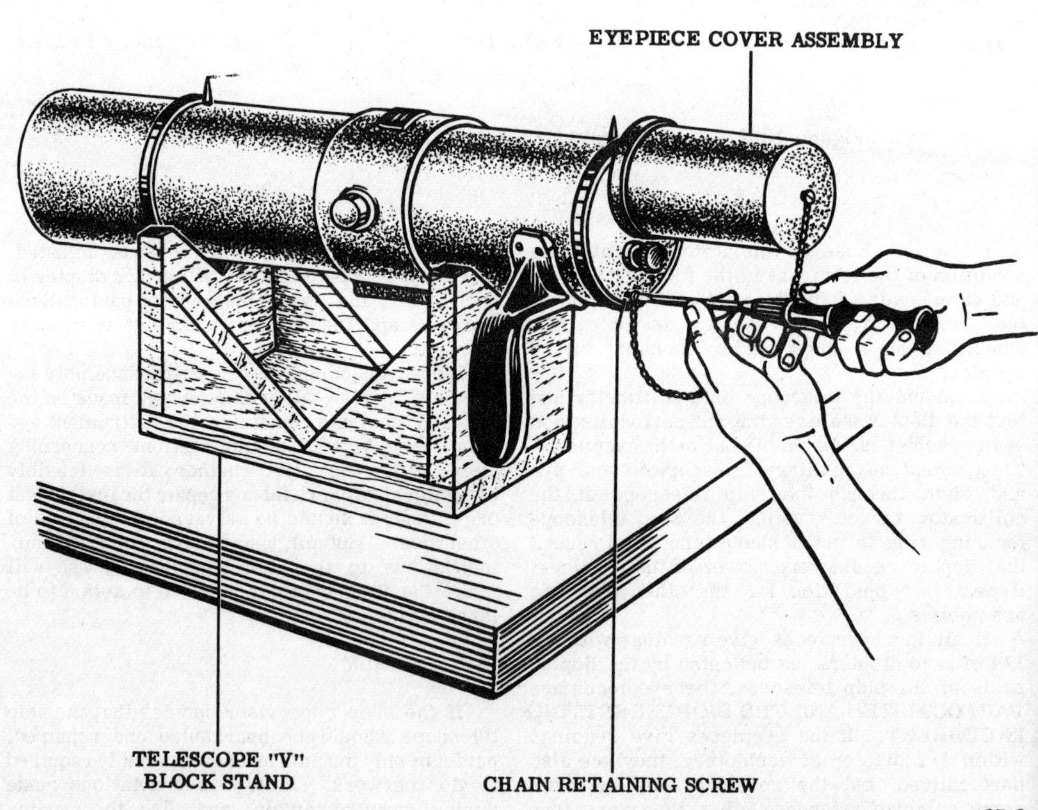

Figure 10-3.—First step in the disassembly of a Mk 1, Mod 0, ship telescope.

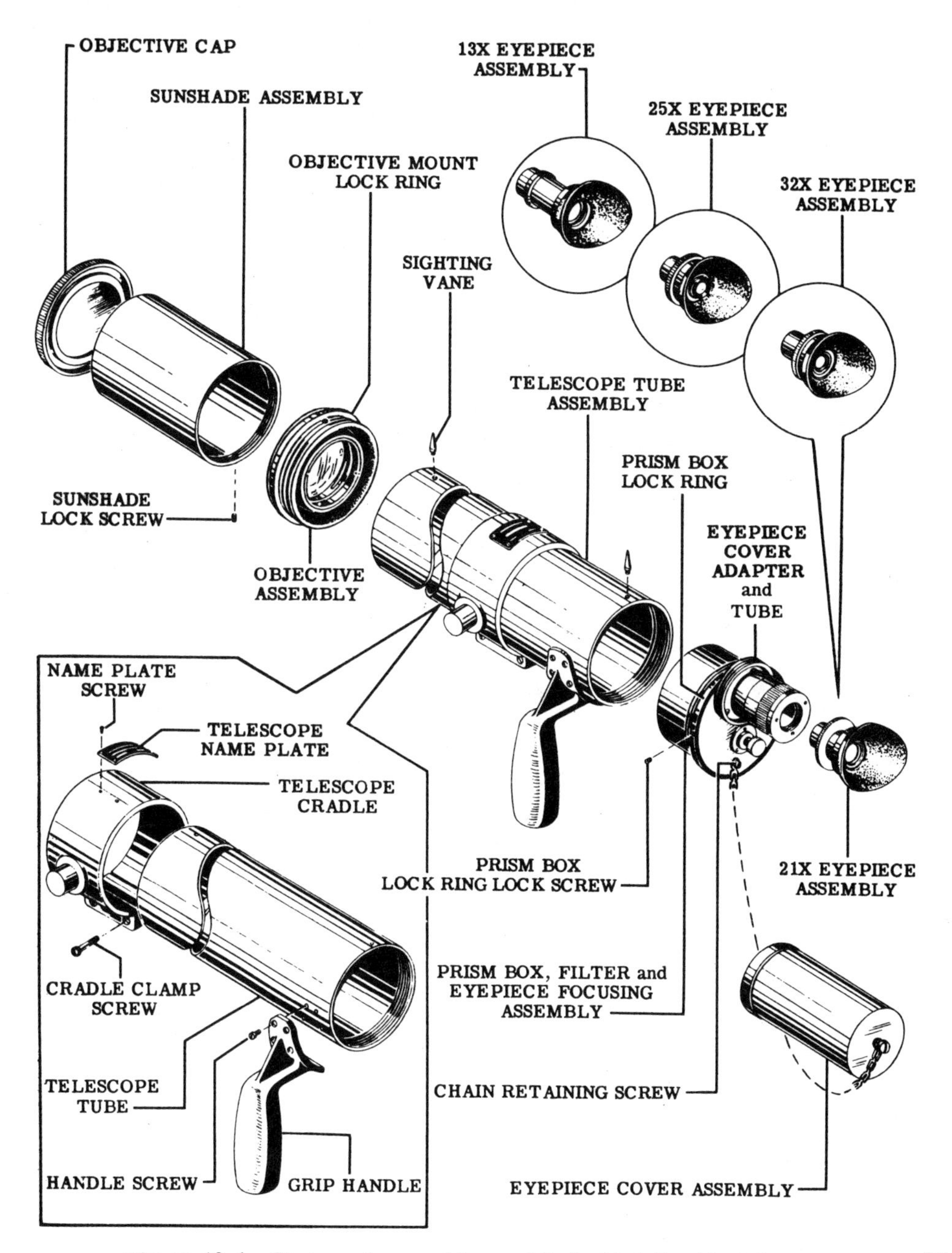

Figure 10-4. Parts and assemblies a Mk 1, Mod 0, ship telescope. 37.2

Eyepiece Assemblies and Sighting Vanes

One eyepiece assembly is mounted in the eyepiece tube of the telescope, and the other three are kept in the eyepiece case. If the eyepiece case is not available in the repair shop, mark the serial number of the telescope on each eyepiece assembly and keep all assemblies together. They are optically matched to each other and their particular ship telescope.

The steps to follow for removing a mounted eyepiece assembly and the sighting vane are:

1. Unscrew the mounted eyepiece assembly from the eyepiece tube. See illustration 10-4.
2. With a sighting vane pin wrench (fig. 10-5), loosen and unscrew the two sighting vanes from the top of the telescope assembly (fig. 10-4).

NOTE: The sighting vanes are also set screws. The front one secures the objective assembly and the rear one secures the prism

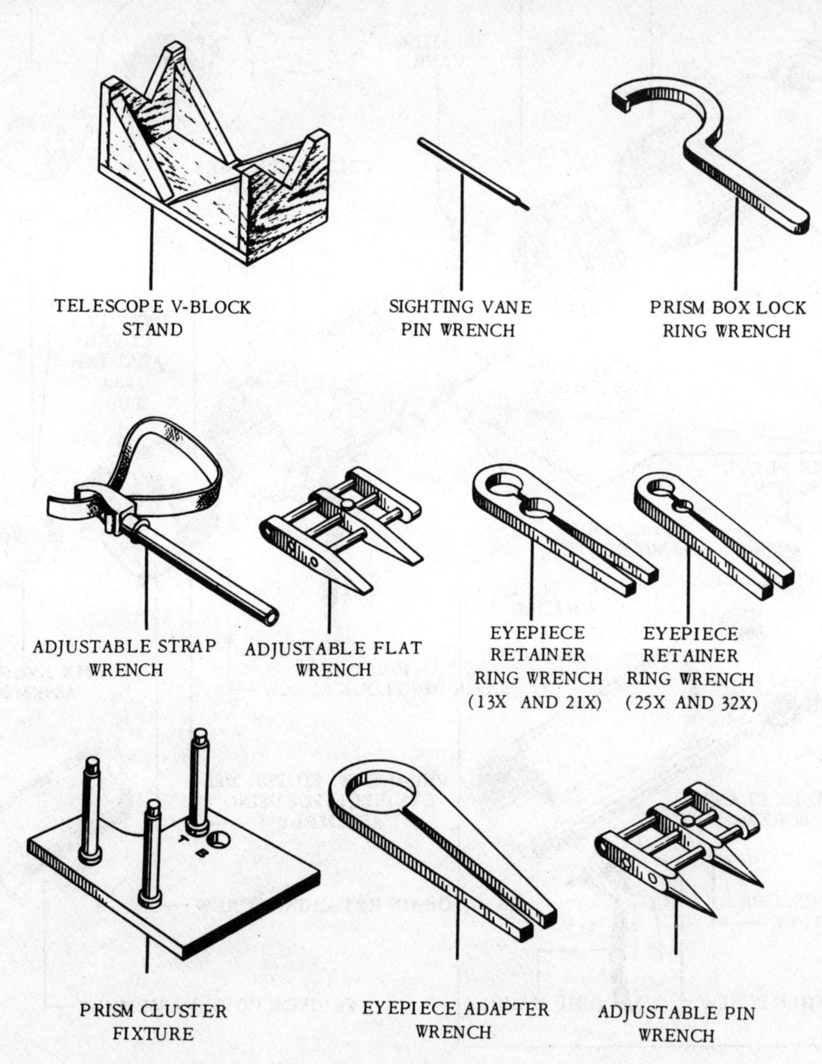

137.209

Figure 10-5.—Ship telescope special service tools.

box, and you cannot remove these assemblies until you remove the sighting vanes.

Prism Box, Filter and Eyepiece Focusing Assembly

The procedure for removing the prism box, filter and eyepiece focusing assembly follows:

1. Remove the prism box ring lock screw from the prism box lock ring (fig. 10-4).
2. With a prism box lock ring wrench (fig. 10-5), loosen the prism box lock ring by turning it one-half to one revolution. Study figure 10-6.
3. With your hand (fig. 10-7), unscrew and remove the prism box and the filter and eyepiece focusing assembly from the telescope

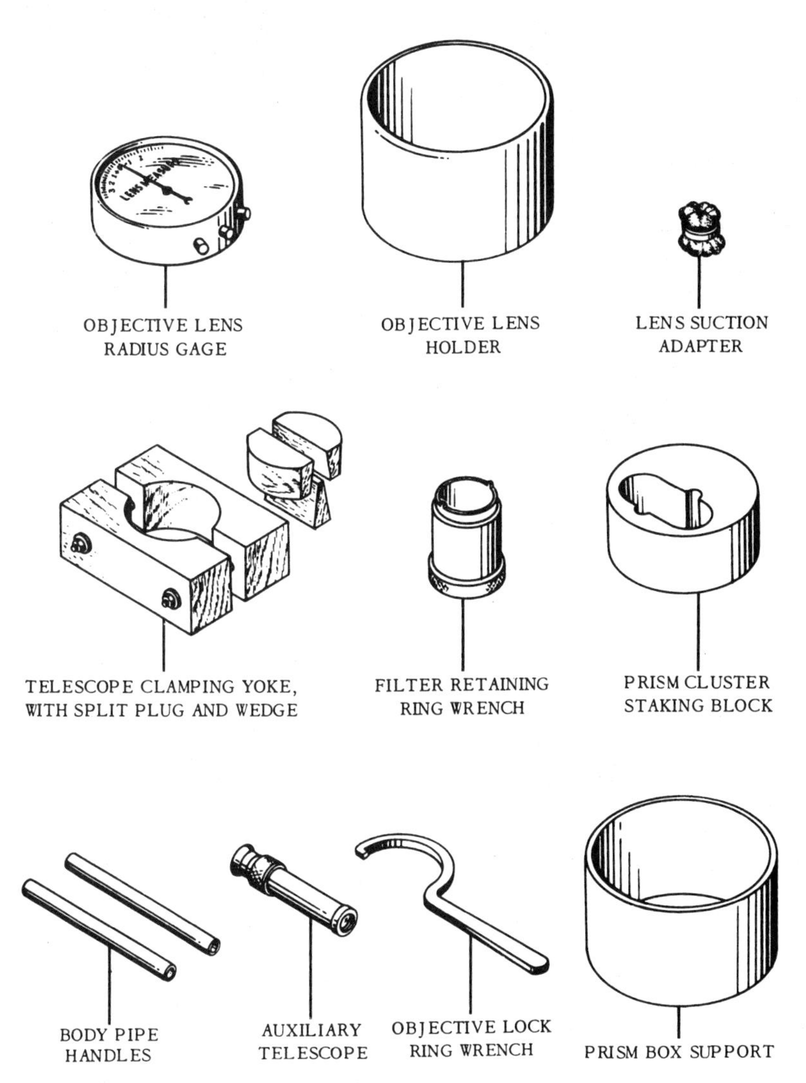

Figure 10-5.—Ship telescope special service tools—continued.

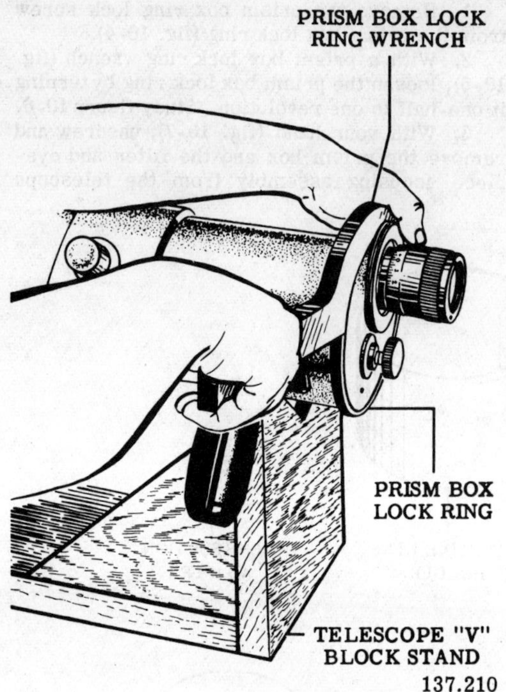

137.210
Figure 10-6.—Removing the prism box lock ring.

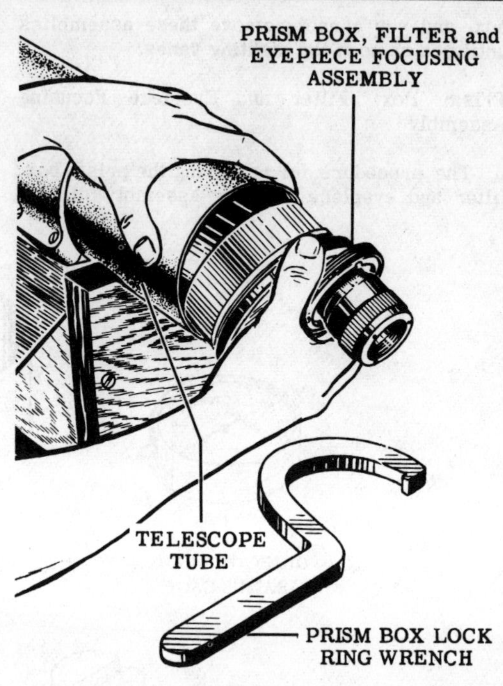

137.211
Figure 10-7.—Removing the prism box and the filter and eyepiece focusing assembly.

tube. NOTE: You may have to heat these assemblies first to soften the sealing compound.

If the prism box is frozen or corroded to the telescope body, do the following: Secure the telescope in a telescope clamping yoke (fig. 10-8) in a bench vise, as illustrated, and use an adjustable strap wrench to remove the prism box. NOTE: Insert the split plug and wedge part of the yoke inside the body of the telescope to prevent crushing.

Sunshade Assembly

To remove the sunshade assembly, remove the sunshade lock screw and unscrew the sunshade assembly from the objective mount (part A, fig. 10-9). If the assembly is frozen, use an adjustable strap wrench to remove it; and if necessary, use a telescope clamping yoke and bench vise, or have an assistant hold the instrument with body pipe handles (part B, fig. 10-9) while you remove the assembly.

Objective Assembly

You can remove the objective assembly from the telescope by loosening the objective mount lock ring (up to one revolution) with an objective lock ring wrench, shown in figure 10-10, and unscrewing the objective assembly from the telescope tube.

NOTE: If the objective assembly is stuck, use an objective mount lock ring, bench vise, and body pipe handles to remove it; but put the yoke on the objective assembly (not on the sunshade) and do not use a split plug (fig. 10-5). Always use an assistant to hold the instrument with body pipe handles when you do not have a telescope clamping yoke.

137.212
Figure 10-8.—Removing the prism box with an adjustable strap wrench.

Always remove the sunshade assembly before you attempt to remove the objective assembly. Before you disassemble the objective assembly clean your hands, your tools, and your workbench. Use lens tissue or a clean cloth to protect a lens from damage.

CAUTION: Use care to prevent scratches on lenses. Place the open side of the objective assembly on your workbench, because the outer convex objective lens bulges out of the front side of the mount and would be scratched if you put this side on the workbench.

Unscrew the objective mount lock ring and remove the objective retaining ring lock screw from the inner rim of the ring (fig. 10-11).

As you know, the two lenses in an objective lens assembly are placed in the position necessary to give the best image fidelity, and their position relative to the objective mount must

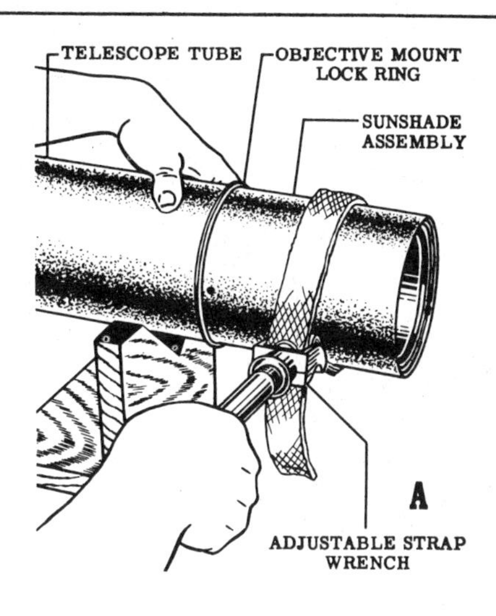

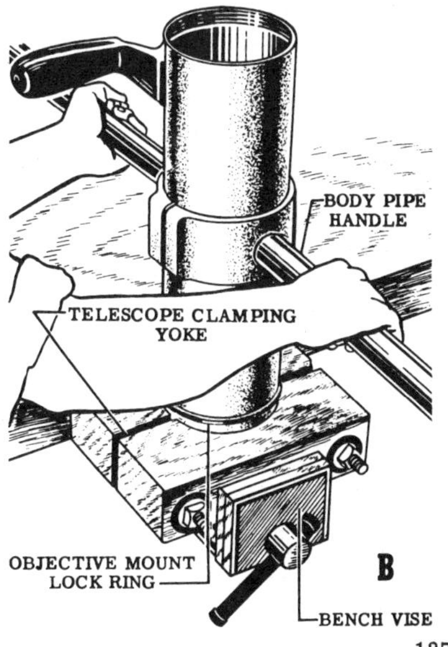

137.213
A. Procedure for holding body pipe handles.
B. Removing the sunshade with an adjustable strap wrench.
Figure 10-9.—Removing the sunshade.

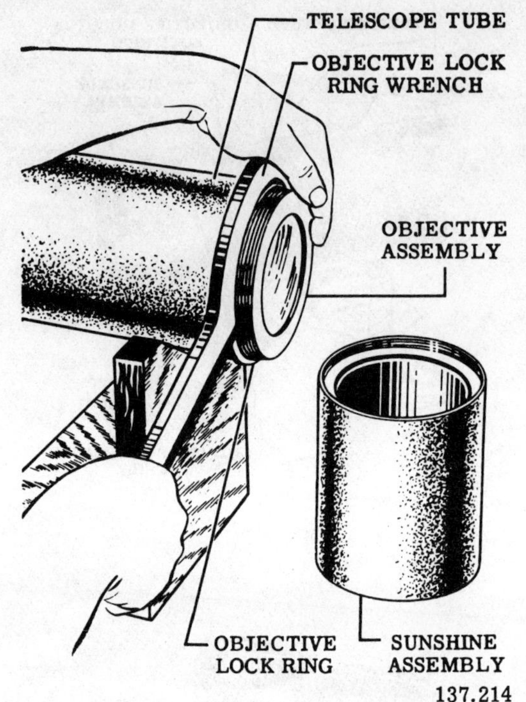

137.214
Figure 10-10.—Removing the objective lock ring.

therefore be maintained. The manufacturer of lenses generally puts guide marks on lenses for use during assembly and disassembly; but if they are not present, use the sighting vane hole in the lens mount as a guide and put a mark on each lens as you disassemble it. See illustration 10-12. It is a good idea to make these marks on the objective lenses as you disassemble them and then look for the manufacturer's marks.

CAUTION: The inner objective lens will be loose when you remove the lens retaining ring; so do not turn the objective mount over until you lift out the objective ring spring.

With an adjustable flat wrench (fig. 10-13), unscrew the objective lens retaining ring and lift out the objective spring ring (fig. 10-11).

Now place the top side of an objective lens holder (fig. 10-5) down against the inner objective lens of the assembly (in your hand). Then turn the objective assembly over, keeping the lens supported with the lens holder, and set the assembly on the workbench. Study illustration 10-14.

Next, cover the surface of the objective lens with lens tissue and press down on the lens as you lift the objective mount slightly to break the sealing compound (fig. 10-14), but do not lift the objective mount off the lens.

Lift the objective mount straight up to show the edge of the objective lens assembly and mark the edges of the lenses (fig. 10-12) to correspond to the sighting vane hole in the objective mount. Then remove the objective mount and wrap the objective lens assembly in tissue to protect it from damage.

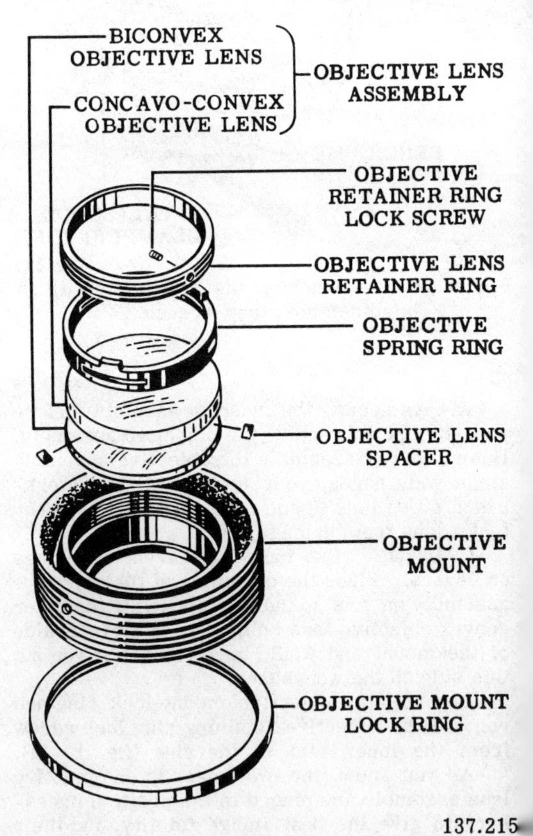

137.215
Figure 10-11.—The objective assembly.

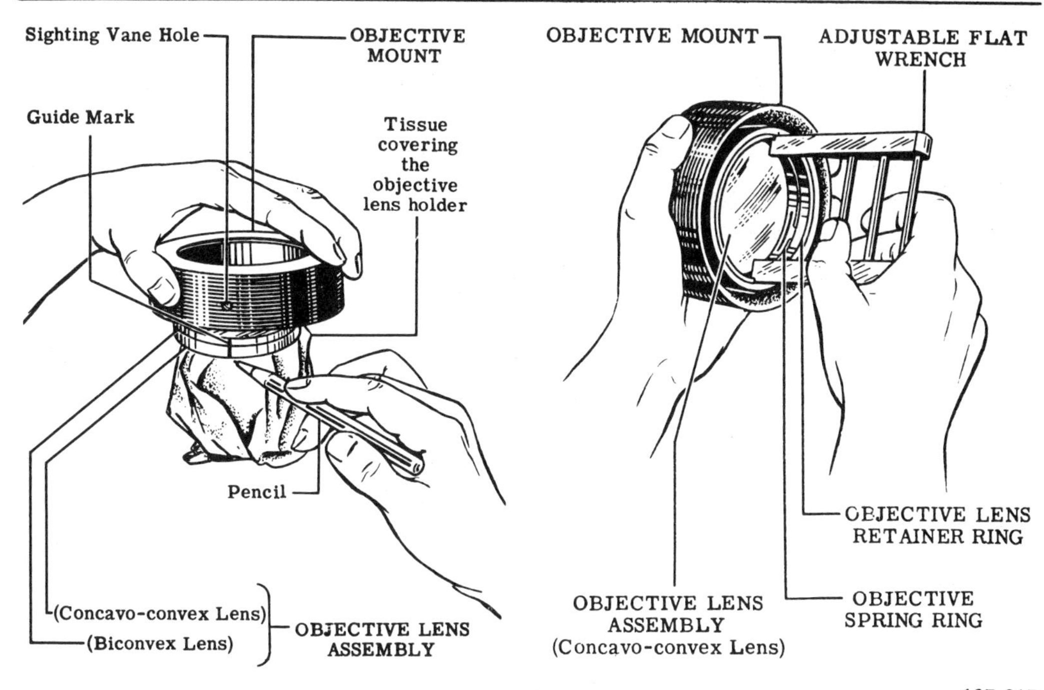

137.216
Figure 10-12.—Placing guide marks on the edges of the elements of the objective lens.

137.217
Figure 10-13.—Unscrewing the objective lens retaining ring.

Telescope Tube Body

When you disassemble the telescope tube body, do not remove the name plate or the grip handle, unless instructed to do so. This handle is secured with six screws and should be removed first. Then remove the cradle clamp screws and slide the telescope cradle off the tube (fig. 10-4).

NOTE: Do not remove the cradle from the telescope tube unless there is a stipulated requirement for repairs to or replacement of cradle parts.

Eyepiece Assemblies

The disassembly procedure for all four eyepiece assemblies (13x, 21x, 25x, and 32x) is essentially the same. The procedure which follows is for the 13x assembly (Kellner eyepiece), with an explanation of differences in procedure for the other assemblies.

CAUTION: Observe the rules given in chapter 8 for handling glass optics when you disassemble an eyepiece assembly. After you disassemble lenses, wrap them in lens tissue and put them in a parts tray in a safe location.

1. Remove the eyeguard retainer screw (fig. 10-15) and take the eyeguard and its retainer off the eyeguard ring.

2. Hold the eyepiece assembly with the eyelens end down and unscrew the eyepiece retaining ring. If the ring is stuck, use an eyepiece retaining ring wrench (fig. 10-16). Observe in illustration 10-15 that the collective lens is loose, so do NOT TURN the eyepiece upside down and permit it to fall out. Mark an arrow on the edge of the collective lens to indicate which side faces into the eyepiece lens mount, to ensure correct mounting.

NOTE: See illustration 10-5 for the tool to use on 25x and 32x eyepiece retainer rings.

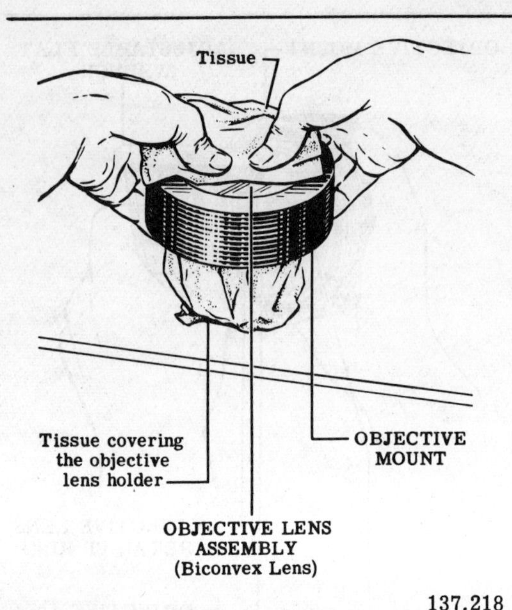

137.218

Figure 10-14.—Breaking the sealing compound from the objective lens and its mount.

3. Slide the collective lens and the eyelens spacer on a piece of lens tissue.

NOTE: The collective lens in a 21x assembly is a cemented triplet, and the eyelens is a single element; the other three eyepiece assemblies (13x, 14x, and 32x) have single collective lenses and cemented doublet eyelenses.

The four eyepiece assemblies on each ship telescope are parfocalized and are not interchangeable.

4. Put a piece of clean lens tissue under the eyepiece lens mount on your workbench, and then use another piece of lens tissue under your thumb as you press down on the doublet eyelens to break it loose from its sealing compound. See illustration 10-17. The bottom piece of lens tissue will prevent damage to the eyelens when it falls out.

5. Separate the eyeguard ring and the eyepiece lens mount by driving two or more wedges between the flanges between the eyeguard ring and the mount. Then remove the eyeguard snap ring (fig. 10-15) from the eyeguard ring and put it in the parts tray.

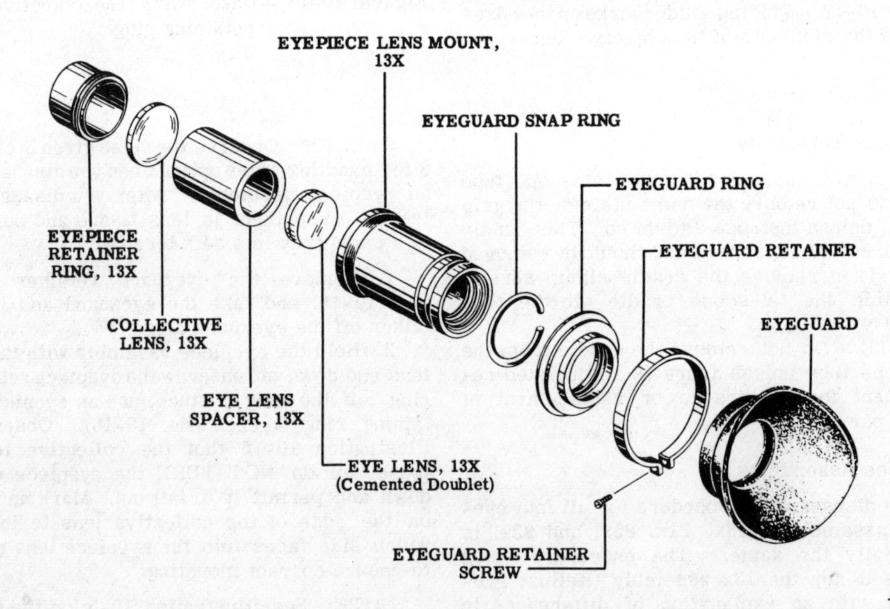

137.219

Figure 10-15.—Eyepiece assembly (13X).

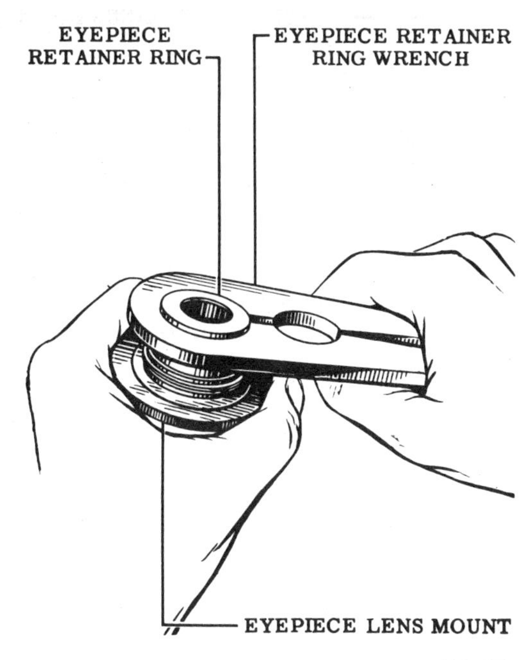

137.220
Figure 10-16.—Unscrewing the eyepiece retaining ring.

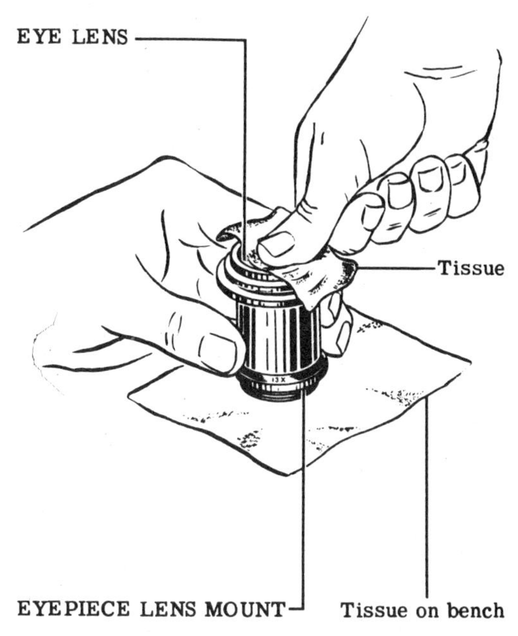

137.221
Figure 10-17.—Pressing the 13x doublet eyelens loose from its sealing compound.

Eyepiece, Filter, and Prism Box Subassemblies

The procedure for disassembling the eyepiece, filter, and prism box subassemblies is as follows:

1. Put the prism box assembly on a prism box support (fig. 10-5), remove the two filter shaft nuts (5/16-inch, open-end wrench), and lift off the filter mount assembly and the mount washer. Study illustrations 10-18 and 10-19 carefully, the last of which gives a top view of the assembly.

NOTE: The ball detent and detent spring button will be pushed, and will fly out when you lift the filter mount off, so hold a piece of flat stock over it to prevent loss, as shown in figure 10-19.

2. Now remove the detent ball and spring button and the detent spring from the hole in the prism box (fig. 10-18).

3. With a filter retaining ring wrench (fig. 10-5), remove the three filter retaining rings from the filter mount assembly. Study illustration 10-20. NOTE: Do NOT forget the order in which the filters are placed in the filter assembly.

4. Remove the filters. They are loose in the filter mount and should come out when you push on them with a few fingers (wrapped with lens tissue to prevent acid from getting on the filters). If the filters do not come out easily, check for shellac on the filter retaining rings. If shellac is present, dissolve it with alcohol or acetone.

5. To disassemble the filter diaphragm and bearing assembly, remove the screws and lift the assembly away from the prism box. Then remove the filter shaft spacer from the filter shaft.

6. Disassemble the filter shaft bearing from the filter diaphragm. See illustrations 10-18 and 10-21.

7. Pull on the knob (fig. 10-22) to slide the filter shaft and knob assembly out. Then remove the filter shaft packing housing and the filter

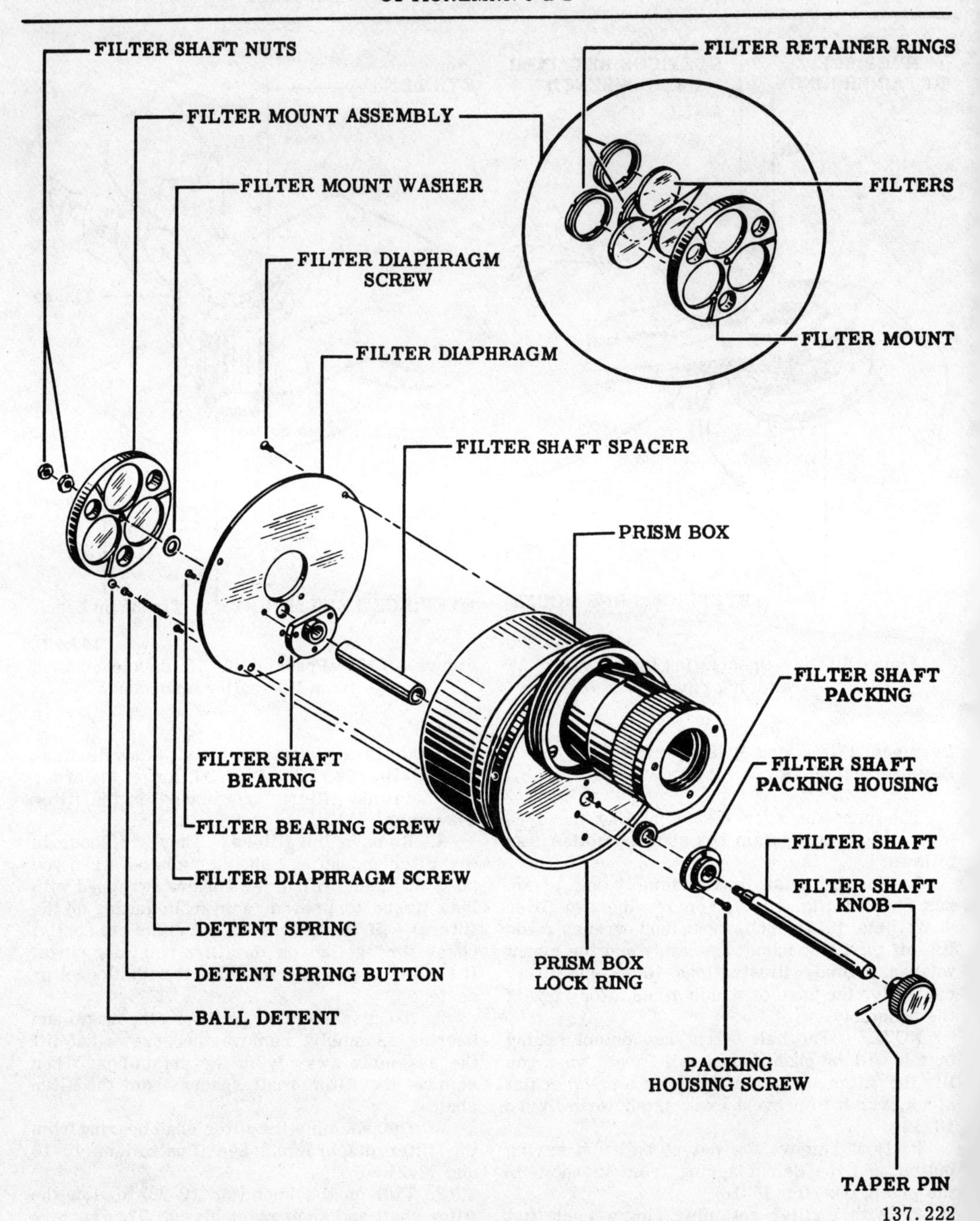

Figure 10-18.–Prism box, filter and eyepiece focusing assembly.

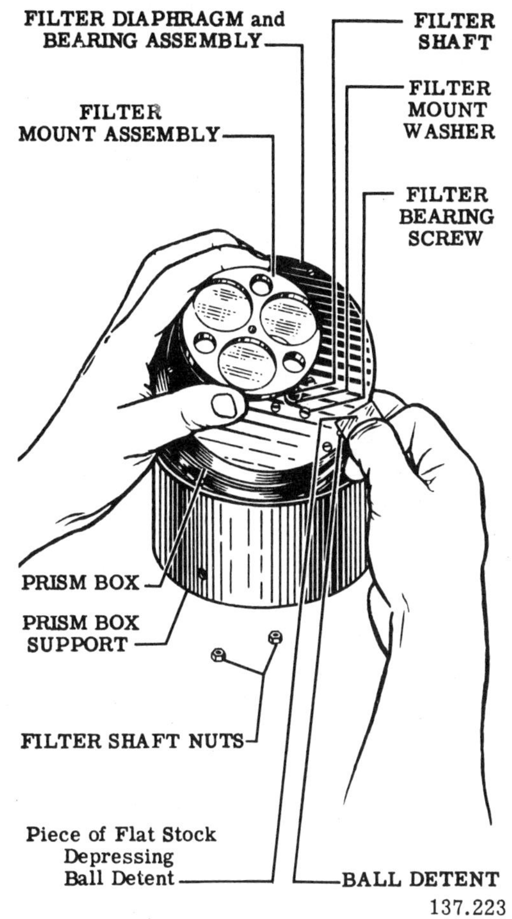

137.223

Figure 10-19.—Removing the filter mount assembly.

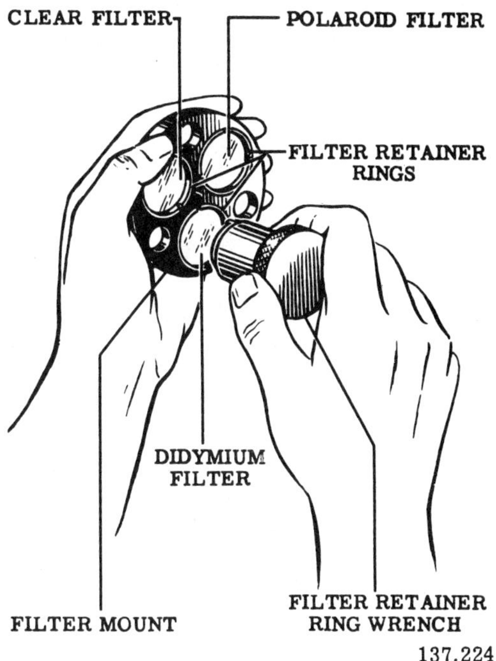

137.224

Figure 10-20.—Removing the filter retaining rings.

shaft packing from the top of the prism box (fig. 10-18).

8. If the filter shaft (or filter shaft knob) requires repairs or replacement, use a drift pin and mallet to knock out the taper pin which secures the knob to the shaft.

9. To disassemble the prism cluster plate from the prism box, remove the 3 screws which secure the plate. CAUTION: Do NOT scratch the prisms. See illustration 10-23. NOTE: The prism cluster plate is aligned with dowel pins to the prism box.

10. Remove the prism cluster by pulling straight out (fig. 10-24) and set it on a prism cluster fixture, with its top prism up. Study figures 10-24 and 10-25.

NOTE: The prism cluster fits on the prism cluster fixture with its prism or side uppermost, as determined by the position of the movable pin on the fixture. Observe in the circled portion of illustration 10-25 that the holes in the prism cluster fixture are coded T (top) and B (bottom). When you assemble the prism cluster in the telescope, the prism nearest the eyepiece is the TOP prism.

CAUTION: The two prisms are matched in the positions they occupy on the prism plate, so put guide marks on them for use during reassembly. Study illustration 10-25. Mark the prisms on the FROSTED SURFACE (T for top and B for bottom) and on the end over the common hole in the prism plate; that is, the one over which both prisms are placed. Keep the two prisms with their plate after disassembly.

11. Remove the two prism clip screws and lift off the prism clip and prism clip pad (fig. 10-23). Then cover the Porro prism with tissue, lift it off the plate, wrap it in lens tissue, and place it in the parts tray.

12. Lift the prism plate off the prism cluster fixture, set the movable pin to the B position, and place the prism plate back on the fixture, with the assembled bottom prism on top (fig. 10-25).

13. Repeat steps 10, 11, and 12 to remove the bottom prism, and mark it with the letter B (bottom).

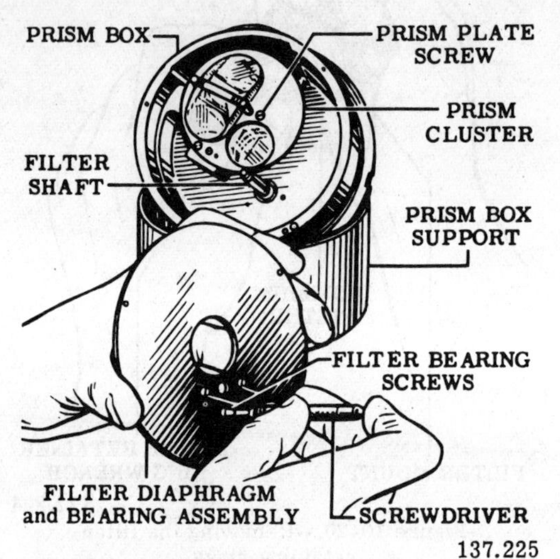

137.225
Figure 10-21.—Disassembling the filter shaft bearing from the filter diaphragm.

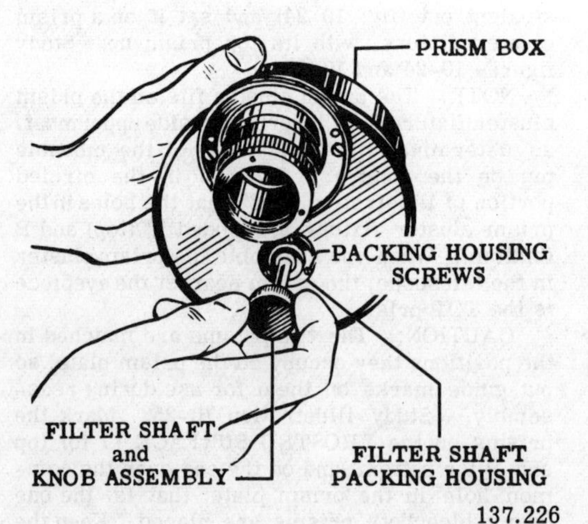

137.226
Figure 10-22.—Filter shaft and knob assembly.

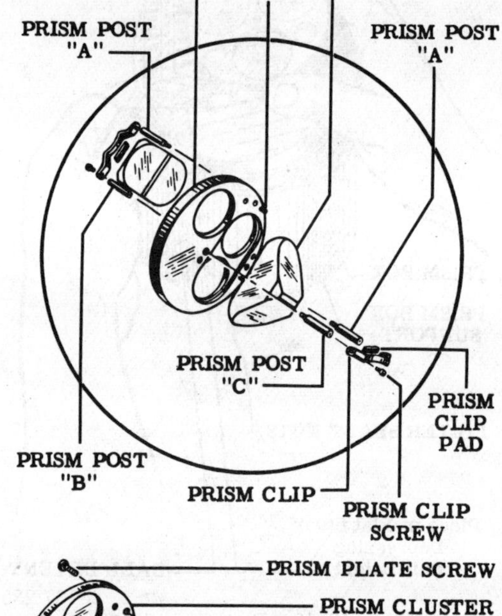

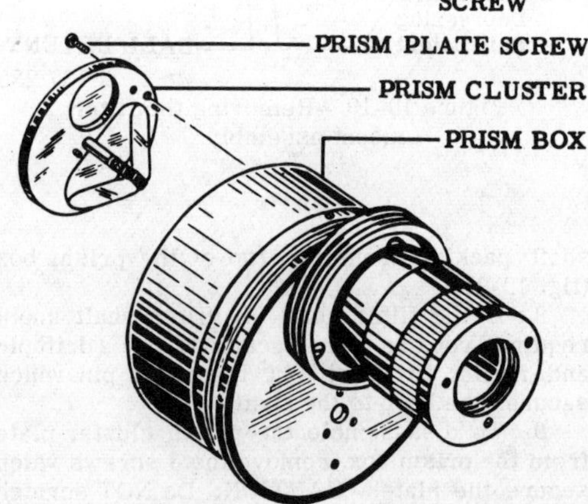

137.227
Figure 10-23.—Prism cluster.

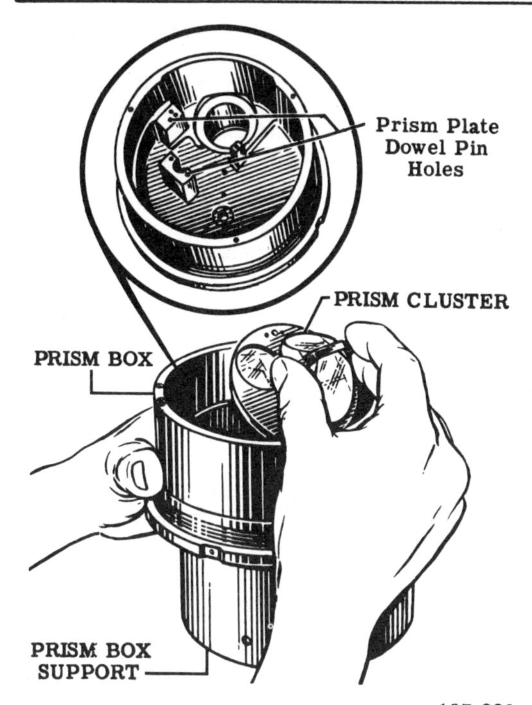

137.228
Figure 10-24.—Removing the prism cluster from the prism box.

14. Turn the knurled focusing ring all the way out (fig. 10-26), and scribe a line across the neck of the prism box and the eyepiece adapter as a guide for use during reassembly.

15. Remove the eyepiece adapter setscrew and unscrew the eyepiece focusing assembly from the prism box (fig. 10-27). If the assembly is stuck, remove the three focusing ring screws from the face of the knurled focusing ring, mark one of the screw holes in it and its matching hole in the eyepiece tube (for correct reassembling), and use an eyepiece adapter wrench, as shown in illustration 10-28.

NOTE: If the knurled focusing ring is not replaced in the position it occupied at the time of disassembly, the diopter scale gives false readings. The eyepiece tube and the eyepiece adapter are lapped together and fitted with multiple-lead threads, are not interchangeable with replacement parts, and must not be interchanged with similar assemblies of other telescopes.

16. Remove the eyepiece stop ring screw from the eyepiece tube stop ring (fig. 10-27); and with an adjustable pin wrench, remove the eyepiece tube stop ring from the end of the eyepiece tube. See illustration 10-29.

17. Turn the eyepiece tube in the eyepiece adapter to bring the bottoms of both parts even and scribe a corresponding mark across their edges, as shown in figure 10-30. Then unscrew the eyepiece tube from the eyepiece adapter.

18. To disassemble the prism box, remove the three eyepiece cover adapter screws and take the cover off the box. See illustration 10-27. Then unscrew and remove the prism box lock ring.

NOTE: Do NOT remove the two adapter dowel pins from the eyepiece cover adapter; except for replacement of parts, this is unnecessary.

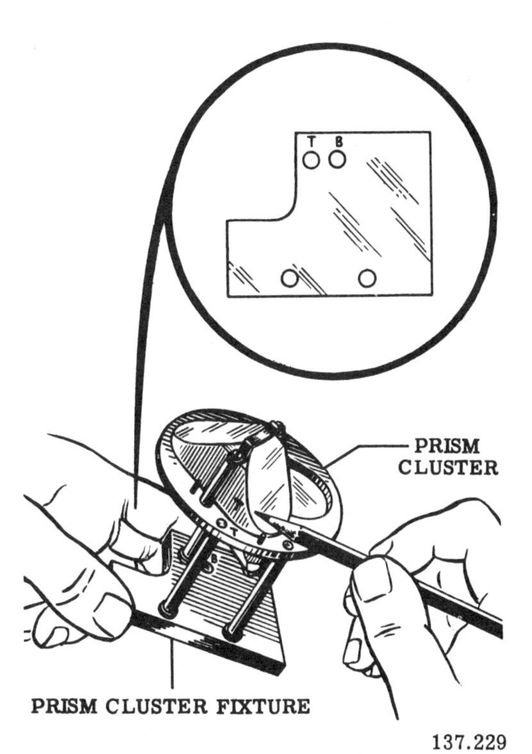

137.229
Figure 10-25.—Making a mark (T, for top) on the side of a prism in the cluster and on the plate of the prism cluster fixture.

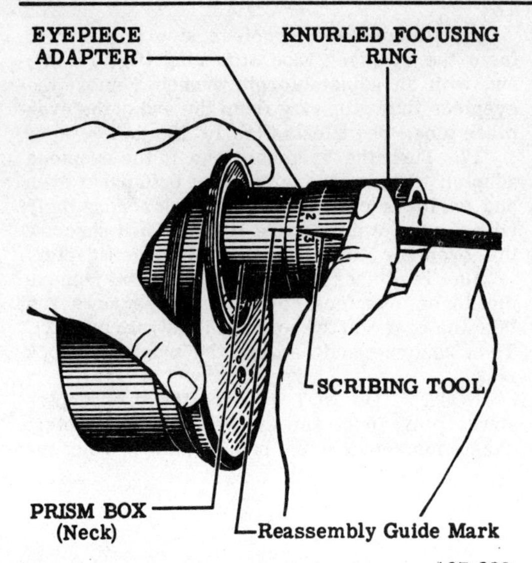

137.230
Figure 10-26.—Scribing a line across the neck of the prism box and the eyepiece adapter.

Yoke Assembly

There is generally no necessity for removing the yoke assembly from a shop telescope; but if you are required to do this work, proceed as follows:

1. Remove the yoke clamp screws (fig. 10-31).
2. With an arbor press and a suitable tool, remove the two yoke clamps by pressing out the yoke clamp pins. Then press out the yoke cap pins and remove the caps.
3. Remove the yoke name plate and unscrew the yoke end cover from the yoke.

Cover Assemblies (Sunshade and Eyepiece)

You can unscrew the plate from the sunshade with an adjustable pin wrench, as illustrated in figure 10-32.

Unless parts must be repaired, refinished, or replaced, do not disassemble the eyepiece cover assembly. The procedure for disassembling this assembly is as follows: Remove the chain screw nut inside the eyepiece cover and remove the screw, the chain, and the chain screw washer.

PARTS INSPECTION AND REPAIR

Review in chapter 8 the procedure for inspecting and repairing optical and mechanical parts of optical instruments.

Inspection of parts after disassembly should include functional defects listed on the job order and also TROUBLESHOOTING—looking for malfunctioning parts. As you find trouble, record it on the casualty analysis sheet for the instrument.

Optical Parts

Review illustration 8-1 and the discussion of various optical defects shown in the illustration.

When you inspect optical parts, do the following:

1. Check the antireflection coating on the eyelenses, prisms, and both surfaces of the two lenses in the objective lens assembly.
2. Inspect the condition of cement between the lenses in the 13x, 25x, and 32x doublet eyelenses, and also the 21x triplet collective lens. Record your recommendations on the casualty analysis sheet.

When authorized, have recementing of optics performed in accordance with procedures listed in chapter 8 of this manual; otherwise, replace defective parts. Some optical instruments are procured with uncoated optics which must be replaced before the instruments are used.

CAUTION: Lenses in an objective lens system are so expensive that you should replace them only when there is an absolute necessity. Defective lenses which impair image formation must be replaced.

3. Check code marks on prisms for correct matching of deviations and prism heights; make replacements when necessary. If you must form prism clusters, refer to the Control Manual (NavShips 250-624-12) for the correct procedure.

Mechanical Parts

Inspection standards for mechanical parts of optical instruments stipulate the amount of time authorized for effecting repairs, and they also authorize repairs and/or replacements of parts in accordance with their conditions. For further information concerning inspection of mechanical parts, see chapter 8.

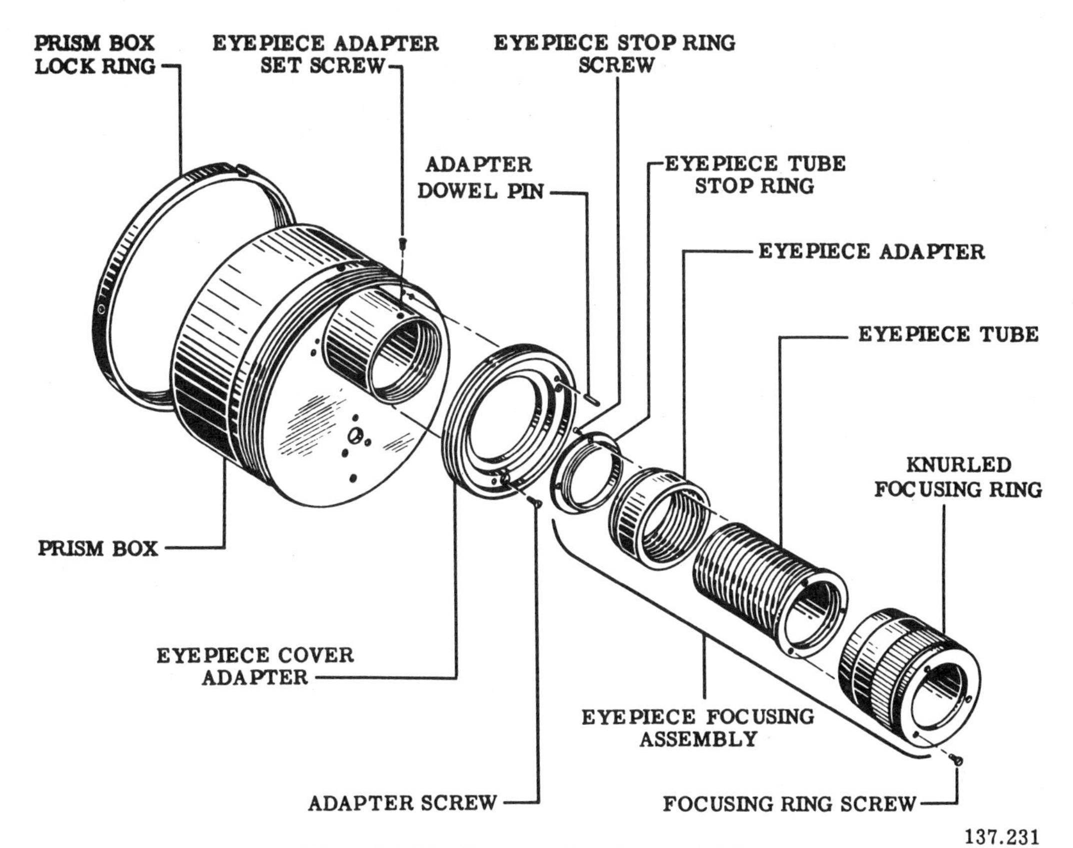

Figure 10-27.—Eyepiece focusing assembly.

Repair Procedure

The rule to follow concerning repairs is this: Repair used parts ONLY when repair of them is economically feasible. When in doubt about the serviceability of a part, check with the leading shop petty officer. Review repair procedures outlined in chapter 8.

REASSEMBLY AND INSPECTION

Reassembly of a ship telescope is essentially the same as disassembly, but in reverse order and with a few adjustments to individual subassemblies. Reassembly of the subassemblies comes first, followed by their optical and mechanical adjustments and their replacement and sealing in the assemblies. Sealing is accomplished as the subassemblies are replaced in the instrument, along with the sealing of all joints, fittings and screws.

Follow the disassembly procedures in reverse order. Review the information on reassembly and ajustment of optical instruments in chapter 8, and then study the following additional information and reminders on reassembly:

1. When you reassemble the Porro prisms onto the prism plate, check the prism system for LEAN and make necessary adjustments, as described in chapter 8. Make certain your reassembly guide marks on the prisms are positioned properly. After reassembly, check the prisms with a polariscope for strain.

2. If a polaroid filter is used in place of a dark smoke filter, there is only one position in which to reassemble it in the filter mount.

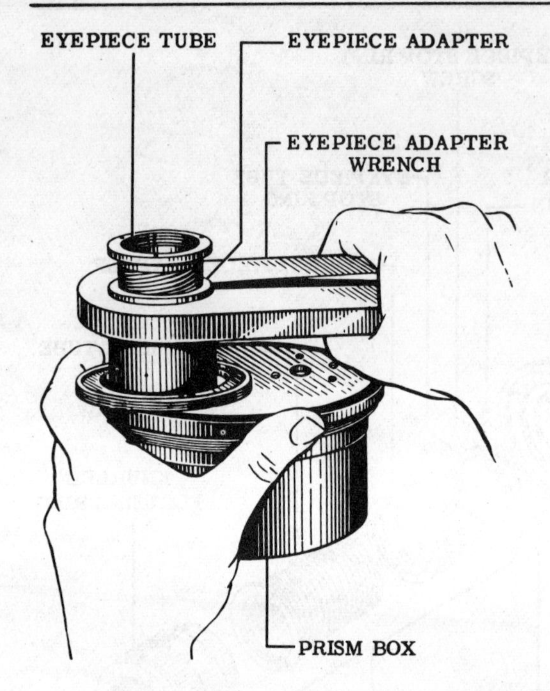

137.232
Figure 10-28.—Removing the eyepiece focusing assembly with an eyepiece adapter wrench.

The polaroid filter has two scratch marks on one of the polished surfaces near the edge. These marks indicate the axis of transmission and MUST BE IN THE VERTICAL POSITION during normal use; that is, when the observer is looking through the completely assembled instrument the marks must be in the vertical plane to reduce the glare a maximum amount (glare from water reflection, etc.). Check the filter mount where the polaroid filter is inserted in its mount for two guide marks which may be on the mount to indicate the filter's position.

3. In the objective lens assembly, positive and negative lenses are replaceable individually; but they must be positioned correctly in relation to each other in order to have them function properly. The positive and negative elements fit together ONLY ONE WAY; that is, only one side of the positive lens fits properly against the concave side of the negative lens, and you can determine which side this is with a Geneva lens measure (chapter 8).

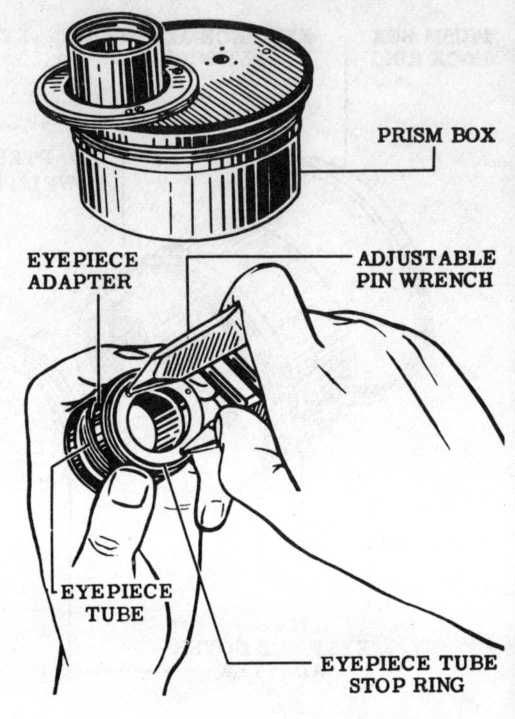

137.233
Figure 10-29.—Removing the eyepiece tube stop ring.

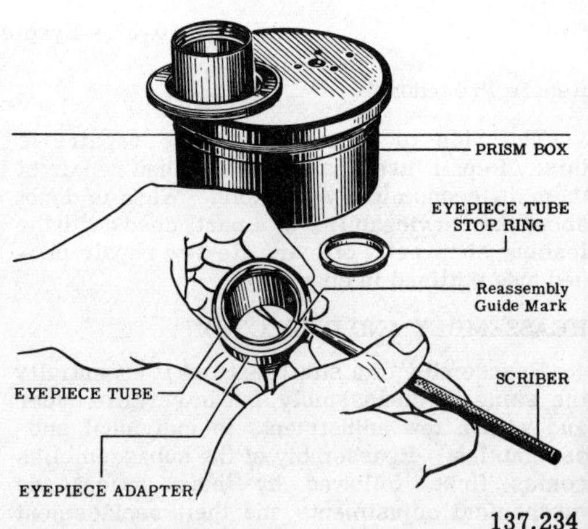

137.234
Figure 10-30.—Scribing reassembly guide marks on the eyepiece tube and the eyepiece adapter.

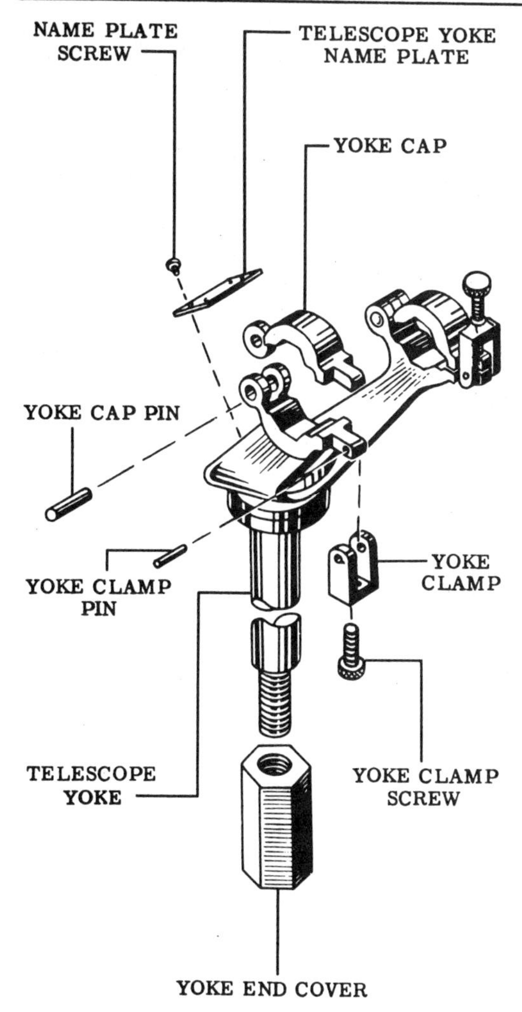

137.235
Figure 10-31.—Yoke assembly.

Place the Geneva lens measure (fig. 10-33) against the concave surface of the negative lens and note the minus reading. One side of the positive lens must have that same reading but of the opposite sign (plus). This side of the positive lens must fit against the concave side of the negative lens.

If you do not have a Geneva lens measure, pick off the three tin foil shims (spacers) on the concave side of the negative lens and place one side of the positive lens against the concave side of the negative lens and check for fit. If you have the correct sides together, the matching surfaces of the two lenses will fit so closely together that they create a suction which makes their removal from each other difficult. If the incorrect side of the positive lens is against the negative lens, there will be no suction. As a safety precaution, try both sides of the positive lens for fit.

You can also tell which side of the positive lens will match the negative lens correctly by placing a drop or two of water on the concave surface of the negative lens. When the correct side of the positive lens is against the negative lens, the water spreads evenly over the surfaces of the two lenses, making it almost impossible to remove the positive lens from the negative lens by lifting straight up. They must be slidden apart.

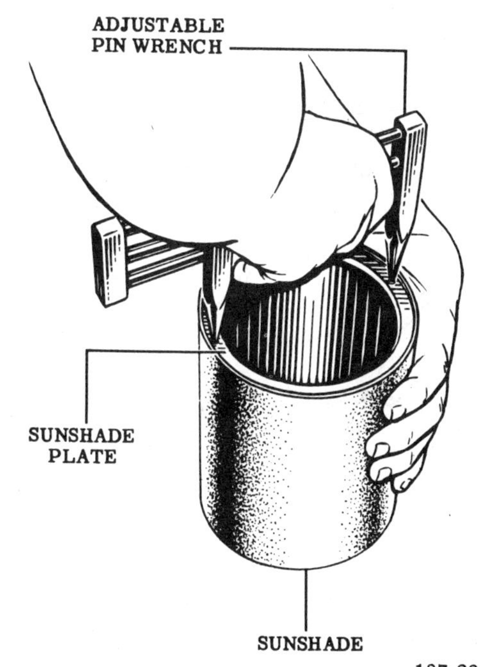

137.236
Figure 10-32.—Unscrewing the plate from the sunshade with an adjustable pin wrench.

When you are sure of the side of the positive lens which should fit against the negative lens, place the positive lens on a clean cloth or lens tissue and then replace the tin foil shims on the negative lens by aligning them in the following manner:

1. Pick up the negative lens of the objective lens by its edges and place its convex surface on the objective spring ring, on the three lugs, and mark on the frosted edges of the negative lens where the lugs touch the lens (fig. 10-34).

2. Now pick up the negative lens and place it on the objective lens holder, convex side down (fig. 10-35), and cement with collodion three pieces of tin foil or aluminum foil, 1/16

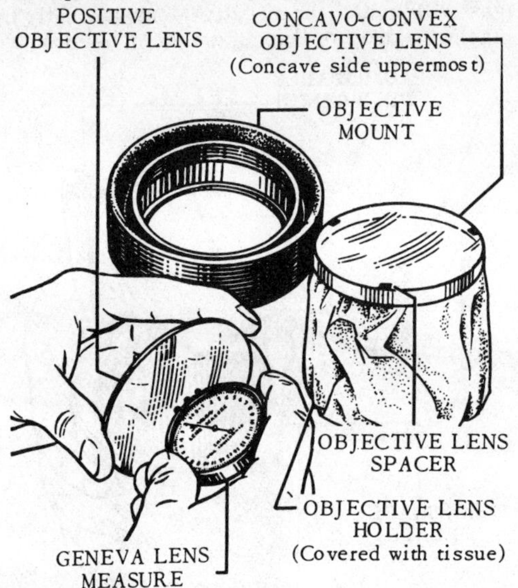

137.237

Figure 10-33.—Taking a reading of the curvature of an objective lens with a Geneva lens measure.

inch by 3/16 inch (.001 or .002 inch thick) to the concave side of the negative lens near the edge and where the three marks are located. All three pieces of foil MUST BE of the same hickness, preferably .001 inch.

3. Place the fitted side of the positive lens against the concave side of the negative lens (both lenses immaculately clean).

4. Place a string of sealing compound on the inside shoulder of the clean objective lens mount and align the guide markings on the two lens elements with the sighting vane hole and insert the objective lens assembly into the objective lens mount. Make certain that these guide marks are aligned and do not shift during insertion.

If, however, a new lens element was replaced during overhaul of the instrument, the objective lens assembly must be retested for image fidelity. To give the objective lens assembly an image fidelity test, insert the two lens elements into the objective lens test fixture and sight a resolution chart (chapter 8) 75 feet away. See illustration 10-36. Rotate one lens against the other; and at the position which gives the clearest and sharpest image of the lines on the chart, make a mark across the frosted edges of both lenses. Then remove the two lenses from the test fixture and insert them in the objective lens mount, positive lens first, against the sealing compound and mount shoulder, with the aid of the lens holder.

Replace the reassembled objective lens mount and objective mount lock ring in the telescope body and realign the sighting vane hole in the objective mount with the sighting vane hole in the telescope body. A little lubrication on the objective lens mount and telescope body threads will help prevent the two from freezing.

Eyepiece Assemblies

No special procedure is required for reassembling eyepiece assemblies. Reassemble them in reverse order to disassembly.

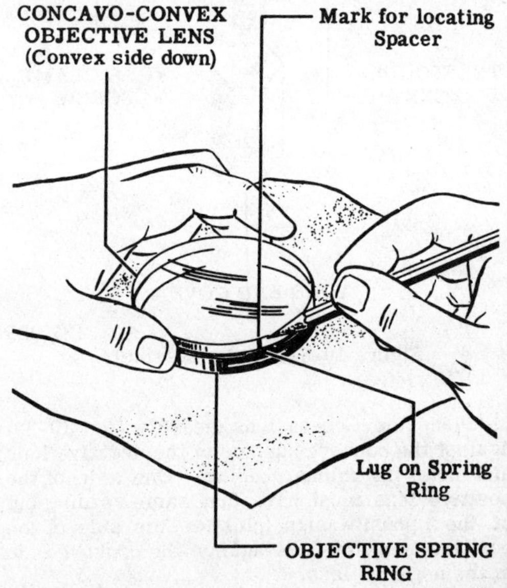

137.238

Figure 10-34.—Putting tin foil lens spacer marks on the edge of an objective lens.

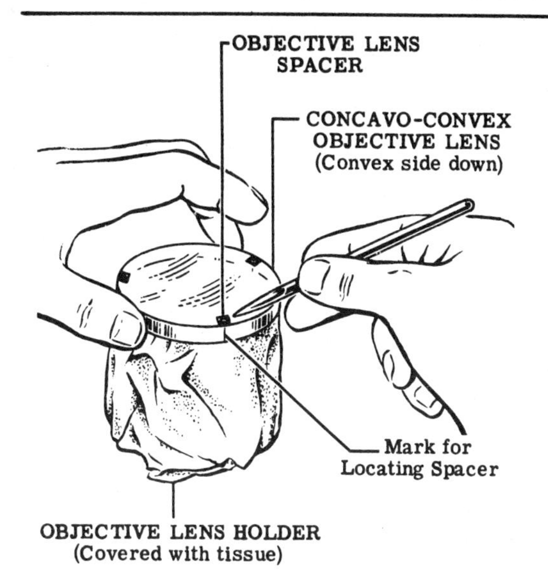

137.239
Figure 10-35.—Cementing tin or aluminum foil to the concave surface of a convex objective lens.

INSPECTION AND COLLIMATION

After you completely reassemble a ships telescope, you must then give it a thorough pre-collimation inspection, collimate it as necessary, and make a final shop inspection. Some collimation procedures were discussed in chapter 8; but unusual collimation standards require special consideration of collimation of a ships telescope in this chapter.

PRE-COLLIMATION INSPECTION

The things you should inspect before you collimate a ships telescope are as follows:

1. Check the optical elements for cleanliness, and recheck them for condition in the manner outlined in chapter 8.
2. Sight the telescope on a distant object and observe the clarity of the field of view. If it is sharply defined in the telescope near its center but badly distorted near its edges, the indication is that a lens is in place backward, usually the positive element of the objective lens assembly, or the collective lens of the eyepiece.
3. Inspect the mechanical action of the eyepieces, the filter shaft, the detent filter plate, and the yoke assembly for freedom of movement.

COLLIMATION PROCEDURE

The primary purpose of collimating a Mk 1, Mod 0, ship telescope is to ensure the accuracy of the diopter scale reading and to make it constant for all four eyepieces. Alignment of the mechanical sighting vanes with the optical axis must also be checked.

During collimation, you must test and adjust the eyepieces as necessary to make all four of them give a focused image for a fixed diopter setting. Each one must have the same distance between its mounting shoulder and its focal plane, which is represented by A in illustration 10-37. In other words, the diopter reading for sharpest focus will not change when the four eyepieces are interchanged. The eyepieces are said to be PARFOCALIZED when they are adjusted in this manner. Each eyepiece must indicate a diopter reading of 0 diopters, plus or minus one quarter, when the telescope is focused on a distant object.

Test equipment required for collimating a ship telescope consists of an auxiliary telescope and a ship telescope collimator. Study figure 10-38.

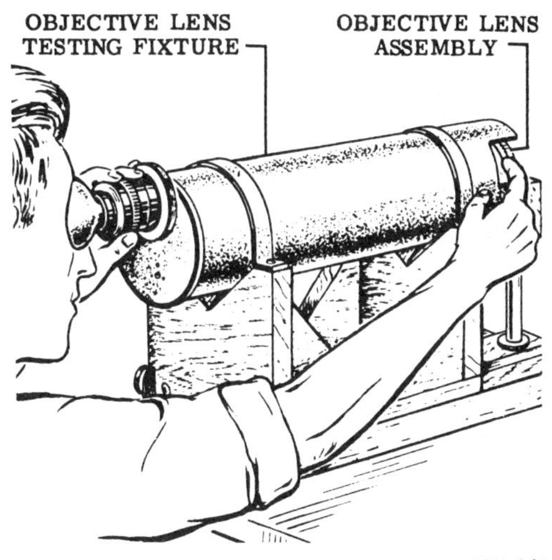

137.240
Figure 10-36.—Sighting a resolution chart with an objective lens testing fixture.

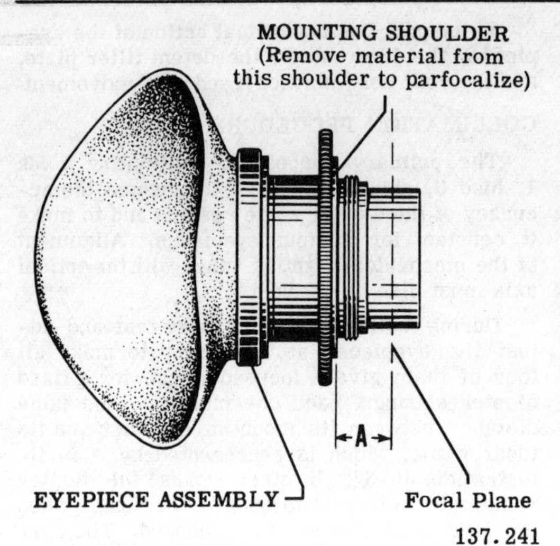

137.241

Figure 10-37.—Correct distance between the mounting shoulder and the focal plane of an eyepiece assembly, represented by A.

Testing Diopter Readings and Parfocalization

The recommended procedure for testing diopter readings and parfocalization for a ship telescope, Mark 1, Mod 0, follows:

1. Sight a distant object to set the auxiliary telescope to your eye (left or right) requirement. Adjust it carefully, so that you have the sharpest focus of the image and the crossline; then, throughout the tests, do not disturb this setting, and always use the same eye for making the tests.
2. Mount the ship telescope on the support of the collimator (fig. 10-38).
3. Assemble one of the eyepiece and set the auxiliary telescope against the eye lens mount. Sight through the instrument at the crossline target in the collimator and line up the crossline in the auxiliary telescope, by shifting the ship telescope. To get the sharpest image of the target, adjust the knurled focusing ring on the ship telescope. Observe and record the diopter reading and the power of the eyepiece.
4. Repeat step 3 for the other three eyepieces and record the diopter reading for each.

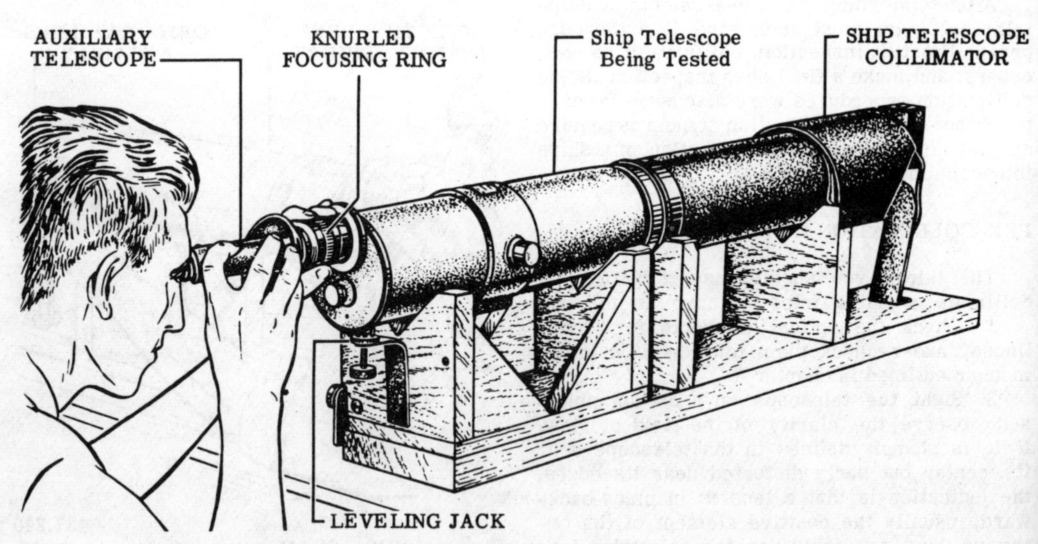

137.242

Figure 10-38.—Test equipment for collimating a ship telescope.

If all four eyepieces read 0 within one-quarter diopter, perform ONLY steps 9 and 10, which follow, to complete the collimation procedure.

If the four eyepieces give readings within one-half diopter of each other, they are parfocalized. The common reference for these readings, however, may be some reference other than 0 (one diopter, for example), in which case the diopter scale must be set. Proceed with all steps which follow, except 11 through 13.

If you find that the four eyepieces give readings which vary by an amount greater than one-half diopter of each other, they are not parfocalized; and the eyepieces and the diopter scale setting must be adjusted. Make the diopter scale correction first; then adjust the diopter scale.

Setting the Diopter Scale

If all four eyepieces are parfocalized, use the 21x eyepiece (chosen to give average magnification) for setting the diopter scale. If all four eyepieces are not parfocalized, use the eyepiece which gave the HIGHEST diopter reading in a positive direction in steps 3 and 4.

5. With the telescope on the collimator, with the aid of the auxiliary telescope, repeat step 3. The diopter scale should indicate 0 diopters, ± one-quarter diopter.

If you learn that the eyepieces are parfocalized but the diopter readings are taken from a reference other than 0, the indication may be that the eyepiece tube was reassembled incorrectly.

If new parts were NOT substituted in the eyepiece focusing assembly, the diopter readings should be about a 0 reference. If the reference is found to be other than 0, it is possible that the eyepiece tube was started in the wrong thread of the multiple threads of the eyepiece adapter. Try a different thread before you proceed with step 6.

6. You can possibly correct a SMALL error (about one diopter) on the PLUS side in the diopter scale reading by screwing out the eyepiece adapter part of a turn. Loosen first the eyepiece adapter set screw; then use a drill (.095 inch) to spot a set point for it in the new position.

You can correct a SMALL error on the NEGATIVE side by moving the objective assembly, but proceed with step 7 first.

7. Try to correct the diopter reading by removing the three focusing ring screws and turning the knurled focusing ring to align with the holes in a different position. Then replace the screws.

NOTE: Perform step 8 ONLY as a last resort. The three methods just described—changing the multiple thread, adjusting the eyepiece adapter, and shifting the knurled focusing ring—will provide in almost all cases a diopter reading within the allowed tolerance. Try them AGAIN before you go to step 8.

8. If the diopter reading is STILL NOT WITHIN one-quarter 0, unscrew the sighting vane which goes into the objective mount and turn the objective assembly IN or OUT in order to get the sharpest focus of the collimation target, as seen through the auxiliary telescope.

9. If during repair the objective assembly was moved, or if a new objective mount was used, use a .136-inch drill to drill and tap a new hole for the sighting vane. Then clean the objective assembly and the telescope tub of foreign matter or lubricant deposited by the drilling.

Aligning the Sighting Vanes

Alignment of the sighting vanes is STEP 10 in the collimation procedure. The vanes should line up with the vertical on the collimator target; and if they do not, loosen the two cradle clamp screws and rotate the telescope as necessary in its cradle without disturbing its position along the tube to line them up with the vertical.

Parfocalizing the Eyepieces

The three steps required to parfocalize the eyepieces of a ships telescope are:

1. With the telescope in the collimator, and with the auxiliary telescope, remove the eyepiece which gave the highest diopter reading and assemble another eyepiece. Sight through the telescope and focus for the sharpest image.

2. Place a Geneva lens gage against the back of the eyepiece and turn the focusing assembly as required to bring it to 0 on the diopter scale. Note and record the distance the eyepiece moved out. This is the amount of material you must remove (face off) from the eyepiece mount. Review illustration 10-37. Then repeat this operation for the other two eyepiece assemblies.

3. Set the eyepiece in a lathe and check the flange face to make certain it is running true.

Then cut off the mounting shoulder to the depth indicated in step 12.

The four eyepieces are parfocalized for their particular ships telescope and the diopter scale should thereafter read 0 diopters, ± one quarter for each. Review step 3.

FINAL SHOP INSPECTION

After you reassemble and seal a ships telescope, it should be in perfect condition and ready for use aboard ship. Before you release the instrument for use, however, give it a final inspection, as follows:

1. Look through both ends of the telescope and the four eyepieces for dirt, grease, and fingerprints on the optics, or fogginess in the optical system.
2. Check the mechanical parts for finish, and tightness in assembly, and check all screws for tightness.
3. Inspect the yoke assembly for freedom of movement of parts.
4. Assemble one eyepiece in the telescope and put the other three in the eyepiece case.
5. Place the eyepiece case, the yoke assembly, the telescope cover, and the telescope in the carrying case.

QUARTERMASTER SPYGLASS

A Quartermaster's spyglass (also called long glass) has 16 power and is generally used by a Signalman or the Quartermaster for reading flags and observing distant objects beyond the range of hand-held binoculars. Illustration 10-39 shows two Quartermaster spyglasses on top of a case, and an OOD spyglass beside the case.

Because the construction of a Quartermaster's spyglass is similar to that of the OOD spyglass, no discussion of its construction and operation is given in this chapter. If you understand thoroughly the following coverage of the OOD spyglass, you will have no difficulty with repair of a Quartermaster's spyglass.

OOD SPYGLASS

The officer-of-the-deck spyglass (OOD) shown in figure 10-39 is a Mk 2, Mod 2, instrument used by the officer of the deck on a ship in port to read flags and signals, to pick up buoys and other markers, and to observe small boats in the harbor.

An OOD spyglass consists of a main tube, an eyepiece assembly, and an objective mount. Its magnification (all marks) is 10x. A Mk 2, Mod 2, spyglass has the erector mount attached to the eyepiece; in some other marks it is mounted in the main tube. Hexagonal flanges at the ends of the main tube keep the instrument from rolling when placed on a smooth surface such as a table.

Optically, QM and OOD spyglasses are identical, except that the objective lens of the QM instrument is approximately twice the diameter and has about one and a half times the focal length as the objective lens of an OOD spyglass. All other optical elements of the two spyglasses are identical and interchangeable. A comparison of the optical elements of the two instruments is given in the following diagram:

	OOD	QM
Magnification......	10x	16x
True Field........	5°30'	3°30'
Apparent Field	55°	56°
Eye Distance......	29.0 mm	28.0 mm
Exit Pupil Diameter	3.5 mm	4.0 mm

The optical elements (system) for the OOD and the QM spyglasses are illustrated in figure 10-40. Study this illustration carefully.

The objective lens is a cemented doublet which refracts the incident rays to the principal focal plane, on or near the plano surface of the collective lens. The image formed by the objective lens is not affected by the collective lens and is therefore real and inverted.

The collective lens, a convexo-plano singlet receives its name from the fact that it collects the extreme principal rays of light (fig. 10-40) which otherwise would be lost and refracts them into the erector lens. Thus, without the collective lens, the center of the field would be well illuminated but the edges of the field would appear quite dark. Because the collective lens is placed within the focal length of the objective lens, it has little or no effect on the focal length of the objective lens, or on the magnifying power of the telescope as a whole. The only purpose of the collective lens is to collect rays and send them into the erector lens to produce a well illuminated field and IMAGE.

The erector lens is a cemented doublet, with its greatest curvature on the exposed surface of the negative lens. The primary purpose of the erector lens is to erect the inverted image formed by the objective lens; hence, the erector is placed two focal lengths from the inverted image in order to produce an erect, real image two focal lengths behind the erector lens. A plano-plano sealing window is placed between the erector lens and the image created by it, and it is used to seal the telescope near

37.3

Figure 10-39.—AN OOD spyglass and two QM spyglasses.

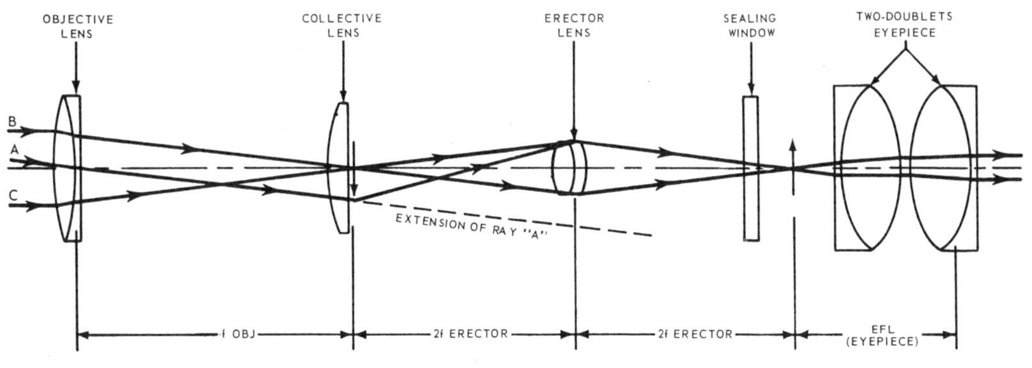

137.243

Figure 10-40.—Optical elements for OOD and QM spyglasses.

the eyepiece end. Because of its position, the sealing window has an effect of lengthening the erector lens' focal distance—NOT the focal length.

The eyepiece of the QM spyglass and the OOD spyglass is TWO-DOUBLET (asymmetrical). The only difference is that the eyelens in the QM spyglass is smaller in diameter, with a longer focal length, as compared with the field lens. A bevel is ground on the rear edge of the eyelens to aid in sealing the eyepiece assembly.

MECHANICAL CHARACTERISTICS

The QM and the OOD spyglasses are essentially the same mechanically, with the following exceptions: The telescope body tube of the QM spyglass is longer than that of the OOD spyglass, because its objective lens has a longer focal length, an hexagonal flange on the eyepiece mount support tube, and another hexagonal shaped part at the objective end of the OOD spyglass. The hexagonal flange and the hexagonal shaped objective end keep the telescope from rolling when it is placed on a smooth surface such as a table top. Other mechanical differences are few, and most all other mechanical parts may therefore be interchanged.

The objective lens mount, with the objective lens spacing rings forward and aft of the lens mount, are slid into the forward end of the body tube against a machined shoulder. The spacing rings and the lens mount are securely held in place by a lock ring threaded into the foremost end of the body tube. Permanently fixed and spaced at intervals along the interior of the body tube are metal diaphragms which aid in controlling aberrations and preventing internal reflections.

The collective and erector lenses are mounted in their respective mounts in a short support tube, one on either end. The support tube slides with a bearing fit into the forward end of the eyepiece mount support tube and is secured with a single screw. The eyepiece mount support tube is in turn threaded into the after end of the telescope body tube and secured with a single set screw. In the opposite end (one which protrudes from the body tube) of the eyepiece mount support tube is threaded the sealing window mount and its lock ring. Threaded onto that end of the eyepiece mount support tube is the eyepiece mount which houses the SPIRAL KEYWAY MECHANISM and the eyepiece drawtube containing the eyepiece lenses. A diopter scale ring is secured to the eyepiece mount with a single screw, and the scale is graduated from -6 diopter to +6 diopters.

In the eyepiece drawtube, just forward of the field lens, is mounted a single metal diaphragm which serves to control chromatic aberrations. A single lock ring secures the eyepiece lenses and their spacer. The lock ring is threaded into the drawtube just aft the diaphragm to lock against the field lens. An eyelens spacing ring between the field lens and the eyelens serves to separate the two lenses the distance required for them to function according to their design.

The eyepiece mount, eyepiece mount support tube, and the collective-erector mount support tube may be threaded into and removed from the after end of the body tube as a single unit. This is possible because all of the above parts are attached and secured to the eyepiece mount support tube which screw into the rear of the body tube. This is an advantage in collimating the telescope.

The QM and the OOD spyglasses are provided with gassing and drying screws which permit drying of the instruments from their sealing windows, to and including their objective lenses. On the OOD spyglass, the inlet gassing screw is located just forward of the hexagonal flange on the eyepiece mount support tube, and the gas outlet screw is located on the hexagonal objective end of the body tube. The inlet and outlet gassing screws are located similarly on the QM spyglass.

PREDISASSEMBLY INSPECTION

Predisassembly inspection procedures for the QM and the OOD spyglasses are checked and inspected in the same manner as for any other optical instrument; therefore, review chapter 8 for inspection procedures of these instruments. Write your findings on an inspection sheet and proceed with the disassembly, or consult your shop supervisor for advice concerning overhaul of the instruments.

DISASSEMBLY

The procedure for disassembling a Mk 2, Mod 2, OOD spyglass is given by steps in the next pages. Proceed as follows:

1. Remove the setscrew which secures the eyepiece mount support tube in the body tube

(fig. 10-41). Then unscrew the eyepiece mount support tube and pull it from the body tube (fig. 10-42).

2. Remove the setscrew which secures the eyepiece mount to the eyepiece mount support tube (fig. 10-43) and unscrew and separate the eyepiece mount from the eyepiece mount support tube (fig. 10-44).

3. Remove the setscrew which secures the actuating ring (commonly called FOCUSING RING) retainer ring lock ring (fig. 10-45) and unscrew the lock ring. Then unscrew the actuating ring retainer ring (fig. 10-46). NOTE: These two rings are NOT identical. The lock ring has a bevel on each side and a lock ring; the actuating ring retainer ring has only one bevel.

4. Remove the eyepiece cap from the rear of the eyepiece drawtube and slip off the actuating retainer ring and its lock ring.

5. Remove the knurled actuating ring first by rotating it counterclockwise to disengage it from the focusing key and then slide it from the eyepiece mount.

6. Remove the focusing key. It is aligned with two dowel pins and secured with two screws. When the screws have been removed (fig. 10-47) with a jeweler's screwdriver, lift the focusing key from the longitudinal slot with a pair of tweezers. The dowel pins should come out with the focusing key; if they do not, remove them from the eyepiece drawtube with a pair of tweezers. The draw tube is now free within the eyepiece mount; remove it by pulling straight out (fig. 10-48).

7. With an adjustable retainer ring wrench, loosen the diaphragm lock ring just enough so that it turns freely. Do not use the retainer wrench to remove the lock ring completely from the drawtube, as the wrench may damage the fine threads on the inner wall of the drawtube. Use a tool such as the one shown in figure 10-49 to remove the lock ring (preferably a pegwood stick). Measure and record the distance the diaphragm is in the drawtube. NOTE: The position of the diaphragm is very

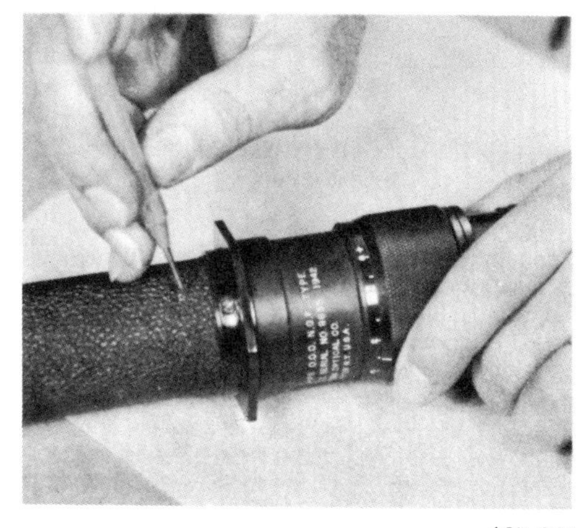

137.244
Figure 10-41.—Releasing the eyepiece mount support tube setscrew.

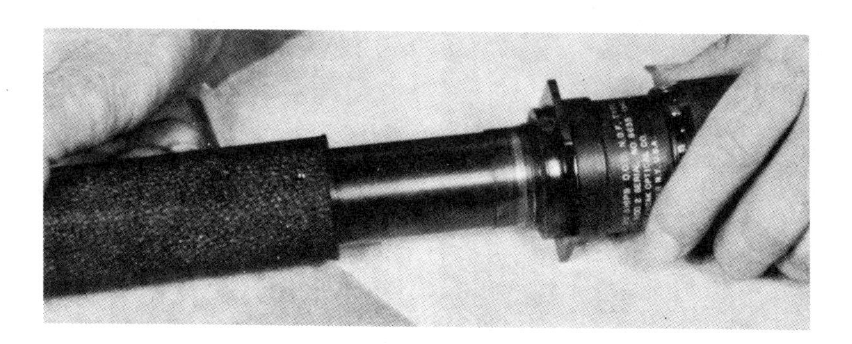

137.245
Figure 10-42.—Removing the eyepiece mount support tube.

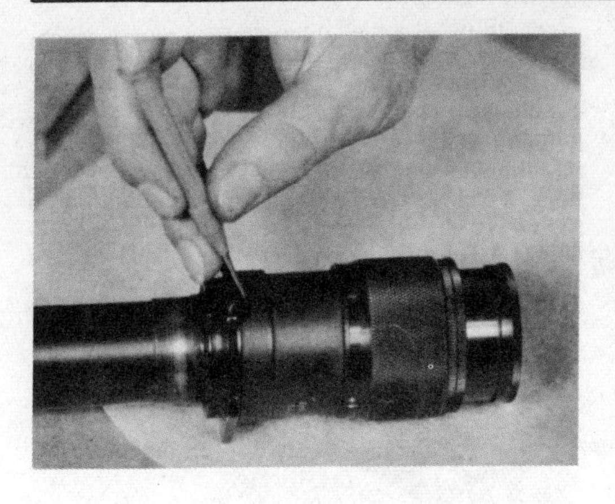

137.246
Figure 10-43.—Removing the eyepiece mount setscrew.

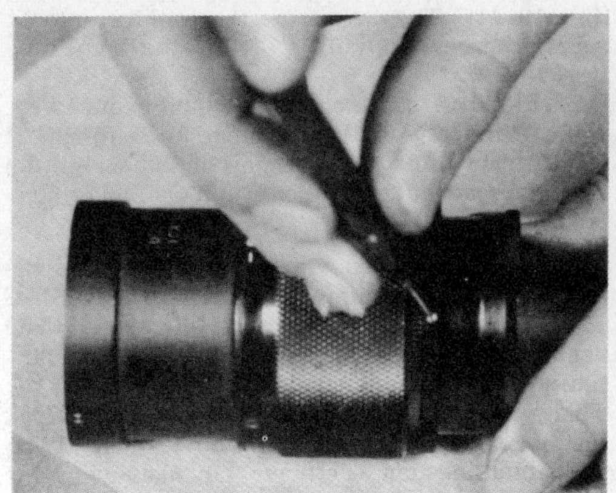

137.248
Figure 10-45.—Removing the actuating ring retaining ring lock ring setscrew.

important, for it controls chromatic aberration. Remove the diaphragm in the same manner you remove its lock ring.

8. Remove the lock ring which secures the eyepiece lenses and their spacer. CAUTION: When you remove the lock ring, the eyepiece lenses and their spacers are loose and they can easily fall out. This lock ring is almost the same diameter as the lock ring diaphragm; do not get these rings mixed.

9. With a piece of lens tissue on the plano surfaces of the field lens and the eyelens, slowly turn the eyepiece drawtube over to allow the spacer and the eyelens to slide out into your hand. The rear surface of the eyelens is sealed, so apply a little pressure with your thumb to break the seal. CAUTION: The clearance between the lenses and the inner wall of the drawtube is so small that the lenses may become cocked; if they do, follow the procedure given in chapter 8 to remove them. When the lenses and spacers are removed from the drawtube,

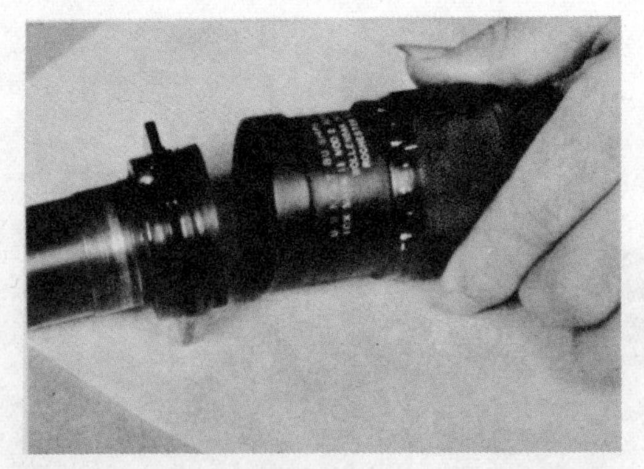

137.247
Figure 10-44.—Removing the eyepiece mount from the eyepiece mount support tube.

137.249
Figure 10-46.—Removing the actuating ring retaining ring.

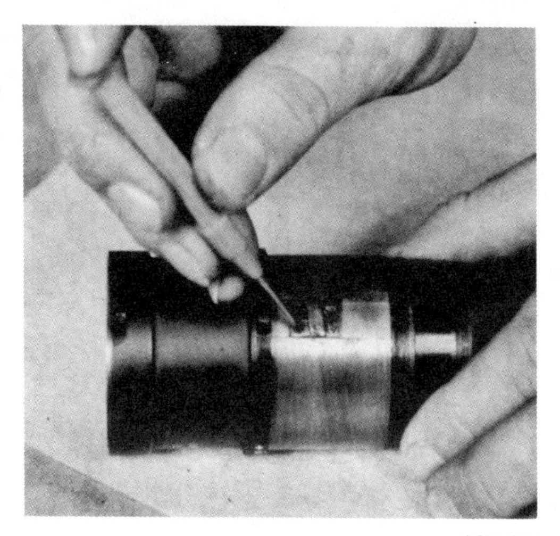

137.250
Figure 10-47.—Removing the focusing key screws.

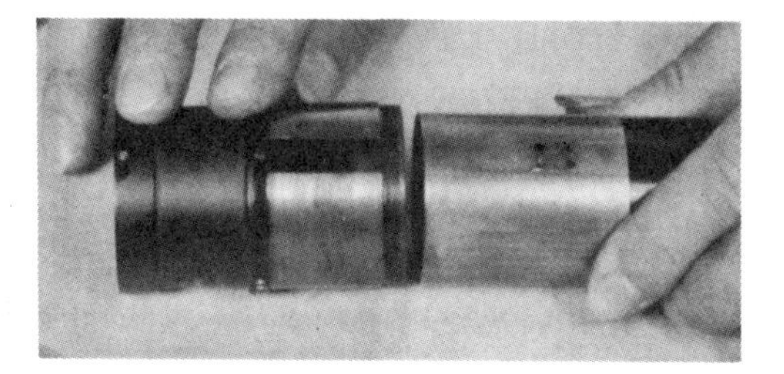

137.251
Figure 10-48.—Removing the eyepiece drawtube from the eyepiece mount.

mark them to indicate the manner in which they fit in the drawtube; there is only one correct way in which they fit when assembled. Wrap the lenses in lens tissue and stow them in a safe place, AWAY FROM THE METAL PARTS OF THE INSTRUMENT.

10. Remove the setscrew which secures the collective-erector mount support tube in the eyepiece mount support tube, and pull STRAIGHT OUT (fig. 10-50) on the tube to remove it from the eyepiece mount.

11. Loosen the collective lens mount lock ring and unscrew the collective lens mount from the support tube. Remove the collective lens lock ring and remove the collective lens. Then wrap it in lens tissue. NOTE: If this lens has pits, scratches, or chips, it must be replaced; because the lens is near the focal plane of the objective lens, any fault of the collective lens is very apparent in the field.

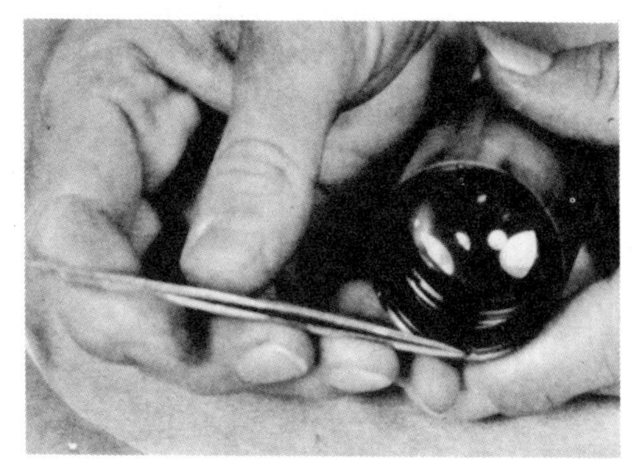

137.252
Figure 10-49.—Removing the diaphragm lock ring.

12. Remove the erector lens mount lock ring from the support tube. Use precaution to prevent damage to the fine threads on the inner wall of the support tube. Remove the erectors lens mount from the support tube. NOTE: The erector lens mount may come out of the support tube in reverse of that shown in figure 10-51; to facilitate collimation, this mount is designed for mounting either way. Remove the erector lens lock ring and then the erector lens. Note that the exposed surface of its negative element has the greatest amount of curvature. Mark the lens and wrap it in lens tissue.

13. Loosen and remove the sealing window mount lock ring in the eyepiece end of the eyepiece mount support tube. Remove the sealing window mount and the sealing window lock ring; then withdraw the sealing window. The window is sealed with sealing compound. If necessary, apply heat to soften the wax and use a suction cup pressed tightly against the window to help break the seal.

14. Loosen the objective mount support with a fiber grip wrench and remove the objective mount support. NOTE: No setscrew secures the objective mount support to the body tube.

137.253
Figure 10-50.—Removing the collective-erector support tube setscrew.

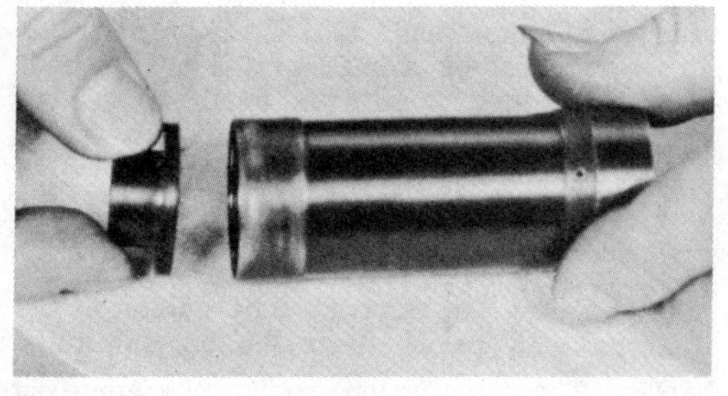

137.254
Figure 10-51.—Removing the erector lens mount from its support tube.

The objective mount support of a QM spyglass is PART OF THE BODY TUBE and therefore cannot be removed.

15. Loosen and remove the lock ring which secures the objective lens spacing rings and the objective lens mount to the interior of the body tube. Remove the front spacing rings, the objective mount, and the rear spacing rings by pulling them straight out of the body tube. Mark each spacer when you remove it. Press on the objective lens to break the seal which secures it in the mount.

16. Remove the two gassing screws, one of which is located just forward of the hexagonal flange on the eyepiece mount support tube, and the other is on the hexagonal section of the body tube. Check both gassing screw orifices for freedom from obstructions.

You have now completed disassembly of the OOD spyglass. Continue with overhaul and repair of the instrument. Follow the procedures listed in chapter 8, and also the additional reassembly procedures discussed in the next paragraphs.

REPAIR AND REASSEMBLY

When you reassemble the collective lens in its mount, mark on the plano surface of the lens near the center with a wax pencil, or pass a thin wire through the gas orifices in the collective lens mount and draw it taut by twisting the ends together. The wax pencil mark or the wire will act as a reference point to aid in the proper positioning of the collective lens during collimation.

Before you reassemble the eyepiece drawtube, check the mechanical 0 diopter setting of the eyepiece assembly at MID-THROW. You can do this by reassembling the complete mechanical action of the eyepiece mechanism, with the exception of the inner lens and parts of the drawtube. When you have the mechanical parts of the eyepiece assembled, turn the eyepiece actuating ring until the drawtube stops on the IN POSITION. Then, with a lead pencil, mark on the drawtube a line where it protrudes from the eyepiece mount. Next, turn the actuating ring counterclockwise until the drawtube stops in the OUT POSITION. Then measure the full amount the drawtube traveled from STOP to STOP and divide the amount by 2 to get the MID-THROW POSITION of the drawtube. Put a mark on the drawtube to indicate its mid-throw position and turn the actuating ring until the drawtube moves in, to the mark for the mid-throw position.

Now observe where the index mark on the actuating ring is pointing; it must point to 0 diopters on the diopter scale ring when the eyepiece is at the mid-throw position. If it does NOT point to 0 diopters, remove the setscrew which secures the diopter scale ring and rotate the ring until the 0 diopter mark is aligned with the index mark of the actuating ring. Then drill and tap a new hole for the diopter scale ring setscrew. When you complete this task, you will have the mechanical 0 diopter setting at MID-THROW.

Disassemble the mechanical parts of the eyepiece and begin reassembly of the lenses and the inner parts of the eyepiece drawtube. Re-position the drawtube diaphragm to its original position in the tube. After you insert the diaphragm, check its position by looking through the eyepiece lenses. When correctly positioned, the diaphragm is sharp and clear; if it is not, screw the diaphragm in or out until it is sharp and clear, and note the bright-yellow fringe which should be around the diaphragm field. You can check the diaphragm later for correct positioning, after you collimate the instrument, by checking the overhauled instrument for chromatic aberration (chapter 8).

Reseal all assemblies during reassembly, except the eyepiece mount support tube, which is withdrawn several times during collimation and must therefore be sealed ONLY after the instrument is collimated.

COLLIMATION

You can collimate this telescope in the same manner as for any telescope with a single erector lens. The procedure for collimating a single erector telescope is outlined in chapter 8 of this manual.

Because spyglasses are hand-held and do not have bearing pads or feet, it is not necessary that you collimate the collimator. All you need is an infinity target such as an outside target or the crossline of a collimator. Then place the telescope on V-blocks in front of the collimator and proceed with collimation.

First, remove the parallax between the instrument's crossline and that of the collimator's. The reference mark (wax pencil mark) on the plano surface of the collective lens or the wire through the collective lens mount serves as a temporary crossline for positioning the collective lens in the optical system. To remove parallax in this system, adjust the collective lens mount by screwing it in or out of the support tube. The eyepiece mount support tube must be removed from the body tube each time the collective lens mount is adjusted; therefore, tighten the eyepiece mount support tube against the shoulder of the body tube after each adjustment to eliminate possible errors in the parallax readings. Furthermore, the setscrew which secures the collective-erector support tube in the eyepiece mount support tube must be in place in order to eliminate such errors.

If you CANNOT remove parallax because of insufficient movement of the collective lens mount in the support tube, re-position one or more of the objective lens spacing rings in order to re-locate the objective lens in the desired direction. When you do this, the collective lens mount is so adjusted that you can remove the final errors in parallax. After you completely remove parallax from the telescope, lock the collective lens mount lock ring against the shoulder of the support tube and make another test for parallax.

Now set the eyepiece to 0 diopters optically by adjusting the erector lens mount. Review the procedure in chapter 8. You may find, however, that there is insufficient movement of the erector lens mount. If this is true, remove the erector lens mount from the support. Then remove the erector lens, turn it over in the mount, and replace the erector mount in the support tube in the opposite direction to what you had it before. The mount is so designed that it may be placed in the support tube in either direction. CAUTION: If you do turn the mount over, remember that the erector lens MUST ALSO BE TURNED OVER in its mount.

When you have optical 0 diopters set on the eyepiece, lock the erector lens mount lock ring and give the instrument a final check for parallax and diopter setting.

For the last time, remove the eyepiece mount support tube from the body tube and clean off the wax pencil mark from the collective lens. Do this by removing the collective lens lock ring without tampering with the collective lens mount itself. If you remove the collective lens mount during removal of the collective lens, you will not have the collective lens in proper position; so be very careful to avoid this.

When you complete the task just explained, place a string of sealing wax around the eyepiece mount support tube and seal and secure it in the body tube. Then give the instrument a final inspection.

GASSING AND DRYING

The procedure for gassing and drying a Mk 2, Mod 2, spyglass is as follows:

1. Remove the inlet and outlet gassing screws from the telescope and connect a gassing hose to the inlet hole.

2. Run dry nitrogen through the instrument and purge it at the same time.

3. After you completely dry the instrument, replace the two gassing screws and seal them.

CHAPTER 11

BORESIGHT TELESCOPES

This chapter contains a general description of boresight telescopes, a discussion of their use, and repair and maintenance procedures pertaining to them.

Because boresight telescopes are designed specifically for use in boresighting, it is important that you have an understanding of boresights and the boresighting procedure.

BORESIGHTING PROCEDURE

Although boresighting is not a duty of an Opticalman, it is explained briefly here to show the importance of the work you do on boresight telescopes used in the operation. If the telescopes you repair for use in the process do not function properly, boresighting is unsatisfactory.

Boresighting is the procedure of aligning the lines of sight of a gunsight telescope so that they intersect the axis of the bore of the gun at a predetermined range, or are parallel to the axis of the bore.

In order for a gun which uses gunsights to fire accurately, the gun barrel and the gunsights must use the same point of aim. If the two are not in proper alignment, the gunsight may be on the target but the gun barrel will be using a point of aim removed from the target by an angle equal to the error in gunsight alignment. Illustration 11-1 shows how these errors may exist in deflection (A) and elevation (B).

The lines of sight for antiaircraft machine guns, and others which fire at short range, must be so adjusted that they converge with and intersect the gun axis at a predetermined mean firing range, as shown in illustration 11-2.

A gun may be boresighted at a particular range by using a sighting target at that range, tightly stretched and so mounted that its lines are vertical and horizontal (fig. 11-1). An object ashore with a good reference point and at the desired distance, however, may be used as a target for boresighting. If the gunsights are to be aligned parallel to the gun bore, the horizon may be used as a target in elevation and an object at a similar or greater distance may be used as a target in deflection. A 6-inch Mk 37 boresight, and also an 8-inch Mk 31 boresight, has illuminated crosslines so that they may be used in boresighting on celestial objects at night.

The following step-by-step instructions for boresighting a gun are typical, but they vary for different gun mounts:

1. Examine the gunsight telescopes for cleanliness, parallax, focusing, and clearness and sharpness of crosslines.
2. Operate the sight mechanisms through their entire area of motion in deflection and elevation to determine whether they operate freely and without excessive lost motion.
3. Open the gun breech and securely lash the plug or breech block open to prevent its swinging against the boresight or boresight telescope.
4. Install and align the boresight in accordance with the procedure required for the boresight being used.
5. Man all stations—pointer's, trainer's, and boresight. Use the regular gun pointer and the regular trainer, and a Gunner's Mate at the boresight, if possible.
6. Adjust sight-setting mechanisms to zero readings.
7. Boresight the gun for deflection or elevation. Suppose you decide to boresight first for elevation, with the horizon as a target. The boresight checker must coach the gun pointer in so aligning the gun that the horizontal crossline of the boresight lies on the horizon; and when it does, he calls ''Mark.'' The pointer then notes the distance his crossline is off the horizon and makes necessary adjustment of the telescope to put it back on the horizon. When he thinks his telescope is adjusted, the pointer also calls ''Mark.''

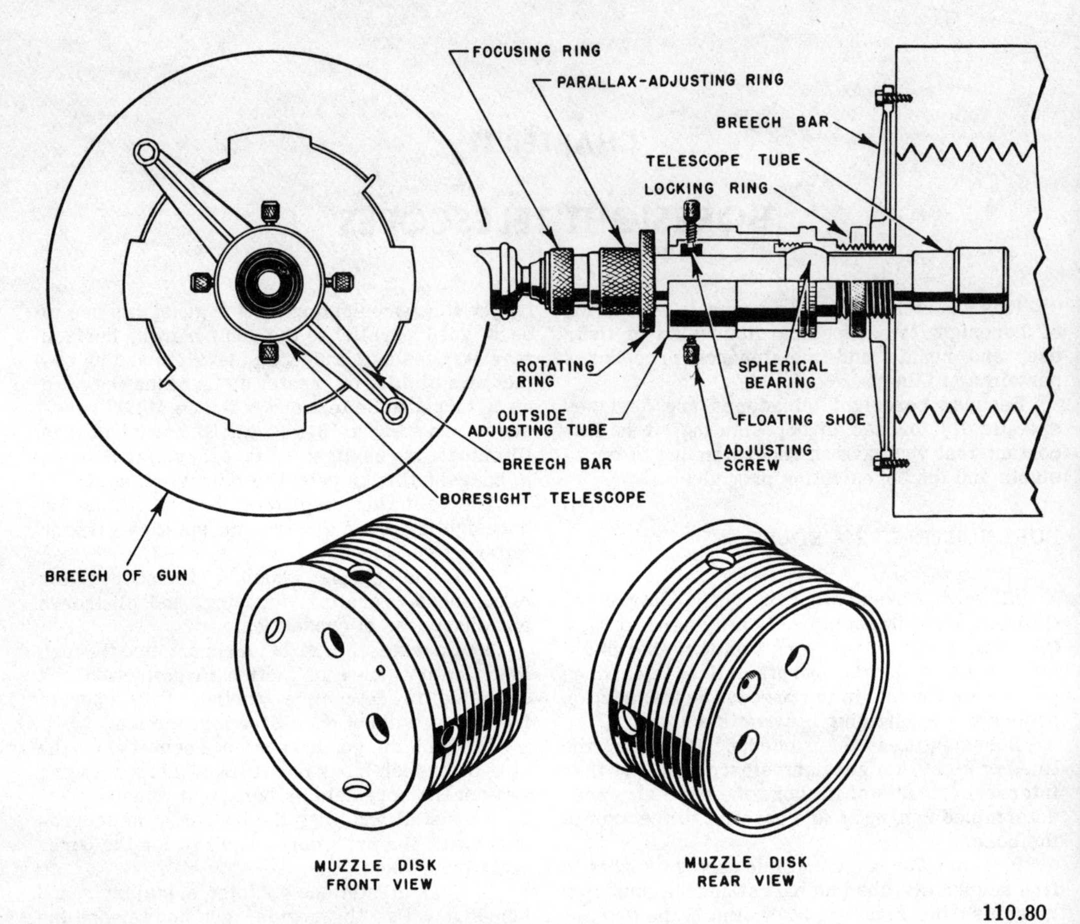

Figure 11-1.—Boresighting equipment.

When the boresight checker and the pointer simultaneously call "Mark" several times, the horizontal crossline is considered adjusted. The adjustment should then be secured tightly and another check made to determine if a shift has occurred. The boresight checker and the gun pointer should then change stations and make a final check.

8. Adjust the trainer's telescope in elevation in the manner just explained for the trainer, with the trainer calling "Mark" to indicate when his horizontal crossline is on the horizon.

9. To complete the check, all three individuals must now "Mark" when on the horizon. This is the final adjustment.

Boresighting in deflection is accomplished in the same manner, except that the vertical crossline of the trainer's sight is adjusted first (on all mounts except those which have control for both train and elevation at the pointer's station).

When boresighting is completed, alignment of the boresight should be rechecked for shift that may have occurred during the procedure.

BORESIGHTS

A boresight is a device used to ESTABLISH A LINE OF SIGHT along the axis of the bore of a gun, or A LINE OF SIGHT PARALLEL WITH

AND CLOSE TO THIS AXIS. Most boresights include optical elements which provide magnification of the field of view, as well as crosslines to fix the line of sight. Some boresights, however, contain only line-of-sight reference points, such as orifices and crosswires.

TYPES OF BORESIGHTS

Boresights are divided into two general classifications: (1) breech bar boresights, and (2) boresights with self-contained optical elements.

Breech Bar Boresights

A breech bar boresight normally consists of a boresight telescope and several mechanical elements used to mount and align the telescope in order to have its line of sight coincide with the axis of the bore of the gun. Telescopes used with boresights are specially designed for boresight service and may be used interchangeably with the mechanical elements of any breech bar boresight. These telescopes provide medium-power magnification and include crosslines to fix the line of sight.

Mechanical elements of breech bar boresights are not interchangeable with other boresights. They generally consist of an adaptor (breech bar) with a central hole for mounting the telescope in the center of the breech of the gun, and a muzzle disk for aligning the telescope.

Study illustration 11-3. Observe the breech of the gun on the left, the breech bar mounted in position on the breech with screws, and the boresight telescope (right) mounted in the center hole of the breech bar. Note the extension of the telescope tube within the breech of the gun, and also the two views of the muzzle disk.

The muzzle disk is machined for exact fitting in the muzzle of a gun and has a small central hole (fig. 11-1) on which the crosslines of the telescope may be adjusted. Scribe marks on the outer surface of the disk ensure correct mounting in the muzzle. Once the telescope is properly aligned, the muzzle disk can be removed to allow a direct view of the target through the telescope.

Boresights with Self-contained Optics

Boresights with self-contained optics use the gun barrel itself as a telescope tube; and the optical elements are mounted in assemblies installed in the breech and muzzle of the gun. Some of these boresights have no lenses to provide magnification; they have a reticle in the muzzle assembly, and an orifice in the breech assembly to fix the line of sight. A mirror is usually provided to permit viewing from above the breech.

Other boresights in this class use individual telescopes which cannot be used with any other boresight; and the telescopes are installed with mandrels fitted into the gun barrel. The telescope may be offset from the barrel, but its line of sight must be parallel with the axis of the bore of the gun.

BORESIGHT TELESCOPES

Boresight telescopes are designed for use in the boresighting procedure just discussed. When used with a breech bar, muzzle disk, and other necessary elements, they comprise a breech bar boresight.

In design, a boresight telescope must be short in length for the amount of magnification required, and its diameter must be correspondingly small. These characteristics are necessary in order to mount the telescope in breech bars and to facilitate handling.

Because of its short length, a boresight telescope must have an objective lens of short focal length; and it must be capable of focusing at distances of approximately 10 feet to infinity. This characteristic makes focusing on the boresight muzzle disk as well as on the target possible. The short objective focal length permits this type of focusing with little adjustment of distance between the objective lens and the crossline plate (plano-plano, with a reticle engraved on one surface).

Since there is very little motion between a boresight telescope and the target during boresighting, and also because of the small size of the target, the instrument need not have great width of field. Because of its short objective length, however, a boresight telescope may have some curvature of the field; but since only the center of the field where the crosslines intersect is used, this effect can be discounted.

MARK 75 BORESIGHT TELESCOPE

Because the Mk 75 boresight telescope is used extensively on 3"/70 and 5"/54 guns, as well as on 16-inch guns, it is discussed in considerable detail in this chapter.

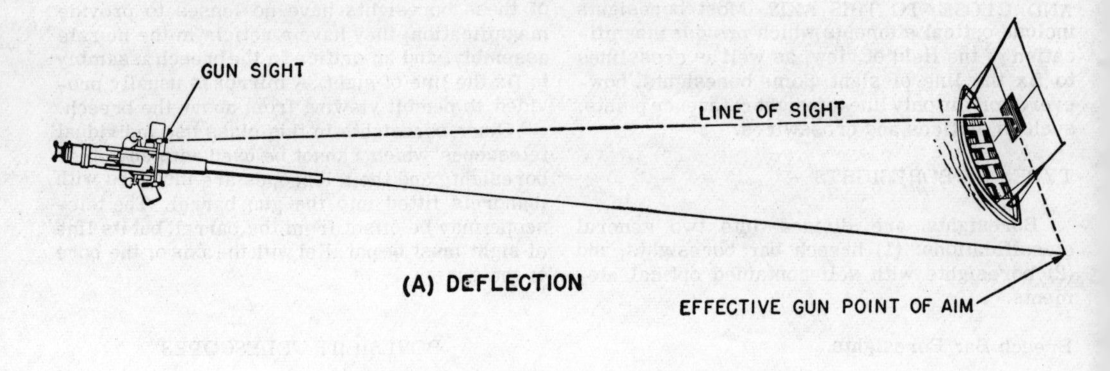

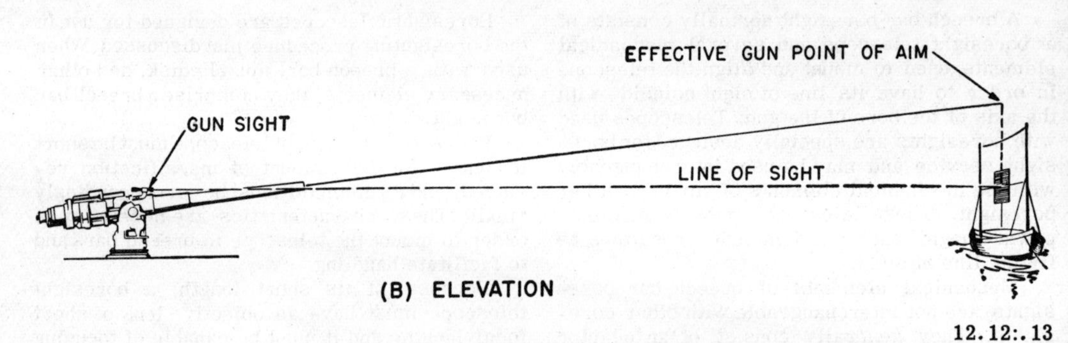

12.12:.13

Figure 11-2.—Effect of sight misalignment.

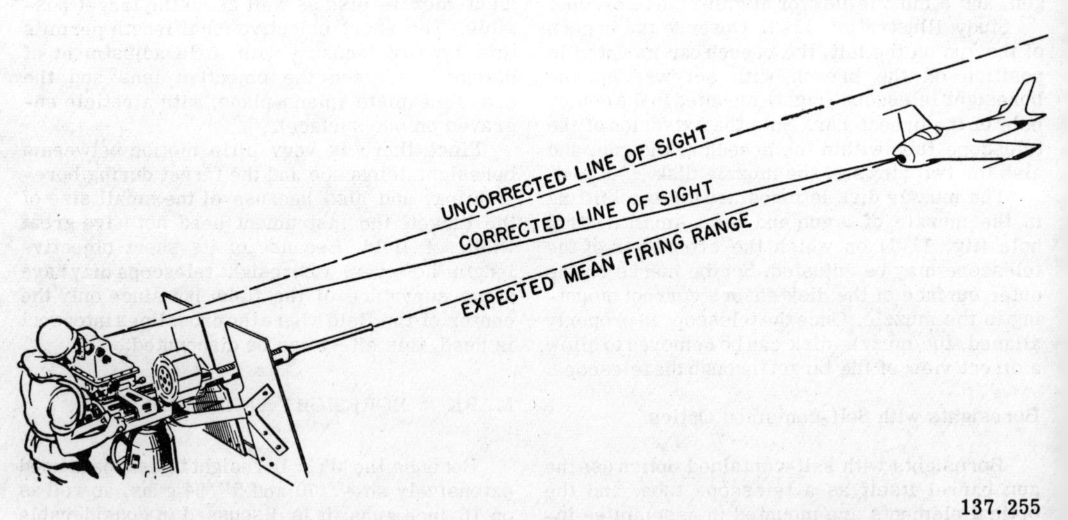

137.255

Figure 11-3.—Antiaircraft machine gun boresighted at mean firing range.

Physical Characteristics

A Mk 75 boresight telescope is illustrated in figure 11-4. Study the nomenclature. It is an 8-power telescope designed specifically for boresighting, and the two Mods (1 & 0) are similar in construction. Their differences in construction and optical characteristics are explained later in this chapter.

The approximate overall dimensions of a Mk 75 boresight telescope are:

Length.	11 1/4 inches
Maximum diameter	3 1/4 inches
Weight.	4 pounds

This boresight telescope consists of the telescope proper, telescope adaptor, and the adjusting-screw housing assembly. Study illustration 11-5, which gives a cutaway view of all parts of the instrument.

The telescope body tube has a spherical journal soldered to it, and the telescope adaptor forms the spherical bearing which supports the telescope about this journal. The clearance between the adaptor and the body tube is adequate for adjustment of the telescope within the bearing. The adaptor carries the external threads with which the telescope may be mounted in a breech bar, and also a lock ring for locking it in any desired position.

The adjusting-screw housing of the telescope surrounds the body tube and is secured to the telescope adaptor. The four adjusting screws (90° apart) contact the four flat surfaces on this square (quadrangular) bearing soldered to the body tube. By manipulating the adjusting screws, you can move the telescope with its spherical journal within its spherical bearing to change its alignment with respect to the adaptor; that is, to line up the telescope with the muzzle disk. This arrangement of adjusting screws, however, does not permit rotation of the telescope within the adaptor housing.

The telescope drawtube slides within the body tube and carries with it all optical parts except the objective lens. The focusing key is attached to the drawtube behind the adjusting-screw housing and extends through a fore-and-aft slot in the body to engage a spiral groove cut in the actuating ring which surrounds the body tube. Rotation of the actuating ring advances the key through the spiral groove as the slot in the body tube restricts movement of the key to a fore-and-aft direction. The key carries the drawtube backward or forward to change the position of the crossline lens relative to the objective lens and thus permits parallax-free focusing on objects from 10 feet to infinity.

Differences in internal construction of a Mk 75, Mod 0, boresight telescope, as compared to Mod 1, are shown in illustration 11-6. Compare the nomenclature in this illustration with that in figure 11-5.

In the Mk 75, Mod 0, the crossline plate mount (fig. 11-6) contains the crossline plate only—not both the crossline plate and the front erecting lens mount, which is secured to the telescope drawtube in the same manner as the rear erecting lens.

The diaphragm in the Mod 1 telescope consists of a flanged bushing which slides into the end of the drawtube; in the Mod 0 telescope it is a ring which is threaded into the drawtube and held in place by a lock ring (fig. 11-6).

Optical Characteristics

Optical characteristics of the Mk 75 boresight telescope (Mods 1 and 0) are:

	Mod 0	Mod 1
Magnification	8x	8x
Field.	3°30'	3°30'
Exit Pupil	2.5 mm	2.5 mm
Eye Distance	18.0 mm	19.4 mm

The image formed by a Mk 75 boresight telescope is erect and normal, which means that it may be viewed exactly as observed by the naked eye (except for magnification provided by the telescope, and such distortion as curvature of the field).

Arrangement of optical elements in a Mk 75 boresight telescope is shown in figure 11-7. Observe the position of all lenses, and the plan view of the crosslines.

OBJECTIVE LENS.—The objective lens is a cemented doublet consisting of a double convex lens of barium-crown glass and a concave-plano lens of dense-flint glass. It is mounted in a fixed position against a seat ring at the end of the body tube and held in position by a retaining ring. The body tube, of course, is the main support tube for all elements in the telescope.

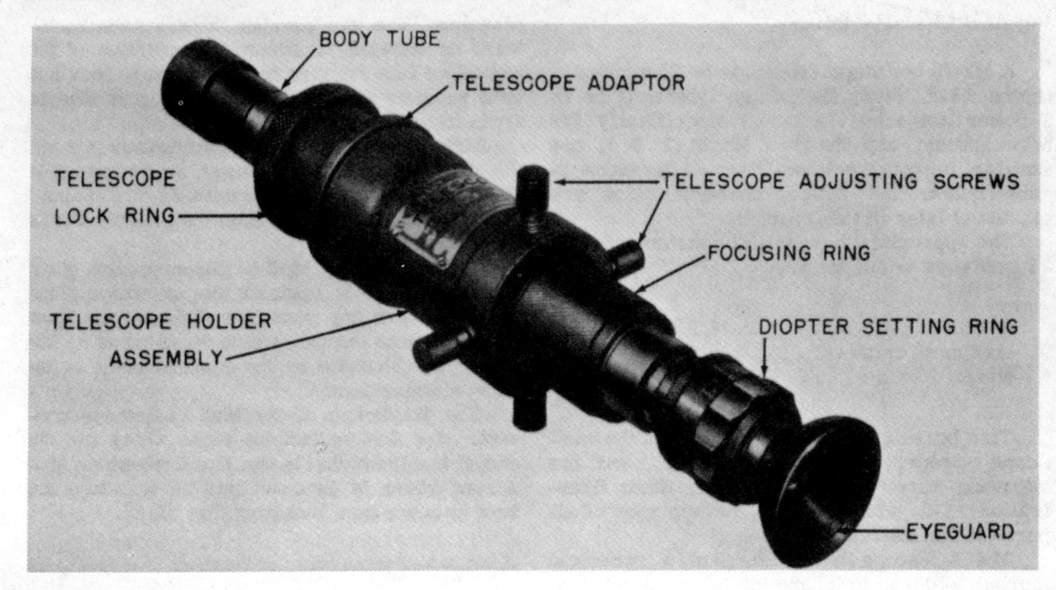

84.207

Figure 11-4.—Mark 75 boresight telescope.

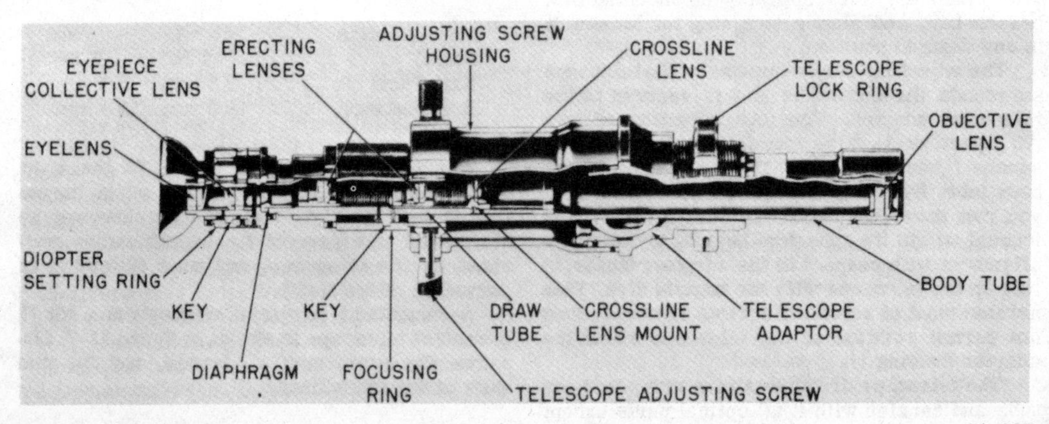

84.207

Figure 11-5.—Cutaway view of a Mk 75, Mod 1, boresight telescope.

CROSSLINE AND ERECTING LENSES.—The crossline plate is plano-plano and made of borosilicate-crown glass, with its crosslines etched into the surface away from the objective lens. It is secured in a mount held in position by an aligning pin which passes through the draw-tube. No means is provided, however, for adjusting the radial position of the crossline plate; but

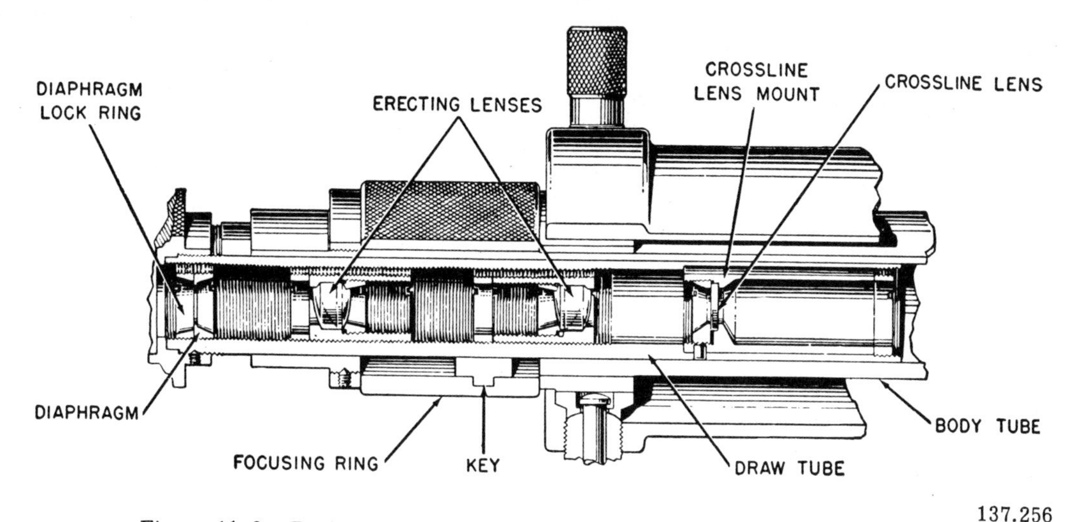

137.256

Figure 11-6.—Partial cutaway view of a Mk 75, Mod 0, boresight telescope.

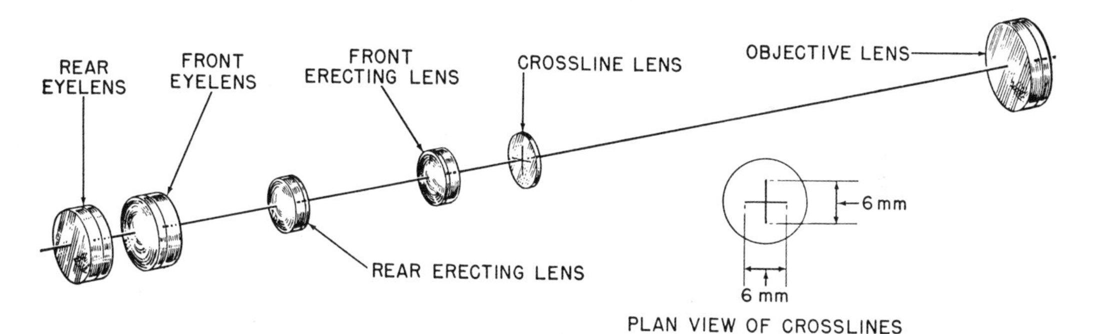

137.257

Figure 11-7.—Optical system of a Mk 75 boresight telescope (Mods 0 & 1).

since the crosslines are located in the geometric center of the plate (concentric with mechanical parts), no radial adjustments should ever be required. The procedure for orienting the crosslines in order to have them located in the same planes as the adjusting screws is explained under collimation at the end of this chapter.

The two symmetrical, cemented doublet erecting lenses consist of a double convex lens of borosilicate-crown glass and a concavoconvex lens of very dense flint glass, and they are mounted in a fixed position in the crossline and erecting system drawtube (double convex lenses facing each other). Both lenses are secured in mounts—front erector lens mount (toward objective lens) secured by a lock ring in the end of the crossline lens mount, the rear erector lens mount secured in the telescope drawtube by a lock ring. Review illustration 11-7.

EYEPIECE LENSES.—The lenses of the eyepiece are symmetrical, cemented doublets with their convex surfaces facing each other. The lenses are secured at opposite ends of a mounting drawtube with the double convex lenses facing each other (fig. 11-7). The mounting drawtube slides within the eyepiece mount screwed onto the end of the telescope drawtube. The diopter setting can be varied by rotating the eyepiece actuating ring around the eyepiece mount. The focusing key attached to the eyelens mounting drawtube fits in a spiral groove in the eyepiece mount; so when you move the actuating ring you also move the focusing key. An index mark on the eyepiece and graduations on the actuating ring indicate the diopter setting.

Predisassembly Inspection

Review the discussion of the predisassembly inspection for telescopes in chapter 8 of this manual. One step in the predisassembly inspection of a Mk 75 boresight, however, which needs mentioning here is the following: Put the telescope in a collimator and superimpose the crosslines on those of the collimator, and then (with the horizontal adjusting screws) move the vertical crossline of the telescope from that of the collimator. If the horizontal crossline of the telescope does not stay on the crossline of the collimator, the crossline lens is incorrectly positioned.

Disassembly Procedure

Refer to illustration 11-8 and study carefully the Mk 75 boresight telescope assembly for Mods 0 and 1. The procedure for disassembling the parts shown in figure 11-8 is as follows:

1. Remove the eyeguard (1) from the eyepiece end of the telescope.
2. Hold the eyepiece assembly to the crossline and erecting system drawtube assembly and remove the eyepiece setscrew (2) and unscrew the eyepiece.
3. Check the focusing ring (5) for ease and smoothness of operation. If the action is not satisfactory, inspect parts for burrs, scratches, and so on, as you remove them. Remove the retaining ring setscrew (3), the retaining ring (4), and the focusing ring (5) from the body tube (24).
4. Remove the two setscrews (6) from the focusing key (8). Remove, next, the focusing key, and then remove the two dowels (7) from the key.
5. Check the operation of the crossline and erecting system drawtube assembly within the body tube for smoothness and ease of motion and then withdraw it. If the action is not smooth, check for burrs, scratches, and so forth.
6. Release (back off) the telescope adjusting screws, remove the telescope holder setscrew (9), and unscrew the telescope holder assembly from the telescope adaptor (20).
7. Remove the objective cap setscrew (15) and unscrew the objective cap (16).
8. Rotate the body tube within the telescope adaptor and check for smoothness of action. If a flaw is apparent, check the bearing surfaces of the adaptor (20), the telescope retainer (18), and the spherical journal (25) for various defects as you disassemble them. Remove the telescope adaptor setscrew (17) and unscrew the telescope retainer (18). Then slide off the telescope adaptor (20) and remove the lock ring (19) from it.
9. Remove the objective lens retaining ring (21) and drop the objective lens (22) out into the palm of your hand.
10. Remove the objective seat (23) from the body tube (24).
11. The spherical journal (25) and the quadrangular bearing (26) are permanently attached to the telescope body tube and damage inflicted upon them cannot be repaired. If these two bearings are damaged, REPLACE THE ENTIRE ASSEMBLY.

This completes the disassembly of the Mk 75 telescope components, and the next step is disassembly of the assemblies (or subassemblies) which comprise the mechanism of the telescope.

EYEPIECE ASSEMBLY.—Proceed as follows to disassemble the eyepiece assembly (fig. 11-9) of a Mk 75, Mods 0 & 1, boresight telescope:

1. Check the actuating ring (5) for ease and smoothness of operation; and if unsatisfactory, inspect surfaces for burrs and other defects as you remove parts. Remove the retaining ring setscrew (1) and the retaining ring (2) from the eyepiece mount (10).
2. Remove the actuating ring (5) from the mount, unscrew the diopter scale setscrew (3), and slip the diopter scale (4) from the ring.
3. Remove the index ring setscrew (6) and slip the index ring (7) from the eyepiece mount.
4. Remove the focusing key screws (8) and the focusing key (9) from the eyelens mount. NOTE: There are no dowel pins in this focusing key.
5. Check the movement of the eyepiece drawtube assembly within the eyepiece mount

(10) before you separate the two parts. If a flaw in the action is evident, check the outside bearing surface of the eyepiece drawtube and the inside bearing surface of the eyepiece mount for burrs, scratches, and so forth.

EYELENS MOUNT ASSEMBLY.—The eyelens mount assembly for the Mk 75 boresight (Mods 0 and 1) is shown in figure 11-10. The two steps in disassembly are:

1. Remove the eyelens retaining ring (1) and drop the field lens (2) out.

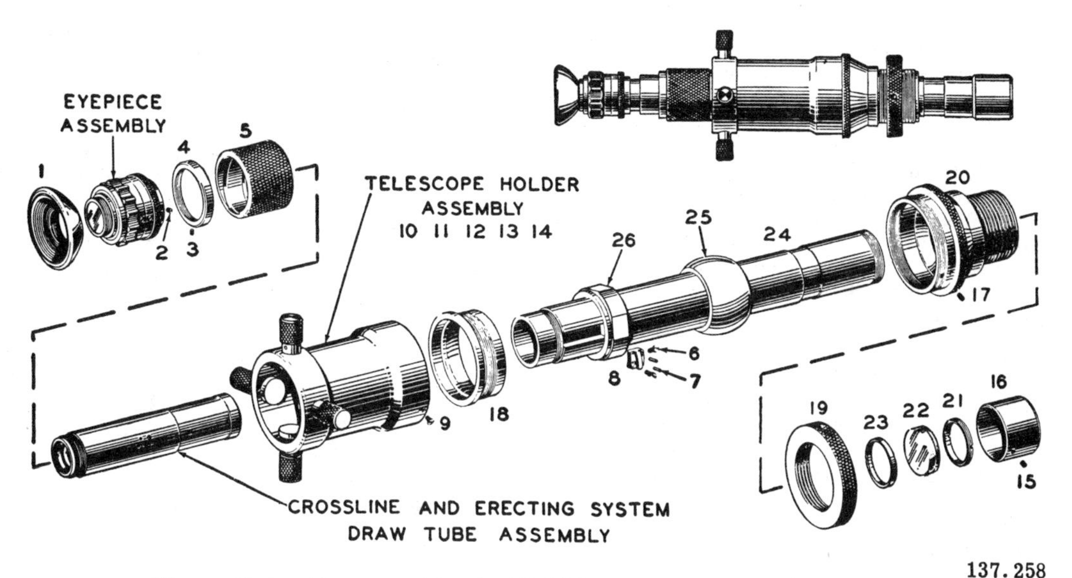

137.258

Figure 11-8.—The assembly of a Mk 75 boresight telescope (Mods 0 & 1).

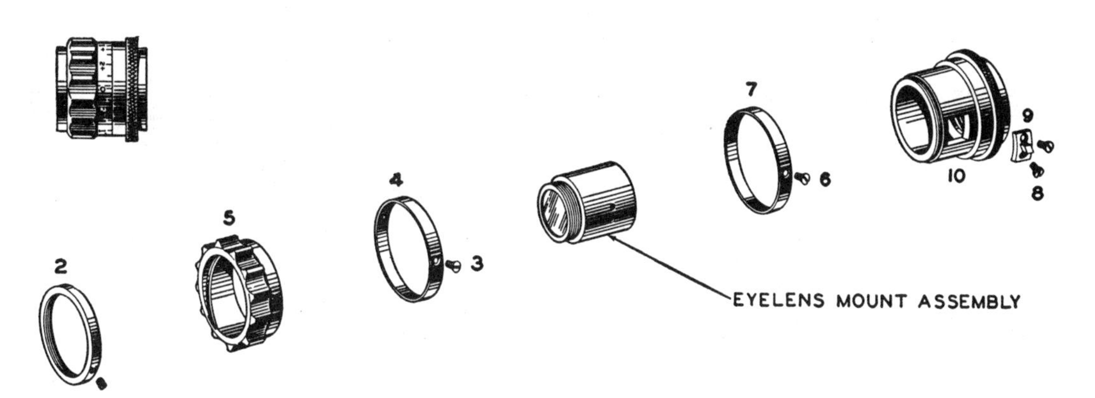

137.259

Figure 11-9.—Eyepiece assembly of a Mk 75 boresight telescope (Mods 0 & 1).

2. Remove the separating ring (3) and drop the eyelens (4) from the eyelens mount (5).

CROSSLINE AND ERECTING SYSTEM DRAWTUBE ASSEMBLY (Mod 0).—Study the components of the crossline and erecting system drawtube assembly of the Mk 75, Mod 0, boresight telescope in illustration 11-11, the disassembly of which is as follows:

1. Remove the diaphragm lock ring (1) and the diaphragm (2) from the drawtube (6).

2. Remove the rear erecting lens mount lock ring (3) from the drawtube and then remove the rear erecting lens mount assembly.

3. Remove the front erecting lens mount lock ring (4) from the drawtube, and then take off the front erecting lens mount assembly.

4. Unscrew the crossline mount lock ring (5) from the other end of the drawtube; then remove the crossline mount assembly (illustrated). The aligning pin (7) is permanently installed in the drawtube (6).

CROSSLINE AND ERECTING SYSTEM DRAWTUBE ASSEMBLY (Mod 1).—The crossline and erecting system drawtube assembly for the Mk 75, Mod 1, boresight telescope is illustrated in figure 11-12. Compare this assembly with the drawtube assembly for the Mk 75, Mod 0, boresight telescope (fig. 11-11). The procedure for disassembling the Mod 1 crossline and erecting system drawtube assembly is as follows:

1. Remove the diaphragm (1) from the drawtube (9). Then unscrew the rear erecting lens mount lock ring (2) and the rear erecting lens mount assembly (illustrated).

2. Next, remove the crossline mount lock ring (3) and the spacer (4). NOTE: This spacer was not on early models of the Mk 75, Mod 1, boresight telescope. It was installed after Serial No. 3664.

3. Slide the crossline mount assembly from the drawtube (9).

4. Remove the crossline retaining ring (6) and the crossline plate from the crossline mount (5).

5. From the other end of the crossline mount, remove the front erecting lens mount retaining ring (8) and the front erecting lens assembly. The aligning pin (10) is permanently installed in the drawtube.

ERECTING LENS MOUNT ASSEMBLIES (Mods 0 & 1).—Study the erecting lens mount

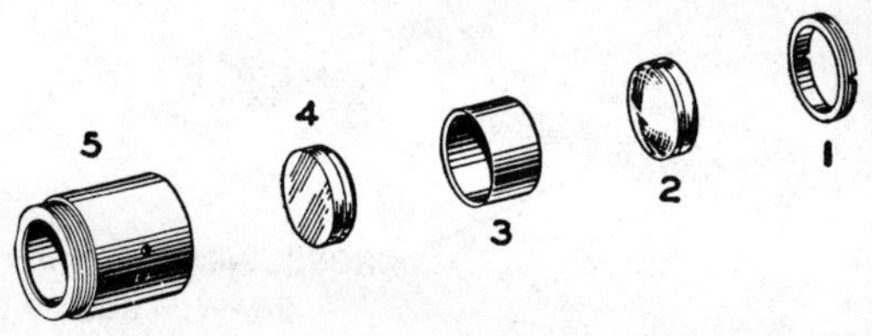

137.260

Figure 11-10.—Eyelens mount assembly of a Mk 75 boresight telescope (Mods 0 & 1).

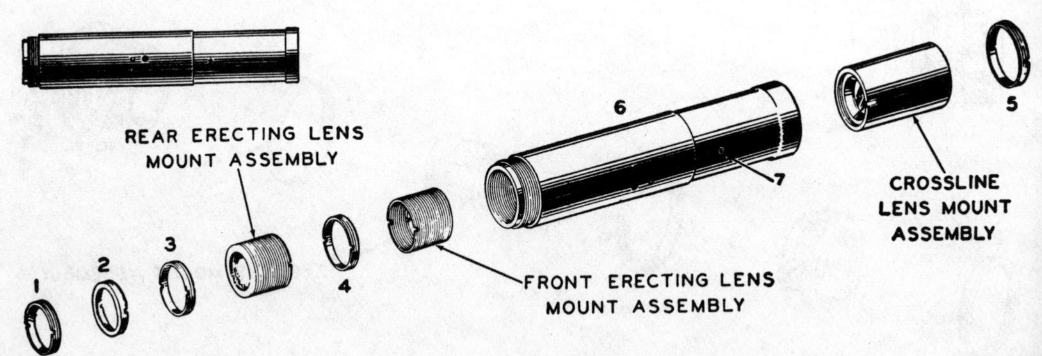

137.261

Figure 11-11.—Crossline and erecting system drawtube assembly of a Mk 75 boresight telescope (Mod 0).

assemblies for the Mk 75 boresight telescope (Mod 0 & 1) in illustration 11-13. The disassembly procedure for these assemblies is as follows:

1. From the front erecting lens mount (3), remove the retaining ring (1) and the front erecting lens (2).
2. Unscrew the lens retaining ring (4) and remove the rear erecting lens (5) from the rear erecting lens mount (6).

CROSSLINE MOUNT ASSEMBLY (Mod 0).—The crossline mount assembly for the Mk 75 boresight telescope (Mod 0) is shown in illustration 11-14. To disassemble it, remove the crossline retaining ring (1) and the crossline plate from the crossline mount (3).

Repairing and Cleaning

Refer to chapter 8 of this manual for maintenance procedures applicable to all optical instruments. Information presented in this section applies specifically to boresight telescopes.

One important thing for you to look for when you repair a boresight telescope is straightness of the tube. Check the tube on a mandrel; and if it is slightly bent, spin it between centers on a lathe as you force a mandrel through it. If there is a pronounced bend in the tube, replace it; because it will crumple if you try to force a mandrel through it.

If you removed burrs from the spherical bearing or the journal, it may be necessary to lap in the spherical bearing. Check carefully, and be thorough.

Check the fit of the crossline and the erecting system drawtube and the eyelens mount for fit, which should be smooth and easy, without lateral play. If these parts do not meet this requirement, lap them in, or take other corrective action, if necessary.

Study notations and recommendations on the casualty analysis sheet for the instrument, and effect essential repairs. When in doubt about procedure, or the action you should take, consult your shop supervisor.

Reassembly

Refer to illustration 11-8 as you study the procedure for reassembling a Mk 75 boresight telescope. Some of the things you should do when reassembling the instrument are:

1. Apply a small amount of APPROVED lubricant on all parts which require it.
2. Lightly lubricate the spherical journal (25, fig. 11-8), the spherical bearing (20), and the telescope retainer (18). Then slip the adaptor over the eye end of the drawtube, and slip the retainer over the other end. Screw these parts together until they fit snugly over the spherical journal, with no lateral or end play but not tight enough to restrict rotation of the body tube. Then secure the adaptor with the setscrew (17).
3. Apply a small amount of lubricant to the flat surfaces of the adjusting screw bearing (26, fig. 11-8) and the swivel heads in the telescope holder assembly. Release all the way the four adjusting screws and slip the holder assembly over the eye end of the body tube. Screw the telescope holder to the telescope retainer and rotate the body tube as necessary to have the four swivel heads bear evenly on the four flat surfaces of the bearing ring. Then turn the adjusting screws in until they bear lightly and evenly on the pads and secure the telescope holder with the setscrew (9).
4. Put the telescope lock ring (19) on the telescope adaptor.
5. Slip the objective lens seat (23) into the unslotted end of the body tube and drop the objective lens (22) into position, with its plane surface bearing against the lens seat. Secure the lens with the objective lens retaining ring (21). Replace the objective cap and secure it with its setscrew (15).
6. Lightly lubricate the exterior of the crossline and the erecting system draw tube assembly and slip it into the body tube, crossline end first. Check the fit by working it forward and backward. The motion should be smooth and easy, without hindrance. Then line the focusing key screw holes up with the slot in the body tube.
7. Put the focusing key (8) in the slot over the two screw holes, insert the dowels (7) through the key and into the drawtube, and secure the key with the two screws (6). Finally, work the key backward and forward to recheck the drawtube for smoothness of action; then screw on the eyepiece assembly.

CROSSLINE AND ERECTING SYSTEM DRAWTUBE ASSEMBLY FOR MOD 0.—Use the

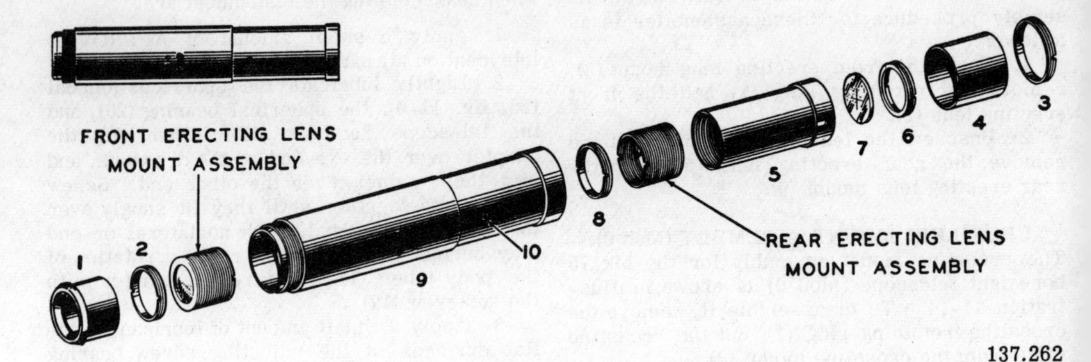

137.262

Figure 11-12.—Crossline and erecting system drawtube assembly of a Mk 75 boresight telescope (Mod 1).

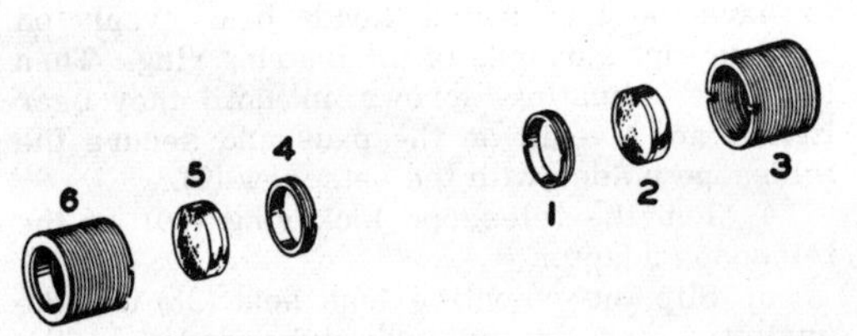

137.263

Figure 11-13.—Erecting lens mount assemblies of a Mk 75 boresight telescope (Mods 0 & 1).

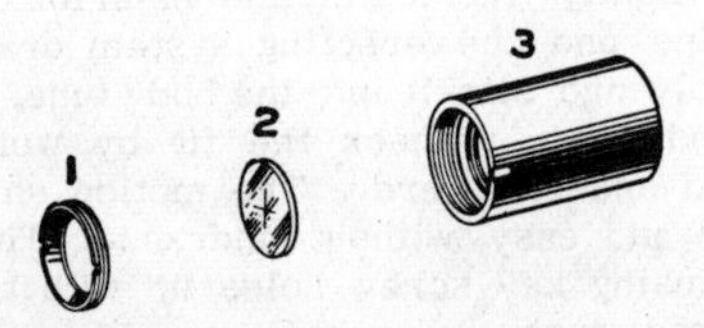

137.264

Figure 11-14.—Crossline lens mount assembly of a Mk 75 boresight telescope (Mods 0).

procedure outlined next for assembling the crossline and erecting system drawtube assembly (fig. 11-11) for Mod 0 of the Mk 75 telescope:

1. Place the crossline mount assembly in the objective end of the erecting system drawtube (6). Aligning pin 7 fits in the keyway in the mount. Secure the assembly with the crossline plate mount lock ring (5).

2. Insert the front erecting lens mount assembly, lens end first, in the eye end of the drawtube and screw it down to within a half inch of the end of the threaded area in the tube.

3. Select an auxiliary eyepiece of sufficient focal length for focusing it on the crossline plate. Place the drawtube on a V-block on a collimator bed with the front erecting lens toward the collimator. With the auxiliary telescope, sight through the drawtube from the crossline lens end and adjust the position of the erecting lens mount until all parallax is removed. Then put in the front erecting lens mount lock ring (4), lock the mount in place, and recheck for parallax. NOTE: Locking of the mount may have disturbed its position.

4. Insert the rear erecting lens mount in the eye end of the drawtube, lens end last, and screw it down to within a half inch of the front lens mount lock ring. Then attach the telescope eyepiece assembly to the drawtube, set the diopter scale on zero, place the auxiliary eyepiece on the telescope eyepiece, and sight on the crossline plate for definition of the crosslines. Remove the auxiliary telescope and eyepiece and adjust the rear erecting lens mount assembly as necessary to obtain maximum definition on the crosslines of the telescope. Recheck after each adjustment. When definition of the crossline plate is sharp and clear, turn the diopter scale off zero and refocus it. If the definition is still not sharper at zero, readjust

the rear erecting lens mount assembly. After you do this, remove the eyepiece and lock the erecting lens mount with the rear erecting lens mount lock ring (3) and recheck the definition.

5. Insert the diaphragm (2), with the slotted side toward the inside of the drawtube, and screw it in far enough to allow the lock ring (1) to screw down tight against the diaphragm, with its rear edge one or two threads within the tube.

CROSSLINE AND ERECTING SYSTEM DRAWTUBE ASSEMBLY FOR MOD 1 OF THE MARK 75 TELESCOPE.—Refer to figure 11-12 as you study the procedure for reassembling the crossline and erecting system drawtube assembly for the Mk 75, Mod 1, telescope. Proceed as follows:

1. Insert the crossline plate (7) into the crossline plate mount, with its etched side against the lens seat. Then center one of the crosslines in the keyway cut in the exterior of the mount and secure the lens with the retaining ring (6).

2. Screw the front erecting lens mount assembly, the end with the lens seat first, into the other end of the crossline plate mount until there is enough room to insert the retaining ring (8). NOTE: Do NOT insert the ring at this point.

3. Select an auxiliary eyepiece of sufficient focal length for focusing it on the crossline plate. Then place the crossline plate mount on a V-block on a collimator bed, with the front erecting lens toward the collimator; and with the auxiliary eyepiece, sight from the crossline end of the mount and adjust the position of the front erecting lens mount until all parallax is removed. Lock the erecting lens mount with the retaining ring (8) and recheck for parallax.

4. Screw the rear erecting lens mount assembly, the end with the lens seat last, into the eye end of the erecting system drawtube (9). Then turn the assembly in enough to permit you to insert the lock ring (2) flush with the threads in the drawtube. Do NOT insert the lock ring at this point.

5. Slide the crossline lens mount into the other end of the drawtube, so that the aligning pin (10) will engage the keyway in the mount. Then insert the spacer (4) and secure it and the mount with the lock ring (3). NOTE: If the serial number of the telescope is below #3666, there will be no spacer.

6. Screw the eyepiece assembly onto the drawtube and set the diopter scale to zero. Then place an auxiliary telescope against the eyepiece, focus on the crosslines, and note the definition of the crosslines. Remove the auxiliary telescope and eyepiece and adjust the rear erecting lens mount for maximum definition of the crosslines. Recheck after each adjustment.

When the definition is sharp and clear, turn the eyepiece diopter scale off zero and refocus it. If the definition is still not sharper at zero, readjust the position of the rear erecting lens mount. After you do this, remove the eyepiece and secure the lens mount with the lock ring (2) and recheck the definition.

7. Insert the diaphragm (1) in the drawtube.

Collimation

Review the collimation procedure for optical instruments in chapter 8 and proceed in the following manner to collimate a Mk 75 boresight telescope (Mods 0 & 1):

1. Place the instrument in a collimator fixture on the collimator bed and screw it up against the lock ring on the telescope adaptor. See illustration 11-15. Line up the adjusting screws vertically and horizontally and lock the telescope in position with the lock ring. Then, with the focusing key, focus on the collimator crosslines; and with the telescope adjusting screws, superimpose the intersection of the telescope crosslines on the intersection of the collimator crosslines. Usually, before the instrument is collimated, the crosslines look like those shown in figure 11-16.

If the crosslines of the telescope are not superimposed on those of the collimator, release the telescope lock ring and rotate the telescope until they are superimposed.

137.265

Figure 11-15.—Mounting a boresight telescope in a collimator fixture.

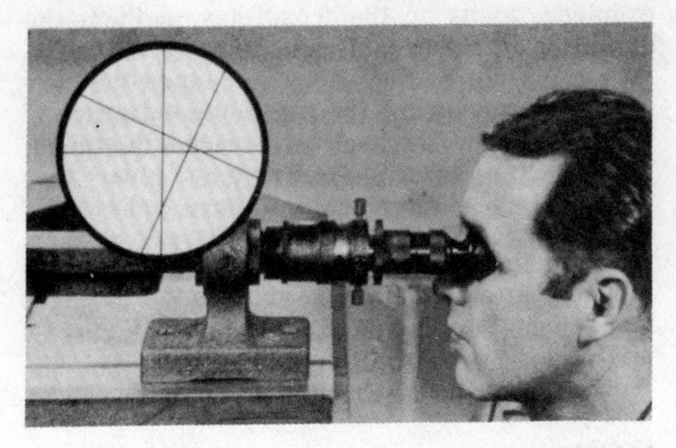

137.266

Figure 11-16.—Typical sight of collimator crosslines before a boresight telescope is collimated.

2. Use the horizontal adjusting screws to move the vertical crossline to the right or to the left (fig. 11-17). The crossline intersection should remain on the horizontal crossline of the collimator; if it does not, adjustment of the crosslines is essential. See illustration 11-18.

Remove the focusing key and slide the drawtube assembly from the body tube (24, fig. 11-8). Then remove the crossline lens mount, loosen the crossline retaining ring, and rotate the crossline in its mount in the direction required. Replace the crossline mount, the drawtube, and the focusing key.

3. Again, superimpose the telescope crossline intersection on that of the collimator crossline and rotate the telescope until the lines themselves are superimposed. Repeat step 2, as necessary.

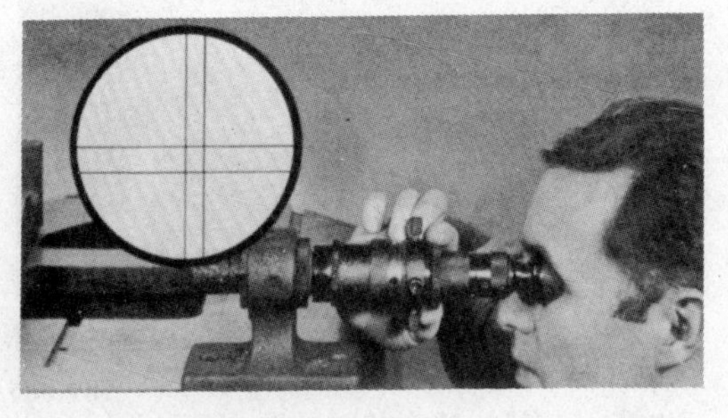

137.267

Figure 11-17.—Squaring the crosslines.

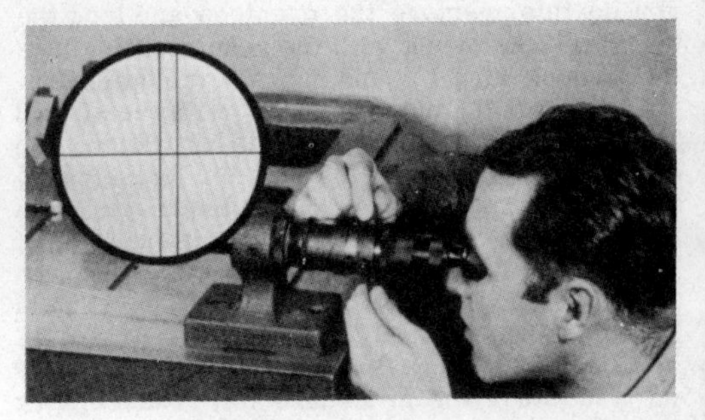

137.268

Figure 11-18.—Adjusting the horizontal crossline.

4. Now use the vertical adjusting screws to check the vertical adjustment, as explained in step 2. If the horizontal adjustment checks and the vertical adjustment does not check, there must be a twist in the body tube. Replace the tube.

5. Remove the eyepiece assembly. Lightly lubricate the interior surfaces of the focusing ring (5, fig. 11-8) and slip the actuating ring over the body tube in the manner required to have its spiral groove (in the ring) engage the focusing key. Then screw on the focusing ring retainer (4) and secure it with the setscrew (3). Rotate the focusing ring to check its action and that of the draw tube.

6. Screw the eyepiece assembly into place and secure it with the setscrew (2), and screw the eyeguard (1) into position.

7. Make a final check on the telescope crosslines by repeating steps 1, 2, and 4. When collimation of the instrument is satisfactory, its crosslines and those of the collimator look like those shown in figure 11-19.

MARK 8 BORESIGHT TELESCOPE

The Mk 8 boresight telescope is similar in appearance to a Mk 75 boresight telescope. It is 11 1/4 inches long, has a maximum diameter of 3 1/4 inches, and has 9.6 power. Its eye distance is 11.0 mm, as compared to 18.0 mm for the Mk 75, Mod 0, and 19.4 mm for the Mk 75, Mod 1, boresight telescope.

A cutaway view of a Mk 8, Mod 6 (only mod of the Mk 8 currently in use), boresight telescope is shown in illustration 11-20. Study it

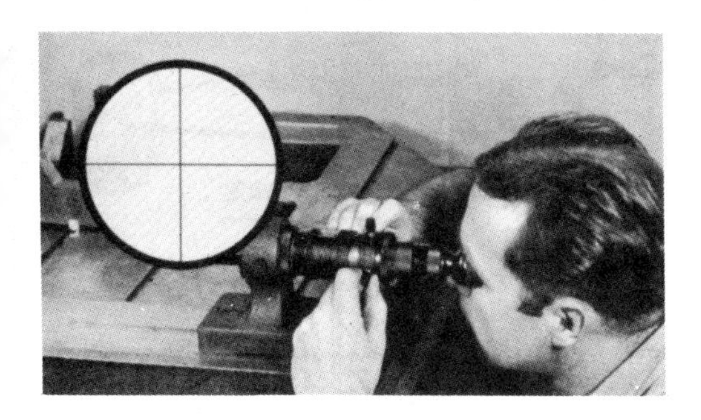

137.269
Figure 11-19.—Position of the crosslines when a boresight telescope is correctly collimated.

carefully and compare the components with those of a Mk 75, Mod 1, boresight telescope (fig. 11-5).

On occasions, you will be required to repair Mk 8 boresight telescopes, because they are still used on some ships.

The crossline lens in a Mark 8 boresight telescope can be adjusted, and the entire telescope can be rotated in the adjusting-screw housing to align the crosslines with the telescope adjusting screw. NOTE: A Mk 75 boresight telescope does not have these features.

The optical system of a Mk 8, Mod 6, boresight telescope is illustrated in figure 11-21. Compare this optical system with that of a Mk 75 boresight telescope (fig. 11-7).

For additional information on Mk 8 boresight telescopes, refer to OP 1449.

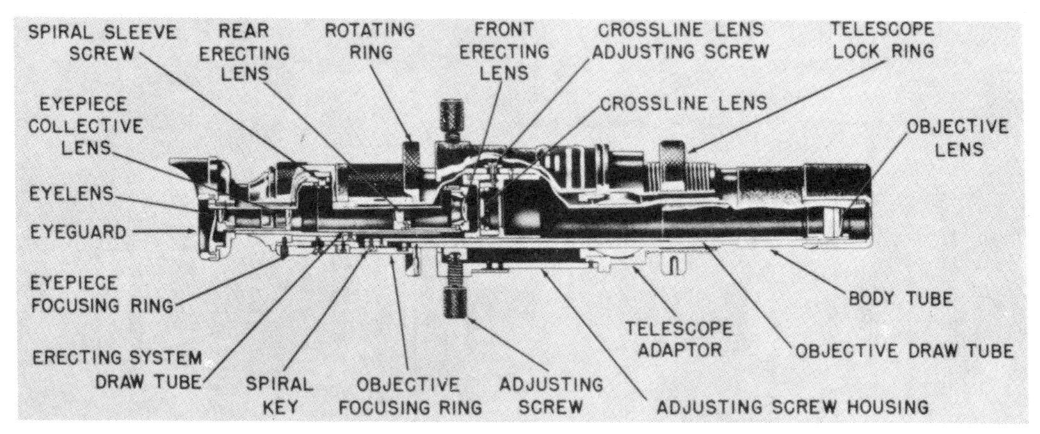

137.270
Figure 11-20.—Cutaway view of a Mark 8, Mod 6, boresight telescope.

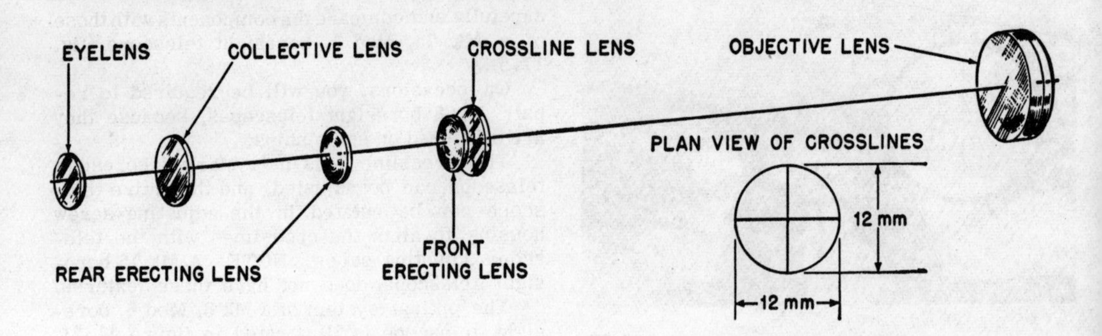

Figure 11-21.—Optical system of a Mark 8, Mod 6, boresight telescope.

CHAPTER 12

FIXED PRISM GUNSIGHT TELESCOPES

Fixed prism antiaircraft gun mount telescopes used by the Navy are designed for 3-inch and 5-inch guns used in antiaircraft fire. These telescopes enable the pointer and the trainer to establish accurate lines of sight from a gun to the target.

Antiaircraft gun mount telescopes considered in this chapter are Mks 74, 77, and 79. A brief summary is given for the Mk 74 telescope; Mks 77 and 79 telescopes are discussed in detail.

MARK 74 TELESCOPES

Two views of a Mk 74 telescope are shown in figure 12-1. All mods of this telescope are fixed-power, single-eyepiece instruments.

Mark 74 telescopes are used on 3-inch 50-caliber gun mount sights. Each mount requires three telescopes: one for the pointer, one for the trainer, and one for the checker.

OPTICAL FEATURES

A diagram of the optical system of a Mk 74 telescope is illustrated in figure 12-2. The objective window is a plano-plano disk which excludes dirt and moisture from the instrument and also seals it. The objective lens is a cemented, achromatic doublet properly positioned to remove parallax from the instrument. The roof prism deviates the line of sight 90° and erects the image.

The reticle is a plano-concavo lens whose plane surface is in the focal plane of the objective lens. A crossline is etched on the plane surface. By diverging the light rays slightly, the reticle increases the eye distance and helps to protect the observer's eye from gunfire shock.

There are four filters: yellow, red, variable-density, and clear. All filters are plano-plano disks. The clear glass, of course, is not actually a filter; but it is included so that the diopter setting will remain constant when you turn one of the filters out of the line of sight.

The eyepiece in the optical system of a Mk 74 telescope is symmetrical.

MECHANICAL FEATURES

Study the mechanical features of the Mk 74 telescope illustrated in figure 12-3. The body tube is a bronze casting which houses the objective lens and the roof prism, and it supports the color filter and eyepiece assemblies. The telescope is mounted on a gun by two bearings: (1) a quadrangular bearing at the rear, and (2) a spherical bearing near the front.

The crossline illuminator in a Mk 74 telescope (fig. 12-4) lights the crossline and thus makes the gunsight usable at night.

Mark 74 telescopes, Mods 0 and 1, have focusing-type eyepieces which have spiral keyway arrangements. Mods 2 and 3 have fixed eyepieces which allow sealing of the entire instrument.

NOTE: Some Mk 74, Mods 0 and 1, telescopes are still in service. When you receive one of them for repairs, perform ORDALT 2039-2 and convert it to a Mk 74, Mod 3, fixed-eyepiece type telescope.

To keep the telescope dry, charge it with dry nitrogen, through the air inlet valve at the back.

Mark 74, Mods 0 and 1, telescopes are gas-tight between the objective window and the crossline, and they are moisture-tight between the crossline and the eyelens. Mark 74, Mods 2 and 3, telescopes are gas-tight between the objective window and the eyelens.

MARKS 77 AND 79 TELESCOPES

Pointers and trainers use Mk 77 and Mk 79 telescopes on 3-inch 50-caliber and 5-inch

137.272
Figure 12-1.—Two views of a Mk 74 telescope.

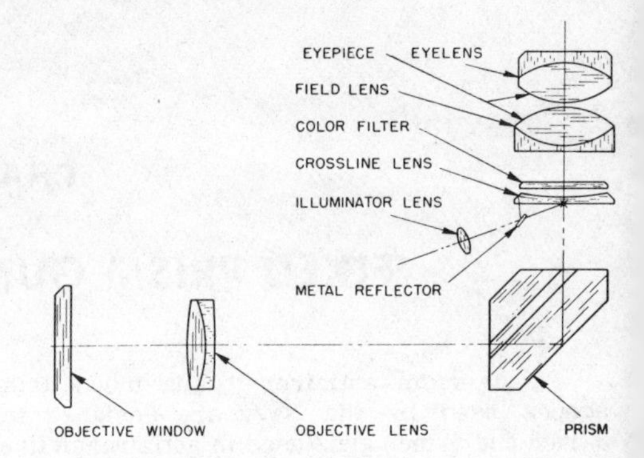

137.273
Figure 12-2.—Diagram of the optical system of a Mk 74 telescope.

38-caliber gun mounts. These telescopes are prismatic, fixed-power, single-eyepiece instruments which offset the line of sight horizontally and vertically. See illustration 12-5.

OPTICAL FEATURES

The optical systems of Mks 77 and 79 telescopes are identical. Study their components in figure 12-6. A Mk 79 telescope has indirect illumination provided to the crossline (fig. 12-7). A Mk 77 telescope has no illumination. All other parts of these optical systems are interchangeable. The illuminator body is sealed in the telescope body by a lead gasket, a brass washer, and a brass lock ring.

The objective window in a Mk 77 or a Mk 79 telescope is a plano-plano disk which excludes dirt and moisture and seals the instrument. The objective lens is a cemented, achromatic doublet so positioned that it removes parallax from the telescope.

The porro prisms in the optical system offset the line of sight and erect the image. Refer to chapter 8 for the correct method of mounting these prisms.

The reticle is a plano-plano disk with the crossline on the surface, facing the prisms. This surface is located in the focal plane of the objective lens, within the focal length of the eyepiece system.

Mk 77 and Mk 79 optical systems have four color filters: red, yellow, variable-density, and clear. The eyepiece is symmetrical.

MECHANICAL FEATURES

Figure 12-8 shows the outlines of Mk 77 and Mk 79 telescopes. The mechanical features of a Mk 77 telescope are illustrated in figure 12-9. The body tube is a bronze casting which houses the objective window, objective lens, porro prism assembly and crossline; and it also supports the color filter and eyepiece assemblies.

A Mk 77 telescope is mounted on its side on a gun, and it can be mounted on either side to serve a pointer or a trainer (fig. 12-10). Because of its mounting feet, a Mk 79 telescope can be mounted in one position only.

Marks 77 and 79 telescopes have fixed eyepieces and they can therefore be completely sealed, from the objective window to the eyelens.

OVERHAUL AND REPAIR

Overhaul and repair procedures discussed in this section are for Mks 77 and 79 telescopes.

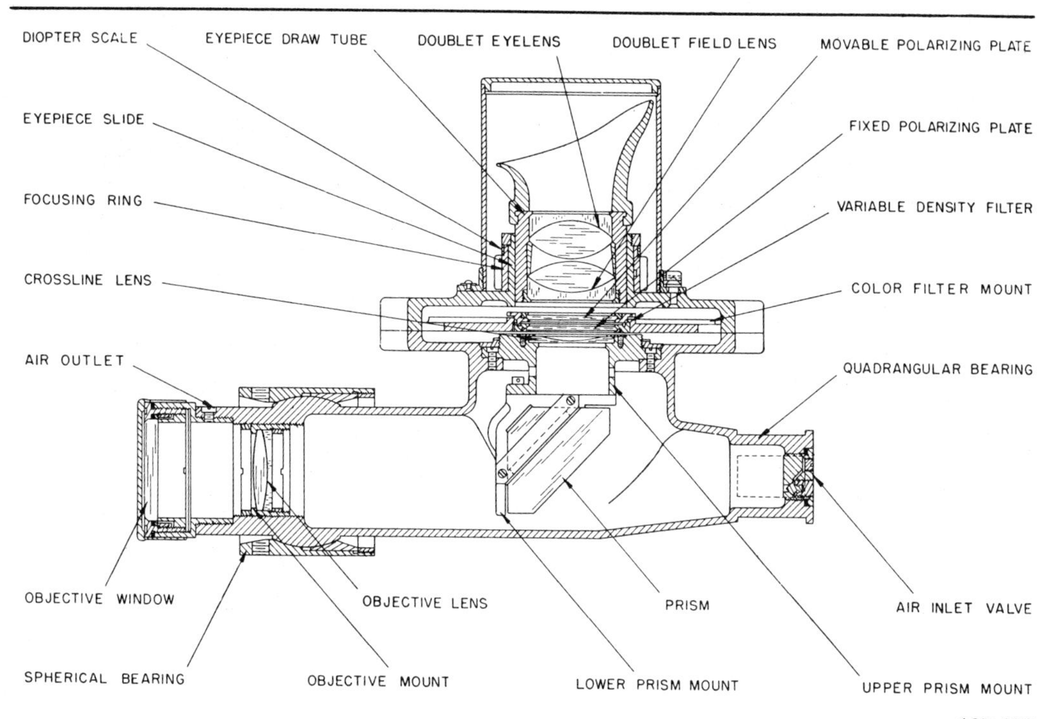

137.179

Figure 12-3.—Mechanical features of a Mk 74 telescope.

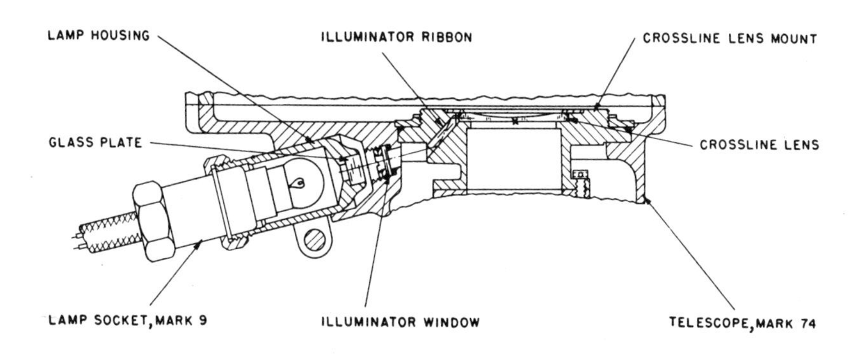

137.274

Figure 12-4.—Crossline illuminator of a Mk 74 telescope.

If you understand the procedure for overhauling and repairing these instruments, you will have no difficulty repairing other fixed prism gunsight telescopes.

Predisassembly Inspection

When you inspect Mk 77 and Mk 79 telescopes, proceed in the following manner:

137.275

A. Mk 77 telescope.
B. Mk 79 telescope.
Figure 12-5.—Fixed prism gunsight telescopes.

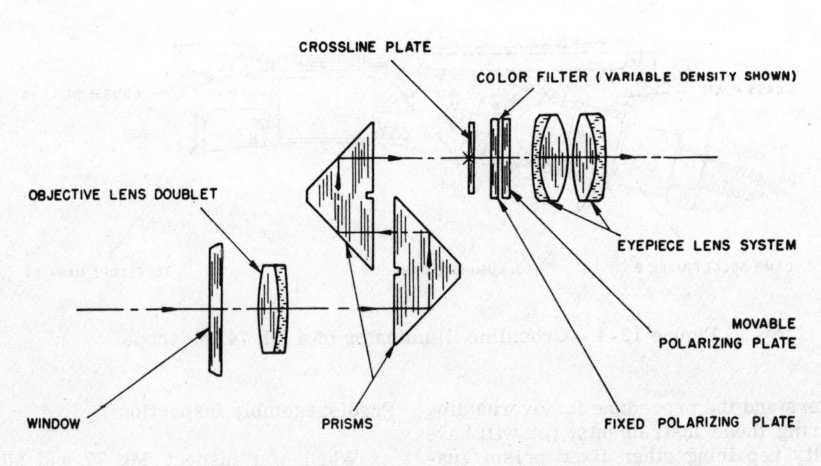

137.276

Figure 12-6.—Diagram of the optical system of Mk 77 and Mk 79 telescopes.

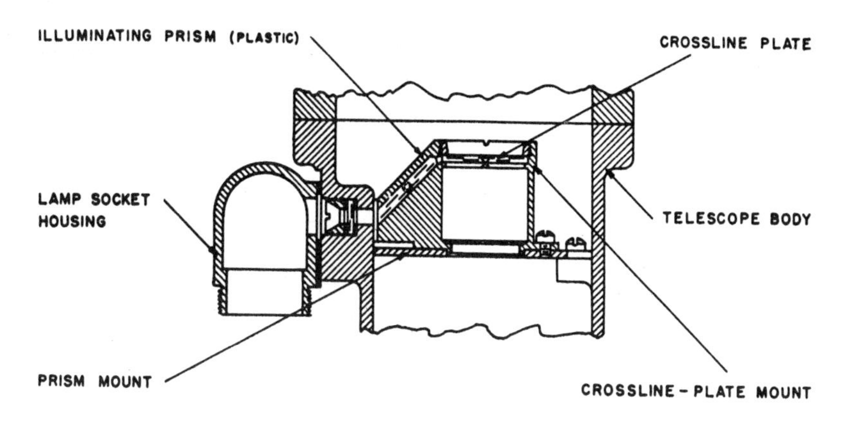

137.277

Figure 12-7.—Crossline illuminator of a Mk 79 telescope.

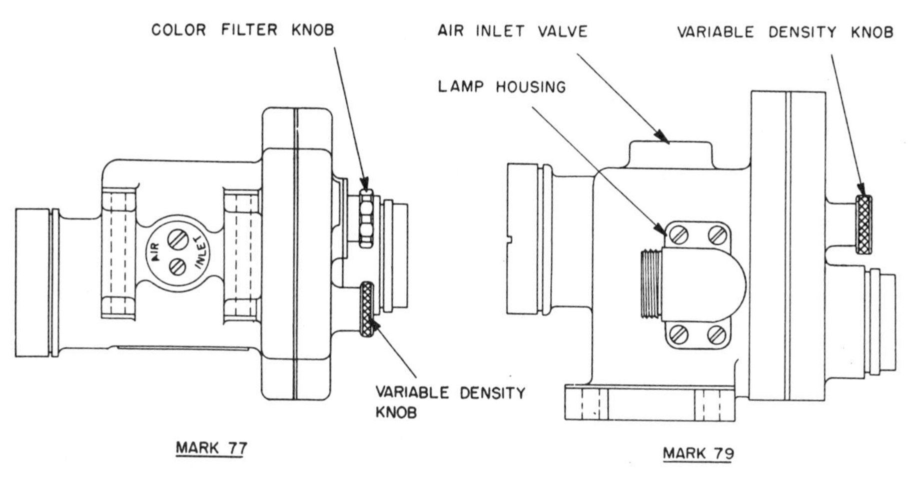

137.278

Figure 12-8.—Outlines of Mk 77 and Mk 79 telescopes.

1. Check bearing surfaces for signs of damage.
2. Test the action of the color filter and variable-density knobs.
3. Move each filter (in turn) into the field of view and examine it for dirt and damage.
4. With the objective cap in place over the objective window, turn on the crossline illuminator (Mk 79) and check through the eyepiece end for even illumination of the crossline. The rest of the field should be dark. If you see bright spots or hazy clouds in the field, the crossline is dirty.
5. Inspect the eyeguard for damage, and for signs of hardening or cracking. The rubber eyeguard is particularly important on gunsight telescopes, because it helps to protect the operator's eye from recoil shock from the gun.

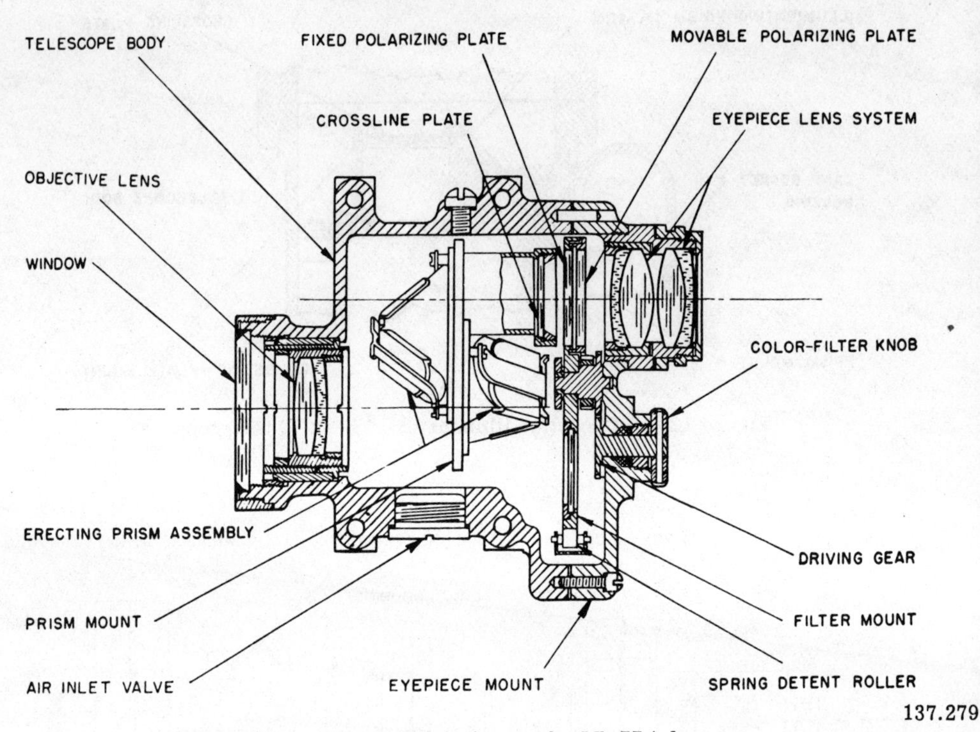

Figure 12-9.—Mechanical features of a Mk 77 telescope.

Disassembly

The Procedure for disassembling a Mk 77 or a Mk 79 telescope is as follows:

1. With a proper checking telescope, collimate the collimator and align the telescope fixture.
2. Completely inspect the instrument.
3. Remove the gas outlet screw and release the pressure.
4. Remove the screws (13) from the ray filter housing.
5. Break the ray filter housing away from the body tube. Lift straight up, and exercise care to prevent bending or damage to the dowel pins in the ray filter housing.
6. To remove the ray filter selection knob, and the density knob, support the knob on your bench block and use a drift punch to drive out the taper pin which secures it. Then remove the knob.

CAUTION: If you bend the shafts, improper sealing and operation will result.

7. Remove the screws from the ray filter mount bearing strip.
8. Press the ray filter shaft and the density shaft inward until they touch the ray filter mount. If you FAIL TO PERFORM THIS STEP before you remove the ray filter mount, damage to the teeth of the gear on the density shaft will follow.
9. Remove the ray filter mount. Lift it carefully, and allow the detent spring and roller to force the ray filter mount toward the eyelens to free it from its lower bearing.
10. Check the glass in the ray filter mount for chips, cracks, dirt, and so forth; and check the condition of the polaroid plates. Clean, wrap, and stow the ray filter mount in a safe place. Do NOT remove the ray filters from the mount unless you must replace them.

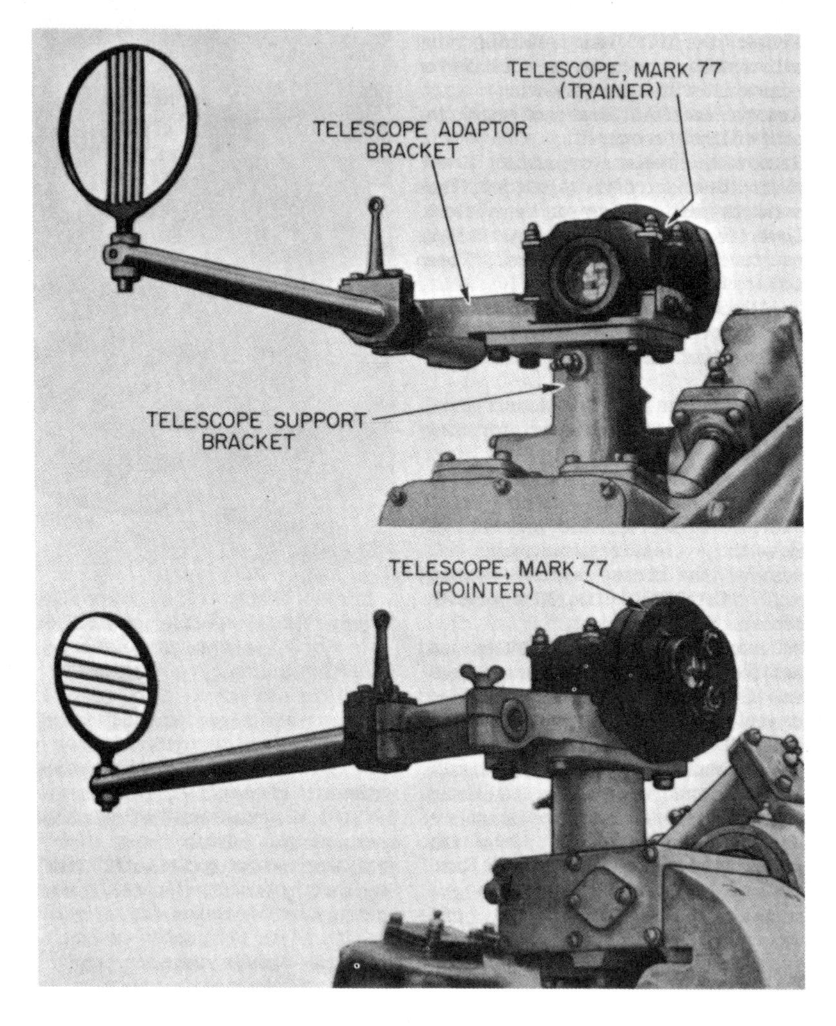

137.280

Figure 12-10.—Mk 77 telescope mounted on a 5-inch sight (Mk 31, Mod 8).

11. Remove the ray filter shaft and gear, and also the density shaft and gear. Check the gear teeth for burrs, and check the shaft for binding.

12. Remove the packing from both glands.

13. Disassemble the Mk 77, Mod 0, eyepiece. Eyepieces for the Mk 77, Mods 0, 1, and 2, and the Mk 79 are basically the same; but there are some minor differences in mechanical design. Use the following procedure to disassemble a Mk 77, Mod 0, telescope eyepiece:

a. With a proper spanner wrench, remove the eyepiece mount from the ray filter housing.
b. Hold the eyepiece mount in a fiber-grip wrench as you use a spanner wrench to remove the eyelens retaining ring.

c. Remove the brass eyelens sealing ring and the lead gasket. Do NOT bend the sealing ring.
d. Protect the field lens retaining ring with a rubber or cardboard disk as you remove it with a pin wrench.
e. Remove the field lens and mark the path of light through it.
f. Remove the eyelens by pushing it out, toward the rear of the telescope. Then mark the path of light on the eyelens. Cover the lens with lens paper before you touch it with your fingers. Clean and stow the lenses.

Although the two lenses in the eyepiece group are identical, it is best to replace them in their original positions.

14. Disassemble the Mk 77, Mods 1 and 2, and Mk 79 eyepieces. The procedure for doing this is as follows:

a. Remove the lock screw from the eyelens retaining ring.
b. With the proper spanner wrench, remove the eyelens retaining ring.
c. Remove the brass eyelens sealing ring. CAUTION: Do NOT bend it.
d. Remove the lead gasket.
e. Put lens paper over the field lens and push both lenses out. When the eyelens is free of the mount, mark and inspect it. Keep pushing on the field lens until the spacer and lens are free of the mount. Inspect and mark the field lens and the spacer. Clean and wrap both lenses and stow them in a safe place. CAUTION: Stow the spacer where it will NOT be bent.
f. Run a copper wire through the gas passage in the lens mount. This passage runs through a boss on the outside, at an angle of about 45° to the mount. CAUTION: If this passage is not kept open, gas cannot reach the eyelens and fogging of the eyepiece will result when the telescope is in service.

15. Remove the screws from the prism mount. See illustration 12-11.

16. Remove the prism mount and the crossline mount as a unit. If you encounter difficulty in freeing the prism mount, release by a few turns the three screw dowel pins.

17. Remove the crossline mount from the prism mount. Take the crossline lens out of its mount; then clean, inspect, wrap, and stow it. Inspect the illuminator ribbon in the Mk 79 crossline mount. Do NOT remove this ribbon unless it is absolutely necessary.

137.281
Figure 12-11.—Prism mount screws in Mk 77 and Mk 79 telescopes.

18. If necessary to clean or to replace them, remove the prisms from their mount. New prisms must be of the SAME THICKNESS as the replaced prisms. Chapter 8 explains the procedure for checking prisms for squareness.

19. With a proper wrench, remove the objective window retaining ring.

20. Remove the objective window sealing ring and the lead gasket.

21. With a suction cup, remove the objective sealing window; then clean and stow it in a safe place. Note the position of the window and mark it (fig. 12-12), so that you can replace it in its original position. This is important, for some windows may be slightly wedge-shaped.

If you CANNOT remove the objective window with a suction cup, do the following:

a. With a pin wrench, reach into the back of the telescope body and remove the objective lens retaining ring.

137.282
Figure 12-12.—Reassembly guide mark on the window.

137.283
Figure 12-13.—Removing the eccentric mount lock ring.

b. Remove the objective lens; then inspect, mark, and properly stow it.
c. Put a rubber disk over the inside of the objective window and, with a wooden stick, push the objective window out of its mount.

22. Place a rubber or cardboard disk over the objective lens and, with a pin wrench, remove the eccentric mount lock ring (fig. 12-13).

23. Pull straight out on the eccentric ring, eccentric mount, and the objective lens mount to remove them as a unit.

24. Check the eccentric ring for smoothness of operation. Remove the eccentric ring and check its surfaces for burrs.

25. Remove the objective lens mount lock ring.

26. Remove the objective lens mount by screwing it out of its eccentric mount. As you do this, check the mount for smoothness of operation. It is important that the objective lens mount move freely during collimation.

27. Remove the objective lens retaining ring.

28. Remove, inspect, clean, and stow the objective lens.

Repair Procedure

After you disassemble the telescope, make required mechanical repairs. If you are not sure about certain procedures, review chapter 8.

Clean all mechanical parts in an approved cleaning solution and rinse them in warm, soapy water. Dry them in a warm oven.

Reassembly and Collimation

Chapter 8 gives the lubrication procedure to follow when you reassemble an optical instrument.

To reassemble a complete unit (eyepiece, ray filter assembly, objective assembly, etc.), follow the disassembly procedure, in reverse order.

1. Reassemble the eyepiece and seal the eyelens with .040-inch lead wire. Resin core

solder is NOT adequate. Use special tools to form and seat the gasket, as illustrated in figure 12-14. Replace the eyelens sealing ring and the eyelens retaining ring. Then tighten the retaining ring with a spanner wrench.

2. Reassemble the ray filter assembly in its housing. Then check the detent action, order of the filters, and operation of the polaroid plates. Replace the HYCAR packing in the packing glands.

NOTE: This packing should be lubricated with GLYDAG (a mixture of glycerine and graphite) to ensure a good seal and proper operation of the filter shafts.

3. Reassemble the objective assembly and put it in the telescope. Then tighten the eccentric mount lock ring sufficiently to seat the assembly properly.

4. Replace the prism mount and the crossline mount in the telescope. The etched surface of the crossline faces the prisms. Make certain the prism mount is seated properly.

5. Secure the ray filter housing to the telescope body with the screws provided.

6. Put the instrument in the collimator and, with an auxiliary telescope, focus on the telescope crossline. Screw the objective lens mount in or out, as necessary, to bring the collimator crossline into focus. When both crosslines are in sharp focus at the same point, the instrument is free of parallax. Tighten the objective lens mount lock ring and recheck to make certain the objective lens mount did not move when you tightened the lock ring.

7. Rotate the crossline in its mount to square the crossline. Then tighten the lock ring and recheck for squareness.

8. Align the optical and mechanical axes of the telescope. The axes of a Mk 77 telescope are aligned when the telescope crossline falls at the same point on the collimator crossline when the telescope is rotated 180 degrees. To get proper alignment, mount the telescope and adjust the collimator until the telescope crossline is superimposed on the collimator crossline. Turn the telescope 180° (on its opposite side) and observe the distance the telescope crossline has moved the collimator crossline, horizontally and vertically. To remove half of this error, rotate the eccentric mount in the telescope.

Return the telescope to the 0° position and superimpose the telescope crossline on the collimator crossline again. Rotate the telescope

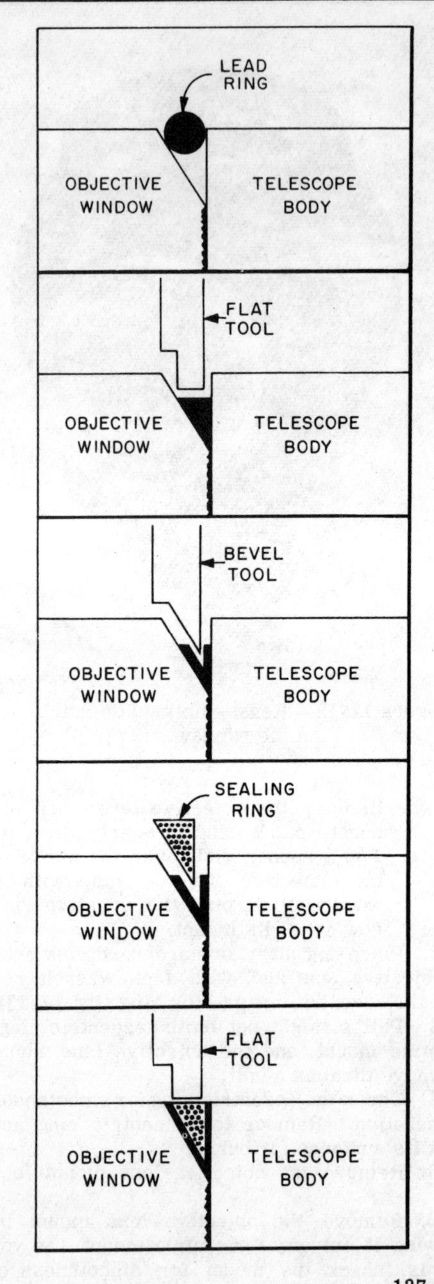

137.284

Figure 12-14.—Sealing the gasket and replacing the eyelens retaining ring.

180° and note the position of the telescope crossline. Again, remove half the error by rotating the eccentric ring and the eccentric mount.

Continue this process until the telescope crossline remains superimposed on the collimator crossline in both the 0° and the 180° position.

You need not change the position of the telescope to align the optical and mechanical axes of a Mk 79 telescope. Rotate the eccentric ring and the eccentric mount until the telescope crossline is superimposed on the collimator crossline.

The allowable tolerance for superimposing telescope crossline is 2 1/2 minutes of field, which is equivalent to the thick part of the crossline.

9. When you have the optical and mechanical axes aligned, secure the eccentric mount in position with the eccentric mount lock ring and recheck the collimation. This check is essential because the eccentric mount could have slipped while you were tightening it.

10. Replace the objective window and make another check of your collimation. If the window is wedge-shaped, it will change the line of sight. If collimation is still within the allowed tolerance, seal the objective window with .040-inch lead wire, in the same manner you sealed the eyelens.

Setting the Fixed Eyepiece

Fixed eyepieces are generally set to minus 3/4 or minus 1 1/2 diopters. This means that rays of light which leave the eyelens are slightly divergent. This value is set to accommodate the average operator of optical instruments, because most operators need a MINUS setting of the eyepiece to use the instrument.

To obtain a minus dioptric setting for a fixed eyepiece, the final, real image produced in the instrument, and/or the reticle, MUST BE LOCATED WITHIN the focal length of the eyepiece system. When an image, or object, is placed within the focal length of any positive lens, the rays emitted by the image are diverging, strike the lens, are refracted, and are still divergent when they emerge.

Because most fixed eyepiece mounts are a part of the telescope housing, the eyepiece mount, or lenses, cannot be adjusted in any manner. How, then, can the crossline of the instrument be positioned within the focal length of the eyepiece if (1) the eyepiece is not adjustable, (2) the crossline itself cannot be adjusted without the introduction of parallax, and (3) the instrument has a prism erecting system?

Most small instrument body housings which contain the fixed eyepieces are cast in two parts, the main body housing, and the ray filter housing. The main body housing contains the objective lens, the prism erecting systems, and the crossline. The ray filter housing contains the ray filters and the eyepiece. The two castings are secured together with screws and sealed with a gasket.

Since the optical system itself cannot be adjusted to obtain proper dioptric setting, without disturbing the previous steps in collimation, mechanical adjustment is the only means left. This can be accomplished by increasing or decreasing the thickness of the gasket between the two housing castings.

A thin gasket moves the ray filter housing closer to the main body housing and thus moves the eyepiece mount and lenses closer to the crossline. The gasket selected must allow the ray filter housing to be so positioned that the crossline is properly located within the focal length of the eyepiece system.

If the dioptric value of the rays leaving the eyepiece must be minus 3/4 diopter (diverging rays), how can you determine when this value is reached? When you use a standard auxiliary telescope, the telescope crossline must come into focus WHEN THE INDEX MARK OF THE AUXILIARY TELESCOPE POINTS TO PLUS 8 DIOPTERS (graduations), plus your own eye correction. One example will explain this.

If the instrument's eyepiece must be set at a value of minus 3/4 diopter, the crossline must come into focus at plus 8 diopters (graduations) on the auxiliary telescope.

NOTE: The ONLY time you focus the eyepiece of the auxiliary telescope to set diopters is when you set a fixed eyepiece; otherwise, you focus the eyepiece of the telescope being collimated.

When you set a fixed eyepiece to minus 1 1/2 diopters, the telescope crossline must come into focus at plus 17 diopters (graduations) on the auxiliary telescope. The auxiliary telescope, however, will not focus out to plus 17 diopters. A special auxiliary telescope must therefore be constructed from a standard auxiliary telescope to allow the eyepiece to conform to the reading. Ask your shop supervisor to demonstrate the use of this special auxiliary telescope.

The rules to follow when you set dioptric value to any fixed eyepiece are as follows:

1. If the dioptric reading (number of graduations) on the auxiliary telescope is PLUS (more than required), use a THICKER gasket between the ray filter housing and the main body housing.

2. If the dioptric reading is MINUS, use a THINNER gasket.

Testing and Gassing

The procedure for testing and gassing fixed prism gunsight telescopes is not discussed in this chapter because the procedure was fully covered in chapter 8.

After you fill your repaired instrument with gas, it is ready for service.

CHAPTER 13

MAGNETIC COMPASSES

A magnetic compass is an instrument which indicates direction. As shown in illustration 13-1, a steersman must constantly watch a compass or a gyro repeater to determine the direction his ship is traveling.

Small ships and boats depend entirely on magnetic compasses for determining their direction, and large ships use them to check their gyro repeaters. When magnetic compasses are damaged or broken, they are sent to the optical shop for repair.

PRINCIPLE OF OPERATION

Somebody in Magnesia (on the coast of the Aegean Sea) discovered a long time ago that certain stones (magnetite or lodestone) could attract iron. Another person learned that when he rubbed an iron bar with a piece of lodestone the bar became a magnet. A Chinaman then learned that when he attached little floats to a magnetized needle and put it in water the needle pointed approximately North and South. Some time later, an Italian navigator balanced the needle on a pivot and learned that its action then was the same as when it was balanced on water.

These people learned through experimentation the principle of operation of a magnetic compass. From the needle on floats or a wooden disk to the compass box and hanging compass, action of the compass needle has always been the same—only the method for holding it has changed.

You can learn how a magnetic compass operates by doing a little experimenting on your own. Hold a small compass level and observe the action of its needle. Regardless of the direction you turn the compass, its needle always points north; and by turning the compass until the N on the card is under the point of the needle, you can determine any direction.

If you take the needle out of the compass, you will find that it is magnetized at both ends. Each end attracts iron; but ONLY ONE end will point North. The reason for this action is that there are TWO KINDS of magnetism (red and blue) and every magnet has both kinds. If we call the points where magnetism in a magnet is strongest the magnetic poles, every magnet has a NORTH-SEEKING pole and a SOUTH-SEEKING pole. In a bar magnet, or in a compass needle, the two poles are at the ends.

The earth itself is a HUGE magnet, with a north magnetic pole in northern Canada and a south magnetic pole in Antarctica. Like poles REPEL each other; unlike poles ATTRACT each other. Put two bar magnets side by side, with both north poles together, and observe what happens. The two magnets repel each other; but if you turn ONE magnet end for end, the two magnets attract each other.

For the reasons just explained, if we say that the earth's north magnetic pole has NORTH magnetism, the SOUTH pole of a compass needle points north; if we say the NORTH pole of a magnet points north, then the north pole of the earth is actually its SOUTH magnetic pole. All of this seems confusing, and mariners have tried to make it a bit simpler by speaking of RED magnetism and BLUE magnetism. The Navy paints one end of its bar magnets RED and the other end BLUE. The RED end points north; the BLUE end points south.

COMPASS VARIATION

The amount a compass needle is offset from true north (caused by attraction to the position of magnetic north) is called variation, because it varies at different points on the earth's surface. Even in the same location it usually does not remain constant—it decreases or increases annually at a certain known rate. Deposits of iron ore tend to pull a compass needle away from its true pole (the end of the imaginary axis on which the earth rotates), and in some parts of

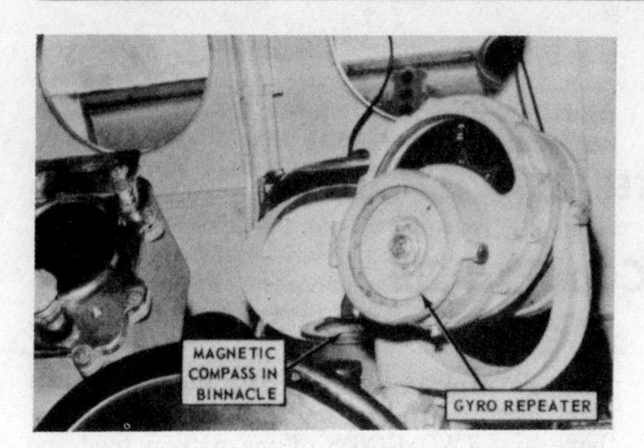

69.1

Figure 13-1.—In the pilot house of a combat ship.

DEVIATION TABLE					
SHIPS HEADING MAGNETIC	DEV.	SHIPS HEADING MAGNETIC	DEV.	SHIPS HEADING MAGNETIC	DEV.
000°	14°W	120°	15°E	240°	4°E
015°	10°W	135°	16°E	255°	1°W
030°	5°W	150°	12°E	270°	7°W
045°	1°W	165°	13°E	285°	12°W
060°	2°E	180°	14°E	300°	15°W
075°	5°E	195°	14°E	315°	19°W
090°	7°E	210°	12°E	330°	19°W
105°	9°E	225°	9°E	345°	17°W
				360°	14°W

69.12

Figure 13-2.—Simple deviation table.

the world the needle may point as far as 60° from true north.

Because all our maps and charts are drawn to true north, readings of a compass for variation must be corrected before it is used. Magnetic compass variation through all navigable waters, however, has been accurately determined and recorded for use by ships.

COMPASS DEVIATION

A compass needle is a magnet, and iron and steel attract magnets. Because ships are made of steel, they affect the action of compass needles. The amount of magnetic deflection of a magnetic compass needle from true north by magnetic material in the ship is called DEVIATION. Deviation is different for different compasses, and also for different parts of a ship.

Although deviation remains a constant amount for any given compass heading, the amount is not the same for all headings. Deviation gradually increases, decreases, increases, and decreases again as the ship goes through an entire swing of 360 degrees. Because the deviation for each heading must be known in order to correct the error, a deviation table is made up for every ship; and this table usually shows the deviation for each 15° of swing. See illustration 13-2.

To find the deviation, swing the ship in 15° increments around to 360° and note the amount the compass points away from each magnetic heading. This is the compass deviation for that particular heading. In using the table, the deviation for the heading nearest the one being checked is selected. If the deviation for a 17° heading is desired, for example, the deviation for 15° (10° W) would be selected.

CORRECTING COMPASS ERROR

Variation and deviation combined constitute magnetic compass error. If you know the true course for a ship, worked out from the chart, you must then know the compass course to steer to make the true course good. This is accomplished by applying compass error, in the shape of variation and deviation, to the true course. On the other hand, if there is a bearing taken by a magnetic compass, variation and deviation must be applied to the compass bearing to obtain the true bearing.

All compass errors, whether caused by variation or deviation, are either easterly or westerly. There are no northerly or southerly errors. Correction for error is made by ADDING easterly error and SUBTRACTING westerly error when correcting compass course to true course, and by SUBTRACTING easterly error and ADDING westerly error when uncorrecting true course to compass course.

TYPES OF MAGNETIC COMPASSES

Most ships carry two or more magnetic compasses like the one illustrated in figure 13-3. This is a U.S. Navy 7 1/2-inch compass.

45.595([illegible])
Figure 13-3.—U. S. Navy 7 1/2-inch magnetic compass.

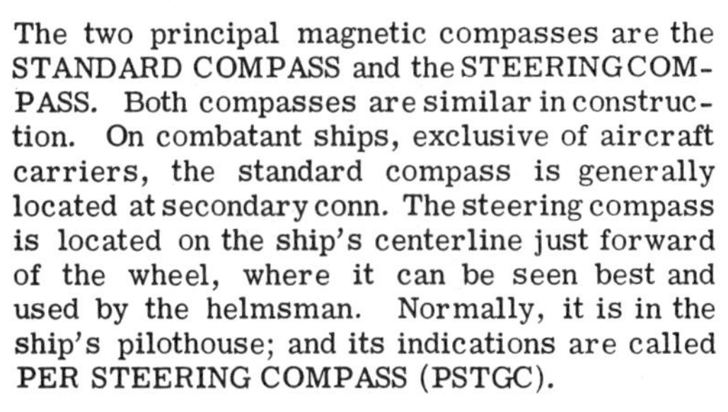

The two principal magnetic compasses are the STANDARD COMPASS and the STEERING COMPASS. Both compasses are similar in construction. On combatant ships, exclusive of aircraft carriers, the standard compass is generally located at secondary conn. The steering compass is located on the ship's centerline just forward of the wheel, where it can be seen best and used by the helmsman. Normally, it is in the ship's pilothouse; and its indications are called PER STEERING COMPASS (PSTGC).

Wet compasses are liquid filled, usually with varsol. Other compasses contain no liquid and are designated as dry compasses, but this type of compass is seldom used in the Navy. A wet compass consists of a bowl filled with liquid, which supports a hollow float to which the compass card and magnets are attached. The liquid steadies the compass against the motion of the ship and the shock of gunfire; and since the liquid supports most of the weight of the magnets, it reduces the pressure and friction on the pivot.

Navy magnetic compasses vary slightly with respect to purpose. Part A of figure 13-4

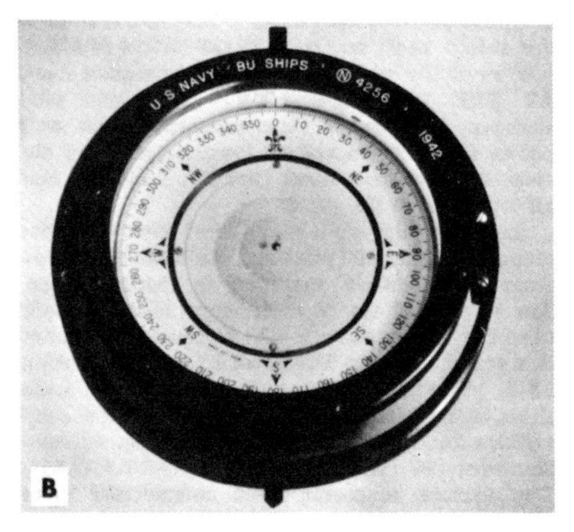

45.23
A. Standard 4-inch boat compass.
B. Steering compass.
Figure 13-4.—Types of magnetic compasses.

shows a standard 4-inch boat compass, and part B of this illustration shows a 6 3/4-inch steering compass.

The size of a compass is designated by the diameter (in inches) of its card. The compass shown in figure 13-3 has a translucent card, or one with perforated markings; and a light in its stand shines up through a ground-glass plate in the bottom of the compass to illuminate the card.

CONSTRUCTION OF A MAGNETIC COMPASS

The compass card of a pocket compass is printed on the bottom of the case, and its needle is not hindered in its motion. A Navy compass, on the other hand, has no needle. The steersman is not interested in knowing the direction of North; all he wants to know is the direction his ship is heading.

The compass card of a magnetic compass is mounted on a pivot and the magnets are attached to the card, so that the card itself will swing and point its zero mark to the north. Observe on the bowl of the compass shown in figure 13-3 the LUBBER'S LINE. The compass is always mounted so that an imaginary line from the compass pivot to the lubber's line is parallel to the ship's keel; so, to read the ship's heading, you read the graduation on the compass card AT THE LUBBER'S LINE. When the ship changes its course, the compass card still points its zero graduation toward north; but the ship, the compass bowl, and the lubber's line all turn—under the card.

In order that it will stay level even when the ship is rolling and pitching, the bowl of a magnetic compass is mounted in gimbal rings. The bottom of the bowl is very heavy, to help keep the compass level. The compass is mounted in a stand, called a BINNACLE. See illustration 13-5. The two hollow soft iron spheres on the sides of the binnacle are called quadrantal correctors for deviation. (Deviation changes direction every 90°, hence the name QUADRANTAL.) The earth's magnetic field magnetizes these spheres by induction. The induced magnetism of the spheres counteracts the induced magnetism of the ship and forces the compass needle to point toward magnetic north. The force exerted by these spheres can be altered by their distance from the compass. The size of a sphere also affects its force.

A cross section of a typical magnetic compass is shown in figure 13-6. Refer to the nomenclature as you study the discussion of the illustration.

The bowl of the compass is filled with a liquid, and there is an expansion chamber in the bottom of the bowl to hold excess liquid created by expansion. The expansion chamber is made of thin, flexible metal; so, when the compass liquid gets warm and expands, the extra liquid is forced by pressure into the expansion chamber and expands the chamber.

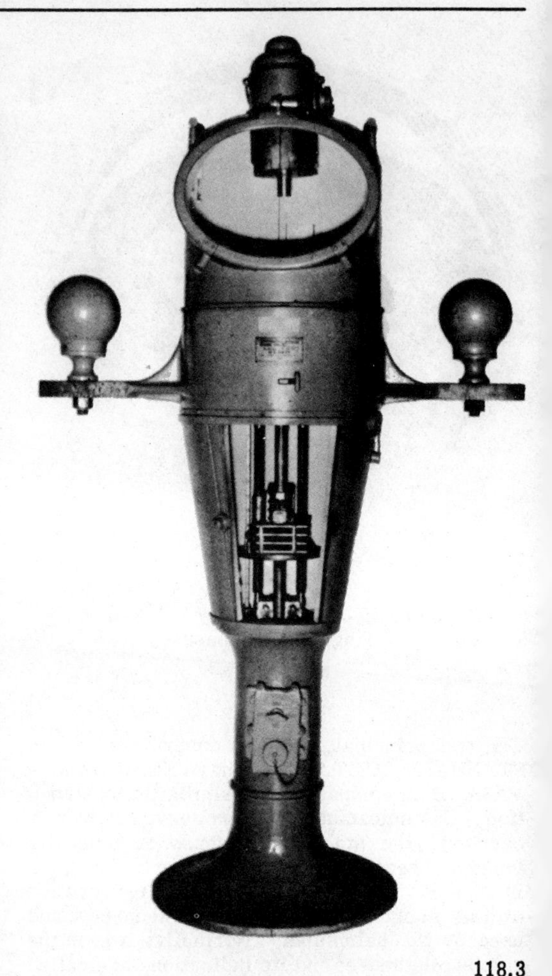

118.3
Figure 13-5.—Magnetic compass and binnacle.

The top of the bowl is covered with a glass plate, secured by a BEZEL ring. A rubber gasket is placed between the ring and the glass to prevent leakage. The pivot which holds the float is secured to the bottom of the bowl, and it has a rather sharp point which fits in a jewel located in the top-middle part of the float. Study illustration 13-7, which shows the pivot tip and the location of the jewel. The pivot tip fits in a cavity in the bottom of the jewel to allow smooth action of the float balanced on the

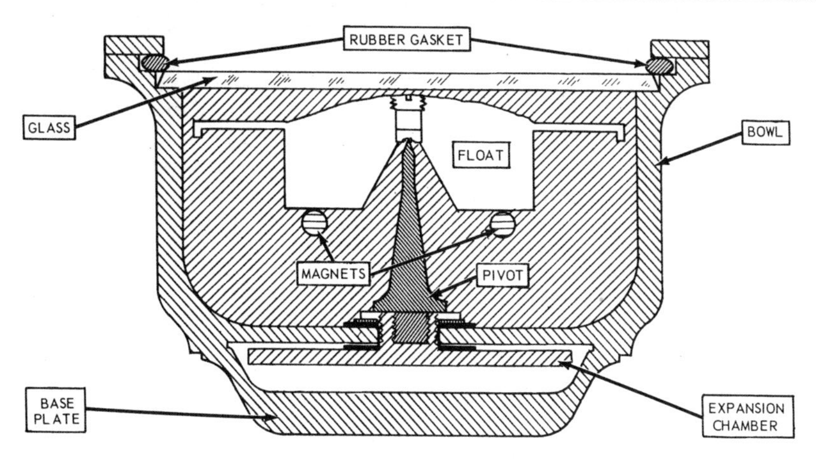

137.285

Figure 13-6.—Cross section of a magnetic compass.

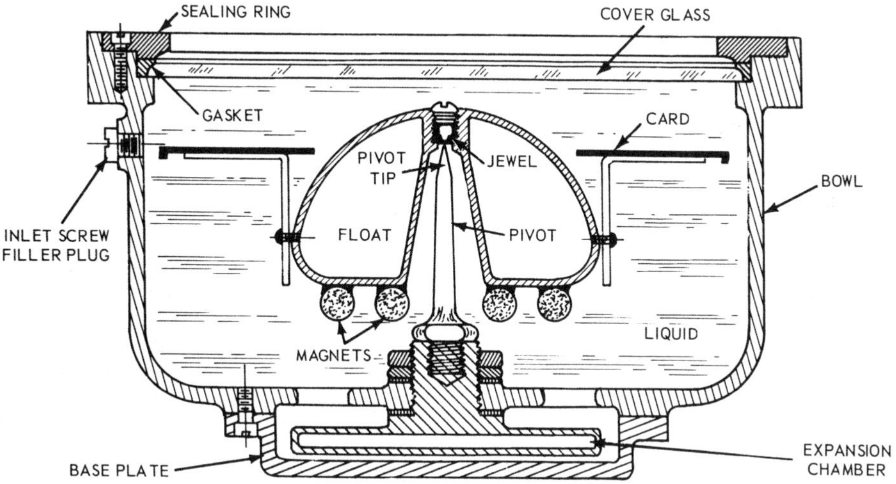

137.286

Figure 13-7.—Nomenclature of a magnetic compass.

pivot. Observe also in figure 13-7 the inlet screw filler plug and the compass card. In this compass, the expansion chamber is filled with air and is surrounded by the compass liquid; when the liquid expands, it compresses the chamber.

The compass card is secured TO THE TOP of the float. The bar magnets are fastened UNDER THE FLOAT, as shown.

The float is a hollow, metal chamber submerged in the liquid, which gives it enough buoyancy to support most of the weight of the magnets. The float assembly of a 7 1/2-inch compass, for example, weighs 3,060 grains in air, but it weighs less than 90 grains when submerged. The float therefore makes the compass more sensitive and more stable.

OVERHAUL AND REPAIR

The following discussion of overhaul and repair of a magnetic compass is for a 4-inch boat compass, whose construction is very similar to that of a 7 1/2-inch compass. If you understand the procedure for overhauling and repairing the 4-inch instrument, you will experience no difficulty in repairing other types of magnetic compasses.

PREDISASSEMBLY INSPECTION

When you receive a defective compass for repairs, give it a careful inspection. You can often determine the trouble with the instrument at this time, before you start disassembly. Look for the following:

1. If the compass card is level and there is a large bubble under the glass cover, liquid is leaking out under the cover glass, or there is a leak in or around the expansion chamber.
2. If the compass card is tilted and a large bubble is under the glass cover, but there is no leak around the cover, there is probably a leak in the float.
3. If there is no bubble under the cover glass but the card is tilted, the magnets have shifted, or the balancing solder has fallen off the float, or the float has jumped off its pivot.
4. Put the compass on a level workbench and turn it until the north point on the card is at the lubber's line. With your magnet, deflect the compass card exactly 11° and then quickly remove the magnet. The compass will then swing back; and as the zero mark crosses the lubber's line, start your stopwatch. The zero mark will reach the end of its swing and start back; and as it crosses the lubber's line the second time, stop your stopwatch and read it. The time you read is THE PERIOD OF THE COMPASS, and it should be 10 seconds or less. If it is longer than 10 seconds, the magnets are weak, or the pivot point is in poor condition.
5. If the float does not swing freely under the influence of a magnet, the pivot point or the jewel is broken.

DISASSEMBLY PROCEDURE

The recommended procedure for disassembling a magnetic compass is as follows:

1. Remove the filler plug and drain out a small quantity of the liquid, to prevent spillage in trying to handle a full compass bowl. Then replace the filler plug. Save the liquid you drew out.
2. Mark the lip of the bowl and the edge of the bezel ring, for you must put the bezel ring back in the same position it occupied before removal.
3. Remove all screws from the bezel ring. CAUTION: Loosen each screw a little at a time, in rotation or opposite each other, to prevent tilting of the bezel ring by the rubber gasket and probable breakage of the glass.
4. Lift off the bezel ring and then remove the rubber gasket. See illustration 13-8. CAUTION: Use care to prevent damage to the gasket.
5. With a suction gripper (fig. 13-9), or a pegwood stick, lift the glass. CAUTION: The glass is beveled to a thin edge and chips easily.
6. Test the float for leaks. Push down on one side of the float, as shown in illustration 13-10, hold it down for several seconds, and then release it. If the float stays down, it contains liquid. Repeat this test at three different points around the card.

137.287

Figure 13-8.—Removing the rubber gasket.

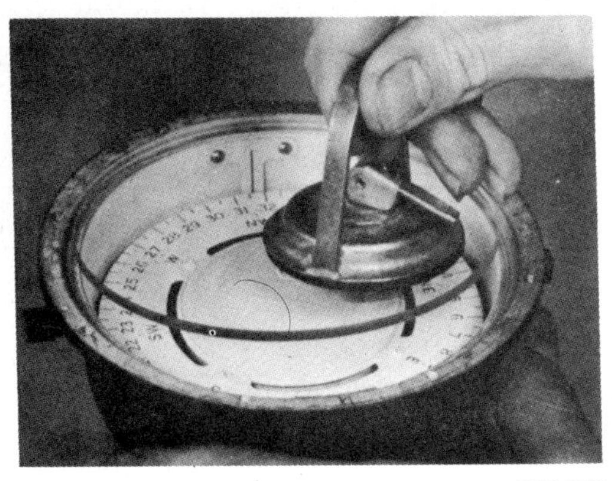

137.288
Figure 13-9.—Removing the cover glass.

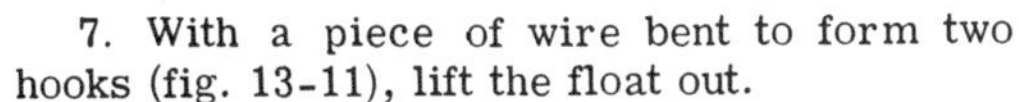

7. With a piece of wire bent to form two hooks (fig. 13-11), lift the float out.

8. Pour the remainder of the liquid from the bowl and filter it through filter paper or absorbent cotton into a clean bottle for future use.

9. To remove the pivot, fit a socket wrench over its hexagonal base and turn counterclockwise. CAUTION: Be sure the center hole of the wrench is deep enough to provide clearance for the pivot point.

10. Turn the bowl over and, with a punch, make light register marks on the bowl and the base plate, to guide you in reassembling the base plate in its original position.

137.289
Figure 13-10.—Testing the float for leaks.

137.290
Figure 13-11.—Removing the float assembly.

11. Remove the screws from the base plate and lift it off. See illustration 13-12. This base is made heavy to help keep the compass on an even keel.

12. Note in illustration 13-12 the bottom of the expansion chamber, and then study figure 13-13 to learn how the chamber is secured through the hole in the bottom of the compass bowl. Beneath the expansion chamber nut is a friction brass washer, and under this washer is a lead washer. Between the chamber and the bottom of the bowl is another lead washer. When these washers are put under pressure, they seal the opening in the bottom of the bowl.

13. Turn the bowl over and remove the expansion chamber lock nut with a socket wrench.

14. Remove the expansion chamber from the bowl and inspect it for leaks or other damage.

REPAIR AND REASSEMBLY

Inspection of parts, repair, and reassembly of a magnetic compass are discussed conjointly, step by step, as follows:

1. If the expansion chamber is in good condition, reassemble it. CAUTION: Do NOT forget the lead washer between the expansion chamber and the bottom of the bowl. If this washer is not in perfect condition, replace it.

2. Replace the second lead washer, inside the bowl and replace the brass friction washer. If necessary, use a new washer. Start the hexagonal lock nut by hand and tighten it with a

137.291
Figure 13-12.—Removing the base plate.

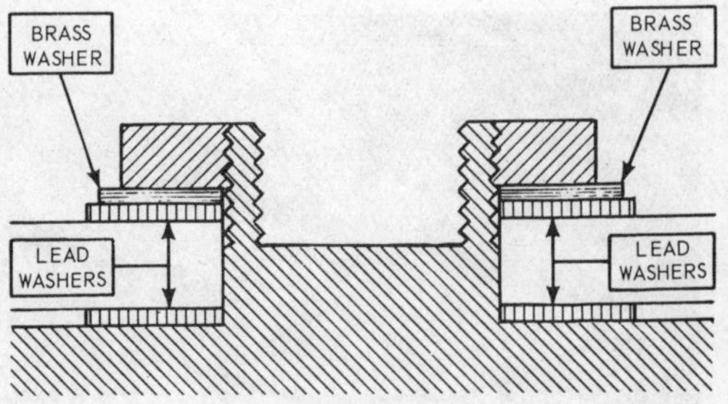

137.292
Figure 13-13.—Expansion chamber secured to bottom of compass bowl.

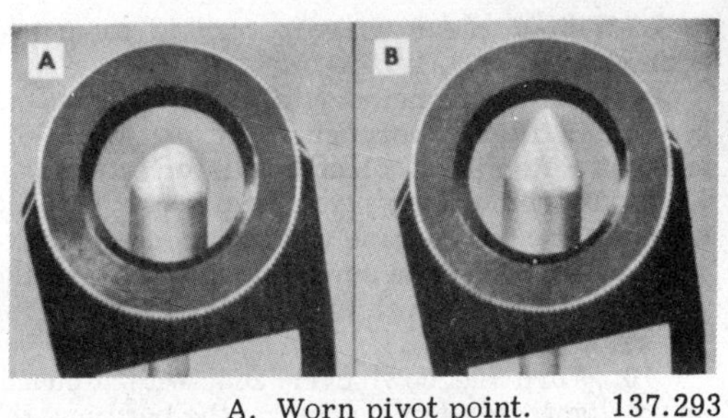

A. Worn pivot point. 137.293
B. Good pivot point.
Figure 13-14.—Pivot points.

socket wrench. NOTE: Use enough tension to make a good seal at the lead washers.

3. Put the base plate back into position; then replace the base plate screws and tighten them. CAUTION: Be sure to line up your two marks you made during disassembly; otherwise, the compass will be out of balance.

4. With a magnifying glass, inspect the pivot point for wear. Study illustration 13-14. The magnified pivot in part A of figure 13-14 is badly worn. Observe the round appearance. The pivot point shown in part B of this illustration has proper shape. NOTE: A badly worn pivot point makes a compass sluggish.

5. If the pivot point is worn, put it in a lathe and reshape it with a fine carborundum ship (fig. 13-15). Then polish it with an Arkansas oil stone and inspect again for correctness of shape. The tip of the pivot should have a radius of .005 inch.

6. Remove the screw from the top of the float and use a piece of pegwood with a rounded end to push the jewel and its spacer out of the float. Study illustration 13-16. Then hone a steel needle to a sharp point on an oil stone and rest it on your finger nail (fig. 13-17). If it slides under its own weight, it is NOT sharp enough; if it catches on your thumb nail, it has correct sharpness. Now slide the needle under its own weight over the whole bearing surface of the jewel, as shown in figure 13-18. If the surface of the jewel has a crack or a pit, it will snag the fine point of the needle. NOTE: If the jewel is defective, replace it.

7. Test the float for leaks by submerging it in warm water (120° F). The heat will expand the air inside the float; and if there are leaks in the float, air will bubble out through them. Use a pencil to mark the position of a leak.

If the float has a leak, drill a small vent hole in it, drain out the liquid, and dry the float

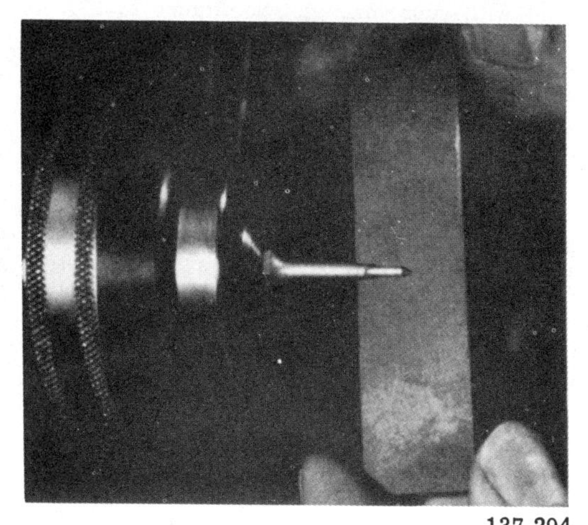

137.294
Figure 13-15.—Shaping a worn pivot point.

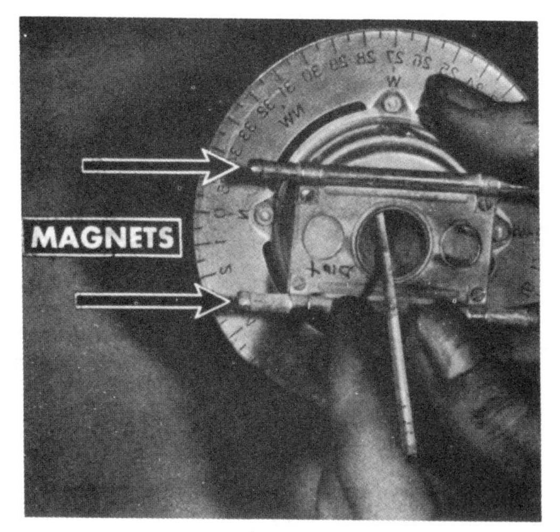

137.295
Figure 13-16.—Removing the jewel from the float.

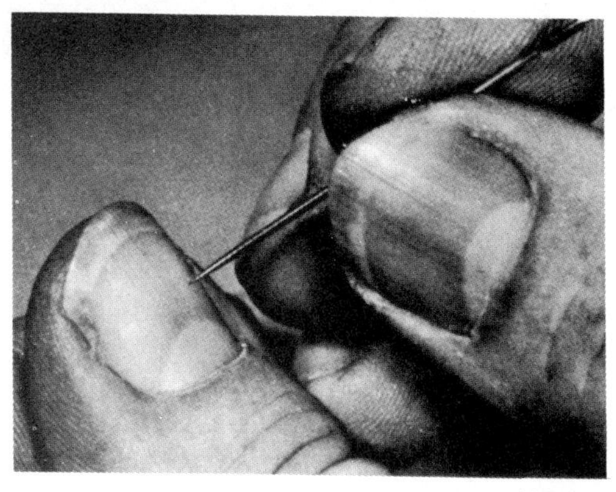

137.296
Figure 13-17.—Testing a needle for sharpness.

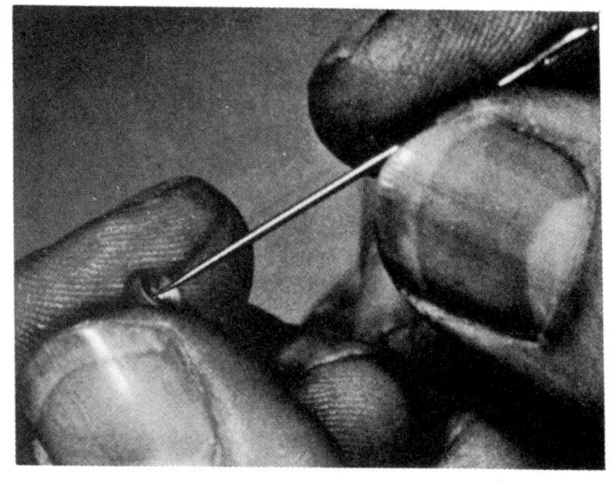

137.297
Figure 13-18.—Testing a pivot jewel with a needle.

in an oven. Then scrape the float down to base metal at each leak, clean the metal, and solder all leaks. Scrape the area around the vent hole and close the hole with solder.

Put the float back into warm water and recheck for leaks. NOTE: Leaks in the cone section of the float are difficult to close; and if you cannot seal them, replace the float.

8. Use a pegwood stick with a flat end to press the jewel and its spacer back into the float, as illustrated in figure 13-19. Then replace the retaining screw in the top of the card and tighten it. CAUTION: Do NOT use force; too much pressure will crack the jewel.

9. When you repair a float or replace a jewel, you generally destroy the balance of the float and must rebalance it. Materials required for making a float balance test are shown in figure 13-20.

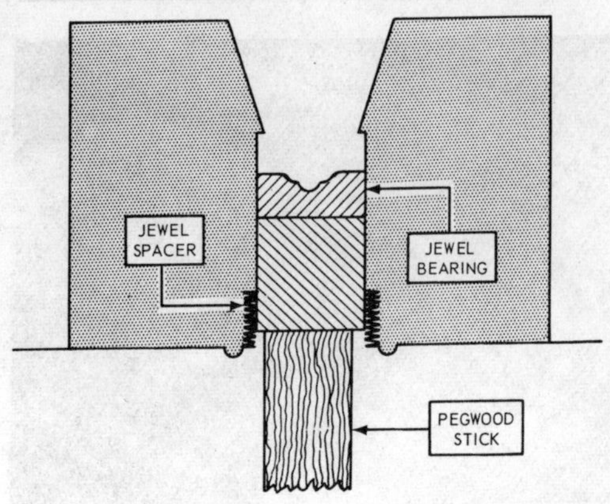

137.298
Figure 13-19.—Replacing pivot jewel and spacer.

10. To get bubbles from under the compass card and out of the cone section of the float, immerse the float edgewise in the compass liquid in the jar, as illustrated in figure 13-21. Then ease the float onto the pivot.

11. Set the point of your sighting rod at the same height as the compass card and spin the float with your magnet. See figure 13-22. As the card spins, compare its level with the sighting rod. If the float is balanced, the card will stay level while it is spinning. If the float is out of balance, you will see a high spot (fig. 13-23).

Remove the float and scrape a clean spot on its edge at the high point. Then apply a small amount of solder at the spot shown in figure 13-24. Put the float back on the pivot and retest for balance, and keep adding solder and retesting until you have the float in perfect balance. NOTE: If you apply too much solder, scrape off some of it with a knife.

12. Inspect the seats for the cover glass and the rubber gasket (fig. 13-25). If they are corroded, scrape them by hand or remove the corrosion on a lathe. Then clean the surfaces thoroughly with an approved cleaner. NOTE: When in doubt about anything, consult your shop supervisor.

13. Inspect the beveled edge of the glass cover. NOTE: The side which seats against the bowl has the larger diameter. If you find chips which would extend beyond the seat, as illustrated in figure 13-26, install a new glass cover.

14. Clean the bowl with a soft-bristle brush.

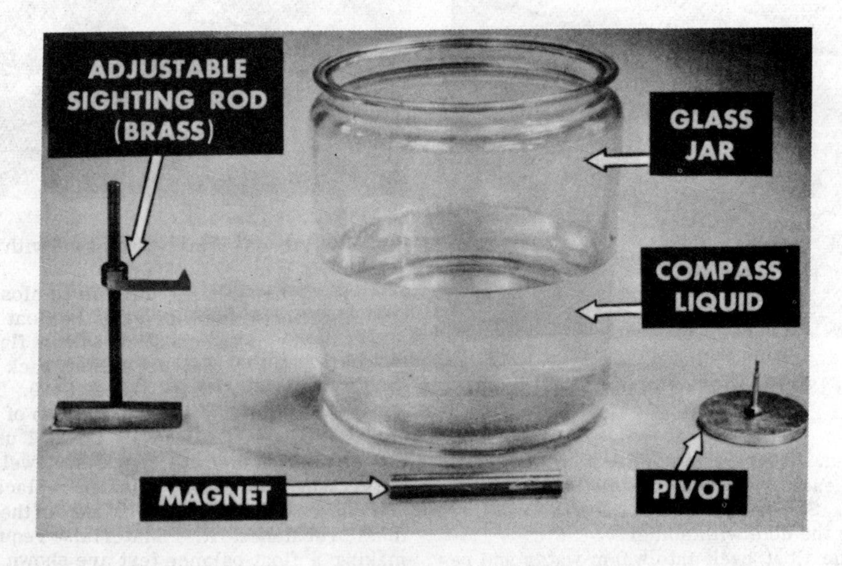

137.299
Figure 13-20.—Equipment for testing float balance.

137.300
Figure 13-21.—Mounting the float for a balance test.

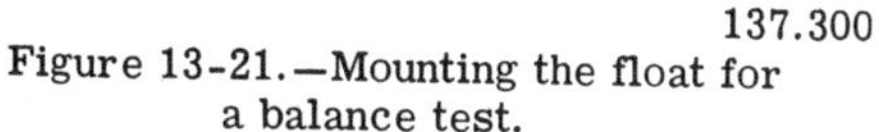

137.301
Figure 13-22.—Making the balance test.

15. Fill the expansion chamber with compass liquid. See illustration 13-27.

16. Replace the pivot and tighten it with a socket wrench.

17. At several points, measure the distance from the rim of the bowl to the tip of the pivot. NOTE: The pivot point should be exactly centered in the bowl. If necessary, adjust the point with a pair of pliers in the manner shown in figure 13-28. Be careful, lest you inflict damage to the point.

18. With wire hooks, lower the float onto the pivot.

19. Measure the distance between the edge of the card and the inner rim of the bowl. If it is not the same all the way around, remove the float and readjust the pivot.

20. Remove the float and fill the bowl with compass liquid to a level one half inch below the cover glass seat.

21. Replace the float and, with a pegwood stick sharpened to a chisel point, carefully place the glass in position.

137.302
Figure 13-23.—A float out of balance.

22. Fit the rubber gasket around the edge of the cover glass (fig. 13-29). The ends of the gasket should meet perfectly. If they overlap, trim them to perfect fit; if the gasket is too short, install a new one.

23. Replace the bezel ring, insert the screws, and turn them tight with your fingers. Then use a screwdriver to tighten all screws, one half turn at a time in rotation, until the ring is secure.

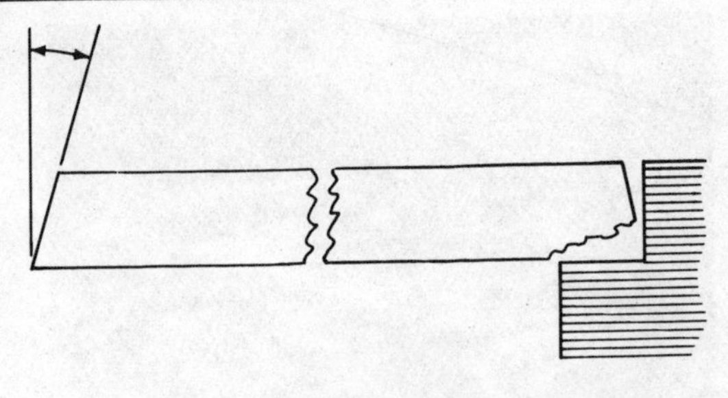

137.305
Figure 13-26.—Inspection of cover glass.

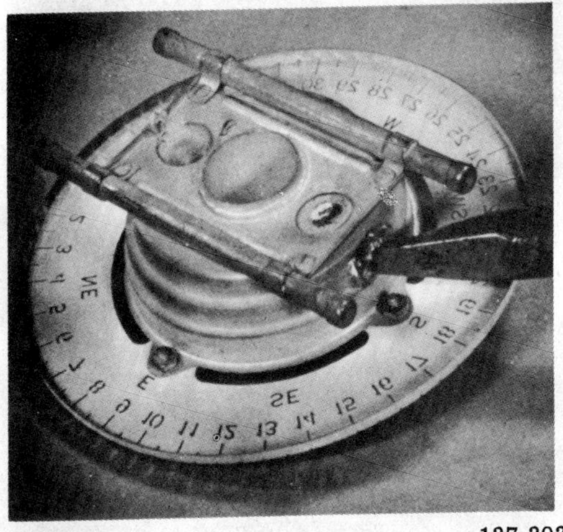

137.303
Figure 13-24.—Applying solder to the float.

137.306
Figure 13-27.—Filling expansion chamber.

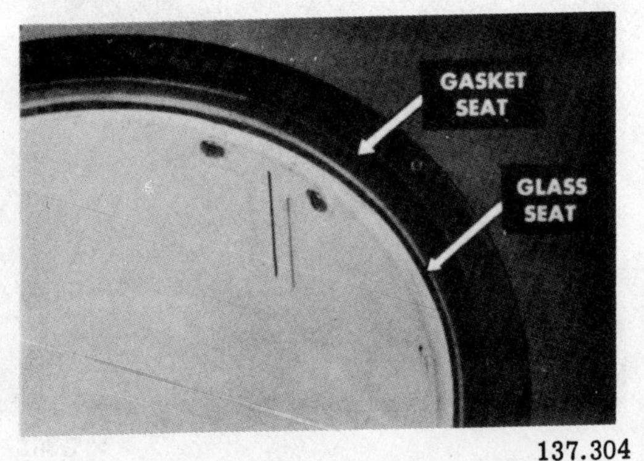

137.304
Figure 13-25.—Cover glass and gasket seats.

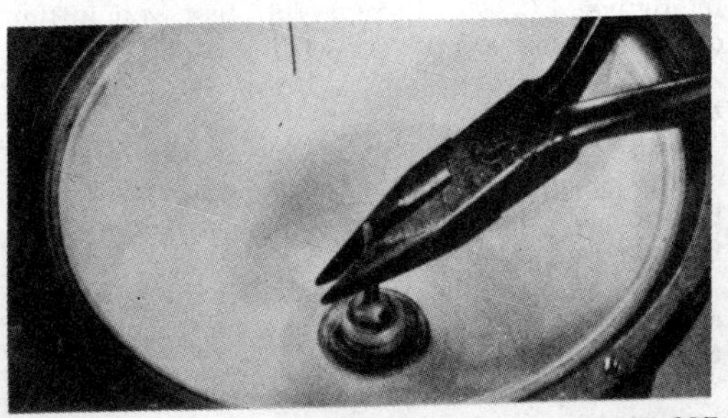

137.307
Figure 13-28.—Adjusting the pivot with pliers.

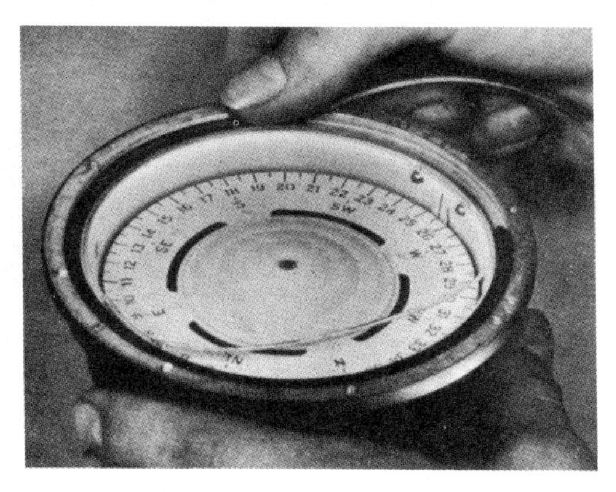

137.308
Figure 13-29.—Replacing the rubber gasket.

137.309
Figure 13-30.—Testing for leaks around the bezel ring.

TESTING AND ADJUSTING

The procedure for testing and adjusting your reassembled compass is as follows:

1. To test for leaks around the bezel ring, make a screw to fit the filler hole and drill a small hole through the center of the screw. Insert the screw in the filler hole and fit a piece of rubber tubing over the screw. Suck on the tube and then pinch it off (fig. 13-30). If there are leaks around the ring, bubbles will rise from them.
2. With a rubber bulb syringe, finish filling the bowl with liquid; then replace the filler vent plug and secure it.
3. Put the compass in a warm place and let it stand for 24 hours, with the filler hole up. This amount of time allows trapped bubbles to rise and dissolved air to come out of the compass liquid. NOTE: Less time is satisfactory if the air is fairly warm. Remove bubbles by adding more liquid and then replace the plug.
4. Retest the period of the compass. See step 5 under DISASSEMBLY.
5. Test the compass for balance. To do this, you need the material shown in illustration 13-31. Mount the compass bowl on the V's and put the level on the glass. NOTE: Be sure to center the level; otherwise, the level itself may unbalance the compass. See illustration 13-32.

If your compass does not balance, file the lugs (projections by which the compass is held) to move the bearing edge over toward the heavy side (fig. 13-33). Make a light cut with your file and test the balance. Repeat this process until the balance is perfect.

137.310
Figure 13-31.—Equipment for testing compass balance.

Now mount the compass in its gimbal ring and mount the ring on the V's (fig. 13-34). Test the balance. If necessary, file the lugs (slight amount each time) of the gimbal ring until you have the balance perfect.

45.23
Figure 13-32.—Testing compass balance.

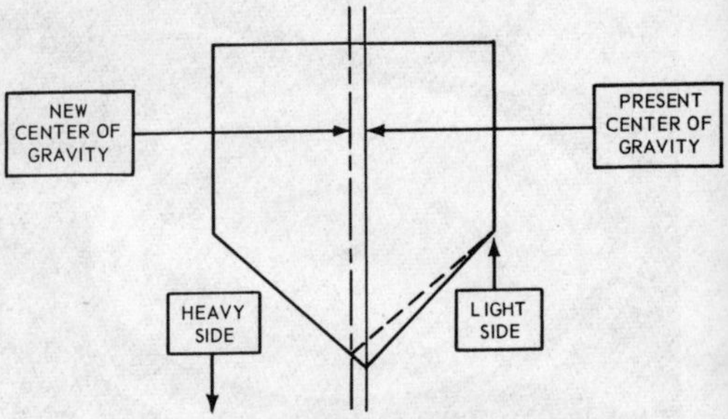

NEW CENTER OF GRAVITY

PRESENT CENTER OF GRAVITY

HEAVY SIDE

LIGHT SIDE

137.312
Figure 13-33.—Restoring compass balance.

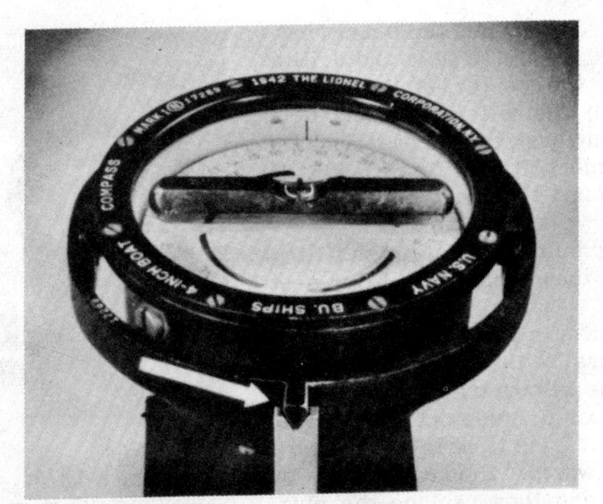

45.23
Figure 13-34.—Final balance test.

CHAPTER 14

AZIMUTH AND BEARING CIRCLES

The meaning of the terms AZIMUTH and TRUE BEARING is the same; namely, the direction of an object from true north (measured clockwise in degrees). In the Navy, however, there is a difference in the use of these two terms—AZIMUTH is used in connection with CELESTIAL BODIES, and TRUE BEARINGS are taken of TERRESTIAL OBJECTS. A Quartermaster, for example, takes a bearing of a lighthouse, but he gets the azimuth of the sun. Relative bearing is the direction of an object relative to the heading of a ship (measured in degrees).

CONSTRUCTION

The instrument used for measuring all bearings (true and relative) is the BEARING CIRCLE, which consists of a balanced, non-magnetic ring made to fit snugly over a magnetic compass or a gyro repeater. Study illustration 14-1.

Mounted on the ring of a bearing circle is a pair of sights which enable an observer to line up a ship or terrestial object and read the compass bearing of the object on a compass card. This pair of sights can also be used for measuring the azimuth of the sun.

When piloting his ship within sight of land, a navigator uses his bearing circle to obtain his ship's position—by taking the bearing of a landmark(s) ashore. When his ship is in a formation at sea, an officer of the deck uses a bearing circle and a stadimeter to keep his ship in proper position relative to the guide ship. (A stadimeter is used aboard ship to measure the range of objects of known height.) The navigator and the officer of the deck of a ship keep a bearing circle in almost constant use; for this reason, it must function accurately.

An azimuth circle is exactly like a bearing circle except that it has an additional pair of sights made especially for measuring the azimuth of the sun. Study illustration 14-1. A navigator uses measurements of the sun's azimuth to check the deviation of his ship's magnetic compass and the accuracy of the gyro compass.

An azimuth circle consists of a balanced, non-magnetic ring which fits over the bowl of a standard 7 1/2-inch Navy compass or compass repeater. To prevent disturbance of the accuracy of magnetic compasses on which they are mounted, azimuth and bearing circles must be made of non-magnetic metals. Parts are made of brass or bronze, except screws, which are made of nickel silver. The full assembly is then balanced. Because the compass bowl is mounted on pivots, the azimuth circle must be accurately balanced to prevent tipping of the compass in its mount.

Illustration 14-1 shows two sets of sights (mirrors) mounted on the azimuth circle ring. The set mounted on the 0° and 180° graduations is the same as the set on the bearing circle; the set mounted on the 90° and 270° graduations is the one made especially for measuring the azimuth of the sun. Each set of mirrors (sights) has a small spirit level to indicate when the circle is in a horizontal plane. See illustration 14-2, which gives an enlarged view of the front sight assembly. Observe the open sight and the prism.

NOTE: If the azimuth circle is out of the horizontal plane when a bearing is taken, the bearing is inaccurate.

A small pentagonal box mounted at 0° on the ring holds all optical elements of the 0° and 180° set of sights. The spirit level is mounted horizontally at the inside edge of the pentagonal box.

Illustration 14-3 shows the path of light rays through the pentagonal prism. As shown, light from the compass card is internally reflected at two different faces of the prism. When you look into the front face, you see a virtual image of part of the compass card. Because the image

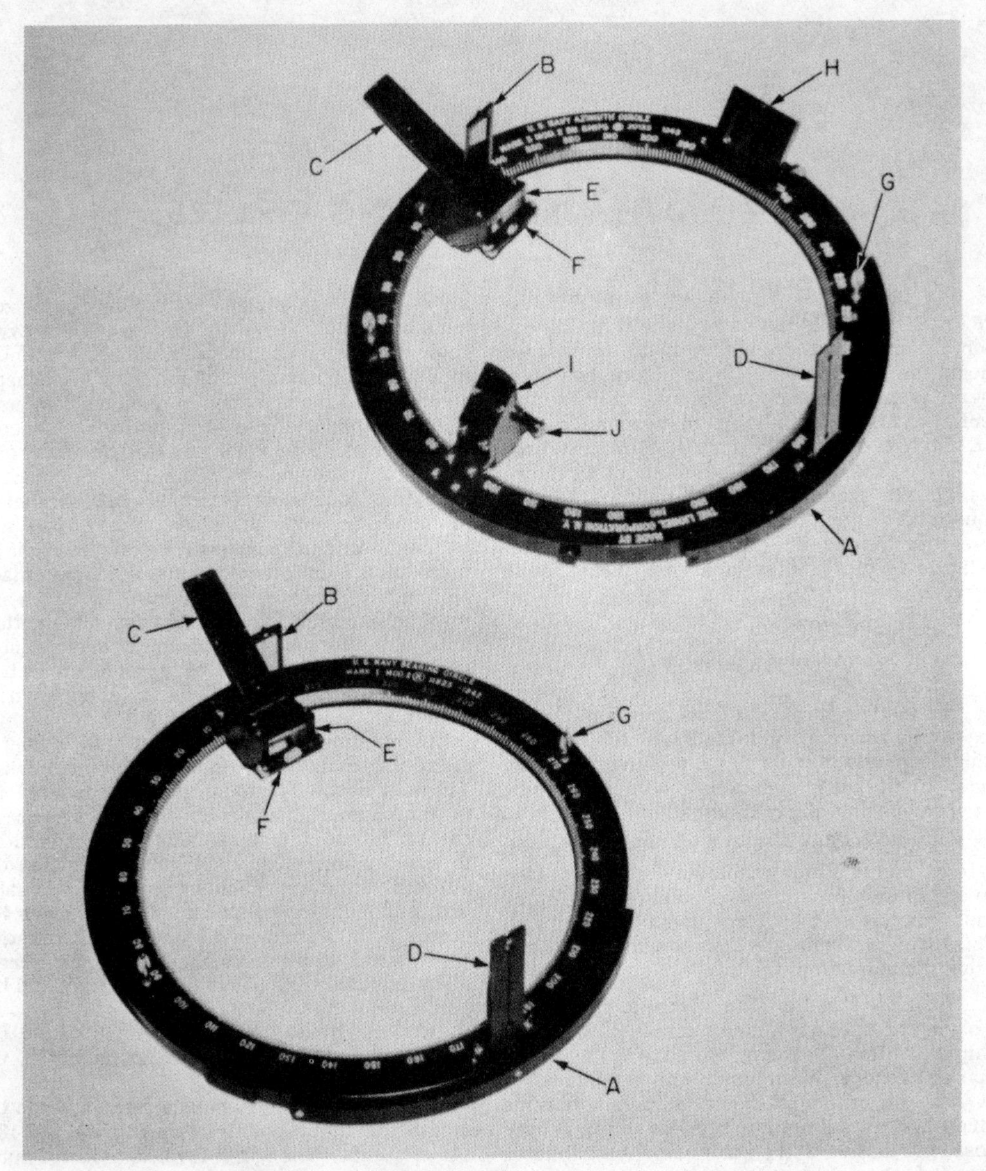

65.122

A. Counterweight.
B. Front sight.
C. Black mirror.
D. Rear sight.
E. Penta prism.
F. Penta spirit level.
G. Hand knot.
H. Curved mirror.
I. Right-angled prism assembly.
J. Right-angled spirit level.

Figure 14-1.—Mark 3, Mod 2, Azimuth circle and Mark 1, Mod 2, bearing circle.

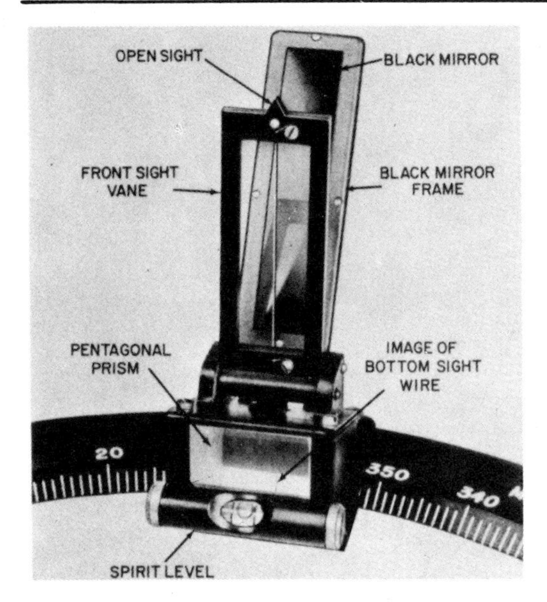

137.313
Figure 14-2.—Enlarged view of the front sight assembly.

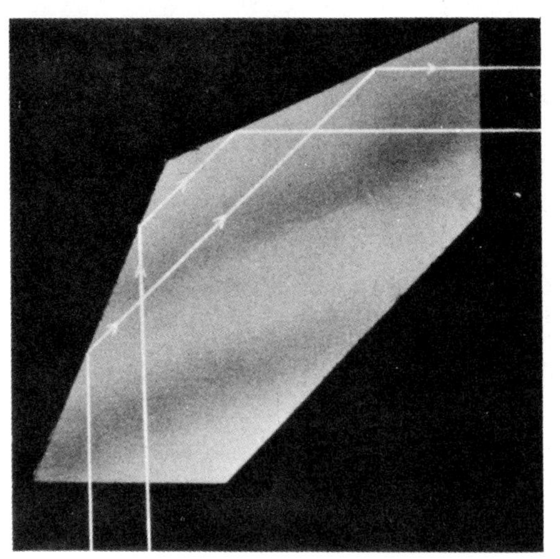

137.314
Figure 14-3.—Path of light through the pentagonal prism.

is reflected twice, it appears erect and normal. A sight wire mounted on the bottom face of the prism housing serves as a reading index of the virtual image of the compass card.

The front sight vane (fig. 14-2), sometimes called the far vane (farthest from your eye), is a rectangular frame with a fine wire stretched down the center of its long dimension. The whole vane (black mirror) moves on a horizontal axis to allow movement down and out of the way when the vane is not in use. The triangular point on the upper edge of the frame is the far point of the open sight.

The rear sight vane is mounted on the 180° graduation directly opposite the pentagonal box. As you can see in figure 14-1, this vane is a thin, rectangular plate with a vertical slot down its center. It swings on a horizontal axis to enable the operator to turn it out of the way when not in use. The V-shaped notch in the top edge (center) of the frame is the rear half of the open sight.

The cylindrical mirror of the 90° and 270° set of sights is mounted over the 270° graduation on the scale, and it reflects the sun's rays to the right-angled prism. This mirror swings on an horizontal axis to enable its operator to adjust it to the angle required to reflect the sun's rays to the right-angled prism. See illustration 14-4.

The right-angled prism in the 90° and 270° set of sights is mounted in a metal housing and located at the 90° graduation on the ring of the azimuth circle. A narrow, vertical slit in the face of the housing (fig. 14-5) and in the focal plane of the cylindrical mirror allows entrance of reflected light from the mirror. The prism receives the light, just like a mirror, and reflects it downward to the cylindrical lens mounted under the prism in the prism housing. This lens then focuses the light on the compass card in the form of a bright, narrow band.

PRINCIPLE OF OPERATION

When a Quartermaster or a navigator desires to take a bearing, he puts a bearing or an azimuth circle on a magnetic compass or ship's course indicator and follows a definite procedure. Suppose, for the sake of illustration, you are piloting a ship within sight of land and spot ashore a lighthouse whose bearing you need. You can get this bearing by using the 0° to 180°

137.315
Figure 14-4.—Front view of the cylindrical mirror.

137.316
Figure 14-5.—Enlarged view of the right-angled prism housing assembly.

set of sights on a bearing or an azimuth circle in the following manner:

1. Put an azimuth circle on a compass and turn the front and rear sight vanes to the vertical position. Then turn the black mirror down and out of your way. The front sight assembly in illustration 14-6 is in position (black mirror down) for taking a bearing.
2. Use the open sight and turn the circle to an approximate bearing on the lighthouse.
3. Move your eye down an inch or more and sight through the slit in the rear vane.
4. Adjust the circle so that the vertical wire on the front sight vane appears to split the lighthouse.

137.317
Figure 14-6.—Front sight assembly in position for taking a bearing.

5. Check the spirit level to determine whether the bearing circle and compass are horizontal; if not, level them and sight again.
6. Look into the prism and read the number of degrees on the compass card, at the point where the bottom sight wire cuts across the image. This is the COMPASS BEARING of the lighthouse.
7. To get the true bearing, correct the compass bearing for variation and deviation.

You can find the relative bearing of the lighthouse by lining up the sights of the azimuth circle as you did to get a true bearing, and by reading at the point on the inner lip of the ring just above the lubber's line the number of degrees on the scale of the azimuth circle.

You can measure the sun's azimuth on the general purpose sights (0° to 180°) on an azimuth or a bearing circle when the sun is partially obscured. The image of the sun is reflected to the observer's eyes from the black mirror (raised for this operation), which fully utilizes available light from the sun and produces a clear, distinct image.

When the sun is too bright to measure its azimuth with the general purpose sights, use the

90° to 270° sights on the azimuth circle, in the following manner:

1. Turn the circle until the cylindrical mirror bracket (fig. 14-4) is toward you and adjust its angle until it reflects a band of sunlight to the prism housing.

2. Now, turn the circle until the light reflected by the cylindrical mirror enters the slit in the housing (fig. 14-5). When it does, you will see a band of light under the prism housing and superimposed on the graduations of the compass card.

3. Check the spirit level to make certain the ring is horizontal.

4. Read the compass card at the point where the band of light intersects the scale.

OVERHAUL AND REPAIR

Because the corresponding parts of azimuth and bearing circles are identical, except for the holes in the circle rings for mounting different parts, a common repair procedure is applicable to both types of instruments. In this section, therefore, basic and complete coverage is given for the Mk 3, Mod 2, azimuth circle, with additional coverage for peculiarities of the Mk 1, Mod 2, bearing circle.

Review again the repair and maintenance procedures discussed in chapter 8.

PREDISASSEMBLY INSPECTION

Before you start to repair or overhaul an azimuth or bearing circle, give it a predisassembly inspection to determine whether it should be repaired or surveyed and salvaged. Record all your findings and recommendations on a casualty analysis sheet.

The things for which you should look when you make a predisassembly inspection are:

1. GRADUATIONS. Are the graduations clear? Can you read them easily?

2. PAINT. Because azimuth and bearing circles generally are exposed to severe weather for long periods of time, their metal parts must be protected by paint. If these parts are worn or chipped, the logical decision is to follow established procedures and repaint them.

3. OPTICAL ELEMENTS. Examine the cylindrical mirror, the black mirror, and the two exposed faces of the pentagonal prism for scratches, cracks, pitting, watermarking, or peeling of silver on the cylindrical mirror.

4. SIGHT WIRES. Check the vertical sight wire in the front vane sights and the bottom wire at the base of the pentagonal prism for straightness and tautness in their frames. If the wire in the front sight vane is loose, it can be tightened; if it is kinked or broken, replace it with .011-inch brass wire. If the bottom sight wire is loose or broken, replace the wire.

5. CIRCLE RING AND PARTS. Inspect the azimuth or bearing circle ring for distortion. It must be a true ring. Check also all parts on the ring.

6. SPIRIT LEVELS. Put the ring on a level surface and look at the bubble in the spirit level. If the bubble does not fall exactly between the two lines on the vial, the level is not correctly adjusted.

7. HINGE MOTION OF SIGHTS AND MIRORS. The motion of hinges on sights and mirrors must be smooth and easy but tight enough to hold the sights and mirrors in any desired position.

8. SCREWS. Inspect the slots of screws for burrs or deformities.

DISASSEMBLY

Unless otherwise indicated, disassembly procedures in this section apply to both azimuth and bearing circle. Illustrations are for the azimuth circle. We explain first how to remove the major assemblies and then give the step-by-step procedure for disassembling the subassemblies.

Removing the Rear Sight Assembly

To remove the rear sight assembly, do the following:

1. Remove the two rear sight bracket screws. See figure 14-7.

2. With a prying tool, lift the rear sight assembly off the azimuth or bearing circle ring assembly. The rear sight bracket is dowelled to the ring, and the two rear sight dowel pins (fig. 14-8) should come out of the ring and remain in the bracket. If they do not, pull them out of the ring with pliers.

Removing the Front Sight and Prism Assembly

The black mirror and a spirit level are also attached to the front sight and prism assembly,

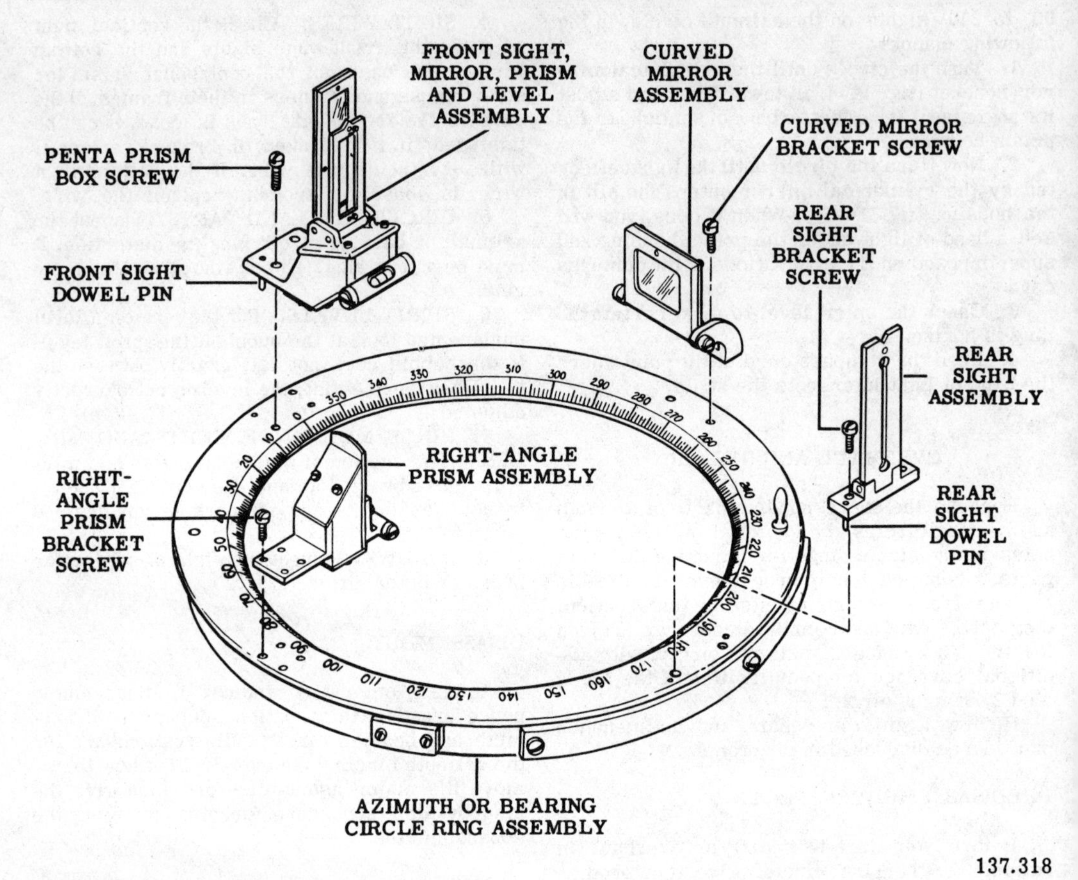

Figure 14-7.—Azimuth or bearing circle ring assembly.

as indicated in illustration 14-7. To remove all parts of this assembly, proceed as follows:

1. Remove the penta prism box screws.
2. Grasp the sides of the penta prism box and lift the front sight, mirror, prism, and level assembly off the azimuth or bearing circle assembly. If the assembly is stuck to the ring, use a prying tool to loosen it.

NOTE: The two front sight dowel pins (fig. 14-7) should come out of the ring and remain in the prism box. If they do not, pull them out with pliers.

Removing the Cylindrical Mirror Assembly

You can remove the cylindrical mirror assembly (fig. 14-7) by taking out the two screws which secure the bracket and lifting the assembly off the azimuth circle ring assembly.

NOTE: The bearing circle assembly (Mk 1, Mod 2) does not include a cylindrical mirror assembly and a right-angled prism assembly.

Removing the Right-Angled Prism Assembly

The right-angled prism assembly is secured to the circle with four screws, and when you remove them you can lift the assembly off. See illustration 14-7.

This step completes the removal of major assemblies from an azimuth or a bearing circle and we are now ready to consider the procedure for disassembling subassemblies.

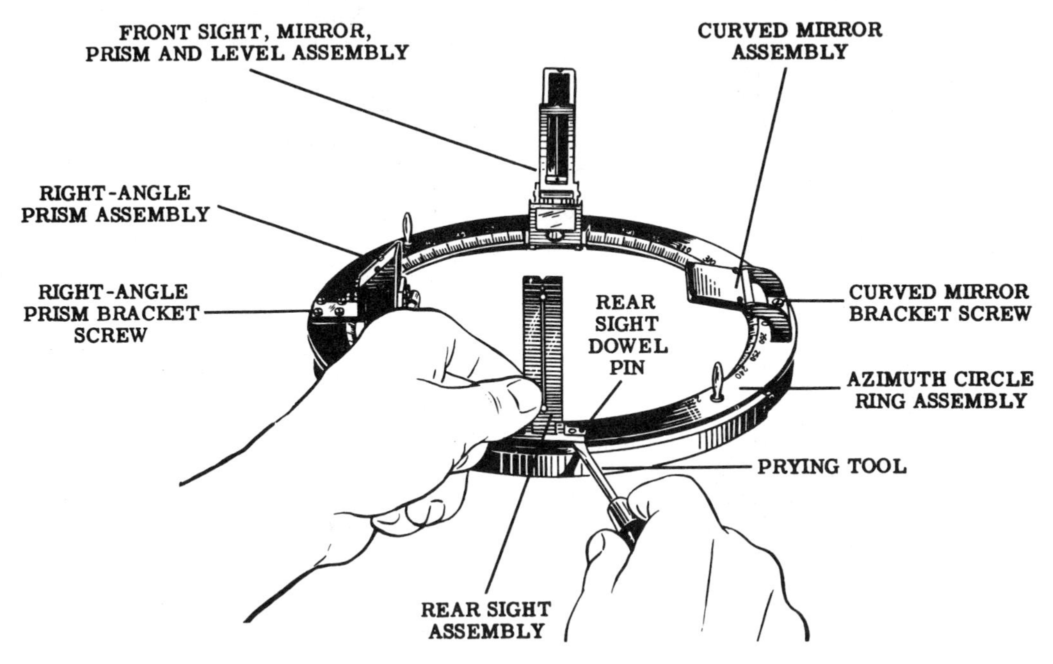

137.319

Figure 14-8.—Removing the rear sight assembly.

Disassembly of the Rear
Sight Assembly

The procedure for disassembling the rear sight assembly is as follows:

1. Set the assembly on an anvil block, with the end of the rear sight bearing pin over a hole in the block. Then (with a hammer and punch) drive the bearing pin out of the rear sight and rear sight bracket. Study figure 14-9.

NOTE: Do NOT remove these pins unless it is absolutely necessary.

2. Use the metal block as a support and drive the rear sight dowel pins out of the rear sight bracket. See illustration 14-10.

Disassembly of the Front Sight
and Prism Assembly

All parts of the front sight, black mirror, penta prism, and spirit level assembly are shown in figure 14-11. Disassemble it in the following manner:

1. Take out the two bottom frame screws and remove the bottom sight assembly (bottom part of fig. 14-11).

2. Remove the four penta prism spirit level mounting screws and the spirit level assembly will come off easily.

3. With a screwdriver, remove the two spirit level caps from the spirit level mount. Use tweezers to pull out the cotton wadding packed over the ends of the spirit level. Study illustration 14-11.

4. Slide the spirit level out of the tube in the penta spirit level mount. Then slide out the white paper level tube liner from the tube and discard the paper liner.

5. Cover the end of the penta prism with lens tissue and push it out of the penta prism box (fig. 14-12). Then grasp the prism by its unpolished sides, wrap it in lens tissue, and set it in the parts tray.

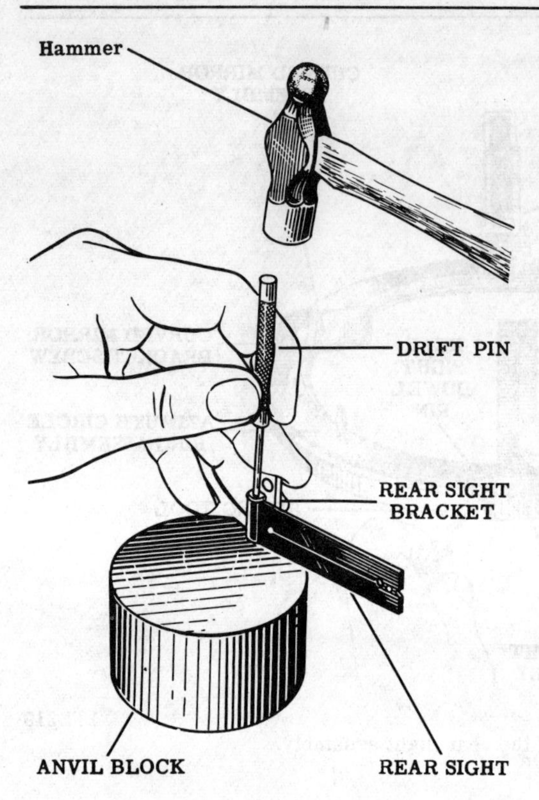

137.320
Figure 14-9.—Driving the bearing pin out of the rear sight bracket.

NOTE: The penta prism may be sealed in its prism box. If so, remove the sealing compound with a hardwood stick.

6. Take out the two penta prism side shims in the penta prism box (fig. 14-11).

7. Now, unscrew the four front sight bracket screws and remove the front sight and black mirror assembly from the penta prism box.

8. Remove the black mirror plate screws and remove the mirror backing plate and the black mirror from the mirror frame. Wrap the mirror in lens paper and set it in a safe place (parts tray).

9. To remove the black mirror frame from the front sight and mirror bracket, drive out the black mirror bearing pin. Study illustration 14-13.

10. To disassemble the front sight assembly from the front sight and mirror bracket, remove the front sight bearing pin.

11. Remove the two front sight wire screws to loosen the sight wire from the front sight frame and pull the wire out of the frame.

12. With a hammer and punch, remove the two front sight dowel pins (fig. 14-11) from the penta prism box.

13. Next, remove the four cylindrical mirror plate screws and take the mirror backing plate off the back of the mirror frame. Then remove the mirror, wrap it in lens tissue, and put it in a safe place. See illustration 14-14.

14. Remove the two cylindrical mirror bearing screws and separate the mirror bracket from the mirror frame.

15. Now, remove the screws from the cover of the right-angled prism box and take the cover off. Refer to figure 14-15.

16. Alwasy spill the right-angled prism out of its box onto several thicknesses of lens tissue. If the prism is sealed in the box remove the sealing compound with a pointed, hardwood stick and pick the prism loose with the stick. Then use a stick padded with lens tissue to push the lens out of the prism box.

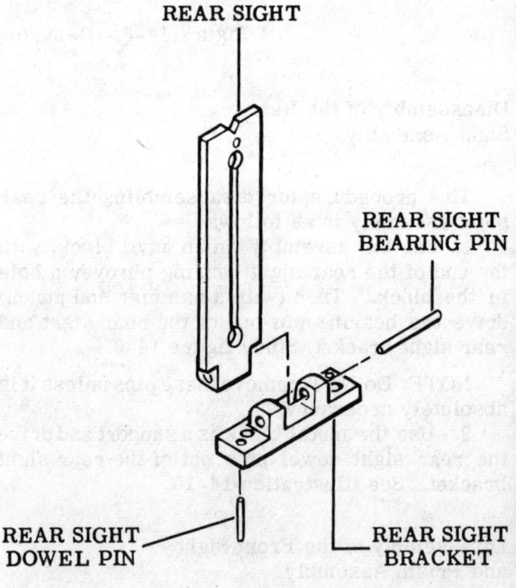

137.321
Figure 14-10.—Disassembled parts of the rear sight.

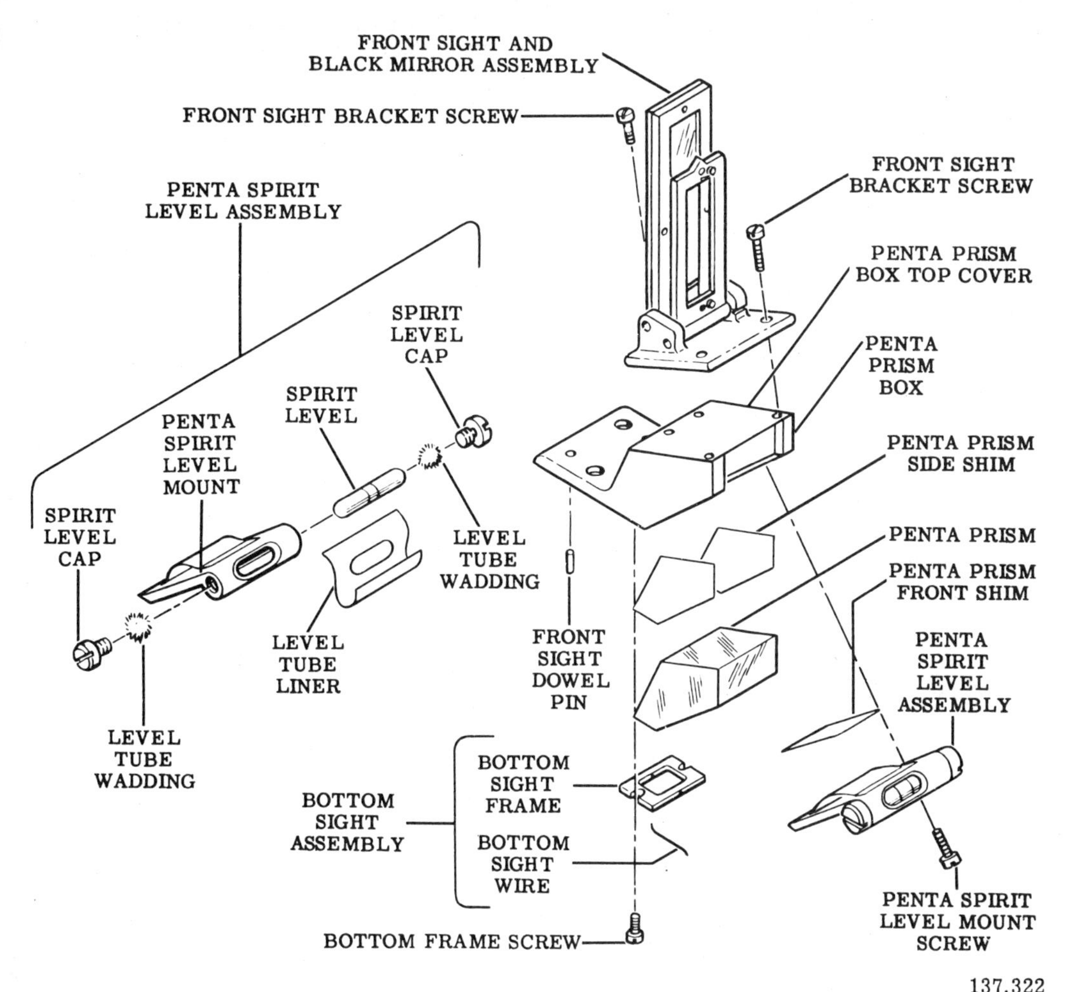

Figure 14-11.—Disassembled front sight and mirror, pentaprism, and penta spirit level assembly.

17. Next, remove the right-angled level mount screws to disassemble the right-angled spirit level assembly from the right-angled prism box. See figure 14-15.

18. To disassemble the right-angled prism bracket from the right-angled prism box, remove the bracket-to-box screws (fig. 14-15).

19. Disassemble the two circle adjusting screws and their adjusting screw lock nuts, and remove the circle spring screw and the circle spring shoulder screw to disassemble the circle adjusting spring. Study figure 14-16.

NOTE: For modified azimuth or bearing circle assemblies which include ball-type detents (fig. 14-17), omit the last step. For unmodified azimuth or bearing circle assemblies (fig. 14-16), omit step 22.

20. To disassemble the three ball detents (fig. 14-17), remove the two detent screws in each. Disassemble the detent spacer and detent spacer shims at the same time. Then wrap each ball detent and its spacer and shims in a piece of paper marked 120°, 240°, and 360°, at which points they were assembled.

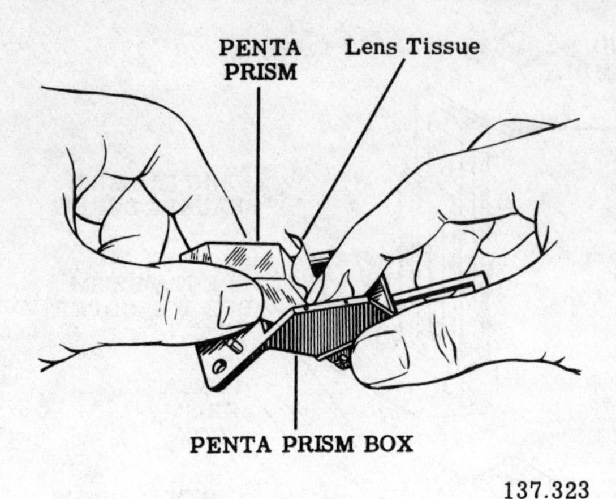

137.323
Figure 14-12.—Pushing the penta prism from its box.

21. Disassemble the counterweight from the circle (figs. 14-16 & 14-17).

22. Wrap the hand knobs with cloth (as a protection) and loosen them with pliers.

This completes the disassembly of an azimuth circle (Mk 3, Mod 2) or a bearing circle (Mk 1, Mod 2). The next step in maintenance is inspection of parts and repair.

PARTS INSPECTION AND REPAIR

This section pertains to the procedure for inspecting and repairing disassembled parts of azimuth and bearing circles in accordance with the procedures prescribed in the Control Manual (NavShips 250-624-12) and the Manual for Overhaul, Repair, and Handling of Azimuth and Bearing Circles (NavShips 250-624-7). When in doubt about any procedure, or the action you should take, consult your shop supervisor.

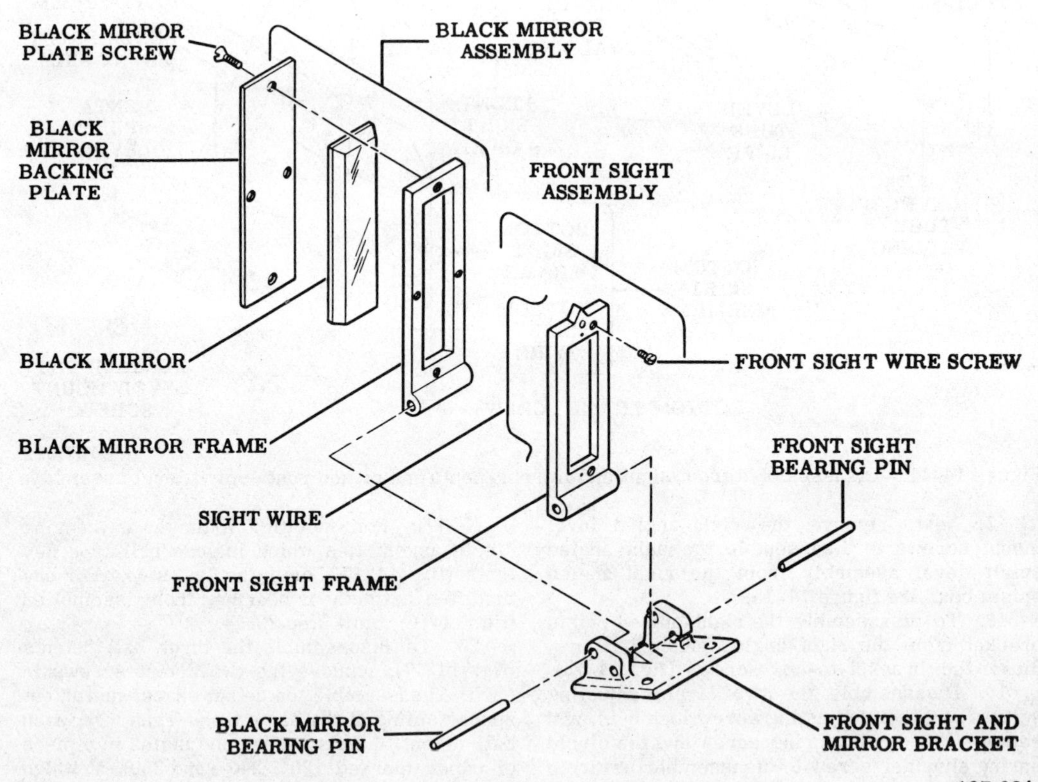

137.324
Figure 14-13.—Black mirror assembly.

Inspection of Parts

On optical parts, inspect the condition of the silver on the cylindrical mirrors. If the reflecting surface of either is defective, replace the mirrors. Proceed as follows to inspect mechanical parts:

1. Check for wear, dents, damaged threads, burrs, and distortions.

NOTE: Judge defects on the basis of their performance in the instrument, not on appearance. Performance of the instrument must be perfect.

2. If the sight wire in the bottom sight frame is missing or damaged, install a new one.
3. The recessed spots on the rear sight and the front sight frame are used for sighting and rough alignment at night. When necessary, refill these spots with monofill.
4. Inspect mechanical parts for appearance and condition of finish. After repairs have been completed on a part, refinish or repaint the part, as required.

Observe in illustration 14-16 the circle adjusting screw and the adjusting screw lock nut, and also the circle adjusting ring. Then look at the ball detent, the detent spacer, and the detent spacer shim on the azimuth or bearing circle ring in figure 14-17. The ball detent adjustment is a replacement for the screw and circle adjusting ring; and if modifications are necessary, drill holes to secure the ball detents. Note in illustration 14-17 the two small screw holes on the sides of the large hole for the ball detent.

Materials required for installing the ball detents consist of three ball detents (with springs and housing), six detent screws, three detent spacers, and an assortment of detent spacer shims. These are available in three thicknesses: .002, .003, and .004 inches.

Observe the engraved markings on the circle ring. If they are indistinct, clean them out with a scriber and refill with monofill. If the circle requires paint, remove the old paint and repaint it before you restore the engraved markings.

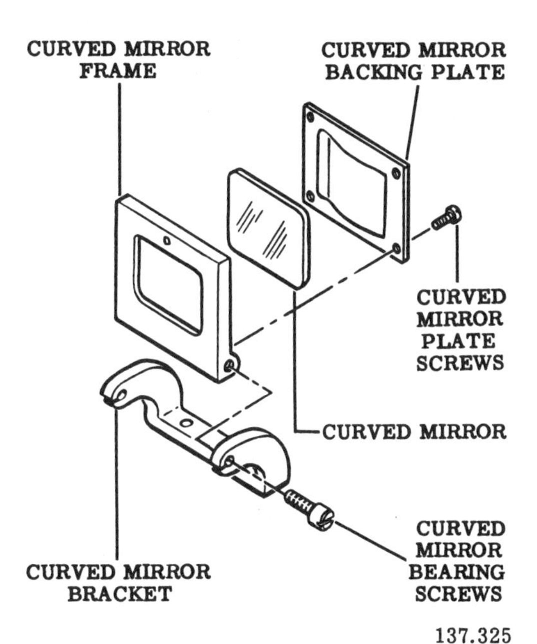

Figure 14-14.—Curved mirror assembly.

Straightening Distorted Rings

If a circle ring is not flat, inspect the underside for burrs, dents, or anything which may prevent even seating. If you find defects, file them down smooth with the surface of the ring.

If a circle ring is bent, put it on a circle testing ring and press down on top of the ring (fig. 14-18) at various points to determine where the ring is rocking. When you determine the location of a bend, put the circle ring on a 2 x 4-inch wood block on a bench, firmly grasp the ring in two places, as shown in figure 14-19, and press down on it hard enough to remove the bend.

When an azimuth or bearing circle is out-of-round, hammer it down over the circle repair ring, as shown in figure 14-20. Use a rawhide mallet and hammer all around the edge to shape the ring. To remove an azimuth or bearing circle ring from a repair ring, place both rings inside the removing ring (part A, fig. 14-21), which will support the azimuth or bearing circle ring, and hammer the repair ring to drive it down and out of the bearing circle ring.

Modification of a Circle Ring

All Mk 3, Mod 2, azimuth circle and Mk 1, Mod 2, bearing circles must be modified to their respective Mod 3 types when they are in the shop for repair. To do this work, proceed as follows:

1. Clamp the circle ring modifying drill jig tool (part B, fig. 14-21) on the azimuth or bearing circle ring, with the jig's guide mark aligned with the 120° line on the circle ring.

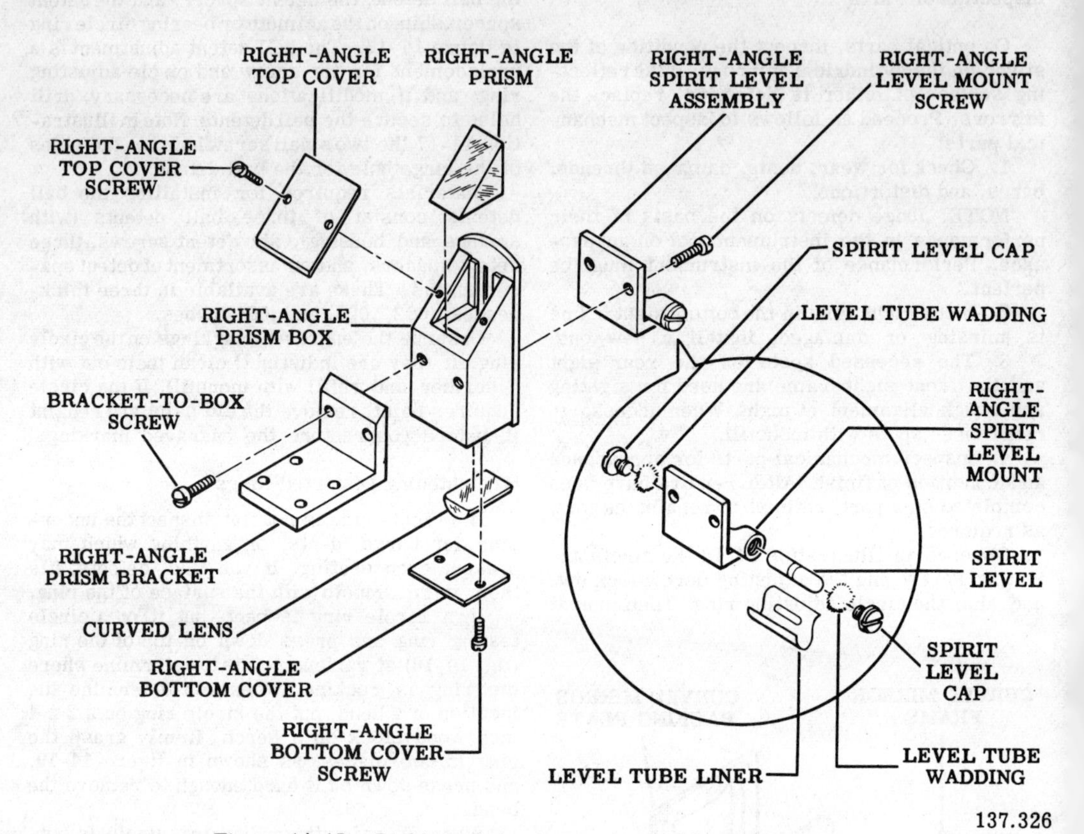

Figure 14-15.—Right-angled prism and spirit level assembly.

2. With a 13/32-inch diameter commercial drill, drill a .406-inch hole in the circle ring, through the center hole of the jig. Then use a .0935-inch drill to drill the two side holes in the positions indicated in illustration 14-22 and tap the holes with a No. 4-48NF-3 tap.

3. Repeat steps 1 and 2 at the 0° and 240° points on the azimuth or bearing circle ring.

4. Change Mod 2 on the data plate of the circle to Mod 3.

5. Turn the assembly as necessary to line up the 0° and 180° cardinal point with the 0° to 180° DUMB LINE on the circle testing ring.

NOTE: The two DUMB LINES on the dummy stand of the collimator are lines drawn to represent the 0° to 180° and 0° and 270° axes of the stand.

6. The 90° and 270° points on the ring should line up (within the width of a marking line) with the 90° to 270° dumb line on the circle testing ring. If they do not, try different spacer shims under the ball detents to get alignment of the 0°, 90°, 180°, and 270° points with the corresponding dumb lines of the testing ring.

7. As a further test of roundness which will give true readings for all points on the azimuth or bearing circle ring assembly, turn the ring to align the 45° and 225° points with the 0° to 180° dumb line on the testing ring. The 135° and 315° points should now line up with the 90° to 270° dumb line. When you complete this test, try several other combinations around the ring.

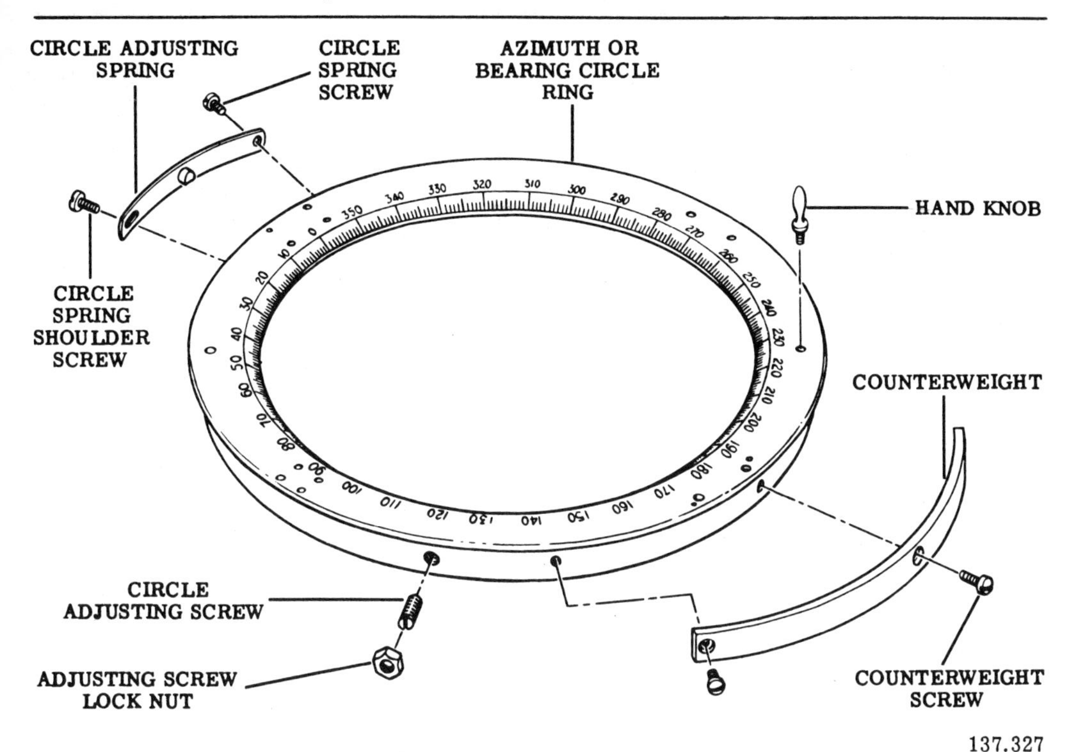

Figure 14-16.—Circle ring with adjusting screw lock nuts and adjusting ring.

Wiring the Bottom Sight Frame

If the bottom sight wire must be replaced in the bottom sight frame, proceed as follows:

1. Heat the solder which holds the sight wire in the grooves in the bottom sight frame and remove the old wire. See illustration 14-23.
2. With a triangular file, remove enough solder from the grooves in the frame to provide space for the new wire.
3. Cut a short length of No. 28 gage phosphorous-bronze wire and attach one end under the binding screw of the bottom sight wiring fixture (fig. 14-23). Wrap the other end of the wire around the tension key of the fixture with the wire pulled around the pins on the fixture.
4. If you use a new frame, tin the wire grooves with solder. There should always be some solder in these grooves, whether the frame is new or old.
5. Place the bottom sight frame under the bottom sight wire and position it to make the wire line up with the grooves. Then tighten the tension key (fig. 14-23) to draw the wire tight across the frame.
6. Hold the wire taut and solder it in the grooves in the frame.

CAUTION: Do NOT release the tension on the wire until the solder has had time to cool and harden. The wire must be tight and straight.

7. Clip the excess wire with a diagonal cutter (fig. 14-23) and file off the excess solder, even with the face of the frame. Then paint the wire black and touch up the solder with black paint.

REASSEMBLY PROCEDURE

The reassembly procedure for the Mk 3, Mod 3, azimuth circle is presented in this section in step-by-step operations. Exceptions in the procedure for a bearing circle are clearly noted.

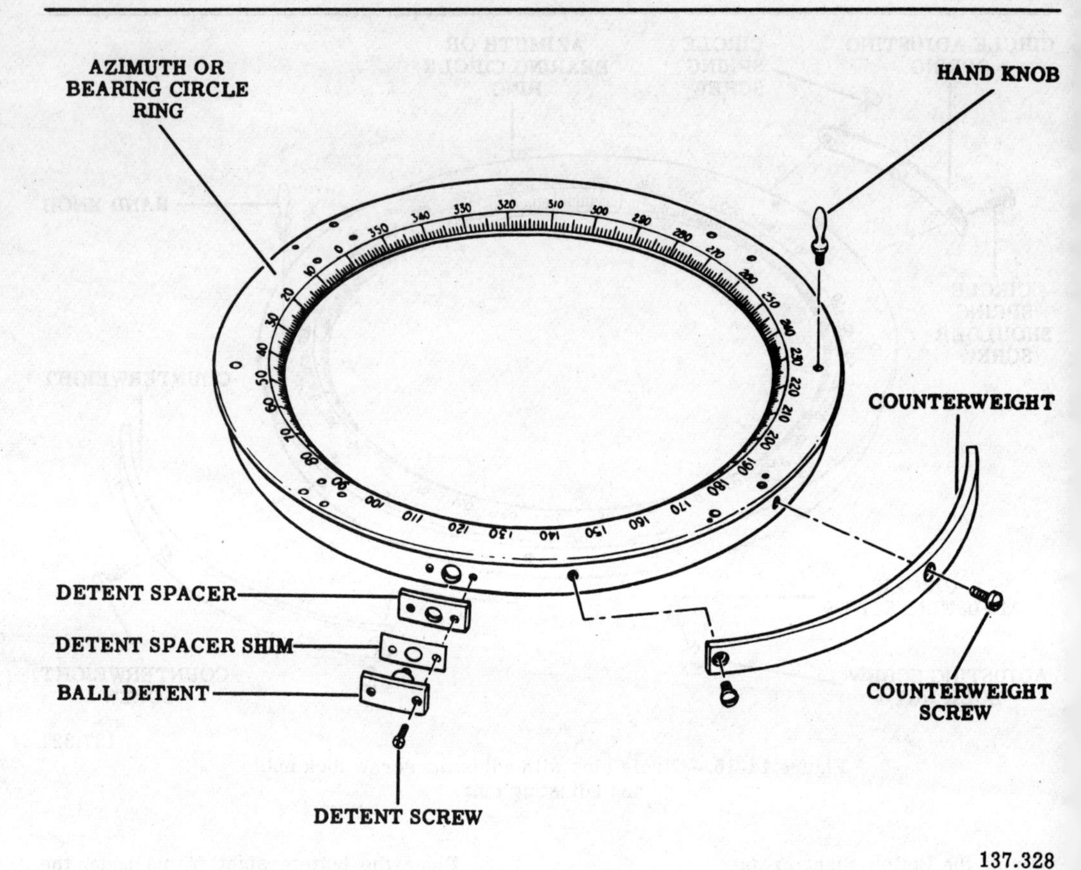

Figure 14-17.—Circle ring with ball detents used for adjusting.

Rear Sight Assembly

To assemble the rear sight assembly, do the following:

1. Fit the rear sight between the bearing lugs of the rear sight bracket, align the holes, and put a rear sight bearing pin in the aligned holes. Support the lugs on an anvil block and tap the pin in with a hammer.

2. If the bearing pin does not hold (by friction) the rear sight in any position you place it, set the lugs on the anvil block and spread the end of the bearing pin with a ball-peen hammer. If the hinge is still too loose, spread the other end of the pin. When this action does not produce enough friction, install a new rear sight and a rear sight bearing pin.

Front Sight and Black Mirror Assembly

Refer to illustration 14-13 as you study the procedure for reassembling the front sight and black mirror assembly, as follows:

1. Fit the black mirror frame between the rear set of lugs of the front sight and mirror bracket, align the holes in both parts, and insert the black mirror bearing pin. Use an anvil block for this operation. If the mirror does not stay in any position you put it, peen (spread) the ends of the bearing pin. If this does not help, replace the frame and the bearing pin.

2. If the sight wire is missing from the front sight frame, insert the front sight wire screws in the frame, and cut a piece of black sight wire about 2 1/2-inches long. Insert one end of the wire

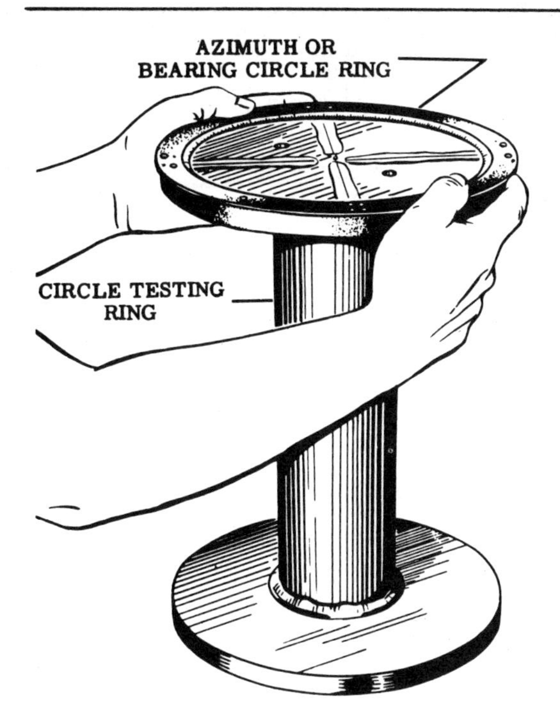

137.329
Figure 14-18.—Checking a circle ring for bends.

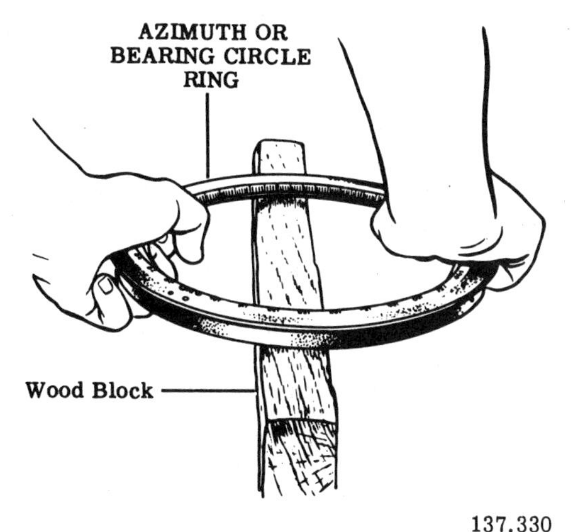

137.330
Figure 14-19.—Removing bends from a circle ring.

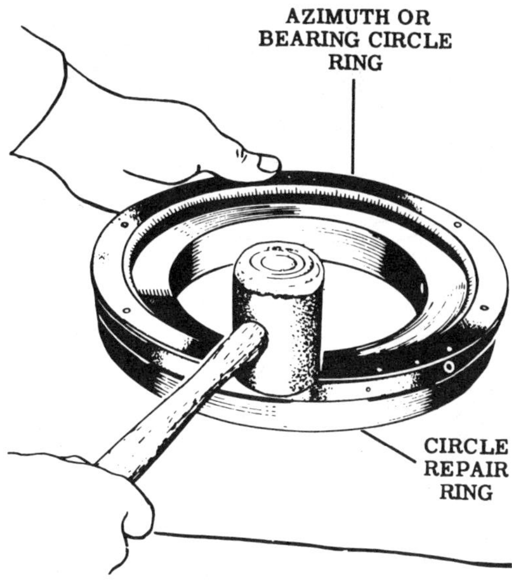

137.331
Figure 14-20.—Shaping a circle ring with a rawhide hammer.

through one of the holes in the frame and wrap it clockwise 1 1/2 turns under the head of the front sight screw. Tighten the screw down on the wire and insert the loose end of the wire in the other hole in the frame. Pull the wire taut across the frame with a small pair of long-nosed pliers and wrap it 1 1/2 turns clockwise under the head of the adjacent front sight wire screw. If any wire remains, cut it off and tighten the screw down on the wire.

3. Put the front sight assembly (fig. 14-13) in the front sight and mirror bracket and insert the front sight bearing pin. Tension exerted by the bearing pin should hold this assembly in any position; and if it is not, peen the ends of the bearing pin. If the friction now is not satisfactory, replace the front sight frame and the bearing pin.

4. Clean the black mirror, place it against the black mirror frame (fig. 14-13), put the black mirror backing plate against it, and insert the screws in the plate and tighten them. If the mirror is loose in the frame, put a fish paper shim (same size as mirror) between the backing plate and the mirror.

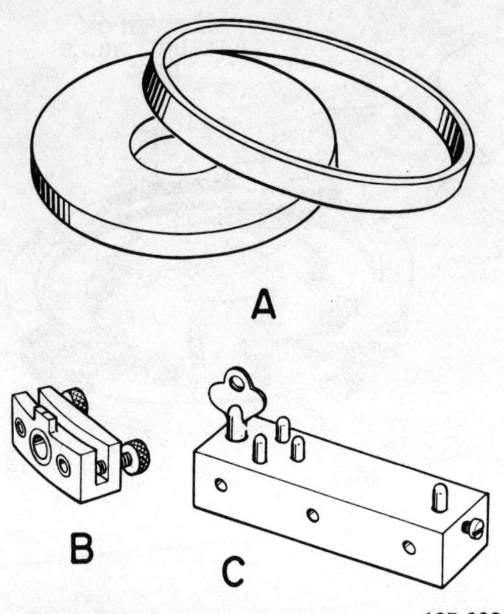

137.332
A. Circle repair ring, with removing ring.
B. Circle ring modifying drill jig.
C. Bottom sight wiring fixture.
Figure 14-21.—Special service tools for azimuth and bearing circles.

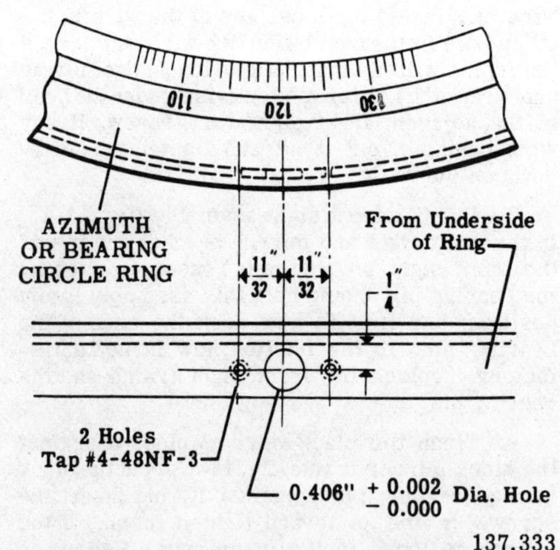

137.333
Figure 14-22.—Measurements for drilling a hole in a circle ring.

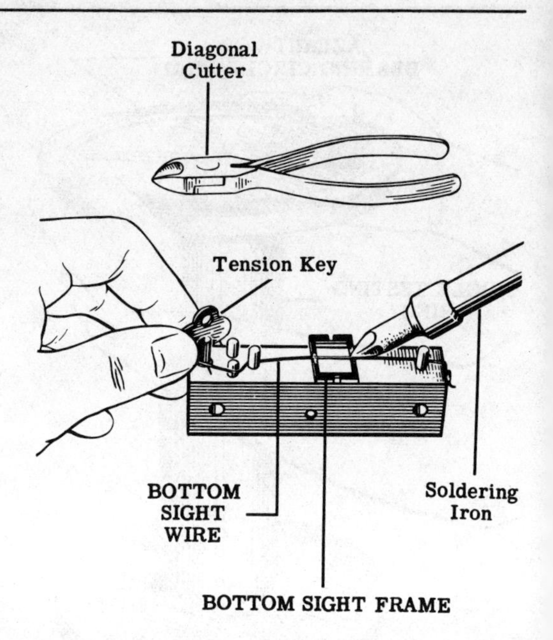

137.334
Figure 14-23.—Installing a bottom sight wire.

5. Put the front sight and black mirror assembly on top of the penta prism box and secure it with screws.

Bottom Sight Assembly

Assembly of the bottom sight assembly is as follows:

1. Screw the bottom sight assembly (fig. 14-24) to the bottom of the penta prism box.
2. Clean the polished surfaces of the penta prism, place a penta prism side shim on each side of the prism, and slide it into its box (fig. 14-24). Use as many shims as required to get a good fit.

NOTE: Cut the penta prism side shims from .01-inch-thick fish paper, a little smaller than the actual dimensions of the side of the prism, so that they will not extend into the sealing bevel on the edges of the box.

3. Apply sealing compound all around the edge of the penta prism to seal it in the prism box. The edges of the box are beveled to

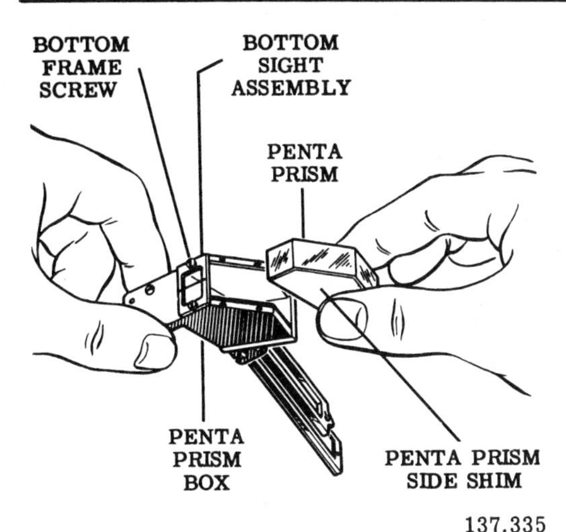

137.335
Figure 14-24.—Putting a penta prism in its box.

provide space for the compound. Use a spatula to smooth the compound flat with the edges of the prism and the box.

Penta Spirit Level Assembly

Refer to illustration 14-11 as you study the following procedure for reassembling the penta spirit level assembly:

1. Wrap enough level tube liner (.005-inch-thick white paper) around the spirit level to make it fit snugly in the tube of the penta spirit level mount, with the spring. The paper must not cover the portion of the level exposed and centered in the opening in the top of the tube (area around the two black leveling lines). See illustration 14-11.

2. Pack level tube wadding (white cotton) in the tube against the ends of the spirit level and screw a spirit level cap into each end of the tube.

NOTE: The caps should compress the wadding lightly, so add or remove wadding as necessary to center the leveling lines of the spirit level in the opening in the tube.

3. Place one penta prism front shim on the long exposed surface of the penta prism and assemble the penta spirit level assembly to the penta prism box with the four penta spirit level mount screws (fig. 14-11).

4. Apply sealing compound to the joint between the top edge of the penta spirit level mount and the exposed face of the prism.

All subassemblies of a Mk 1, Mod 3, bearing circle have now been reassembled. Reassembly operations from this point to REASSEMBLY OF ASSEMBLIES pertains only to the Mk 3, Mod 3, azimuth circle.

Cylindrical Mirror Assembly

Refer to illustration 14-14 as you study the procedure for reassembling the curved mirror assembly. Proceed as follows:

1. Assemble the cylindrical mirror frame to the cylindrical mirror bracket with the two cylindrical mirror bearing screws. Tighten the screws and test the friction of the frame in the bracket by moving the frame back and forth. If it is loose, recheck the bearing screws for tightness. If it is still loose, disassemble the parts and squeeze the split lugs of the bracket together.

NOTE: Put cardboard on the lugs to protect the finish when you squeeze them together.

2. Clean the cylindrical mirror and place it in its frame, uncoated side in.

3. Screw the cylindrical mirror backing plate (with spring) onto the back of the frame.

Right-Angled Prism Assembly

The procedure for reassembling the right-angled spirit level assembly is illustrated in figure 14-15 and is the same as for the penta prism, previously explained under reassembly of subassemblies.

Other steps in the reassembly of the right-angled prism assembly are:

1. Attach the right-angled spirit level assembly to the right-angled prism box with the mounting screws.

2. Clean the right-angled prism with lens tissue and place it in the right-angled prism box, as shown in illustration 14-25.

3. Apply sealing compound all around the edges of the prism and the box, in the beveled space provided for the compound.

4. Assemble the top cover to the prism box with the two top cover screws. See figure 14-15.

5. Clean the cylindrical lens with lens paper and place it flat side down in the opening in the bottom of the right-angled prism box. NOTE: The curved side of the lens faces out.

6. Apply sealing compound all around the edges of the lens and box, in the beveled space provided.

7. Screw the bottom cover to the prism box.

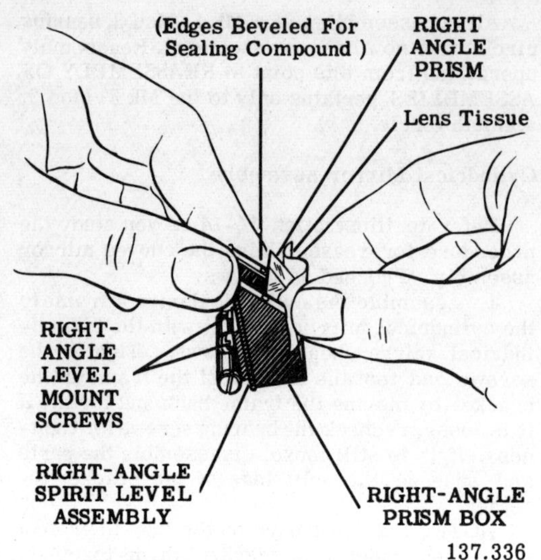

137.336
Figure 14-25.—Putting a right-angled prism in its box.

NOTE: The front cover of the prism box is soldered on.

8. Assemble the right-angled prism bracket to the prism box.

The subassemblies of a Mk 3, Mod 3, azimuth circle have now been assembled.

Reassembly of Major Assemblies

The procedure for assembling the major assemblies on the circle ring is given in the following steps. Refer to illustration 14-7.

NOTE: Reassemble the counterweights and all other parts to the bearing ring before you start to install the major assemblies. See illustrations 14-16 and 14-17.

REAR SIGHT ASSEMBLY.—Attach the rear sight assembly to the bearing circle ring assembly. If the original parts are being used, align the holes for the rear sight dowel pins and tap them in.

FRONT SIGHT AND PRISM ASSEMBLY.—Attach the front sight, mirror, prism, and level assembly to the ring assembly with the screws provide. If you are using original parts, align the holes for the front sight dowel pins and tap the pins in.

This last step completes the reassembly of a Mk 1, Mod 3, bearing circle; but there are two more steps in the assembly of a Mk 3, Mod 3, azimuth circle, and they are explained next.

CYLINDRICAL MIRROR ASSEMBLY.—Study figure 14-7. Screw the curved mirror assembly to the azimuth circle ring assembly with the two curved mirror bracket screws.

RIGHT-ANGLED PRISM ASSEMBLY.—Put the right-angled prism assembly in position on the azimuth circle ring assembly and secure it with the four screws provided.

The Mk 3, Mod 3, azimuth circle is now ready for final inspection, testing, and adjusting.

COLLIMATION

When you complete repairs on an azimuth or bearing circle, give the assembled instrument a careful inspection. Test the ring assembly for flatness and trueness; and if inspection results are satisfactory, collimate the instrument.

Collimation of an azimuth or bearing circle is performed on a collimator which simulates a gyrocompass repeater or a standard ship magnetic compass with the sun at a known azimuth. Study the azimuth and bearing circle collimator shown in figure 14-26. This collimator consists of a dummy stand, representing a cyrocompass repeater, and an artificial sun aligned with the 0° and 180° axis of the stand. Observe the nomenclature in figure 14-26. Study particularly the enlarged view in the circle.

The following discussion on collimation is for both azimuth and bearing circles. Procedures inapplicable for a bearing circle are so designated.

ALIGNING THE REAR SIGHT

The procedure for aligning the rear sight of an azimuth or bearing circle is as follows:

1. Put the circle assembly on the collimator stand, as shown in illustration 14-26.

2. Turn the circle assembly as necessary to align the 0°, 180°, 90°, and 270° marks, called cardinal points, of the azimuth circle ring with the four cardinal points on the collimator stand, with the 180° mark set to the zero point. The rear sight will then be toward the artificial sun of the collimator.

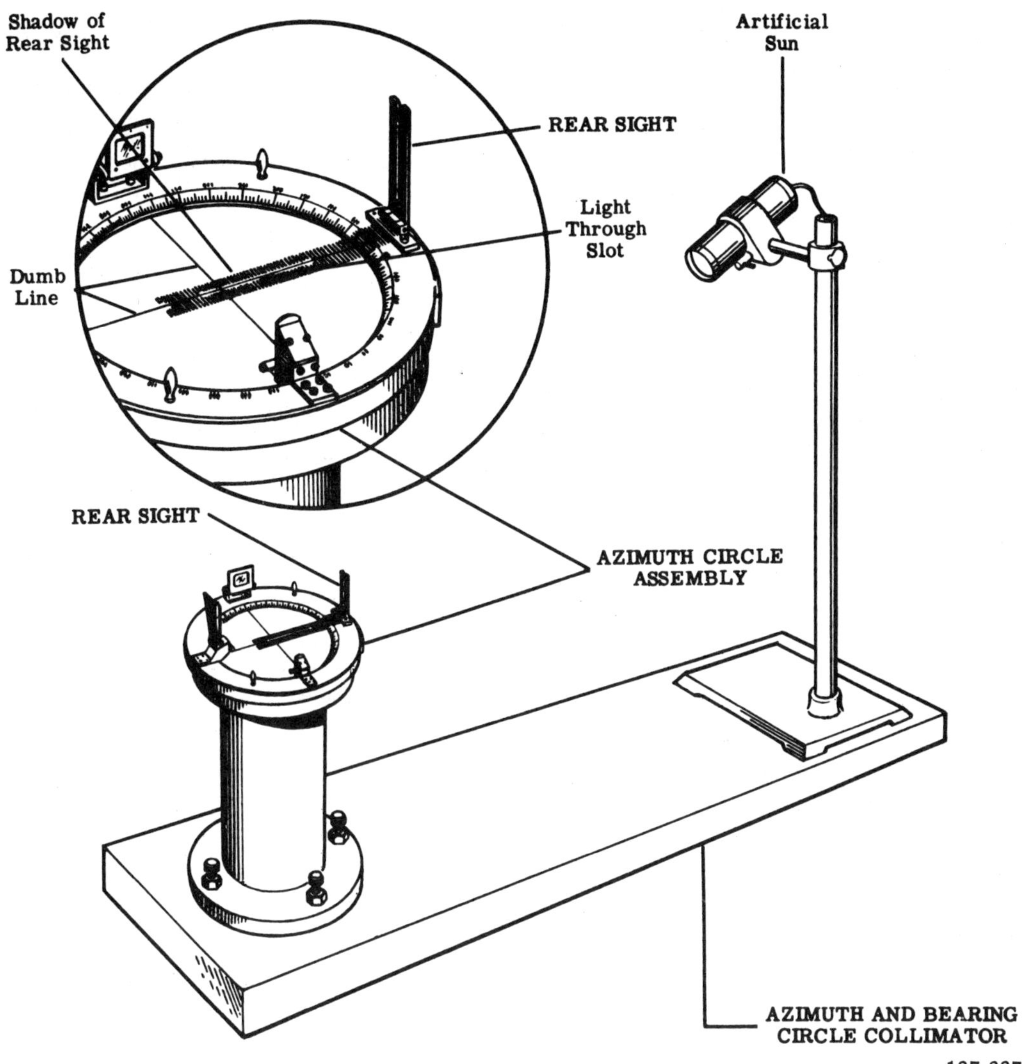

Figure 14-26.—Azimuth and bearing circle collimator.

3. Turn the rear sight to a vertical position and note the shadow of the sight cast by the artificial sun on the collimator stand. Light which passes through the slot should be centered on the 0° to 180° dumb line on the stand. If the light is not aligned at one end or the other, use a shim (fish paper, for example) on one side in order to get alignment.

If the light is not even all along the length, loosen the rear sight bracket screws and shift the rear sight assembly sideways. Turn the sight from the vertical position to the horizontal position and observe the light through the slot. If should still be centered on the dumb line; and if it is out at one end, loosen the rear sight bracket screws and turn the whole rear sight assembly. Then check vertical and horizontal alignment in all positions.

NOTE: If the rear sight dowel pins were assembled and the rear sight assembly needs adjustment, drive out the pins and effect necessary adjustment. Then replace the dowel pins.

FRONT SIGHT, BLACK MIRROR, AND BOTTOM SIGHT

To align the front sight, the black mirror, and the bottom sight, proceed as follows:

1. Align the 0°, 90°, 180°, and 270° marks on the circle ring with the corresponding points on the collimator stand; that is, at the ends of the dumb lines.
2. Look through the rear sight at the face of the penta prism for the reflection of the bottom sight wire and the dumb line on the collimator stand. The reflected dumb line should line up with the front sight wire (with the front sight assembly vertical).
3. Raise both front and rear sights to the vertical position and sight through the front sight at the rear sight. The front sight wire should be exactly centered along the slot in the rear sight. If each sight was properly aligned, the front and rear sight will align with each other.
4. Look through the sights and raise the black mirror to reflect the artificial sun. The front sight wire should appear to split the image of the artificial sun for all positions of the mirror which reflects the sun into the sights. If the sun is displaced to one side throughout the travel, disassemble the mirror and place a thin strip of shim paper between the long edge of the mirror and its frame on the opposite side to which the image was displaced. If the image is off at one end of the mirror only, shim the opposite corner across from the direction of displacement.
5. Look through the rear sight into the face of the penta prism at the image of the bottom sight wire and the 0° to 180° dumb line on the collimator stand. The bottom sight wire should coincide with the dumb line; and if it does not, shift the bottom sight assembly as required to bring the bottom sight wire into coincidence with the dumb line.

LEVELING THE PENTA SPIRIT LEVEL

To level the penta spirit level, do the following:

1. Since the collimator stand is level, the bubble in the spirit level should be centered between the leveling lines on the level. To make a SMALL adjustment, loosen the penta level mounting screws and adjust the spirit level assembly as necessary to center the bubble. When a GREATER AMOUNT of adjustment is necessary to center the bubble, remove the spirit level caps and shim the level with cotton wadding and paper liner.

At this point, a Mk 1, Mod 3, bearing circle is completely aligned. The next step in collimation for this instrument is "Dowelling the Front and Rear Sights," which is explained later. For a Mk 3, Mod 3, azimuth circle you must collimate the azimuth elements.

CYLINDRICAL MIRROR AND RIGHT-ANGLED PRISM ALIGNMENT

The procedure for aligning the cylindrical mirror and the right-angled prism follows:

Turn the azimuth circle assembly to bring the face of the cylindrical mirror toward the artificial sun. The 90° and 270° marks on the azimuth circle ring must be aligned with the 0° to 180° dumb line on the collimator stand. The 90° mark should be set at the 0° end of the dumb line.

LEVELING THE RIGHT-ANGLED SPIRIT LEVEL

The right-angled spirit level bubble should be centered between the two leveling lines on the level. Make small adjustments by loosening the right-angled level mounting screws and shifting the right-angled spirit level assembly. For larger adjustments, remove the two spirit level caps and shim the right-angled spirit level.

This last step completes the collimation procedure for a Mk 3, Mod 3, azimuth circle.

DOWELLING THE FRONT AND REAR SIGHTS

The front sight, mirror, prism and level assembly, and the rear sight assembly must be dowelled to an azimuth or bearing circle ring. To do this, proceed in the following manner:

1. If the sights need realignment, the original holes for the dowel pins may be slightly misplace. When this is the case, use a tapered bottom reamer to enlarge and form a straight tapered hole and install a tapered dowel pin.

2. If a new azimuth or bearing circle ring was assembled, use the original holes in the rear sight bracket and the penta prism box. With a .086-inch drill, drill holes 1/8-inch deep in the azimuth or bearing circle. Then assemble the rear and front sight dowel pins. See illustration 14-7.

3. If you must use a new rear sight bracket or penta prism box, plug the original holes and use an .086-inch drill to make new holes 1/8-inch deep in the ring and assemble the front and rear sight dowel pins.

FINAL SHOP INSPECTION

After you collimate an azimuth or bearing circle, give it a final inspection to make certain everything concerning repair and adjustment of the instrument is satisfactory. Inspect as follows:

1. Check the circle for completeness of parts.

2. Inspect the general appearance, finish of parts, tightness in the assembly, legibility of engravings, and tightness of screws.

3. Examine optical parts for defects and cleanliness.

4. Inspect the pivot tension of the front and rear sight vanes, and the black and cylindrical mirror assemblies.

Make final notations (if any) on the casualty analysis sheet for the instrument and put it in the case.

CHAPTER 15

SEXTANTS

A sextant is an instrument used for measuring the angle between two objects. The arc on which the scale for reading angles is engraved is approximately one-sixth of a circle; hence the name of the instrument, SEXTANT.

When a ship is at sea and away from visible landmarks, the navigator must use celestial navigation to determine his ship's position. Celestial navigation is possible because the navigator can use a sextant to determine the angle which the sun or another celestial body makes with the visible horizon. He can then determine the position of the particular celestial body at the time he took the sight by referring to the NAUTICAL ALMANAC. If he knows the position of the celestial body at the TIME OF SIGHT and the angle it makes with the horizon, he can (after applying certain correction factors) ascertain the exact position of his ship.

A sextant is well adapted for measuring angles at sea for three reasons:

1. It is small, light, and can be held easily in one hand.
2. It does not need a stable mounting.
3. It measures angles accurately to the nearest tenth of a minute.

Two of the earliest types of sextants are shown in figure 15-1. The ASTROLABE was a round, wooden disk with graduations from 0° to 359 degrees. A movable wooden pointer was fastened to the center of the disk. The instrument was suspended from a plumbline, supposedly to keep the horizon line level. When a navigator desired to measure the altitude of a star with this sextant, he sighted along the pointer to aim at a star and then read the scale on the disk at the end of the pointer.

The CROSS STAFF had a little more accuracy than an astrolabe but it had one big disadvantage—the navigator had to look in two directions at the same time. A cross staff was made of two wooden boards at right angles to each other, as illustrated, and the vertical board could be moved back and forth along the horizontal board. To take a sight, the navigator sighted from the end of the horizontal board to the celestial body and then moved the vertical board until its tip was on his line of sight. All the while, he had to keep the horizontal board pointed at the horizon. A scale on the horizontal board at the point where the vertical board crossed it gave him the angle formed by the celestial body with the horizon.

Although Sir Isaac Newton was probably the first man to put in writing the idea of the modern sextant, a man by the name of Hadley was perhaps the first man who actually made one (1731). Since that time many improvements have been made to sextants, with the result that those in use today are very accurate.

TYPES OF SEXTANTS

The only sextants currently in use in the Navy are the PIONEER and the DAVID WHITE, and both of these are designated as ENDLESS TANGENT SCREW TYPES. During your study of the next few pages you will learn the meaning of "endless tangent screw."

This chapter is limited in scope to a discussion of David White sextants, with references to differences in the pioneer sextant. The David White sextant is used much more extensively than the Pioneer.

CONSTRUCTION

At this point, study illustration 15-2, particularly the nomenclature and location of parts. Refer to this illustration as you study the construction of the instrument.

The principal parts of a sextant are:

1. ARC, OR LIMB. The arc of a sextant is the lower curved part of the frame, with a scale graduated in degrees engraved on it (fig. 15-2).

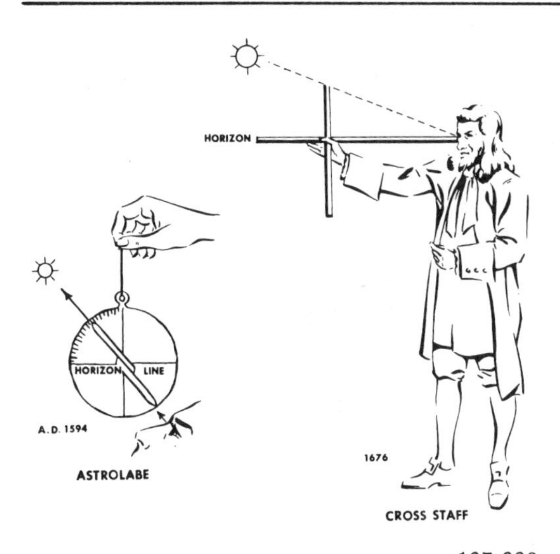

137.338
Figure 15-1.—Early types of sextants.

Gear teeth (one tooth to each degree on the scale) are cut in the lower edge of the arc.

2. INDEX ARM. The index arm moves on a pivot mounted at the geometric center of the arc, and the index mirror is attached to the upper part of the arm (at the pivot). The index mirror is silvered plate glass and moves with the index arm. Its plane is perpendicular to the plane of the index arm.

Near the lower end of the index arm is the index mark, where you read the plane of the arc. The endless tangent screw (fig. 15-3), attached to the lower part of the index arm, engages the gear teeth on the arc.

If you turn the tangent screw through one revolution, you advance the index mark one degree. A micrometer drum and vernier are mounted on the shaft of the tangent screw to enable an observer to read an angle accurately to a small part of a degree. Up to and including 90 degrees, the maximum permissible error for declination and inclination readings is plus or minus 30 seconds of arc. Above 90 degrees, the

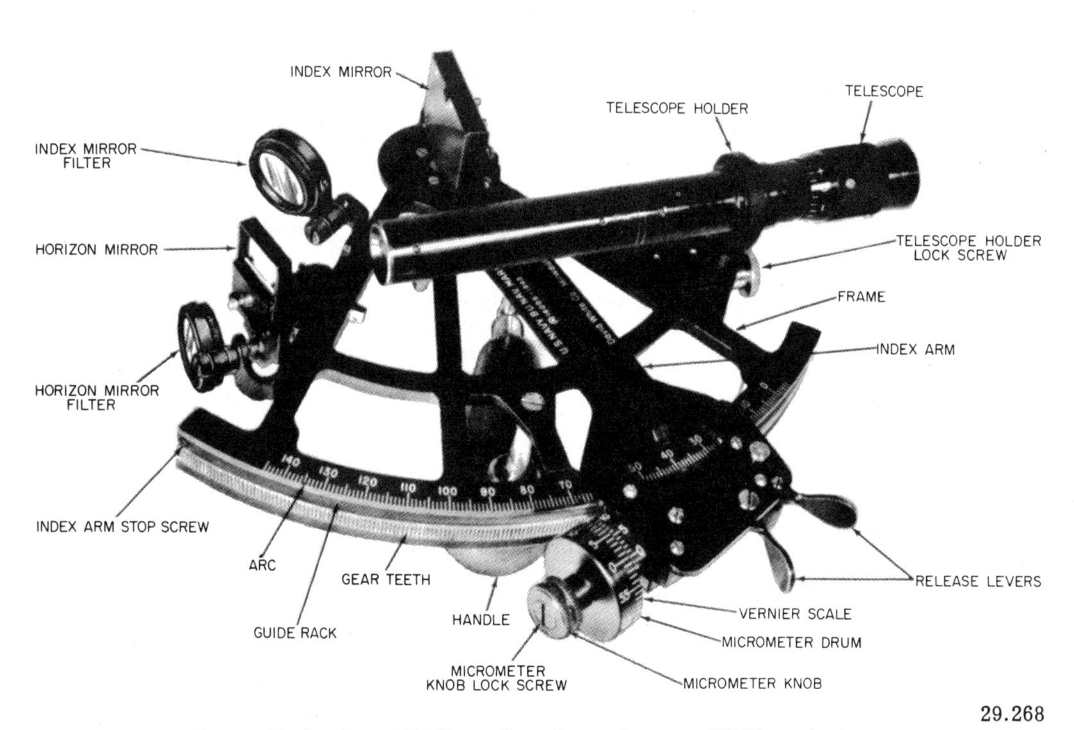

29.268
Figure 15-2.—David White endless tangent screw (ETS) sextant.

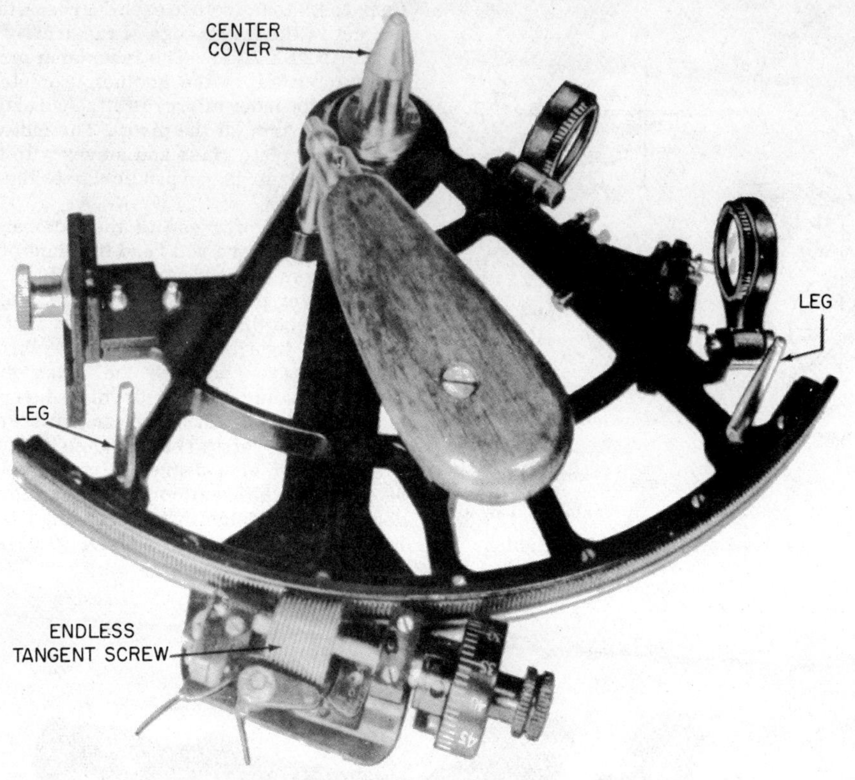

137.339

Figure 15-3.—Bottom view of a David White sextant.

maximum permissible error is plus or minus 35 seconds of arc.

NOTE: DECLINATION readings are taken by going one degree ABOVE the required reading, and then by rotating the micrometer drum to set the index mark at the required reading. INCLINATION readings are taken by starting one degree BELOW the required reading, and then by setting the index mark to the required reading with the micrometer drum.

When you press the two release levers together (fig. 15-2), you disengage the tangent screw from the gear teeth of the arc and obtain freedom of movement of the index arm.

3. HORIZON GLASS. The horizon glass (fig. 15-2) is attached to the frame and does NOT move. The half of the glass closest to the frame is silvered so that it will reflect images of celestial bodies to the eye of an observer looking through the sextant telescope. The image is reflected from the index mirror to the horizon mirror.

The outer half of the horizon glass is clear to enable an observer to see the horizon through it. The horizon glass is perpendicular to the plane of the arc; and when the index mark is at 0° on the scale, the horizon glass is parallel to the index mirror.

4. TELESCOPE. A sextant telescope enables an observer to see objects (images) more clearly, and it helps him to direct his line of sight to the horizon glass. The telescope has a magnification of 3, and its resolving power must be 18 seconds of arc in the center of the field.

5. POLAROID FILTERS. There are two sets of polaroid filters. When an observer looks through the clear part of the horizon glass, he

should use the filters with circular frames to reduce glare from the horizon. He should use the shades with square frames to eliminate or reduce glare produced by the reflected image.

PRINCIPLE OF OPERATION

A sextant consists basically of two optical systems, one rotatable and one fixed, and a sextant must have both to perform satisfactorily. If you understand how these optical systems function, you will have little difficulty in understanding the principle of operation of a sextant.

FIXED OPTICAL SYSTEM

The components of the fixed optical system of a David White or Pioneer sextant are: (1) a horizon mirror, and (2) a telescope. Study part X of illustration 15-4. You already learned in this chapter the function of these components.

ROTATABLE OPTICAL SYSTEM

The rotatable optical system of a sextant is composed of the index arm and the index mirror, which is mounted on the index arm. See part X of figure 15-4. The index arm rotates around a center point (top), and it indicates on the sextant arc scale (by means of an index mark) the angle in degrees a celestial body makes with the visible horizon.

OPERATION OF THE OPTICAL SYSTEMS

Study part Y of illustration 15-4, which is a schematic diagram of a sextant. Compare this part with part X. The letter C represents a celestial body whose angular altitude you must know. Your eye is at point 0, which would be next to the eyepiece of the telescope shown in part X. Line OD is your direct line of sight to the horizon. This means that angle COD is the one you must determine, because it represents the angle of the celestial body above the horizon.

The horizon glass is represented by H, and I is the index mirror, attached to index arm IV. When you swing the index arm along arc AB, you change the angle of the index mirror. When you move the arm to the point where the reflected image of the celestial body appears to lie on the horizon, rays from the body travel from C to I, from I to H, and from H to O. As you can see, these rays enter your eye along the same line of sight as rays from the horizon. Your next step, therefore, is to read the angle on the graduated scale at point V.

Because one degree on the arc is marked as two degrees on the scale, when angle VIZ is 15° the pointer at V shows exactly 30° on the scale. If the sextant is to give a true reading, angle COD must therefore be twice angle VIZ. How can you prove that it is of this size?

Line FE in the diagram is the normal to the index mirror and HE is the normal to the horizon glass. Angle CIF is therefore the angle of incidence on the index mirror, and FIH is the angle of reflection. Because both of these angles are equal, as you learned in chapter 3, we can designate both as a; and since the angles of incidence and reflection at the horizon glass are also equal, we can call both of them angle b.

Line IZ goes from the geometrical center of the arc to the zero mark on the scale, and the horizon glass is always parallel to this line. Since line HE is perpendicular to the horizon glass, it is also perpendicular to line IZ. Line IV lies along the reflecting surface of the index mirror, so it is also perpendicular to the normal (FE).

A theorem in plane geometry states that: "If the two arms of an angle are respectively perpendicular to the two arms of another angle, the two angles are equal." Angles VIZ and IEH are therefore equal. A principle of operation of a sextant also states that: "The angle between the first and last directions of a ray of light that has suffered two reflections in the same plane is equal to twice the angle that the two reflecting surfaces make with each other." The reflecting surfaces in this case are the index mirror and the silvered section of the horizon mirror.

It follows in reverse order, then, that if the celestial body you have under observation is 60° above the visible horizon, the angle which the index mirror and the silvered section of the horizon mirror must make with each other to bring the celestial body tangent to the visible horizon is 30 degrees. This is just half the angular height of the celestial body.

READING A SEXTANT

The scale on the arc of a sextant is graduated in degrees. From this scale, therefore, you can read with accuracy ONLY TO the nearest degree. Look at the index mark and read the number of degrees on the arc. Then use the micrometer to get a more accurate reading.

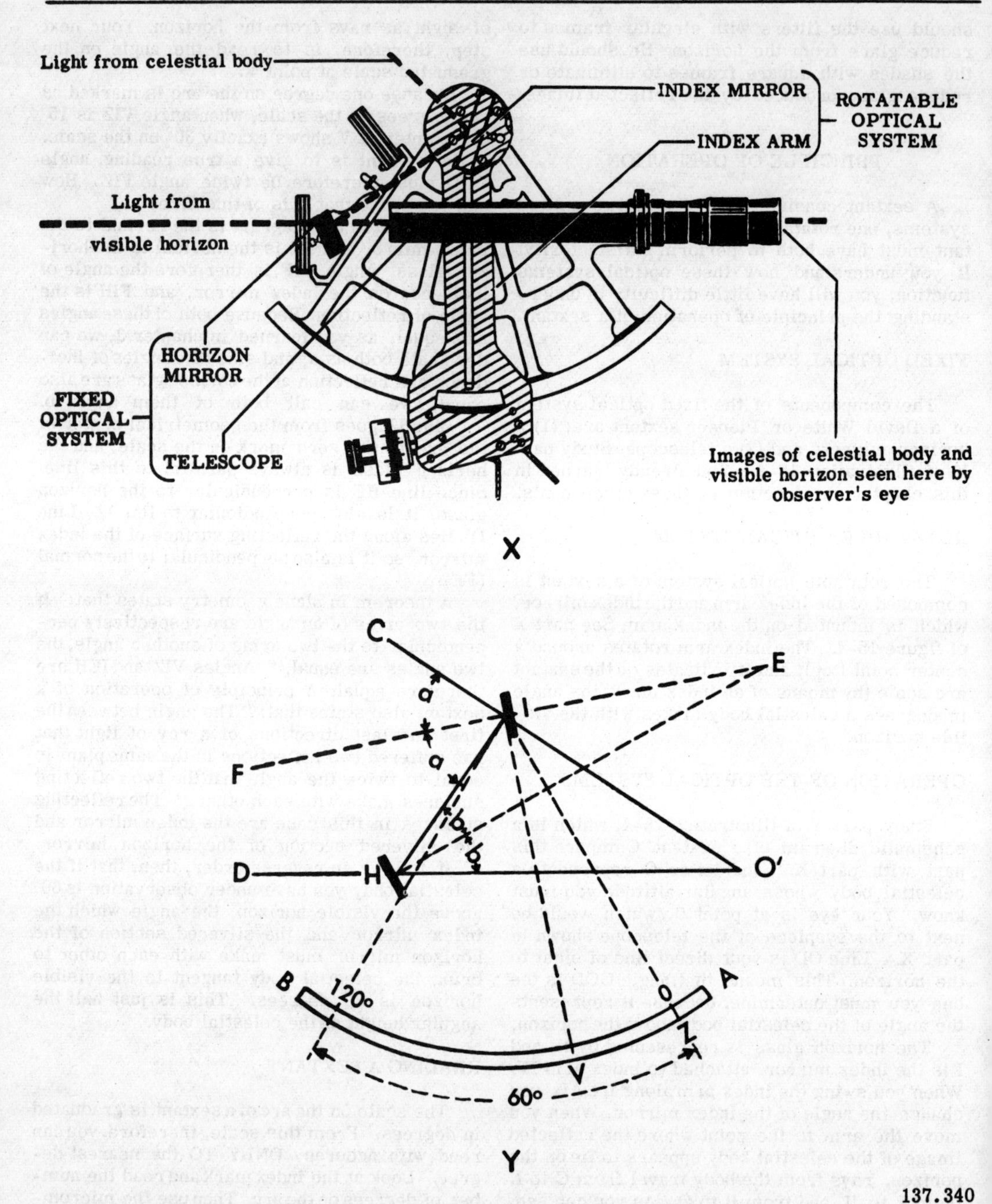

137.340

Figure 15-4.—Schematic drawing of the principle of operation of a sextant.

The micrometer drum has a scale with sixty divisions, and each division represents one minute of arc. To increase further the accuracy of the reading, use the vernier scale located alongside the micrometer drum. This scale has ten divisions and enables you to determine the angle being measured to one-tenth of a minute, or to the nearest six seconds or arc.

To read a sextant, therefore, you read degrees on the arc at the index mark. Then you add the number of minutes read on the micrometer drum, and also the number of tenths of a minute you read on the vernier scale.

Study illustration 15-5, which gives two sample sextant readings. In part A of this illustration, the reading on the arc is 13 plus (at the index mark), the 0 mark on the vernier scale is between 16 and 17, and the first mark on the vernier which coincides with a mark on the drum is 7 on the vernier scale; so the reading is 13°16.7'. The reading in part B of figure 15-5 is 55°25.2'.

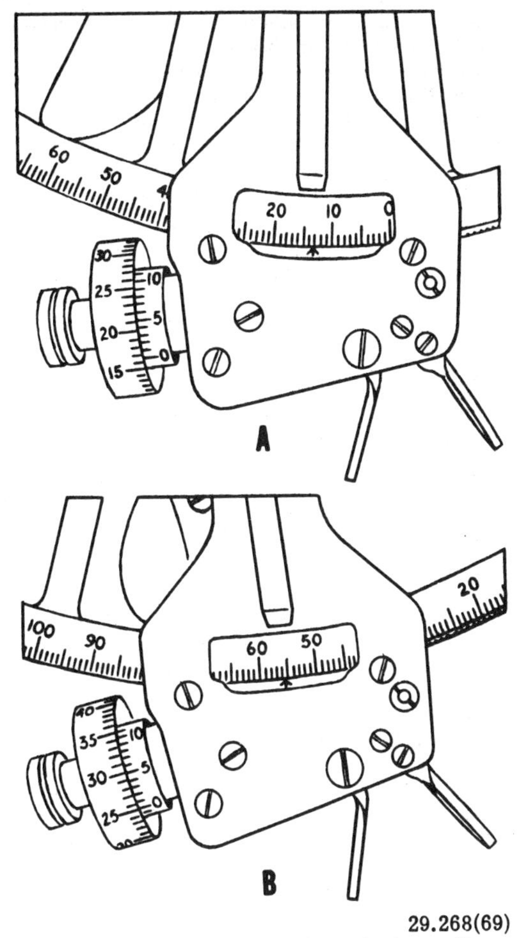

29.268(69)

Figure 15-5.—Examples of sextant readings.

OVERHAUL AND REPAIR

Overhaul and repair procedures discussed in this section are for Mk 2 endless tangent screw David White sextants. Space does not permit a detailed discussion of Pioneer sextants; but to the degree possible, differences in procedure for the Pioneer instrument are explained.

Go back at this time and review the repair procedures given in chapter 8.

PREDISASSEMBLY INSPECTION

A preliminary decision must always be made concerning the feasibility of repair of an instrument. This is the purpose of a predisassembly inspection, to determine whether the instrument should be repaired or surveyed and salvaged; and if repair is the decision, the extent of disassembly required.

Some of the things to check when giving a sextant a predisassembly inspection include:

1. Condition of silver on mirrors.
2. Corrosion, and failure of protective finishes.
3. Evidence of unauthorized tampering and disassembly.
4. Appearance, finish, and condition of parts in the sextant assembly. Examine scale markings for legibility.
5. Cleanliness and physical condition of the telescope assembly. If mounted, remove the telescope from its sliding bracket before you make this test.

NOTE: Be certain the diopter scale reference mark is at the top when you mount the sextant telescope in the sliding bracket.

6. Action of the diopter focusing ring. It should be smooth over the entire diopter scale range, but it should be fairly tight.
7. Polaroid filter assemblies. There should be no cracks or chips, cloudiness, or dark spots

caused by dirt or moisture between the individual glasses of each filter.

NOTE: Polaroid filters must have a protective coating on their edges.

8. Rack teeth. Check with an eye loupe for wear, bends, and chipping. Clean the rack teeth, the endless tangent worm and worm gear thread, and the guide slot with a suction line or a nylon brush.

DISASSEMBLY PROCEDURE

To prevent confusion of terms, removal of major assemblies from the sextant is considered first. Disassembly of the subassemblies then follows. Before proceeding, study the special service tools for sextants (fig. 15-6).

Removal of Mirror Filter Assemblies

Remove the mirror filter assemblies first to eliminate unnecessary chances of damaging them. The parts of these assemblies are shown in figure 15-7. Remove the screws which secure these assemblies, along with the washers and spring, and then carefully remove the assemblies and protect them in a parts tray.

Horizon Mirror Mounting, Bracket, and Filter Bracket Assembly

With a capstan head screw pin wrench (fig. 15-6), remove the horizon mirror and filter bracket capstan head screw. See illustration 15-8. Then remove the horizon mirror and filter bracket screw and the assembly.

Removal of Index Mirror Mounting

Remove the index mirror mounting screws which secure the mounting to the index arm and worm frame assembly (fig. 15-8).

CAUTION: Exercise extreme care to prevent damage to the index arm. Rough treatment will damage it beyond repair.

Removal of Telescope Bracket Assembly

To remove the telescope bracket assembly, grasp the sextant by its handle and hold it close to the bench. Then, with an adjusting-screw wrench (fig. 15-6), remove the telescope bracket assembly, which secures the telescope to the frame (fig. 15-2).

Index Arm and Worm Frame Assembly

The index arm and worm frame assembly (fig. 15-9) are attached to the male center, which is critically important in the instrument. Remove first the center lock screw and washer before removing the arm.

Remove the index arm stop screw from the 0-degree end of the arc. Then press the release levers together and swing the index arm free of the arc, as shown in illustration 15-10.

The last step completes the removal of major assemblies from the sextant, with the exception of male and female centers, which should NEVER be removed.

Disassembly of the Worm Frame Assembly

Before you can disassemble the worm frame assembly, you must first remove it from the index arm:

1. Remove the retainer block screw from the back of the index arm to free the worm frame retainer block (fig. 15-11).

2. With an end thrust nut wrench (fig. 15-6), remove the disengaging lever end thrust nut and lift the disengaging lever and shoe off the pivot screw bushing. Then remove the disengaging lever pivot screw (fig. 15-11) from the back of the index arm.

NOTE: If the disengaging lever shoe remains in its recess in the worm frame during this operation, remove the shoe from its recess and place it with the disengaging lever.

3. Remove the end-thrust spring screws to release the spring from the worm frame (fig. 15-11).

4. With a forked screwdriver (fig. 15-6), remove the end-thrust screw from the end of the shaft of the endless tangent worm and shaft worm gear. See figure 15-11.

5. Remove the spring support block screw from the back of the index arm (fig. 15-11).

6. With a pivot screw lock nut screwdriver (fig. 15-6), remove the worm frame pivot screw lock nut (fig. 15-11); then remove the pivot screw and tension washer.

Refer to illustrations 15-11 and 15-12 as you study the procedure for disassembling the worm frame assembly.

1. Remove the worm frame holding spring stop screw and the spring from inside the worm frame.

2. Remove the drum lock screw and the washer and slide the micrometer drum off its adapter.

3. Remove the adapter lock pin by pressing it out. If necessary, drive it out with a pin and mallet. Then slide the micrometer drum adapter off the shaft of the endless tangent worm and shaft worm gear. Be careful NOT to bend the worm gear shaft.
4. Remove the vernier yoke clamp screw and slip the yoke off the front worm bearing. To remove the front bearing cap, remove the screws which secure it.

The front bearing cap is lapped to the worm frame. If these parts were not marked by the manufacturer to ensure their being kept together, scribe a mark on one face of the front bearing cap and on the worm frame. See illustration 15-13.

If you must remove the front bearing cap locating pin (fig. 15-12), support the bearing cap (protected by pads) between the jaws of a vise and press on the pin, or tap it lightly with a flat-end pin and mallet (fig. 15-6). Then remove the rear worm bearing clamping screw and push the bearing out of the frame.

Slide the endless tangent worm and shaft worm gear out of the worm frame (fig. 15-12) and place the frame with the front bearing cap. Be very careful NOT to damage bearing surfaces. Slide the front worm bearing off the worm gear shaft and wrap it in tissue paper.

To remove the vernier yoke locating pin (fig. 15-12) in the front worm bearing, grasp the bearing firmly between the index finger and the thumb of one hand and remove the pin with a pair of flat-nosed pliers.

Disassembly of the Index Arm

The index arm guide lugs, stationary lever, and index arm stiffening rib (fig. 15-14) are generally left assembled to the index arm. If these parts must be removed, however, do the following:
1. Remove the stationary lever screws from the back of the index arm to free the lever.
2. Remove the index arm guide lug (fig. 15-14). Then turn the index arm guide lug over and scribe the letter R beside the hole in which the lug fits. Scribe the letter L on the left hand index arm guide lug and screw.
3. Remove the stiffening rib from the back of the index arm and put a reassembly guide mark on the top of the rib, the part near the male center (removed in this illustration).

Horizon Mirror Mounting, Bracket, and Filter Bracket Assembly

Proceed as follows to disassembly the parts named in this heading:
1. Remove the horizon mirror mounting coupling screw (fig. 15-15) and the top adjusting screw directly below the mounting coupling screw hole. The mirror mounting will then be free. Then remove the bottom mounting adjusting screw.
2. Remove the mirror bracket coupling screw and the inner bracket adjusting screw. Then remove the outer mirror bracket adjusting screw.

Disassembly of the Polaroid Filter Assembly

To disassemble either the index or the horizon mirror polaroid filter assembly, proceed as follows:
1. Remove the filter yoke screw (fig. 15-16) and slide the yoke off the assembly.
2. Put the filter assembly on a bench with the filter holder facing up and rotate the holder counterclockwise until it is stopped by the pin in the holder. Then press down on the filter with your index finger with a sliding counterclockwise motion and push the rotating filter holder out of the fixed filter holder.
3. Disassemble the filter spacer, the filters, and the filter spring from the holders.

Disassembly of the Telescope Bracket Assembly

Use the procedure outlined next to disassemble the telescope bracket assembly:
1. Remove the clamp nut screw from the telescope sliding bracket clamp nut. See illustration 15-17.
2. Unscrew the bracket clamp nut by hand and remove the parts secured to it. Then remove the bracket retainer screw from the bracket support. If necessary, remove the fulcrum pin with a pin punch and a light hammer.

Disassembly of the Sextant Frame Assembly

At this stage of disassembly, you have removed everything from the sextant frame assembly except the: (1) rack, (2) handle assembly, (3) arc support studs, (4) index filter stop stud, and (5) the index arm center assembly. The rack

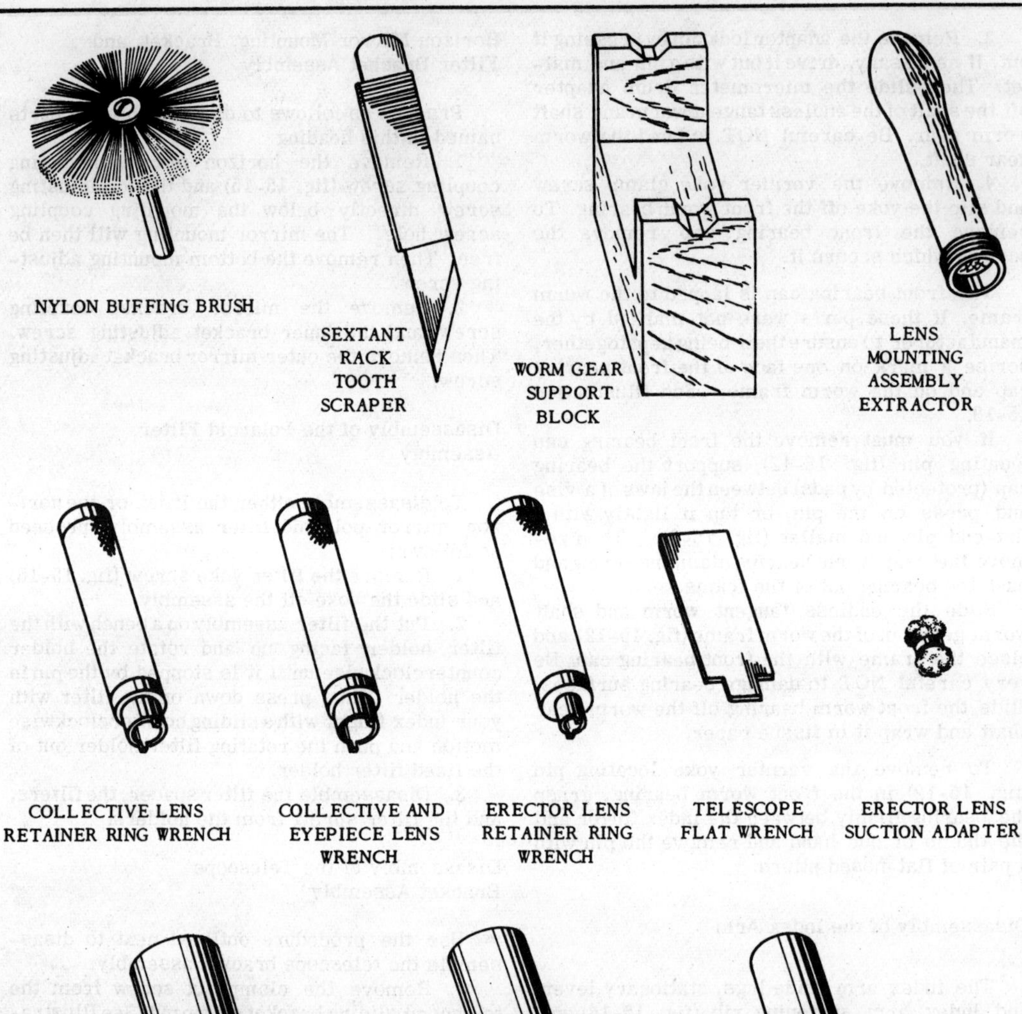

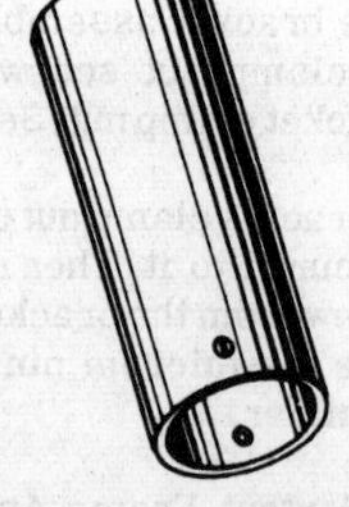

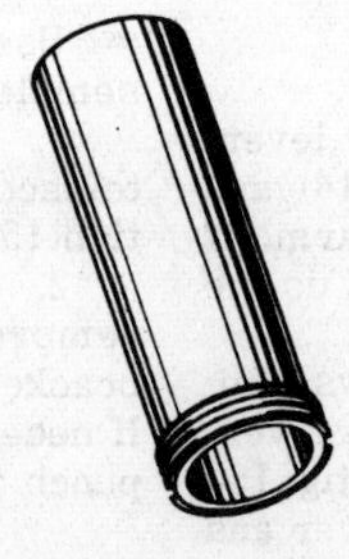

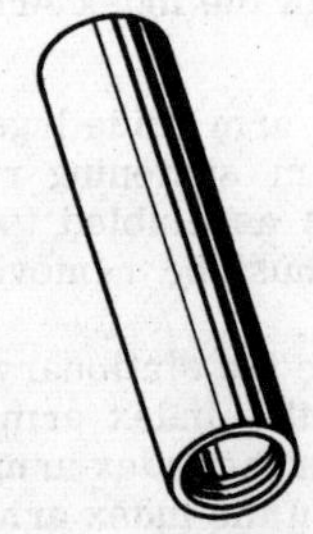

Figure 15-6.—Special service tools. 137.341

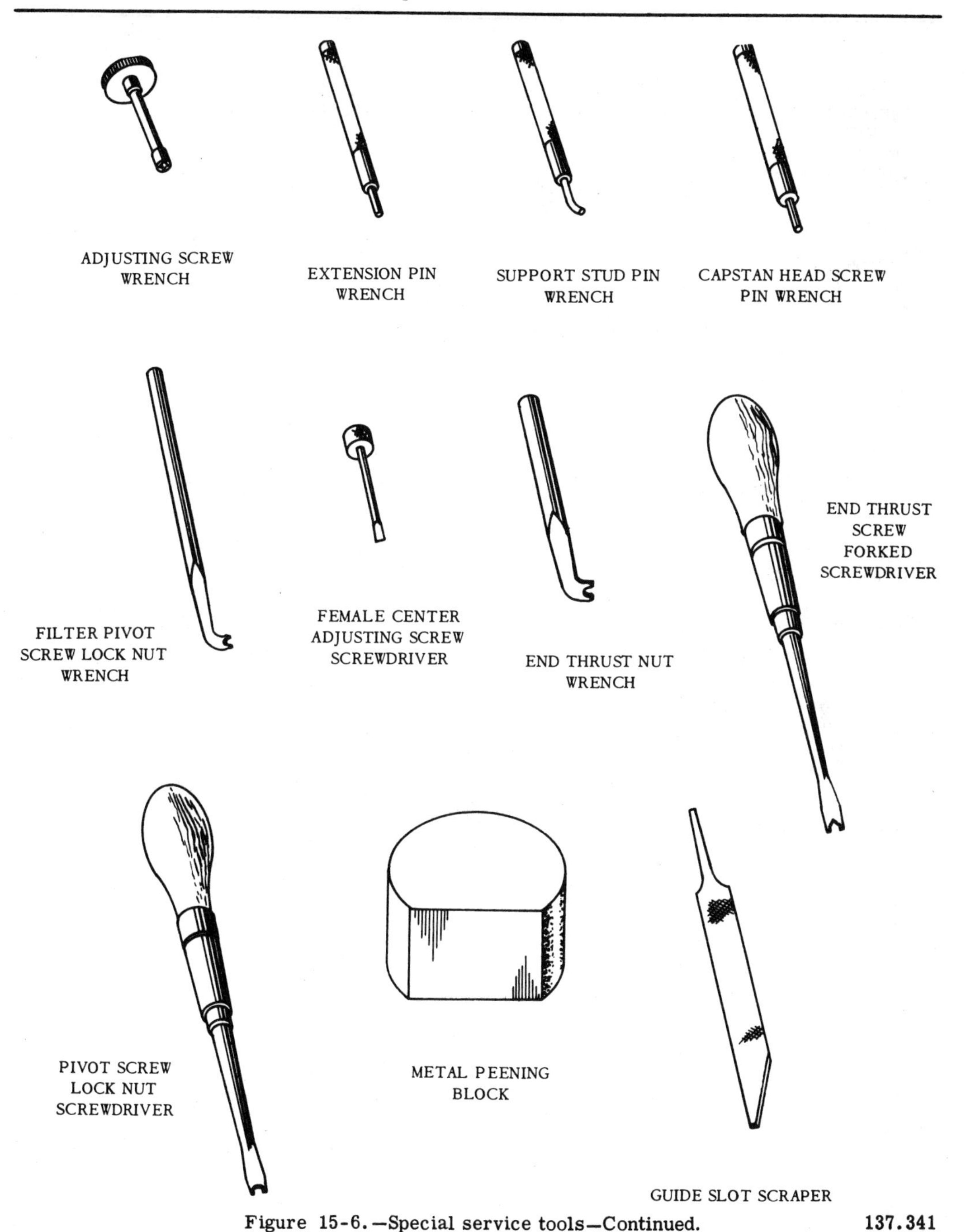

Figure 15-6.—Special service tools—Continued. **137.341**

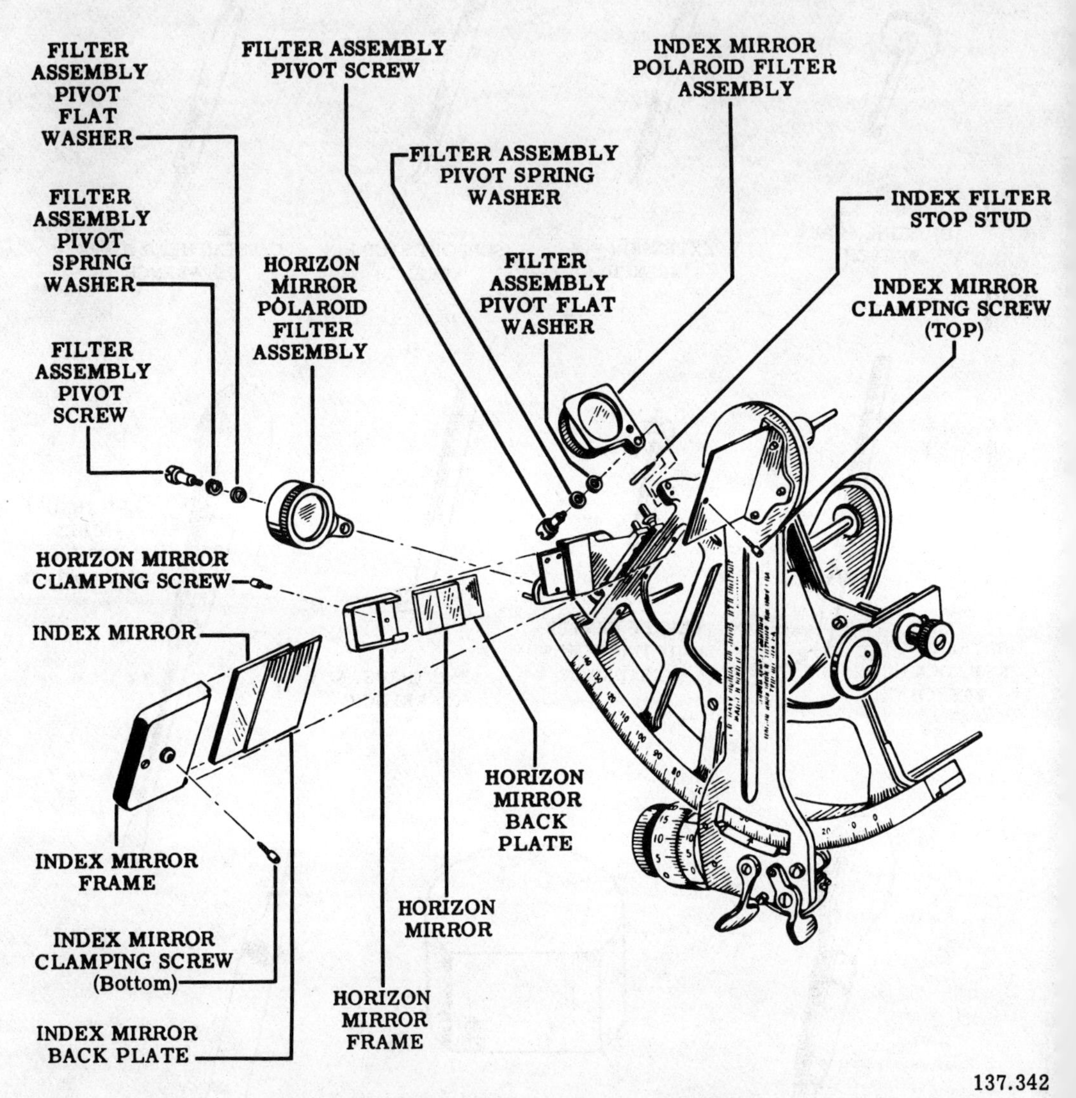

Figure 15-7.—Sextant mirror filter assemblies.

and frame are considered as one unit, and the rack must NEVER be removed from the sextant frame.

Remove the grip assembly screws (fig. 15-18), the hand grip cap-stud end support, the hand grip retainer screw (to remove hand grip support), and the cap-stud screw from the hand grip. Then disassemble all parts of the grip handle, as shown.

Telescope Eyepiece and Field Lens Assembly

To disassemble the eyepiece and field lens, do the following:

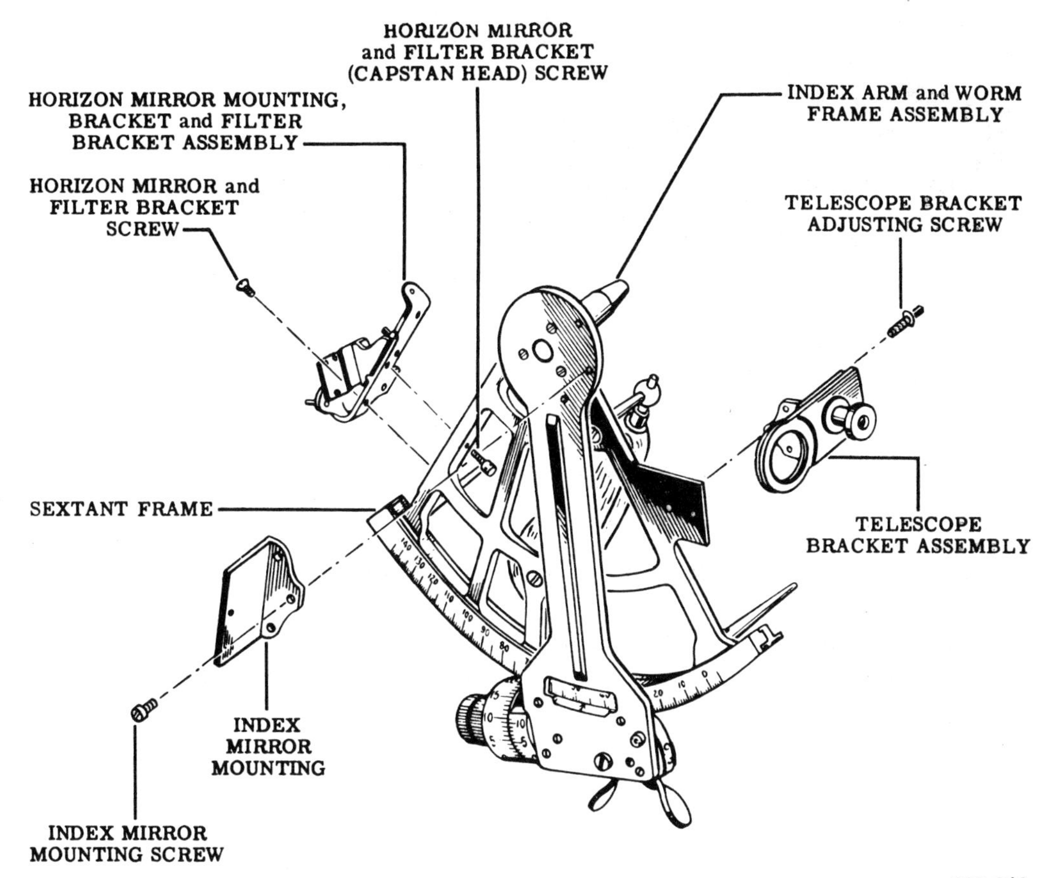

Figure 15-8.—Horizon mirror mounting, bracket, and filter bracket assembly.

1. Remove the key screw and slide the eyepiece and field lens assembly off the end of the telescope body. See illustration 15-19. When you have the assembly moved far enough off the body to expose the eyepiece movement shoe (illustrated) pick out the shoe from its slot in the body. Scribe reassembly guide marks on the eyepiece cap, diopter focusing ring, and the eyepiece lens mount assembly.

2. Remove the lock screw in the diopter focusing ring (top part, fig. 15-19) and unscrew the eyepiece cap and the eyepiece lens mount assembly. If necessary, use the eyepiece cap wrench and the eyepiece lens mount assembly wrench.

3. To disassemble the eyepiece lens mount assembly, remove the retainer ring lock screw in the side of the eyepiece lens mount.

4. With a flat telescope wrench, unscrew and remove retaining ring for lenses F and G.

5. Tip the eyepiece lens mount for lenses F and G over a pad of lens tissue on the workbench and carefully slide the lenses and spacers out of the mount and onto the cloth. With a BLACK lead pencil, mark the first lens which comes out with an F, on its edge; and mark the last lens out at the same place with a G.

29.268
Figure 15-9.—Removing the index arm.

Disassembly of the Telescope Body

Before you disassemble the telescope body, scribe a reassembly line across the two halves and the erector lens mount (fig. 15-19).

The telescope body is composed of two halves threaded onto the erector lens mount assembly and locked in place with two lock screws (fig. 15-19). Remove the lock screws and unscrew the two halves of the body from the erector lens mount assembly. If necessary, use a telescope body and objective lens mount wrench.

NOTE: The telescope securing ring is force-fitted to the eyepiece end of the telescope body and must not be disassembled from the body. Objective lenses A and B are in the objective end of the body and comprise the body tube and objective lens assembly. Set this aside for further disassembly.

Disassembly of the Erector Lens Mount Assembly

Scribe a reassembly guide mark on the erector lens retaining housing and on the mount before you disassemble an erector lens mount assembly. Then continue as follows:

1. Unscrew the retaining housing for lens C from the erector lens mount and mark the housing with a C.
2. Put lens tissue on the workbench and gently spill erector lenses D and E (fig. 15-19) from the lens mount onto the tissue. With a black lead pencil, put the letter D on the first lens out; mark the second lens with an E.
3. Remove the lock screw in the retaining housing assembly for lens C; then remove the retaining ring and mark it with a C.
4. Tip the retaining housing for lens C and slide the lens onto lens tissue and put the letter C on the lens.

Body Tube and Objective Lens Assembly

Scribe reassembly guide marks on all parts of the body tube and the objective lens assembly; mark also the objective end of the telescope body. Then do the following:

1. Remove the lock screw on the objective end of the telescope body and unscrew the objective lens mount. The retaining housing assembly for lens B also comes out at this time.
2. Grasp the objective lens mount by its knurled edge and unscrew the retaining housing assembly for lens B from the mount. If it is stuck or frozen, use a retainer housing wrench.
3. Remove objective lens A from its mount and put the letter A on its edge.
4. Follow the same procedure as for lens C to disassemble the retainer housing assembly for lens B.

This completes the amount of disassembly you will perform on sextants in optical shops. The next step in the repair procedure is inspection of disassembled parts.

INSPECTION OF DISASSEMBLED PARTS

During the predisassembly inspection of the sextant you made recommendations concerning some parts on the casualty analysis sheet for the instrument. After disassembly, inspect all disassembled parts and make a decision concerning their usability. If they are still good, clean them in the approved manner and protect them until

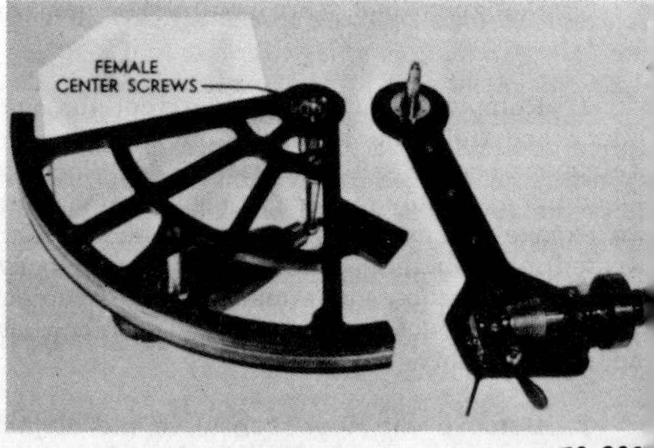

29.268
Figure 15-10.—Index arm removed from the frame.

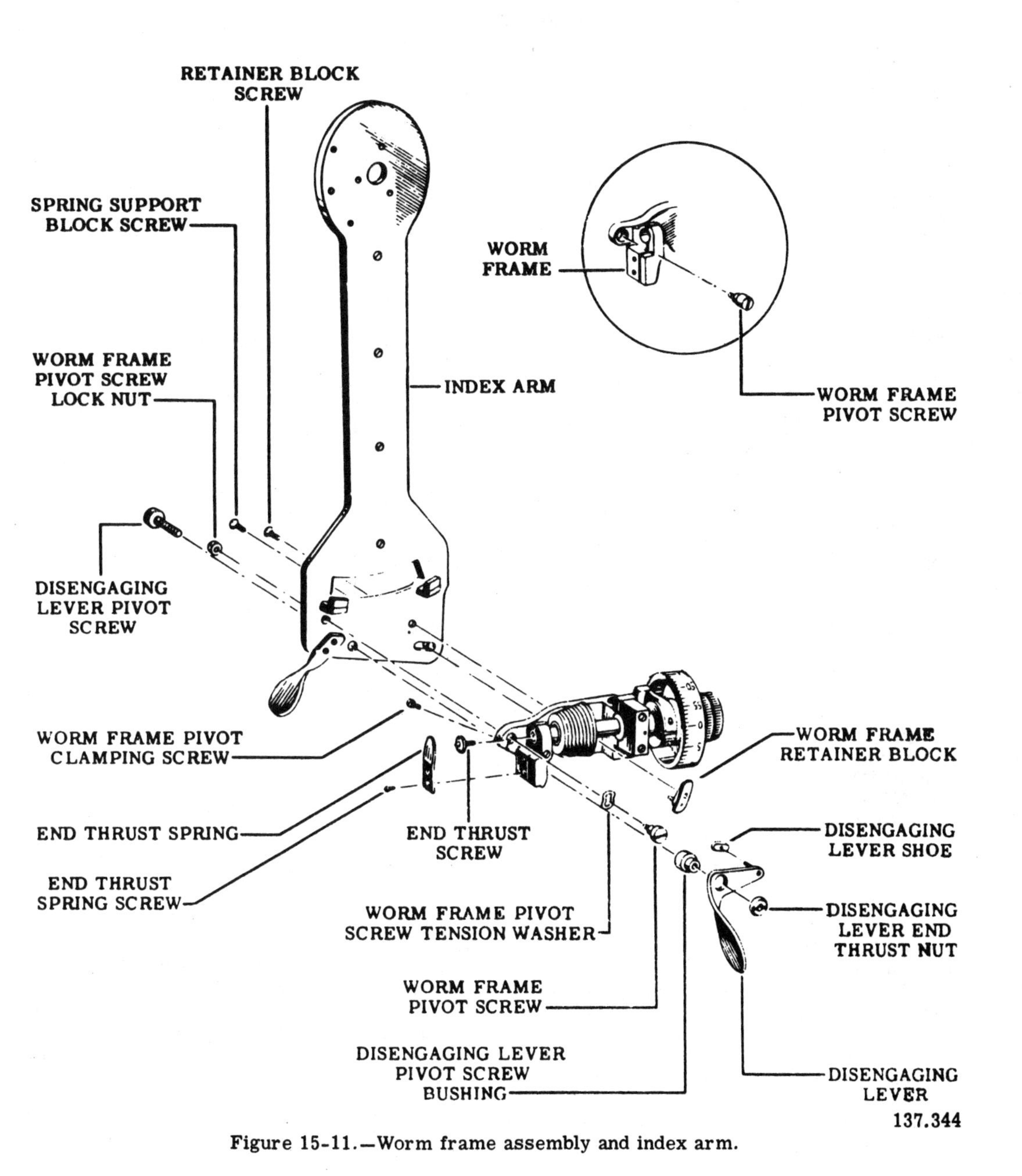

Figure 15-11.—Worm frame assembly and index arm.

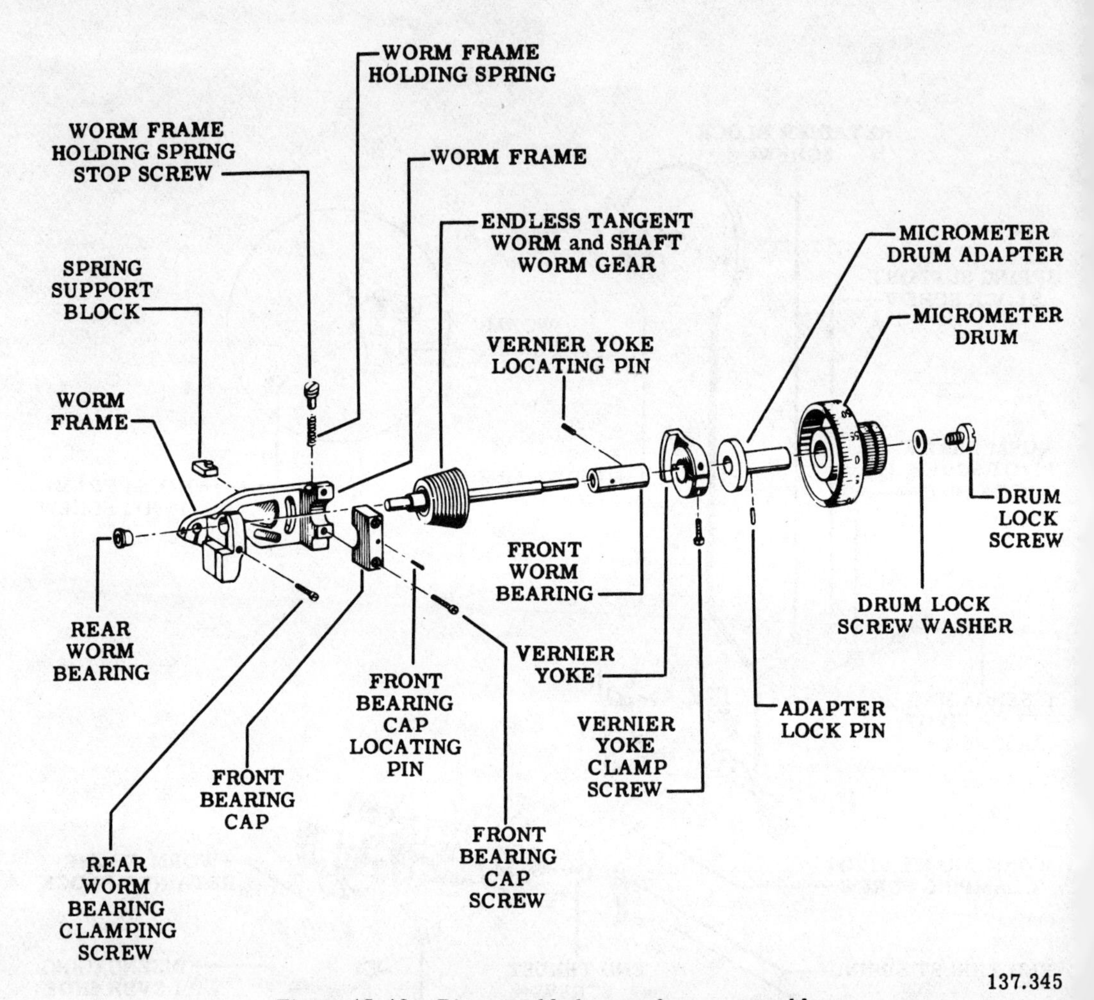

Figure 15-12.—Disassembled worm frame assembly.

needed. Put parts which can be repaired in a special tray; discard parts which have no further usefulness.

At this time, review the discussion of maintenance in chapter 8.

A sextant's accuracy is dependent upon accurate engagement of the endless tangent worm and shaft worm gear in the sextant rack teeth. Nicks or burrs on the rack teeth, or high spots on the worm gear thread, will cause large errors in readings.

Components of the worm frame assembly are listed in the parts list for the David White sextant in the repair manual for the instrument (NavShips 250-624-10). All parts are replaceable individually except the worm frame and the front bearing cap. If either of these parts must be replaced, get the worm frame and front bearing cap assembly.

The index arm and its stiffening rib are fitted together to form the index arm assembly and are

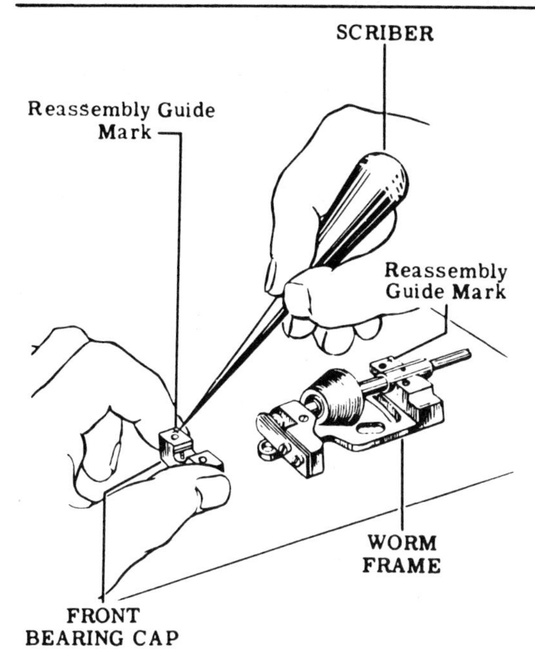

137.346
Figure 15-13.—Scribing reassembly guide marks on the worm frame and the front bearing cap.

not available separately. If either part is damaged, procure a complete index arm assembly.

Inspect the condition of cement between lenses, and the condition of the silver on the horizon and index mirrors. Inspect each polaroid filter for the condition of the cement between the glass elements, and also the condition of the protective coating around the edge of the filter.

REPAIR PROCEDURE

The following discussion of repair of a David White sextant pertains to some of the repairs you will be required to perform on the instrument. Repair involves everything necessary to put the sextant in excellent working condition, in accordance with prescribed standards.

Replacing the Diopter Focusing Ring

Put the eyepiece lens mount in the diopter focusing ring, with the flange on the eyepiece lens mount against the end of the diopter focusing ring. Then place the eyepiece cap in the diopter focusing ring.

With a drill .0595 inch in diameter, drill a hole approximately 1/16-inch deep in the diopter focusing ring and into the eyepiece lens mount and eyepiece cap. This hole should be 9/64 inch from the eyepiece cap end of the ring, at the point where the two parts butt together in the diopter focusing ring. Be careful, lest you drill completely through the mount and cap and damage the surface of the eyepiece lens.

To provide room for the head of the lock screw, counterbore the hole in the diopter focusing ring a depth of .040 inch and with a diameter of .140 inch. Tap the drilled hole for a No. 1-72NF thread and assemble the lock screw. Then put an assembly mark on the eyepiece cap and also on the mount.

Scraping the Rack Teeth

Examine the rack teeth carefully with an eye loupe to locate defective teeth. Then hold the

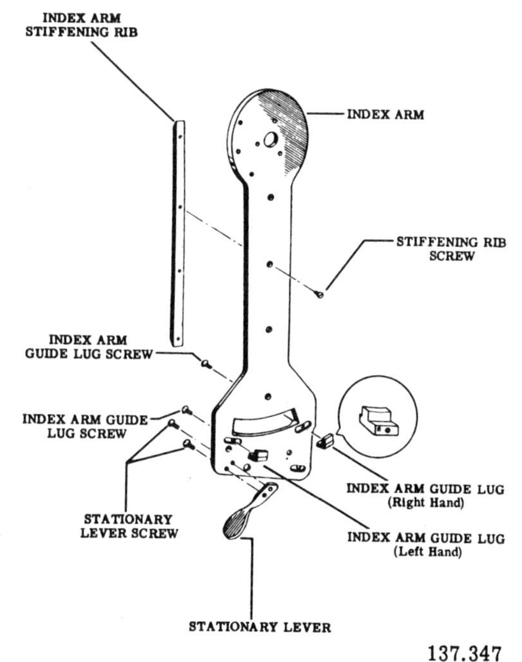

137.347
Figure 15-14.—Disassembled index arm.

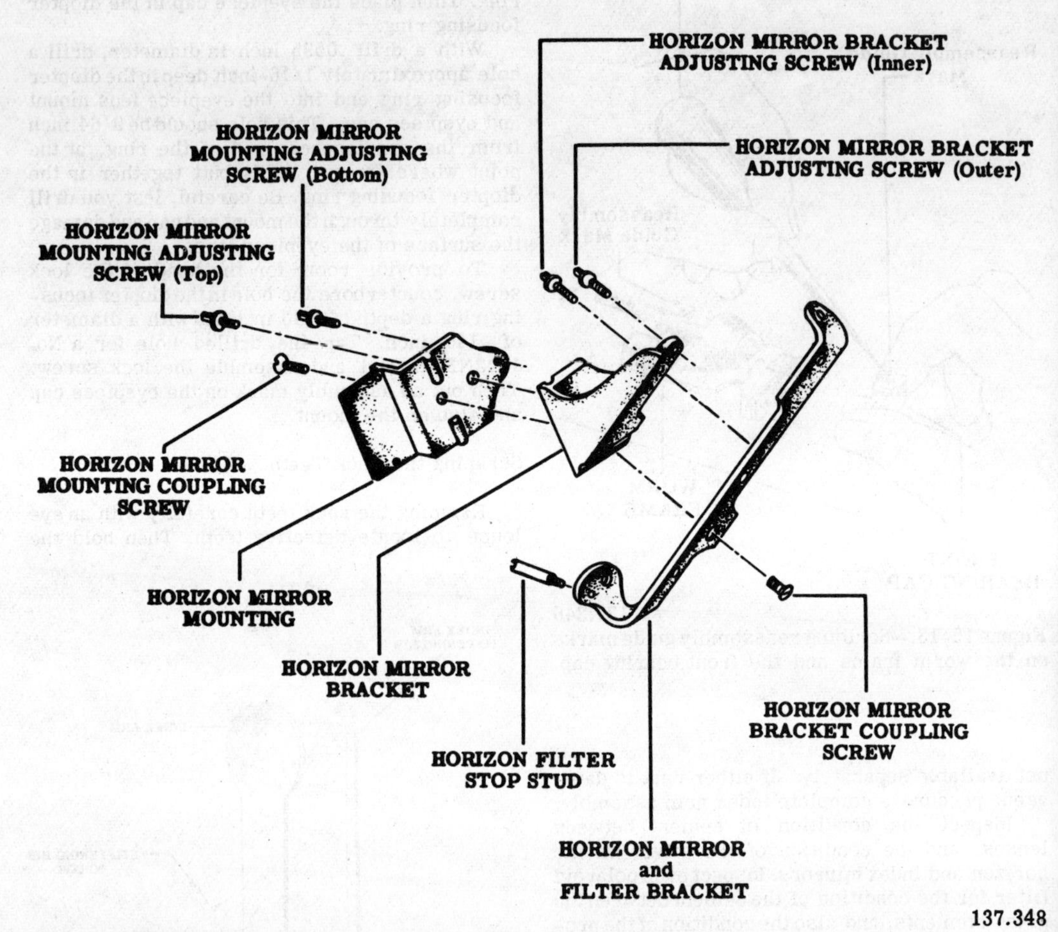

Figure 15-15.—Horizon mirror mounting, bracket, and filter bracket assembly.

sextant assembly in the manner shown in illustration 15-20 while you remove burrs or nicks on the teeth with a sextant rack tooth scraper. The purpose of this operation is to bring the teeth back to almost perfect condition without removing TOO MUCH metal.

Endless Tangent Worm and Shaft Worm Gear Thread

The endless tangent worm and shaft worm gear thread must fit the rack teeth perfectly in order to attain the degree of accuracy required of sextants. For this reason, you must remove high spots or burrs on the worm gear thread. Examine the worm gear thread with an eye loupe for burrs or other defects. High spots will show up as small, highly polished areas on the surface of the thread.

Place a worm gear support block in a bench vise and hold the endless tangent worm and shaft worm gear on the block as you remove with a triangular stone existing burrs or high spots on the worm gear thread.

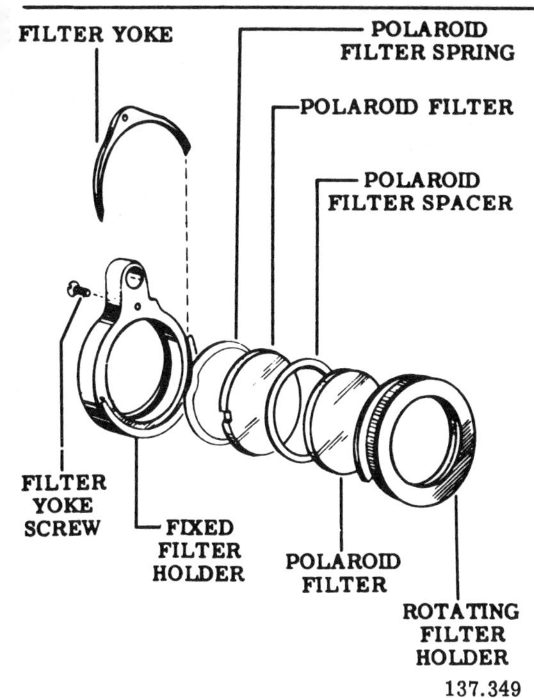

Figure 15-16.—Disassembled polaroid filter assembly.

CAUTION: If you damage the worm gear thread, you will be required to procure a new worm frame assembly, complete with pivot.

Lapping the Worm and Shaft to the Rack Teeth

Run the endless tangent worm and shaft worm gear along the rack teeth and FEEL for spots which indicate binding. If a high spot or a defect is on the worm gear thread, binding will occur at regular intervals along the rack. If defects exist on the rack teeth, binding will probably occur at erratic points where such defects exist along the rack. Mark the spots where binding is indicated and then repeat the process outlined under "Scraping the Rack Teeth." Do NOT allow metal dust to enter bearing surfaces of the worm frame.

Check the movement of the worm gear thread in the sextant rack teeth a second time for binding or lifting of the worm gear. If this still occurs, apply a fine film of sextant rack and worm gear lapping compound to the teeth with a small, soft-bristle brush. Then run the worm gear back and forth over the entire length of the rack. Do NOT press down on the worm gear. With an approved solvent and a nylon brush, clean the lapping compound from the teeth and thread and recheck the movement of the worm gear thread in the rack teeth.

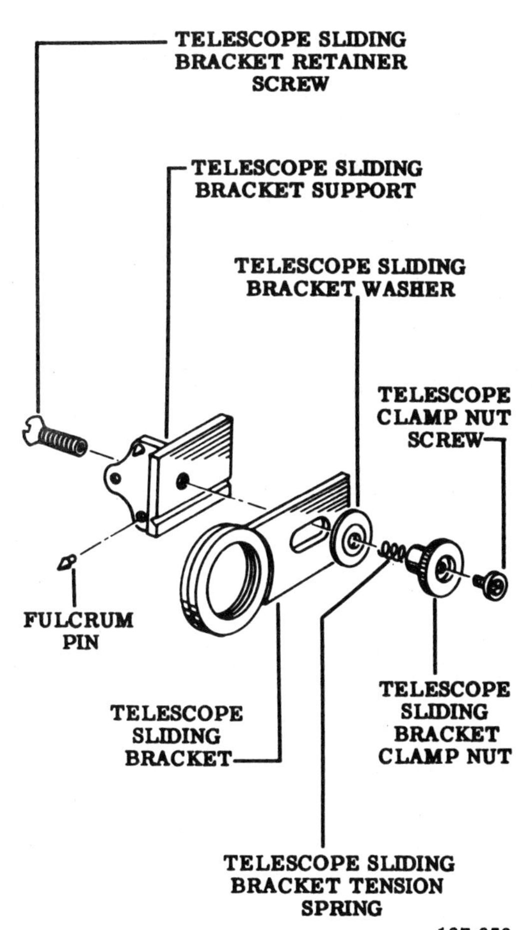

Figure 15-17.—Telescope bracket assembly.

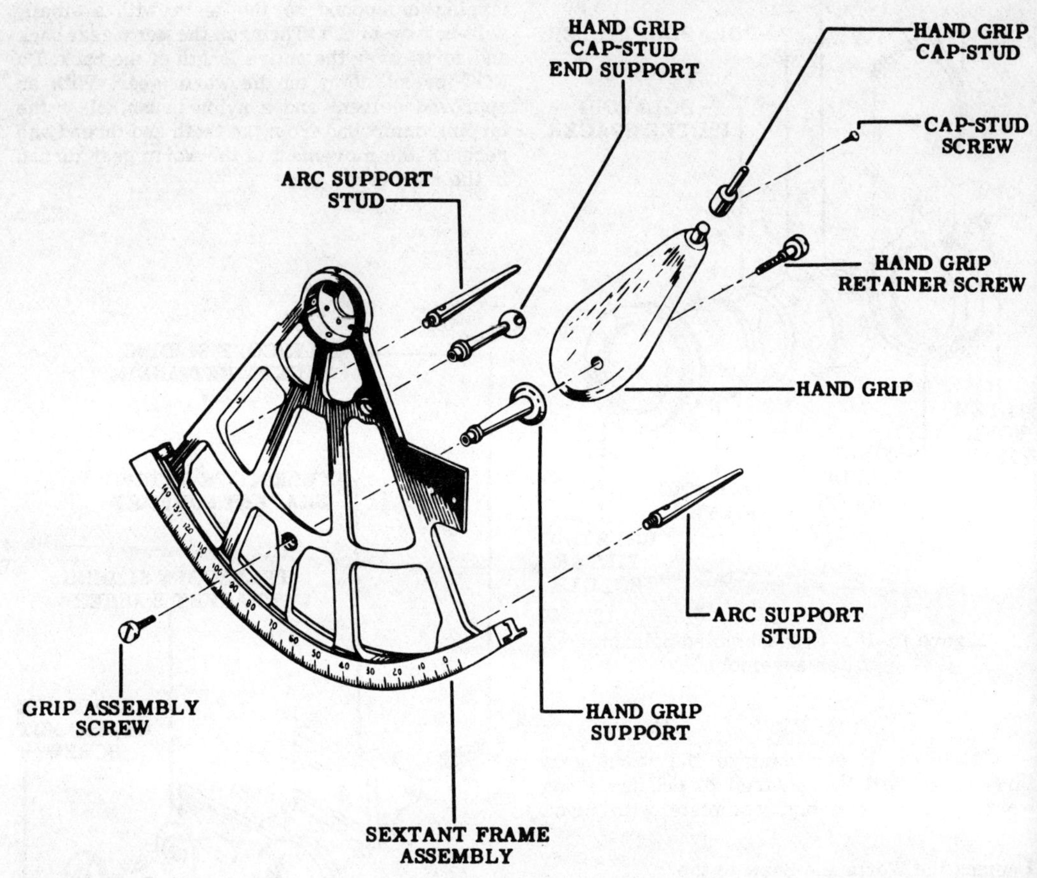

Figure 15-18.—Sextant frame assembly.

Lapping the Worm Frame Pivot Screw to the Frame

If you replace the worm frame pivot screw, you must lap the screw to the worm frame to provide smooth movement of the frame. Proceed as follows:

1. Apply a coat of worm frame pivot lapping compound to the surface of the worm frame pivot screw.

2. Lubricate the surface of the worm frame pivot lapping plate with an approved oil. Put the worm frame pivot screw tension washer on the pivot screw, and then secure the worm frame to the pivot lapping plate with the pivot screw. Put the pivot screw in the No. 2 hole in the plate. Then put the lock nut on the pivot screw, and put the worm frame pivot clamping screw in its frame.

3. Tighten the clamping screw sufficiently to have the frame move stiffly on the pivot screw and lap the two parts together. As you do this, tighten the clamping screw gradually until you obtain the best fit.

4. Clean all parts and replace the worm frame on the pivot lapping plate.

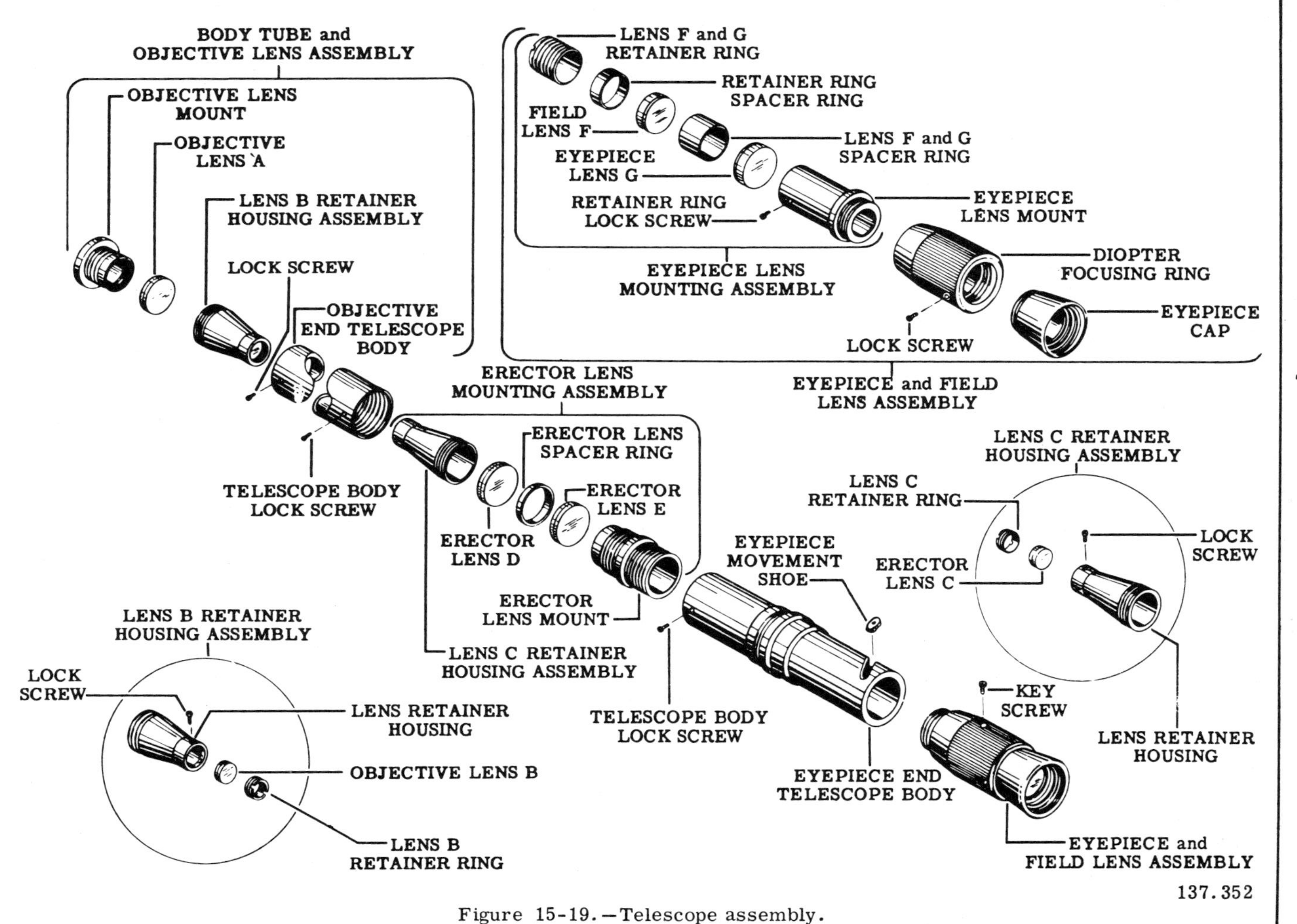

Figure 15-19.—Telescope assembly.

Replacing a Worm Bearing or a Worm Frame

Front and rear worm bearings must have a smooth-running fit on the shaft of the endless tangent worm and shaft worm gear. If either bearing (or both) is tight or has been replaced, do the following:

1. Put the front or rear worm bearing on the shaft and check for tightness.
2. Apply a thin coat of worm bearing lapping compound to the bearing surface of the shaft, replace the bearing, and lap it to the worm gear shaft.
3. Clean the worm bearing and the worm gear shaft.
4. Apply a thin strip of worm gear shaft oil to the worm gear shaft and replace the worm bearing. Then check the action of the bearing on the shaft. It should rotate smoothly, without play. Clean the parts.

NOTE: If fitted, the front bearing cap locating pin must be removed before you perform step No. 5.

5. Put the front worm bearing in the frame, flush with the inner face of the worm frame seat. Then assemble the front bearing cap and check its fit to the worm frame. It should seat squarely and hold the front worm bearing firmly. If the cap does not seat on the frame when light pressure is applied by the cap screws, remove enough metal from the outer surface of the front worm bearing to get a correct fit.

NOTE: Perform step No. 6 only when the ORIGINAL front bearing cap is used; perform step No. 7 only when a new cap is used.

6. Replace the front bearing and drill a hole 1/16 inch in diameter and 180° from the vernier yoke locating pin hole in the bearing for the front bearing cap locating pin. Use the existing hole in the bearing as a guide. The hole should be deep enough to accommodate the end of the locating pin, but it should not pierce the inner bearing surface of the front worm bearing.
7. Assemble the front worm bearing and drill a hole 1/16 inch in diameter and 3/32 inch from the inner face of the front bearing cap in the cap and also in the bearing for the cap locating pin. The hole must be deep enough to accommodate the end of the locating pin.

CLEANING OPERATIONS

Cleaning operations considered in this section are peculiar to sextants.

Cleaning the Rack Teeth

Proceed in the following manner to clean the rack teeth of a sextant:

1. Apply a coating of approved sextant rack polish to the teeth and use a small, soft brush to rub it well into the teeth.
2. Hold the sextant frame horizontally and press it lightly against a nylon buffing brush, mounted on a suitable buffing motor. Then move the rack continuously across the brush.

CAUTION: If you allow the rack to remain stationary at a spot, it will cause uneven wear of the teeth.

3. After you remove dirt and corrosion from the teeth, blow off excess polish with an air hose or remove it with a nylon brush.

Cleaning the Worm and Shaft Worm Gear

To clean the endless tangent worm and the shaft worm gear, do the following:

1. Apply a coat of approved worm gear polish to the thread of the worm gear and polish it well into the thread with a small, soft brush.
2. Hold the worm gear by its shaft at each end and press it lightly against a nylon buffing brush. Rotate the worm gear in an even manner, to ensure uniform buffing everywhere on the gear.
3. After removing dirt and/or corrosion from the gear, remove excess polish with a soft nylon brush or an air hose.

REASSEMBLY

The reassembly procedure discussed in this chapter is for a David White sextant. During reassembly, oil and/or grease moving parts with approved lubricants. Clock oil is approved by the Navy for sextant centers, worm gear bearings, and pivots; lubriplate is approved for the shoe; and petroleum jelly is approved for the arc groove. Apply anti-seize lubricant to screws before you replace them. No lubricant should be applied to the rack teeth or the endless tangent worm and worm gear thread.

Subassemblies are reassembled first, in reverse order of disassembly; reassembly of major assemblies and the telescope assembly follows.

Grip Assembly and Arc Support Studs

Steps in the reassembly of the grip assembly and arc support studs are:

1. Put the hand grip cap-stud on the grip and replace the cap-stud screw. Review figure 15-18.
2. Replace the hand grip support and secure it with the retainer screw.
3. Slide the hand grip cap-stud end support on the hand grip cap-stud into the hole in the support. Then assemble the two supports in the sextant frame assembly.
4. Replace the grip assembly screws and tighten them.
5. Replace and tighten the two arc support studs in the sextant frame assembly. Use a pin wrench.

Reassembly of the Index Arm

Refer to illustration 15-14 as you reassemble the index arm. The procedure follows:

1. Replace the index arm stiffening rib and secure it. Follow the reassembly guide mark.
2. Put the index arm on a bench (rib beneath), with the worm frame end of the arm near you, and assemble the index arm guide lug marked L in the recess on the left-hand side of the arm guide lug. Secure the lug with its screw. Replace the index arm guide lug marked R in the same manner.

NOTE: At this point, check the fit of the index arm guide lugs in the guide slot of the sextant rack.

3. Move the index arm along the rack and check the movement of the index arm guide lugs along the guide slot. There should be no binding or slackness. If you feel binding, look for dirt or corrosion in the guide slot. If the guide slot is slightly damaged, remove existing high spots or burrs with a guide slot scraper.

If a guide lug is tight in the slot, file some metal from the face of the guide lug, just enough to make it fit correctly. If the guide lug is slack, support it on a metal peening block and peen the top of the guide lug with a small hammer and chisel. Then recheck the fit. Continue with this procedure until the fit is correct.

When the index arm guide lugs are correctly fitted and aligned, the index arc is just clear of, and parallel to, the sextant arc scale. Check the amount of clearance with a feeler gage. The endless tangent worm and shaft worm gear teeth should also be centered in the sextant rack teeth.

If you replace an index arm guide lug, fit it to the guide slot by filing metal from both sides of the lug. Use a good lug as a guide when you do this.

5. Put the stationary lever on the index arm (same side as the guide lugs) and secure it with the screws provided.
6. Replace the spring support block in its recess in the index arm.

Reassembly of the Worm Frame Assembly

Proceed as follows to reassemble the worm frame assembly:

1. If you removed it, replace the vernier yoke locating pin in the front worm bearing (fig. 15-12) and tap it lightly into position with a small hammer. The flat sides of the pin should be in line with the front worm bearing, so that it will slide easily into the vernier yoke slot when you assemble the yoke.
2. Apply a thin stripe of worm gear shaft oil to the shaft of the endless tangent worm and shaft worm gear. Apply the oil along the shaft where the front and rear worm bearings will be located.
3. Replace the front worm bearing on the shaft of the endless tangent worm and shaft worm gear and slide the shaft of the worm gear into the end of the worm frame in which the rear worm bearing is assembled. At the same time, bring the front worm bearing down to rest on its seat on the worm frame.
4. Slide the rear worm bearing over the worm gear shaft and into the worm frame; then assemble the rear worm bearing clamping screw.

CAUTION: Do NOT exert enough pressure on the rear worm bearing clamp screw or the bearing to SQUEEZE them out of the round. Check the rotation of the worm and shaft for freedom of fit and action.

5. If removed, reassemble the front bearing cap locating pin and tap it lightly into place with a small hammer, from the flat side of the bearing cap.
6. Rotate and position the front worm bearing until the hole drilled in the bearing to accommodate the front bearing cap locating pin is correctly positioned. Place the front worm bearing cap over the bearing and insert the locating pin in the cap in the corresponding hole in the front worm bearing. Follow the reassembly guide marks.

7. Secure the front bearing cap to the worm frame. Then check the rotation of the worm gear in its bearings for freedom and smoothness of action.

8. Assemble the vernier yoke on the front worm bearing. Slide it into position to allow the yoke locating pin to enter the slot in the yoke. Then insert the vernier yoke clamp screw but do not tighten it.

9. Slip the micrometer drum adapter onto the shaft of the endless tangent worm and shaft worm gear and pin it in position with the adapter lock pin.

When you assemble this pin, support the worm gear shaft firmly and squarely on a wooden block, and use care to prevent bending of or damage to the shaft. If it is necessary to fit a new drum adapter, drill a hole for the adapter lock pin with a .10-inch diameter drill.

10. Put the micrometer drum on the adapter and lock it in position with its lock screw and washer.

11. Lightly lubricate the worm frame holding spring with approved grease and place the spring in the frame. Then replace the frame holding spring stop screw, with its head flush with the frame.

Reassembly of the Worm Frame Assembly to the Index Arm

Some early types of David White sextants (Mk 2) have a worm frame pivot screw which does not have assembled with it a pivot screw tension washer, a pivot screw lock nut, or a worm frame pivot clamping screw. In this case, omit step No. 1 in the reassembly procedure.

1. Apply a thin coat of worm frame grease to the three bearing surfaces on the bottom side of the worm frame, and a thin film of pivot oil over the bearing surface of the worm frame pivot screw. Then assemble the worm frame assembly to the index arm (fig. 15-11). With a pair of tweezers, engage the frame holding spring on the stud of the spring support block. See illustration 15-21. Then lower the frame assembly onto the index arm and replace the pivot screw and its tension washer. See figure 15-11.

2. With a pivot screw lock nut screwdriver, replace the worm frame pivot screw lock nut.

3. Insert and tighten the worm frame pivot clamping screw, just tight enough to eliminate play or slackness. There must be smooth action of the worm frame around the pivot.

4. Lightly coat the two bearing surfaces on the underside of the worm frame retainer block flange with retainer block grease and replace the block. Secure it with its screw.

5. Move the worm frame back and forth to spread the lubricant on the worm frame retainer block, along the top sides of the slot in the worm frame. Make certain the worm frame assembly moves smoothly and freely, without binding or play.

With a suitable feeler gage (fig. 15-22), measure the clearance between the bottom bearing surfaces of the worm frame and the index arm. The maximum clearances should be .001 inch when measured at the worm frame retainer block end of the frame. The frame and index arm should be in direct contact at the pivot end of the frame.

If the frame retainer block is tight, use a fine file to remove a little metal from the two bearing surfaces on the underside of the flange, an equal amount from each bearing surface. If the block is too slack, remove metal from its base with a fine file.

6. Adjust the tension of the worm frame holding spring with its stop screw. Tension of the spring should be enough to hold the endless tangent worm and shaft worm gear firmly against the sextant rack teeth; and when you have it adjusted correctly, lock the worm frame holding spring stop screw by spreading the slot at right angles to the screwdriver slot. Use a screwdriver with a sharp blade. Study illustration 15-23. The worm frame should move smoothly and freely.

7. Put a drop of worm gear shaft oil on the exposed end of the rear worm bearing.

8. Replace and tighten the screw in the end of the worm gear shaft. Its head must bear evenly against the rear worm bearing flange.

9. Bend the end thrust spring (fig. 15-11) as necessary to provide ample thrust when assembled to the worm frame. Apply a drop of end thrust screw oil to the tip of the end thrust screws and replace the end thrust spring.

NOTE: Tension of the end thrust spring should be such that no movement of the endless tangent worm and shaft worm gear is felt when the micrometer drum is lightly pushed or pulled. Test it by turning the worm gear. There should be slight resistance, but the worm gear should rotate smoothly, without binding. With a screwdriver blade, bend the end thrust spring outward to adjust its tension.

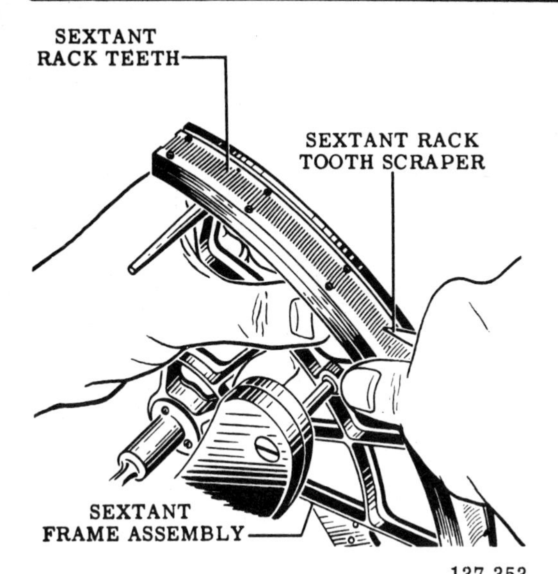

137.353
Figure 15-20.—Scraping the rack teeth.

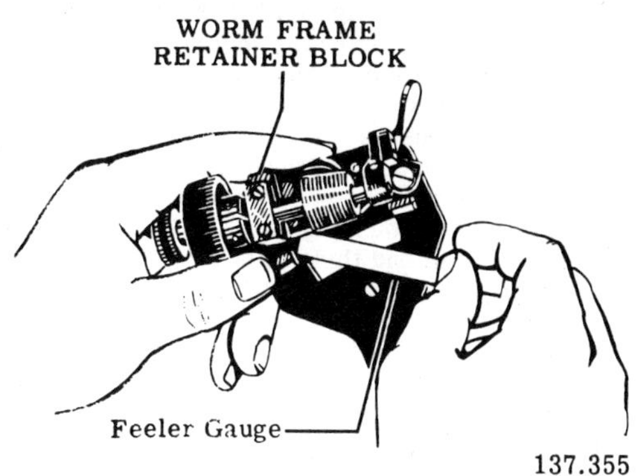

137.355
Figure 15-22.—Measuring the clearance between the bottom bearing surfaces of the worm frame and the index arm.

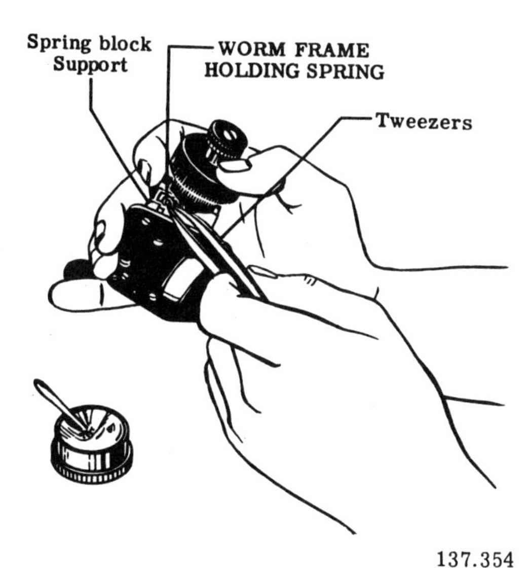

137.354
Figure 15-21.—Engaging the worm frame holding spring on the stud of the spring support block.

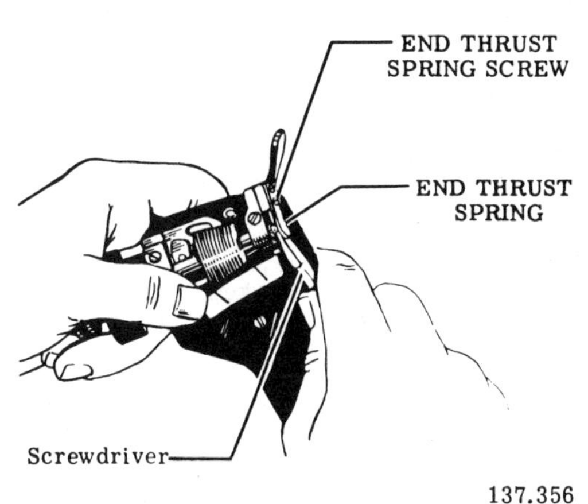

137.356
Figure 15-23.—Adjusting tension of the end thrust spring.

10. Position the vernier yoke and move it along the front worm bearing as required to have the edge of its scale located .001 to .0015 inch from the edge of the micrometer drum scale. Measure with a suitable feeler gage. Tighten the vernier yoke clamp screw and slack off slightly on the micrometer drum lock screw, because the micrometer drum will be adjusted to zero during testing and adjusting.

11. Insert and tighten the disengaging lever pivot screw bushing to the index arm.

12. Apply disengaging lever grease to the following parts: bearing surface of the lever pivot screw bushing, track in the worm frame for the lever shoe, and the pin in the disengaging lever. Then put the shoe on the lever and put the lever on the bushing (fig. 15-11).

13. Use an end thrust nut wrench to replace the disengaging lever end thrust nut on the lever pivot screw.

This assembly is now reassembled. Wipe off excess lubricant with a dry cloth. Check the assembly for smoothness of action when you depress the disengaging lever, and also the freedom of action under spring tension.

Horizon Mirror Mounting, Bracket, and Filter Bracket Assembly

To assemble all of these parts, proceed as follows:

1. Assemble the horizon mirror bracket to the mirror and filter bracket. Use an adjusting screw wrench. See figure 15-15. If you must tap a hole for a new bracket, use a No. 5-44NF thread.
2. Replace the outer mirror bracket adjusting screw.
3. Assemble the horizon mirror mounting to its bracket, and replace the bottom adjusting screw.
4. Replace the filter pivot screw extension in the mirror and filter bracket.
5. If removed, replace and tighten the horizon filter stop stud to the mirror and filter bracket.

Telescope Sliding Bracket Support

Replace the telescope sliding bracket retaining screw (fig. 15-17). Tighten the screw. If you disassembled the fulcrum pins from the telescope sliding bracket support, press-fit them into place. CAUTION: Do NOT damage the conical heads of the fulcrum pins.

Index Arm and Worm Frame Assembly

The index arm, endless tangent worm, and the shaft worm gear thread (or rack teeth) can be damaged easily; handle them with care. Proceed as follows:

1. Apply a thin coat of guide slot grease to the slot in the sextant rack (fig. 15-24), but do

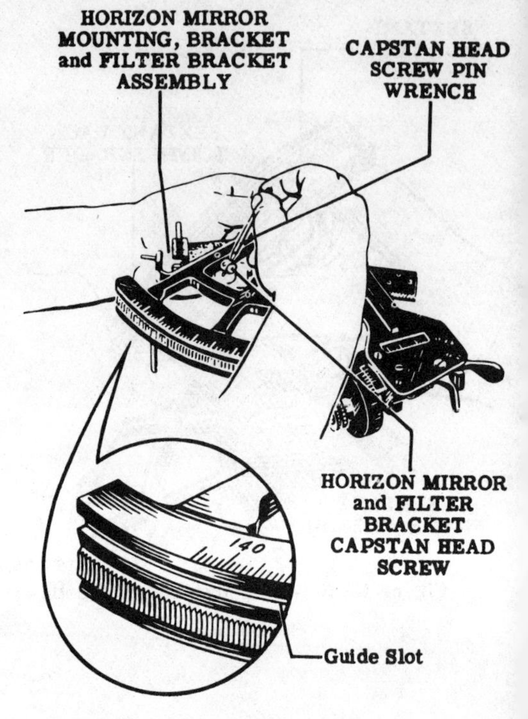

137.357

Figure 15-24.—Replacing the horizon mirror and filter bracket capstan head screw.

NOT allow grease to get on the surface of the rack or teeth.

2. Press the disengaging lever to free the endless tangent worm and shaft worm gear from the sextant rack and swing the index arm over the end of the rack. Make certain the index arm guide lugs (fig. 15-14) are engaged in the guide slot as you move the index arm onto the rack.

The endless tangent worm and shaft worm gear should be centered in the rack teeth and in line with the rack. Check with your eye. If they are not centered and in line with the rack, the index arm guide lugs were not adjusted correctly. Disengage the worm gear and move the index arm back and forth along the arc scale. The movement should be smooth.

3. Rotate the worm gear along the full length of the rack and check the action. It should be smooth, but there should be a slight, constant drag on the worm gear.

4. Put the index arm stop stud in the end of the sextant rack nearest the location of the horizon mirror.

Horizon and Index Mirror Polaroid Filter Assemblies

Put the filter assembly pivot spring washer, flat washer, and the horizon mirror polaroid filter assembly on the filter assembly pivot screw. Then mount them on the mirror and filter bracket and tighten the screw. See illustration 15-23.

NOTE: The horizon mirror polaroid filter assembly must be mounted with the rotating filter facing the mirror mounting.

Reassemble the filter assembly pivot spring washer, pivot flat washer, and index mirror polaroid filter assembly on the filter assembly pivot screw. Then mount them to the horizon mirror and filter bracket.

NOTE: The index mirror polaroid filter assembly must be mounted with the rotating filter away from the direction of the mirror.

Replace the index mirror mount on the index arm; then insert and tighten the screws.

The sextant assembly, less the telescope, is now ready for testing, adjusting, and final inspection.

Reassembly of the Telescope Assembly

All optical parts and components of the telescope assembly must be perfectly clean when you reassemble them. Refer to figure 15-19 as you study the reassembly procedure.

EYEPIECE AND FIELD LENS ASSEMBLY.— Do the following to reassemble this assembly:

1. Set the eyepiece lens mount on the bench (threaded collar down) and, with a lens suction adapter, pick up the eyepiece lens (G) by its convex surface and place it flat side down in the eyepiece lens mount.

2. Place the F and G lens spacer ring in the eyepiece lens mount, on eyepiece lens G.

3. With a lens suction adapter, put field lens F in the eyepiece lens mount, flat surface up.

4. Place the retainer ring spacer ring in eyepiece lens mount, on field lens F.

5. Screw the F and G retainer ring into the eyepiece lens mount and tighten the retainer ring down with a telescope flat wrench.

6. Insert the retainer ring lock screw into the side of the eyepiece lens mount and into the hole in the F and G lens retainer ring. If the holes do not line up, turn the retainer ring to line them up. The lenses must be held tight in the eyepiece lens mount.

NOTE: If you use a new eyepiece lens mount, drill a hole .0460 inch in diameter and 7/32 inch from the retainer-ring end of the eyepiece lens mount when assembled. Tap the mount and ring for a No. 0-80NF screw. Countersink the hole to provide space for the screw head.

7. Put the eyepiece lens mount assembly in the diopter focusing ring, with the flange of the lens mount against the end of the diopter focusing ring. Tighten the assembly with an eyepiece lens mount assembly wrench.

8. Screw the eyepiece cap in the diopter focusing ring until the reassembly guide marks on the eyepiece and the diopter focusing ring align.

9. Insert and secure the lock screw in the diopter focusing ring to lock the assembled parts together.

ERECTOR LENS MOUNT ASSEMBLY.— Assemble the erector lens mount assembly in the following manner:

1. With a lens suction adapter, pick up erector lens E (fig. 15-19) by its side of greatest curvature and place it flat side down in the erector lens mount.

2. Put the erector lens spacer ring in the erector lens mount, on erector lens E.

3. Pick up lens D by its flat side with a lens suction adapter and assemble it (convex side first) in the erector lens mount.

4. Put the lens retaining housing for lens C on the bench; then put lens C in the housing, with the side of greatest curvature of the lens down.

5. Replace the retaining ring for lens C and tighten it with a telescope flat wrench.

6. Replace the lock screw in the retaining housing for lens C and tighten it.

7. Put the lens C retaining housing assembly in the erector lens mount and align the reassembly marks on the lens retaining housing and erector lens mount you made during disassembly. Tighten the assembly with a retaining housing wrench.

BODY TUBE AND OBJECTIVE LENS ASSEMBLY.—The procedure for reassembling the body tube and objective lens assembly in a David White sextant telescope follows:

1. Put the objective lens mount on the bench, knurled edge down, and place the objective lens

(A) in it, convex side down. Use a lens suction adapter.

2. Assemble lens B retaining housing assembly, including the lens, retaining ring, and lock screw.

3. Place the retaining housing assembly of lens B in the objective lens mount and tighten the assembly with a retaining housing wrench. Align the reassembly guide marks.

NOTE: If you replaced the retaining housing, drill a hole .0469 inch in diameter in the housing. Use the hole in the objective lens mount as a guide.

4. Screw the objective lens mount assembly into the objective end of the telescope body, which you marked during disassembly. If required, use the telescope body and objective lens mount wrench to tighten the assembly. Align the reassembly guide marks.

5. Replace the lock screw of the objective lens mount.

NOTE: If you replace the objective end body of the telescope or the objective lens mount, loosely assemble the objective lens mount assembly. Replace the lock screw during testing and adjusting.

EYEPIECE, TELESCOPE BODY, AND ERECTOR LENS MOUNT ASSEMBLY.—The procedure for assembling these parts is as follows:

1. Assemble the erector lens mount assembly to the body tube and the objective lens assembly by screwing the mount assembly into the objective end of the telescope body. Align the reassembly guide marks.

2. Replace the telescope body lock screw to lock the erector lens mount assembly and body tube and objective lens assembly together.

3. Screw the eyepiece end of the telescope body onto the erector lens mount assembly and align the reassembly guide marks. If required, use the telescope body and objective lens mount wrench.

4. Replace the telescope body lock screw to lock the eyepiece end telescope body to the erector lens mount assembly.

EYEPIECE AND FIELD LENS ASSEMBLY.—Use the procedure which follows to replace the eyepiece and field lens assembly.

1. Put a thin coat of eyepiece grease on the inside of the slotted end of the eyepiece end of the telescope body. Extend the coating approximately 3/4 inch inside the body to lubricate the eyepiece lens mount as it slides in the body.

2. Slide the eyepiece and field lens assembly onto the eyepiece end of the telescope body far enough to bring the edge of the diopter focusing ring just over the edge of the diagonal slot in the body.

3. Apply a light coat of approved grease to the eyepiece movement shoe and place the shoe, convex side up, in the diagonal slot in the eyepiece end of the telescope body. It must slide smoothly in the slot.

4. Slide the eyepiece and field lens assembly forward to bring the hole for the key screw in the diopter focusing ring over the hole in the eyepiece movement shoe; then insert the key screw.

FINAL REPAIR INSPECTION

Final repair inspection for an overhauled David White sextant should be made from the standpoint of (1) the sextant assembly and (2) the telescope assembly.

Sextant Assembly

After you reassemble a sextant telescope, inspect the following:

1. TENSION OF THE END THRUST SPRING. Disengage the endless tangent worm and shaft worm gear from the rack teeth. Then pull and push lightly on the micrometer drum to check for lateral movement of the endless tangent worm and shaft worm gear relative to the worm frame. There must be no lateral movement. If there is lateral movement, backlash will occur during collimation. Proper tension on the spring eliminates backlash.

2. Disengage the endless tangent worm and shaft worm gear from the sextant rack and move the index arm back and forth along the arc scale. The arm should move freely without binding of the index arm guide lugs, or the index arm guide stud and index arm guide lug on a Pioneer sextant.

Check the guides for slackness. Disengage the worm gear from the rack teeth, grasp the worm frame end of the index arm between the thumb and forefinger, and apply up-and-down pressure to the index arm. There should be NO movement of the index arm up or down.

Check for slackness in the worm frame. Disengage the endless tangent worm and shaft worm gear from the rack and move the micrometer drum up and down at right angles to the plane of the index arm surface. There should be no movement as a result of:

a. Loose worm frame bearings (Pioneer).

b. Loose worm bearings, front or rear (David White).

c. Loose worm frame pivot (Pioneer) or worm frame pivot screw (David White).

d. Loose worm frame retainer block (David White), or guide screw (Pioneer).

4. Use a feeler gage to check the worm frame on a Pioneer sextant for parallelism with the index arm. The frame should be parallel to within .001 inch when tested at each end. In David White sextant, the clearance between the bottom bearing surfaces of the worm frame and the index arm must not be greater than .001 inch when measured at the worm frame retaining block end of the worm frame. The worm frame and the index arm must be in direct contact at the pivot end of the frame.

5. Check the tension of the worm frame inner and outer springs (Pioneer), and the worm frame holding spring (David White). Disengage and re-engage the worm gear thread in the rack teeth, and apply pressure with your fingers to the micrometer drum. There should be no further movement of the worm gear thread into the rack teeth. Check the worm frame assembly for freedom of movement on the index arm. Erratic action caused by the following must not be tolerated:

a. Over-tight worm frame pivot (Pioneer and David White).

b. Over-tight worm frame retaining block (David White).

c. Over-tight guide screw (Pioneer).

d. Binding between the base of the worm frame and the surface of the index arm (Pioneer and David White).

6. Check moving parts for freedom of movement and smoothness of action. Run the index arm back and forth along the arc scale by rotating the endless tangent worm and shaft worm gear, which should rotate smoothly, without sticking or knocking resulting from high spots on the worm gear thread.

7. The polaroid filter assemblies should rotate smoothly. The assemblies should be firm on their respective pivots.

Telescope Assembly

Check the telescope assembly in the following manner:

1. Inspect optics for cleanliness and physical condition.

2. Make a visual inspection of the appearance and finish of mechanical parts and legibility of engravings.

3. Check the action of the diopter focusing ring, which should move evenly and smoothly over the entire diopter scale range.

Diopter Setting

The procedure for adjusting and checking the diopter setting of a David White sextant is as follows:

1. Set the diopter setting auxiliary telescope to your eye requirements by sighting the collimator target and setting the diopter ring on the auxiliary telescope to give the sharpest focus.

2. Mount the sextant telescope assembly on a suitable stand in front of the collimator tube.

3. Place the diopter setting telescope against the sextant telescope eyepiece cap and, with the diopter focusing ring on the sextant telescope, bring the collimator target into sharp focus. The diopter reading from the sextant telescope assembly should be 0 within plus or minus one-quarter diopter. Each mark on the diopter scale represents one diopter. If the telescope does not meet this performance requirement, proceed with the following steps; if the requirements are met, uses steps No. 7 and No. 8 only.

4. If assembled, remove the lock screw for the objective lens mount assembly and loosen the assembly in the objective end of the telescope body. See figure 15-25. Set the diopter scale of the sextant telescope at 0, place the diopter setting auxiliary telescope against the sextant telescope eyepiece cap, and sight the collimator target. Then slowly rotate the objective lens mount to move the mount assembly out of the objective end of the telescope body. If the collimator target comes into sharp focus, measure the distance (A, fig. 15-25) from the inner edge of the objective lens mount to the nearest edge of the mount. Replace the existing objective end of the telescope body with one of the length of A in figure 15-25.

If the collimator target does not come into focus when you move the objective lens mount

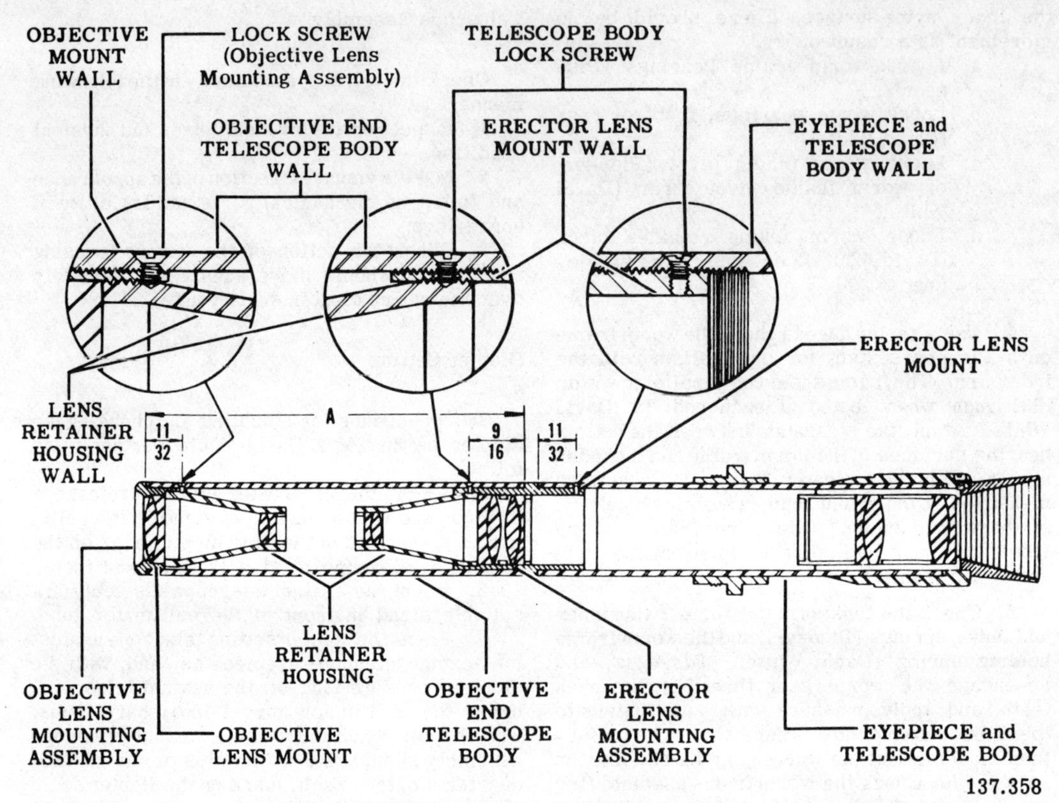

Figure 15-25.—Adjusting the diopter setting.

assembly out of the telescope body, proceed with the operations which follow:

NOTE: If the objective end of the telescope body is too long, turn it down in a lathe to the length required. Be sure to take an equal amount of metal from each end of the body. Leave the body 1/64 inch longer than the length of A in figure 15-25.

5. Remove the objective lens mount assembly from the objective end of the telescope body, and disassemble the objective end of the telescope body from the erector lens mount. Then assemble the objective adjusting tube loosely on the erector lens mount and put the mount assembly in the objective adjusting tube.

6. Repeat step No. 3. If the sextant telescope still does not meet performance requirements, disassemble the objective end of the telescope body and remove a SMALL amount of metal from one end. Then reassemble the objective end of the telescope body and the lens mount assembly and check the performance. If it is still unsatisfactory, remove more metal; but each time remove a smaller amount than you removed the previous time. Continue this procedure until performance is satisfactory.

7. To drill and tap holes for the telescope body lock screws, or the lock for the objective lens mount assembly, locate the holes as indicated in the top portion of figure 15-25. Use a drill .0409 inch in diameter, and drill each hole to the depth shown in the circles in illustration 15-25. Be sure to clean the telescope of metallic particles after you finish drilling; you will generally have to disassemble the telescope to do this.

8. Screw the telescope into the sliding bracket on the sextant and check the reference

mark for the diopter scale, which should be at the top when the sliding bracket is assembled.

Adjusting the Mirrors

You can adjust the index and horizon mirrors by sighting natural targets, but check your adjustments with an artificial target. These mirrors get out of adjustment easily, even during normal use. One procedure for checking the mirrors before taking a sight is discussed next.

PERPENDICULAR ADJUSTMENT OF THE INDEX MIRROR.—The reflecting surface of the index mirror must be perpendicular to the sextant frame in order for the sextant to function properly. To make proper adjustment, do the following:

1. Place the sextant on a stand (with the arc scale farthest from you) and rest it on the arc support studs and center cap. Then release the endless tangent worm and shaft worm gear from the rack teeth and shift the index arm around the arc scale until the index mark is about on 35 degrees.

2. With one eye, sight the 0° mark on the arc scale. Then bring your line of sight down until you can see simultaneously the reflected image of that section of the arc scale in the vicinity of the 140° mark in the index mirror. You should see the reflected image level with the directly-viewed scale, as shown in part C of figure 15-26, provided the index mirror is adjusted properly. If the mirror is not adjusted correctly, you will view the conditions shown in part A or part B of illustration 15-26. When this is true, proceed with the adjustments which follow.

3. Adjust the index mirror clamping screws to bring the reflected image level (in coincidence) with the directly-viewed scale (part C, fig. 15-26). If the reflected image is high, loosen the top clamping screw and tighten the bottom clamping screw; if the reflected image is low, reverse this procedure. ALWAYS loosen one of the index mirror clamping screws before you tighten the other, to prevent cracks in the mirror. Tighten both screws LIGHTLY and EQUALLY when adjustment is correct.

4. Check the index mirror adjustment by moving the index arm along the arc scale from the 0° mark to about the 55° mark and, at the same time, view the arc scale and its reflected image. The directly-viewed arc scale and its reflected image should appear continuous as the index arm is moved along the arc scale. If there is an apparent shaking or wiggle of the line formed by the edge of the arc scale, the mirror is not adjusted correctly.

PARALLEL ADJUSTMENT OF THE HORIZON MIRROR.—The horizon mirror must be so adjusted that the reflecting surface of the silvered section of the mirror is parallel to that of the index mirror when the sextant is set at zero. Assemble the sextant telescope in the sliding bracket and proceed as follows:

1. Set the index arm at 1° on the arc scale and then make the setting 0 with the micrometer drum knob (Pioneer), or micrometer drum (David White).

2. Hold the sextant so that its frame is horizontal and sight through the telescope at a suitable vertical object such as a spire, crane boom, or flagpole with a ball on top. Study figure 15-27. The object should be at least 300 yards distant. If the mirrors are parallel, the directly-viewed object and its reflected image will be in line with each other in the vertical plane (part C, fig. 15-27). If they are not, proceed with the next steps.

3. If the reflected image is to the left of the directly-viewed object, the left-hand edge of the horizon mirror is toward you (part A, fig. 15-27). Adjust the horizon mirror as required to bring the directly-viewed object and its reflected image into line. Loosen the outer horizon mirror bracket adjusting screw and tighten the inner one so that the left-hand edge of the mirror moves away from you. See figure 15-28 for the position of the adjusting screws. If the reflected image is to the right of the directly-viewed object (part B, fig. 15-27), reverse this procedure. When properly adjusted, the directly-viewed object and the reflected image line up and are centralized in the mirror.

To prevent stripping of threads, loosen one of the bracket adjusting screws before you tighten the other. If the reflected image of the vertical object observed is far to the left or to the right of the directly-viewed object when the outer horizon mirror bracket adjusting screw is about at its halfway point of adjustment, the index mirror mounting should be adjusted. With the index arm set at 0°, loosen the index mirror mounting screws and move the mirror as necessary to have the reflected image near coincidence with the directly-viewed object and also in the center of the horizon mirror. Then tighten the screws and check the perpendicular adjustment. If necessary, use the horizon mirror bracket

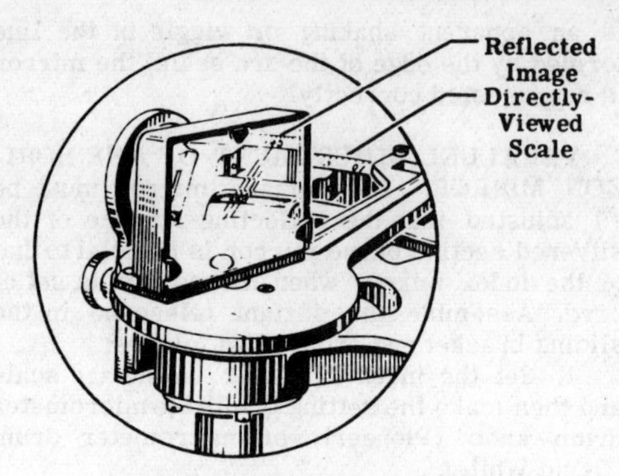

A. Index Mirror leaning forward
(Reflected image high)

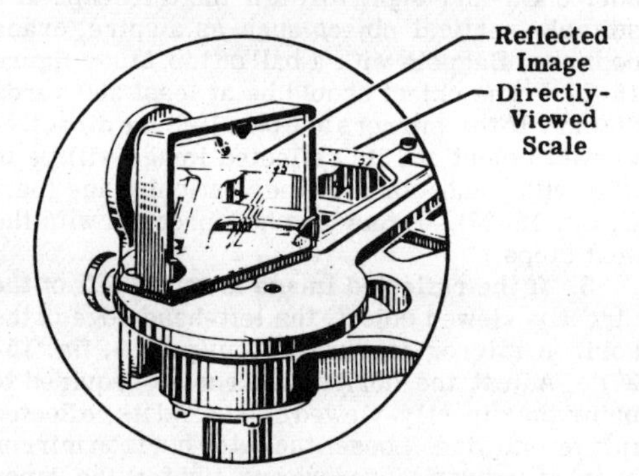

B. Index Mirror leaning backward
(Reflected image low)

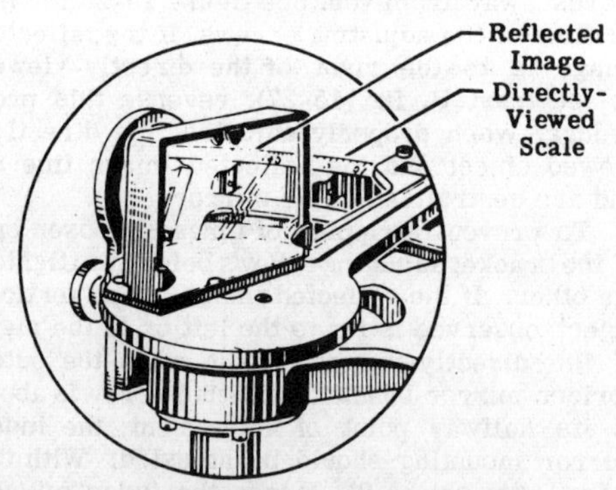

C. Index Mirror perpendicular to
the plane of the Sextant Frame
(Reflected image level with
directly-viewed scale)

137.359

Figure 15-26.—Adjusting the index mirror.

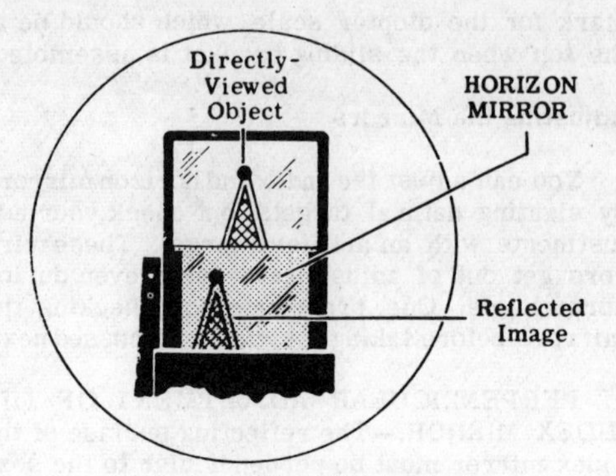

A. Horizon Mirror left hand edge
toward you

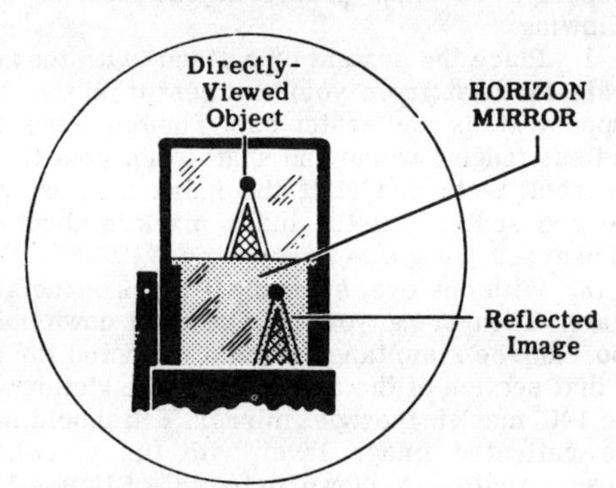

B. Horizon Mirror left hand edge
away from you

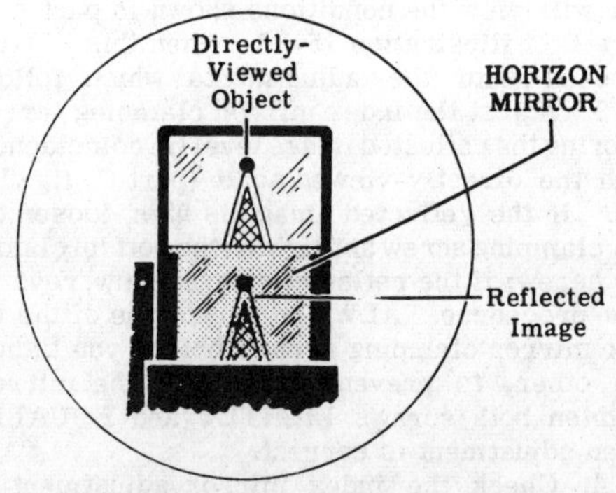

C. Horizon Mirror parallel to Index
Mirror but not perpendicular
to the Sextant Frame

137.360

Figure 15-27.—Parallel adjustment of
the horizon mirror.

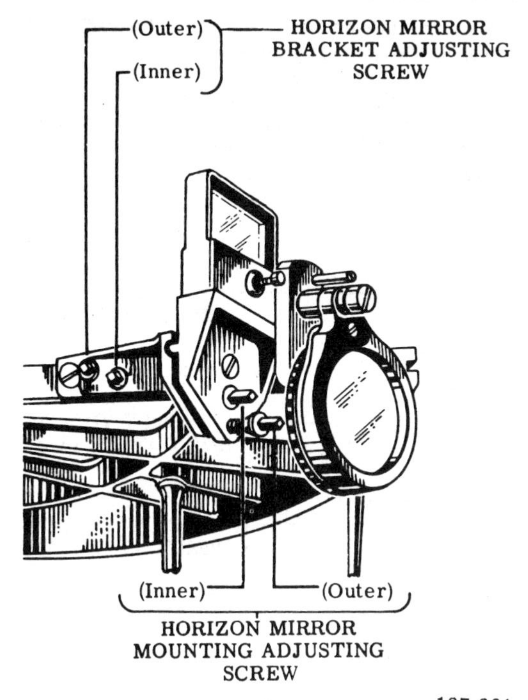

137.361
Figure 15-28.—Horizon mirror mounting and adjusting screws.

adjusting screws to bring the directly-viewed object and its reflected image into line.

The horizon mirror is now parallel with the index mirror. The next adjustment is to make the horizon mirror perpendicular to the sextant frame.

PERPENDICULR ADJUSTMENT OF THE HORIZON MIRROR.—You can adjust the horizon mirror perpendicularly in two steps:

1. With the index arm set at 0 and the sextant held horizontally, sight the same vertical object and check for coincidence of the reflected image with the directly-viewed object. If the mirror is adjusted correctly, the object and the image will be in coincidence. If this is not true, proceed with step No. 2.

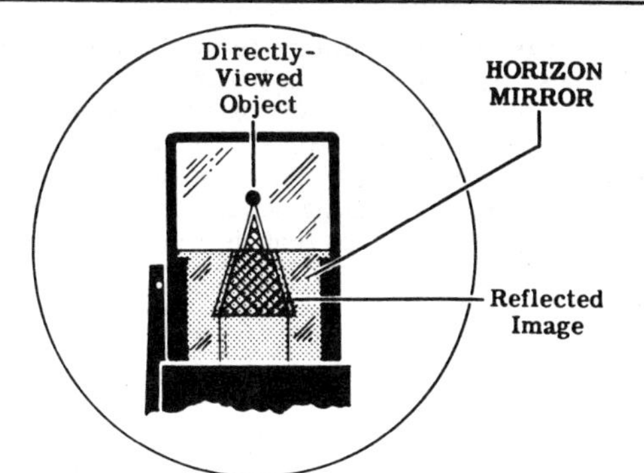

A. Horizon Mirror top edge leaning toward you

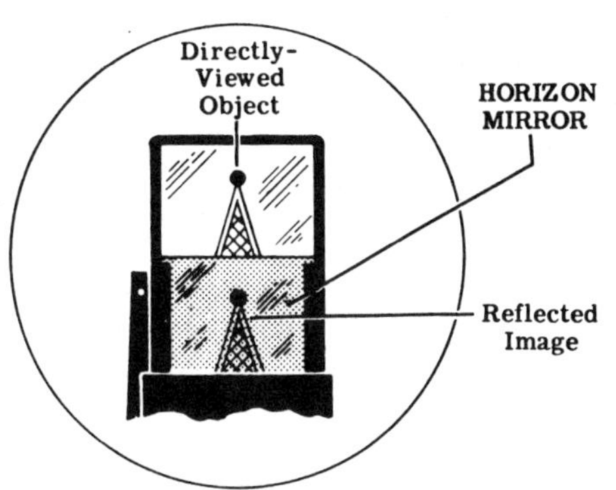

B. Horizon Mirror top edge leaning away from you

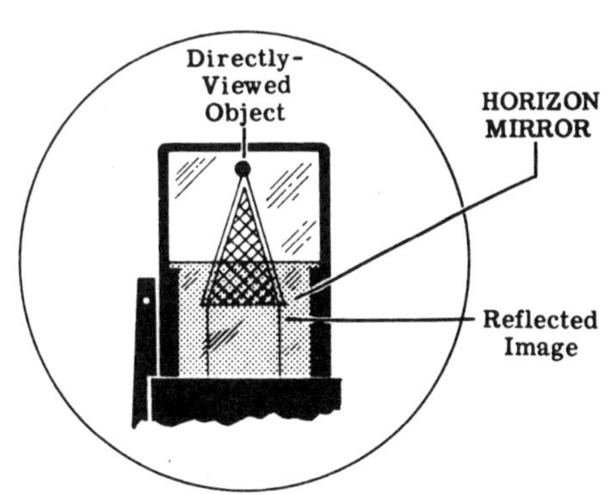

C. Horizon Mirror perpendicular to the Sextant Frame and parallel to the Index Mirror

137.362
Figure 15-29.—Perpendicular adjustment of the horizon mirror.

2. If the reflected image is high, the horizon mirror's top edge will lean toward you (part A, fig. 15-29). Adjust the mirror as necessary to bring the directly-viewed object and its reflected image into coincidence. Then loosen the outer horizon mirror mounting adjusting screw and tighten the inner one, so that the top of the horizon mirror moves away from you. If the reflected image is low, reverse this procedure.

After you complete the perpendicular adjustment of the mirror, recheck for parallelism with the index mirror. Correct adjustment is shown in figure 15-30.

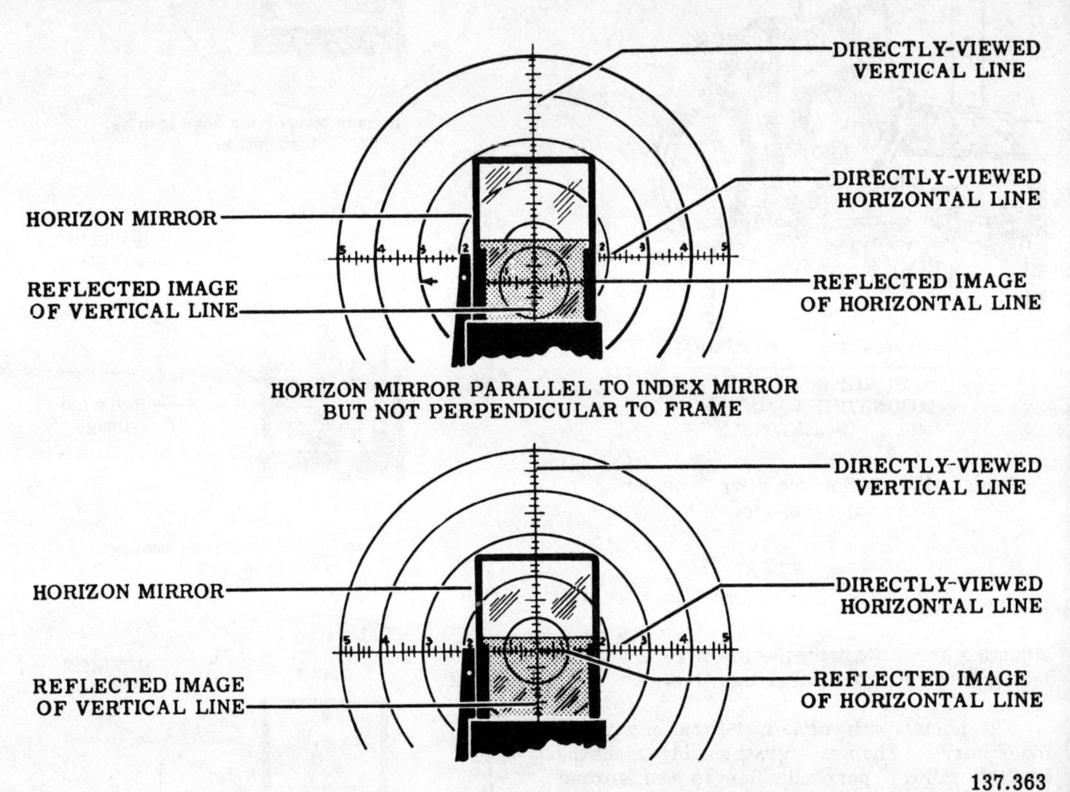

Figure 15-30.—Correct adjustment of the horizon mirror.

CHAPTER 16

STADIMETERS

A stadimeter is an instrument used to measure the range of objects of known height. Aboard ship, the officer of the deck uses a stadimeter to maintain his position in a formation by sighting on the guide ship for range. The height scale is calibrated in feet for objects from 50 to 200 feet in height. The range scale is calibrated in yards for readings from 200 to 10,000 yards and infinity.

TYPES OF STADIMETERS

The Navy uses two types of stadimeters: (1) sextant, which looks like a sextant (part A, fig. 16-1) and (2) Fiske, which has a rectangular frame, as shown in part B of figure 16-1. The sextant-type is a Mk 5, Mod 0; the Fiske stadimeter is made in two types, I and II, but type II is used much more extensively than type I.

Study the nomenclature of the two stadimeters illustrated in figure 16-1, and note the differences in construction.

CONSTRUCTION

Refer to illustrations 16-1 and 16-2 as you study the construction of a Fiske stadimeter, the principal parts of which are:

1. FRAME. The frame is the rectangular base on which all other parts of the instrument are mounted.
2. INDEX ARM. This arm carries the height scale; and it swings on a pivot at one corner of the frame.
3. INDEX MIRROR TABLE. The index mirror table is an adjustable platform mounted on the index arm (directly over the pivot) to carry the index mirror and its frame.
4. HORIZON MIRROR TABLE. This table is an adjustable platform which supports the horizon mirror and its frame.
5. CARRIAGE SCREW. The carriage screw moves the carriage block back and forth on the frame, and it is used to set the carriage index mark to the proper height on the height scale.
6. MICROMETER DRUM AND SCREW. The micrometer drum shows in yards the range of an object. For any given position on the height scale, the position of the drum controls the angle of the index arm and the index mirror.
7. CARRIAGE BLOCK. The carriage block carries the micrometer drum and screw (on a track) along the length of the frame.
8. INDEX MIRROR. The index mirror receives rays of light from the target and reflects them to the horizon mirror.
9. HORIZON MIRROR. This mirror, as in a sextant, enables an observer to see two images of the target, a direct image and an image reflected from the index mirror. Unlike the horizon glass of a sextant, however, the horizon mirror of a stadimeter has no clear glass on one side. It is merely a half-sized mirror in half of the mirror frame.
10. TELESCOPE. The telescope of a Fiske stadimeter is a low-power Galilean telescope which directs the line of sight toward the horizon mirror and gives a magnified image of the target.
11. MAGNIFYING GLASS. The magnifying glass is mounted on a bracket above the micrometer drum to give the observer a magnified image of the range scale. The bracket is adjustable, to allow movement of the glass up and down for focusing on the scale.

PRINCIPLE OF OPERATION

A stadimeter operates in the following manner: It elevates the observer's line of sight in order that he may view the top of an object which moves AWAY FROM or TOWARD him. If the object moves away, the angle of elevation of the observer's line of sight becomes smaller; if the

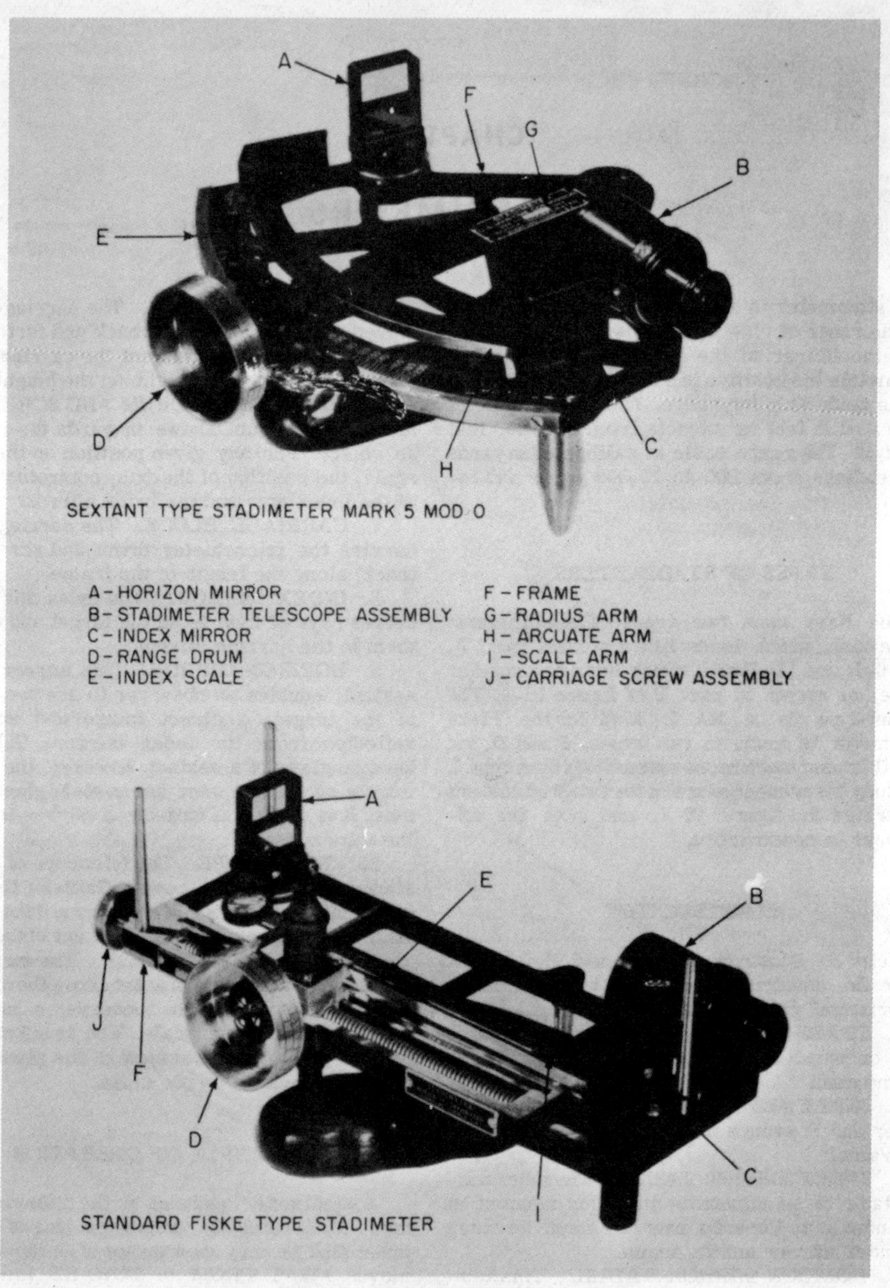

58.78

Figure 16-1.—Types of stadimeters.

object moves toward the observer, the angle becomes larger. Study the top portion of illustration 16-3. This means that if the object is distant (1, fig. 16-3), the angle of elevation is small; if the object is near, the angle is large.

Heights of objects at sea, such as the masts of various classes of ships, are generally known; and by means of its optical system a stadimeter measures the angle to which an observer's line of sight is elevated to permit him to view the top of an object. Then the calibrated range drum of the stadimeter converts the angular measurement to a range reading in yards.

This calibration of the range drum would be all right if all observed objects had the same height. Observe in illustration 16-3 that the object in the bottom portion is at the same distance from the observer as the object in the top portion; but the extra height of the object increases the angle of the observer's line of sight. To compensate for this, a sextant-type stadimeter has an arcuate arm and the Fiske stadimeter has a scale arm (fig. 16-1) which gives a greater or lesser displacement of the index mirror (in accordance with the height of the object observed) for the same movement of the range drum. The distance of the object (regardless of height) is therefore measured correctly.

The sighting triangle is not actually a right triangle, because the observer is generally above the waterline of the ship he is observing (bottom part, fig. 16-4). However, because of the distance involved, the triangle can be considered right-angled in actual practice.

When the officer of the deck desires to determine the range of a ship, he swings the range drum (fig. 16-1) on a sextant-type stadimeter over to the graduation on the index scale which represents the known height of the object. On a Fiske stadimeter, he runs it along on the carriage screw assembly to the proper-height setting on the index scale. Then he:

1. Holds the stadimeter by the handle with his right hand.
2. Screws the telescope into its mount.
3. Brings the instrument up to his eye (fig. 16-4), with the frame in a vertical plane, sights through the telescope toward the object, and focuses the telescope.
4. Views the bottom of the object directly through the open half of the horizon mirror, and at the same time views the top of the object as a reflected image in the index mirror and the mirror half of the horizon mirror. See the circled portion of figure 16-4.
5. Turns the index mirror by rotating the range drum until the reflection of the object coincides with the directly-viewed bottom of the object.
6. Reads on the range drum in yards the angle between the horizon mirror and the index mirror. Actually, the angle (subtended by the object) measured is twice the angle between the two mirrors.

OVERHAUL AND REPAIR

Overhaul and repair procedures considered in this section are for a Fiske stadimeter, because it is used more extensively than a sextant-type stadimeter. Before you study the following discussion, however, review the repair procedures outlined in chapter 8.

Some special tools required for repairing a stadimeter are shown in figure 16-5.

PREDISASSEMBLY INSPECTION

When you start to work on a stadimeter, prepare a casualty analysis sheet for the instrument and make a predisassembly inspection to determine whether it should be repaired or surveyed and salvaged. Record your findings on the inspection sheet.

During the predisassembly inspection of a stadimeter, look for:

1. Excessive play of the drum screw in the carriage.
2. Too much play of the carriage to the frame.
3. Excessive play in the center assemblies.
4. Condition of lubricants.
5. Excessive errors resulting from the condition of the scale arm.
6. Condition of silver on the mirrors.
7. Corrosion and/or failure of protective finishes.
8. Evidence of unauthorized tampering and disassembly.

DISASSEMBLY

The disassembly procedure considered in this section for a Fiske stadimeter is divided into two parts: (1) REMOVAL of assemblies, and (2) DISASSEMBLY of the subassemblies. The complete assembly of the instrument consists of the stadimeter assembly plus the telescope assembly.

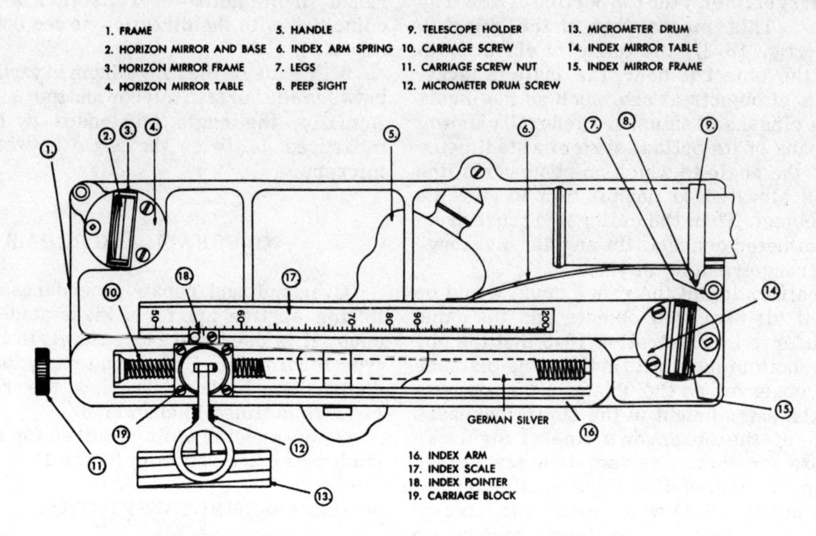

137.364

Figure 16-2.—Construction of a Fiske stadimeter.

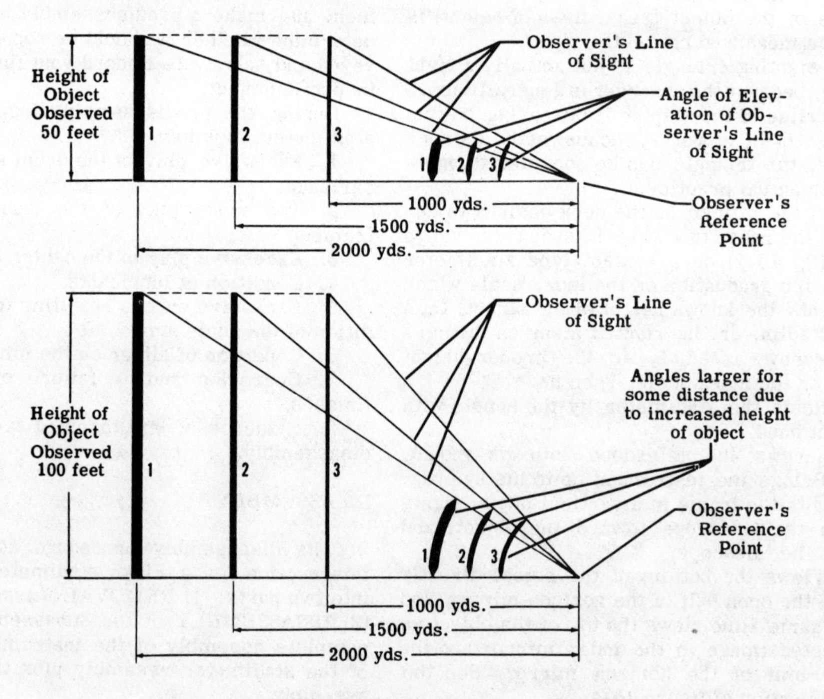

137.365

Figure 16-3.—Stadimeter principle of operation.

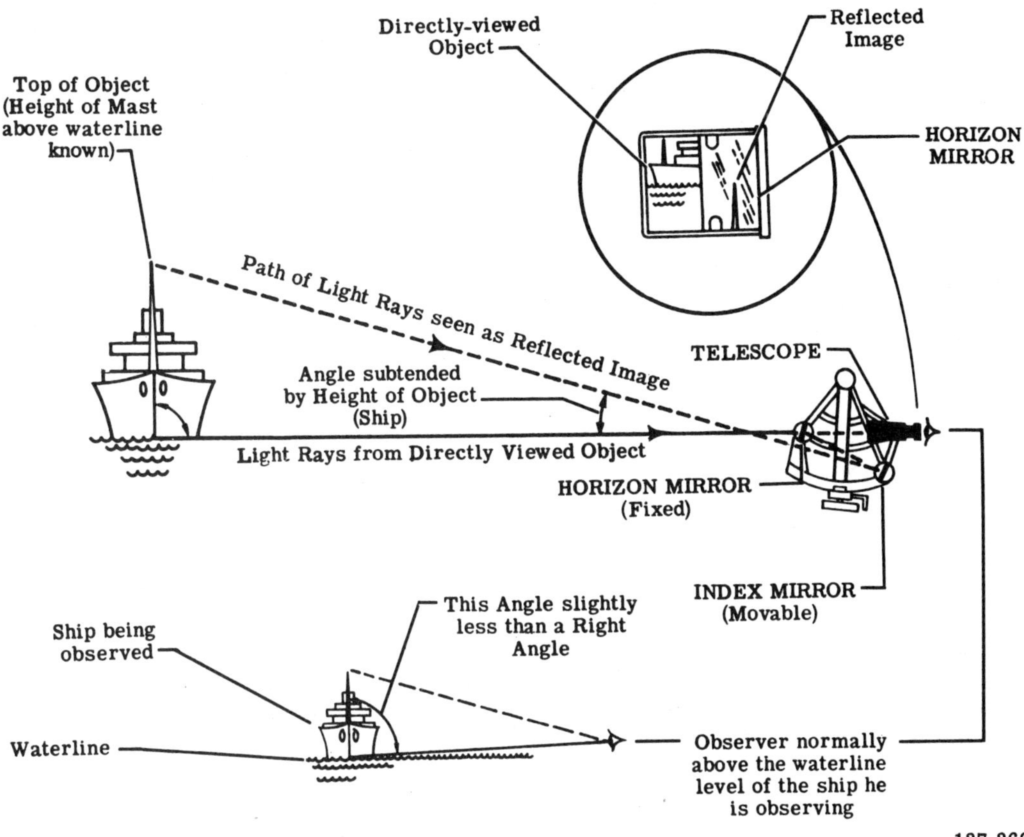

Figure 16-4.—Determining the range of an object.

Removal of the Telescope Assembly

To remove the telescope assembly from the stadimeter, unscrew it from the telescope holder. The position of the peephole in the telescope holder is critical; so, do NOT BEND or TWIST the holder when you remove the telescope.

Removal of the Horizon and Index Mirrors and Range Drum Magnifier

To remove the horizon and index mirrors, with their frames and the range drum magnifier, do the following:

1. Remove the mirror clamping screws from the backs of the horizon and index mirror, as shown in figure 16-6.

2. Cover each mirror with lens tissue (as a protection) and lift the mirror and frame from the stadimeter.

Study the enlarged view of the horizon and index mirror assemblies in figure 16-7.

3. Hold the range drum magnifier by its frame and remove the magnifier retaining screw (figs. 16-6 and fig. 16-7).

NOTE: Do NOT disassemble the magnifier lens from its frame, because the range drum magnifier must be replaced as a unit if either part is defective.

4. With an adjusting screw wrench (fig. 16-5), remove the two horizon mirror radial adjusting screws (figs. 16-7 & 16-8).

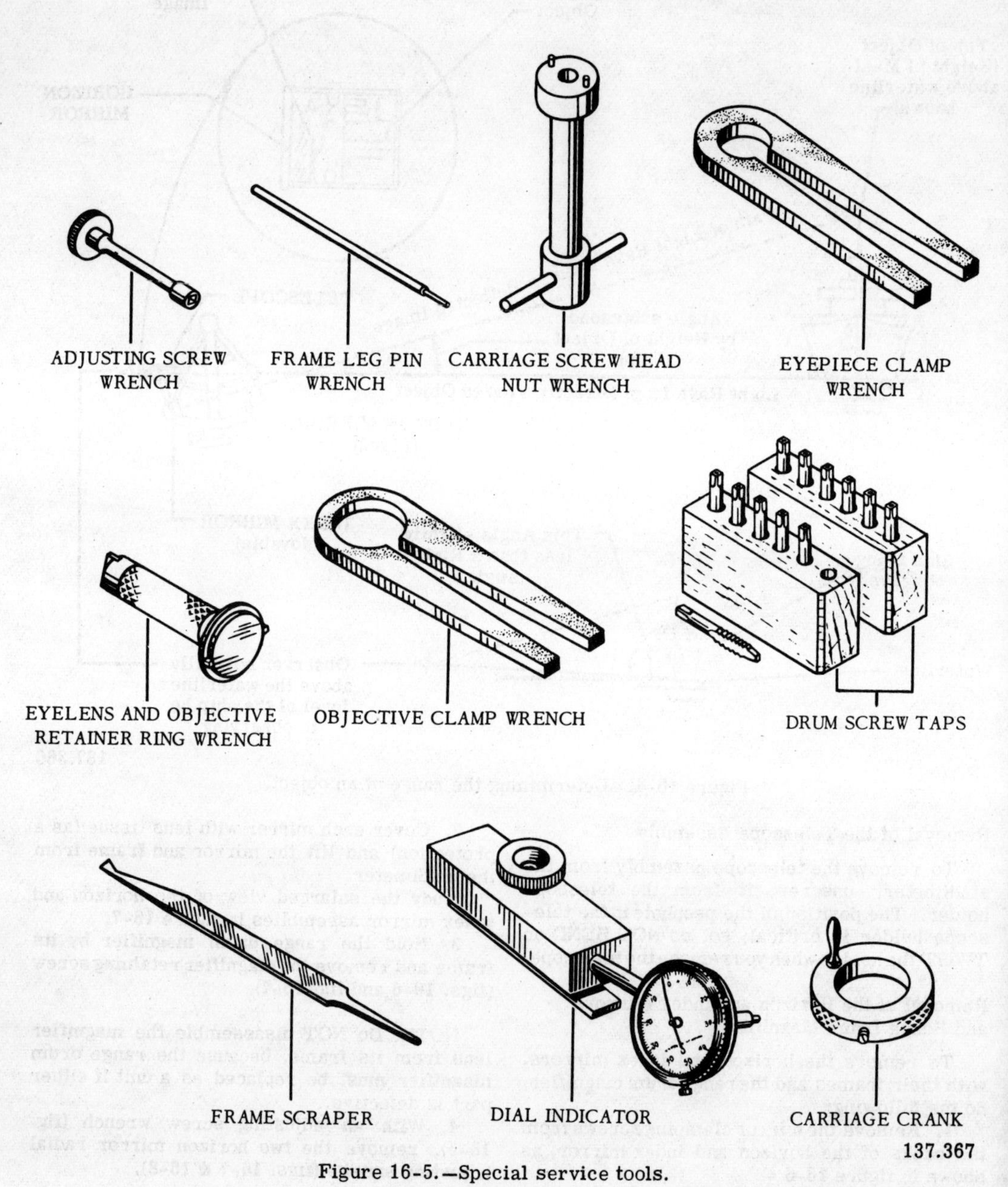

137.367

Figure 16-5.—Special service tools.

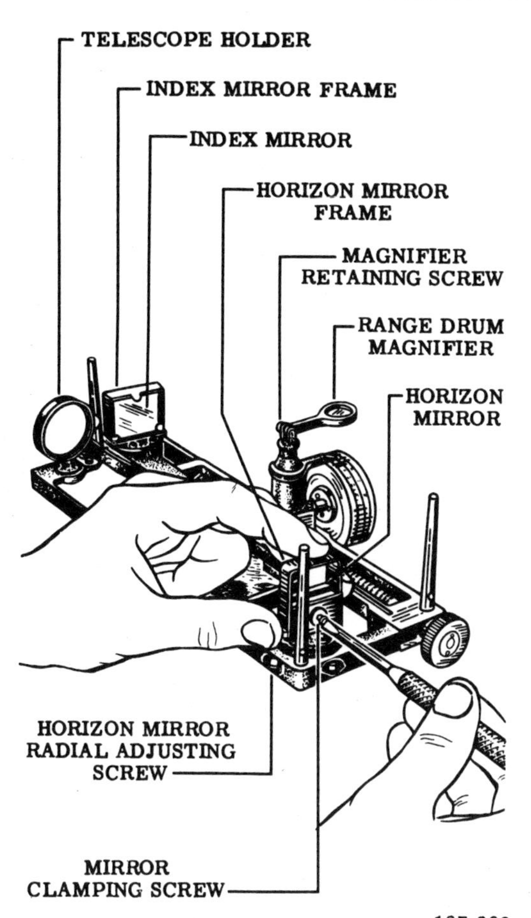

Figure 16-6.—Removing the horizon mirror clamping screw.

5. Remove the horizon turntable mounting screw and washer. Then remove the horizon mirror base and turntable assembly.

6. Turn the stadimeter over and let it rest on its legs. Then remove the scale arm guide screw. See figure 16-9.

7. Remove the center cap (fig. 16-7).

8. Support the scale arm to keep the male center (part of the index mirror base and scale arm assembly) in the female center as you remove the male center screw (fig. 16-7).

9. Scribe a reassembly guide mark across the end of the male center and the male center washer, as illustrated in figure 16-10. Note the enlarged portion in the circle.

10. Turn the stadimeter over and lift the index mirror base and scale arm assembly out of the female center, as illustrated in figure 16-11. Observe the position of the left hand, which should hold the scale arm spring to prevent pressure on the scale arm.

Removal of Handle Assembly

Remove the handle and handle-to-frame post screws and lift the post off, as shown in figure 16-12. Then remove the handle post and bracket (fig. 16-13).

NOTE: Replace the handle and stud as a complete unit whenever necessary. Do NOT disassemble it.

Removal of Carriage and Drum Screw Assemblies

Run the carriage and drum screw assembly all the way down on the carriage screw and scribe a reassembly guide mark across the carriage guide plate and the carriage, as illustrated in figure 16-14. Then remove the bearing cap screws and lift off the carriage screw bearing cap and guide plate. Be sure to reassemble together the bearing cap and screw. They are parts fitted to the frame.

Lift the carriage screw assembly from its open bearing (fig. 16-15), withdraw it from the end bearing in the frame, and unscrew it from the carriage and drum screw assembly.

Removal of Frame Legs and Scale Arm Assembly

Do NOT remove the frame legs and the scale arm assembly spring unless you must do so on effect repairs, make a replacement, or refinish the frame. The procedure for removing these parts follows:

1. Unscrew the three frame legs (fig. 16-16).

2. Unscrew the spring assembly screws and remove the scale arm spring. Do NOT disassemble the spring assembly. The two springs are riveted together and must be replaced as a unit.

Removal of Telescope Holder

The telescope holder is secured to the frame with a screw, and it is doweled to correct

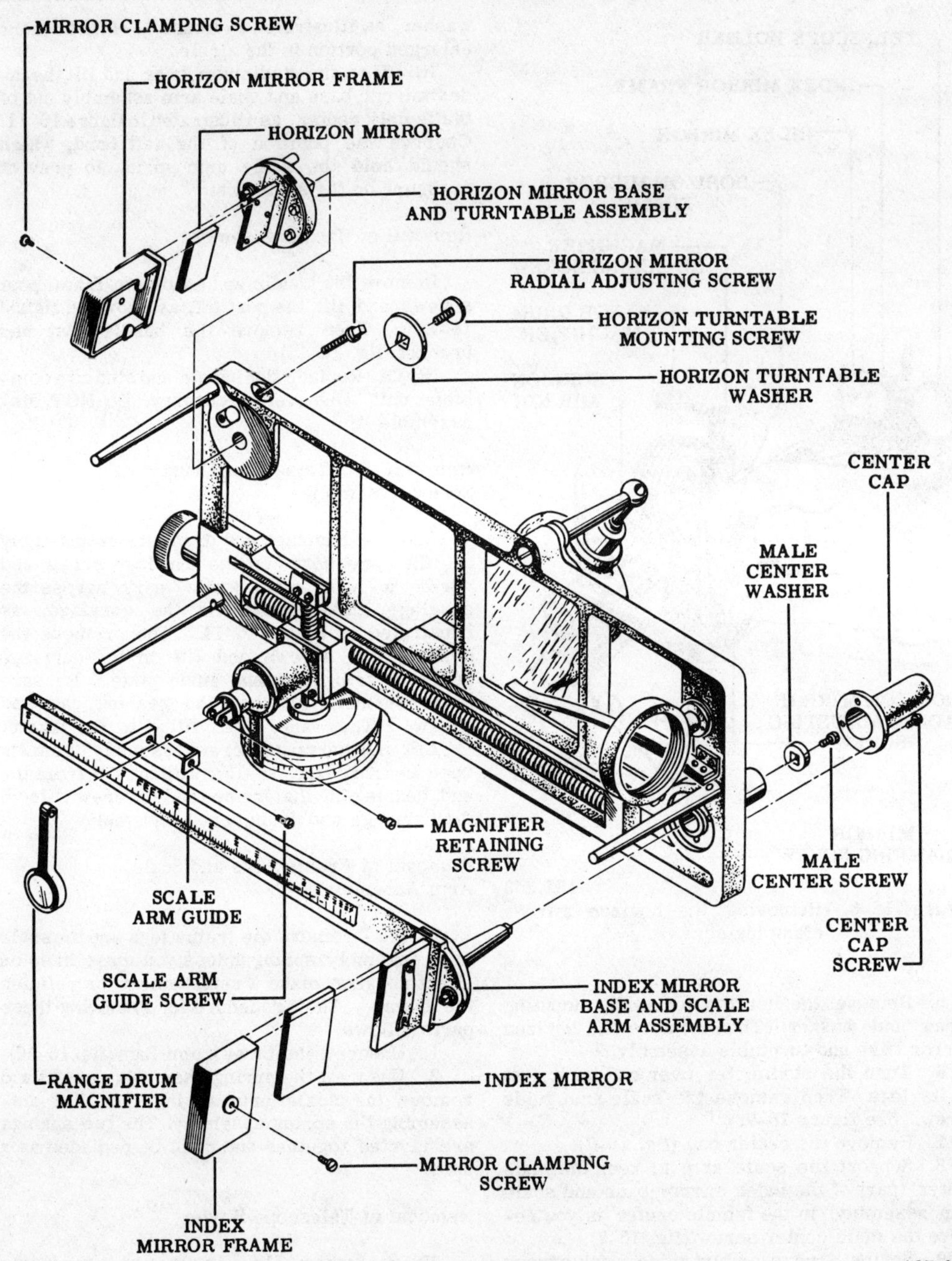

Figure 16-7.—Horizon and index mirror assemblies. 137.369

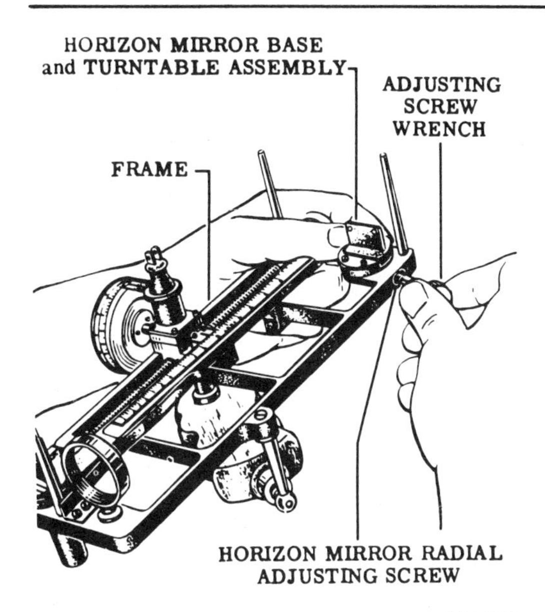

137.370
Figure 16-8.—Removing the horizon mirror radial adjusting screw.

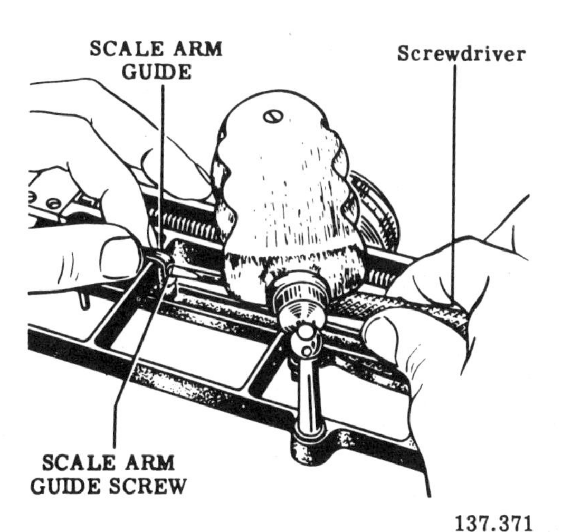

137.371
Figure 16-9.—Removing the scale arm guide screw.

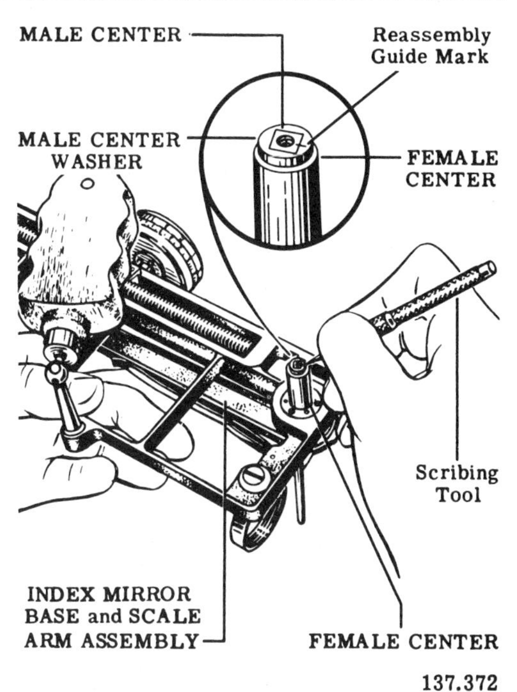

137.372
Figure 16-10.—Scribing reassembly guide marks on the male center and washer.

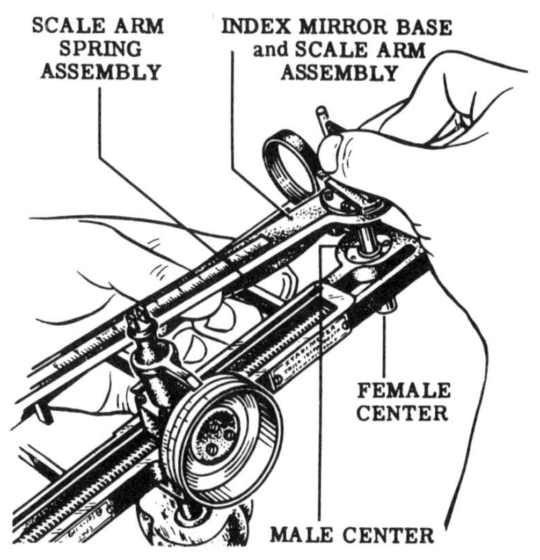

137.373
Figure 16-11.—Removing the index mirror base and scale arm assembly.

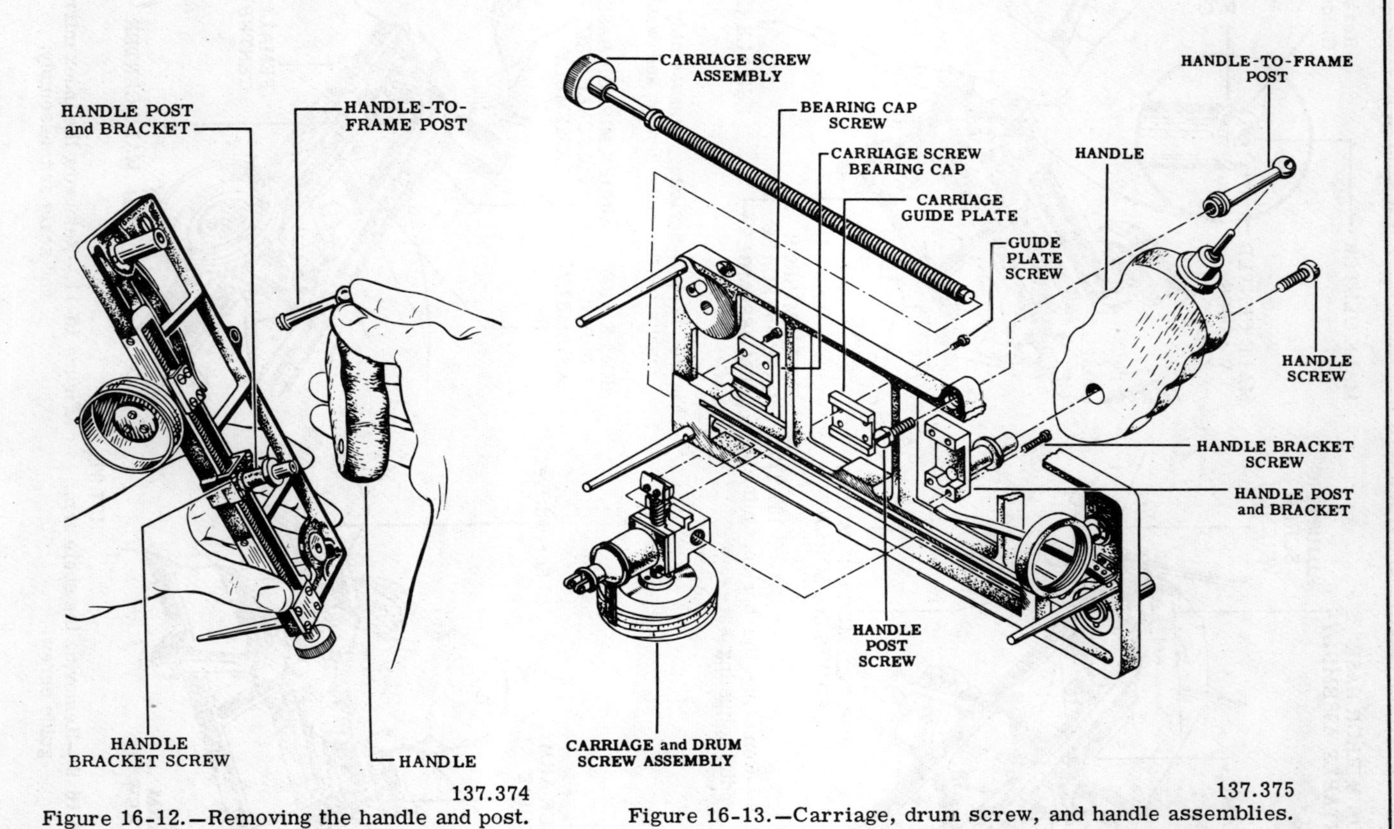

Figure 16-12.—Removing the handle and post.

Figure 16-13.—Carriage, drum screw, and handle assemblies.

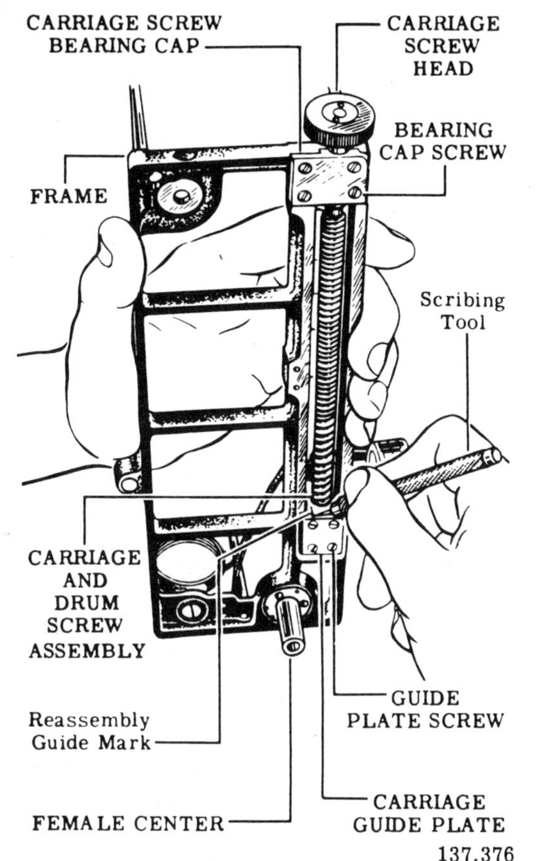

137.376
Figure 16-14.—Scribing reassembly guide marks across the carriage plate and carriage.

position for the peephole (small hole in the car) with the telescope holder pin. Remove the screw and pry the holder off the pin (fig. 16-16). Keep the holder with the frame for correct reassembly.

Disassembly of Carriage Screw Assembly

With a carriage screw head nut wrench (fig. 16-5), remove the carriage screw head nut (fig. 16-17). Then unscrew the carriage screw head from the screw, as illustrated.

Disassembly of Carriage and Drum Screw Assembly

Refer to illustrations 16-18 and 16-19 as you study the procedure for disassembling the carriage and drum screw assembly, as follows:

1. Remove the index plate screws, the index plate, and the two parts of the drum screw socket from the end of the drum screw. These parts are fitted with dowel pins. Put reassembly guide marks on them to ensure proper reassembly. The drum screw has a double thread fitted to a corresponding thread in the carriage.
2. Turn the drum screw as required to bring the end of the thread flush with the surface of the carriage and scribe a reassembly guide mark across the end of the screw and the face of the carriage. See illustration 16-20.
3. Remove the drum screw from the carriage.
4. Scribe a reassembly guide mark across the side of the carriage (fig. 16-21). The drum screw and carriage must be reassembled in their original positions.
5. Remove the four drum index arm screws (fig. 16-19).

NOTE: Keep the carriage and the drum index arm together, because they were drilled and tapped as an assembly for the drum screw.

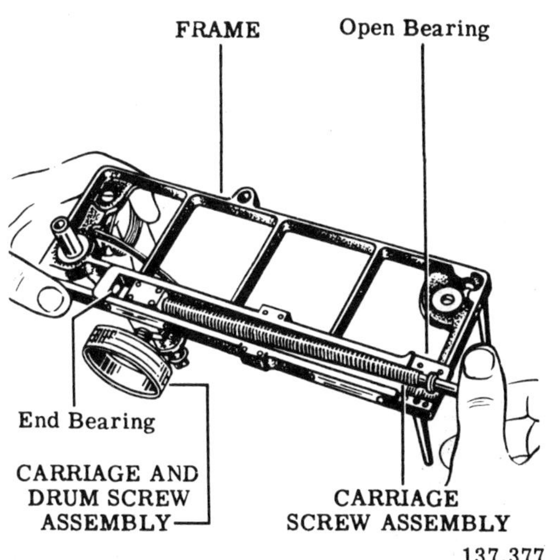

137.377
Figure 16-15.—Removing the carriage screw assembly.

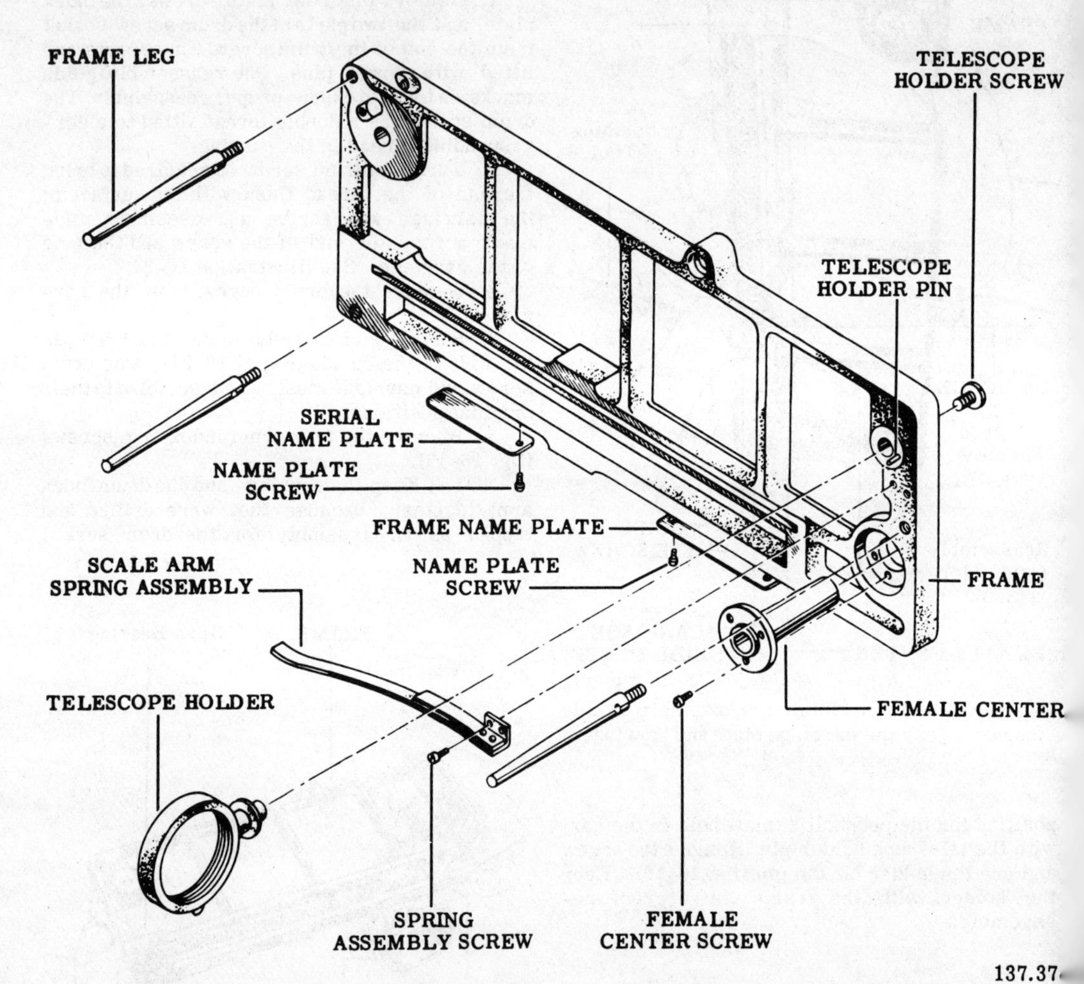

137.37

Figure 16-16.—Removal of the female center, legs, scale arm spring assembly, and telescope holder.

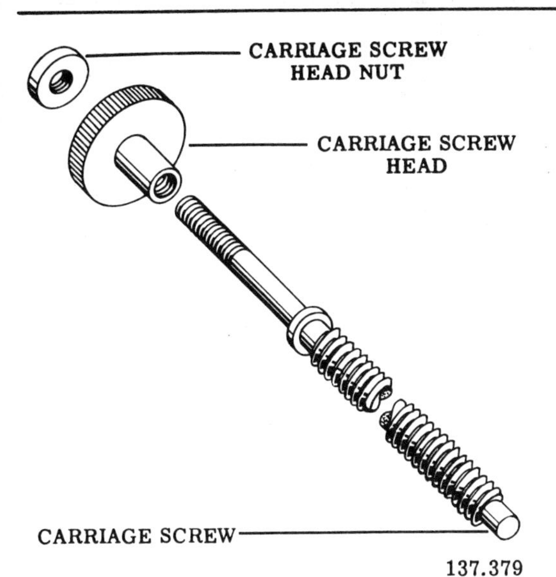

137.379
Figure 16-17.—Disassembly of the carriage screw assembly.

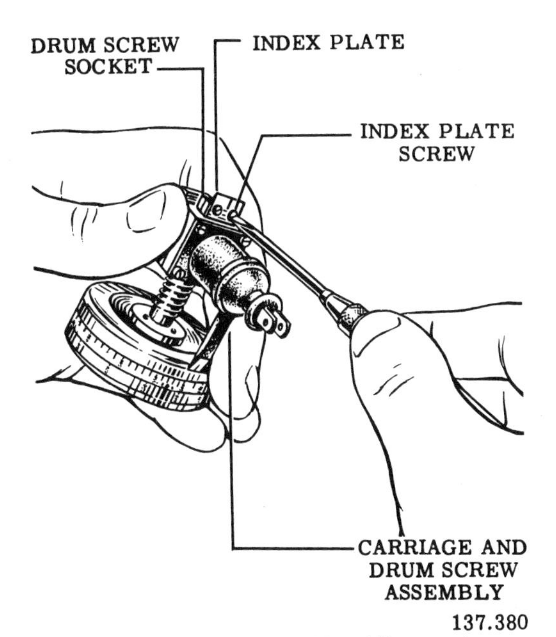

137.380
Figure 16-18.—Disassembly of the carriage and drum screw assembly.

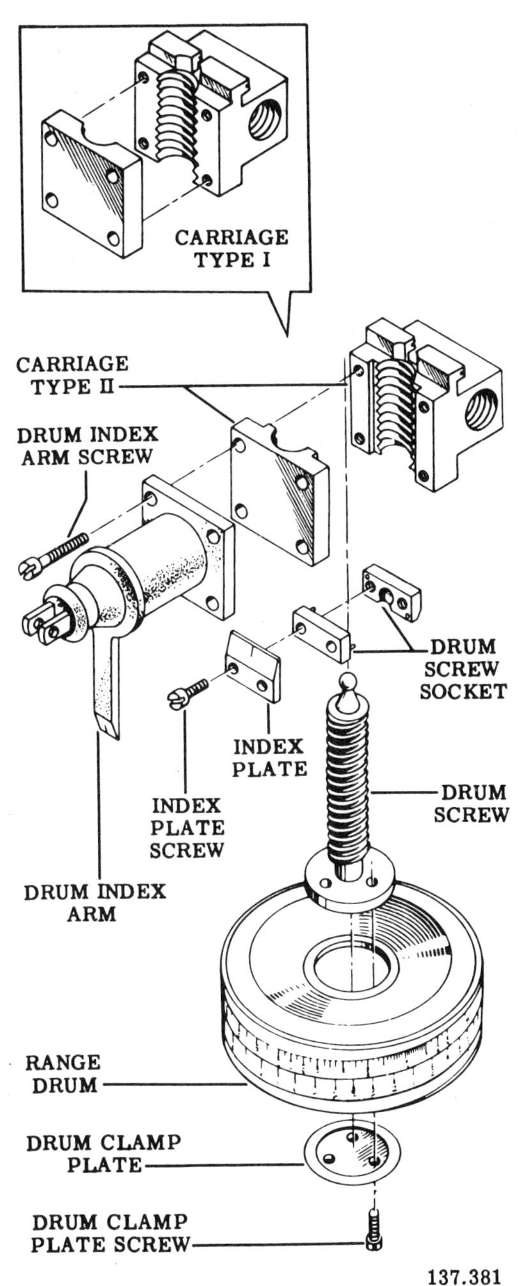

137.381
Figure 16-19.—Components of the carriage and drum screw assembly.

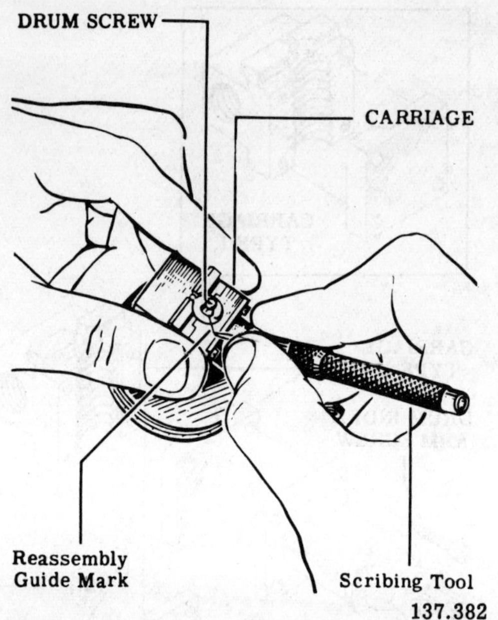

137.382
Figure 16-20.—Scribing reassembly guide marks on the drum screw and the face of the carriage.

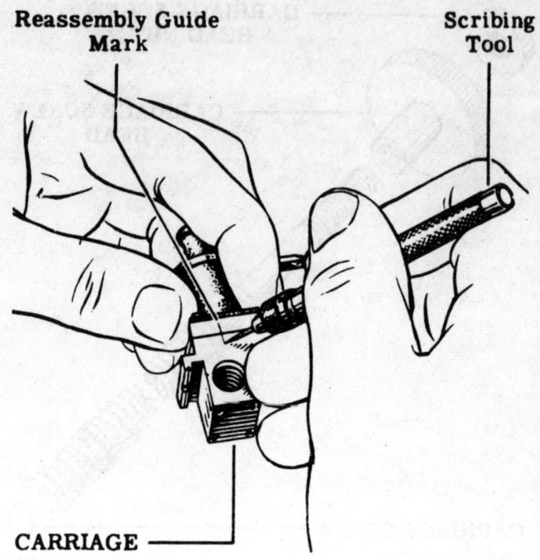

137.383
Figure 16-21.—Scribing a reassembly guide mark across the side of the carriage.

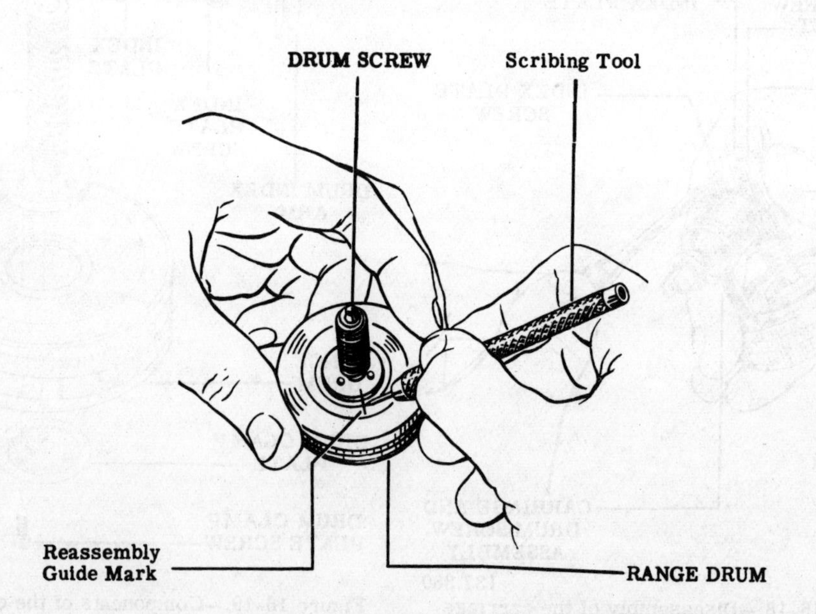

137.384
Figure 16-22.—Scribing reassembly guide marks on the drum screw and the range drum.

Scribe a reassembly guide mark on the drum screw and the range drum, as shown in figure 16-22. Then remove the clamp plate screws and plate and disassemble the range drum from the screw.

Disassembly of Index Mirror Base and Scale Arm Assembly

Do not bend or damage the scale arm when you disassemble the index mirror base and scale arm assembly. Bumps or depressions on the finished edge WILL CAUSE ERRORS in range readings. Proceed as follows to disassemble the assembly:

1. Remove the scale setscrew from the bottom of the scale arm (fig. 16-23).
2. Hold the scale arm flat on the bench, as shown in figure 16-24. Then press down on the slot in the end of the index scale with a screwdriver and pull the scale arm straight toward you. Do NOT bend or twist the arm.
3. To disassemble the index mirror base and clamp spring assembly from the scale arm, remove the index mirror vertical adjusting screw (fig. 16-25).
4. Remove the mirror clamping spring from the index mirror base and clamping spring assembly. See illustration 16-23.

Disassembly of Horizon Mirror Base and Turntable Assembly

The procedure for disassembling the horizon mirror base and turntable assembly follows.

1. Use an adjusting screw wrench to remove the vertical adjusting screw (fig. 16-26). Note the differences in the assemblies illustrated.
2. To disassemble the clamping spring, remove the screw which secures it.

Disassembly of the Telescope Assembly

The following disassembly procedure is applicable to telescopes on Fiske and sextant-type stadimeters:

1. Grasp the knurled shoulder of the eyepiece lens mount and pull the eyepiece drawtube (fig. 16-27) from the telescope body. If it is stuck, use an eyepiece clamp wrench.
2. Unscrew the eyelens mount from the eyepiece drawtube. See illustration 16-28.
3. With an eyelens and objective retainer ring wrench (fig. 16-5), unscrew the eyelens retainer ring and remove the eyelens. Study figure 16-29. If necessary, use an eyepiece clamp wrench to hold the eyelens mount when you do this task.
4. The eyelens is loose in its mount (fig. 16-28); so use lens tissue as a protection when you remove it.
5. Unscrew the objective lens mount from the telescope body (fig. 16-30).
6. With an eyelens and objective retainer ring wrench, unscrew the objective lens retainer ring, as shown in illustration 16-31.
7. Use lens tissue to protect the objective lens as you remove it from the mount. Then wrap it in the tissue.

REPAIR PROCEDURE

After disassembling the stadimeter, inspect mechanical and optical parts, particularly those parts which were not easily accessible during the predisassembly inspection. Repair those parts which still have useful life; replace broken and missing parts. Clean usable parts in the prescribed manner and protect them until needed.

The repair procedures discussed herein are peculiar to stadimeters.

Fitting the Drum Screw to the Carriage

The procedure for fitting the drum screw to the carriage is the same for sextant-type and Fiske stadimeters.

The drum screw of a Fiske stadimeter fits in the two sections of the carriage in a 3/8-inch-12 National Special, right-hand, double thread. It is a Class 4 fit, which is a lapped fit. A sextant-type stadimeter has a 3/8-inch American National Acme thread, which is a Class 3 fit. A fit as close as either of these is essential to eliminate play in the drum screw. A sideshake or backlash (endshake) of .001 inch will cause an appreciable error in range readings.

The thread wears from use and causes play in the drum screw. Proceed as follows to fit the parts:

1. Remove old grease from the parts with an approved cleaning solvent and test the fit of the drum screw in the carriage for sideshake and endshake. Study illustrations 16-32 (Fiske) and 16-33 (sextant-type). If the screw is looser at some points than at others, uneven wear has taken place and the screw must be replaced.

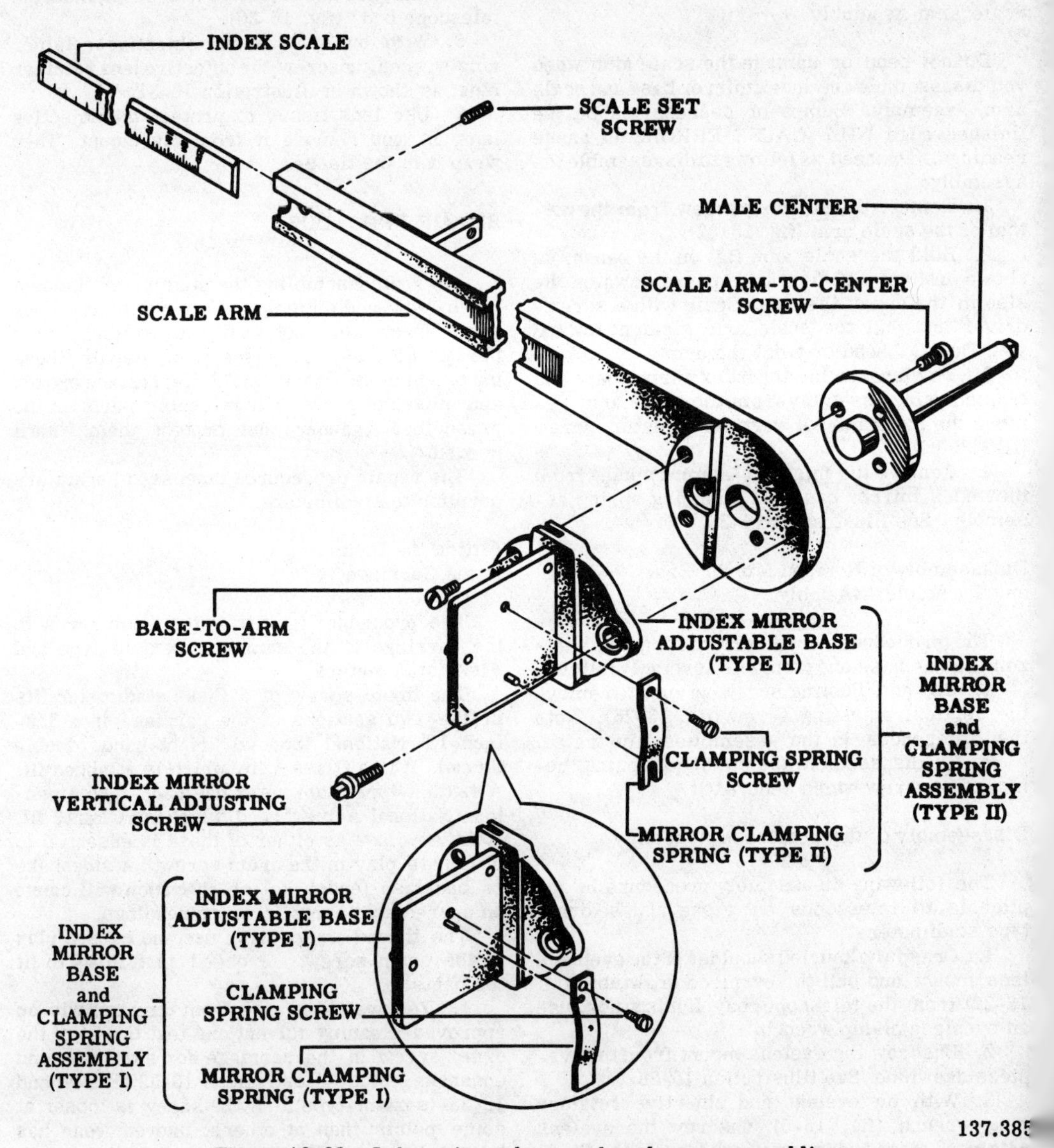

Figure 16-23.—Index mirror base and scale arm assemblies.

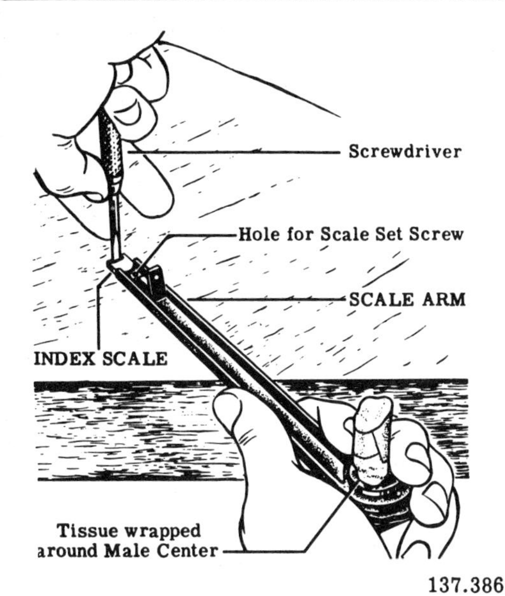

137.386
Figure 16-24.—Pulling the scale arm out.

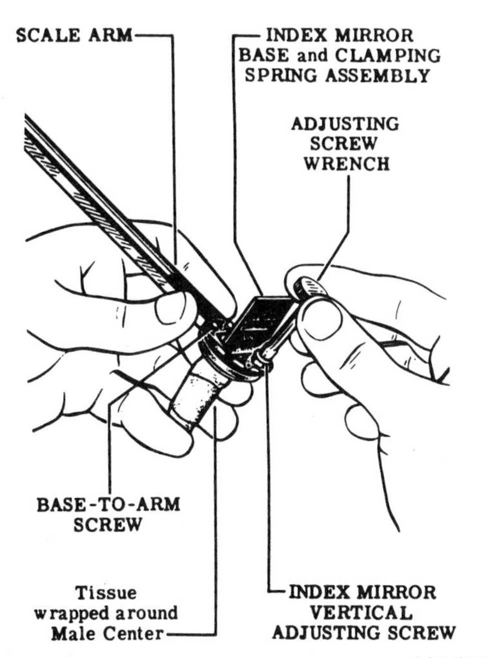

137.387
Figure 16-25.—Removing the index mirror vertical adjusting screw.

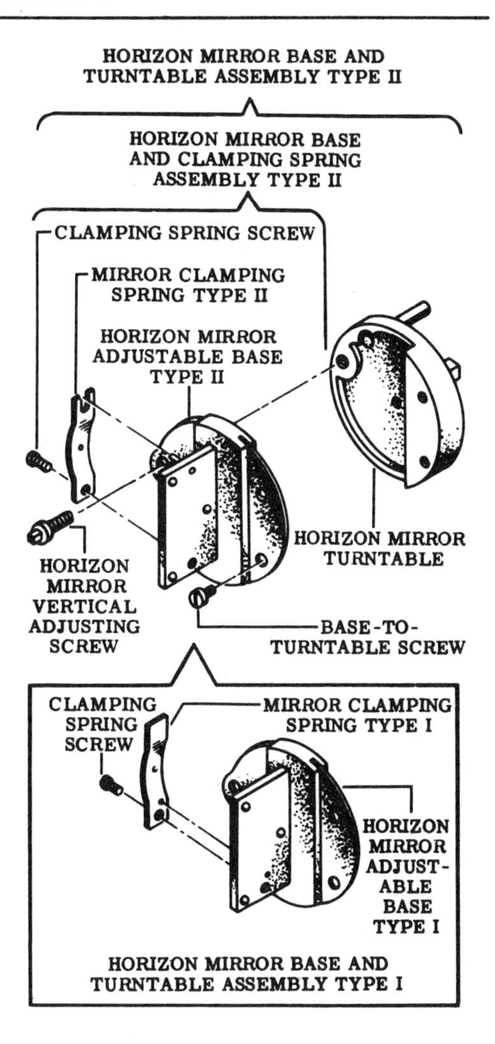

137.388
Figure 16-26.—Disassembled horizon mirror base and turntable assembly.

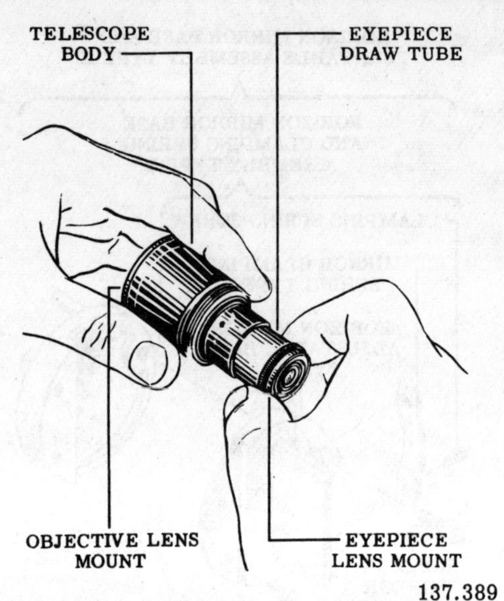

137.389

Figure 16-27.—Pulling the eyepiece drawtube from the telescope body.

2. If a drum screw is loose, partially release the four drum index arm screws (Fiske) and shift the upper part of the carriage as necessary to eliminate looseness (fig. 16-34). Then tighten the screws.

The aligning step in a Type II Fiske or a sextant-type stadimeter keeps the threaded portion of the upper and lower parts of the carriage aligned. Study illustration 16-34. If necessary, use step No. 3 to eliminate play.

3. On a Fiske stadimeter, separate the two parts from the carriage and file their contact faces with a fine file to close up the threaded hole for the drum screw. Then reassemble the parts and try the fit of the drum screw.

CAUTION: Remove a MINIMUM amount of metal, because this adjustment makes the hole egg-shaped and causes rapid wear on the threads as a result of increased pressure on them.

4. If a drum screw does not fit after you file it down, use the set of six drum taps provided and tap the carriage. These taps range from .002 inch below the .0045 inch above the normal of 3/8 inch provided for tapping the carriage. Start with the smallest tap and try the drum screw. Then try the next larger size and

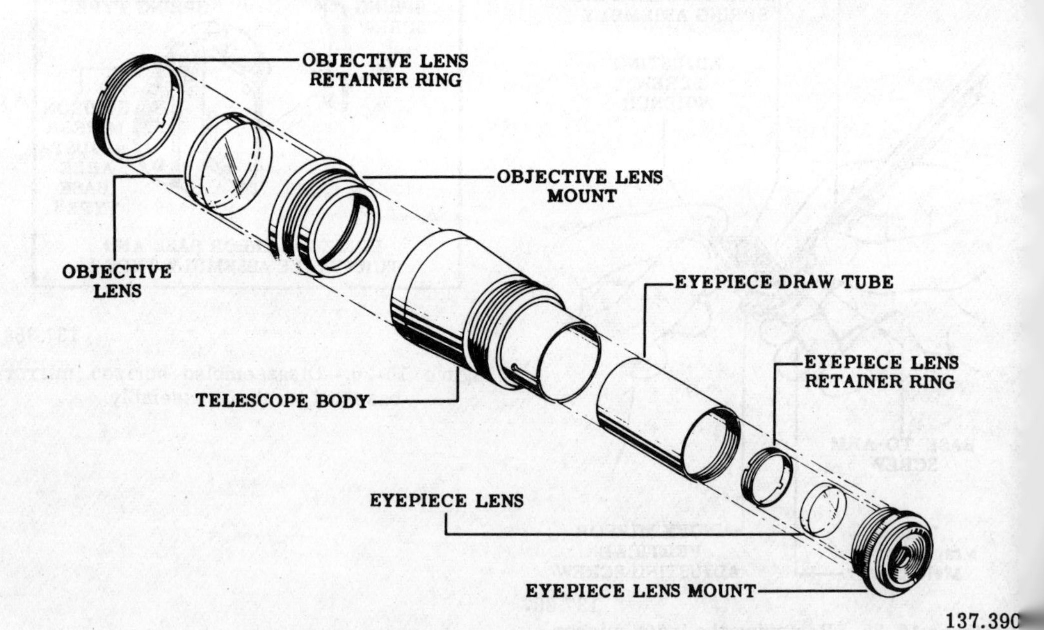

137.390

Figure 16-28.—Components of a stadimeter telescope.

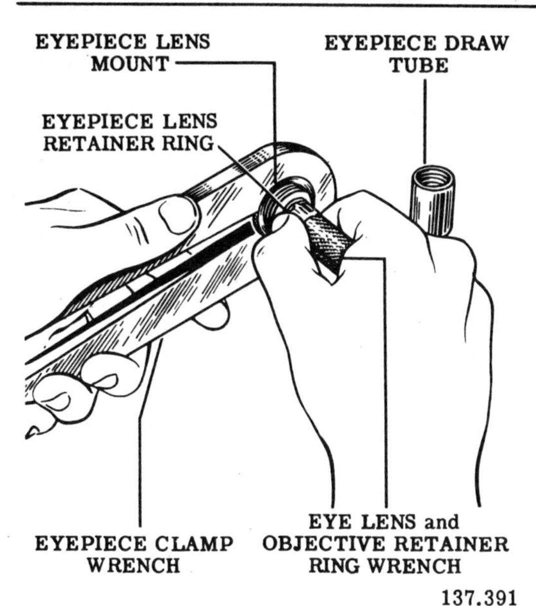

Figure 16-29.—Unscrewing the eyepiece lens retainer ring.

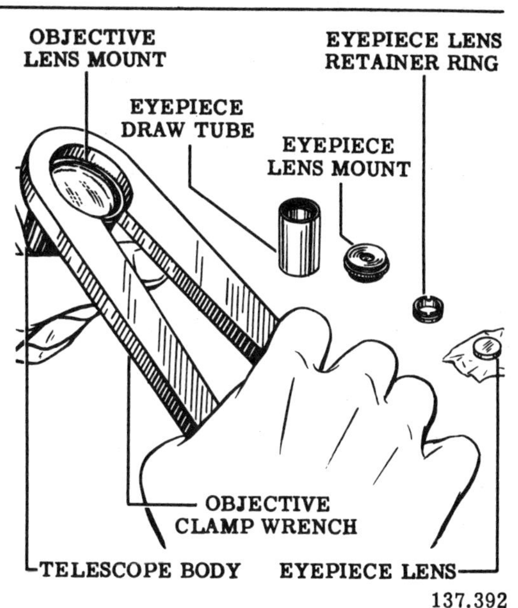

Figure 16-30.—Removing the objective lens mount from the telescope body.

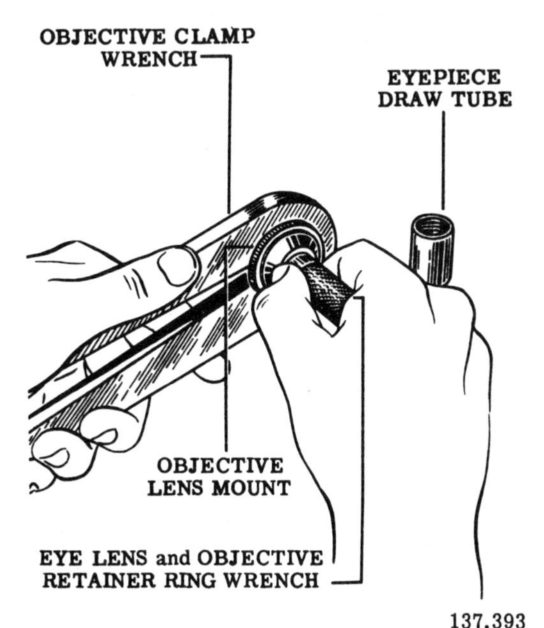

Figure 16-31.—Unscrewing the objective lens retainer ring.

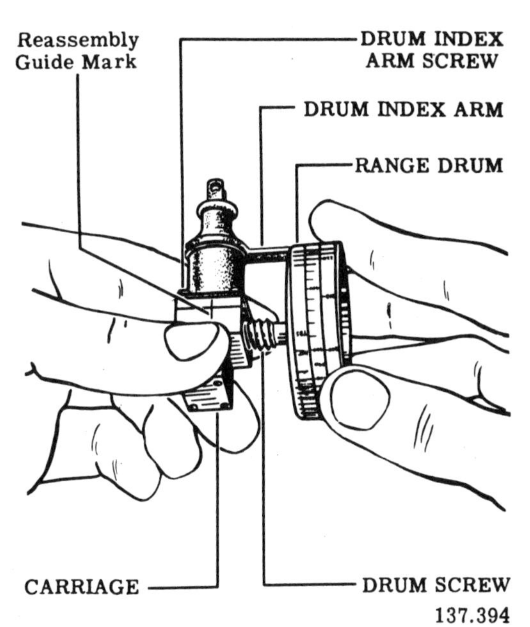

Figure 16-32.—Testing the fit of a Fiske drum screw in the carriage.

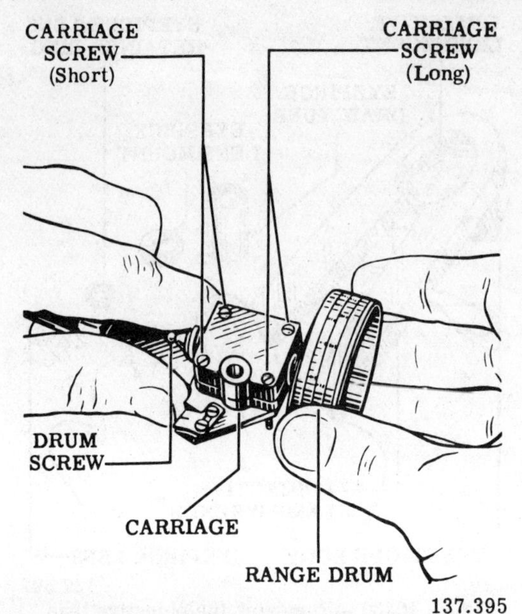

137.395
Figure 16-33.—Testing the fit of the drum screw in the carriage of a sextant-type stadimeter.

as many others as necessary until a proper fit is obtained.

Fitting the Carriage to the Frame

To prevent play of the carriage, its tongue (fig. 16-35) must fit the groove in the frame over its entire length; and the carriage guide plate must fit the carriage and the frame in order to hold the carriage in the groove for a close fit. The fitting procedure outlined next will eliminate play in original and new parts.

1. Slide the lower part of the carriage on the frame, as shown in illustration 16-35. Carriages for Types I and II Fiske stadimeters are shown in figure 16-36.

2. If the lower part of the carriage is loose over the entire length of the frame groove, the tongue is too narrow. Set the tongue of the carriage (carriage upside down) on a metal block and spread the tongue by hitting each section on its bottom edge with a chisel, as illustrated in figure 16-37.

CAUTION: To prevent damage to the carriage during this operation, hold it securely on the metal block.

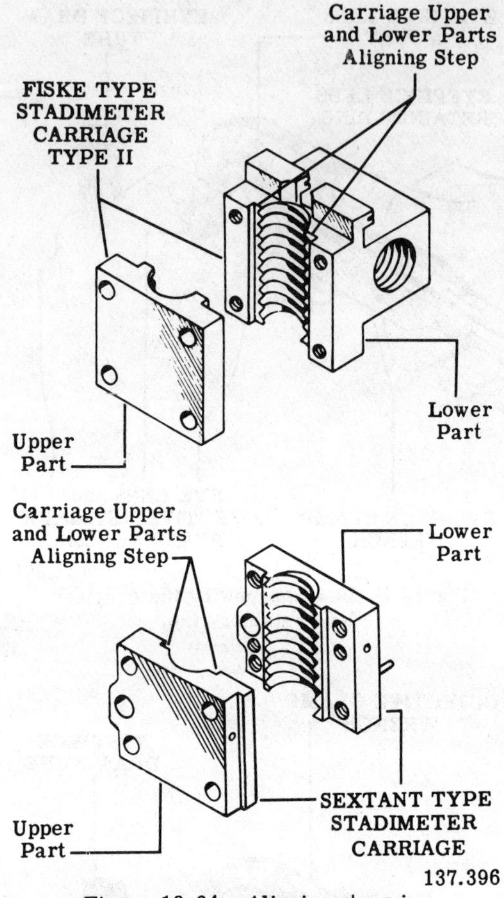

137.396
Figure 16-34.—Aligning steps in stadimeter carriage.

3. Try the tongue for fit in the widest point in the frame groove. If the tongue is too wide, remove enough material from its side with fine emery cloth to make it fit. This action is sometimes necessary when a new carriage is being fitted.

4. Start at the widest point to which you fitted the lower part of the carriage and move the carriage along the frame to try it for fit. If necessary, use a frame scraper to scrape the side of the groove to the extent required to make it fit the tongue of the carriage. Study illustration 16-3. Arrow No. 1 in the illustration points

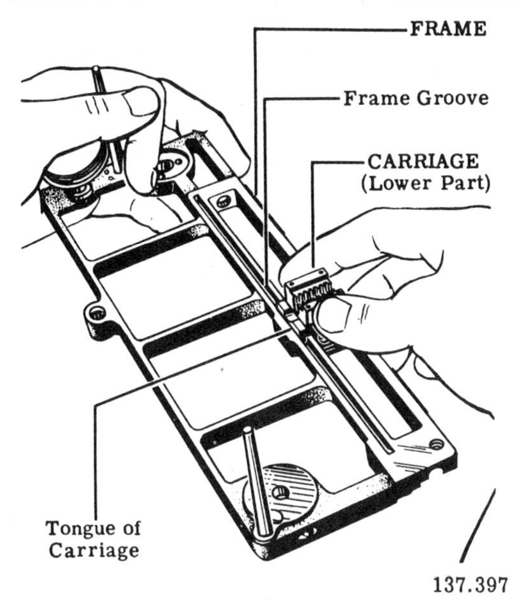

Figure 16-35.—Fitting the carriage and carriage guide plate to the frame.

to the side which should be scraped, because the tongue bears more heavily against it and therefore causes greater wear than on the other side (arrow No. 2). Check the unscraped side for burrs or other defects.

5. When the lower part of the carriage slides over the length of the groove without play or binding, replace the carriage guide plate on the carriage (fig. 16-39). Assemble original parts in accordance with guide marks. If you fit new parts, scribe reassembly guide marks on them.

6. Slide the lower part of the carriage along the frame and locate the position where the carriage has the most play. Then remove the carriage guide plate and file the bottom of the lower part of the carriage to make the guide plate come down and hold the carriage without play.

If the lower part of the carriage (with guide plate attached) is tight all along the frame, use fine emery cloth to remove a little material from the two bottom bearing surfaces of the carriage guide plate.

7. Start with a fit at the widest (loosest) point, move the carriage along the frame, and scrape the frame in the manner shown in figure 16-39. If a smooth, close fit cannot be obtained all along the frame, the frame is warped and must be replaced.

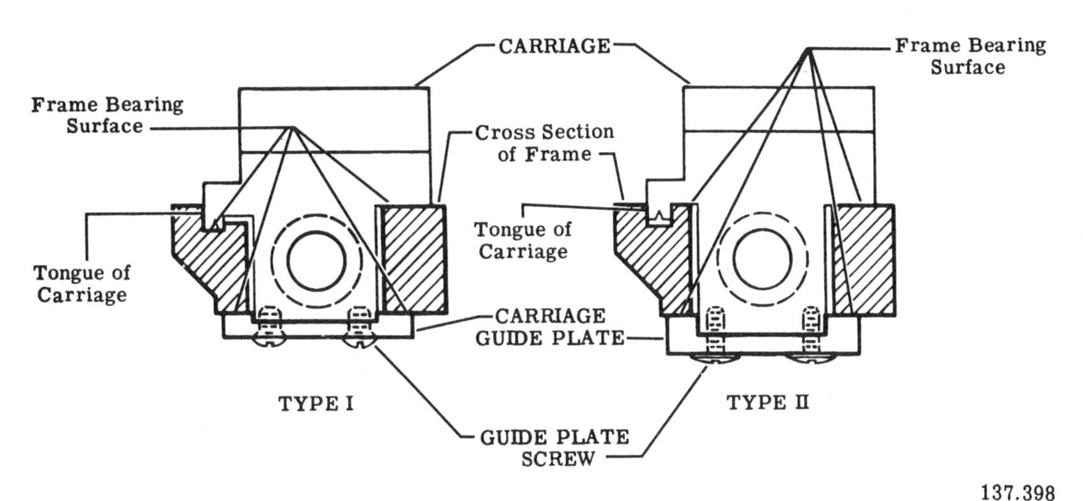

Figure 16-36.—Carriage for Type I and Type II Fiske stadimeters.

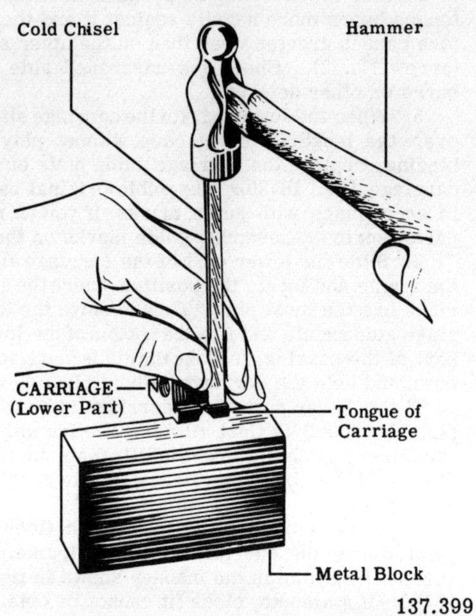

Figure 16-37.—Spreading the tongue of the carriage.

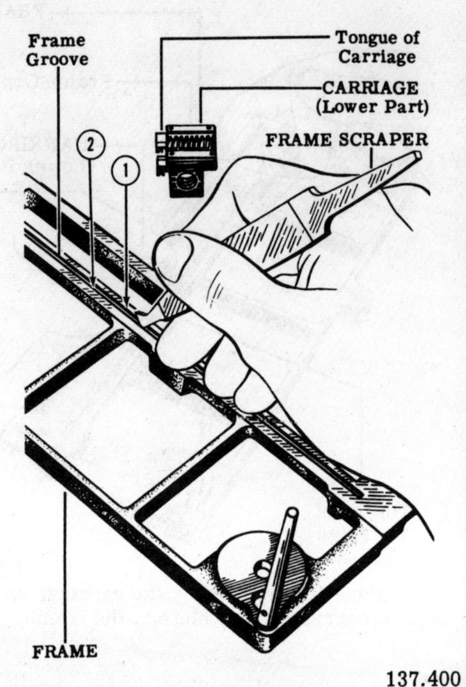

Figure 16-38.—Removing metal from the frame groove.

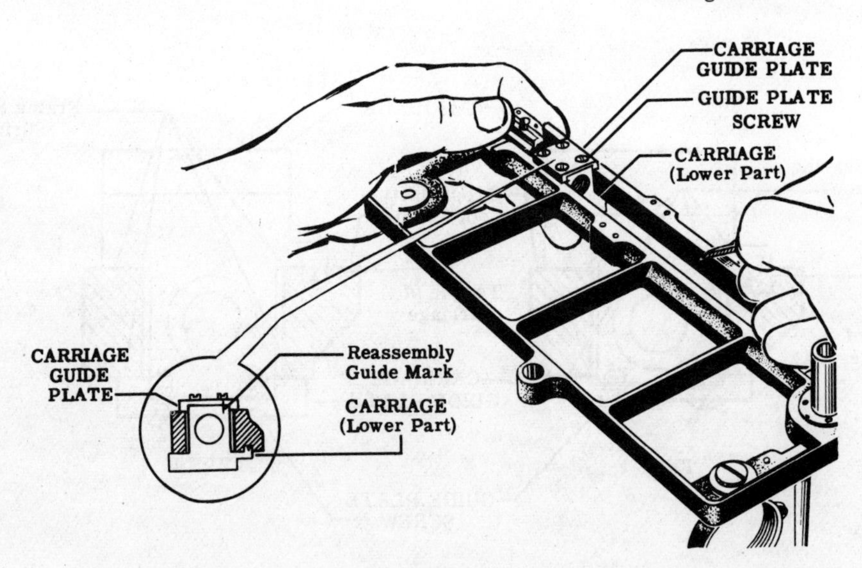

Figure 16-39.—Assembly of the carriage guide plate.

Positioning and Pinning the Telescope Holder

A telescope holder must be so positioned on the frame that the optical center of a telescope mounted in it is in line with the pivotal point of the horizon mirror. See illustration 16-40. The telescope holder pin fits tight in the frame and in a sliding fit in the telescope holder. See the circled portions at the bottom of illustration 16-40.

The procedure for drilling holes for a telescope holder is somewhat detailed and is not discussed here. You may on occasions be required to drill these holes, however, and the instructions in NavShips 250-624-6 will be most helpful.

Repair of Sextant-Type Stadimeters

If you understand the procedure for disassembling and repairing a Fiske stadimeter,

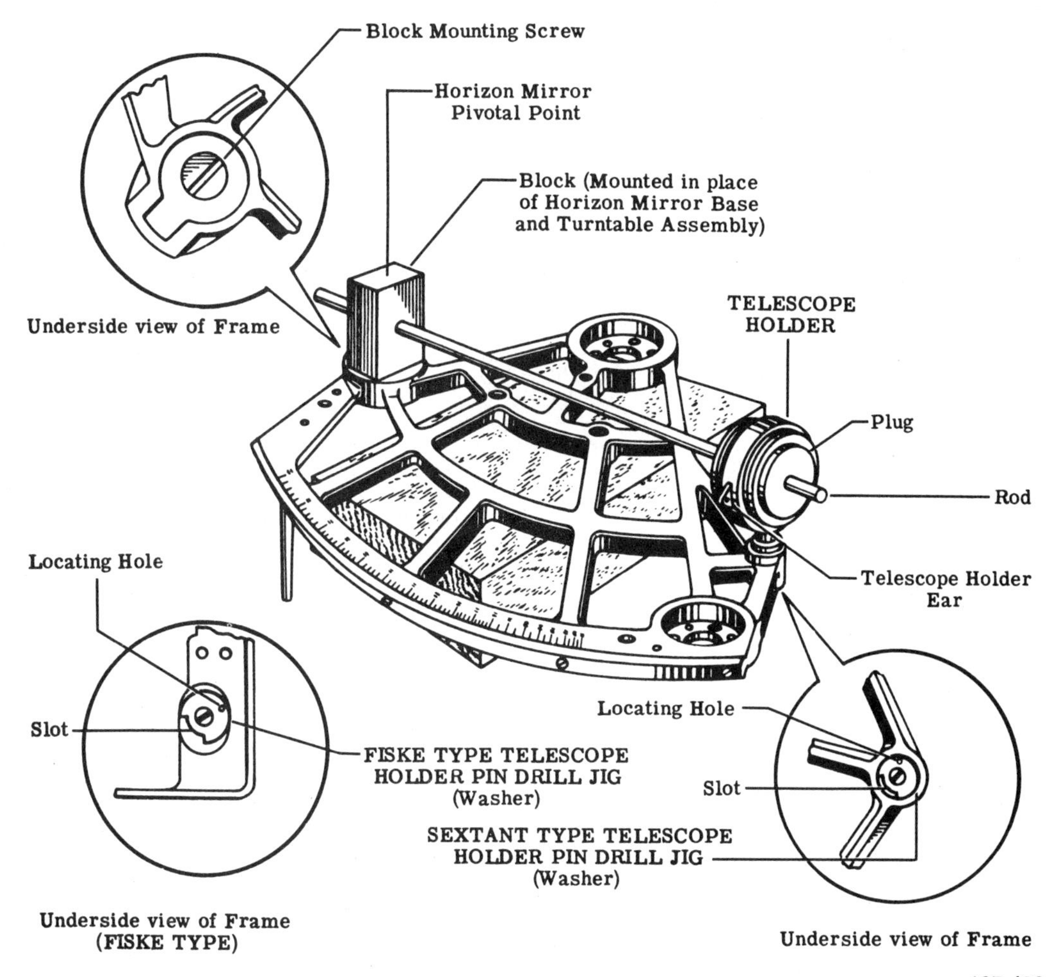

Figure 16-40.—Positioning and pinning the telescope holder.

you will experience little or no difficulty in repairing a sextant-type stadimeter. There are some differences in the procedure, however, which should be emphasized, as follows:

1. After removing the telescope and mirrors from a sextant-type stadimeter, remove the screws from the arcuate arm guard (fig. 16-41) and then the guard (fig. 16-42).

2. Remove the center covers from the two moving centers.

3. Remove the locking screw and bushing from the center assembly of the arcuate arm (fig. 16-43).

4. Screw the index drum out and lift the arcuate arm from its pivot (fig. 16-44).

5. The arcuate arm tension spring will push the plunger out of the spring housing (fig. 16-45). Remove the spring and the plunger.

6. Drive the dowel pin out of the retaining ring on the end of the lock spindle screw (fig. 16-46) to free the lock nut assembly spring (fig. 16-47).

7. Withdraw the lock spindle from the other end of the assembly (fig. 16-48).

8. The drum index marker (fig. 16-49) is adjustable. When you reassemble the instrument, adjust the pointer so that the circular lines on the index drum are visible and line up the index mark with the range marks on the drum.

9. Before replacing the arcuate arm pivot in its center assembly, turn the drum screw out enough to give room for working. Then depress the spring and plunger with the left thumb while you slip the arm in place (fig. 16-50).

REASSEMBLY

This section contains the standardized procedure for reassembling Fiske stadimeters. Subassemblies are assembled first, in reverse order to disassembly, and they are then placed in the assemblies for final installation on the frame.

Horizon Mirror Base and Turntable Assembly

Proceed as follows to reassemble the turntable assembly:

1. Replace the mirror clamping spring on the back of the adjusting base. These parts comprise the horizon mirror base and clamping spring assembly. Review illustration 16-26. The slot in the clamping spring fits around the clamping spring guide pin. If you removed it, replace the guide pin in the adjustable base. Use care as you tap the pin into position.

2. With an adjusting screw wrench, replace the horizon mirror base and clamping spring assembly on the horizon mirror turntable.

Index Mirror Base and Scale Arm Assembly

The first two steps in the reassembly of the index mirror base and scale arm assembly are the same as for the horizon mirror. After completing these steps, slide the index scale into the groove in the scale arm, high end of scale (marked 200) first. The parts are shown in illustration 16-23. Replace the scale set screw in the underside of the scale arm, but do not tighten it until you collimate the instrument.

Carriage and Drum Screw Assembly

When reassembling the carriage and drum screw assembly, proceed as follows:

1. Put the upper part of the carriage on the lower part and line up the reassembly guide marks. Review illustrations 16-19 and 16-21.

2. Place the drum index arm on top of the upper part of the carriage, with the index arm facing away from the tongue on the carriage, and replace the four drum index arm screws.

3. Replace the range drum and clamp plate to the drum screw (fig. 16-19), but do not tighten the screws. Turn the drum screw assembly over, look for the reassembly guide mark on the drum screw and range drum (fig. 16-22), and then realign them. Tighten the drum clamp plate screws.

4. Apply a thin coat of approved drum screw grease on the thread of the drum screw and screw the assembly into the hole formed by the two parts of the carriage. The shoulder of the drum screw must be flush with the other side of the carriage. If the reassembly guide marks on these parts (fig. 16-20) coincide, the double thread is reassembled in its matched position.

5. If the drum screw assembly runs smoothly throughout its travel in the carriage, the parts are assembled properly. Wipe off any drum screw grease which was squeezed out during the operation. There must be NO play of the drum screw in the carriage. Shift the upper part of the carriage to adjust for fit.

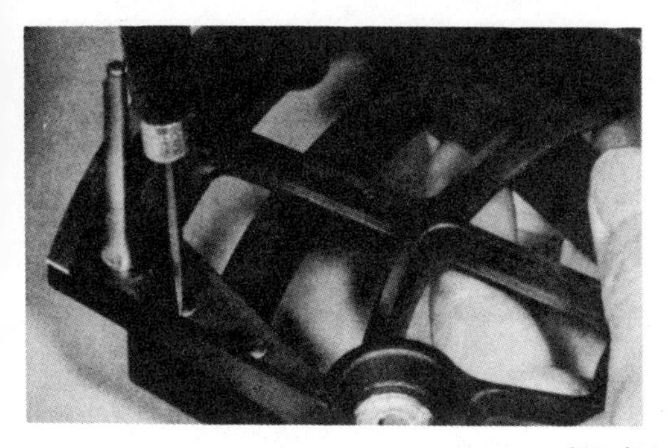

137.403
Figure 16-41—Removing screws from the arcuate arm guard.

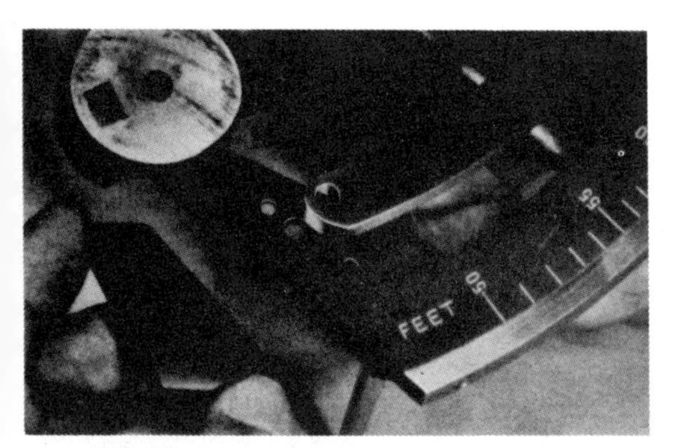

137.404
Figure 16-42.—Removing the arcuate arm guard.

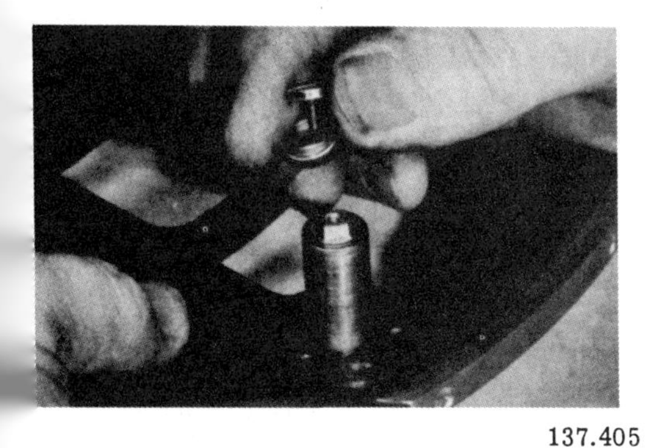

137.405
Figure 16-43.—Releasing the arcuate arm centers.

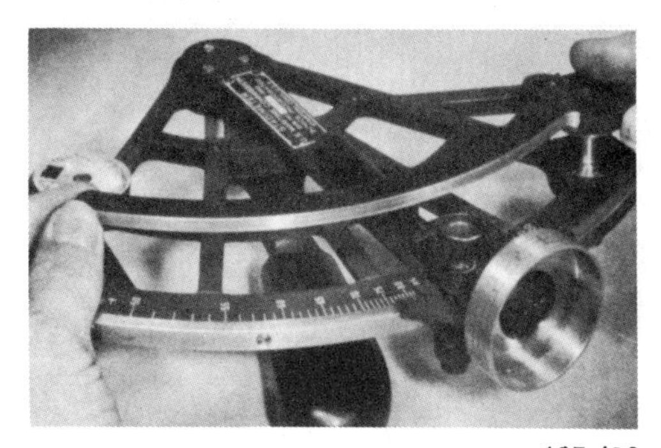

137.406
Figure 16-44.—Removing the arcuate arm.

137.407
Figure 16-45.—Removing the tension spring and plunger.

137.408
Figure 16-46.—Removing the dowel pin.

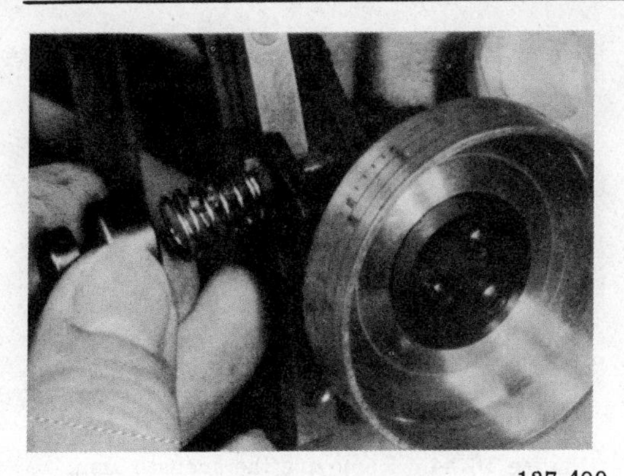

137.409
Figure 16-47.—Removing the lock nut assembly spring.

137.411
Figure 16-49.—Index drum and pointer.

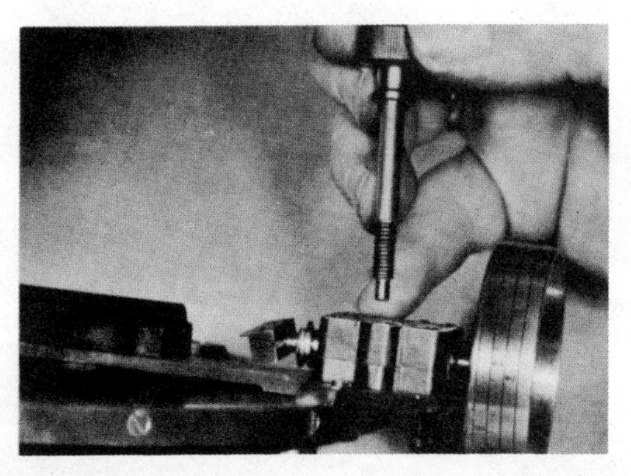

137.410
Figure 16-48.—Removing the lock spindle.

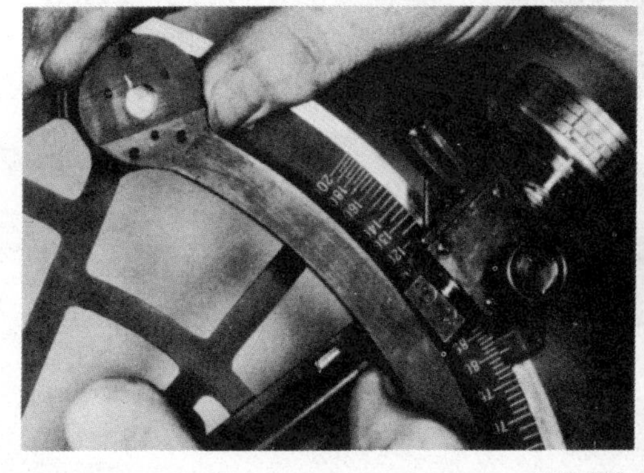

137.412
Figure 16-50.—Remounting the arcuate arm.

6. Apply a thin film of drum screw socket oil on the ball end of the drum screw assembly and put the two parts of the drum screw socket on the end of the screw. These parts are fitted with two dowel pins permanently fixed in the upper section of the socket.

7. Screw the index plate (figs. 16-18 & 16-19) to the top of the drum screw socket and set the carriage and the assembly aside until you replace it on the frame.

Thread the carriage screw head (fig. 16-51) onto the threaded end of the carriage screw, and screw the carriage screw head nut on. Do NOT tighten these parts until the carriage screw is properly fitted in the frame.

Telescope Assembly

The procedure for assembling the telescope assembly for Fiske and sextant-type stadimeters follows.

1. Clean the eyepiece lens in the manner explained in chapter 8 and use tweezers to set its concave side down in the eyepiece lens mount (fig. 16-28).

2. Replace the eyepiece lens retainer ring and then tighten it (fig. 16-29), but not tight enough to put strain in the lens.

3. Screw the eyepiece lens mount into the drawtube (fig. 16-28).

4. Apply a thin coat of eyepiece grease on the outside surface of the drawtube and slide it into the telescope body (fig. 16-31). Wipe off excess grease.

5. Clean the objective lens and use tweezers to set it in the mount, flat side down (fig. 16-28).

6. Replace the objective lens retainer ring.

7. Screw the objective lens mount into the large end of the telescope body.

This completes reassembly of subassemblies of the stadimeter. The next step is the replacement of subassemblies in their proper positions in the instrument.

Legs and Spring Arm Assembly

Screw the three frame legs into the frame and tighten them with a frame leg pin wrench. See figure 16-16.

Screw the scale arm spring assembly to the frame and insert the study of the telescope holder in its position in the frame. Then position the holder with its projecting ear towards the adjacent frame leg, and line up the hole in the holder for the telescope holder pin in the frame. Replace and tighten the holder screw.

Carriage Screw and Drum Screw Assemblies

Follow the procedure outlined next for replacing these assemblies on the stadimeter frame.

1. Spread a film of carriage screw grease over the thread on the screw (bearing end) and on its bearing surface which fits the open bearing in the frame (fig. 16-15).

2. Set the frame on its legs and hold with your left hand the carriage and drum screw assembly in place in the frame. Pick up the carriage screw assembly with your right hand and screw it into the carriage far enough to make its bearing end protrude. Then push the assembly along the frame so that you can insert it in the end bearing and put it down in the open bearing.

3. Secure the carriage screw bearing cap to the frame with the screws provided (figs. 16-13 & 16-14).

4. Put the carriage guide plate in position on the bottom of the carriage and secure it with its four screws. Realign the assembly guide mark.

5. Apply a thin film of grease along the length of the frame where the carriage guide plate rides and screw the carriage up to the end near the screw head.

6. Check for end play in the carriage screw head by pushing and pulling it. To adjust for play, turn the head up against the frame and the bearing cap. Then tighten the screw head nut and lock it in position with a carriage screw head nut wrench. See illustration 16-52.

Handle Assembly

Review illustrations 16-12 and 16-13. Reassembly of the handle assembly is the same as disassembly, in reverse order. Because of the simplicity of the task, the procedure is not repeated here.

Index Mirror Base and Scale Arm Assembly

Do the following to assemble the index mirror base and scale arm assembly to the frame:

1. As illustrated in figure 16-53, oil with a thin film of oil the full length of the male center, on both sides. The oil will spread when the centers are replaced.

2. Pick up the index mirror base and scale arm assembly and insert the male center in the female center (fig. 16-11). Hold the scale arm spring assembly clear of the scale arm; and when the scale arm is resting on the frame, release the arm spring assembly. The arm should be pushed against the drum screw socket and the index plate should be over the edge of the arm. Work the scale arm back and forth and wipe off excess oil (squeezed out).

3. Turn the frame over and support the index mirror base and scale arm assembly with your left hand. Place the male center washer over the square lug on the end of the male center and align it with the reassembly guide mark on the center. Review illustrations 16-7 and 16-10.

4. Replace the male center screw (fig. 16-7) and also the center cap over the female center.

5. Put the scale arm guide in position against the projecting lug beneath the scale arm. Then insert and tighten the screw which secures it. See illustrations 16-7 and 16-9.

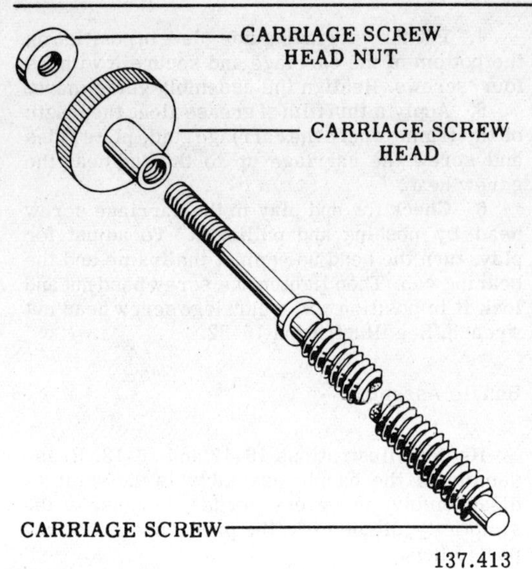

Figure 16-51.—Carriage screw assembly.

Horizon Mirror Base, Turntable, and Range Drum Assembly

To replace the horizon mirror base and turntable assembly, do the following:

1. Lubricate with turntable grease the underside of the horizon mirror base and turntable assembly and the flat side of the turntable washer (fig. 16-7).
2. Secure the assembly to the frame.
3. With an adjusting screw wrench, screw the two horizon mirror radial adjusting screws into the frame, but leave them LOOSE.
4. Screw the range drum magnifier to the drum index arm.

The overhauled stadimeter is now reassembled as completely as possible prior to final tests and adjustments.

POST REASSEMBLY INSPECTION

After reassembling the stadimeter, make a final repair inspection to detect flaws in workmanship before testing and adjusting. Inspect the following:

1. Finish of each part, and legibility of engravings. Then examine the general appearance of the instrument for defects which may affect its useful life.
2. Check the drum screw for play in the carriage: and on a Fiske stadimeter, check the carriage for play on the frame.
3. Check all moving parts for freedom of action and smoothness of motion.

When the instrument passes this inspection satisfactorily, proceed with the tests and adjustments.

TESTS AND ADJUSTMENTS

The range drum is calibrated at infinity because at infinite range the index and horizon mirrors remain parallel to each other, regardless of the height of the object under observation. At infinity setting, therefore, provided the stadimeter is varied along the index scale, the index mirror (arcuate arm or scale arm) remains stationary and provides a convenient method for setting the range drum to infinity and a simple check on the trueness of the arcuate or scale arm.

Test and adjustment procedures for Fiske and sextant-type stadimeters are slightly different, but lack of space in this chapter prohibits a detailed discussion for the sextant-type. If necessary, refer to NavShips 250-624-6 for additional information on this instrument.

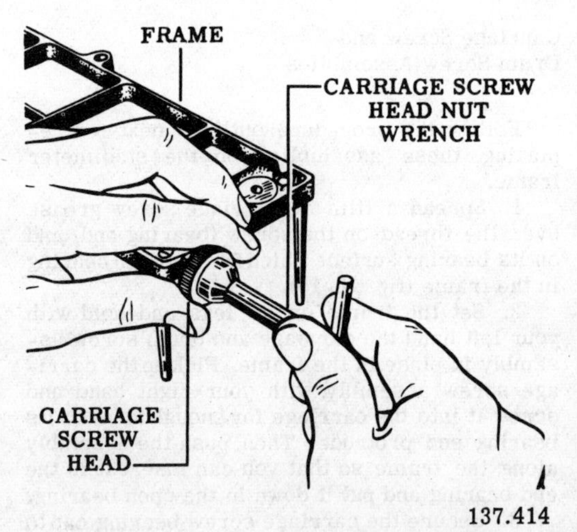

Figure 16-52.—Locking the carriage screw head.

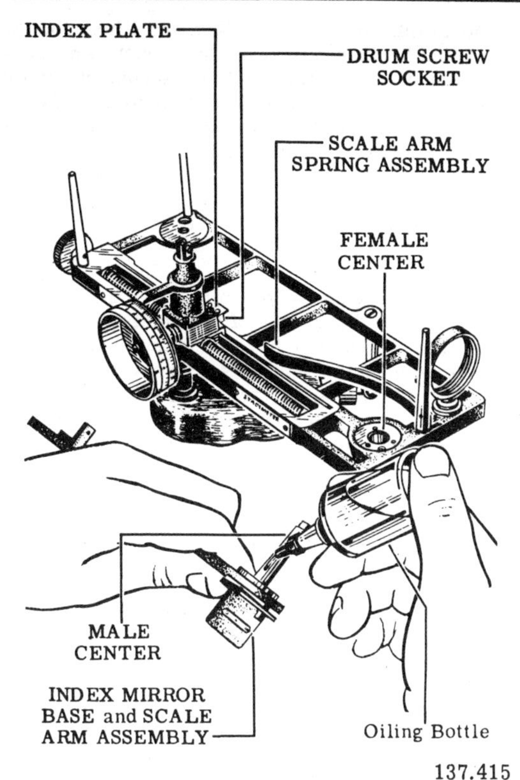

137.415
Figure 16-53.—Applying oil to the male center.

Setting the Range Drum at Infinity

The following procedure for setting the range drum at infinity is applicable for Types I and II Fiske stadimeters:

1. Clamp the dial indicator (with bracket) to the frame and position it as necessary to bring the indicator arm against the back of the scale arm. See figure 16-54.
2. Slide the carriage crank onto the carriage head and tighten its setscrew. Then turn the range drum to bring the scale arm parallel with the groove in the frame in which the tongue of the carriage slides. If the drum screw and range drum are correctly assembled, the drum will read "INF".
3. Crank the carriage to the 50-foot mark on the index arm and set the indicator dial to zero, Then run the carriage back to the 200-foot mark and read the indicator. Correct the reading to zero by turning the range drum, run the carriage back to the 50-foot mark, and read the deflection on the dial. Then turn the range drum to add one half of the deflection reading to the present deflection. Reset the dial to zero.

Repeat this adjustment until the indicator shows a deflection of no more than plus or minus .0005 inch at the 50-foot mark and at the 200-foot mark. As a rule, four or five repetitions of

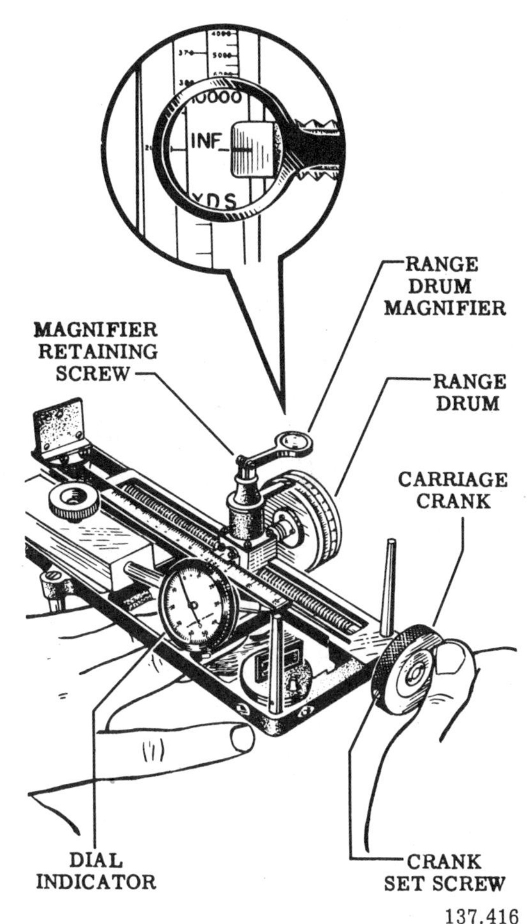

137.416
Figure 16-54.—Setting the range drum at infinity.

this operation are required to get correct adjustment.

4. The range drum must be set at infinity (INF) on its scale, without changing the position of the drum screw. To do this, loosen the three drum clamp plate screws which secure the range drum on the screw. Then set the drum at infinity, tighten one of the screws, and read the indicator dial to check the infinity setting. It should read 0 within plus or minus .0005 inch (circled portion, fig. 16-54). If the setting is correct, tighten the other screws.

Checking Arcuate and Scale Arms

After setting the range drum at infinity, check the arcuate arm (sextant-type) or the scale arm (Fiske) on a stadimeter to make certain the index mirror will not be moved at any intermediate point on the index scale more than the allowable deflection of plus or minus .0005 inch. The procedure for doing this is identical for both types of stadimeters.

NOTE: A true check of the arcuate or scale arm is impossible when slack and backlash are PRESENT in the instrument.

1. Set the range drum at infinity, position the carriage to read 50 feet on the index scale, and set the indicator at zero.

2. Constantly read the dial while moving the carriage in small steps from the 500-foot mark to the 200-foot mark. If the deflection exceeds .0005 inch, record the index reading and the deflection with its plus or minus sign. For a sextant-type stadimeter, minus deflections are caused by high spots on the arm, and plus deflections are caused by low spots. For a Fiske stadimeter, these readings are reversed—plus for high spots and minus for low spots.

3. With fine emery cloth, remove the high spots on the arm, polish them with croecus cloth, and wash off abrasive residue—to prevent unnecessary wear. To prevent FLAT spots, use a sweeping motion when you remove high spots on the arm. If there are FLATS, rub the scale arm down with emery cloth; but remove only the amount of material necessary to make the arm even.

When a scale or arcuate arm is bent or uneven to a considerable extent, it is generally more economical to replace it.

When the indicator shows NO MORE THAN a plus of minus .0005-inch deflection at infinity setting of the range drum and all across the arm, the stadimeter is ready for collimation. Before you do this, however, replace and adjust the index and horizon mirrors. They must be perpendicular to the frame of the instrument and (at infinity settings) parallel with each other. These mirrors become loose and get out of adjustment easily; and they must be adjusted each time the stadimeter is used.

Perpendicular Adjustment of the Horizon Mirror

The top edge of the silver on the horizon mirror and the center of the small peephole in the telescope holder have the same height above the frame of the stadimeter; so if the horizon mirror is perpendicular to the plane of the instrument, you can see the reflection of one half of the peephole in the silvered portion of the mirror.

If you cannot see one half of the peephole in the mirror, turn the vertical adjusting screw in the direction necessary to enable you to see it. Study illustration 16-55.

If you cannot see the peephole and telescope holder in the horizon mirror, the mirror may be at an incorrect angle. Adjust it by turning the radial adjusting screws.

Perpendicular Adjustment of the Index Mirror

If the index mirror of a stadimeter is not perpendicular to the frame, the directly-viewed object and its reflected image will not be aligned.

Hold the stadimeter in the manner shown in figure 15-56 (frame vertical) and look through the telescope holder at a small vertical object (mast or flagpole) to determine whether the directly-viewed object and the reflected image coincide, as illustrated in the left part of figure 16-57. If they do not coincide, use an adjusting screw wrench to turn the index mirror adjustable base as necessary to make them coincide. Wobble the stadimeter while holding it in a vertical plane and check to determine whether the directly-viewed object appears to wiggle. If it does not, the mirror is not properly adjusted. If adjustment of the mirror does not remove the wiggle, repeat the procedure for perpendicular adjustment.

Parallel Adjustment of Mirrors

With the range drum set at infinity, make the horizon and index mirrors perpendicular to the frame and parallel to each other, in the following manner:

1. Set the range drum at infinity. The index or height scale can be set at any position.

2. Hold the stadimeter in a vertical plane and look through the telescope holder at a distant object. If the reflected image and directly-viewed horizon are not continuous, as illustrated in the left portion of figure 16-58, loosen one of the horizon mirror radial adjusting screws and tighten the other enough to make the horizontal line appear continuous.

When you hold the stadimeter horizontally, a distant vertical object such as a smoke stack can serve as the horizon.

3. Wobble the instrument along the horizon to check the alignment of the object and its reflected image. If the horizontal line appears to wiggle, the mirrors are not exactly parallel.

4. As an overall check on the adjustment of the mirrors, hold the stadimeter diagonally at 45° and wobble the stadimeter as you sight the horizon. If there is a wiggle, the mirrors are not vertical to the frame or parallel to each other. Make necessary adjustments.

Now that the mirrors are correctly adjusted and aligned, the instrument is ready for collimation.

COLLIMATION PROCEDURE

The purpose of collimating a stadimeter is to calibrate the index scale and check the range readings.

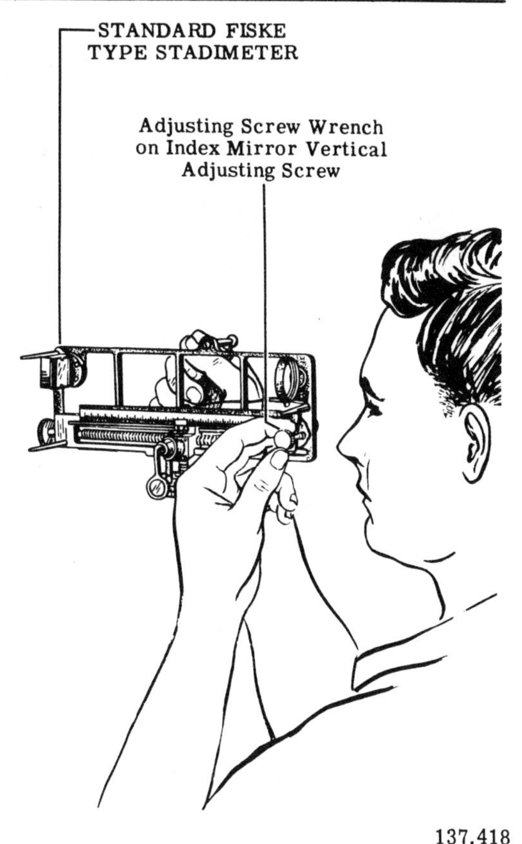

137.418
Figure 16-56.—Perpendicular adjustment of the index mirror.

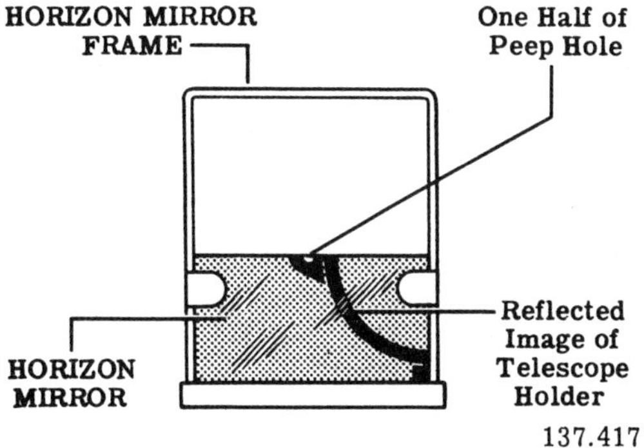

137.417
Figure 16-55.—Perpendicular adjustment of the horizon mirror.

A Mk 4, Mod 5, collimator (fig. 16-59) or a Mk 5 collimator may be used for collimating sextant-type and Fiske stadimeters. The target in the Mk 4, Mod 5, collimator, however, is more suited for collimating stadimeters because it has two flagstaffs. See illustration 16-60.

The radial lines of the collimator target cross the coordinate axis at one degree intervals. Each intersection is marked with the number of degrees which it is displaced from the target center. The Mk 4, Mod 5, target is used in this discussion. The flagstaffs on this target are spaced to subtend an angle of 3°48'48". The reason for this will be clear after you study the next few paragraphs.

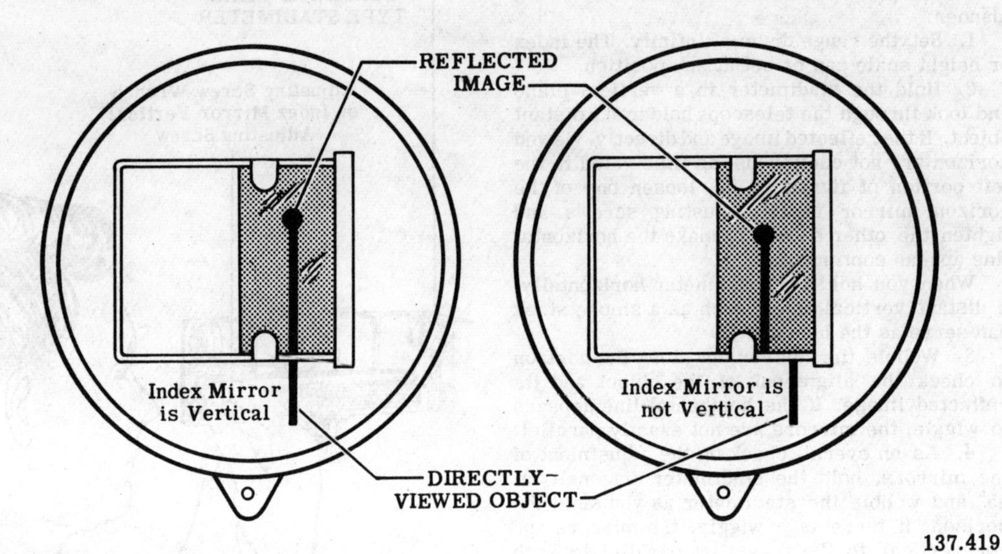

Figure 16-57.—Directly-viewed object and reflected image in perpendicular alignment (coincidence).

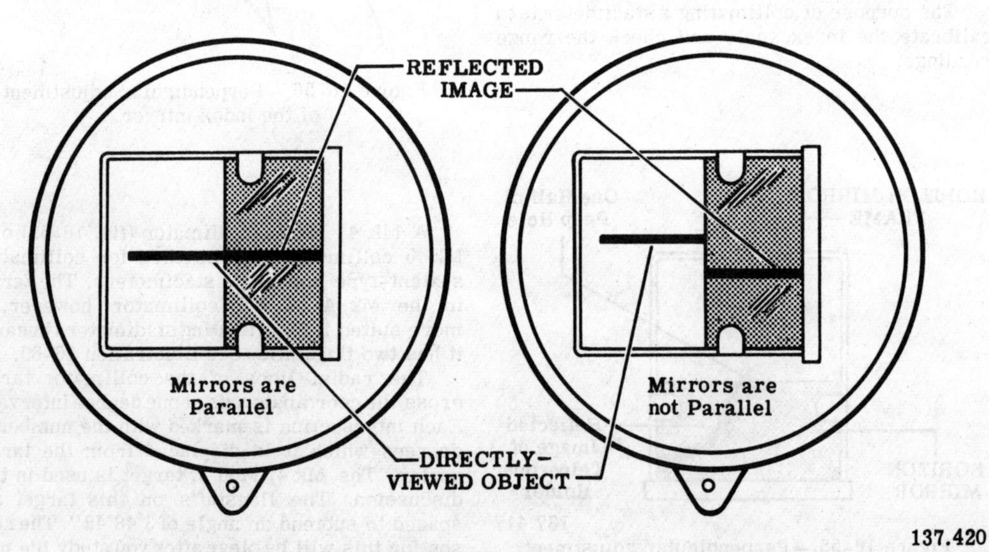

Figure 16-58.—Parallel adjustment of horizontal and index mirrors.

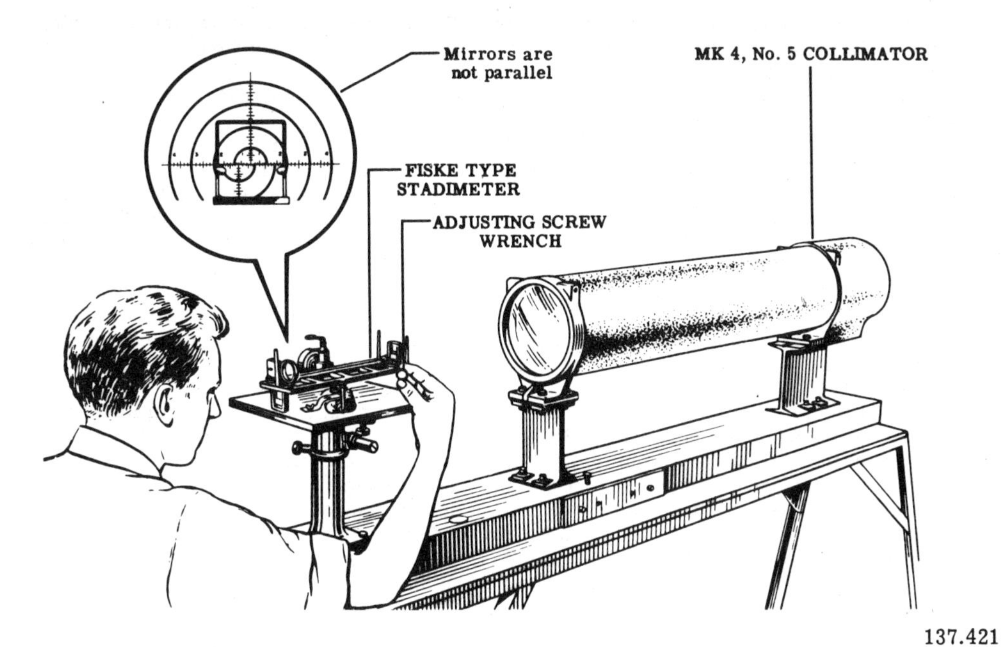

137.421

Figure 16-59.—Stadimeter collimator.

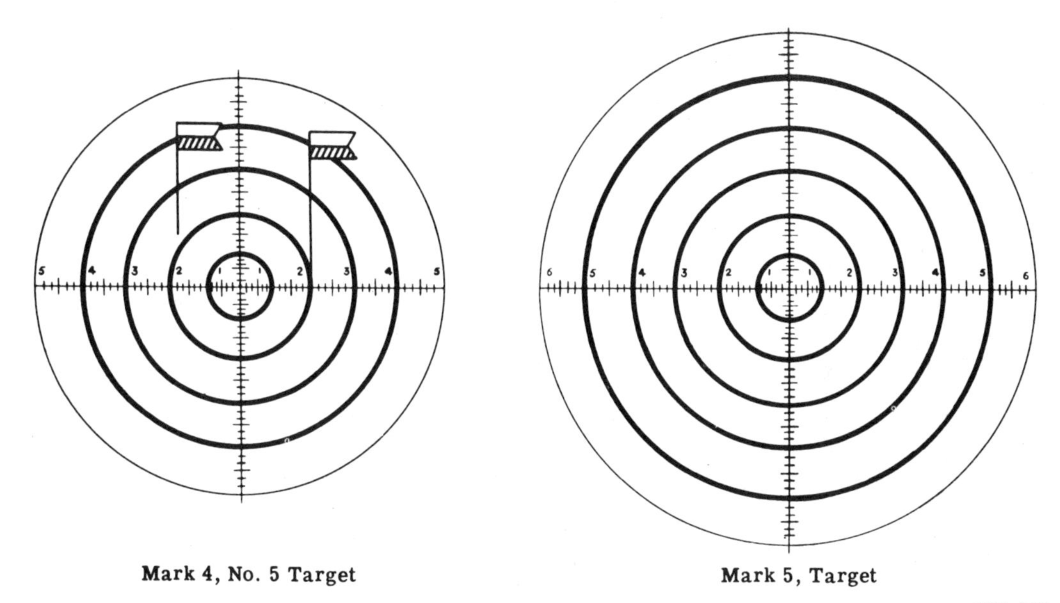

137.422

Figure 16-60.—Collimator targets.

A stadimeter determines ranges of objects of known heights by sighting the angle subtended by the object. This can be accomplished by varying the index mirror as required to superimpose the directly-viewed and reflected images. If the angular height of the object is known, the ratio of the height to the range will be equal to the tangent of the angle. Study illustration 16-61. The formula for determining the tangent of the angle subtended is:

$$\text{Tangent of Angle} = \frac{\text{Height}}{\text{Range}}$$

The tangent of the angle subtended in illustration is 3°48'48", which equals .0666, or $\frac{1}{15}$. So the ratio of height to range is 1 to 15 if they are expressed in the same distance units. Because stadimeter scales are calibrated in feet for the height (index) scale and in yards for range (on the range drum), the ratio becomes 1 to 5. For this reason, if the stadimeter scales are set to the 1-to-5 ratio (50 ft. & 250 yds, 100 ft. & 500 yds., etc.), the flagstaffs in the collimator will be superimposed.

Any other convenient ratio can be set up and targets made to suit it. If you use a collimator such as the Mk 5, you must select two graduations on the target which subtend a known angle. Targets such as flagstaffs are desirable because they are easily identified.

Mirror Adjustment and Infinity Setting

To check the mirror adjustment and the infinity setting, proceed as follows:

1. Set the stadimeter on the collimator stand, as shown in illustration 16-59. The telescope is not on the instrument because it is not required for this operation.

2. Sight through the peephole in the telescope holder and check the vertical position of the horizon mirror. If the mirror is not vertical to the frame, turn its adjusting screw as necessary to make it vertical.

3. Set the range drum at infinity and sight the target to check the vertical position of the index mirror. If the mirror is not vertical to the frame, the reflected portion of the target in the horizon mirror will be displaced up or down. To bring the reflected image into position, adjust the index mirror vertical adjusting screw. Shake your head up and down and look at the top edge of the horizon mirror. If the image jumps, the index mirror is not exactly vertical.

Horizontal displacement of the reflected image indicates that the infinity setting of the range drum is incorrect, or that the horizontal mirror

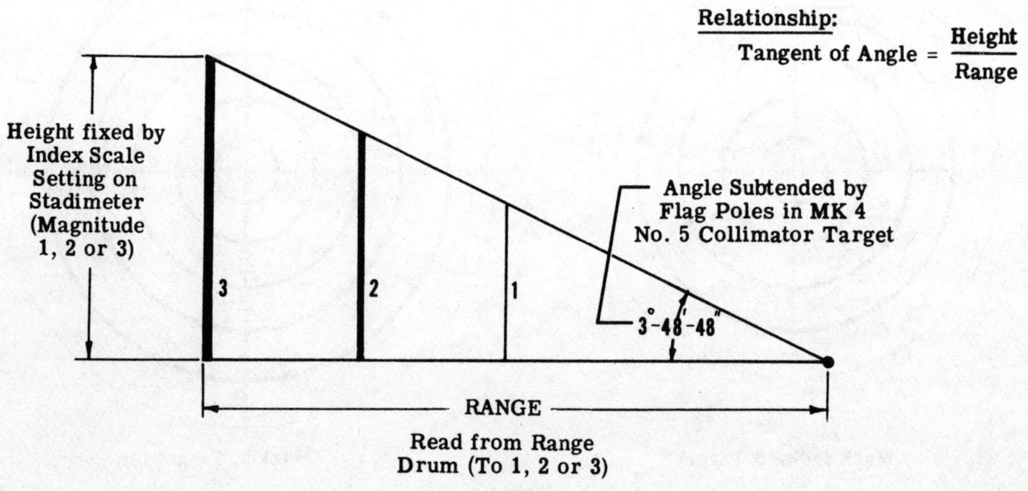

Figure 16-61.—Ratio of height to range of an object.

is not parallel to the index mirror. Adjust the range drum setting to eliminate this.

4. Set the range drum at infinity and sight the target to determine whether the horizon mirror is parallel to the index mirror. If the reflected image is displaced horizontally, adjust it as necessary with the horizon mirror radial adjusting screws. See the circled portion of figure 16-59. For an accurate check, look at the vertical line on the target between the directly-viewed portion of the line and its reflected image and shake your head up and down. If the line appears to wiggle (snake), the horizon mirror is not adjusted properly.

5. To check the infinity setting, set the range drum at infinity, run the carriage to the 50-foot end of the scale, and check for horizontal displacement in the manner explained in step four. Then check again at the 200-foot mark. If displacement is evident at either or both ends, the infinity setting is incorrect. Set the range drums at infinity and collimate again.

Setting the Index Scale

A Fiske stadimeter has an index scale which slides in the index arm. To set the scale, do the following:

1. Set the range drum to read 250 yards, sight the collimator target, and superimpose the right flagstaff on the left flagstaff by moving the carriage with the carriage screw. To check the superimposition, shake your head up and down and look for a wiggle.

2. Partially release the index scale setscrew on the underside of the scale arm and slide the scale to make the 50-foot mark coincide with the line on the index plate. Tighten the setscrew and fill the space (if any) between the end of the scale (at the 200-foot end) and the scale arm slot with sealing compound.

Checking Range Readings

Fiske and sextant-type stadimeters must be accurate enough to determine the range of objects of known height to within 5 percent. If the adjustment was made properly, all range readings should be correct. This test procedure is intended as a final check on the work accomplished previously on the instrument during overhaul. If the stadimeter does not give accurate readings, diagnose the cause and correct it.

The test of range readings is for 200 yards to 1000 yards for objects of heights between the index scale limits of 50 to 100 feet. Perform the test in the following manner:

1. Set the index scale at 50 feet and sight the collimator target (fig. 16-60). Turn the range drum to make the right flagstaff move and superimpose on the left flagstaff. Shake your head up and down and look for a wiggle. Align the flagstaff accurately and read the range drum. It should read 250 yards.

At a range of 200 yards, a 5 percent error (allowed) is 10 yards. This is a measurement of almost 5/8 inch on the range drum scale; however, 5/8 inch along the scale at the high end is almost 700 yards. Misalignment or play in the stadimeter, therefore, has the greatest effect at long ranges.

Allowed tolerance at 10,000 yards is 500 yards, which is equal to approximately 1/64 inch on the range drum at the 10,000-yard mark. At any range for an object of any height, therefore, provided the reading is within plus or minus 1/64 inch, the instrument will be within tolerance for the 10,000 yard range and ALL lesser ranges. The range drum magnifier gives a magnification of about twice the length; so with a little experience, you will be able to judge the tolerance in the magnifier.

2. Repeat step No. 1 at 55, 60, 70, 80, 90, 100, 110, 120, 140, 160, 180, and 200 feet on the index scale; and in each case multiply the index scale reading by 5 to determine the range reading expected.

Illustration 16-62 gives some defects in range readings and the probable causes.

At this point, the stadimeter is fully tested and adjusted, ready for the final inspection.

FINAL SHOP INSPECTION

Final inspection of the overhauled stadimeter should reveal details previously overlooked during the repair procedure. Inspect as follows:

1. Check for completeness of component parts.

2. Inspect parts for finish, tightness in assembly, and legibility of engravings. Check all screws for tightness.

3. Look at an image through the telescope assembly and check for brightness, clearness, and distortion (shape and color).

4. Examine both mirrors for cleanliness and freedom from visible defects.

5. When you are satisfied with the condition of the instrument, put it in its carrying case, which should have an attached nameplate, and an instruction sheet pasted on the underside of the cover. Include a spare set of horizon and index mirrors and an adjusting screw wrench. For a sextant-type stadimeter, include two spare radius arm springs; for a Fiske stadimeter, include two scale arm spring assemblies.

Wrong Range Reading for Particular Index Setting	High or Low Range Readings over the Entire Scale	Inconsistent Readings
Probable Cause	Probable Cause	Probable Cause
1. Bumps or low spots on the arcuate or scale arm. 2. Play in the drum screw. 3. Looseness of carriage block in the frame groove (Fiske).	1. Range drum infinity setting is incorrect. 2. Index scale setting is incorrect. 3. Defective mirrors.	1. Play in the drum screw. 2. Play in the carriage block (Fiske). 3. Loose mirror mountings.

137.424

Figure 16-62.—Defects in range readings.

CHAPTER 17

TELESCOPIC ALIDADES

Telescopic alidades were designed for use aboard ship to enable personnel to sight ACCURATELY and CLEARLY distant objects and to get precise bearings of those objects. A telescopic alidade must indicate a bearing with an accuracy of less than two minutes of arc, relative to the bearing circle assembly.

CONSTRUCTION

A telescopic alidade consists of a telescope, with a reticle for sighting purposes, mounted on a bearing circle. An auxiliary telescope system which permits simultaneous sighting of a portion of a compass card and the object under observation is located in the body of the telescope. The entire assembly is mounted on a gyrocompass repeater or a standard ship magnetic compass to permit rapid taking of bearings. A Mk 2, Mod 3, alidade is equipped with a prismatic altitude head which permits the taking of bearings of celestial objects.

Study the telescopic alidades shown in figure 17-1. Note the nomenclature, and observe the external similarities and differences in construction.

OPTICAL SYSTEM

The complete optical system of a basic telescopic alidade is shown in figure 17-2. Mk 3, Mods 0 and 1, and Mk 4, Mods 0 and 1, have this system. The optical system of a Mk 2, Mod 3, alidade comprises the basic alidade optical system, but it also has an ALTITUDE prism head assembly and a RETICLE illuminator window.

The basic optical system of an alidade is essentially a terrestial telescope with a reticle (K, fig. 17-2). In a Mk 2, Mod 3, alidade, the reticle is illuminated through the illuminator window by a small lamp attached to the telescope body.

The OBJECTIVE LENS (E, fig. 17-2) of a telescopic alidade gathers reflected light from a distant object and forms an inverted and reverted image within the optical system, at the rhomboid collective lens.

The RHOMBOID COLLECTIVE LENS (I, fig. 17-2) collects the extreme rays of the object image and the auxiliary field image (explained later) and diverts them into the erector lens. This lens is called a rhomboid collective lens because it is next to the auxiliary rhomboid prism. Without the rhomboid collective lens, extreme principal rays of light would miss the small erector lens.

The ERECTOR lens (J, fig. 17-2), located between the rhomboid collective lens and the crossline plate, transforms the image of the object from the rhomboid collective lens into a normal, erect image in the plane of the crossline plate (reticle side).

The CROSSLINE PLATE, commonly called reticle, is a piece of plane-parallel optical glass with a fine vertical line engraved across its diameter (on the side toward the eyepiece lenses). The image of the object formed by the objective lens and inverted and reverted by the erector lens comes to focus in the plane of the reticle line. When viewed through the eyepiece, the reticle appears superimposed in sharp focus on the image of the object under observation. This plate (reticle) is an accurate guide for aligning an alidade on a bearing point.

The EYELENS and the EYEPIECE COLLECTIVE LENS form an enlarged image of the object and the reticle line. See L and M in figure 17-2. These lenses are mounted in a drawtube which can be moved in or out to obtain a sharply focused image.

AUXILIARY TELESCOPE SYSTEM

The auxiliary telescope system of a telescopic alidade is a second optical system consisting of a

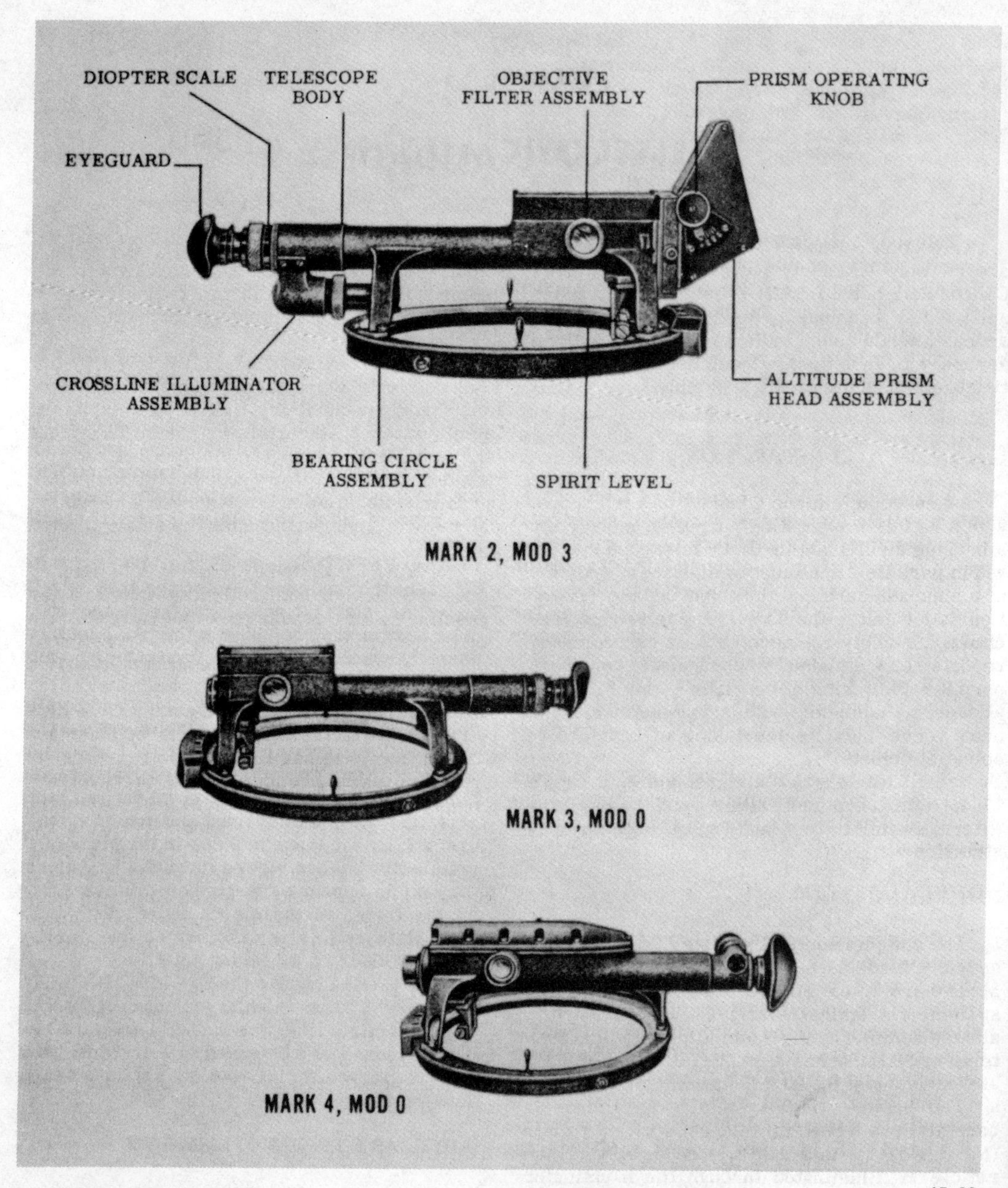

45.39

Figure 17-1.—Mks 2, 3, and 4 telescopic alidades

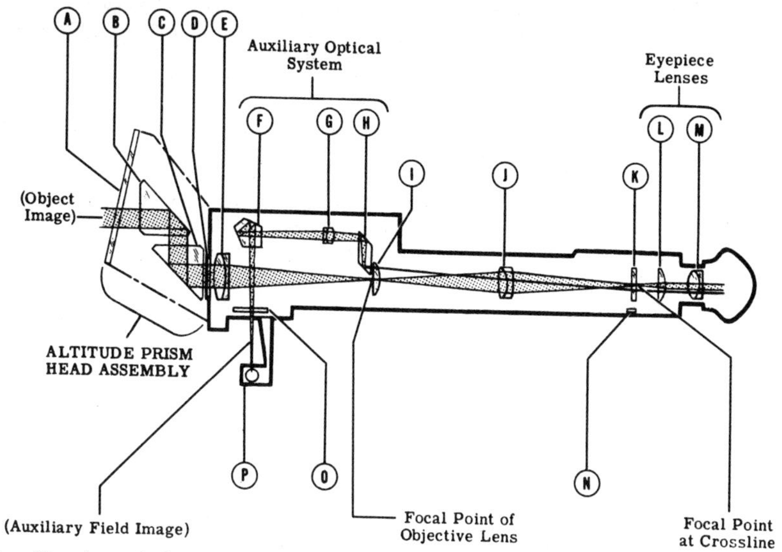

A. Housing window
B. Altitude head movable prism
C. Altitude head fixed prism
D. Objective filter
E. Objective lens
F. Auxiliary penta prism
G. Auxiliary objective lens
H. Auxiliary rhomboid prism
I. Rhomboid collective lens
J. Erector lens
K. Crossline plate (reticle)
L. Eyepiece collective lens
M. Eyelens
N. Reticle illuminator window
O. Auxiliary window
P. Spirit Level

137.425

Figure 17-2.—Cross section view of a telescopic alidade.

penta prism, a small objective lens, and a rhomboid prism. See F, G, and H, figure 17-2. This optical system sights a portion of the compass card through the auxiliary window in the bottom of the main telescope body when the alidade is mounted on a gyro compass repeater or on a ship magnetic compass. At the same time, the spirit level of the alidade is visible.

The purpose of the auxiliary system is to project the image of the compass card and the spirit level bubble from the auxiliary rhomboid prism into the main telescope system, which covers about half of the rhomboid collective lens. The object being sighted, the instrument's crossline, a portion of the compass card, and the bubble are visible at the same time. (This is illustrated and explained under adjustments at the end of the chapter.) The bearing of an object is read on the compass card, as indicated by the portion of the crossline in the auxiliary field.

BEARING CIRCLE ASSEMBLY

As shown in figure 17-1, the telescope body of a telescopic alidade is rigidly mounted on a bearing circle assembly which fits over the bezel of a ship magnetic compass or a gyrocompass repeater. The bearing circle assembly is used to indicate relative bearings.

ALTITUDE PRISM HEAD ASSEMBLY

A Mk 2, Mod 3, telescopic alidade (and some others) is equipped with a detachable altitude prism head assembly which fits over the objectives end of the telescope, as shown in illustration 17-1. This assembly enables an observer to sight objects which are at an angular elevation, by means of two prisms within the head: (1) a fixed prism (C, fig. 17-2), and (2) a movable prism (B, fig. 17-2), so that objects from 0° to 60° of angular elevation may be sighted.

A glass housing window (A, fig. 17-2) completes the altitude head housing, which protects the altitude head prisms from dirt and accidental damage.

TYPES OF TELESCOPIC ALIDADES

Five different marks of telescopic alidades are currently used by the Navy: Mks 2, 3, 4, 6, and 7. Mk 4, Mods 0 and 1, telescopic alidades are used more extensively by the Navy than other alidades and are discussed in detail in this chapter. Mks 6 and 7, however, will also be used rather extensively and they are discussed at the end of the chapter in as much detail as space and material permit.

MARK 4 TELESCOPIC ALIDADES

A Mk 4 telescopic alidade does not have a crossline illuminator assembly or an altitude prism head assembly. The principal design feature of this alidade is that an attempt was made to make it water- and moisture-proof by adding rubber gaskets to the seats of the eyelens and the objective lens, a rubber seal over the objective mount, and a sealing ring on the eyepiece.

The telescope body of a Mk 4 alidade was also made in one piece (instead of three) in an effort to eliminate joints through which moisture could enter the interior. The eyepiece mount moves in and out of the end of the body; it does not have a separate adapter. Focusing is accomplished by turning a pinion which engages a rack cut on the eyepiece lens mount. Instead of being in a fixed plane, the crossline plate is mounted in its own mount and can be positioned IN or OUT in relation to the eyepiece.

The whole auxiliary telescope system is mounted on a plate so that it can be made up as a subassembly before it is placed in the telescope body. To absorb moisture which DOES ENTER the interior, a container of dessicant (silica gel) is placed within the body of the telescope. The body cover is made watertight by a rubber gasket between it and the telescope body.

DISASSEMBLY PROCEDURE

The disasembly procedure considered in this section is for the Mk 4, Mods 0 and 1, telescopic alidade. To prevent misunderstanding, the first section of disassembly is designated as REMOVAL for the major assemblies. The procedure for disassembling the subassemblies follows.

Removal of Objective Filter Assemblies

Pull the two objective filter assemblies off the filter carriers, which are on each side of the telescope body. Study illustration 17-3. Then unscrew the two filter carrier screws from the telescope body and remove the filter carrier. Follow the same procedure to remove the filter carrier on the other side of the body.

Removal of the Objective Assembly

Refer to illustration 17-3 as you study the following procedure for removing the objective assembly:

1. Slip the sunshade off and remove the seal ring screws. Then take off the objective mount seal ring.
2. With a prying tool (fig. 17-4), remove the objective mount gasket from the recess in the telescope body.
3. Unscrew the main objective assembly from the telescope body.

Removal of the Eyepiece Assembly

To remove the eyepiece assembly, proceed as follows:

1. Remove the diopter ring screws with an Allen screw wrench and slide the ring off (fig. 17-3).
2. With a 3/64-inch drift pin and a hammer, remove the adjusting pinion taper pin from the adjusting pinion collar.

CAUTION: Excessive pressure will bend the adjusting pinion; ALWAYS support the pinion collar on a block.

3. With a pinion collar and packing ring wrench, unscrew the adjusting pinion collar, as shown in illustration 17-5. Then unscrew the adjusting pinion packing ring (fig. 17-3).
4. As illustrated in figure 17-6, pull the eyepiece assembly out of the telescope body as you turn and remove the adjusting pinion. Slide the adjusting pinion packing and bushing off the pinion and discard the packing.
5. Unscrew the bushing cap screws from the eccentric bushing cap on the left side of the telescope body. Then remove the bushing cap and the pinion eccentric bushing, which is loose in the

adjusting pinion cylinder. See illustrations 17-3 and 17-6.

Auxiliary Telescope Assembly and Desiccator

Remove the cover screws and lift the cover and gasket off. Then remove the body desiccator and pull off the front and rear grommets. Discard the desiccator. Study illustration 17-7.

Unscrew the auxiliary telescope plate screws. Then push the end of the telescope plate with your left thumb and lift the assembly out of the telescope body; but be careful NOT to chip the lower edge of the auxiliary rhomboid prism which projects from the bottom of the telescope assembly (fig. 17-8).

Removal of the Crossline Assembly

The position of the crossline along the optical axis is critical. To save time during reassembly and collimation, measure and record the depth of the crossline assembly in the body. Insert a scale inside the body, as illustrated in figure 17-9, and measure the depth to the nearest 1/64 inch at the face of the body.

CAUTION: Do NOT touch the crossline plate with the scale. If you scratch the glass, it must be discarded.

Insert the outer element (fig. 17-10) of a crossline adjusting wrench through the eyepiece end of the body and unscrew the crossline assembly lock ring.

Use the outer element of a crossline adjusting wrench as a guide and insert the inner element of this wrench through the eyepiece end of the tube to engage the slots in the crossline plate mount. See figure 17-11. Then unscrew the crossline assembly, tilt the body slightly, and catch the crossline assembly in one hand when it falls out. Refer to illustration 17-7.

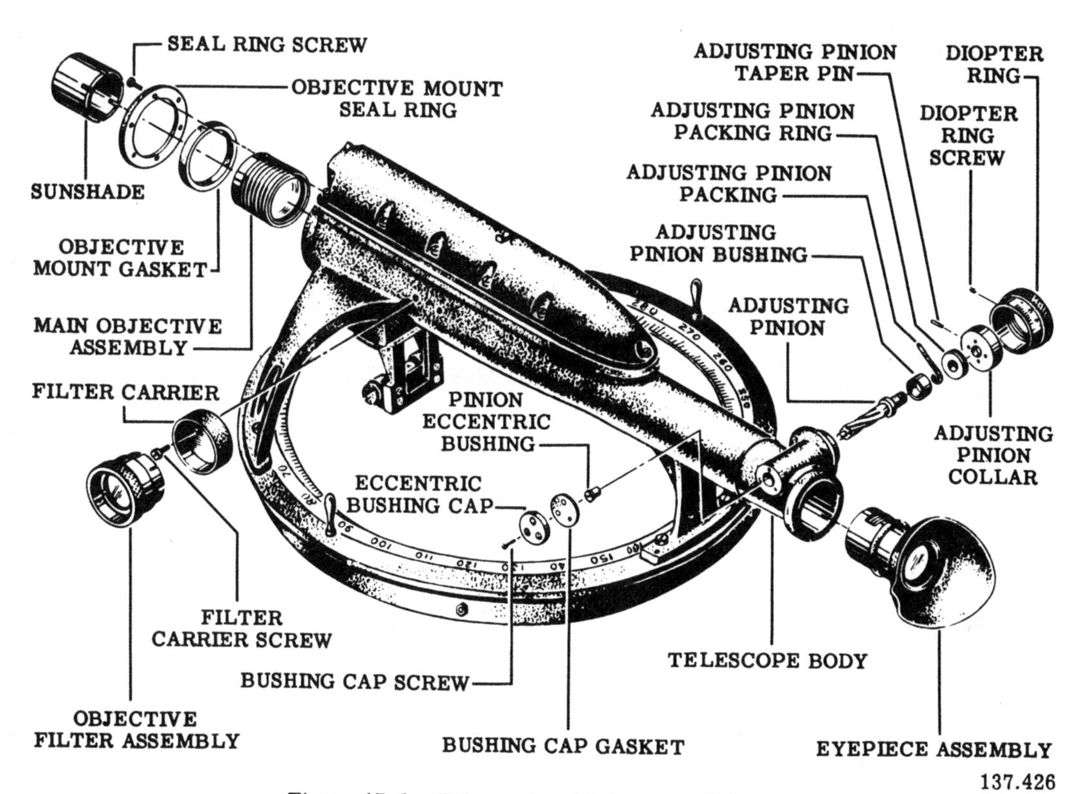

Figure 17-3.—Telescopic alidade assemblies.

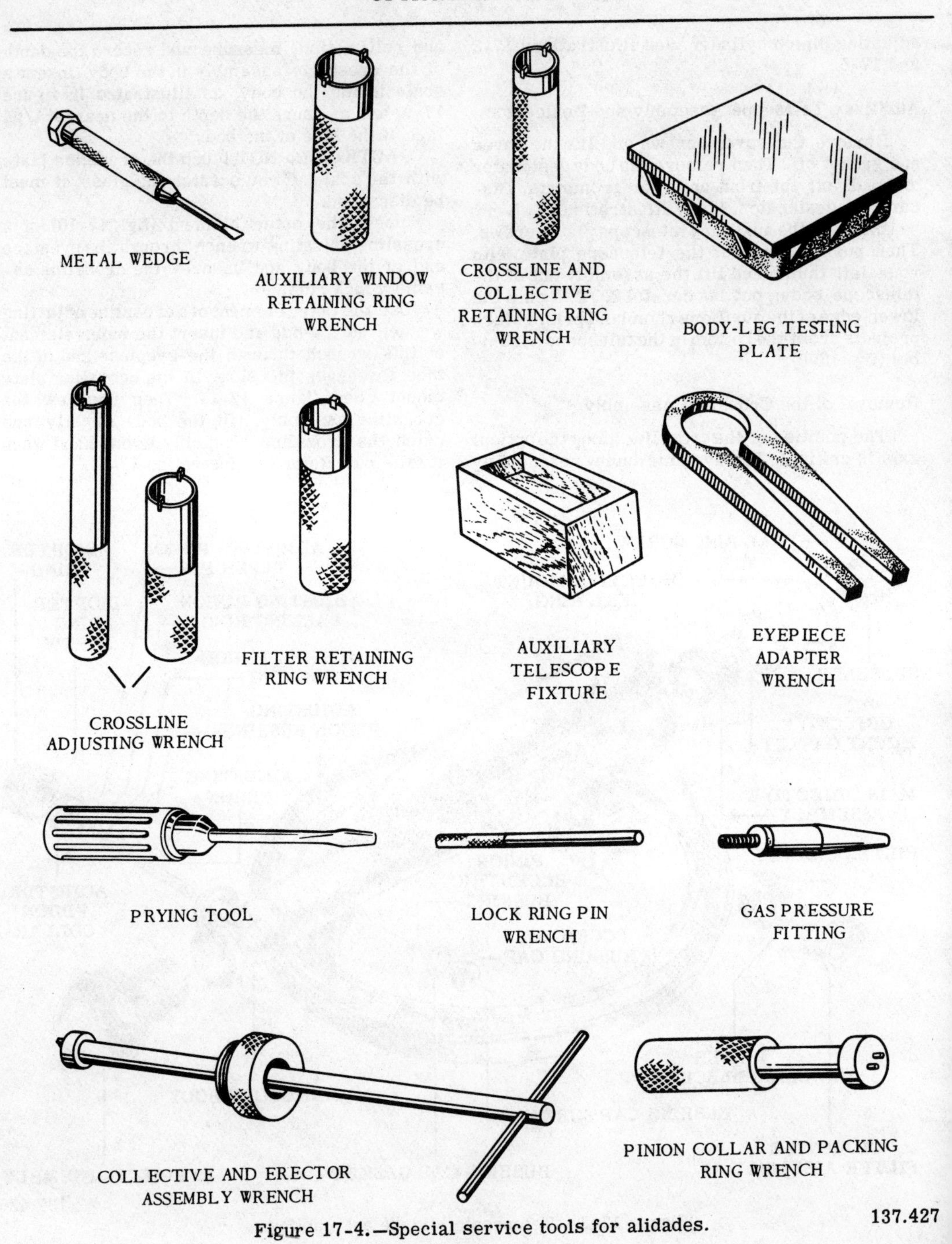

Figure 17-4.—Special service tools for alidades.

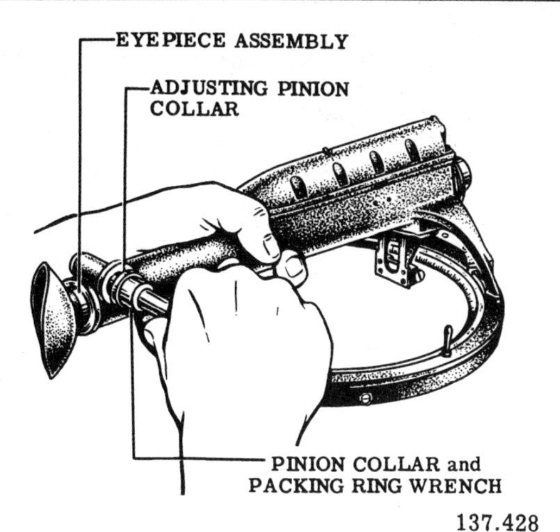

137.428
Figure 17-5.—Unscrewing the adjusting pinion collar.

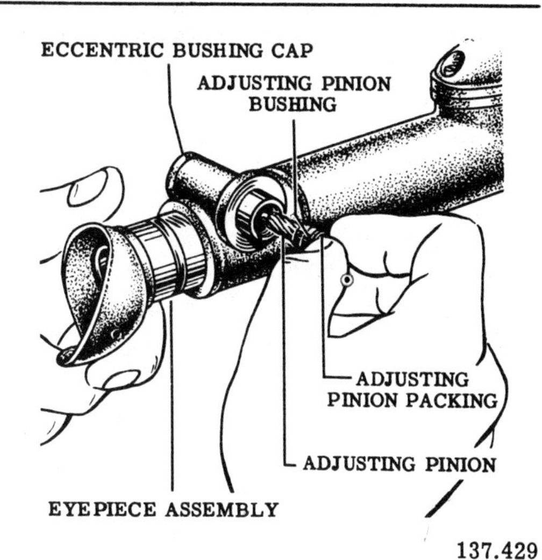

137.429
Figure 17-6.—Removing the eyepiece assembly and the adjusting pinion.

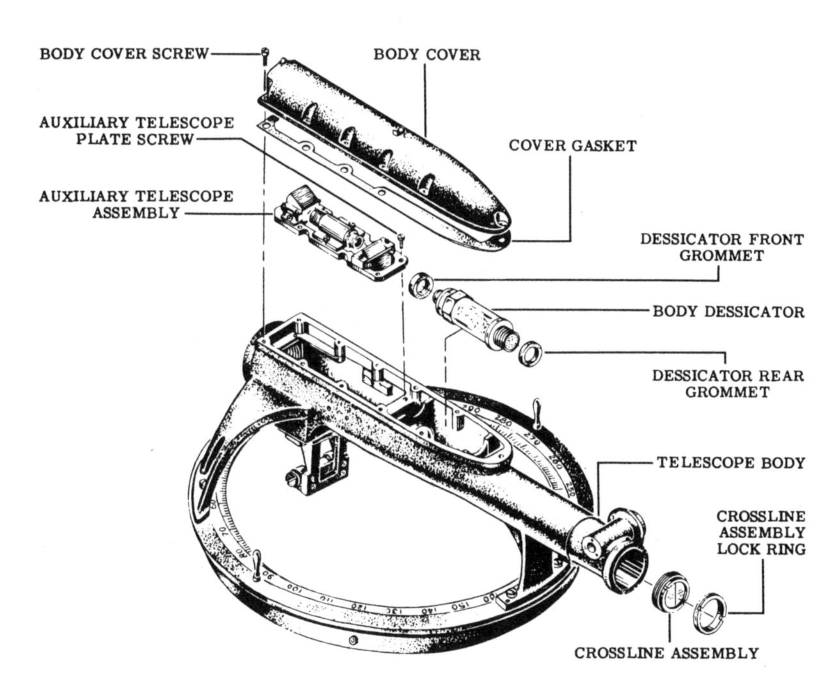

137.430
Figure 17-7.—Crossline and auxiliary telescope assemblies and desiccator.

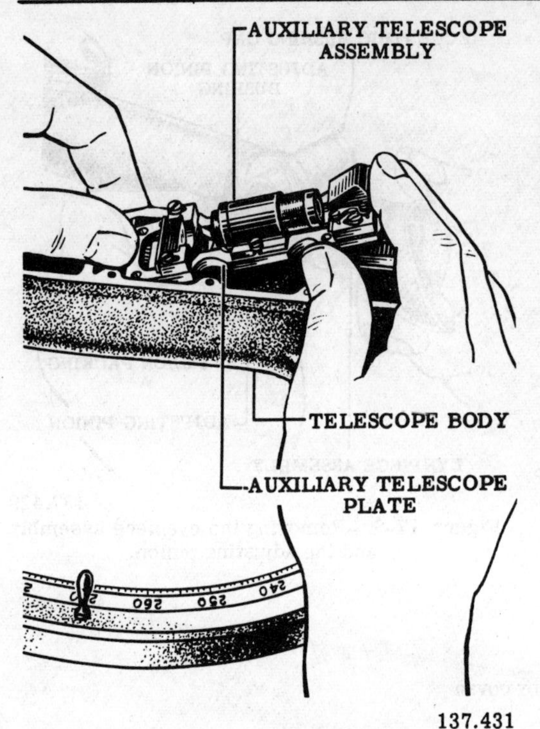

137.431

Figure 17-8.—Removing the auxiliary telescope assembly.

CAUTION: Wrap the crossline assembly in lens tissue to protect the crossline plate from scratches.

Removal of Rhomboid Collective and Erector Assemblies

As is true for the crossline assembly, the rhomboid collective assembly and the erector assembly must be placed in definite positions along the optical axis in the body. Scribe reassembly guide marks on them, and scribe a longitudinal mark along the length of the rhomboid collective mount and on the inside rib of the body. Then scribe a circumferential mark around the mount where it enters the opening in the body. Study illustration 17-12. Then (in the same manner) put marks on the erector mount and on the collective assembly lock ring.

Now do the following:

1. Insert the long rod of a collective and erector assembly wrench into the objective end of the body, as shown in figure 17-13. Engage the slots in the rhomboid collective mount and unscrew the rhomboid collective assembly. Then remove the assembly with tweezers.
2. With a lock ring pin wrench, unscrew the rhomboid collective assembly lock ring.
3. Remove the erector assembly lock ring and unscrew the erector assembly. Then use tweezers to remove the assembly. Both assemblies are shown in illustration 17-14.

Removal of the Bearing Circle Assembly

Unscrew the body leg screws. Then use a metal wedge and a hammer to separate the telescope body from the bearing circle assembly.

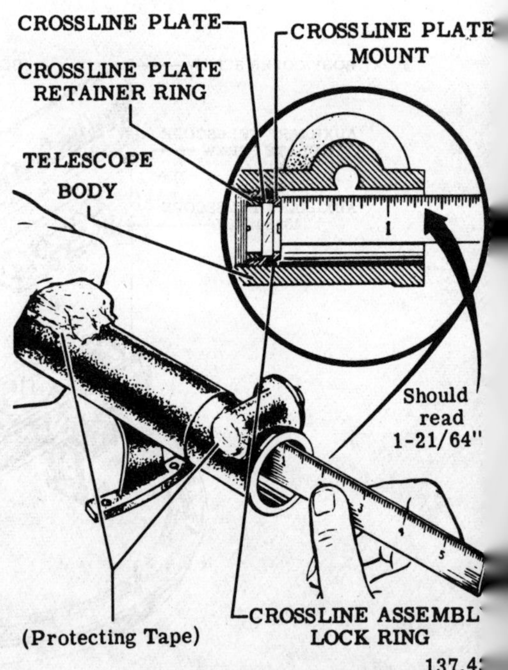

137.4

Figure 17-9.—Measuring the distance of the crossline assembly in the body.

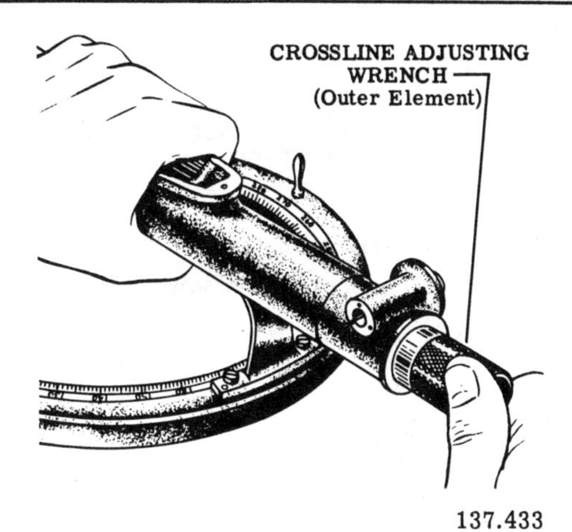

Figure 17-10.—Removing the crossline assembly lock ring.

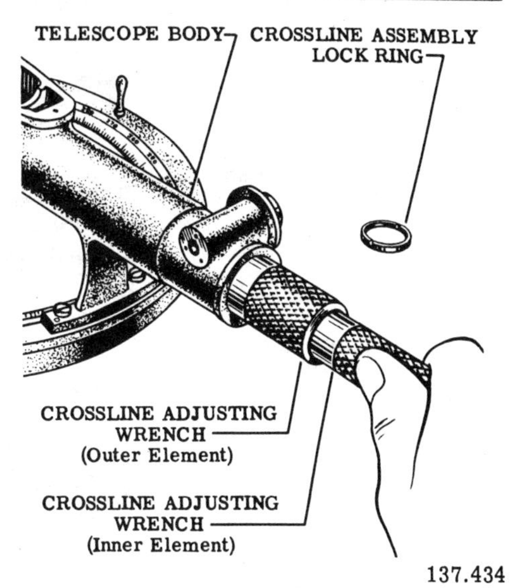

Figure 17-11.—Unscrewing the crossline assembly.

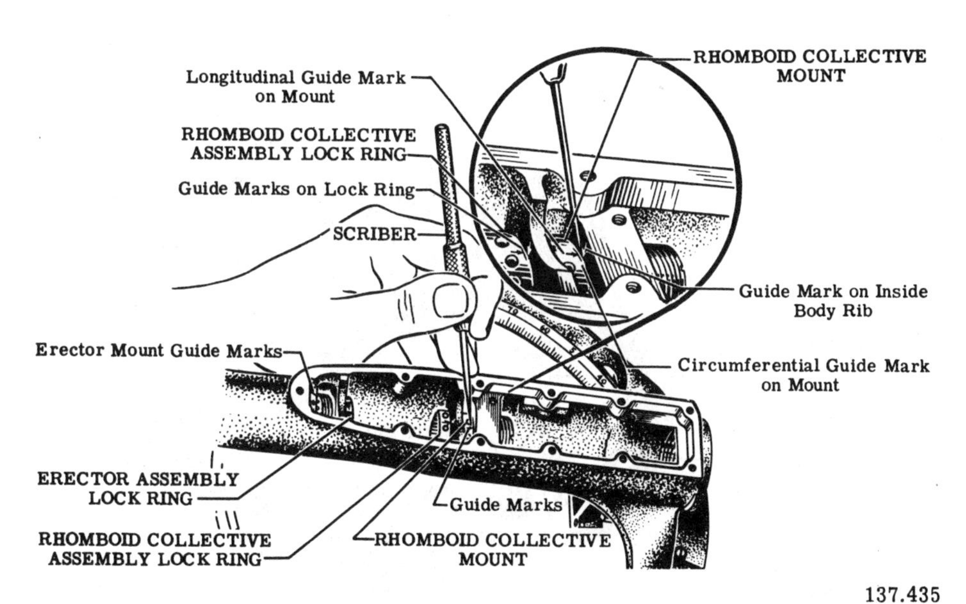

Figure 17-12.—Scribing reassembly guide marks on the rhomboid collective and erector assemblies.

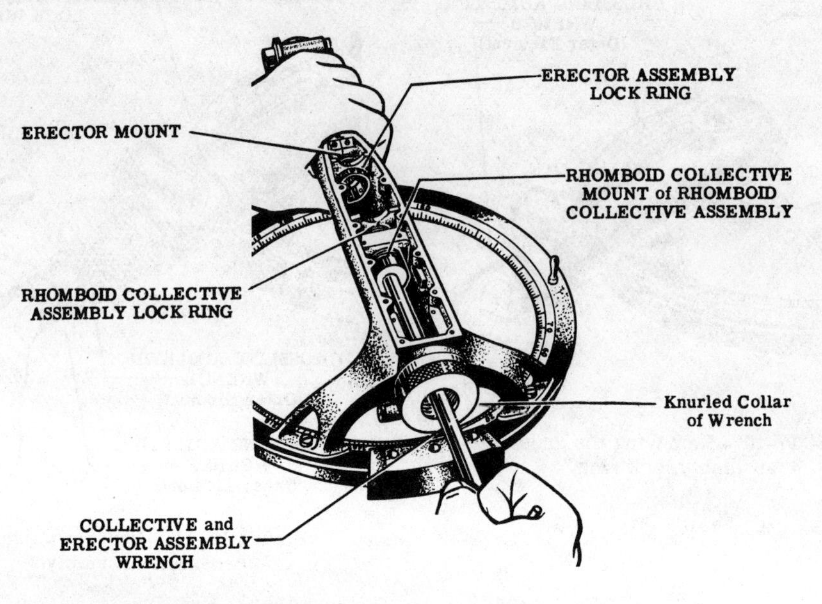

137.436

Figure 17-13.—Unscrewing the rhomboid collective assembly.

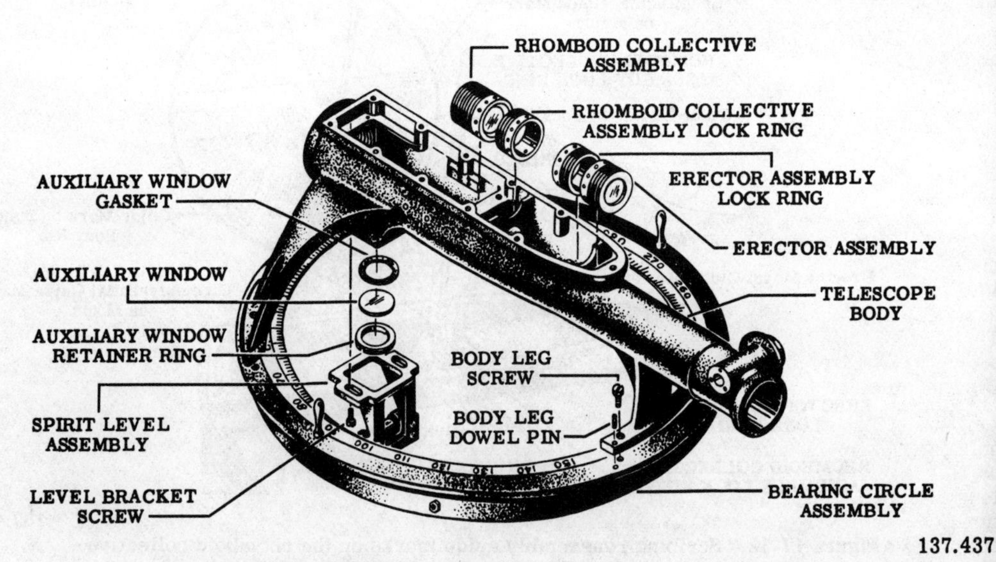

137.437

Figure 17-14.—Rhomboid collective and erector assemblies.

Drive the metal wedge in at the spot shown in figure 17-15, adjacent to the body leg dowel pin.

With a pair of pliers, pull the two body leg dowel pins out of the bearing circle assembly. If they remain in the body leg or support, use a hammer and a 5/64-inch drift pin to remove them.

Spirit Level Assembly and Auxiliary Window

Remove the spirit level assembly by unscrewing the screws which secure the bracket to the main body (fig. 17-14). To remove the auxiliary window, use an auxiliary window retaining ring wrench (fig. 17-16) to unscrew the ring from the body. If the auxiliary window does not drop out with the retaining ring, put a piece of lens tissue or clean cloth over some fingers to protect the glass and gently push the window out into your hand. See figure 17-14.

This last stop completes the removal of the assemblies and the next stop in the repair process is disassembly of the subassemblies.

Disassembly of the Spirit Level Assembly

To disassemble the spirit level assembly, unscrew the caps from the level mount (fig. 17-17) and slide the spirit level out of its mount. Then, to disassemble the level mount from the level bracket, remove the level mount screws.

NOTE: The spirit level is sealed in the mount. If the level caps are stuck, or the spirit level will not slide out, gently heat the parts on a hot plate to soften the sealing compound.

Disassembly of the Objective Filter Assemblies

The two objective filter assemblies (light and dark) in a telescopic alidade are made up of identical parts (except for the filters) and the disassembly procedure is the same for both.

With a filter retaining ring wrench, unscrew the filter retaining ring, as illustrated in figure 17-18. If the ring is tight, use a clamp wrench to hold the filter mount during this operation. Then lift the objective filter out.

Disassembly of the Main Objective Assembly

To disassemble the main objective assembly, hold the assembly in one hand and unscrew the objective lens retaining ring from the objective lens mount. Use lens paper to protect the objective lens from fingerprints and gently press it out of the mount. The assembly parts are shown in figure 17-19.

Disassembly of the Crossline Assembly

To disassemble the crossline assembly, use the following procedure:

1. Place the assembly on a bench and (with a straightedge as a guide) scribe two guide marks on the crossline plate mount, in line with the line in the crossline plate. See illustration 17-20.
2. With a crossline plate retaining ring wrench, unscrew the crossline plate retaining ring.
3. The crossline plate is loose in its mount; protect it with lens paper and gently push it out. Then wrap it in clean lens paper.

Disassembly of the Rhomboid Collective Assembly

To disassemble the rhomboid collective assembly, unscrew the rhomboid collective lens retaining ring from the rhomboid collective mount (fig. 17-21).

Disassembly of the Erector Assembly

Hold the erector assembly in one hand and unscrew the lens retaining ring. Then drop the lens out of its mount onto a clean cloth in your other hand. If the lens does not drop out, wrap a finger with lens paper and gently push it out. The parts of the assembly are shown in figure 17-22.

Disassembly of the Eyepiece Assembly

Proceed as follows to disassemble the eyepiece assembly:

1. Unscrew the eyepiece collective lens retaining ring from the mount. Study illustration 17-23.
2. Turn the assembly over and let the eyepiece collective lens fall out on clean lens paper or cloth.
3. Remove the eyelens retaining ring. Then protect the lens with clean lens paper and gently press it from the mount.
4. With a prying tool, remove the eyeguard snap ring (fig. 17-23). Then slide off the rubber eye guard.

Disassembly of the Auxiliary Telescope Assembly

The procedure for disassembling the auxiliary telescope assembly is as follows:

1. Put the assembly on an auxiliary telescope fixture, which protects the bottom edge of the

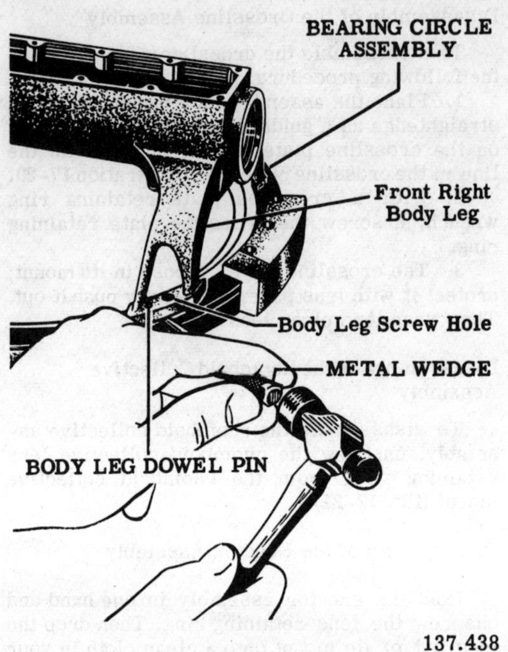

137.438
Figure 17-15.—Removing the bearing circle assembly.

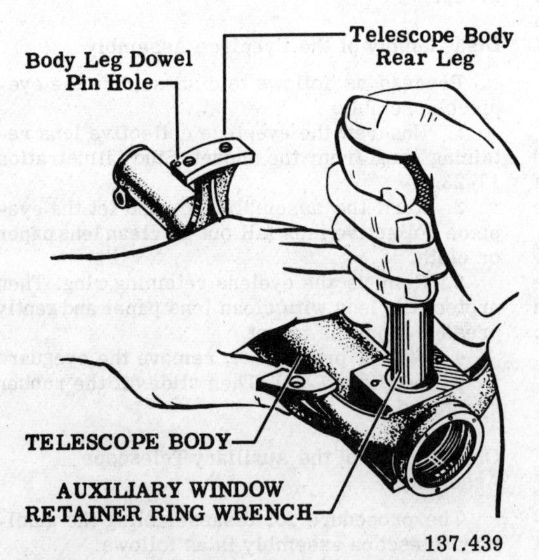

137.439
Figure 17-16.—Removing the auxiliary window retaining ring.

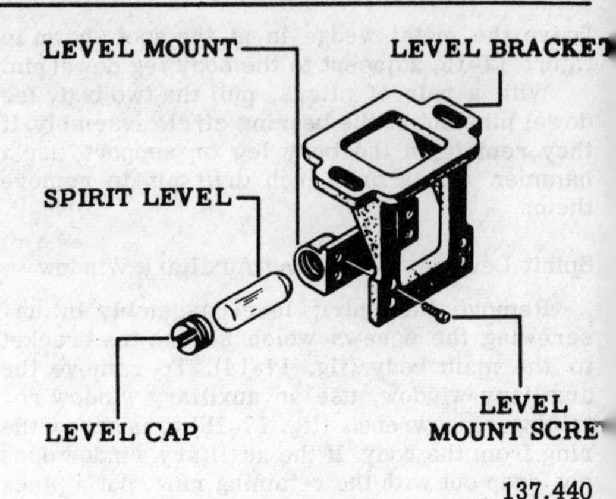

137.440
Figure 17-17.—Spirit level assembly.

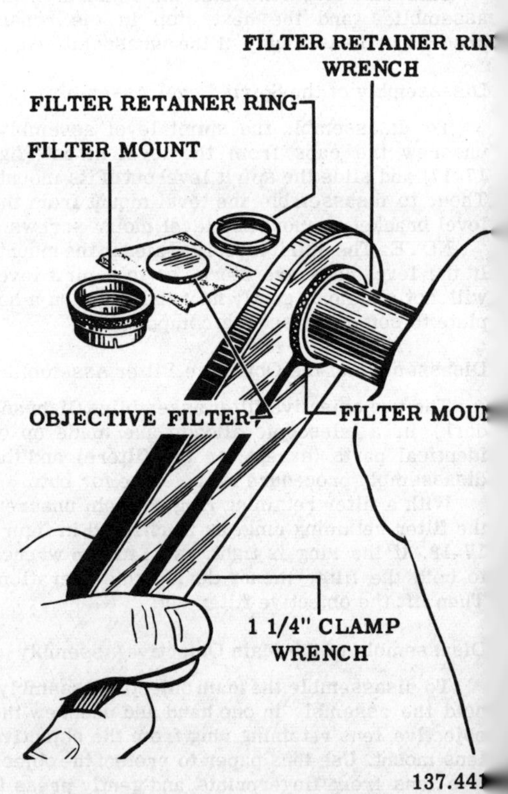

137.441
Figure 17-18.—Unscrewing the filter retaining ring.

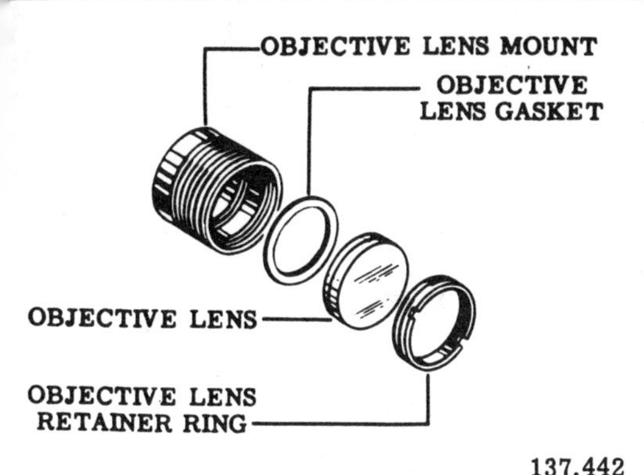

137.442
Figure 17-19.—Main objective assembly.

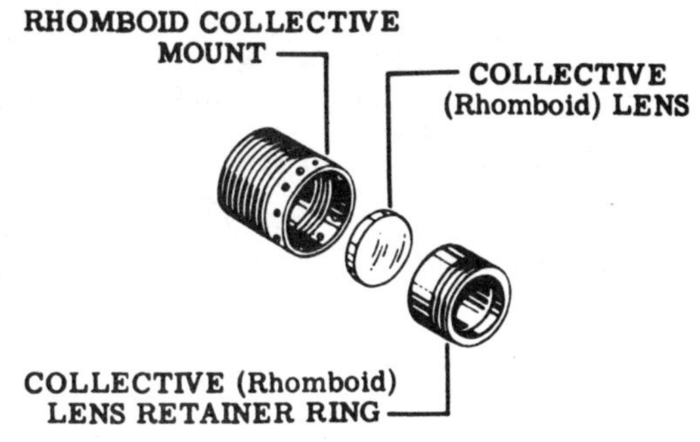

137.444
Figure 17-21.—Rhomboid collective assembly.

auxiliary rhomboid prism, and scribe reassembly guide marks for the auxiliary rhomboid prism assembly on the telescope plate, as shown in figure 17-24. Scribe marks also on the plate along the sides of the auxiliary objective assembly, and along the back of the auxiliary penta prism assembly.

2. Unscrew the auxiliary rhomboid prism assembly screws (fig. 17-25) and lift the auxiliary telescope assembly up in one hand. Use the other hand to turn the auxiliary rhomboid prism assembly one-fourth turn (fig. 17-26), and then pull it back and rotate it out of the opening in the auxiliary telescope plate.

3. Remove the rhomboid prism strap screws and lift the prism strap and pad off.

4. With tweezers, grasp the loose rhomboid prism by its unpolished sides and lift it out.

5. Remove the rhomboid prism shield and pad from the prism mount.

6. Unscrew the screws which secure the auxiliary objective assembly and remove the assembly (fig. 17-25).

NOTE: Do NOT loosen the auxiliary objective mount lock ring or remove the mount from the bracket. Do NOT remove the diaphragm or its lock ring unless it or the auxiliary objective mount must be replaced.

7. With an alidade flat wrench, remove the auxiliary objective lens retaining ring and allow the lens to fall out gently onto a clean cloth. If

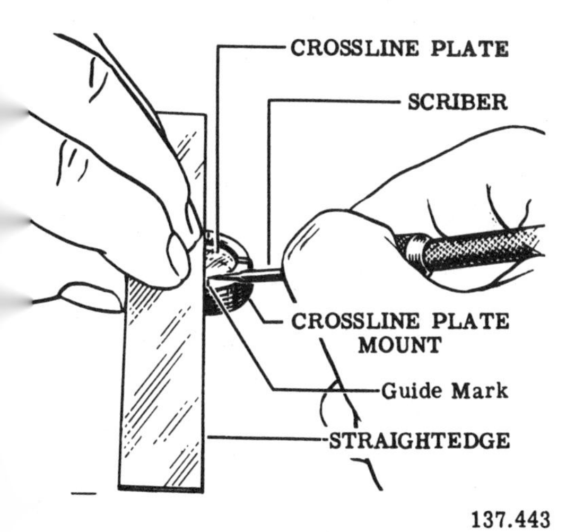

137.443
Figure 17-20.—Scribing guide marks on the crossline plate mount.

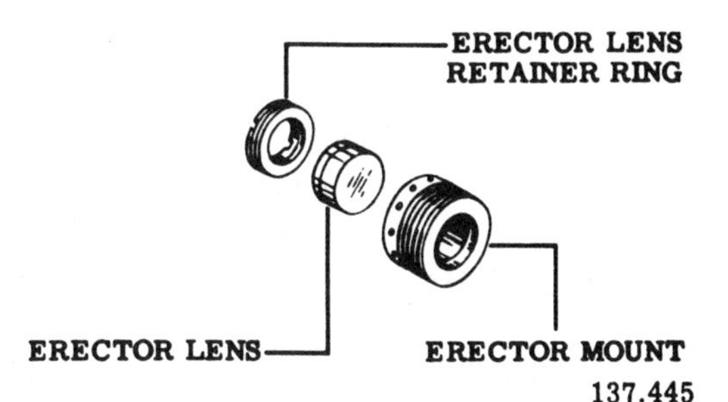

137.445
Figure 17-22.—Erector assembly.

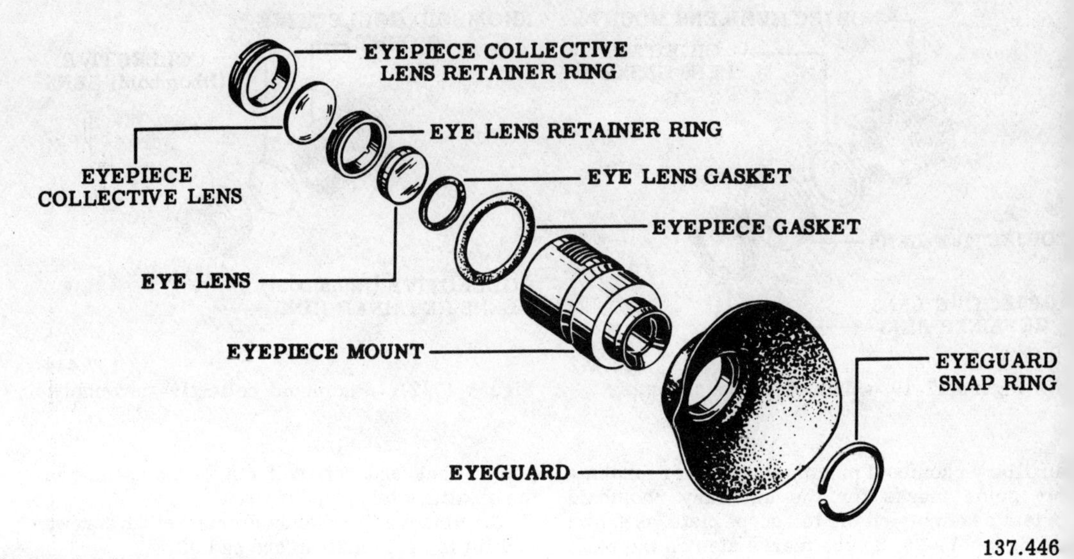

137.446

Figure 17-23.—Eyepiece assembly.

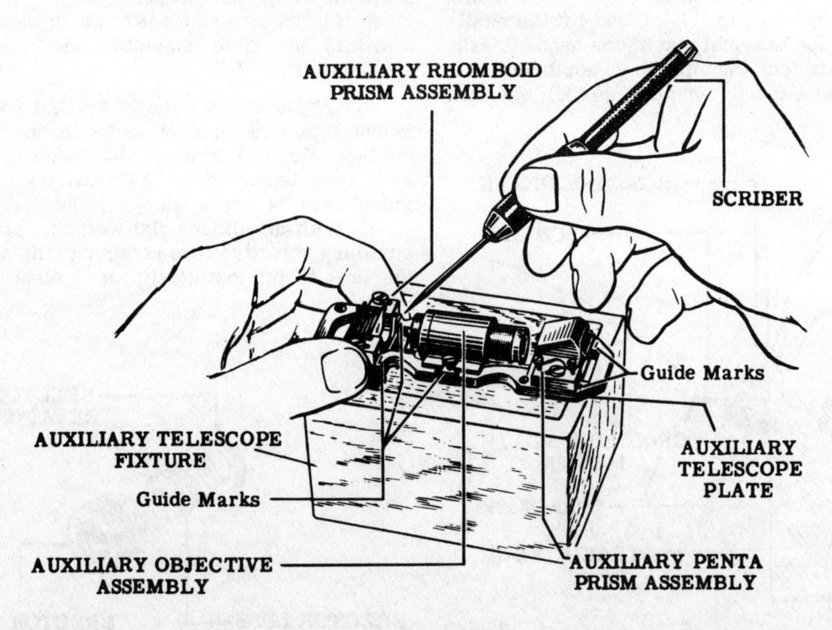

137.44'

Figure 17-24.—Scribing reassembly guide marks on the telescope plate for the auxiliary rhomboid prism assembly.

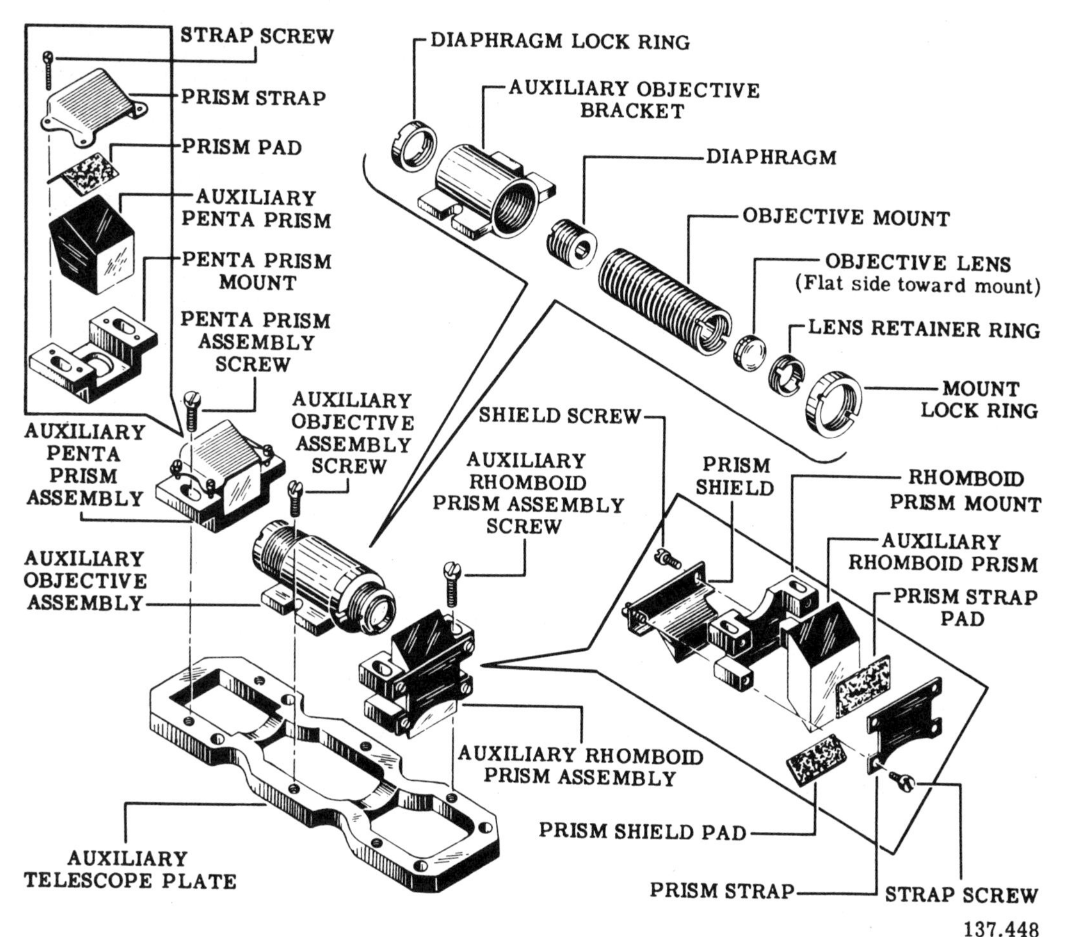

Figure 17-25.—Components of the auxiliary penta prism, auxiliary objective, and rhomboid prism assemblies.

the lens is stuck, wrap a small stick with lens paper and push it out.

8. Measure and record the distance the auxiliary objective mount extends from the bracket. This distance is indicated by A in illustration 17-27. The distance the auxiliary objective diaphragm extends from the mount is indicated by B.

9. With an alidade flat wrench, unscrew the diaphragm lock ring on the auxiliary objective diaphragm and remove the diaphragm.

10. Loosen the auxiliary objective mount lock ring and unscrew the mount from the objective bracket.

11. Remove the auxiliary penta prism assembly screws and then the prism assembly from the telescope plate (fig. 17-25).

12. Remove the penta prism strap screws and lift the penta prism strap off.

13. Remove the penta prism pad, lift the prism out of its mount, and lay it on clean cloth or lens paper.

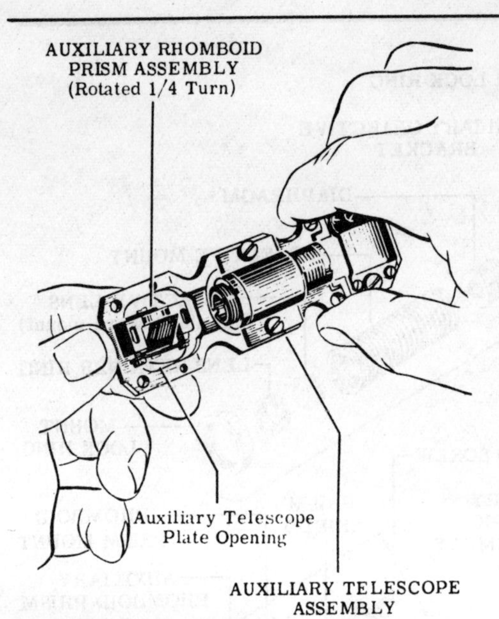

137.449
Figure 17-26.—Removing the auxiliary rhomboid prism assembly.

This last step completes the disassembly of a Mk 4, Mods 0 and 1, telescopic alidade. Refer to chapter 14 for the disassembly procedure for a hearing circle.

OVERHAUL AND REPAIR

After disassembling the instrument, inspect all parts for defects. Some parts may have to be replaced, as is also true for complete subassemblies.

Inspection of Optical Parts

Inspect optical parts of an alidade for the following:

1. ANTI-REFLECTION COATING. Check the anti-reflection coating on the objectives (main and auxiliary), on the erector eyepiece collective, and on the eyelenses by viewing the light reflected from their surfaces.

2. DEFECTS IN CEMENT. Inspect the main and auxiliary objective lenses, the erector lens, and the eyelens for defects in cement between the elements. If necessary, replace the lenses.

3. SILVER PLATING. Inspect the silver plating on the mirror surfaces of the auxiliary penta and auxiliary rhomboid prisms by viewing the mirrors through the polished surfaces of the prisms. Deteriorated mirror surfaces may cause a lack of brightness or poor quality of the auxiliary image.

4. ANGLES FORMED BY RHOMBOID PRISM. Examine both edges formed by the surfaces of the acute angles of the auxiliary rhomboid prism. If one edge is chipped, reverse the prism and use the other side. NOTE: Some auxiliary rhomboid prisms have one beveled edge and one sharp edge. The beveled edge cannot be used.

Inspection of Mechanical Parts

Inspect mechanical parts of the alidade for the following:

1. FUNCTIONAL DEFECTS. Look for dents, burrs, distortion, damaged threads, and wear on mechanical parts.

2. BEARING CIRCLE. Inspect the bearing circle for legibility of compass figures, degree markings, and distortion. If the markings are not clearly legible, they must be refilled.

3. LEGS OF TELESCOPE BODY. The legs of the telescope body must lie in a flat plane; but they are easily bent, especially when the body is removed from the bearing circle. Put the body

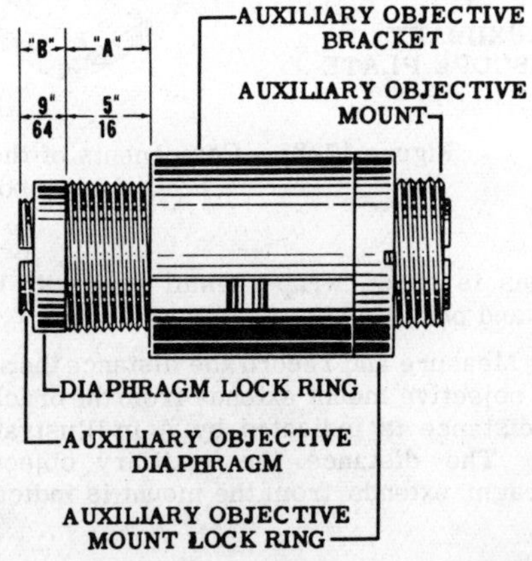

137.450
Figure 17-27.—Distance the auxiliary objective assembly extends from the mount.

on a body-leg testing plate (fig. 17-4) and test for flatness of the legs by pressing down on them with your thumbs. If rocking occurs, the legs are uneven.

4. EYEPIECE PARTS. Inspect the eyepiece parts for backlash or binding and, if present, repair or replace the parts.

5. GENERAL APPEARANCE. Check the finish of all parts and take the action recommended in chapter 8 of this manual, or as directed by the shop supervisor.

Refer to chapter 14 for repair procedures for bearing circles. Some other repairs which may be required on the alidades are explained next.

Objective Lens Mount

Illustration 17-28 shows the location and size of slots which must be put in the objective lens mount of a Mk 4, Mods 0 and 1, telescopic alidade. Use a good file to do this job, and remove burrs formed by the file.

Telescope Body Feet

If the feet of the telescope body are not flat and do not lie in the same plane, the body and the bearing circle will be distorted when tightened together with the body leg screws. File or machine the legs as necessary to have them flat and of the same length. Use emery cloth to finish the job; that is, to remove small amounts of metal at a time from the legs until all are level and have the same length.

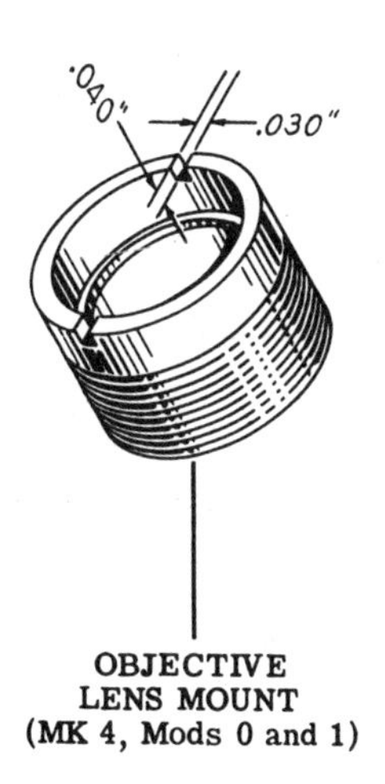

137.451
Figure 17-28.—Position of slots in the objective lens mount.

Assembly Lock Rings

If slots are needed in the erector assembly or rhomboid collective assembly lock rings, put them in the locations shown in figure 17-29. Be sure to use exact measurements.

When repairs on the alidade are completed, clean all parts in the manner explained in chapter 8.

REASSEMBLY

The procedure for reassembling subassemblies of your instrument is explained first. The procedure for replacing the subassemblies follows:

Replace all old gaskets. Lubricate moving parts with approved oils and greases.

Reassembly of the Main Objective Assembly

Study the optical layout of a Mk 4, Mod 1, telescopic alidade in figure 17-30. Then proceed

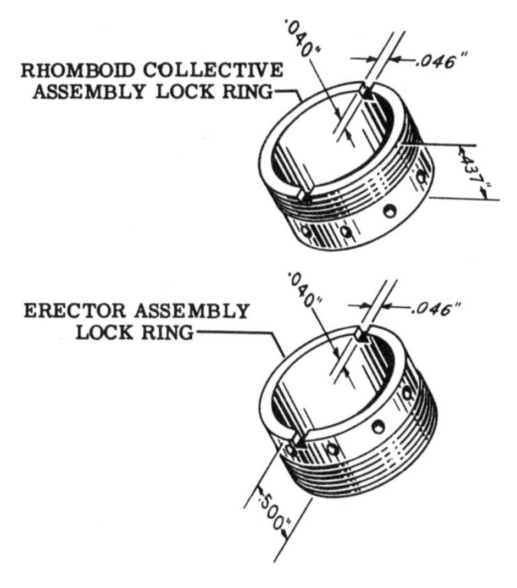

137.452
Figure 17-29.—Location of slots in the erector assembly lock ring and the rhomboid collective assembly lock ring.

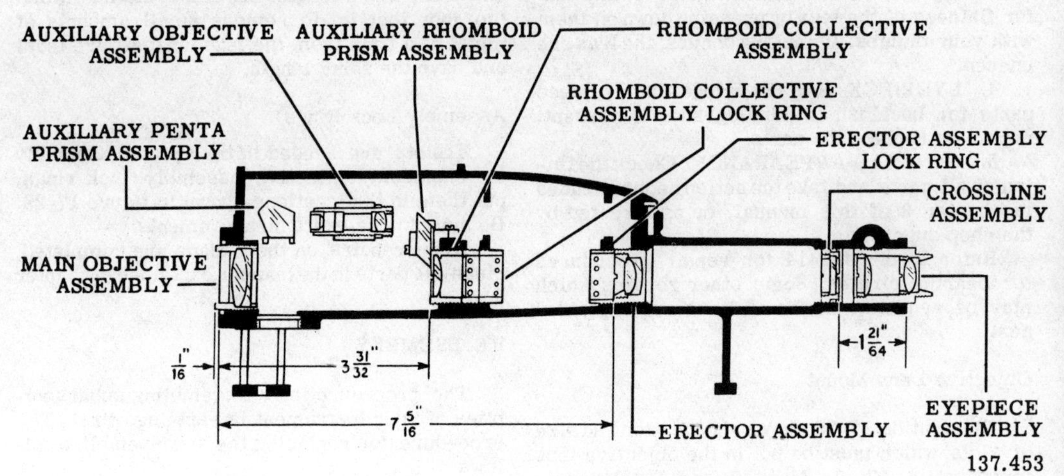

Figure 17-30.—Optical layout of a Mk 4, Mod 1, alidade.

as follows to reassemble the main objective assembly:

1. Put the objective lens gasket on the lens seat of the lens mount (fig. 17-19). Make certain it lies flat.

2. Clean the objective lens and use a lens suction adapter to put it in the objective lens mount, convex side down.

3. Replace and tighten the objective lens retaining ring.

Reassembly of the Rhomboid Collective and Erector Assemblies

Clean the rhomboid collective lens and use a lens suction adapter to place it flat-side down in its mount. Then clean the erector lens and replace it in its mount. Replace and tighten the collective lens retaining ring. Review illustration 17-22.

Reassembly of the Objective Filter Assemblies

Put a filter gasket (fig. 17-31) on the seat of the filter mount and make it flat and even all around. Then place the clean objective filter on top of the gasket in the filter mount. Replace and secure the filter retaining ring.

Reassembly of the Crossline Assembly

Clean the crossline plate and use a lens suction adapter to set it in its mount, engraved crossline face down. See illustration 17-32. Screw the crossline plate retaining ring into the mount but do not tighten it. Then cover your fingers with lens paper and turn the plate until you have it aligned with the reassembly guide marks on the back of the mount.

Reassembly of the Eyepiece Assembly

Clean the eyelens and the interior of its mount (fig. 17-23). Then put an eyelens gasket in the mount and replace the eyelens, flat-side down. Replace and tighten the lens retaining ring.

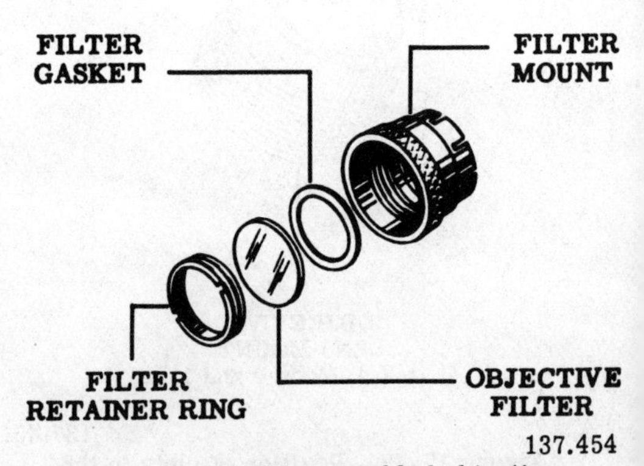

Figure 17-31.—Disassembled objective filter assembly.

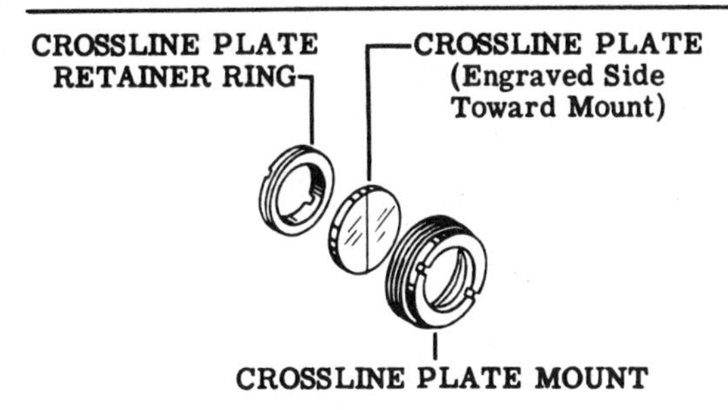

137.455
Figure 17-32.—Crossline assembly.

Put the clean eyepiece collective lens (fig. 17-23) in the eyepiece mount, with its convex side facing the eyelens. Then replace and tighten the collective lens retaining ring. Put the eyepiece gasket over the eyepiece mount and seat it in the groove. Slide the eyeguard onto the end of the mount and insert the eyeguard snap ring into its groove.

Reassembly of the Level Assembly

Assemble the spirit level mount to its bracket (fig. 17-17) and then fit the level in it. Replace one level cap and fill the assembled level cap with sealing compound. Use a sealing compound tool and an alcohol lamp, as illustrated in figure 17-33.

Now slide the level into its mount and position it in the soft compound, with its leveling line on top and centered in the holder. Then fill the space on top of the level with compound and replace the other level cap.

Reassembly of the Auxiliary Telescope Assembly

Refer to illustration 17-25 while studying the procedure for reassembling components of the auxiliary telescope assembly.

RHOMBOID PRISM ASSEMBLY.—Attach the rhomboid prism shield to its mount and clean the polished surfaces of the auxiliary rhomboid prism. Replace the prism shield pad and use tweezers to pick up the prism by its unpolished sides and put it in the mount, parallel to the vertical sides and centered between them. Then use tweezers to replace the prism strap pad on top of the prism. Next, replace the prism strap and tighten the screws evenly and just enough to hold the prism. Excessive pressure will cause strain in the prism.

PENTA PRISM ASSEMBLY.—Clean the auxiliary penta prism, pick it up with tweezers, and place it in its mount. See illustration 17-25 for the position of each part. Next, pick up the penta prism pad with tweezers and put it on top of the prism. Then replace the prism strap. Make certain the vertical face of the prism is even and square with the side of the mount; then tighten the screws to hold the prism firmly in position. Excessive pressure will cause strain.

OBJECTIVE ASSEMBLY.—Screw the auxiliary objective mount into the auxiliary objective bracket. The distance the mount should project from the bracket is represented by A in illustration 17-27.

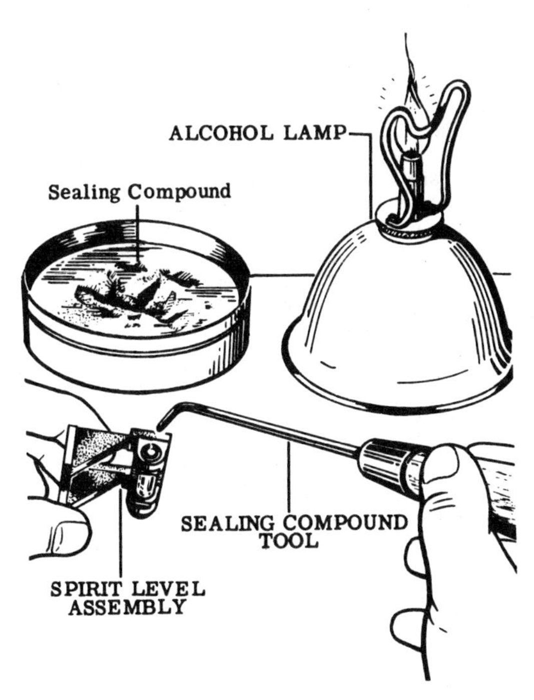

137.456
Figure 17-33.—Filling a spirit level cap with sealing compound.

Replace and tighten the mount lock ring. Then screw the objective diaphragm into the other end of the mount and replace the lock ring. Position the ring to the dimension indicated by B in figure 17-27 and then tighten it.

Clean the objective lens, pick it up with tweezers, and place it flat-side down in its mount.

ATTACHING THE AUXILIARY ASSEMBLIES.—Place the auxiliary telescope plate on the telescope fixture (fig. 17-24). This illustration shows the fully assembled auxiliary telescope, with the reassembly guide marks on the plate.

Replace the penta prism assembly on the plate and position it to the guide marks. (If there are no guide marks on a plate, center the assembly on the plate.) Then assemble the auxiliary objective assembly on the plate and assemble it in accordance with the guide marks, with the diaphragm facing the penta prism assembly.

Pick up the partially assembled auxiliary telescope assembly with one hand, and the rhomboid assembly with the other hand (fig. 17-26), and insert the telescope assembly in the plate. Then rotate the rhomboid prism assembly to bring it into correct position.

Put the assembly on the telescope fixture and position it to the guide marks. Then insert and tighten the screws. (If there are no guide marks, pull the prism assembly toward the objective assembly and tighten the screws.)

This completes the reassembly of subassemblies and you are now ready to replace them in their proper positions in the telescopic alidade.

Assembly of the Auxiliary Window

Use the following procedure to assemble the auxiliary window to the telescope body.

1. Turn the telescope body over on its back and put the window gasket in the seat of the body opening.
2. Pick up the clean window with tweezers and place it on the gasket.
3. Screw in and tighten the window retaining ring (fig. 17-16).

Assembling the Crossline Assembly

Slide the crossline assembly into the eyepiece end of the telescope body, crossline plate retaining ring facing in (fig. 17-34), and screw it into the threads in the body with the inner element of a crossline adjusting wrench. The outer element of the wrench acts as a guide to keep the inner element from slipping.

Reset the crossline assembly to its depth in the telescope tube assembly recorded during disassembly (fig. 17-9), by actual measurement with a rule. If the depth varies more than 1/32 inch from 1-21/64 inch, which is accurate, set it to this depth.

Make certain the crossline is vertical and at the correct depth. Then insert the assembly lock ring, slotted side out, into the body and tighten it against the crossline assembly. Hold the crossline assembly at the measured depth with the inner element (fig. 17-11) of the crossline adjusting wrench and use the outer element to engage the slots in the lock ring to tighten it. Recheck the depth to be sure it has not changed.

Reassembling the Erector Assembly

Screw a collective and erector assembly wrench into the objective end of the body. Then pick up the erector assembly (fig. 17-35) with tweezers and put it down in the body. Screw the assembly into the body and turn the erector mount to align the guide marks with those made on the body during disassembly. Replace the assembly lock ring and screw it in against the erector assembly.

Assembling the Rhomboid Collective Assembly

Use tweezers to put the rhomboid collective assembly lock ring in the body. Then screw the ring into the threaded rib of the body, and position it to align the reassembly guide marks on the lock ring and the rib. See figures 17-34 and 17-36.

Position the rhomboid collective assembly in accordance with the reassembly guide marks on the telescope body and also on the mount (fig. 17-36). When in its aligned position, the assembly should be tight against the lock ring.

Replacing the Main Objective Assembly

Screw the main objective assembly (fig. 17-30) into the end of the telescope body. Turn it in until the objective mount protrudes about 1/16 inch from the end of the body.

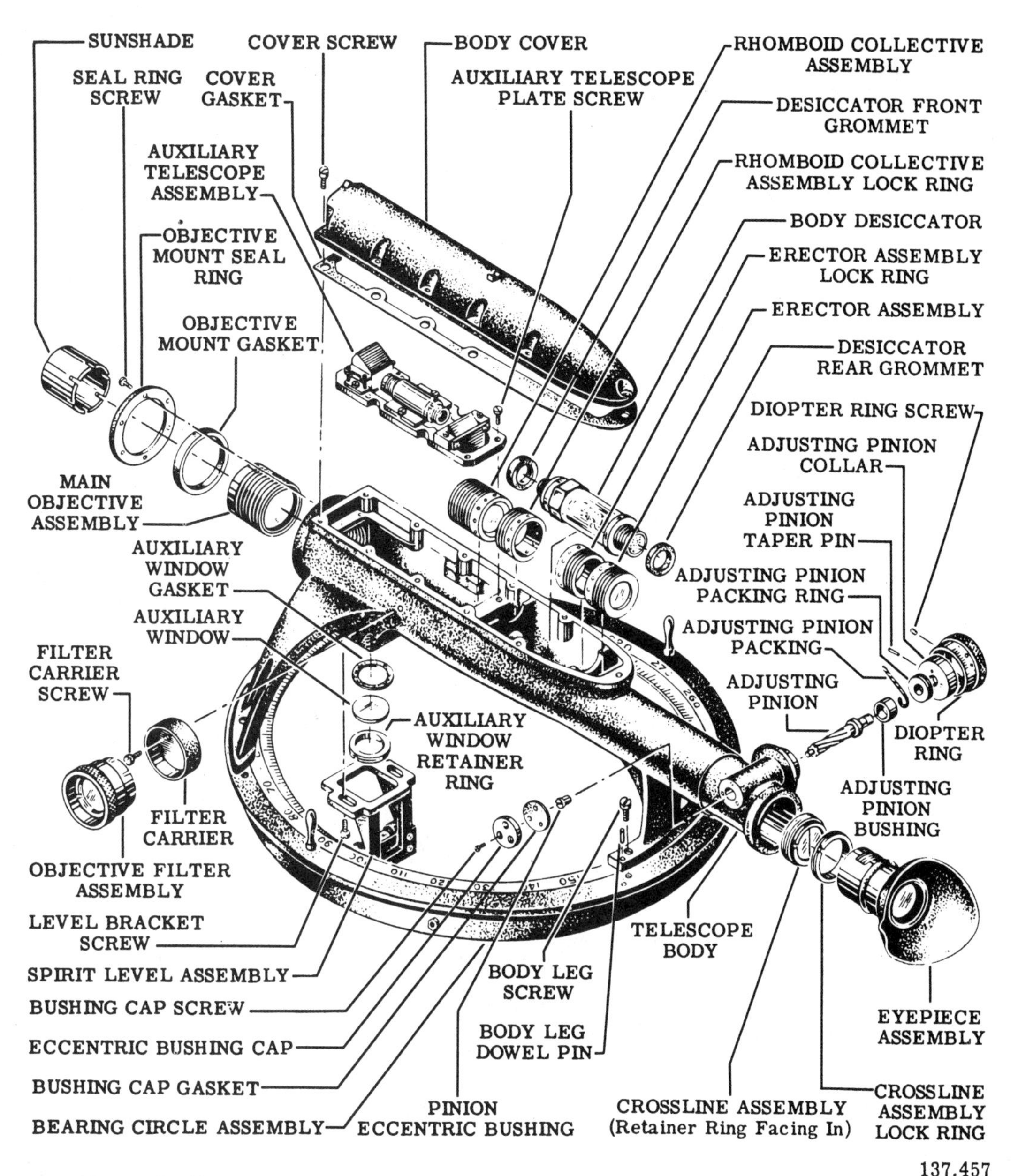

137.457

Figure 17-34.—Location of components and parts of a Mk 4, Mod 1, telescopic alidade.

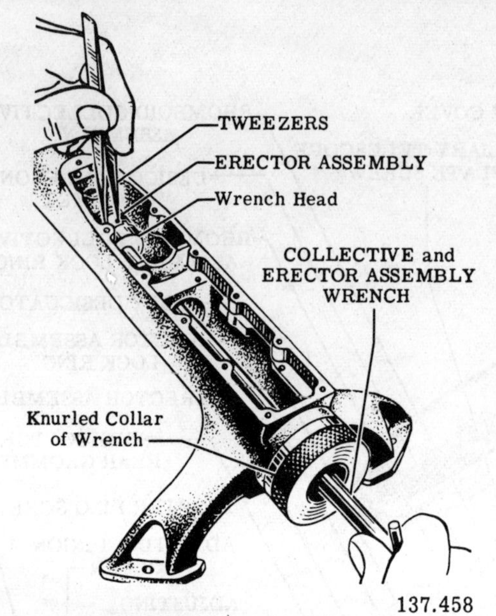

Figure 17-35.—Reassembling the erector assembly.

Do NOT put the objective mount gasket and seal ring in at this time, because you may have to change the position of the assembly during testing to eliminate parallax.

Replacing the Auxiliary Telescope Assembly

Put the auxiliary telescope assembly back into the body. Be sure the assembly is clean; and use a suction line to pick up loose particles of dirt within the body before replacing the assembly.

Pick up the telescope assembly and set it down in the body. Then replace and tighten the auxiliary telescope plate screws. Slide the assembly on the auxiliary telescope plate until the rhomboid prism is against its collective lens mount. Then tighten the prism mount screws.

Put the body cover over the opening in the body and secure it temporarily with two body cover screws. After collimating the instrument, replace the body desiccator, the cover gasket, body cover and all eleven screws which secure the cover.

At this point, put an eyepiece assembly in the eyepiece end of the body and look through

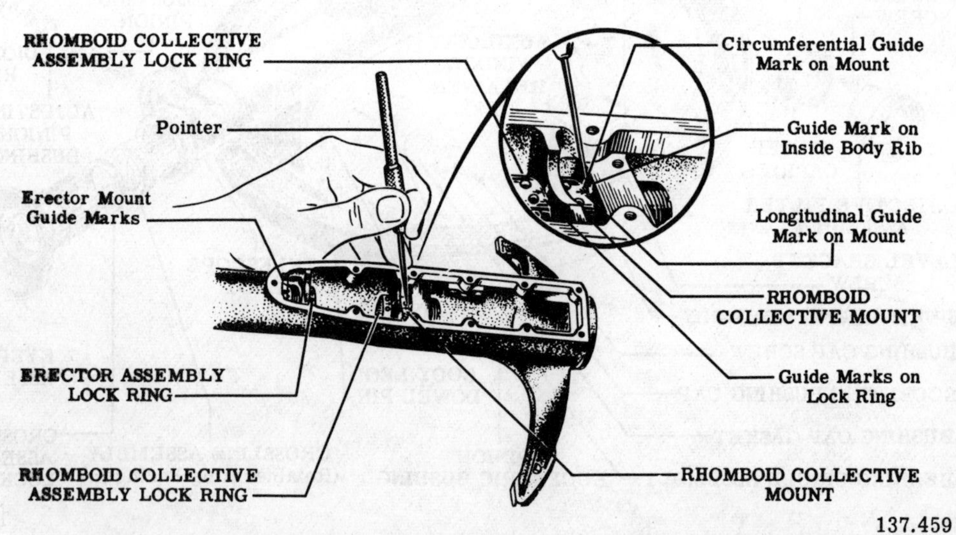

Figure 17-36.—Reassembling the rhomboid collective assembly.

the telescope at a light source to check for cleanliness and position of the optical elements. The edge of the auxiliary rhomboid prism should appear as a sharply focused horizontal line. If it is not in focus with the crossline, the crossline plate may be in backwards or in the wrong position, or the erector assembly and the rhomboid collective assembly may be in the wrong positions.

Replacing the Eyepiece Assembly

Before you replace the eyepiece assembly, put the objective filter assemblies (one on each side) on the body. See figure 17-3.

Proceed as follows to replace the eyepiece assembly:

1. If you removed the eyepiece gasket, stretch it over the eyepiece mount and roll it into the groove around the mount.
2. Make a pencil guide mark on the eyepiece mount, near the eyepiece guard and gear teeth, to indicate the position of the teeth on the eyepiece assembly when it is in the body. See illustration 17-37.
3. Smear a thin coating of lubricant over the eyepiece teeth and the adjusting pinion.
4. Insert the eyepiece assembly in the eyepiece end of the body and push it in until the eyepiece mount is protruding about 3/16 inch from the body. Turn it to bring the teeth on top, as indicated by the pencil mark you put on the mount.
5. Insert the adjusting pinion into the body, as indicated, and turn it slightly until you FEEL it slide through the teeth on the eyepiece mount. Turn the pinion as necessary to engage the teeth.
6. Cut a piece of adjusting pinion packing 3 1/4 inch in length. Then push the end in against the adjusting pinion bushing and use a stock to wind it CLOCKWISE around the pinion, as shown in figure 17-38.
7. Replace and tighten the adjusting pinion packing ring. This ring should feel snug against the packing when it is just inside the body; and if it does not snug up securely, use a longer piece of packing.
8. Replace the adjusting pinion collar on the threaded end of the pinion and tighten it enough to make it flush with the end of the pinion. Then insert the pinion taper pin.
9. Assemble and secure the diopter ring over the adjusting pinion collar.
10. Push the pinion eccentric bushing into the opening on the opposite side of the body and hold the bushing with one finger as you turn the diopter ring to test for backlash in the eyepiece. There should be no movement of the pinion without movement of the eyepiece.
11. Replace the bushing cap gasket and the cap.

COLLIMATION

Before collimating the alidade, inspect all optical and mechanical parts; and check the image fidelity (chapter 8).

Collimation of a telescopic alidade is a step-by-step procedure of interdependent adjustments. The main optical system and also the auxiliary optical system must be collimated to the mechanical axis of the bearing circle. The first step is mechanical alignment of the bearing circle with the axis of the collimator. The second step is alignment of the main optical system with the collimator. The third step is alignment of the auxiliary optical system with the main optical system.

The collimator generally used for collimating a telescopic alidade is a Mk 4, Mod 0, instrument, illustrated and discussed in chapter 8 (fig. 8-22). The collimation procedure discussed in this section is for a Mk 4, Mod 1, telescopic alidade.

Positioning the Crossline

The crossline of an alidade must be vertical when the instrument is in the position of normal use. Test and adjustment procedures for positioning the crossline are as follows:

1. Mount the telescopic alidade on the collimator stand, look through the eyepiece, and turn the alidade on the stand to make its crossline image coincide with the vertical line of the collimator target. If the alidade crossline is not in exact vertical coincidence with the target, proceed to step No. 2.
2. Note the depth to which the crossline assembly is set in the telescope body. If uncertain about the depth, remove the eyepiece assembly and measure the length with a scale. This depth must be maintained; and the crossline must be made vertical at this depth.

Observe the error in the position of the crossline relative to the vertical line of the target. Record the error in terms of the angle it appears to be off the vertical. If it is a small error, remove the eyepiece assembly and use the two concentric elements of the crossline adjusting wrench to turn the crossline assembly. The outer element of the wrench grips the crossline assembly lock ring and the inner element turns the crossline assembly. Then remove the wrenches and check with the eyepiece. Repeat this procedure until the crossline is in vertical coincidence with the collimator target.

If the bottom edge of the auxiliary rhomboid prism is in sharp focus for the same eyepiece setting as the crossline, the crossline is in correct position along the optical axis of the telescope. If the crossline is not in correct position, or if the angular error was large in the vertical position of the crossline, do the following: Disassemble the crossline assembly from the telescope body. Then loosen the crossline plate retaining ring and turn the crossline plate to correct the error noted in the original position. Reassemble the crossline assembly and its lock ring, at the depth measured originally.

Positioning the Objective Lens

The image formed by the objective lens must focus in the plane of the crossline. If it does not, the crossline image will not be in focus at the same time (same eyepiece setting) as the object.

Properly position the objective lens by checking for parallax.

Sight through the alidade and focus on its crossline and move your eye from side to side to check for parallax between the crossline plate and collimator image. If the image of the collimator target formed by the objective lens falls in the plane of the crossline plate (crossline side), no parallax will be observed. If the images are not in focus in the same plane, there will be parallax, which can be removed by screwing the main objective assembly in or out.

Put the objective mount gasket over the objective lens mount of the main objective assembly and press the gasket into its groove in the face of the telescope body. Then replace the objective mount seal ring.

Aligning and Checking the Bearing Circle Assembly

Because the bearing circle assembly indicates the direction in which the alidade is pointing, it must be round and true. Turn the alidade to align the 180° and 360° cardinal points on the bearing circle assembly with the corresponding points on the collimator stand. The 90° and 270° cardinal points should be aligned within the width of a line at the same time. If they are out of alignment, proceed as explained in the next paragraph.

Remove the two detent housing screws to disassemble one of the ball detents (with spring and housing), its detent housing spacer, and spacer shims (if any). If shims must be used to get proper alignment, experiment with several of them under one (or more) of the ball detents until proper alignment is obtained.

When the bearing circle is aligned, clamp it to the collimator stand to keep it aligned.

Aligning the Main Optical System

Telescopic alidades must indicate bearings within two minutes of arc. The collimator target is marked with tolerance limits, and the bearing circle assembly has been aligned with the collimator. Use the procedure discussed next to align the main optical system with the bearing circle assembly.

Sight through the alidade (with the bearing circle clamped on the collimator stand and aligned with the collimator) and note the deviation of the crossline from the vertical line in the center of the target. See illustration 17-39. If it is within the tolerance limits of two minutes of arc, tighten the four body leg screws to fix the body to the bearing circle assembly. If the two vertical lines do not coincide within the tolerance limits, shift the telescope body on the bearing circle assembly as necessary to obtain coincidence.

Loosen the body leg screws of the alidade and withdraw the body leg dowel pins. Then sight the collimator target and gently tap the telescope body legs with a mallet to shift the body on the bearing circle assembly enough to get coincidence of the vertical crosslines. If necessary, elongate the screw holes in the body legs. Then tighten the body leg screws and recheck the vertical lines for authorized coincidence. If necessary, repeat the whole operation.

After tightening the body leg screws, tap the dowel pins in. If you shifted the body on the bearing circle assembly and the dowel pin holes are not in line, drill new holes.

Adjusting the Auxiliary Telescope System

The auxiliary telescope system projects a segment of the compass or repeater into the eyepiece view so that bearing readings may be taken during the sighting of an object. It also

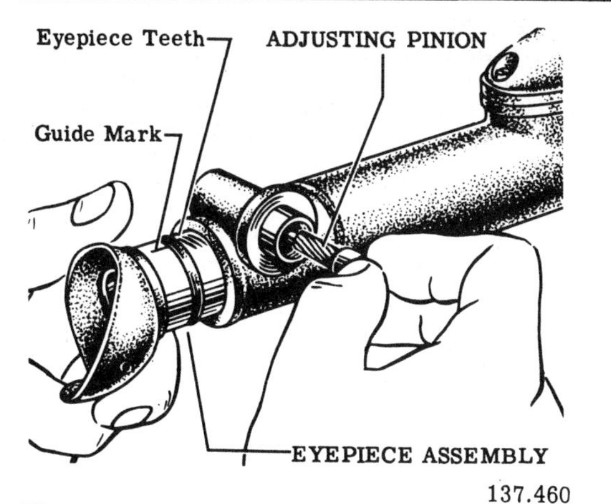

137.460
Figure 17-37.—Location of pencil guide mark on eyepiece mount to indicate position of teeth on the assembled eyepiece.

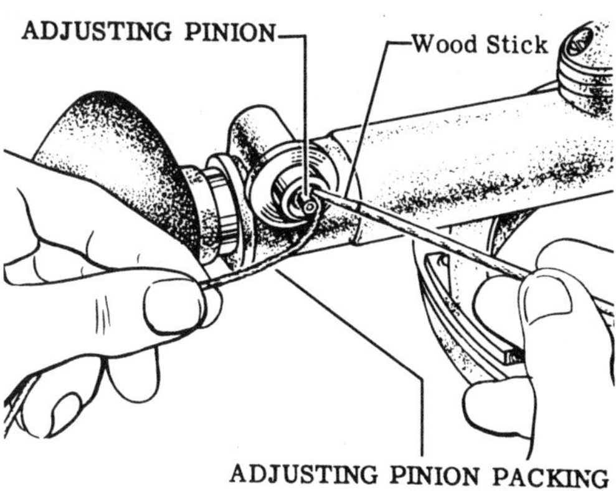

137.461
Figure 17-38.—Inserting adjusting pinion packing.

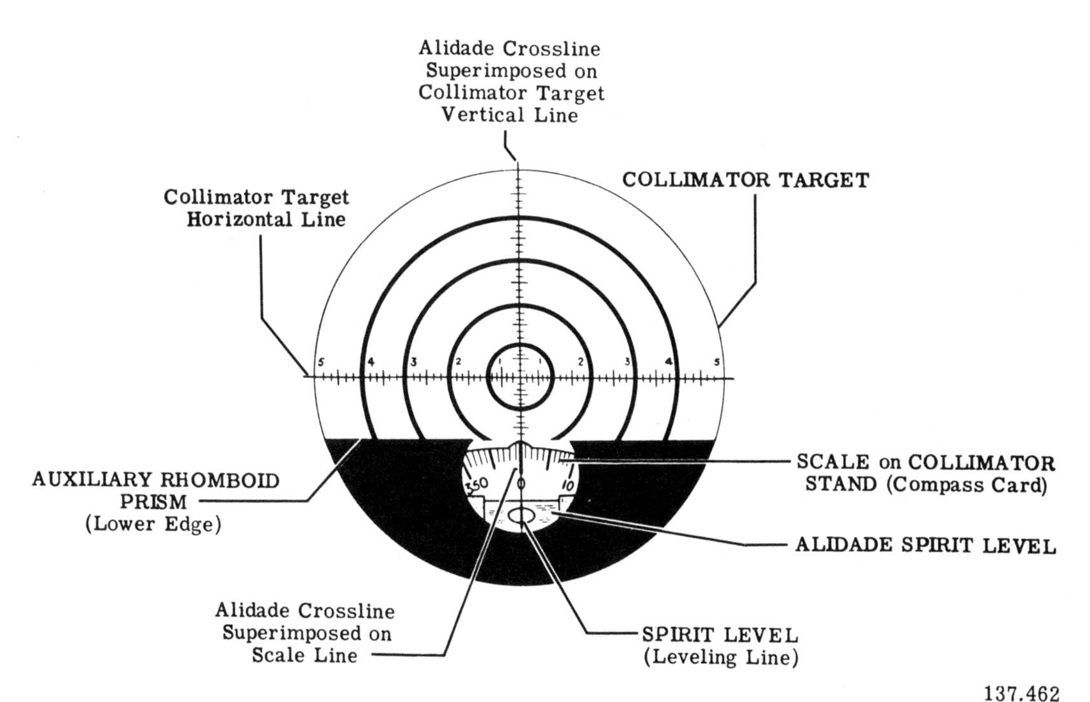

137.462
Figure 17-39.—Collimator target viewed through a telescopic alidade.

provides a view of the spirit level at the same time.

Sight the collimator target and observe the position of the horizontal dividing line in the telescope (formed by the bottom edge of the auxiliary rhomboid prism) relative to the horizontal crossline of the collimator target. It should be parallel to the crossline (fig. 17-39).

The auxiliary field should appear as shown in illustration 17-39. Disregard the position of the spirit level for the present. The 0° line on the collimator stand coincides with the crossline in the telescope—about 10° are visible on each side of the 0° line.

Loosen the penta prism mount or auxiliary objective assembly screws and shift the auxiliary penta prism assembly to shift the auxiliary field up or down. Twist the auxiliary penta prism assembly slightly to eliminate lean of the 0° line on the collimator stand. Make this line parallel to the crossline in the telescope.

To center the auxiliary field and make the 0° line coincide with the crossline, loosen the auxiliary objective bracket or auxiliary objective assembly screws and shift the assembly to one side or the other. Exclusive of the spirit level, the field of view should appear as shown in figure 17-39.

Tighten the screws which hold the auxiliary telescope assemblies in place to make certain they will not shift. Then check the auxiliary field to determine whether the image of the scale on the collimator stand in the auxiliary field is in focus in the same plane as the crossline plate (fig. 17-39). Focus on the crossline and check for parallax between the crossline in the telescope and the image of the lines and figures on the collimator stand.

To remove parallax, loosen the auxiliary objective mount lock ring and screw the mount in and out in the auxiliary objective bracket. After you eliminate parallax, tighten the objective mount lock ring.

Now, look through the telescope to check the position of the bubble in the spirit level. The level in its mount holder is shown in the correct position in illustration 17-39. If the level should be shifted up or down, or straightened parallel to the horizontal, loosen the level bracket screws and shift the entire spirit level assembly. To get to these screws, you must remove the alidade from the collimator stand; so when you replace the alidade, be sure to align the compass points.

The black leveling line on the spirit level glass tube should coincide with the crossline and the 0° line on the collimator. The leveling bubble should also appear centered on the leveling line, as shown in figure 17-39. This is the final adjustment, a slight one, which can be made by shifting the level holder or level mount screws and shifting the level holder.

Replace the body cover to keep the interior clean and secure it with two cover screws. The desiccator, cover gasket, and the other cover screws must be assembled before the Mk 4 alidade is tested for waterproofness.

Setting the Diopter Scale

To set the diopter scale on a Mk 4 telescopic alidade, do the following:

1. Look through the auxiliary telescope into a Mk 4 collimator and sight the target of the collimator. Bring it into sharp focus.
2. Put the auxiliary telescope in front of the eyelens of the alidade and look through the alidade into the collimator. Bring the collimator into sharp focus and observe the reading of the alidade diopter scale. It should be 0 within 1/4 diopter.
3. If the diopter scale error is greater than 1/4 diopter, shift the diopter ring. Loosen the two diopter ring screws and turn the ring as required to align the 0 with the index line. Then tighten the diopter ring screws.

FINAL ASSEMBLY AND WATERPROOFING

At this point in the repair procedure, replace the desiccator in the body. Put a cover gasket over the opening in the body and replace the cover. Replace and tighten the screws, but do not put a screw in the pressure hole. This hole was drilled through the casting for making the waterproof test. See illustration 17-40. Review the procedure for waterproofing and sealing optical instruments in chapter 8.

MARK 6 AND MARK 7 TELESCOPIC ALIDADES

Mark 6, Mod 1, and Mark 7, Mod 0, telescopic alidades are portable navigational instruments. Each type consists of a housing, bearing ring, handle, supports, level vial, main optical system, and an auxiliary optical system.

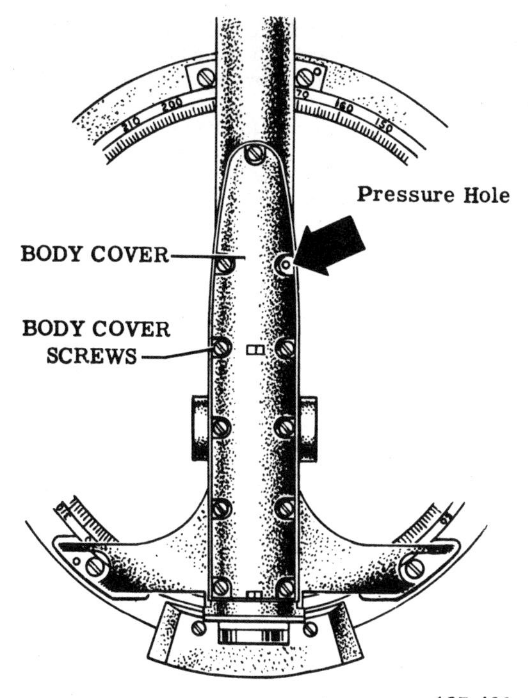

137.463
Figure 17-40.—Location of pressure hole in body cover.

A Mk 6, Mod 1, telescopic alidade (fig. 7-41) is for use only on a 6-inch (card diameter) gyrocompass indicator or a magnetic compass. This alidade does not require adapter rings. When it is mounted on an indicator or a compass, the auxiliary optical system forms an image of the level vial, the reticle, and 25° of the indicator or compass card. The main optical system forms an image of the object or target in the field of view of the instrument.

A Mk 7, Mod 0, telescopic with adapter rings is illustrated in figure 17-42. One adapter ring (type A) fits on a Navy No. 1 magnetic compass. The other adapter ring (type B) fits on a 7 1/2-inch gyrocompass indicator.

MAIN OPTICAL SYSTEM

The main optical system of a Mk 6 or a Mk 7 alidade is a terrestial telescope consisting of an objective lens, polarizing filters, a compensator lens, an Amici prism, a reticle, and the eyepiece elements. The objective lens receives light from an object and forms an image which is magnified by the eyepiece. The two polarizing filters may be rotated in or out of the line of sight to meet varying light conditions. One filter may be rotated independently to vary the intensity of light.

The compensator lens is mounted in the filter assembly, but it is put in the line of sight only when the polarizing filters are not in use. It converges the path of light and maintains the required focus. Light from this lens, or the polarizing filters, is then focused through a fixed stop aperture and on the eyepiece side of the Amici prism.

The Amici prism inverts and reverts the image and deviates the line of sight through a 45° angle. The reticle wire is superimposed on the image and the eyepiece elements (field lens, center lens, and the eyelens) then produce an enlarged, virtual image of the distant object at the eyepoint of the alidade.

AUXILIARY OPTICAL SYSTEM

The auxiliary optical system of a Mk 6 or Mk 7 alidade contains a sealing window, front surface mirror, outer objective lens, inner objective lens, erector lenses, and an auxiliary optical system prism.

45.39
Figure 17-41.—Mark 6, Mod 1, telescopic alidade.

45.39

Figure 17-42.—Mark 7, Mod 0, telescopic alidade with adapter rings.

The image of the compass or indicator card and level vial is transmitted through the window and reflected into the auxiliary optical system by the front surface mirror.

The inner and outer objective lenses converge the image from the front surface mirror, through the stop in the erector lens cell, into the auxiliary optical system prism. The erector lens inverts the image of the inner and outer objective lenses and reticle wire; the auxiliary optical system prism re-inverts the image in the plane of the mask. The reticle wire is superimposed on the compass card and the bubble in the level vial, and an image is formed at the eyepoint of the alidade.

The complete image formed at the eyepoint of the alidade consists of the distant object with the reticle wire superimposed (as viewed through the main optical system), and the image of the compass card, bubble, and reticle wire superimposed (transmitted through the auxiliary optical system).

FOCUSING ASSEMBLY

The function of the focusing assembly is to accommodate for visual variations between different observers. The assembly includes a focusing knob and shaft, diopter scale, and a stuffing box. When an observer rotates the focusing knob, he can adjust the eyelenses to his desired focus. The diopter scale on the focusing knob may be aligned with the white line on the stuffing box to obtain a diopter setting.

FILTER ASSEMBLY

The filter assembly enables an observer to control, by means of two polarizing filters and a compensator lens, the light intensity and glare within the alidade. When an observer turns the knob on the filter drive shaft he can increase or decrease light intensity. Another knob (larger) is provided for turning the polarizing filters out of the line of sight and inserting the compensator lens.

For additional information on Mk 6 and 7 telescopic alidades, refer to NavShips 324-0654 and NavShips 0924-001-6000.

CHAPTER 18

BINOCULARS

Binoculars are the eyes of the Fleet. They magnify distant objects and appear to bring them closer to the viewer for better observation. As the name implies, a binocular pertains to both eyes. There are two binocular optical systems, one for each eye; and in this discussion, each side of a binocular is considered as a binocular optical system, or body. See illustration 18-1.

The magnification, or power, of a binocular is expressed as 7x, 9x, or whatever times larger an object is when observed with a binocular as compared to the size of the object when observed by the naked eye. A 7 x 50 binocular, for example, has 7 power (7x) and an objective lens whose free aperture (usable diameter) is 50 millimeters. This is the code system commonly used for all hand-held binoculars.

You studied stereoscopic vision in chapter 6 and understand its importance. A binocular, by magnifying an image, increases the convergence angles and therefore increases the apparent divergence of these angles. In prism binoculars, the distance between the centers of the two objectives is greater than the distance between the pupils of the observer's eyes. Thus, a binocular increases the effective interpupillary distance of the eyes and therefore increases the convergence angles of objects in the field of view. For this reason, a binocular can extend the range of stereoscopic vision far beyond the 500-yard limit of the unaided eyes and make it easier to detect small differences in the brightness of objects. This is very important when an observer is scanning the sea for a periscope, for example.

This chapter provides information concerning the construction, performance, repair, and collimation of 7 x 50 prismatic binoculars, the type generally used by the Navy. When a binocular leaves your shop, it should meet all performance requirements—users of the instrument should be able to see distant objects CLEARLY without EYESTRAIN.

MECHANICAL FEATURES

Prismatic binoculars considered in this chapter are manufactured in accordance with Navy specifications. The bodies are made of aluminum, as are all other mechanical parts, with the exception of gaskets and some of the hinge parts. The weight of each binocular must be less than 46 ounces—carrying case and strap excluded.

The hinge (fig. 18-2) joins the two optical systems and provides a means for interpupillary adjustment. The design of the hinge is such that it gives smooth action, with sufficient tension to maintain proper spacing between the systems, without play or looseness.

The tapered hinge axle (fig. 18-2) is held firmly in the hinge lugs of the left body and rides freely in the matching taper of the hinge tube, which is set permanently in the hinge lugs of the right body. Hinge tension is controlled by the lower axle screw. When this screw is tightened, the left body hinge lugs are SQUEEZED against the right body hinge lugs. Since the right body lugs are held rigid by the hinge tube, friction developed between the faces of the outer and inner hinge lugs gives hinge tension. The .010-inch cellulose acetate hinge washers between each pair of upper and lower hinge lugs receive the wear resulting from hinge motion. These washers fit firmly in their positions against the hinge lugs and provide friction for hinge tension.

The hinge on a Mk 45, Mod 0, binocular has a different design. The hinge axle fits tightly in the straight hinge tube, and the axle is splined to the top of the tube. The hinge tube is held in the left body hinge lugs. A split hinge expanding bearing is threaded onto a modified Buttress-type thread on each end of the axle. The right body lugs swing on the expanded bearings.

When a split-hinge expanding bearing is tightened against the hinge bearing thrust washer (between the bearing and the bottom of the

37.1

Figure 18-1.—Mark 28, Mod 0, binocular.

threaded axle shoulder), the thrust forces the bearing up on the sloping sides of the threads to cause it to expand and develop friction with the right body hinge lug. Hinge tension is adjusted by tightening the hinge expanding bearing. Hinge locks fit in notches in the ends of the axle and keep the bearing from loosening on the axle. Lubrication is forced into the hinge through a grease fitting, and up through holes in the center of the axle to the top bearing.

OPTICAL SYSTEM

The optical systems in both bodies of a binocular are identical, and the optical axes of the two systems must be parallel and LOOK at the same point on a distant object. This is necessary to prevent eyestrain.

OBJECTIVE LENSES

The objective lens in a binocular is a large, cemented doublet which collects light from an object under observation. Its size, as you learned previously in this manual, determines the amount of light it can gather. Night glasses, for example, have EXTRA LARGE objective lenses which give greater detail to dim objects.

The objective lenses receive the light from a distant object and form a real, inverted and reverted, image in the system at the focal plane of the objectives. A pair of Porro prisms in the path of light in the system receives the image

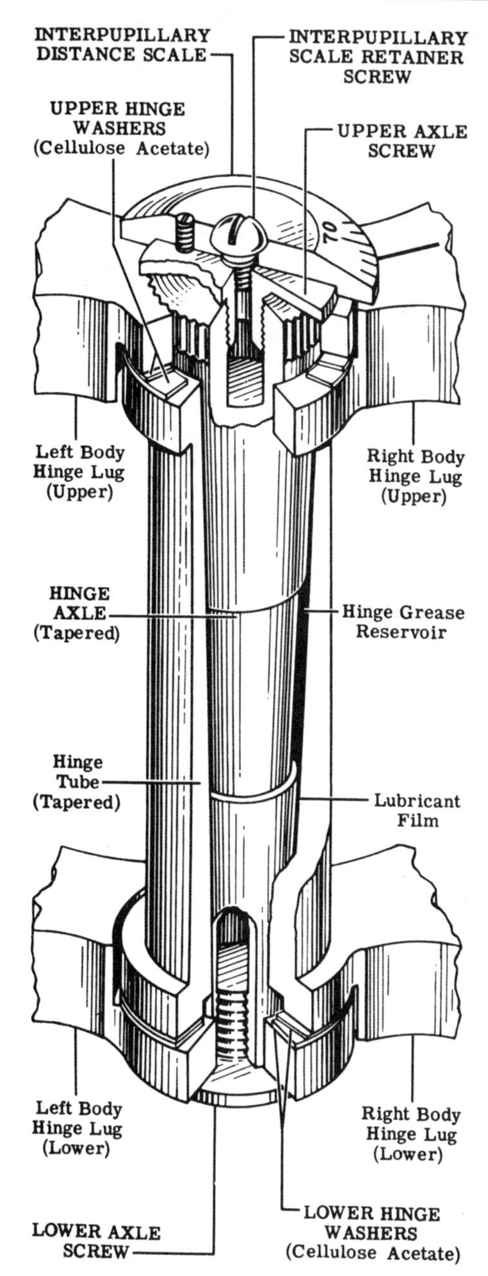

137.464
Figure 18-2.—Hinge mechanism (Mks 28, 32, & 39 binoculars).

and erects it in the form in which the observer sees it in the eyepieces.

PORRO PRISM CLUSTERS

The Porro prisms used as an erecting system in each body are mounted at right angles between the objective lenses and the eyepiece clusters. The prism clusters serve three important purposes:

1. They revert and invert the image formed by the objective lens, to the right-side-up or true position.
2. They increase stereoscopic effect and thus give better depth perception. This is true because they permit greater distance between the left and right objective lenses than would be allowed by the user's interpupillary distance alone.
3. They decrease the physical distance (the length of the body) between the objective lens and the eyepiece by FOLDING UP the path of light, as shown in illustration 18-3. Much longer tubes would be required if an erector lens were used to erect the image.

EYEPIECES AND DIOPTER SCALES

A user of prismatic binoculars sees the image formed by the objective lenses and inverted and reverted by the prism clusters. An eyepiece consists of a collective lens and a doublet eyelens (Kellner type) and is used like an ordinary magnifying glass, except that the object viewed in the binocular is the real image formed by the objective lenses.

To adjust for differences in vision between his right and left eyes, and to focus the image so that he can see it sharply and clearly, the user of a binocular can move the eyepieces in or out.

The scale on each eyepiece (fig. 18-1), called the diopter scale, enables a person who regularly uses a binocular to set the eyepieces to compensate for visual correction required by his eyes. This feature enables him to use the binocular quickly without delay for focusing.

A person who has normal vision should be able to see distant objects in perfect focus with both eyepieces set at the 0 mark, or slightly less.

OVERHAUL AND REPAIR

Overhaul and repair procedures discussed in this chapter are for Mk 28, Mod 0; Mk 32, Mod 7; and Mk 39, Mod 1, binoculars.

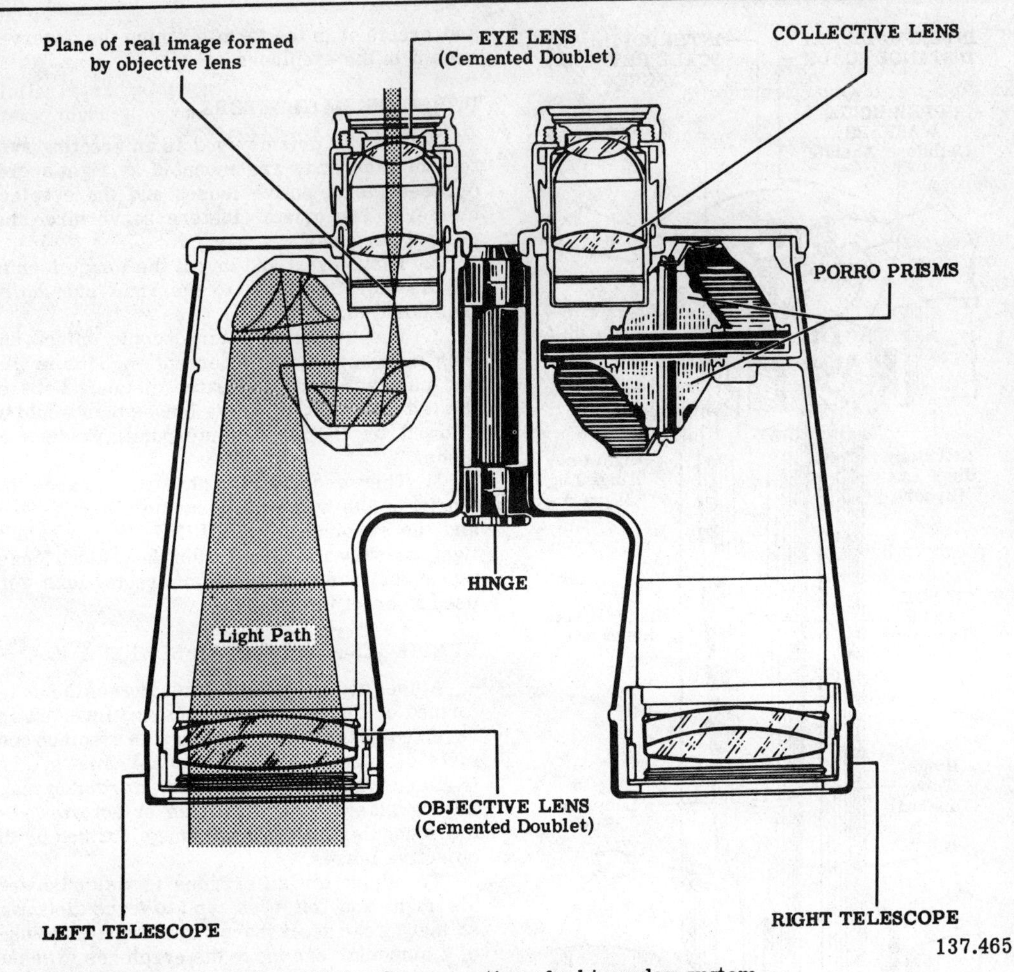

Figure 18-3.—Cross section of a binocular system.

The Mk 28, Mod 0, binocular is the standard, basic design of several marks and mods of 7 x 50 prismatic binoculars. Mod 0 was made waterproof by adding a rubber gasket between the cover and the body, together with a channel of wax. A rubber gasket was also added between the objective mount and the objective lock ring. The eyelens and the objective lens are set in a wax seal.

A Mk 32, Mod 7, binocular differs from a Mk 28, Mod 0, in two ways: (1) the objective lens, mount, and rings are made up as a subassembly with an adapter which screws into the objective end of the body; (2) it has a plastic eyelens gasket, a molded rubber eyepiece sealing ring; and a plastic stop ring gasket in the eyepiece assembly to waterproof the eyepiece assembly.

A Mk 39, Mod 1, binocular is the same as a Mk 28, Mod 0, except that it has a CROSSLINE reticle attached to the top of the right-hand prism cluster.

A Mk 45, Mod 0, binocular is made of drawn aluminum and waterproofed by rubber gaskets to make it capable of withstanding water pressure, which other binoculars cannot do. It is used on board submarines and by demolition personnel.

PREDISASSEMBLY INSPECTION

You already learned in other chapters of this manual the function and importance of predisassembly inspections. Prepare an inspection sheet for your binocular, make the inspection, and record your findings on it. Inspect the following:

1. General appearance and physical condition, including metal finishes, plastic body covering, and legibility of white-filled engravings.
2. Hinge action, for smoothness and tension. There should be NO shake in the hinge.
3. Eyepiece focusing motion, for wear and condition of lubricant. The tension should be fairly tight, and smooth action indicates that the condition of the lubricant in the eyepiece focusing sextuple thread is good.
4. Condition of cement in lenses.
5. Image fidelity, brightness, and contrast.
6. Interpupillary distance. Use an interpupillary distance spacing bar to check this.
7. Collimation and diopter settings. Failure of the binocular to collimate indicates that the prisms are cocked or tilted. If the diopter settings are correct, set the eyepieces at 0 diopters and place a straightedge across the eye caps. If they are not even within 1/16 inch, the prisms, objective lenses, or objective mounts in both telescopes may be mismatched.

DISASSEMBLY PROCEDURE

To prevent confusion of terms, the procedure for removing major assemblies from a binocular is considered first. Then the details of disassembly of the major assemblies are discussed. Some special tools used in the repair of binoculars are shown in figure 18-4.

Keep your workbench clean. Keep unprotected (unwrapped) fingers off polished optical surfaces. Wrap optical elements in clean tissue and protect them until needed. Work with care to prevent damage to parts.

Removal of Eyepiece and Cover Assembly

To remove this assembly, take out the cover screws and discard the fiber washers. See illustration 18-5. Then remove the synthetic rubber cover gasket and the desiccator (fig. 18-6). Discard the desicator. Mk 28, Mod 0, and Mk 39, Mod 1, binoculars have nickel-silver washers which may be cleaned and replaced on the instrument.

Removal of the Reticle Assembly

A Mk 39, Mod 1, binocular has a reticle assembly mounted on a post on the right prism plate. See illustration 18-7. Other binoculars do not have reticles.

Removal of the Prism Cluster

Use a long screwdriver to remove the prism plate screws on Mks 28, Mod 0, and 32, Mod 7, binoculars. On a Mk 39, Mod 1, binocular, remove the reticle mount post and the prism plate screws from the right body (fig. 18-7) and the three plate screws from the left body. Then tilt the cluster upward and rotate it to remove it from the body (fig. 18-8).

Scribe an L (left) or an R (right) on the TOP SIDE OF THE PRISM PLATE to indicate the side of the binocular from which you removed the prism cluster. These clusters are NOT interchangeable.

Removal of the Objective and Mount Assembly

The procedure for removing the objective and mount assembly (figs. 18-9 and 18-10) is as follows:

1. Unscrew and remove the objective cap. If it is frozen, warm it over a hot plate. CAUTION: If you overheat the cap, you will crack the objective.
2. Remove the sealing compound over the objective lock ring setscrew and the objective ring lock screw and then remove the screws.
3. Remove the objective lock ring, objective gasket ring, and the rubber gasket.
4. With a suction cup, pull out the objective and mount assembly.
5. A Mk 32, Mod 7, binocular has an objective adapter (fig. 18-10) screwed into the body. Use an objective adapter clamp wrench to remove it.

Removal of the Interpupillary Distance Scale

Use an appropriate screwdriver to remove the interpupillary scale lock screw (not on Mk 32, Mod 7, binoculars). Then remove the scale retainer screw and the scale. Study illustration 18-11.

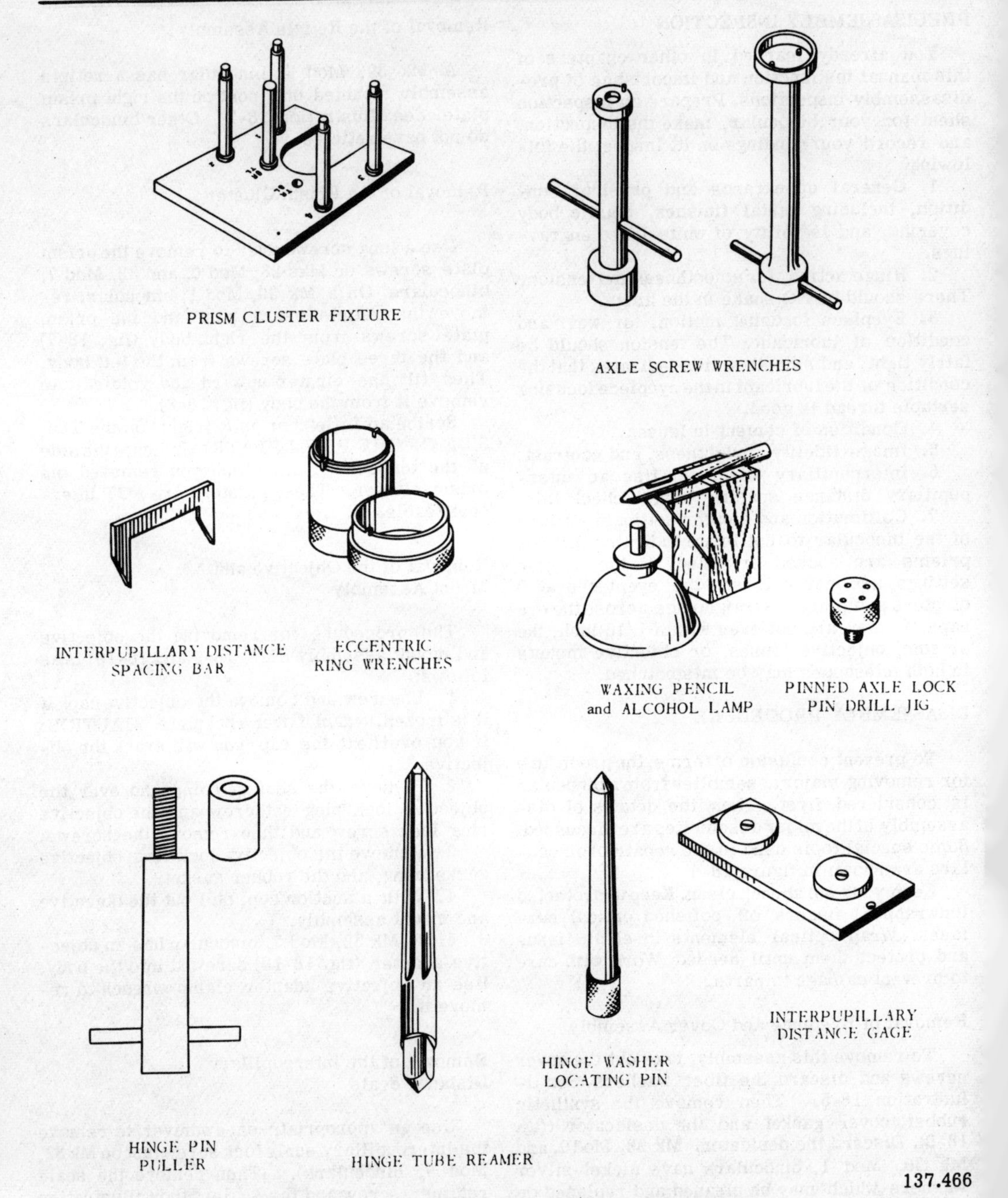

Figure 18-4.—Special service tools for binoculars.

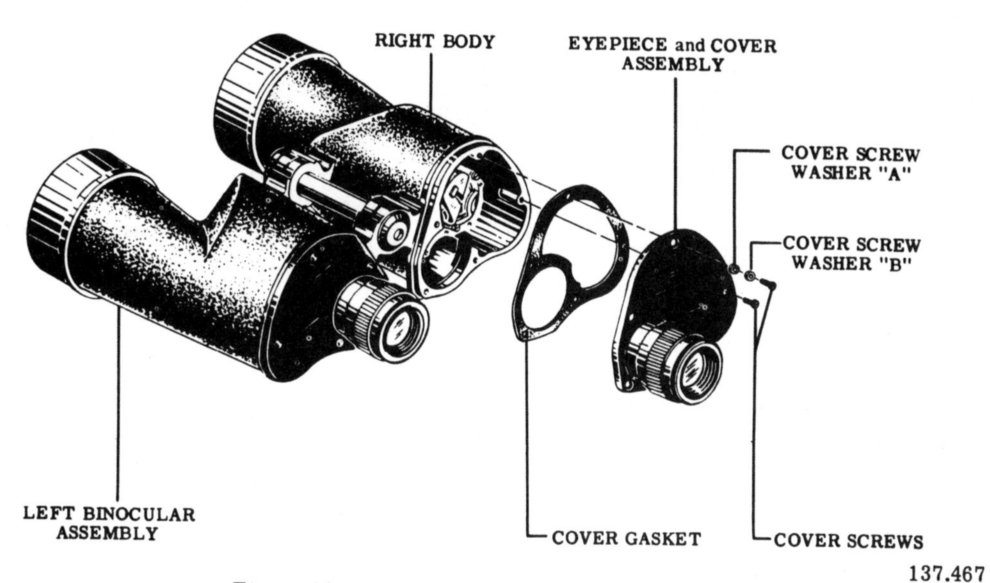

137.467

Figure 18-5.—Eyepiece and cover assembly.

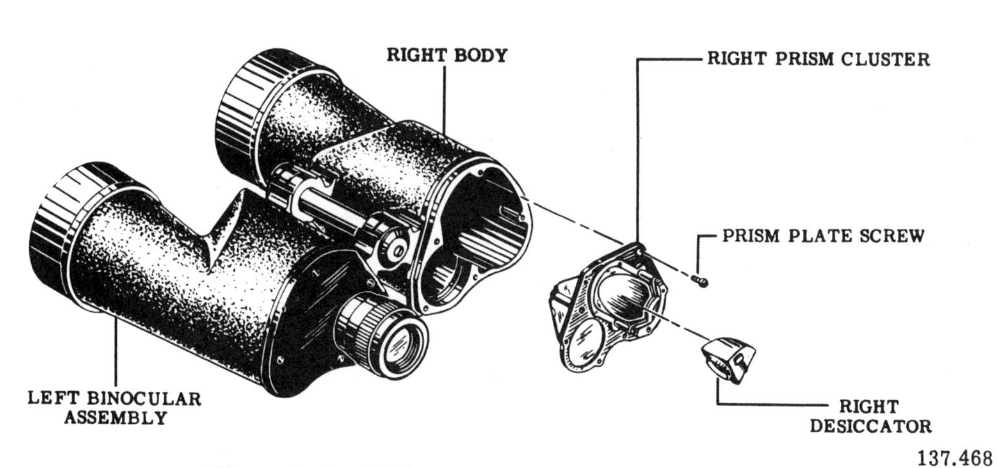

137.468

Figure 18-6.—Right prism cluster and desiccator.

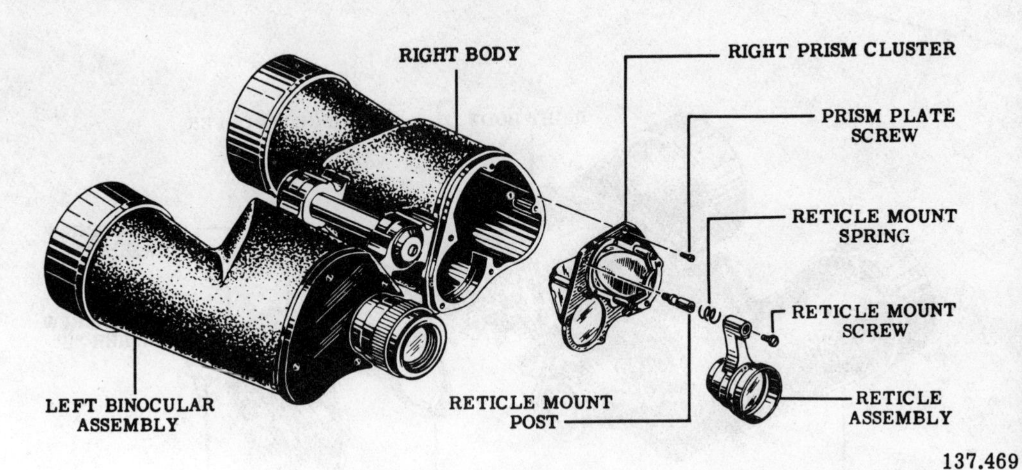

137.469

Figure 18-7.—Reticle assembly.

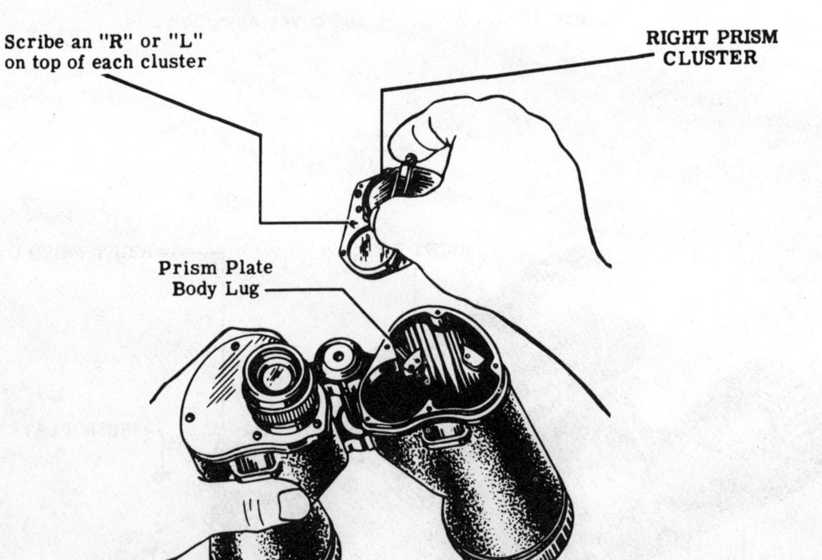

137.470

Figure 18-8.—Identification mark on a prism cluster.

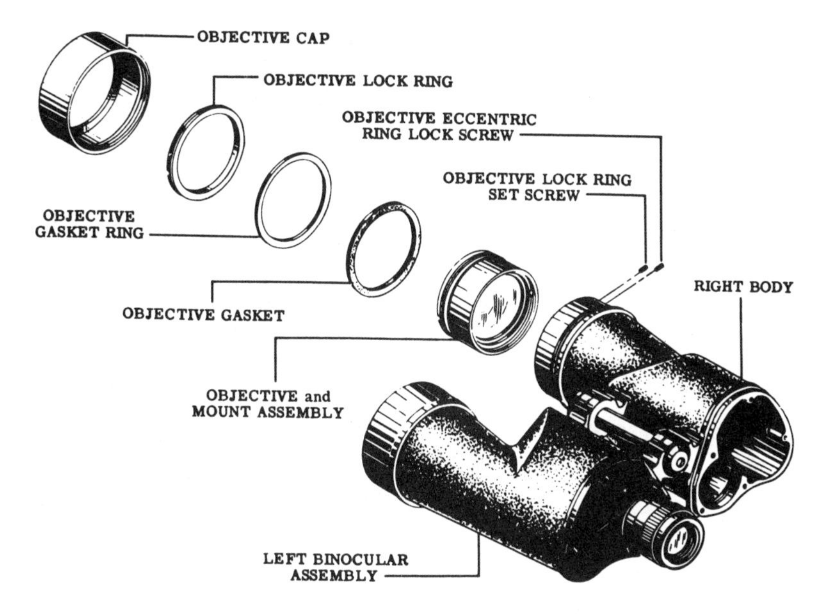

137.471
Figure 18-9.—Binocular objective and mount assembly (Mks 28, Mod 0, & 39, Mod 1).

Removal of the Bodies

To remove the right and left telescope bodies, do the following:

1. Put the binocular assembly on the bench, objective end up, and remove the lower axle lock screw (fig. 18-11) and the axle screw.

2. Turn the binocular assembly around on the bench and use an axle screw wrench to remove the upper axle screw.

3. Scribe a reassembly guide mark across the top of the hinge axle and the hinge lug, as shown in figure 18-12. Note the enlarged part in the circle.

4. Observe the hinge puller in figure 18-4. The long part is the puller and the small part is the collar which fits over the puller. Screw the puller onto the upper end of the hinge pin and keep screwing it down until it pulls the pin out.

5. Discard the cellulose hinge washer. On a Mk 32, Mod 7, binocular, keep the bronze hinge washers (one upper and one lower).

Disassembly of the Eyepiece and Cover Assembly

Use the following procedure to disassemble the eyepiece and cover assembly of a Mk 28, Mod 0, or a Mk 32, Mod 7, binocular:

Hold the assembly in one hand and unscrew the collective lens retaining ring. Then remove the lens. Illustration 18-13 shows the assembly for a Mk 28, Mod 0, binocular and figure 18-14 shows the assembly for a Mk 32, Mod 7, binocular.

2. Reach inside the eyepiece lens mount and slide out the eyepiece lens spacer. This is not on a Mk 32, Mod 7, or a Mk 45, Mod 0, instrument.

3. Cover the open end of the eyepiece and cover assembly with lens tissue and push the eyelens out of its waxed seat. (A Mk 32, Mod 7, or a Mk 45, Mod 0, binocular has a lens retaining ring which must be removed first.)

4. Remove the clamp ring lock screw and the eyepiece clamp; then lift off the knurled focusing ring (fig. 18-14).

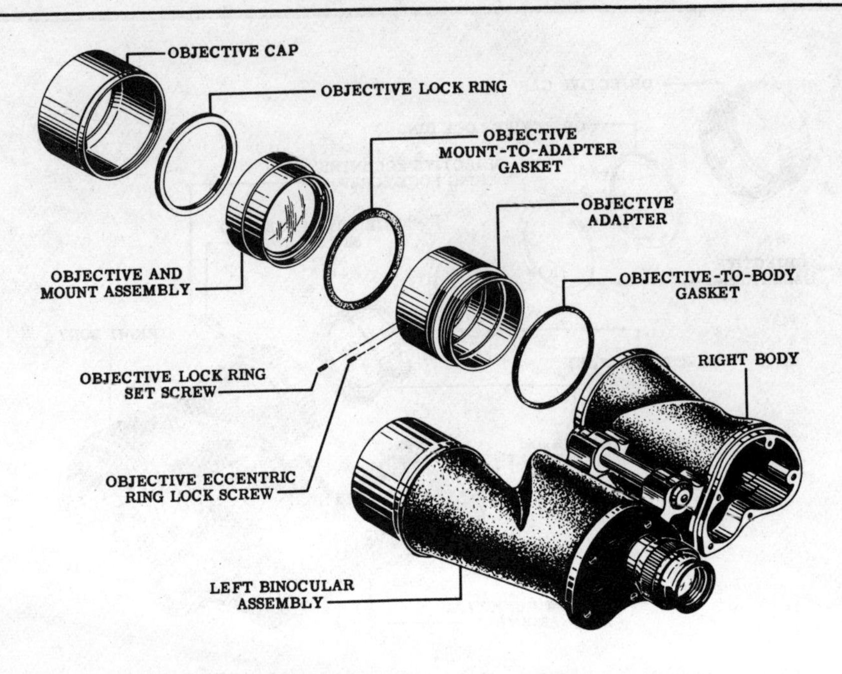

137.472

Figure 18-10.—Objective and mount assembly for a Mk 32, Mod 7, binocular.

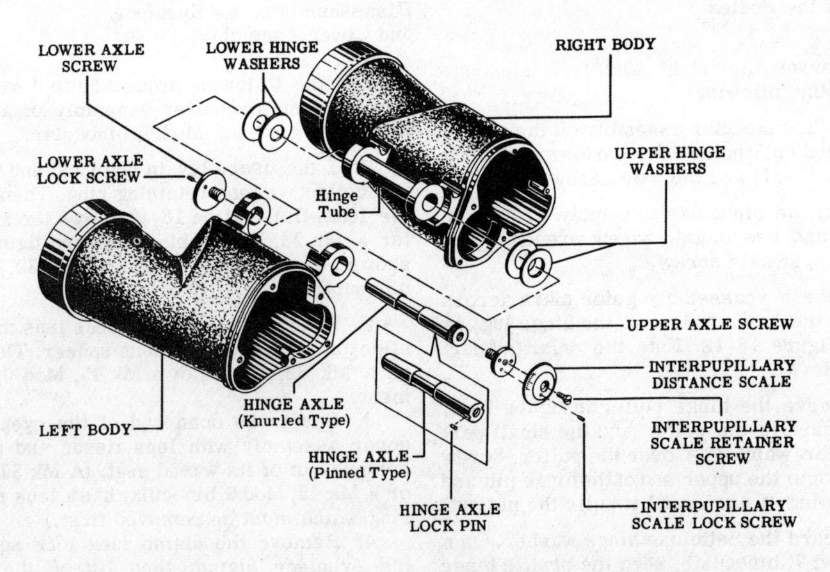

137.473

Figure 18-11.—Components of the interpupillary distance scale.

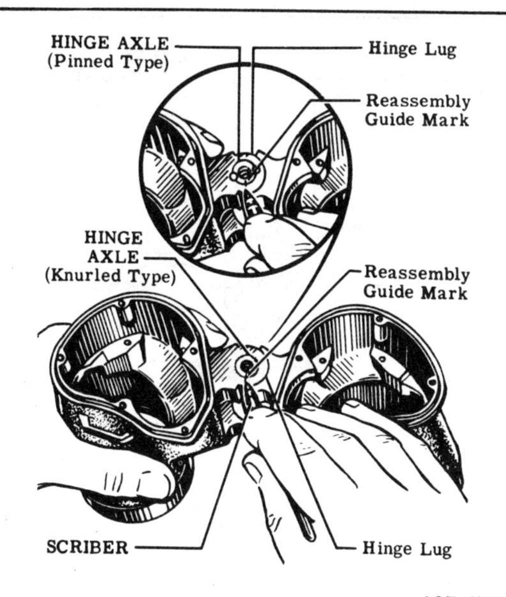

137.474
Figure 18-12.—Scribing a reassembly guide mark across the hinge axle and lug.

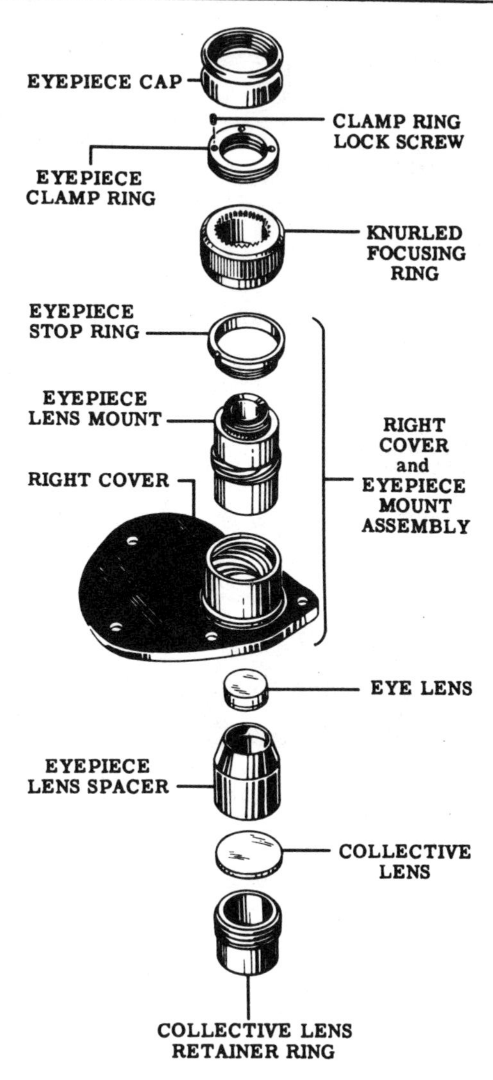

137.175
Figure 18-13.—Eyepiece and cover assembly for a Mk 28, Mod 0, binocular.

5. Unscrew and remove the eyepiece stop ring. Then unscrew the eyepiece lens mount from the cover.

6. Scribe a reassembly guide mark (fig. 18-15) on the top of the eyepiece lens mount corresponding to the diopter line on the neck of the cover. Do NOT interchange the parts of the two bodies.

Disassembly of the Reticle Assembly

On a Mk 39, Mod 1, binocular, remove the lower and upper reticle shields by pulling them off the reticle mount. Then unscrew the reticle retaining ring from the mount and slide the loose reticle off onto a clean cloth. See figure 18-16.

Disassembly of the Prism Cluster

Set the right prism cluster of a Mk 28, Mod 0, or a Mk 39, Mod 1, prism cluster on the prism cluster fixture in the manner shown in figure 18-17.

CAUTION: The two prisms in a cluster are matched. Mark each prism to indicate the plate from which it was removed and its position on the plate—bottom (eyepiece) or top (objective), and right and left.

With a lead or wax pencil, mark each prism on its unpolished surface before you remove it from the prism plate. Make the mark as shown

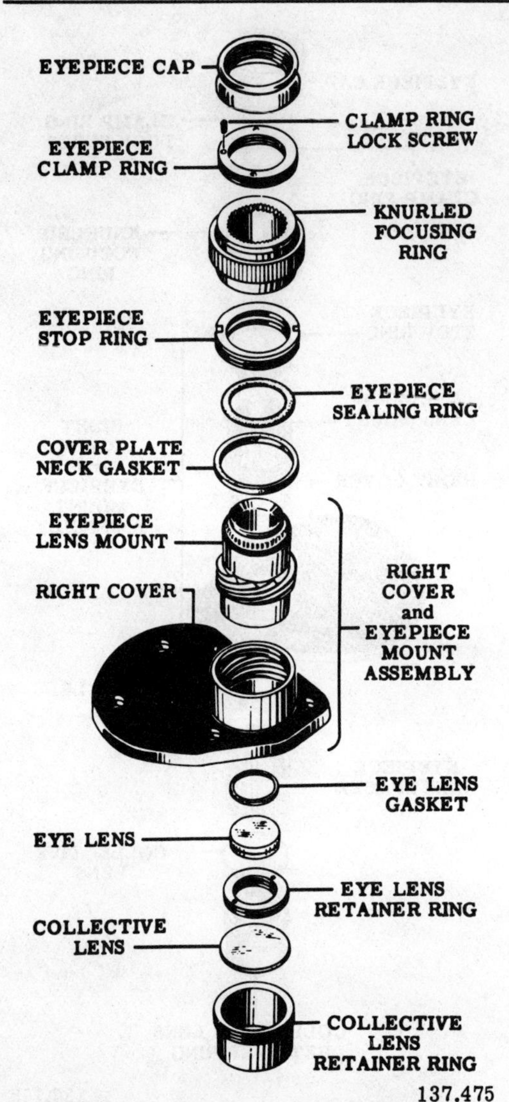

137.475
Figure 18-14.—Eyepiece and cover assembly for a Mk 32, Mod 7, binocular.

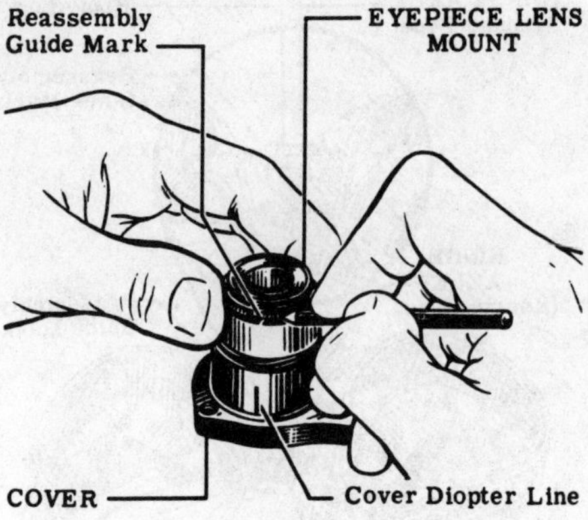

Reassembly Guide Mark — EYEPIECE LENS MOUNT — COVER — Cover Diopter Line

137.476
Figure 18-15.—Scribing a reassembly guide mark on an eyepiece lens mount.

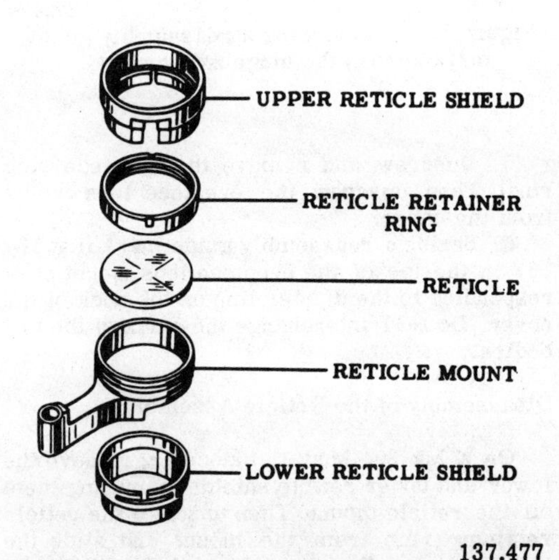

137.477
Figure 18-16.—Reticle assembly for a Mk 39, Mod 1, binocular.

in figure 18-17—TR for the top-right prism cluster, for example. B represents bottom and L represents the left position.

Remove the prism clip screw and the clip. Observe the parts in figure 18-18. Then remove the prism shield and the pad. Discard the pad. Remove next the Porro prism by picking it up by its sides. Then set the prism on its side on a clean cloth. Follow this procedure to remove each prism.

Remove the four prism posts from the prism plate, but do NOT remove the two prism plate

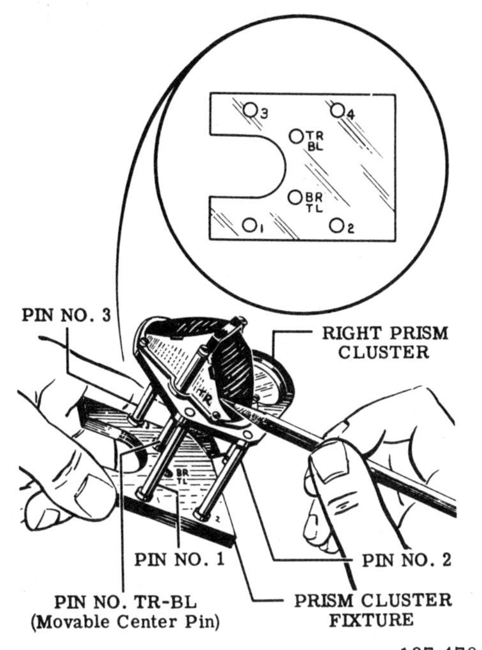

137.478

Figure 18-17.—Disassembly of a right prism cluster.

dowel pins unless they are defective and must be replaced.

Disassembly of the Objective and Mount Assembly

The objective eccentric ring, the objective mount, and the objective lens in each objective and mount assembly are matched to each other. The objective lens in each left and right binocular assembly may also be matched to its respective prism cluster. For this reason, keep the left and right side objective and mount assembly parts separate.

With a rotary motion, slide the objective eccentric ring off the objective mount. Do NOT squeeze the objective eccentric ring out of shape. See illustration 18-19. Then remove the objective lens retaining ring.

Place a clean tissue over the objective lens, hold the objective mount in both hands, and remove the lens by pressing it out with the thumbs onto a clean cloth. The procedure for doing this is shown in figure 18-20.

This completes the disassembly of subassemblies of the binocular. The next step in the repair procedure is inspection, repair, and cleaning of parts.

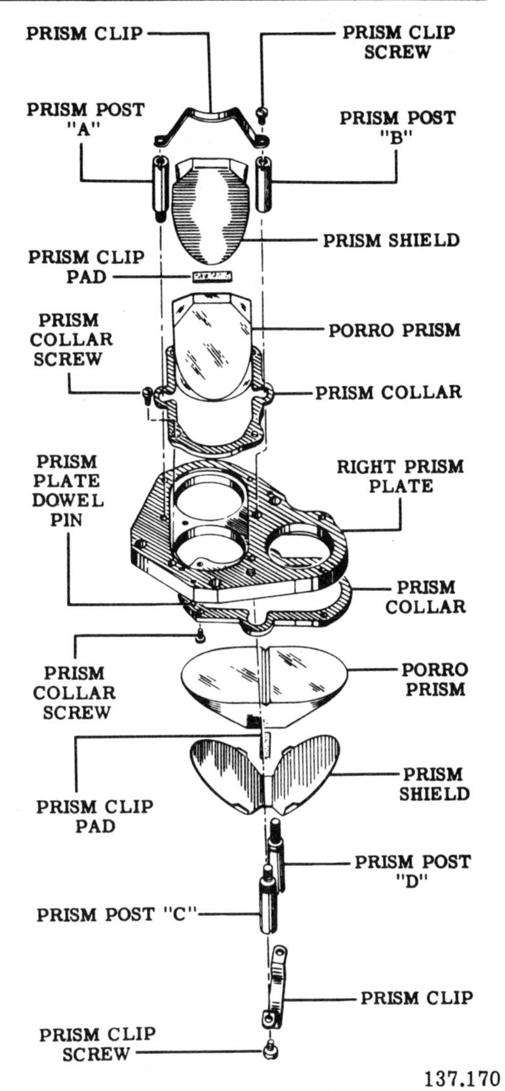

137.170

Figure 18-18.—Disassembled prism cluster.

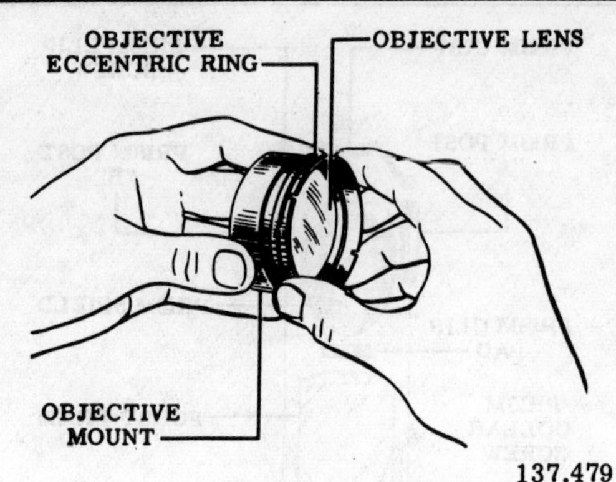

137.479

Figure 18-19.—Disassembling the objective assembly.

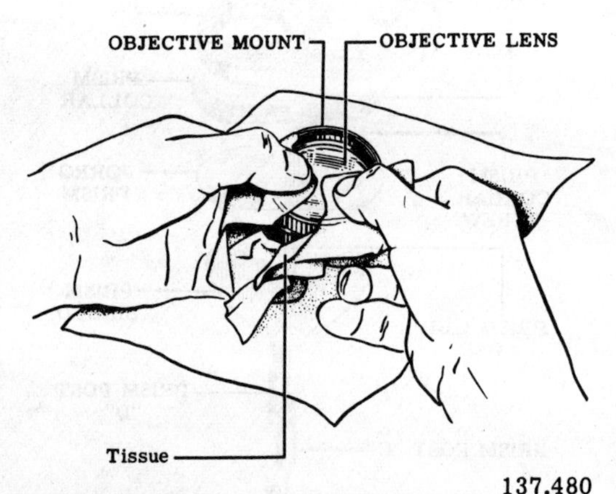

137.480

Figure 18-20.—Pressing the objective lens from its mount.

REPAIR PROCEDURE

After you disassemble a binocular, inspect all parts for defects. This inspection sometimes will reveal defects overlooked or undetected during the predisassembly inspection; and you can give disassembled parts a more thorough inspection.

Inspection of Optical Parts

Post disassembly inspection of optical parts should include:

1. ANTI-REFLECTION COATING. Check the anti-reflection coating on the objective, eye and collective lenses, and the LONG faces of the prisms. The coating must comply with Navy standards.

2. CONDITION OF CEMENT IN LENSES. Inspect the condition of cement between the doublet eyelenses and the objective lenses.

3. CODE MARKS ON PRISMS. Inspect the code marks on the pairs of prisms for correct matching of deviations and prism heights. If there are no code marks, or if the prisms do not match, test them and make necessary replacements.

Inspection of Mechanical Parts

Inspect mechanical parts of your disassembled binocular for the following:

1. FUNCTIONAL DEFECTS. These defects include: distortion, dents, burrs, damaged threads, wear, and so forth. In accordance with prescribed standards, effect repairs or replace the parts.

2. APPEARANCE AND FINISH. Inspect appearance defects and correct them in accordance with prescribed procedures, or the recommendation of your shop supervisor.

Fitting a New Hinge Axle

Hinge tension and collimation of a hand-held binocular are dependent upon the action of the hinge. In order to get proper tension and smooth action in a binocular hinge, you may be required to install an oversized hinge axle on Mks 28, 32, and 39 binoculars. Weak and faulty hinge tension in Mk 45, Mod 0, binoculars can be corrected by using oversized hinge expanding bearings.

An oversized hinge axle must be fitted to the tube in the right body of the binocular, in the following manner:

1. Fit the bodies together and place wedge shims between the lower hinge lugs to push the right body (hinge tube) up. Make certain the shims are clear of the hole for the hinge axle. Study illustration 18-21.

2. With a hinge tube reamer (fig. 18-4), when required, ream the left body hinge lugs and the hinge tube to get a new bearing surface and to help fit the axle. Ream the tapered hole as necessary to make the new axle fit, with its top end approximately flush with or slightly above the top surface of the left body top hinge lug. To draw the axle down, you may have to replace and tighten the lower axle screw.

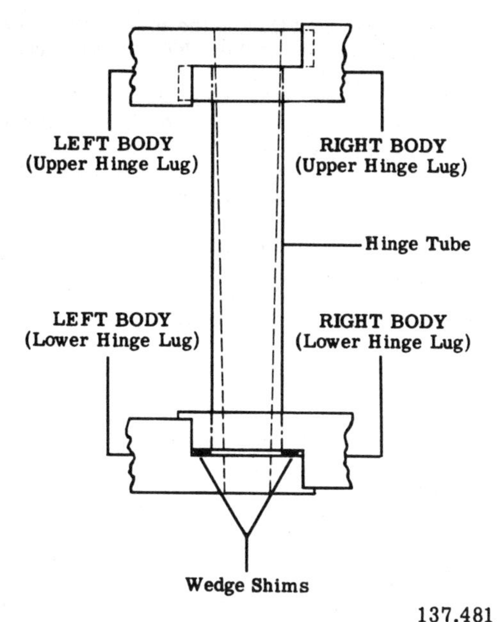

137.481
Figure 18-21.—Wedge shims in position between the lower hinge lugs.

3. If needed, screw the pinned axle lock pin drill jig (fig. 18-4) into the top of the hinge axle and drill two holes 5/64 inch in diameter and 1/8 inch deep. Use two opposite holes in the jig. Then disassemble the axle and the bodies. See figure 18-22.

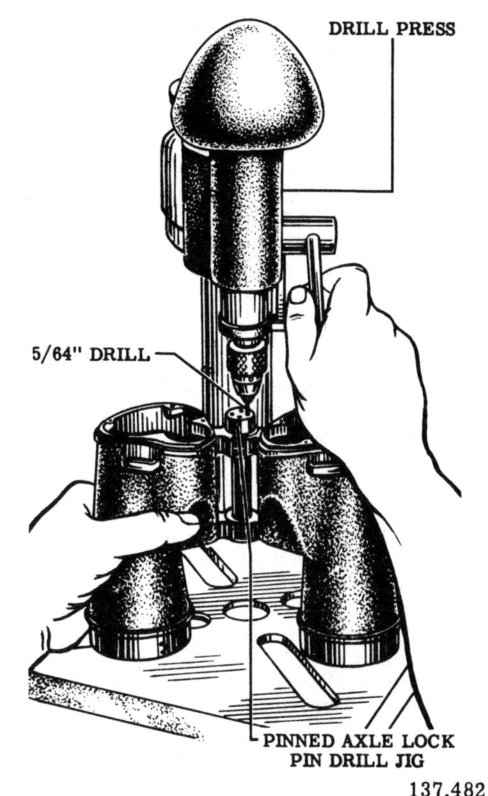

137.482
Figure 18-22.—Drilling a hole in the hinge axle and hinge barrel.

Cleaning Operations

In accordance with procedures discussed in chapter 8, clean the mechanical parts of each binocular optical system and protect them in separate trays. Clean the threads in holes and on all parts to remove paint and corrosion. Use commercial inside and outside thread chasers and taps and dies. If paint on surfaces interferes with reassembly or spoils the appearance, remove it.

Clean optical parts in the prescribed manner just before you replace them.

REASSEMBLY

The reassembly procedure discussed in this section is for Mk 28, Mod 0; Mk 32, Mod 7; and Mk 39, Mod 1, binoculars. Instructions are for reassembling the right half and, except wherever noted, also the left half.

The order of reassembly is as follows: Parts are reassembled to form their particular subassemblies; the subassemblies are then attached to each other to form the major assembly, the complete binocular.

Reassemble all parts in their original positions, for both bodies. Follow your reassembly guide marks. Apply approved lubricants to parts which require them, and in the proper places—hinge grease on hinges only, for example. Apply anti-seize compound on threads and wherever specified.

Joining the Bodies

The best way to join the bodies of a binocular is outlined next.

1. Put cellulose acetate hinge washers between the hinge lugs (upper and lower) of the right and left bodies of Mk 28 and Mk 39 binoculars. These washers are .010 inch thick; but when you must replace them, use washers .0125 inch thick. Mk 32, Mod 7, binoculars have two bronze washers, and the one with the largest hole goes on the top hinge.

2. Slide the two hinges together and check with a straightedge to determine whether the two bodies are even, as shown in figure 18-23. Use different combinations of washers, as necessary, until you have the tops even. The fit must be tight—no shaking or looseness.

3. Separate the bodies and coat the tapered hinge axle, the inside of the hinge tube, and the hinge washers (faces away from the hinge lugs) with hinge grease.

4. Replace the washers on the hinge shoulders of the two bodies and push the two halves together.

5. Use a hinge washer locating pin to align the washers with the hinge lug holes; then turn the bodies to allow the locating pin to work (down) into place. Remove the locating pin and insert the hinge axle in the hinge holes, and re-align the axle with the reassembly guide mark.

Tops of binocular bodies must be even

STRAIGHTEDGE

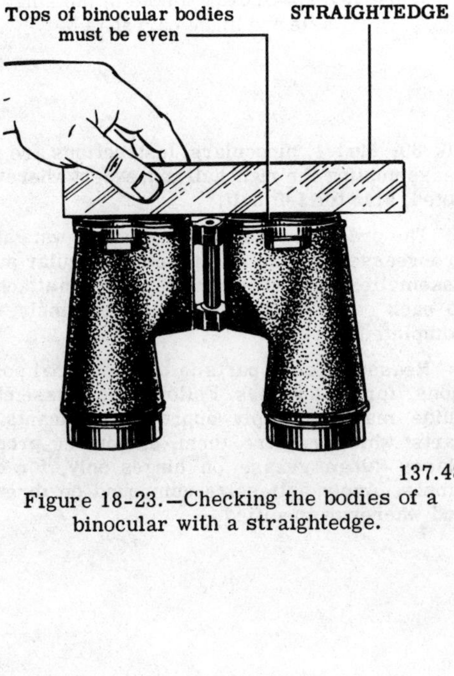

137.483

Figure 18-23.—Checking the bodies of a binocular with a straightedge.

6. Replace and tighten the lower axle screw.

NOTE: Steps 7 and 8 do NOT apply to Mk 32, Mod 7, binoculars.

7. Use tweezers to insert the two hinge axle lock pins. See illustration 18-24. Then complete the job with a rivet set and a hammer.

8. Replace the upper axle screw and screw it down just enough to bring the head against the hinge lug.

9. Test the hinge tension by grasping the bodies and flexing them. Make proper adjustment by loosening or tightening the lower axle screw. Do NOT change the position of the upper axle screw; it keeps the axle at the correct depth in the hinge tube for lubrication clearance.

10. Put an objective-to-body gasket down over the outside threaded end of the objective adapter and seat it on the shoulder. Then screw the objective adapter into the end of the body and tighten it.

11. If you have an interpupillary distance gage, put the assembled bodies on it, as shown in figure 18-25. If you do not have the gage, use an interpupillary distance spacing bar (fig. 18-4). Then replace the interpupillary distance scale and secure it with retainer screw, just

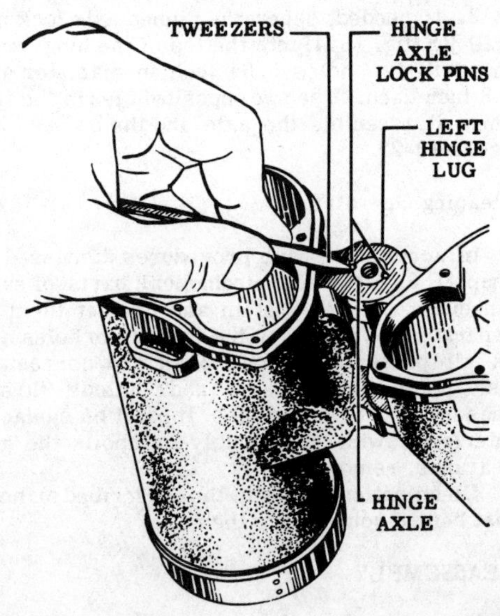

137.484

Figure 18-24.—Replacing hinge axle lock pins.

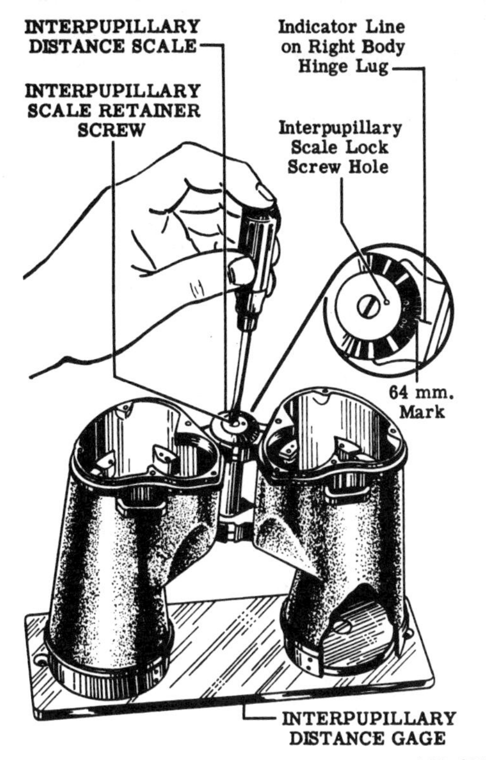

137.485

Figure 18-25.—Setting the interpupillary distance scale.

enough to keep the scale from shifting. Too much pressure will crack the plastic scale.

12. If the interpupillary scale lock screw hole is lined up, replace the screw (Mk 28, Mod 0, & Mk 39, Mod 1). If you did not knock off the lug on the scale during step No. 11 (on a Mk 32, Mod 7, binocular), the scale is properly set and will be held in position by the lug.

Cover the assembly with a clean cloth and put it in a safe place.

Reassembly of the Objective and Mount Assembly

The procedure for reassembling the objective and mount assembly is as follows:

1. Place a string of sealing compound on the inside shoulder of the clean objective mount (Mks 28 & 39 binoculars). On a Mk 32, Mod 7, binocular, place the objective lens gasket on the lens seating shoulder in the objective mount.

2. Put anti-seize compound on the thread of the objective lens retaining ring.

3. Clean the objective lens, pick it up by its edges, and put it in the objective mount, with the positive side (with greatest curvature) down.

4. Replace and tighten the objective lens retaining ring. To relieve possible strain present, unscrew the ring about 1/4 turn; then screw it tight enough to hold the lens in place without shaking.

5. Put a drop of shellac on the threads of the objective mount next to the objective lens retaining ring and run it all the way around. This prevents the ring from becoming loose. Then remove excess shellac and clean the lens.

6. Apply anti-seize compound to the inside surface of the objective eccentric ring. Then slide the ring on the mount and turn it to distribute the compound. NOTE: The slotted end of the ring goes on the outside.

Reassembly of the Prism Cluster

Proceed as follows to reassemble the prism cluster:

1. Clean the prisms and assemble them to their collars. Be sure the code mark on the prism is near the code mark on the collar. Work gently as you do this. The four pads on the collars which bear against the sides of the prisms have been fitted to the prisms. Make certain that you assemble the RIGHT prism to each collar. Do not turn the prisms around in their collars. CAUTION: If the collar pads exert too much pressure on the prisms, strain occurs and must be eliminated. See chapter 8.

2. If you removed them, replace the two prism plate dowel pins by lightly tapping them into the holes in the prism plate. Do the same thing to replace the reticle mount pin in a Mk 39, Mod 1, binocular. Do NOT distort the plate. Study illustration 18-26. If necessary, use a drill to enlarge the holes as required to have a perfect fit of the pins. If the pins still do not fit, try another prism plate.

3. Replace and secure the prism posts to the prism plate. As shown in figure 18-26, Mk 32, Mod 7, binoculars do not have prism posts; they have prism straps.

4. Place the right prism plate on the prism cluster fixture (fig. 18-27) on the pins set for the prism at the top-right position. Move the center

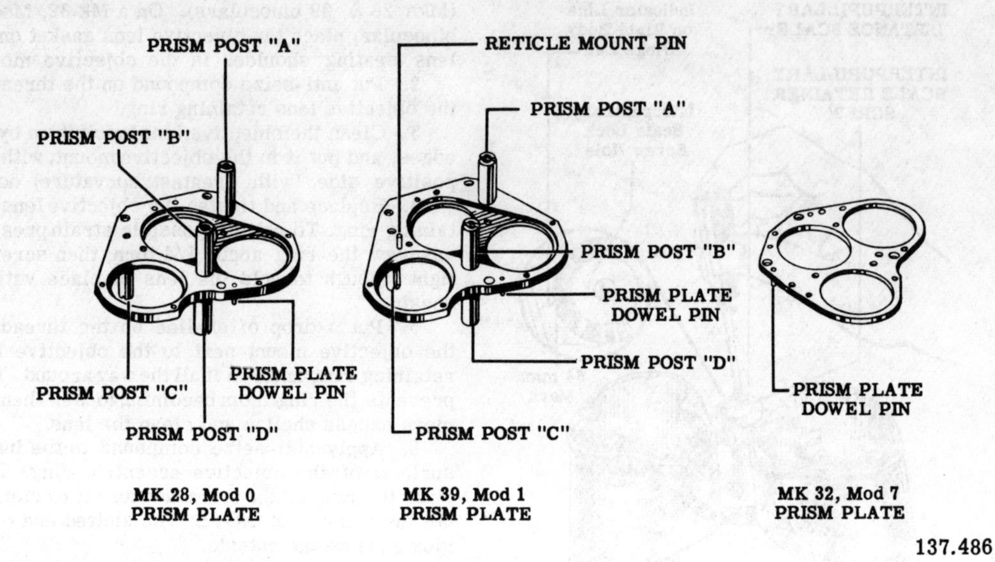

Figure 18-26.—Replacing dowel pins in the prism plate.

pin (fig. 18-27) of the prism cluster fixture to the TR-BL hole. When positioned so that it will receive its T (top) prism on top, the right prism plate will fit on pins Nos. 1 and 2 and the movable center TR-BL. The left prism plate will fit these same pins when so positioned that it will receive its B (bottom) prism on top.

5. Set the top-right prism (with its collar assembled) on the right prism plate on the prism cluster fixture. The prism should fit in the recess in the prism plate, and the two prism posts will fit in the spaces on each side between the prism and the collar.

NOTE: On a Mk 28, Mod 0, or a Mk 39, Mod 1, binocular, place a cork prism clip pad on the apex of the prism, put a prism shield over the prism, and place a prism clip on the tops of the prism posts. These parts are illustrated. On a Mk 32, Mod 7, binocular, put a cork prism pad and a prism shield on the prism and secure the prism with a clip.

6. Reassemble the bottom-left prism in the same manner as explained for reassembling the top-right prism.

The procedure for squaring a prism cluster is explained in chapter 8.

Reassembly of the Eyepiece and Cover Assembly

Proceed as follows to reassemble the eyepiece and cover assembly:

1. Apply a thin coat of eyepiece grease to the threads on the eyepiece lens mount. Then hold the cover and mount on the bench (with guide marks aligned) and screw the mount into the cover (fig. 18-16).

2. Screw the eyepiece stop ring into the neck of the cover and tighten it. Mk 32, Mod 7, binoculars have sealing rings and a cover plate neck gasket in the eyepiece stop ring to keep moisture and dirt out of the body. Stretch the gasket over the threads and into the external groove in the stop ring and insert the sealing ring in the groove. Tighten the stop ring.

3. Place the knurled focusing ring over the eyepiece lens mount and screw on the eyepiece clamp ring. Replace the clamp ring lock screw but do not tighten it.

4. Clean the eyelens and the collective lens.

5. Put a string of sealing compound on the inside shoulder of the eyepiece lens mount and replace the lens. A Mk 32, Mod 7, binocular uses an eyelens gasket instead of sealing compound to seal the lens.

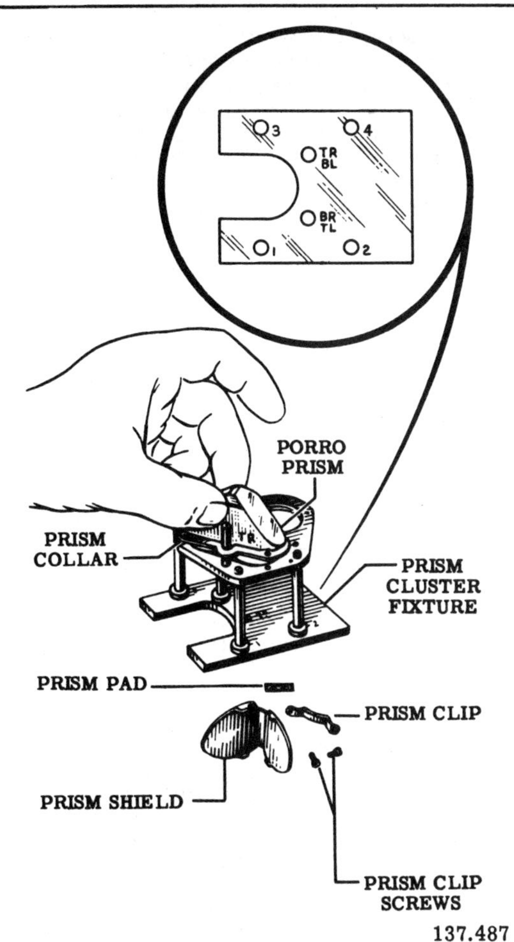

137.487

Figure 18-27.—Replacing prisms on the prism plate.

6. Place the eyepiece lens spacer on the eyelens (side of greatest curvature). If available, use an eyelens assembly fixture when you do this. Then pick up the assembled eyepiece parts and slide the eyepiece lens mount down over the eyelens and eyepiece spacer. Press it down to seat the eyelens in the sealing compound.

NOTE: For a Mk 32, Mod 7, binocular, use a lens suction adapter to set the eyelens in the eyepiece lens mount. Then apply anti-seize compound to the threads on the eyelens retaining ring and replace the ring.

7. Apply anti-seize compound to the threads on the collective lens retaining ring and replace the lens (side of least curvature down) on the ring.

8. Pick up the collective lens and retaining ring with one hand (fig. 18-28). Hold the lens by its edges. Then pick up with your other hand the assembled eyepiece parts and assemble the collective lens in the eyepiece lens mount. Screw in the retaining ring.

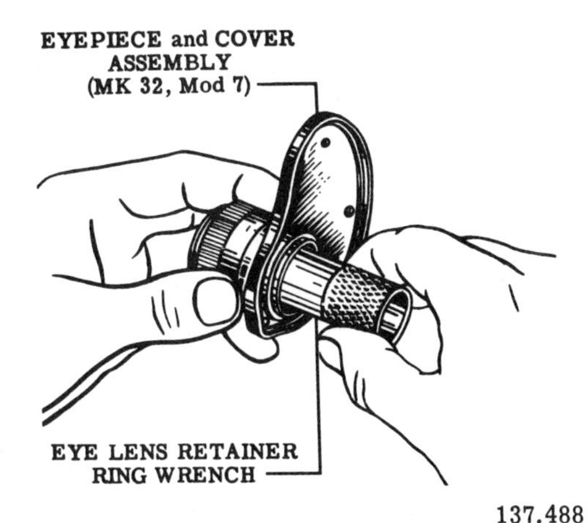

137.488

Figure 18-28.—Assembling the collective lens in the eyepiece lens mount.

9. Screw an eyepiece cap onto the eyepiece clamp ring but leave it loose. Place a drop of shellac on the threads between the mount and the collective lens retaining ring and spread it around. When the shellac hardens, it prevents the retaining ring from becoming loose.

10. On a Mk 39, Mod 1, binocular, clean the reticle carefully and reassemble it, in reverse order of disassembly.

This completes the reassembly of the binocular and it is ready for the final tests and adjustments.

TESTS AND ADJUSTMENTS

This section is concerned with the tests and adjustments necessary to bring your overhauled binocular up to required performance standards. Make certain the instrument is in perfect operating condition before giving it your final approval. Tests for image fidelity and the procedure for waterproofing and sealing are explained in chapter 8. Tests for hinge tension and interpupillary distance were covered previously in this chapter.

Setting the Diopter Scales

The first adjustment you must make in order to bring this binocular up to performance requirements is setting the diopter scales on each eyepiece. They must read 0 diopters plus or minus 1/4 diopter on their scales to get sharp focus of a distant object.

You can set the diopter scales of the binocular with a Mk 5 binocular collimator and a separate auxiliary telescope. See illustration 18-29. Proceed as follows:

1. Sight through the auxiliary telescope and properly adjust it to your eye.
2. Look through the auxiliary telescope and adjust the knurled diopter ring on the binocular eyepiece as necessary to bring the crossline into sharp focus.
3. Unscrew the eyepiece cap and the eyepiece clamp ring but do NOT change the diopter setting.
4. Remove the knurled focusing ring and then set it back on the position necessary to have the mark on the neck of the cover read 0 diopters.
5. Screw the eyepiece clamp ring onto the eyepiece lens mount and tighten it.
6. Check the setting again by repeating step No. 2. Each line on the scale is 1/4 diopter.
7. With a .015-inch drill, make a set point for the lock screw. Then insert and tighten the clamp ring lock screw.
8. Screw on the eyepiece caps.

Checking the Position of the Eyepieces

To check the position of the eyepieces, set both eyepieces to 0 diopters and check with a straightedge across the top of both eyepiece caps to determine whether they are even. They should be even within 1/16 inch.

COLLIMATION

When a binocular is thoroughly overhauled and properly assembled, collimation with a Mk 5 collimator is not difficult. The TAIL-OF-ARC method of collimating is recommended. The step-by-step procedure for using this method is as follows:

1. Turn the binocular upside down and so mount it on the collimator fixture that the left barrel (now on your right) swings freely.

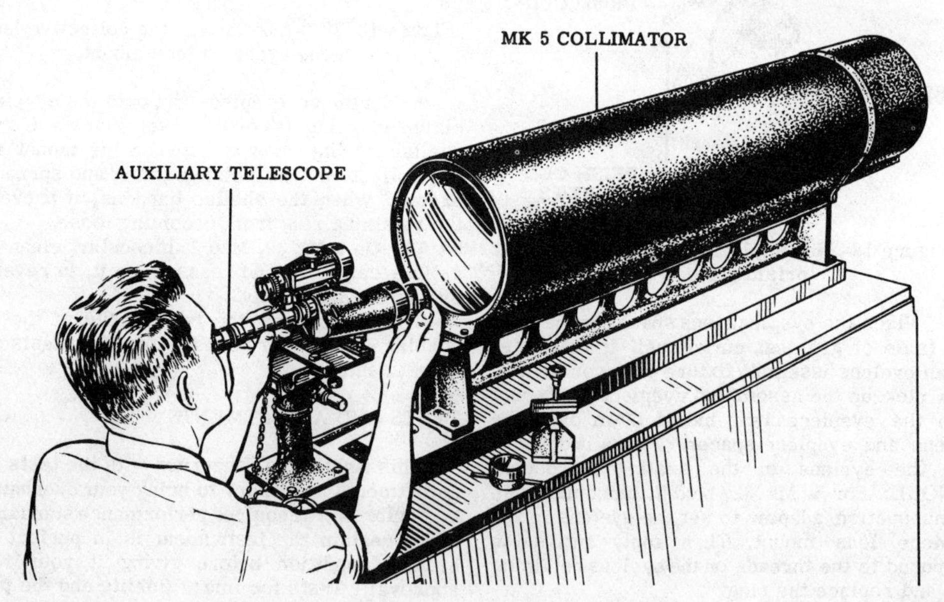

Figure 18-29.—Mark 5 binocular collimator.

2. Look through an auxiliary telescope (with the binocular attachment in place) and the swinging barrel of the binocular. Two images of the collimator crossline should be visible, and one image should be more magnified than the other.

3. With the two adjusting screws on the collimator fixture, superimpose the two crosslines at 58 mm interpupillary distance. Study illustration 18-30.

4. Move the swinging barrel down to 74 mm interpupillary distance and observe the position of the larger crossline, as shown in figure 18-31. Then sketch on graph paper the two crosslines as they appear in the field of view, and construct an equilateral triangle, as explained in the next step. NOTE: ALWAYS figure the displacement of the larger crossline on the scale of the smaller crossline.

5. Use point A in illustration 18-32 as the vertex of a compass and the distance from A to B as a radius and draw an arc CLOCKWISE from B. NOTE: Point A is the intersection point of the smaller crossline and B is the point of intersection of the larger crossline. Next, use point B as the vertex of your compass (with the same radius) and draw another arc which crosses the first arc. This is represented by C in figure

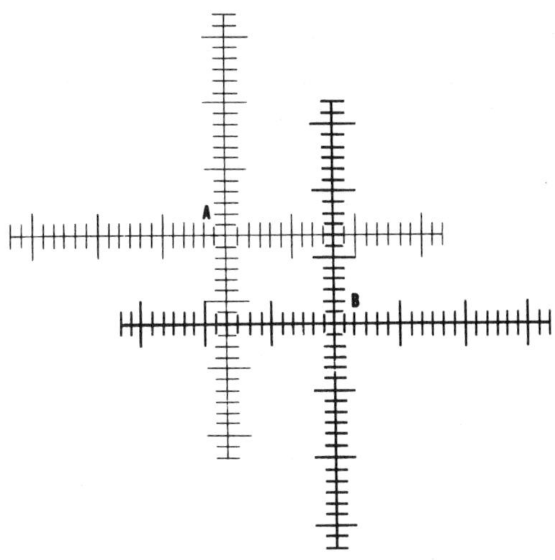

137.491

Figure 18-31.—Position of binocular crosslines (A & B) when the interpupillary distance is at 74 mm.

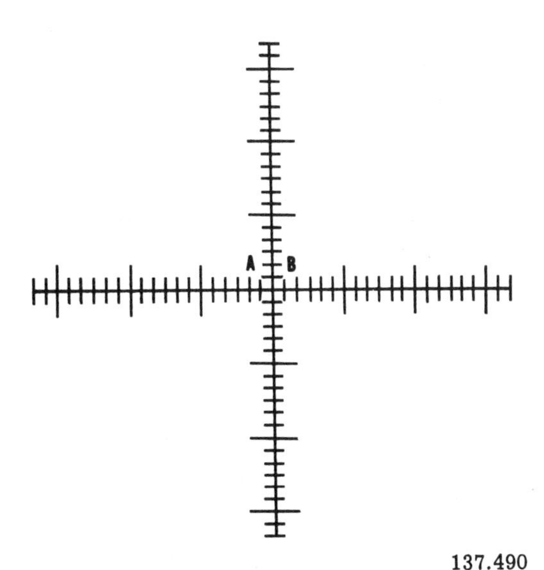

137.490

Figure 18-30.—Binocular crosslines superimposed at 58 mm interpupillary distance.

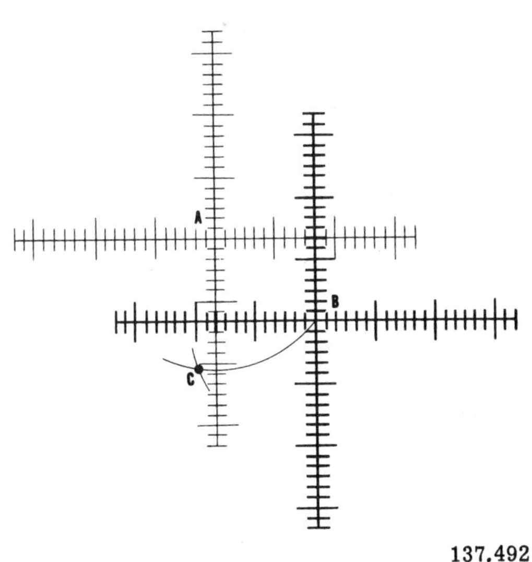

137.492

Figure 18-32.—Locating the mechanical axis of a binocular.

18-33, and C is the mechanical axis of the binocular. The distance from one letter to the other is the same, and the triangle formed by connecting points A, B, and C is therefore equilateral.

NOTE: ALWAYS draw the FIRST arc clockwise from point B.

6. Observe the vertical and horizontal displacement of point C.

7. Look through the auxiliary telescope and the swinging barrel of the binocular and mentally transfer the triangle to your field of view. Then manipulate the objective eccentric rings until the point of intersection of the larger crossline is at point C.

8. Repeat steps 4 through 7 until the crosslines remain superimposed at all settings when the swinging barrel is moved from 58 mm to 74 mm interpupillary distance.

9. Exercise care to prevent throwing the barrel out of collimation and lock the eccentric rings in place with the objective ring setscrew.

10. Leave the mechanical adjustments to the collimator fixture intact and look through the stationary barrel with the auxiliary telescope. Then manipulate the objective eccentric rings until the crosslines are superimposed.

11. With the objective ring setscrew, lock the objective eccentric rings in place. CAUTION: Do NOT throw the barrel out of collimation.

12. Replace the objective gaskets, the objective gasket rings, and the objective lock rings.

13. Check both barrels again to make certain the instrument was not thrown out of collimation. Then replace the objective lock ring setscrews.

14. Put a drop of sealing compound over the objective ring and the objective lock ring setscrews. Then replace the objective caps.

15. Give the instrument another check to be sure it is still collimated.

TOLERANCE AND PERFORMANCE REQUIREMENTS

Binoculars must meet specific tolerances and performance requirements before they are returned to service, as follows:

1. The optical axis of the two barrels must be parallel within: (a) 2 minutes' step (vertical alignment of the two axes); (b) 4 minutes' divergence (spreading apart of the two axes); and (c) 2 minutes' convergence (coming together of the two axes). CAUTION: Failure of the axis to stay within these tolerances causes eyestrain, sometimes severe and accompanied by nausea.

2. When both eyepieces are set to the same diopter reading, they should be even within 1/16 inch. Deviation from this tolerance causes eyestrain.

3. The images of a distant, vertical, straight line formed in the two telescopes (barrels) must be parallel to each other within 1 degree. This is a very liberal tolerance. Failure of the images to stay within this tolerance results in eyestrain.

4. Hinge tension of a binocular is most important. At 70° F, plus or minus 5°, the unsupported side must support a load of 1.80 to 3 pounds at a distance of five inches from the center of the hinge (9 inch pounds minimum and 15 inch pounds maximum). Part of the weight is to the unsupported side of the binocular. If the hinge is TOO tight, it will NOT permit adjustment over the interpupillary distance range. If the hinge is TOO loose, it will not maintain any interpupillary setting.

5. Interpupillary scale readings must be accurate within 1 millimeter. Deviation from this tolerance causes eyestrain.

6. The diopter scales must be set to give readings accurate within 1/4 diopter. If this tolerance is not maintained, a blurred image and eyestrain result.

7. Loss of image fidelity results when optical elements in a binocular are improperly mounted. The image is distorted. Tests of two characteristics of optical performance (central astigmatism and central resolution) provide an overall check on both design and service defects. NOTE: Most defects are eliminated by design of the instrument.

Objects in the center of the FIELD OF VIEW that subtend an angle of 4 seconds of arc must be clearly resolved. In terms of the test to be made, equal-width lines equally spaced .018 inches on centers must be clearly visible as separate and distinct when viewed at 77 feet. Failure to stay within this tolerance results in BLURRED and/or DISTORTED images and eyestrain.

8. Each barrel of a binocular must transmit at least 75 percent of the white light being viewed, and the transmission of the two barrels may not differ by more than 3 percent. Give the binocular a functional test and inspect the optics for proper coating. Loss of light, image brightness, and glare, result when this tolerance is not maintained.

9. When submerged in water, a binocular must withstand a pressure of 3 pounds per square inch; otherwise, moisture will collect in the instrument during service. This means that sealing of the instrument must be thorough.

10. To test a binocular's resistance to shock, drop it from a height of six feet into a box containing six inches of sand and then recheck alignment on the collimator. If the binocular cannot withstand this test, it will be knocked out of collimation during normal usage.

APPENDIX I

TRAINING FILM LIST

Certain training films that are directly related to the information presented in this training course are listed under the appropriate chapter numbers and titles. Unless otherwise specified, all films listed are black and white with sound, and are unclassified. For a description of these and other training films that may be of interest, see the United States Navy Film Catalog, NavPers 10000 (revised).

Chapter 3

CHARACTERISTICS OF LIGHT

MN-2449A Optical Craftsmanship—Introduction to Optics. (17 min.—1945.)

MN-5383 Fundamentals of Photography—The Basic Camera. (15 min.—1948.)

MN-5384 Fundamentals of Photography—Elementary Optics in Photography. (18 min.—Color—1948.)

MN-5385 Fundamentals of Photography—Light-Sensitive Materials. (20 min.—Color—1948.)

Chapter 9

MACHINING OPERATIONS

ME-243A The Engine Lathe—Rough Turning between Centers. (15 min.—1942.)

ME-243B The Engine Lathe—Turning Work of Two Diameters. (11 min.—1942.)

ME-243C The Engine Lathe—Cutting a Taper with a Compound Rest. (11 min.—1942.)

ME-243D The Engine Lathe—Drilling, Boring, and Reaming Work Held in a Chuck. (11 min.—1942.)

ME-243F Operation on the Lathe—Turning a Taper with the Tailstock Set Over. (17 min.—1942).

MC-365B How to Run a Lathe—Plain Turning. (20 min.—Color—1942.)

INDEX

SOME DOVER SCIENCE BOOKS

CHANCE, LUCK AND STATISTICS: THE SCIENCE OF CHANCE,
Horace C. Levinson

Theory of probability and science of statistics in simple, non-technical language. Part I deals with theory of probability, covering odd superstitions in regard to "luck," the meaning of betting odds, the law of mathematical expectation, gambling, and applications in poker, roulette, lotteries, dice, bridge, and other games of chance. Part II discusses the misuse of statistics, the concept of statistical probabilities, normal and skew frequency distributions, and statistics applied to various fields—birth rates, stock speculation, insurance rates, advertising, etc. "Presented in an easy humorous style which I consider the best kind of expository writing," Prof. A. C. Cohen, Industry Quality Control. Enlarged revised edition. Formerly titled *The Science of Chance*. Preface and two new appendices by the author. Index. xiv + 365pp. 5⅜ x 8. Paperbound $2.00

BASIC ELECTRONICS,
prepared by the U.S. Navy Training Publications Center

A thorough and comprehensive manual on the fundamentals of electronics. Written clearly, it is equally useful for self-study or course work for those with a knowledge of the principles of basic electricity. Partial contents: Operating Principles of the Electron Tube; Introduction to Transistors; Power Supplies for Electronic Equipment; Tuned Circuits; Electron-Tube Amplifiers; Audio Power Amplifiers; Oscillators; Transmitters; Transmission Lines; Antennas and Propagation; Introduction to Computers; and related topics. Appendix. Index. Hundreds of illustrations and diagrams. vi + 471pp. 6½ x 9¼.
Paperbound $2.75

BASIC THEORY AND APPLICATION OF TRANSISTORS,
prepared by the U.S. Department of the Army

An introductory manual prepared for an army training program. One of the finest available surveys of theory and application of transistor design and operation. Minimal knowledge of physics and theory of electron tubes required. Suitable for textbook use, course supplement, or home study. Chapters: Introduction; fundamental theory of transistors; transistor amplifier fundamentals; parameters, equivalent circuits, and characteristic curves; bias stabilization; transistor analysis and comparison using characteristic curves and charts; audio amplifiers; tuned amplifiers; wide-band amplifiers; oscillators; pulse and switching circuits; modulation, mixing, and demodulation; and additional semiconductor devices. Unabridged, corrected edition. 240 schematic drawings, photographs, wiring diagrams, etc. 2 Appendices. Glossary. Index. 263pp. 6½ x 9¼. Paperbound $1.25

GUIDE TO THE LITERATURE OF MATHEMATICS AND PHYSICS,
N. G. Parke III

Over 5000 entries included under approximately 120 major subject headings of selected most important books, monographs, periodicals, articles in English, plus important works in German, French, Italian, Spanish, Russian (many recently available works). Covers every branch of physics, math, related engineering. Includes author, title, edition, publisher, place, date, number of volumes, number of pages. A 40-page introduction on the basic problems of research and study provides useful information on the organization and use of libraries, the psychology of learning, etc. This reference work will save you hours of time. 2nd revised edition. Indices of authors, subjects, 464pp. 5⅜ x 8.
Paperbound $2.75

THE RISE OF THE NEW PHYSICS (formerly THE DECLINE OF MECHANISM), *A. d'Abro*

This authoritative and comprehensive 2-volume exposition is unique in scientific publishing. Written for intelligent readers not familiar with higher mathematics, it is the only thorough explanation in non-technical language of modern mathematical-physical theory. Combining both history and exposition, it ranges from classical Newtonian concepts up through the electronic theories of Dirac and Heisenberg, the statistical mechanics of Fermi, and Einstein's relativity theories. "A must for anyone doing serious study in the physical sciences," *J. of Franklin Inst.* 97 illustrations. 991pp. 2 volumes.

T3, T4 Two volume set, paperbound $5.50

THE STRANGE STORY OF THE QUANTUM, AN ACCOUNT FOR THE GENERAL READER OF THE GROWTH OF IDEAS UNDERLYING OUR PRESENT ATOMIC KNOWLEDGE, *B. Hoffmann*

Presents lucidly and expertly, with barest amount of mathematics, the problems and theories which led to modern quantum physics. Dr. Hoffmann begins with the closing years of the 19th century, when certain trifling discrepancies were noticed, and with illuminating analogies and examples takes you through the brilliant concepts of Planck, Einstein, Pauli, de Broglie, Bohr, Schroedinger, Heisenberg, Dirac, Sommerfeld, Feynman, etc. This edition includes a new, long postscript carrying the story through 1958. "Of the books attempting an account of the history and contents of our modern atomic physics which have come to my attention, this is the best," H. Margenau, Yale University, in *American Journal of Physics.* 32 tables and line illustrations. Index. 275pp. 5⅜ x 8.

T518 Paperbound $2.00

GREAT IDEAS AND THEORIES OF MODERN COSMOLOGY, *Jagjit Singh*

The theories of Jeans, Eddington, Milne, Kant, Bondi, Gold, Newton, Einstein, Gamow, Hoyle, Dirac, Kuiper, Hubble, Weizsäcker and many others on such cosmological questions as the origin of the universe, space and time, planet formation, "continuous creation," the birth, life, and death of the stars, the origin of the galaxies, etc. By the author of the popular *Great Ideas of Modern Mathematics.* A gifted popularizer of science, he makes the most difficult abstractions crystal-clear even to the most non-mathematical reader. Index. xii + 276pp. 5⅜ x 8½ T925 Paperbound $2.00

GREAT IDEAS OF MODERN MATHEMATICS: THEIR NATURE AND USE, *Jagjit Singh*

Reader with only high school math will understand main mathematical ideas of modern physics, astronomy, genetics, psychology, evolution, etc., better than many who use them as tools, but comprehend little of their basic structure. Author uses his wide knowledge of non-mathematical fields in brilliant exposition of differential equations, matrices, group theory, logic, statistics, problems of mathematical foundations, imaginary numbers, vectors, etc. Original publications, appendices. indexes. 65 illustr. 322pp. 5⅜ x 8. T587 Paperbound $2.00

THE MATHEMATICS OF GREAT AMATEURS, *Julian L. Coolidge*

Great discoveries made by poets, theologians, philosophers, artists and other non-mathematicians: Omar Khayyam, Leonardo da Vinci, Albrecht Dürer, John Napier, Pascal, Diderot, Bolzano, etc. Surprising accounts of what can result from a non-professional preoccupation with the oldest of sciences. 56 figures. viii + 211pp. 5⅜ x 8½. S1009 Paperbound $2.00